T0318980

WHITE WINE TECHNOLOGY

WHITE WINE TECHNOLOGY

Edited by

ANTONIO MORATA

UPM, Chemistry and Food Technology Department, Escuela Técnica Superior de Ingeniería Agronómica,
Alimentaria y de Biosistemas, Universidad Politécnica de Madrid, Madrid, Spain

ELSEVIER

ACADEMIC PRESS
An imprint of Elsevier

Academic Press is an imprint of Elsevier
125 London Wall, London EC2Y 5AS, United Kingdom
525 B Street, Suite 1650, San Diego, CA 92101, United States
50 Hampshire Street, 5th Floor, Cambridge, MA 02139, United States
The Boulevard, Langford Lane, Kidlington, Oxford OX5 1GB, United Kingdom

Copyright © 2022 Elsevier Inc. All rights reserved.

No part of this publication may be reproduced or transmitted in any form or by any means, electronic or mechanical, including photocopying, recording, or any information storage and retrieval system, without permission in writing from the publisher. Details on how to seek permission, further information about the Publisher's permissions policies and our arrangements with organizations such as the Copyright Clearance Center and the Copyright Licensing Agency, can be found at our website: www.elsevier.com/permissions.

This book and the individual contributions contained in it are protected under copyright by the Publisher (other than as may be noted herein).

Notices
Knowledge and best practice in this field are constantly changing. As new research and experience broaden our understanding, changes in research methods, professional practices, or medical treatment may become necessary.

Practitioners and researchers must always rely on their own experience and knowledge in evaluating and using any information, methods, compounds, or experiments described herein. In using such information or methods they should be mindful of their own safety and the safety of others, including parties for whom they have a professional responsibility.

To the fullest extent of the law, neither the Publisher nor the authors, contributors, or editors, assume any liability for any injury and/or damage to persons or property as a matter of products liability, negligence or otherwise, or from any use or operation of any methods, products, instructions, or ideas contained in the material herein.

British Library Cataloguing-in-Publication Data
A catalogue record for this book is available from the British Library

Library of Congress Cataloging-in-Publication Data
A catalog record for this book is available from the Library of Congress

ISBN: 978-0-12-823497-6

For Information on all Academic Press publications
visit our website at https://www.elsevier.com/books-and-journals

Publisher: Charlotte Cockle
Acquisitions Editor: Patricia Osborn
Editorial Project Manager: Devlin Person
Production Project Manager: Joy Christel Neumarin Honest Thangiah
Cover Designer: Mark Rogers
Cover Pictures: Courtesy of Rafael Cuerda García-Junceda Winemaker

Typeset by MPS Limited, Chennai, India

Working together
to grow libraries in
developing countries

www.elsevier.com • www.bookaid.org

Contents

List of contributors

Diederich Aguilar-Machado Department of Animal Production and Food Science, Food Technology, Faculty of Veterinary, Instituto Agroalimentario de Aragón — IA2— (University of Zaragoza-CITA), Zaragoza, Spain

Ignacio Álvarez Department of Animal Production and Food Science, Food Technology, Faculty of Veterinary, Instituto Agroalimentario de Aragón — IA2— (University of Zaragoza-CITA), Zaragoza, Spain

Jenny Andrea-Silva Chemistry Research Centre-Vila Real (CQ-VR), Food and Wine Chemistry Laboratory, University of Trás-os-Montes and Alto Douro, Vila Real, Portugal; Adega Cooperativa de Figueira de Castelo Rodrigo, Portugal

Antonio Álamo Aroca Oenobrands SAS, Montferrier sur Lez, France

María Antonia Bañuelos enotecUPM, Biotechnology-Plant Biology Department, Universidad Politécnica de Madrid, Campus Ciudad Universitaria, Madrid, Spain; enotecUPM, Biotechnology-Plant Biology Department, Technical University of Madrid, Madrid, Spain

Ana Belén Bautista-Ortín Department of Food Science and Technology, University of Murcia, Campus de Espinardo, Murcia, Spain

Eduardo Boido Oenology and Fermentation Biotechnology Unit, Food Science and Technology Department, School of Chemistry, Universidad de la Republica, Montevideo, Uruguay

Andreea Botezatu The Horticulture Department, Texas A&M AgriLife Extension Service, Texas A&M University, College Station, TX, United States

Laura Canonico Department of Life and Environmental Sciences, Polytechnic University of Marche, Ancona, Italy

Mateo Cardona Winery Designer and Consultant, Krean S. Coop, Gipuzkoa, Spain

Francisco Carrau Oenology and Fermentation Biotechnology Unit, Food Science and Technology Department, School of Chemistry, Universidad de la Republica, Montevideo, Uruguay; Center for Biomedical Research (CEINBIO), School of Medicine, Universidad de la Republica, Montevideo, Uruguay

Hannah Charnock Department of Biological Sciences, Faculty of Mathematics and Science, Brock University, St. Catharines, ON, Canada

Gianella Chávez-Segura Agroindustrial Engineering Program, Universidad Privada del Norte (UPN), Trujillo, Peru

K. Chen College of Food Science & Nutritional Engineering, China Agricultural University, Beijing, P. R. China

Monika Christmann Institute of Oenologie, Hochschule Geisenheim University, Geisenheim, Germany

Maurizio Ciani Department of Life and Environmental Sciences, Polytechnic University of Marche, Ancona, Italy

Francesca Comitini Department of Life and Environmental Sciences, Polytechnic University of Marche, Ancona, Italy

Piergiorgio Comuzzo Department of Agricultural, Food, Environmental and Animal Sciences (Di4A), University of Udine, Udine, Italy

Fernanda Cosme Chemistry Research Centre-Vila Real (CQ-VR), Food and Wine Chemistry Laboratory, University of Trás-os-Montes and Alto Douro, Vila Real, Portugal; Biology and Environment Department, School of Life Sciences and Environment, University of Trás-os-Montes and Alto Douro, Vila Real, Portugal

D. Cozzolino Centre for Nutrition and Food Sciences, Queensland Alliance for Agriculture and Food Innovation (QAAFI), The University of Queensland, Brisbane, QLD, Australia

Rafael Cuerda Bodegas Comenge, Valladolid, Spain

Juan Manuel Del Fresno enotecUPM, Chemistry and Food Technology Department, Technical University of Madrid, Madrid, Spain

Eduardo Dellacassa Laboratory of Aroma Biotechnology, School of Chemistry, Universidad de la Republica, Montevideo, Uruguay

Carlota Delso Department of Animal Production and Food Science, Food Technology, Faculty of Veterinary, Instituto Agroalimentario de Aragón — IA2— (University of Zaragoza-CITA), Zaragoza, Spain

Carlos Escott enotecUPM, Chemistry and Food Technology Department, Technical University of Madrid, Madrid, Spain

Sergi Ferrer Institute of Biotechnology and Biomedicine (BioTecMed) and Department of Microbiology and Ecology, University of Valencia, Valencia, Spain

Luís Filipe-Ribeiro Chemistry Research Centre-Vila Real (CQ-VR), Food and Wine Chemistry Laboratory, University of Trás-os-Montes and Alto Douro, Vila Real, Portugal

Vincenzo Gerbi Department of Agricultural, Forest and Food Sciences, University of Turin, Turin, Italy

Simone Giacosa Department of Agricultural, Forest and Food Sciences, University of Turin, Turin, Italy

Encarna Gómez-Plaza Department of Food Science and Technology, University of Murcia, Campus de Espinardo, Murcia, Spain

Carmen González enotecUPM, Chemistry and Food Technology Department, Universidad Politécnica de Madrid, Campus Ciudad Universitaria, Madrid, Spain; enotecUPM, Chemistry and Food Technology Department, Technical University of Madrid, Madrid, Spain

Lucía González-Arenzana Institute of Grapevine and Wine Science (Government of La Rioja, University of La Rioja and Superior Council of Scientific Investigations), La Rioja, Logroño, Spain

Susana González-Manzano Polyphenols Research Group (GIP-USAL), University of Salamanca, Campus Miguel de Unamuno, Salamanca, Spain; Unit of Excellence Agricultural Production and Environment (AGRIENVIRONMENT), Scientific Park, University of Salamanca, Salamanca, Spain

Ana M. González-Paramás Polyphenols Research Group (GIP-USAL), University of Salamanca, Campus Miguel de Unamuno, Salamanca, Spain; Unit of Excellence Agricultural Production and Environment (AGRIENVIRONMENT), Scientific Park, University of Salamanca, Salamanca, Spain

Buenaventura Guamis Centre d'Innovació, Recerca i Transferència en Tecnologia dels Aliments (CIRTTA), TECNIO, XaRTA, Departament de Ciència Animal i dels Aliments, Facultat de Veterinària, Autonomous University of Barcelona, Bellaterra, Spain

Shunyu Han College of Food Science and Engineering, Gansu Agricultural University, Lanzhou, P.R. China; Gansu Key Laboratory of Viticulture and Enology, Lanzhou, P.R. China; China Wine Industry Technology Institute, Yinchuan, P.R. China

José María Heras Lallemand Ibérica SL., Madrid, Spain

L. Iacumin Department of Agricultural, Food, Environmental and Animal Sciences (Di4A), University of Udine, Udine, Italy

Debra Inglis Department of Biological Sciences, Faculty of Mathematics and Science, Brock University, St. Catharines, ON, Canada; Cool Climate Oenology & Viticulture Institute (CCOVI), Brock University, St. Catharines, ON, Canada

António M. Jordão Polytechnic Institute of Viseu, Agrarian Higher School, Viseu, Portugal.Chemistry Research Centre (CQ-VR) - Food and Wine Chemistry Lab, Vila Real, Portugal

Belinda Kemp Department of Biological Sciences, Faculty of Mathematics and Science, Brock University, St. Catharines, ON, Canada; Cool Climate Oenology & Viticulture Institute (CCOVI), Brock University, St. Catharines, ON, Canada

Paul Kilmartin School of Chemical Sciences, University of Auckland, Auckland, New Zealand

J. Li College of Food Science & Nutritional Engineering, China Agricultural University, Beijing, P. R. China

Wei Li College of Food Science and Engineering, Gansu Agricultural University, Lanzhou, P.R. China; Gansu Key Laboratory of Viticulture and Enology, Lanzhou, P.R. China

Iris Loira enotecUPM, Chemistry and Food Technology Department, Technical University of Madrid, Madrid, Spain; enotecUPM, Chemistry and Food Technology Department, Universidad Politécnica de Madrid, Campus Ciudad Universitaria, Madrid, Spain

Tengzhen Ma College of Food Science and Engineering, Gansu Agricultural University, Lanzhou, P.R. China; Gansu Key Laboratory of Viticulture and Enology, Lanzhou, P.R. China

Manuel Malfeito-Ferreira Linking Landscape Environment Agriculture and Food Research Center (LEAF), Instituto Superior de Agronomia, Universidade de Lisboa, Lisbon, Portugal

Richard Marchal Laboratoire d'Oenologie de Chimie Appliquée, Unité de Recherche Vignes et Vins de Champagne (URVVC), Reims, France

Valentina Martin Oenology and Fermentation Biotechnology Unit, Food Science and Technology Department, School of Chemistry, Universidad de la Republica, Montevideo, Uruguay

Juan Manuel Martínez Department of Animal Production and Food Science, Food Technology, Faculty of Veterinary, Instituto Agroalimentario de Aragón — IA2— (University of Zaragoza-CITA), Zaragoza, Spain

Juana Martínez Institute of Grapevine and Wine Science (Government of La Rioja, University of La Rioja and Superior Council of Scientific Investigations), La Rioja, Logroño, Spain

Marcos Maza National University of Cuyo, Faculty of Agricultural Sciences, Department of Oenological and Food Sciences, Mendoza, Argentina

Karina Medina Oenology and Fermentation Biotechnology Unit, Food Science and Technology Department, School of Chemistry, Universidad de la Republica, Montevideo, Uruguay

Antonio Morata enotecUPM, Chemistry and Food Technology Department, Technical University of Madrid, Madrid, Spain; enotecUPM, Chemistry and Food Technology Department, Universidad Politécnica de Madrid, Campus Ciudad Universitaria, Madrid, Spain

Jean-Roch Mouret SPO-Sciences For Enology, Univ Montpellier, Montpellier SupAgro, Montpellier, France

Fernando M. Nunes Chemistry Research Centre-Vila Real (CQ-VR), Food and Wine Chemistry Laboratory, University of Trás-os-Montes and Alto Douro, Vila Real, Portugal; Chemistry Department, School of Life Sciences and Environment, University of Trás-os-Montes and Alto Douro, Vila Real, Portugal

Pablo Ossorio Winemaker and Oenology Consultant, Hispano-Suizas Cellar, Valencia, Spain

Maria Alessandra Paissoni Department of Agricultural, Forest and Food Sciences, University of Turin, Turin, Italy

Isabel Pardo Institute of Biotechnology and Biomedicine (BioTecMed) and Department of Microbiology and Ecology, University of Valencia, Valencia, Spain

Paula Pérez-Porras Department of Food Science and Technology, University of Murcia, Campus de Espinardo, Murcia, Spain

Luis Javier Pérez-Prieto Department of Food Science and Technology, University of Murcia, Campus de Espinardo, Murcia, Spain

Gary Pickering Department of Biological Sciences, Faculty of Mathematics and Science, Brock University, St. Catharines, ON, Canada; Cool Climate Oenology & Viticulture Institute (CCOVI), Brock University, St. Catharines, ON, Canada; National Wine and Grape Industry Centre, Charles Sturt University, Wagga Wagga, NSW, Australia

A. Power CREST Technology Gateway of TU Dublin, Dublin, Ireland

Isak S. Pretorius ARC Centre of Excellence in Synthetic Biology, Macquarie University, Sydney, NSW, Australia

Javier Raso Department of Animal Production and Food Science, Food Technology, Faculty of Veterinary, Instituto Agroalimentario de Aragón — IA2— (University of Zaragoza-CITA), Zaragoza, Spain

Jorge M. Ricardo-da-Silva LEAF - Linking Landscape, Environment, Agriculture and Food, Universidade de Lisboa, Instituto Superior de Agronomia, Lisboa, Portugal

Susana Río Segade Department of Agricultural, Forest and Food Sciences, University of Turin, Turin, Italy

Luca Rolle Department of Agricultural, Forest and Food Sciences, University of Turin, Turin, Italy

Jean-Marie Sablayrolles SPO-Sciences For Enology, Univ Montpellier, Montpellier SupAgro, Montpellier, France

Faisal Eudes Sam College of Food Science and Engineering, Gansu Agricultural University, Lanzhou, P.R. China; Gansu Key Laboratory of Viticulture and Enology, Lanzhou, P.R. China

Celestino Santos-Buelga Polyphenols Research Group (GIP-USAL), University of Salamanca, Campus Miguel de Unamuno, Salamanca, Spain; Unit of Excellence Agricultural Production and Environment (AGRIENVIRONMENT), Scientific Park, University of Salamanca, Salamanca, Spain

Matthias Schmitt Institute of Oenologie, Hochschule Geisenheim University, Geisenheim, Germany

Remi Schneider Oenobrands SAS, Montpellier sur Lez, France

Mark Strobl Hochschule Geisenheim University, Geisenheim, Germany

José Antonio Suárez-Lepe enotecUPM, Chemistry and Food Technology Department, Technical University of Madrid, Madrid, Spain; enotecUPM, Chemistry and Food Technology Department, Universidad Politécnica de Madrid, Campus Ciudad Universitaria, Madrid, Spain

Maria Jose Valera Oenology and Fermentation Biotechnology Unit, Food Science and Technology Department, School of Chemistry, Universidad de la Republica, Montevideo, Uruguay

Isabelle van Rolleghem Oenobrands SAS, Montferrier sur Lez, France

Cristian Vaquero enotecUPM, Chemistry and Food Technology Department, Universidad Politécnica de Madrid, Campus Ciudad Universitaria, Madrid, Spain

Ricardo Vejarano Dirección de Investigación y Desarrollo, Universidad Privada del Norte (UPN), Trujillo, Peru

Aude Vernhet SPO, Univ Montpellier, INRAE, Institut Agro, Montpellier, France

Sabrina Voce Department of Agricultural, Food, Environmental and Animal Sciences (Di4A), University of Udine, Udine, Italy

James Willwerth Department of Biological Sciences, Faculty of Mathematics and Science, Brock University, St. Catharines, ON, Canada; Cool Climate Oenology & Viticulture Institute (CCOVI), Brock University, St. Catharines, ON, Canada

Fei Yang Cool Climate Oenology & Viticulture Institute (CCOVI), Brock University, St. Catharines, ON, Canada

Fernando Zamora Department of Biochemistry and Biotechnology, Faculty of Oenology, University Rovira i Virgili, Tarragona, Spain

Bo Zhang College of Food Science and Engineering, Gansu Agricultural University, Lanzhou, P.R. China; Gansu Key Laboratory of Viticulture and Enology, Lanzhou, P.R. China

Junxiang Zhang China Wine Industry Technology Institute, Yinchuan, P.R. China; Wine School, Ningxia University, Yinchuan, P.R. China

Prologue

White Wine Technology (WWT) is the logical continuation and the evolution of the previous and successful Red Wine Technology (RWT). Like RWT, it covers a gap in the specialized literature on enology by being highly specialized in the enology, microbiology, and chemistry of white wines with special concern for new trends in white winemaking and the use of emerging technologies and advanced biotechnologies. The elaboration of white wines has too many specificities and thus constitutes an independent discipline. The gentle extraction of the juices is essential to reduce the content of phenolic compounds while avoiding astringency and bitterness. Additionally, the delicate phenolic constitution of white wines makes them especially sensitive to oxidative processes, and they must therefore be protected during winemaking, aging, and storage.

WWT includes not only the advanced knowledge needed to make high-quality white wines but also specific topics that are currently trending in white winemaking. The book is structured into 29 chapters written by knowledgeable experts from prestigious universities, research centers, and wineries from 11 countries and four continents from traditional and emerging growing areas. Most of them are located in wine areas because most of the world's vineyard surface is located in warm regions, but the specifics of cold regions are also included, covered by experts from Canada and northern China.

The assessment of grape quality is described including the main parameters and the evaluation of composition, health, and potential contaminants describing current analytical techniques and sensory evaluation. The use of hyperspectral imaging (HSI) is also extensively detailed, explaining the technique and the applications to control not only vineyards but also the quality of the grape. HSI is increasingly used due to the high-throughput and noninvasive applications that can be developed to control grape composition and quality. Alternative antioxidant products to reduce SO_2 levels are now a trend due to their allergenic effects. Among them, glutathione is being widely used to control oxidations in white grapes and wines.

Preprocessing techniques of grapes are addressed, including traditional systems as well as new perspectives of emerging technologies. The conventional but discontinuous extraction of the must by crushing and pressing the grapes with high-quality inert pneumatic presses produces high-quality juices, but it is a slow step and a bottleneck in the process. Traditionally, continuous extractions using dejuicers and continuous screw presses produced low-quality juices, but now the use of horizontal decanters is a gentle technology used to continuously separate high-quality must from the crushed grapes. Innovative processing of white grapes and juices includes the use of emerging technologies such as ultra high pressure homogenization (UHPH), pulsed electric fields (PEF), or ultrasounds (US) that open new perspectives to obtaining high-quality juices in continuous processes and, in some of them, with the possibility to control microorganisms or oxidative enzymes. After obtaining the raw juice, the design of suitable settling processes is also essential for the quality of white wine. Flotation, cold settling, and other techniques are described in a specific chapter.

New fermentation biotechnologies are covered by describing the use of non-*Saccharomyces* that are now extensively used in the fermentation of white wines. The use of *Hanseniaspora vineae* is specifically discussed because of its great impact on the aroma and body of wines from neutral varieties. The general use of the main non-*Saccharomyces* in fermentation is also described in a specific chapter. The use of *Lachancea thermotolerans* is described in a specific chapter for its applications to control pH and improve freshness in wines from warm areas. Nitrogen management in fermentation is essential to produce high-quality wines and to avoid sluggish or stuck fermentations, especially in reductive conditions at low temperatures. A chapter describes in depth nitrogen nutrition and its impact on fermentation. Synthetic biology and the genetic engineering of wine yeasts, including advanced innovative molecular techniques such as CRISPRCas9 are specifically addressed. And finally, concerning white wine biotechnology, and although not always used, the main applications of malolactic fermentation with an effect on sensory improvement or acidification are also studied.

Pinking is a hot topic in some white wines and the formation or revealing of anthocyanins in some white varieties has recently been studied. Updated information explaining the varieties affected and the influence of the process conditions is included in this topic. The prevention of the light-struck taste in white wine is also a typical concern. The management of this alteration is clearly described. White wines have lower levels of polyphenols, compounds with antioxidant properties and a high impact on health, than red wines. Even considering their lower content, the amount and effect of white wine polyphenols have been analyzed in an updated review.

Enzymes are powerful biotools in many winemaking processes, the use of which encompasses the pretreatment of grapes, extraction and release of aroma compounds, extraction of polyphenols, facilitation of technological processes, aging, and minimization of alterations. Their production, specifications, and wine utilities are extensively studied. The use of near-infrared (NIR) technologies is increasing nowadays due to the simplicity, robustness, noninvasive analysis, and many other features that make it a fast and efficient technique to assess grape and wine quality and alterations.

White wines are usually consumed as young wines, but many high-quality whites are produced by biological (lees contact) or physicochemical (barrel) aging. Aging on lees (AOL) increases wine body and palatability and, at the same time, improves aroma and protects the varietal smell as it is a reductive process. The applications of AOL have been described, especially the impact on quality. Additionally, the use of barrel aging and the advantages and drawbacks on wine sensory profile and quality have been assessed. Alternative wood species to oak have recently been incorporated into winemaking, and their applications and advantages are described in a specific chapter.

Aroma is a key perception in the sensory profile of wine with a clear impact on quality. Emerging techniques to protect and improve wine aroma using phenolics and other alternative antioxidants have been evaluated in depth. Furthermore, the impact of aroma compounds on wine quality has been described taking into account their origin and properties. An essential technological process to preserve wine aroma is the use of inertization technologies and bottling methods to preserve wine aroma and color at an optimal level.

Although wine production is mainly extended in warm areas, wines from cool regions have really interesting properties and specific sensory profiles. The specificities of winemaking in cold areas are included in two chapters focusing on the northern regions of Canada and China. The special management of viticulture, winemaking, and the ways to control the unsuitable maturity of grapes are reviewed. The use of ancient practices such as buried viticulture in extremely cold regions to ensure the survival of the vine during winter is also described.

Alcohol content in wines is always a controversial aspect for considering healthy beverages. The use of dealcoholization technologies can produce alternative wine products that keep the healthy properties together with a suitable sensory profile. The current technological possibilities to dealcoholize wines have been thoroughly described.

Lastly, wine tasting and sensory perception are key aspects of wine quality. The neurophysiological basis of the sensory perception and all the considerations affecting sensory sensation and emotion are extensively detailed in an indepth review chapter.

An edited book is a really complex work of design, selection of key topics, finding the right experts, coordination, and management; more than 500 emails are in my inbox regarding this book. I would like to acknowledge all the prestigious and knowledgeable professors and experts who have contributed because they have made chapters of the highest quality and some of them in very stressful situations. The quality of this book is due to their work and expertise. Special thanks to my colleagues in the Food Technology Department of the Universidad Politécnica de Madrid for their help and understanding during the edition of the book and for their useful advice, especially to professors José Antonio Suárez, Karen González, Iris Loira, Felipe Palomero, María Jesús Callejo, María Antonia Bañuelos, and Carmen López; to postdoctoral researchers Carlos Escott and Juan Manuel del Fresno; and to Ph.D. students Cristian Vaquero and Natalia Gutiérrez. I am grateful to Patricia Osborn, Senior Editor at Elsevier, for her confidence in me again as the editor of WWT and for her support throughout the entire project in the elaboration of this amazing new book. My gratitude also to Devlin Person, Senior Editorial Project Manager at Elsevier, for her full support and kindness in helping me to manage the editing of all the chapters. And finally, but especially, to my family: Cari, Jaime, and María; most of the time in this book belongs to them.

CHAPTER

1

Assessment and control of grape maturity and quality

Luca Rolle, Susana Río Segade, Maria Alessandra Paissoni, Simone Giacosa and Vincenzo Gerbi

Department of Agricultural, Forest and Food Sciences, University of Turin, Turin, Italy

1.1 White grape quality parameters

To achieve high quality in white wine production, it is necessary to obtain grapes presenting optimal compositional parameters. In this sense, vineyard management is of primary importance together with the choice of harvest time. In fact, the basis of wine characteristics is intrinsically confined in the grape quality, and the wide diversity in white wine grape chemical composition is of major technological importance, as each cultivar requires dedicated enological adaptation of the winemaking technique. Depending on the white grape variety to be vinified, the grape chemical-physical characteristics at the harvest, and the style of the desired "wine" type, it is possible to carry out different enological strategies with specific technological variants of classic winemaking (i.e., without the solid parts of the grapes), such as cryomaceration and pellicular maceration or even by carrying out a more or less prolonged maceration (Orange wines, Kwevri wines). Moreover, the vinification can be carried out with specific oxygen management of the must (reductive winemaking, partial oxygen protection, hyperoxygenation), possible variants in the fermentation strategies, *sur lies* aging of wines, and/or the use of wood. The possibility of diversification of the enological technique allows the full valorization of the acquired peculiarities of the grapes from each specific vineyard.

Nowadays, in an ever more competitive and demanding white wine market, even in globally enologically recognized and established areas, producers are always seeking new and original products. Indeed, against the associated effects of climate change that strongly impacts the performances of the grapes at harvest (Mira de Orduña, 2010), the viticulture and enology sector is asked to perpetually reconsider the choice of the ampelographic basis, define the most suitable grape level of ripeness to reach depending on the prefixed enological goals, and choose the most suitable vineyard management strategies in order to obtain the desired grape qualitative characteristics. In particular, referring to the varietal choice to make, it follows that companies often need to turn to imported cultivars, in the scope of singular national enologies. Nevertheless, in more recent times and more particularly for some specific geographical areas of Europe, the rediscovery and valorization of autochthonous or lesser known cultivars have recently gained traction, drawing from white grape's germplasms of national *Vitis vinifera* L., whose valorization cannot however leave aside the knowledge of respective chemical-physical characteristics or relative enological attitudes. In fact, diversity within the grape species is expressed in thousands of vegetatively propagated genotypes differing in the concentration of the various classes of primary and secondary metabolites and in their chemical profiles (i.e., the relative concentration of individual metabolites).

Over the decades, the concept of grape quality has evolved, emphasizing its multidisciplinary nature and that the same "desired quality" might correspond to even strikingly different compositional patterns (Poni et al., 2018). In particular, white grape "quality" at harvest derives from the compositive balance of several primary and secondary metabolites: sugars, organic acids, nitrogenous and mineral substances, polyphenols, and aroma

White Wine Technology.
DOI: https://doi.org/10.1016/B978-0-12-823497-6.00001-6

1

© 2022 Elsevier Inc. All rights reserved.

compounds and their precursors. Alongside these chemical qualitative factors of grapes, recently, many studies have also drawn attention to the physical characteristics of grapes, and more specifically to the so-called mechanical properties of grapes (Rolle, Siret, et al., 2012). In fact, grape berries undergo numerous physiological and biochemical changes during ripening, which induce texture modifications. Hardness and thickness of the berry skin may differentiate the response of individual varieties to climate adversities and to phytopathologies during their ripening period on the vine. These mechanical properties are the expression of the variety and the level of ripeness reached (Río Segade, Orriols, Giacosa, & Rolle, 2011). However, the influence of annual climate variations on skin hardness is also a consequence of the genotype—environment interaction (Rolle, Gerbi, Schneider, Spanna, & Río Segade, 2011). Moreover, abiotic and biotic factors and elicitors may induce differences in the berry skin tissues. The water regimes in vineyard management (Giordano et al., 2013), the use of specific inactivated yeast extract distribution during grape *véraison* (Giacosa et al., 2019b), the presence of *Botrytis cinerea* in noble rot form (Rolle, Giordano, et al., 2012), the use of the cane-cut system in the on-vine withering process (Giacosa et al., 2019a), and the application of an ozone-modified atmosphere for off-vine postharvest treatment (Laureano et al., 2016; Río Segade et al., 2019) are examples of how the viticultural and enological technique can have an impact on the physical—mechanical characteristics of the grapes before their vinification, and consequently, it can change and/or regulate their enological performances and the eventual relative maceration strategy.

1.2 Chemical parameters related to the quality of white wine grapes

As exhaustively reported in the scientific literature, the vine synthesis of metabolites and their accumulation during ripening in the berry are highly dependent on the cultivar genotype (Kuhn et al., 2014; Teixeira, Eiras-Dias, Castellarin, & Gerós, 2013). The accumulation of these compounds is influenced by the grapevine response to growing conditions and particular treatments as a result of the interaction among genetic characteristics, environmental conditions, and cultural practices (Poni et al., 2018). Therefore ripeness grade, soil type, seasonal climatic conditions, agronomical—cultural practices, and growing location are important factors that affect the chemical composition of grapes, including some aspects whose importance is often underestimated by viticulturists such as protein content and enzymatic activities (Fortea, López-Miranda, Serrano-Martínez, Carreño, & Núñez-Delicado, 2009; López-Miranda et al., 2011) or antioxidant activities (Samotica, Jara-Palacios, Hernández-Hierro, Heredia, & Wojdylo, 2018). To improve grape quality chemical characteristics, different canopy techniques, such as training systems, bunch thinning, and defoliation, have been developed in the vineyard to modify the reducing sugar and organic acid composition and to increase the content of secondary metabolites in grapes at harvest (Alem, Rigou, Schneider, Ojeda, & Torregrosa, 2019; Poni et al., 2018).

1.2.1 Grape technological ripeness: primary metabolites

Although, as we previously saw, the quality factors of the grape at the harvest can be multiple, among the operators of the viticulture-winemaking sector the parameters of the so-called "technological ripeness" (i.e., sugars and titratable acidity) are still the main (and in many cases the only) parameters monitored during grape ripening. In particular, the concentration of total soluble solids (TSS) is still the most used parameter to assess ripeness and, in many cases, to determine the economic value of grapes for winegrowers (Poni et al., 2018).

Sugar accumulation is the main driver of TSS parameter changes in grapes. During ripeness, the increase in sugar content in the berries is accompanied by its softening and size increase (Coombe & McCarthy, 2000). In the case of white grapes, there is not a defined sugar or TSS value for the grape harvest. In fact, in production winemaking the desired soluble solids concentration at harvest takes into account a complex range of parameters such as terroir, cultivar adaptation, climate, harvest choices, and equilibrium within all berry metabolites, as well as legal regulations (e.g., minimum grape requirements for wine production), price range, and quality sought by wine producers. As an example, interesting patterns can be deduced by observing the temporal trends of the ripeness degree at harvest in specific wine regions, which are strongly influenced by the abovementioned factors.

Fig. 1.1 shows the grape must TSS (as Brix degree) at grape crush surveyed on Californian (United States) white grape productions, separated by grape variety (USDA, 2020). It may be observed that the whole average of Brix degree (black points and smoothed black line) in the analyzed region accounted for about 22 Brix in the monitored period, and although some vintage variability was registered, it generally increased until the 2014

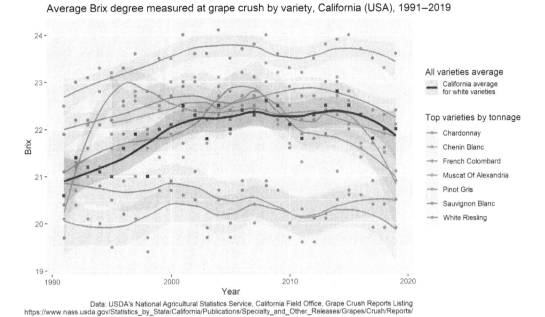

FIGURE 1.1 Evolution of the average total soluble solids content (Brix degree) in California grape productions, detected at grape crushing. Source: *Data from USDA (2020).*

vintage. Apart from the general trend, by comparing the values at harvest for the top seven varieties by tonnage (in the year 2019), it is interesting to note that each variety tended to position itself in a defined range. For instance, Chardonnay showed the highest Brix values of this group in all years surveyed (except 1994 and 1995), followed by Sauvignon blanc and Muscat of Alexandria (the latter included in the report from the year 1995). Two varieties, namely Pinot gris and White Riesling, showed variable values in the monitored period. Lastly, other varieties such as Chenin blanc and French Colombard accounted for values mainly in the range of 19.5–21 Brix. Indeed, it might be hypothesized that the large number of factors (environmental, viticultural, economic, etc.) involved in the ripeness degree choice for harvest had a great role in shaping this registered trend.

Concerning other primary berry metabolites, in order to have balanced wines from a tasting perspective, it becomes evidently important to harvest the grapes with the correct amount of organic acids, with the right balance between tartaric and malic acid contents. These two acids represent the major share of the titratable acidity (TA) of grape musts and wines, and their concentrations undergo significant changes during grape ripening due to the increase in juice volume and, in the case of malic acid, also due to degradation phenomena (Ruffner, 1982a, 1982b). Furthermore, they are instrumental in determining the grape juice, must, and wine pH, together with the concentration of cations such as potassium (Mpelasoka, Schachtman, Treeby, & Thomas, 2003).

In the recent climate change context with very different climatic conditions from one harvest to another in terms of temperatures and water regimes, in order to find the correct balance between these "technological parameters" for each variety and growing location, it is necessary to have high knowledge about the behaviors that bind the factors of genotype-environment-vine management together (Poni et al., 2018). Regarding these parameters, as it has become clearer since the eighties, the sugar—acidity balance remains of great importance for the grape ripeness monitoring, but other characteristics such as aroma may not be on the same ripeness evolution and need the right consideration in determining the harvest date (Coombe, Dundon, & Short, 1980).

1.2.2 Nitrogenous compounds

Despite almost never being the object of direct monitoring during the ripening, amino acids (for example, methionine), peptides, and proteins contribute significantly to the quality of wine because they affect its taste, clarity, and stability. Haze formation is a complex phenomenon involving colloids of different origins (i.e., grape proteins), high molecular phenolic compounds, and non-proteinaceous factors, such as sulfate ions, wine ionic strength, organic acids, and pH level (Colangelo et al., 2019). Therefore total protein content and related composition (i.e., relative contents among thaumatin-like proteins, chitinases, β-glucanases, and others) present in the grapes at harvest are important because they determine the dose and type of enological processing aids used for reducing the colloidal content of must/wine. In any case, removing proteins, clarifying, and fining treatments, particularly with bentonite, can lead to both direct and indirect removal of aroma molecules originating from the grapes (Vincenzi, Panighel, Gazzola, Flamini, & Curioni, 2015).

Among the vitamins present in grapes at harvest, riboflavin (RF) acts as a photosensitizer in many beverages, including white wine. The RF level in the grape is usually lower than few tens of $\mu g/L$ of grape juice (Ribéreau-Gayon, Glories, Maujean, & Dubourdieu, 2006), but it can increase during winemaking mainly due to the metabolic activity of the different *Saccharomyces cerevisiae* strains (Fracassetti, Limbo, Pellegrino, & Tirelli, 2019). In bottled wines, RF can be involved in light-induced reactions, affecting changes in volatile compounds, color, and flavor, known as light-struck taste (LST). The LST is a fault occurring in light-exposed white wines containing methionine and a high concentration of RF, bottled in clear or transparent bottles. These conditions induce the formation of methanethiol and dimethyl disulfide, which are responsible for the defect resulting in a "cabbage-like" aroma (Fracassetti et al., 2017).

1.2.3 Phenolic compounds

The berries of white grape varieties are rich in flavonoids, such as flavonols and flavan-3-ols, as well as non-flavonoid phenols such as hydroxycinnamoyl tartrates (HCTs) like red grapes, from which they differ by the lack of anthocyanins. In the production of common white wines, apart from those produced with more or less intense maceration, the total concentration of polyphenols extracted from the grapes is very low (about 50−200 mg/L). Although present in modest quantities, these compounds are important to the final quality of the obtained wine. In particular, flavan-3-ols and their polymerized forms (proanthocyanidins) are of certain importance because, as previously seen, they can be involved in the phenomena of protein instability and oxidation; the latter are particularly possible in wines with low content of sulfites (Pati, Crupi, Savastano, Benucci, & Esti, 2020). Moreover, HCTs (caffeoyltartaric [caftaric] acid, p-coumaroyltartaric [coutaric] acid, and feruloyltartaric [fertaric] acid) are known to be involved in the browning reactions of must and wine (Romeyer, Macheix, Goiffon, Reminiac, & Sapis, 1983) and can be precursors of volatile phenols in white wines subjected to aging. Besides, they have been shown to be of great significance in the taxonomy of young single-variety wines and can be considered varietal markers (Ferrandino, Carra, Rolle, Schneider, & Schubert, 2012).

1.2.4 Volatile compounds

Aroma is a key contributor to the perception of white wine's quality features, including sensory typicality, perceived diversity, and overall preference. In the wine, more than 800 olfactory-active volatile compounds have been identified with contents ranging from fractions of ng/L to several mg/L (Ferreira et al., 2016). Their presence also permits the discrimination of the wines in terms of authenticity, thanks to primary aroma markers present in the grapes. These aspects are even more relevant for a varietal enology (i.e., winemaking of a single grape variety, which is often an expression of a determined territory). In fact, geographical typicality is being exploited to increase the economic value of wines having distinctive and recognizable aroma characteristics. Interestingly, the differences are not very evident in the volatile composition of grapes from individual parcels of a single vineyard, but they are significant in the resulting wines (Slaghenaufi, Guardini, Tedeschi, & Ugliano, 2019). It is thus particularly important to obtain high-quality raw material because, in this case, there is not the possibility of finding the right sensorial balance from the blend of grapes from different cultivars. The assessment of grape-originated aroma can give important knowledge about the varietal features of the final wine flavor. Therefore the evolution of volatile organic compounds (VOCs) profile and concentration during grape ripening should be monitored to select the harvest date, providing predictive information on grape-derived wine aroma. This would allow winemakers to determine a more targeted harvest date based on the desired aroma compounds.

Technological and aromatic maturity do not occur at the same time. Although each TSS concentration is generally associated with a different grape VOCs profile, the evolution of aroma compounds in the grape berry during ripening is not strictly related to sugar accumulation (Coombe & Iland, 2004; Torchio et al., 2016). In addition, the synthesis of VOCs can be promoted in the advanced stages of ripening, once the sugar increase per berry has slowed. Nevertheless, several factors strongly affect the accumulation of grape VOCs during ripening, such as variety, edaphoclimatic conditions, and cultural practices (Alem et al., 2019; Poni et al., 2018). It has been demonstrated that water deficit activates the carotenoid, isoprenoid, and fatty acid metabolic pathways, increasing the concentrations of many VOCs contributing to the aroma (Deluc et al., 2009; Wang et al., 2019) but reducing the volatile thiol precursors at severe water deficit (Pons et al., 2017); sun exposure usually increases the accumulation of monoterpenes and C_{13}-norisoprenoids (Lee et al., 2007; Zhang et al., 2014) excepting β-damascenone (Kwasniewski, Vanden Heuvel, Pan, & Sacks, 2010), whereas high temperatures and light exposure degrade methoxypyrazines, leading to the formation of much less odoriferous compounds (Pons et al., 2017); in addition, jasmonate treatments induce sesquiterpene biosynthesis in grape cell cultures (D'Onofrio, Cox, Davies, & Boss, 2009). The treatments in vineyards by using biostimulants and/or elicitors are often used in order to increase VOCs concentration in white grapes (Gutiérrez-Gamboa et al., 2020), although they are more generally used to increase other secondary metabolites (Narayani & Srivastava, 2017). Regarding biological elicitors, yeast extracts can induce secondary biosynthetic pathways as a result of plant defense responses stimulated by their content in chitin, β-glucan, N-acetylglucosamine oligomers, glycopeptides, and ergosterol (Granado, Felix, & Boller, 1995). The use of specific inactivated dry yeasts in foliar treatments can also influence norisoprenoid profile and concentration in some wine grape varieties (Crupi et al., 2020).

As mentioned above, the assessment of the grape aromatic maturity level is difficult because the evolution of VOCs during ripening is not only variety dependent, but it is also influenced by the chemical family to which they belong. The main classes of varietal VOCs accumulated in the grapes of *Vitis vinifera* are terpenes, C_{13}-norisoprenoids, C_6 alcohols and aldehydes, benzenoids, volatile thiols, and methoxypyrazines (Ribéreau-Gayon, Glories, Maujean, & Dubourdieu, 2006). These compounds are present in grape berries in free and glycosidically bound forms. The latter ones are odorless sugar-conjugated compounds (aglycone bound directly to a β-D-glucopyranosyl moiety) and can undergo acid or enzyme hydrolysis during the winemaking and aging process, releasing important free volatiles and potentially enhancing the wine aroma (Günata, Bayonove, Baumes, & Cordonnier, 1985). Many of these free compounds are produced from non-volatile precursors such as fatty acids, amino acids, and carotenoids through complex metabolic reactions, which can begin during grape ripening and continue throughout fermentation, aging, and bottling (González-Barreiro, Rial-Otero, Cancho-Grande, & Simal-Gándara, 2015; Swiegers, Bartowsky, Henschke, & Pretorius, 2005), but many grape varieties contain mainly VOC glycoside precursors at harvest and the corresponding odorous compounds are released in the wine during fermentation (Englezos et al., 2018).

Terpenes are secondary metabolites that play a key role in the assessment of grape quality. Some aromatic varieties (having free VOCs in concentrations above their olfactory threshold), such as Muscat, Malvasia, Gewürztraminer, and Riesling, mainly depend on these compounds for giving fruity and floral sensory nuances. Monoterpene alcohols are some of the most odoriferous compounds, particularly linalool, geraniol, nerol, citronellol, and α-terpineol in Muscat varieties, as a consequence of their quite low sensory thresholds ranging from tens to hundreds of micrograms per liter (Ferreira & Lopez, 2019). Regarding Gewürztraminer, (Z)-rose oxide is responsible for the characteristic litchi- or rose-like aroma of wines (Ong & Acree, 1999), even though this terpenic compound has been also detected in Muscat and Riesling varieties. In neutral grape varieties, one of the most interesting sesquiterpene compounds is rotundone. This compound has a very low odor detection threshold (8 ng/L in water) and a distinctive black pepper note. It is found in Gruener Veltliner grapes among white varieties (Caputi et al., 2011). Moreover, α-muurolene was also detected in Riesling grapes (Kalua & Boss, 2010). The accumulation of free mono- and sesquiterpenic compounds starts at *véraison* and increases progressively during grape ripening until it reaches the highest value at maturity (19–20 Brix). Then, a decrease is observed in free forms, particularly for aromatic varieties such as Malvasia (Perestrelo, Silva, Silva, & Câmara, 2018) and Moscato bianco (Torchio et al., 2016; Fig. 1.2), whereas different trends are observed at the last ripening stages for non-aromatic white wine grapes (Caputi et al., 2011; Coelho, Rocha, Barros, Delgadillo, & Coimbra, 2007; Sollazzo, Baccelloni, D'Onofrio, & Bellincontro, 2018; Vilanova, Genisheva, Bescansa, Masa, & Oliveira, 2012). This seems to confirm that the evolution of terpenes during grape ripening is not strictly related to the sugar accumulation. Nevertheless, Torchio et al. (2016) have density sorted Moscato bianco grape berries by flotation at different sampling dates and they have highlighted that free as well as glycosylated terpenes are more affected by grape density than by sampling date, which makes it possible to obtain a different aroma profile on each harvest date based on berry density (Fig. 1.3).

Glycosylated forms are generally more abundant than free terpenes (from 3 to 10 times), although the relative proportion depends on the grape variety. In fact, Muscat varieties are the richest in terpene glycosides and also those in free forms, and some authors have found higher concentrations of free linalool than of the bound form in ripe grapes (Fenoll, Manso, Hellín, Ruiz, & Flores, 2009). The three predominant types of monoterpene glycosides are monoterpenol hexose-pentose, malonylated monoterpenol glucosides, and monoterpendiol hexose-pentoses (Godshaw, Hjelmeland, Zweigenbaum, & Ebeler, 2019). Regarding monoterpenol glycosides, their concentration remains stable at the early stages of berry development and increases progressively and

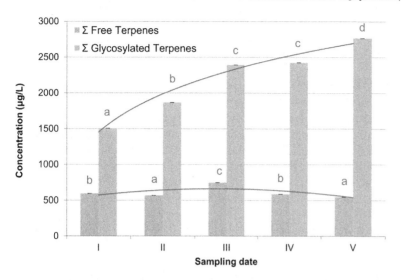

FIGURE 1.2 Total free and glycosylated terpene concentration of Moscato bianco grapes harvested on different dates. Source: *Modified from Torchio, F., Giacosa, S., Vilanova, M., Río Segade, S., Gerbi, V., Giordano, M., Rolle, L. (2016). Use of response surface methodology for the assessment of changes in the volatile composition of Moscato Bianco* (Vitis vinifera L.) *grape berries during ripening.* Food Chemistry, 212, 576−584.

significantly from *véraison* up to the last ripening stages (>20 Brix) (Sollazzo et al., 2018; Torchio et al., 2016; Fig. 1.2), even though there is a variety effect (Vilanova et al., 2012).

C_{13}-norisoprenoids derive from the oxidative degradation of grape carotenoids and contribute significantly to the aroma of many white wine grape varieties, such as Chardonnay, Sauvignon blanc, Chenin blanc, and Sémillon (Gambetta, Bastian, Cozzolino, & Jeffery, 2014). Particularly, β-damascenone and β-ionone are important aroma compounds due to their low sensory thresholds (2 and 120 ng/L for β-damascenone and β-ionone in water, respectively) (Darriet & Pons, 2017). β-Damascenone is described as having applesauce and rose notes, and it is particularly relevant in neutral varieties (Ferreira & Lopez, 2019). Nevertheless, glycoside precursors of l,1,6-trimethyl-1,2-dihydronaphthalene (TDN) have been also detected in grapes. Kerosene or petrol notes of this compound, having a sensory threshold of 20 μg/L in wine, are closely associated with the aroma of aged Riesling wines (Kwasniewski et al., 2010). The evolution of free norisoprenoids follows an inverse trend to monoterpenes during grape ripening probably due to the fact that norisoprenoid glycosylation occurs, which reduces the corresponding free forms (Perestrelo et al., 2018). Particularly, the concentration of β-damascenone increases in the last stages of maturation (late harvest), giving plum-like nuances (Ferreira & Lopez, 2019).

C_6 alcohols and aldehydes are mainly present in free form and they are formed by peroxidation of fatty acids through nearly instantaneous enzymatic/catalytic processes triggered during the disruption of fruit tissues (Ferreira & Lopez, 2019). The most powerfully odorous C_6 compounds are the aldehydes with sensory thresholds in water ranging from 0.25 μg/L for (Z)-3-hexenal to 60 μg/L for (E,E)-2,4-hexadienal (hundreds of times smaller than those of the corresponding alcohols ranging from 70 μg/L for (Z)-3-hexenol to 2500 μg/L for 1-hexanol) (Ferreira & Lopez, 2019). Aldehyde concentration increases significantly during ripening and then decreases toward the end of this phase, whereas alcohols reach significant quantities only at the end of ripening. The prevalence of alcohols during late berry development, with respect to aldehydes in early stages, permits the use of the alcohols-to-aldehydes ratio for the prediction of harvest date in order to enhance grape and wine aroma in non-Muscat varieties (Kalua & Boss, 2009). Therefore early grape harvests can contribute to increased concentrations of C_6 compounds, particularly aldehydes, giving undesirable herbaceous notes as a consequence of their lower odor threshold when compared to alcohols. Moreover, their concentrations are strongly influenced by grape variety (Oliveira, Faria, Sa, Barros, & Araujo, 2006) and also by the berry position into the grape cluster (Noguerol-Pato et al., 2012), with the berries positioned in the shoulders being richer. For these reasons, an important criterion to define the harvest date is to achieve an adequate balance between the concentration of terpenes and norisoprenoids and the concentration of C_6 compounds.

Other important groups of grape VOCs are benzenoids, including benzyl and phenyl derivatives such as benzaldehyde, benzoic acid, benzyl alcohol, and 2-phenylethanol, homovanillic alcohol, and 4-vinyl guaiacol. They are characterized by walnut, almond, spice, and rose nuances. Their concentration increases slightly during grape ripening until reaching a maximum value at approximately $17-20°$ Brix (Martin, Chiang, Lund, & Bohlmann, 2012; Sollazzo et al., 2018). In many neutral grape varieties, these compounds are the major constituents of the glycosidic aroma fraction but their odor detection thresholds are relatively high ($350-10,000$ μg/L in water, Noguerol-Pato et al., 2012). Therefore grape-derived benzenoids contribute marginally to the wine aroma. The health status of grapes strongly affects the aromatic quality of the resulting wines. Particularly, the metabolic

(A)

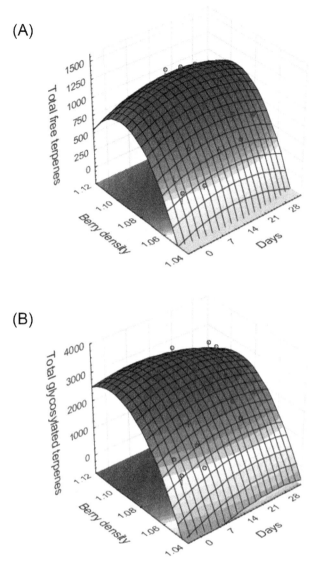

(B)

FIGURE 1.3 Response surface plot showing the effect of sampling date and berry density on the concentration of terpenes (μg/L) in Moscato bianco grapes: (A) total free terpenes; (B) total glycosylated terpenes. Source: *Modified from Torchio, F., Giacosa, S., Vilanova, M., Río Segade, S., Gerbi, V., Giordano, M., Rolle, L. (2016). Use of response surface methodology for the assessment of changes in the volatile composition of Moscato Bianco (Vitis vinifera L.) grape berries during ripening. Food Chemistry, 212, 576–584.*

activity of *Botrytis cinerea* results in an increase in benzaldehyde (Genovese, Gambuti, Piombino, & Moio, 2007), whereas phenylacetaldehyde is formed in sun-exposed botrytized grapes from the Strecker degradation of amino acids (Sarrazin, Dubourdieu, & Darriet, 2007).

Varietal thiol precursors are particularly important for Sauvignon blanc grapes, mainly S-glutathionyl and S-cysteinyl conjugated forms. Thus 3-S-cysteinylhexan-1-ol, 3-S-cysteineglycine-3MH, 3-S-glutathionylhexan-1-ol, and (E)-2-hexenal are precursors of 3-mercaptohexan-1-ol (3MH) and 4-mercapto-4-methylpentan-2-one (4MMP), which have low sensory thresholds (0.8–60 ng/L in model wine) and give grapefruit- and passion fruit–like aromatic notes (Pinu et al., 2019). Nevertheless, other varieties including Gewürztraminer, Riesling, Muscat, Sémillon, Chenin, Pinot blanc, Colombard, and Petit Manseng also contain 3MH. Particularly for Sauvignon blanc grapes, it is of great relevance to monitor the evolution of S-cysteinyl and S-glutathionyl thiol precursors during ripening (Cerreti et al., 2015). The concentrations of cysteine and glutathione conjugates (Cys-3-MH and Glut-3-MH, respectively) increase as ripening progresses in white grape varieties in the last 14 days (Capone, Sefton, & Jeffery, 2011), probably due to the increase in (E)-2-hexenal and glutathione. In addition, moderate vine water deficit is related to the higher accumulation of S-conjugate precursors in the grape berry, whereas severe water deficit causes the lowering of volatile thiol precursor contents (Peyrot Des Gachons et al., 2005; Pons et al., 2017). A particularly interesting study published highlights that harvesting affects varietal thiols in wines (Herbst-Johnstone et al., 2013). Higher concentrations of these compounds (3MH and its acetate) were found in Sauvignon blanc wines made from machine-harvested grapes when compared to hand-picked grapes. Nevertheless, this same effect was also observed for C_6 alcohols, such as 1-hexanol and (Z)-3-hexenol, and their associated acetate esters.

Varietal methoxypyrazines derive from the metabolism of amino acids. They are usually present in free form and are responsible for herbaceous and vegetal sensory characteristics. These compounds have low olfactory detection thresholds (1–16 ng/L in water and white wines). They characterize the Cabernet family, even though they could be present in Sauvignon blanc, Chardonnay, Riesling, and Sémillon varieties at acceptable levels (Zhao et al., 2019). In grape berries, 3-isobutyl-2-methoxypyrazine (IBMP), 3-isopropyl-2-methoxypirazine (IPMP), and 2-methoxy-3-sec-butylpirazine (SBMP) can be present. Nevertheless, 3-isobutyl-2-methoxypyrazine (IBMP) is the most abundant and gives green pepper and pea pod notes. It is accumulated rapidly from the fruit set and reaches maximum concentrations at 2–3 weeks before the onset of *véraison*, but then decreases markedly during grape ripening as a consequence of demethylation to 3-isobutyl-2-hydroxypyrazine (Ryona, Leclerc, & Sacks, 2010).

Other compounds such as lactones, particularly γ-non-alactone, have been found in low contents in botrytized and late harvest grapes. It is important to consider that the aroma of neutral grapes derives from the contribution of a relatively large number of volatile compounds present in low concentrations that are often singularly undetectable by a sensory point of view. Finally, VOC profile patterns permit the characterization and discrimination of white wines by variety and even by region. More recently, the enantiomeric composition of monoterpenes has been already proposed for wine differentiation based on variety, region, and style, showing a high potential for wine authentication (Song, Fuentes, Loos, & Tomasino, 2018).

1.3 Monitoring grape characteristics through analytical control and sensory and remote assessments

White wine grapes quality assessment during ripening and at harvest is clearly based on the evaluation of technological parameters and the concentration of secondary metabolites, in particular the aroma-related compounds. Nevertheless, other important factors must be taken into consideration during quality evaluation, such as the grapes' health status and their conditions at the receiving point. These parameters can be summarized as follows:

- grape yield;
- grape technological maturity (sugars, TA, and pH);
- secondary metabolites concentration (polyphenols and aroma-related compounds);
- grape conditions (temperature, hand- or machine-harvested, damaged berries);
- health status (presence of molds and rots);
- contamination and presence of material other than grapes (MOG).

These evaluations should be referred to parameters that can be quantified and ascribed to a value through analytical measurements if possible or by visual or tasting inspection by trained and experienced staff.

1.3.1 Analytical control of basic metabolites

Considering technological maturity, measurements can be easily made through instrumental methods. Sugars are usually assessed as TSS by refractometry and the value is expressed in Brix degree. This value represents the number of grams of soluble solids per 100 g of solution, and it also takes into account other soluble solids, such as pigments, acids, and glycerol (Zoecklein, Fugelsang, & Gump, 2010). Generally, the fermentable sugars account for 90%–95% of TSS, mostly being represented by glucose and fructose. In winery conditions, Brix degree is commonly used and sometimes converted to Baumé (1.8 Brix is equivalent to 1 Baumé) since this latter expression is used as a rough estimation of the potential alcohol content of the wine to be produced. Together with TSS, TA (expressed in g/L equivalents of tartaric acid) by titration with sodium hydroxide (OIV, 2019) and pH (expressed in pH units), evaluated using a pH meter, are used. Since these parameters (TSS, TA, and pH) are easy to be evaluated instrumentally and provide fundamental information on the grapes' quality, they are often used as specification parameters to follow the grape ripening stage in the vineyards, to identify different stages of maturity inside a vineyard, to determine the final grape quality for harvest, and to establish the grapes' price.

A more precise determination of individual sugars and organic acids can be achieved by using the enzymatic method through spectrophotometric measurement, or with high-performance liquid chromatography (HPLC) coupled with UV and refractive index detectors (Giordano, Rolle, Zeppa, & Gerbi, 2009).

Nondestructive and fast analysis of grapes evaluating technological maturity parameters have been proposed with several in-field, online, and laboratory applications, based on the interaction of the berries or the produced juice and several kinds of sensors used for reflectance, light energy, irradiance, fluorescence, optic, and acoustic measurements (Ferrandino et al., 2017; Poni et al., 2018). The spectra acquired by the sensors are then cleaned from uninformative data, combined (using multivariate statistic approaches), and correlated to measured values from a conventional reference method in order to develop a model (using different regression methods) with predictive scope (Tahir et al., 2019). In this way, it is possible to estimate a parameter in intact samples or in the produced juice in a rapid, fast, and cost-effective way. In this sense, the main techniques used are based on vibrational spectroscopy such as Raman and infrared (IR) spectroscopy. Considering the latter, near-infrared (NIR, 750−2500 nm), often coupled with visible (VIS, 400−750 nm), and mid-infrared (MIR) are the most common options in the applied instrumentation (Poni et al., 2018; Tahir et al., 2019). Several reviews can be found on their application (Dambergs, Gishen, & Cozzolino, 2015; Li, Lecourt, & Bishop, 2018; Poni et al., 2018; Tahir et al., 2019; Walsh, Blasco, Zude-Sasse, & Sun, 2020); technological parameters (TSS, TA, and pH), and also other relevant ones such as yeast assimilable nitrogen (YAN), can be predicted with great accuracy (Dambergs et al., 2015; Gishen, Dambergs, & Cozzolino, 2005). A limitation of these techniques is that the model construction requires a high number of samples analyzed with a referenced method and owns a low intra-laboratory and inter-harvest reproducibility (Walsh et al., 2020). The common methodology is referred to as "point" analysis, as opposed to multi- and hyperspectral imaging (HSI). The HSI generates a three-dimension cube with the image at a range of continuous wavelengths, enabling the use of up to thousands of bands in different spectral regions (of VIS-NIR, MIR, and even Raman) for each pixel (Li et al., 2018). This method provides higher spectral and spatial resolutions, combining spectroscopic and computer vision techniques (Tahir et al., 2019).

1.3.2 Analytical control of aromatic potential

For aroma compounds, the correlation between the classic technological parameters and their concentration are poor and compound- and variety-dependent (Alem et al., 2019; Poni et al., 2018). According to Ruiz et al. (2019), for VOCs analytical determination a classification can be done by separating sulfur-containing compounds (in grapes represented by the thiol-related aromas, TRACs) from non-sulfurous compounds (the other classes, NSCs). These wine VOCs are usually present in grapes as bound precursors, the first as cysteinyl- and glutathionyl- derivatives, whereas the latter as glyco-conjugates, and both are released as free aromas by enzymatic or chemical reactions during the winemaking process. At the research level, gas chromatography coupled with mass spectrometry (GC-MS) of free volatiles is the most common technique for both classes. Given the low concentration (ng/kg to mg/kg depending mainly on the variety and the aroma class), purification, extraction, and concentration steps are required. The most common techniques applied are liquid-liquid extraction (LLE), solid-phase extraction (SPE), solid-phase microextraction (SPME), and stir bar sorptive extraction (SBSE) methodology, or a combination of them (Arcari, Caliari, Sganzerla, & Godoy, 2017; Ruiz et al., 2019). All these steps imply the risk of sample loss or the formation of reaction product as artifacts: the use of internal standards (deuterated analogous or a similarly behaving compound) and, in the case of TRACs, derivatizing agent, is usually applied to improve stability, sensitivity, and accuracy (Fracassetti & Vigentini, 2018; Roland, Schneider, Razungles, & Cavelier, 2011; Ruiz et al., 2019).

In contrast, different analytical techniques can be found in the scientific literature for both TRACs and NSCs corresponding precursors. Concerning the assessment of NSC precursors, it usually involves the release of the free aroma in its volatile form by enzymatic reaction by a glycosidase enzyme. The GC-MS system allows the evaluation of the aromatic profile of grapes, and the quantification of individual VOCs is achieved, thus giving accurate information on the potential aroma depending also on the specific odor active threshold of each aroma. Another option besides GC-MS determination is the "G-G method" (glycosil-glucose, Williams et al., 1995). Through this technique the enzymatic assay of the glucose released by hydrolysis allows the quantification of total glyco-conjugates. It is an unspecific technique since it does not take into consideration the nature of the

VOCs released, and the final wine aroma depends on their individual odor thresholds. Anyway, as an advantage, this method was adopted for winery quality control since it does not require expensive instrumentation (Arévalo Villena, Pérez, Úbeda, Navascués, & Briones, 2006).

Regarding conventional analytical methods for determining the precursors of sulfur-containing compounds, the GC-MS of free volatiles can be applied after enzymatic hydrolysis through β-lipase activity. The most common technique is the direct analysis of the precursors by liquid chromatography combined with mass spectrometry (LC-MS, Roland et al., 2011). SPE purification is often performed to clean samples and concentrate the analytes. Quantification is achieved with labeled compounds, i.e., the deuterated analogous of the investigated precursors. Recently, the direct analysis of non-sulfurous precursors by LC-MS/MS or LC-NMR techniques has been proposed as well (Barnaba et al., 2018; Schievano et al., 2013), giving the advantage to reduce sample manipulation and therefore the risk of sample deterioration and the creation of artifacts. As a limitation, since it is still uncommon, only a few databases of glyco-conjugates aroma in MS/MS detection are available.

Although specificity and accurate results are achieved by these techniques, they are time-consuming, laborious, expensive, and complex for a routine winery control process, and aromatic potential determination remains a challenge. A future perspective can be given by rapid analysis techniques such as the vibrational spectroscopy—based method for aroma potential identification and follow-up during grape ripening. In scientific research, the correlation between spectra from Fourier transform—NIR (FT-NIR) and HSI-UV-NIR has been attempted with the referenced method of free aroma compounds and their precursors. The first attempt was by Gishen & Dambergs, 1998 who proposed that NIR spectroscopy could be used for rapid G-G determination in white grapes. Later, Cynkar, Cozzolino, Dambergs, Janik, and Gishen (2007) assessed this possibility on Chardonnay, Riesling, and Sauvignon blanc clarified juice samples coming from different wineries and vintages in Australia. The combination of NIR spectroscopy and partial least square (PLS) regression led to interesting results, although the precision of prediction did not allow for quantification but discrimination of grape juice depending on quality range, which may help winemakers in separating low, medium, and high aroma potential juices. Further experiments of rapid determination were performed on individual aroma detected by GC-MS. Schneider, Charrier, Moutounet, and Baumes (2004) attempted the prediction of aroma divided by classes with FT-NIR in grape homogenates ($n = 39$) of Melon blanc wine grapes after SPE purification. A PLS regression was done between the GC-MS determination of aglycone for each class and the corresponding NIR spectra. A good prediction of the levels of C_{13}-norisprenoidic and monoterpenic glyco-conjugates was achieved and a fair prediction of alcohol, C_6-compounds, and phenol glyco-conjugates was obtained, but in this case, the sample preparation still required a purification step. Concerning the free aromatic compounds, UV-VIS-NIR determination in white Albariño wine grapes was done ($n = 52$) with GC-MS as the reference method, and PLS regression was chosen to fit the predictive models for individual aromas (Ripoll, Vazquez, & Vilanova, 2017). Analyses were performed on filtered juices (5 mL), and good predictions of C_6-compounds [(E)-2-hexanal, 1-hexanol, (Z)-2-hexanol], aldehydes (benzaldehydes and phenylethanal), one terpene (cis-pyran-linalool oxide), and one alcohol (2-phenylethanol) were achieved. However, only volatiles (free fraction) were analyzed. The authors underlined the limitations of these techniques in the determination of aromatic maturity since several compounds with different odor thresholds contribute to the varietal aroma, and the application of fast screening is often subjected to a high coefficient of variation given by the variety, the geographical area, and the vintage, suggesting that several calibration steps of the NIR technique are necessary for obtaining good predictive models. Another concern is related to the grape sample type considered, which may be produced as homogenate or juice and often subjected to purification treatment before analysis. An interesting method may be the use of HSI techniques on intact berries; in this direction, prediction of the most important free volatiles in Albariño wine grape samples ($n = 12$) were found by Álvarez-Cid, García-Díaz, Rodríguez-Araújo, Asensio-Campazas, and de la Torre (2015) using monochromatic green light as the illuminant, and multispectral cube data were correlated to GC-MS aroma analysis with support vector regression (SVR).

1.3.3 Sensory and visual assessment of grape quality

For aroma compounds and technological parameters (pH, acidity, sugars), and visual and texture parameters, sensory analysis of grape berries can be performed. Rigorous sensory analysis on defined sensory attributes by a panel trained on anchored scales is described in Quantified Descriptive Sensory Analysis (QDSA). The panel should be composed of as many assessors as possible (4—12), and the assessors should be trained on selected attributes (17–30). The samples need to be tasted in a standardized and controlled environment, taking care of serving berries representative of the vineyards (3–5 berries for samples), at a defined temperature, and with a randomized presentation order to minimize fatigue and carry-over effect, with the help of pause and palate cleanser (Olarte Mantilla, Collins, Iland, Johnson, & Bastian, 2012). A contrast sample (possibly grape berries conserved at −20°C from the previous point) will help in sensory assessment during ripening (Zoecklein, Fugelsang, & Gump, 2010), and the adoption of standard references for each attribute helps in discrimination (Olarte Mantilla et al., 2012). Berry sensory analysis (BSA) procedure was firstly described by Rousseau (2001) on several wine grapes including Chardonnay, and it is composed of visual, texture, and gustatory assessments. The evaluated parameters differed for whole berry, bunch, stem, and pedicel, and skin, pulp, and seeds examinations were performed separately on a four-point category scale, where "one" represents the lowest maturity level and "four" the highest. Lohitnavy, Bastian, and Collins (2010) used the BSA on cv. Sémillon evaluation for differentiating berries basing on different vine treatments, using 11 trained judges and 23 agreed attributes divided for skin, pulp, and seed characteristics, all evaluated on 15-cm unstructured scales. Several correlations between the grapes parameters and final wines produced were established (Olarte Mantilla et al., 2012) and allowed the differentiation of grape samples coming from different vineyard treatments as well (Lohitnavy et al., 2010), enforcing the suitability of this technique.

Together with compositional parameter evaluations, grapes' health status and the presence of MOG can be evaluated by visual inspection and evaluation as percentages of infected berries for clusters or clusters impacted in total, or percentages of MOG (Allan, 2003). For the presence of molds and rots, another option can be the evaluation of secondary metabolites produced by the contaminant such as acetic acid, ethanol, glycerol, gluconic acid, ethyl acetate, and laccase (Steel, Blackman, & Schmidtke, 2013; Zoecklein, Fugelsang, & Gump, 2010). The evaluation of these metabolites allows for understanding their potential impact on final wine and can be instrumentally determined. Nevertheless, a spectrophotometer is necessary for enzymatic method determination, or a liquid chromatography system for these individual compounds' quantification. Future perspective of nondestructive determination in this case may be represented by the application of HSI imaging of whole bunches for the estimation of various infection levels, such as in the case of powdery mildew in Chardonnay grapes performed by Knauer et al. (2017) (Fig. 1.4).

1.3.4 Remote sensing assessment of grape quality and yield

In addition to the other techniques previously discussed, in recent years, remote sensing techniques for the evaluation of vineyard traits throughout the vegetative season have been developed (Matese & Di

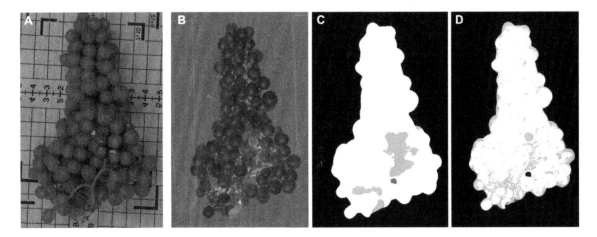

FIGURE 1.4 A data analysis approach for powdery mildew detection applied on a Chardonnay grape bunch: (A) manual annotation of the infection sites; (B) disease-specific visible and near-infrared—hyperspectral image (VNIR-HSI); (C) detection based on spatial-spectral approach; (D) detection results on HSI signatures. Source: *From Knauer, U., Matros, A., Petrovic, T., Zanker, T., Scott, E.S., Seiffert, U. (2017). Improved classification accuracy of powdery mildew infection levels of wine grapes by spatial-spectral analysis of hyperspectral images.* Plant Methods, 13 (1), 47.

Gennaro, 2015). These non-invasive techniques may be useful for the estimation of ripeness and the grape yield and to underline the heterogeneity in the vineyard or among different parcels for specific grape parameters. However, often an appropriate model and actual measurements on "sentinel" vines are necessary to develop the technique and obtain meaningful data. In the case of white wine grapes production, vegetation indices gathered from airborne images were able to be used to monitor the vineyard vine vigor and allowed to delimitate vineyard subzones presenting different grape characteristics (Marciniak, Brown, Reynolds, & Jollineau, 2015). In another study, a slightly different approach using multispectral imaging gathered from unmanned aerial vehicles (UAVs) and associated with appropriate prediction models confirmed the ability to obtain vineyard zones presenting grapes with higher soluble solids and aromatic traits such as terpene compounds (Fig. 1.5; Iatrou et al., 2017).

Specific applications in remote sensing were also proposed to tackle some recent issues in grape production, such as the evaluation of possible vineyard damage by smoke contamination, a growing issue in viticulture due to bushfires and a threat for wine production due to smoke taint defects. Remote (UAV) and proximal sensing (IR thermal imagery and NIR) techniques were studied on the evaluation of smoke-contaminated vineyards and allowed to identify the affected vineyard portions (Fig. 1.6) and to monitor the vegetative responses (Brunori et al., 2020), while proximal sensing was tested to develop a model for the estimation of the levels of compounds related to smoke taint (Fuentes et al., 2019).

To conclude, although the analytical possibilities of grape monitoring are plenty, these tools are useful only in assisting grape and wine industry professionals in their harvest decisions. In the future, the research and development of advanced analytical systems and prediction models will improve the ability to evaluate and monitor grape ripeness, a fundamental factor for the production of high-quality white wines, together with the vineyard management and the enological operations.

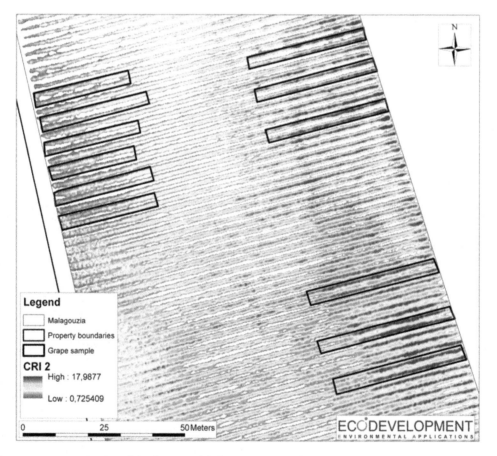

FIGURE 1.5   Remote sensing evaluation of the Carotenoid Reflectance Index 2 (CRI2) in a cv. Malagouisa vineyard. Low CRI2 values were associated with high grape TSS and terpene aroma. Source: *From Iatrou, G., Mourelatos, S., Gewehr, S., Kalaitzopoulou, S., Iatrou, M., Zartaloudis, Z., 2017. Using multispectral imaging to improve berry harvest for wine making grapes.* Ciência e Técnica Vitivinícola, 32, 33–41.

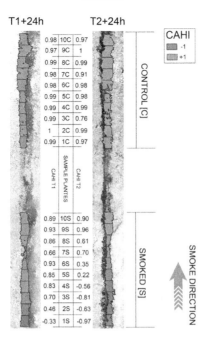

FIGURE 1.6 Application of remote sensing techniques in the evaluation of smoke contamination damage. Source: *Reproduced with permission from Brunori, E., Maesano, M., Moresi, F.V., Antolini, A., Bellincontro, A., Forniti, et al. (2020). Using UAV-based remote sensing to assess grapevine canopy damage due to fire smoke.* Journal of the Science of Food and Agriculture, 100, 4531–4539.

References

Alem, H., Rigou, P., Schneider, R., Ojeda, H., & Torregrosa, L. (2019). Impact of agronomic practices on grape aroma composition: A review. *Journal of the Science of Food and Agriculture, 99,* 975–985.

Allan, W. (2003). Winegrape assessment in the vineyard at the winery. *Australian Viticulture, 7*(6), 20–43.

Álvarez-Cid, M.X., García-Díaz, A., Rodríguez-Araújo, J., Asensio-Campazas, A., de la Torre, M.V. (2015). Goal-driven phenotyping through spectral imaging for grape aromatic ripeness assessment. In *Iberian conference on pattern recognition and image analysis* (pp. 272–280). Springer, Cham.

Arcari, S. G., Caliari, V., Sganzerla, M., & Godoy, H. T. (2017). Volatile composition of Merlot red wine and its contribution to the aroma: Optimization and validation of analytical method. *Talanta, 174,* 752–766.

Arévalo Villena, M., Pérez, J. D., Úbeda, J. F., Navascués, E., & Briones, A. I. (2006). A rapid method for quantifying aroma precursors: Application to grape extract, musts and wines made from several varieties. *Food Chemistry, 99*(1), 183–190.

Barnaba, C., Dellacassa, E., Nicolini, G., Nardin, T., Serra, M., & Larcher, R. (2018). Non-targeted glycosidic profiling of international wines using neutral loss-high resolution mass spectrometry. *Journal of Chromatography. A, 1557,* 75–89.

Brunori, E., Maesano, M., Moresi, F. V., Antolini, A., Bellincontro, A., Forniti., et al. (2020). Using UAV-based remote sensing to assess grapevine canopy damage due to fire smoke. *Journal of the Science of Food and Agriculture, 100,* 4531–4539.

Capone, D. L., Sefton, M. A., & Jeffery, D. W. (2011). Application of a modified method for 3-mercaptohexan-1-ol determination to investigate the relationship between free thiol and related conjugates in grape juice and wine. *Journal of Agricultural and Food Chemistry, 59,* 4649–4658.

Caputi, L., Carlin, S., Ghiglieno, I., Stefanini, M., Valenti, L., Vrhovsek, U., et al. (2011). Relationship of changes in rotundone content during grape ripening and winemaking to manipulation of the 'peppery' character of wine. *Journal of Agricultural and Food Chemistry, 59,* 5565–5571.

Cerreti, M., Esti, M., Benucci, I., Liburdi, K., de Simone, C., & Ferranti, P. (2015). Evolution of S-cysteinylated and S-glutathionylated thiol precursors during grape ripening of *Vitis vinifera* L. cvs Grechetto, Malvasia del Lazio and Sauvignon Blanc. *Australian Journal of Grape and Wine Research, 21*(3), 411–416.

Coelho, E., Rocha, S. M., Barros, A. S., Delgadillo, I., & Coimbra, M. A. (2007). Screening of variety- and pre-fermentation-related volatile compounds during ripening of white grapes to define their evolution profile. *Analytica Chimica Acta, 597,* 257–264.

Colangelo, D., Torchio, F., Rolle, L., Gerbi, V., De Faveri, M., & Lambri, M. (2019). Modelling by Response surface methodology of the clarifying process of Muscat blanc must for the production of a sweet sparkling wine. *American Journal of Enololgy and Viticulture, 70,* 42–49.

Coombe, B. G., & Iland, P. (2004). Grape berry development and winegrape quality. In P. R. Dry, & B. G. Coombe (Eds.), *Viticulture volume 1 – Resources* (2nd, pp. 210–248). Adelaide, South Australia: Winetitles.

Coombe, B. G., Dundon, R. J., & Short, A. W. S. (1980). Indices of sugar-acidity as ripeness criteria for winegrapes. *Journal of the Science of Food and Agriculture, 31,* 495–502.

Coombe, B. G., & McCarthy, M. G. (2000). Dynamics of grape berry growth and physiology of ripening. *Australian Journal of Grape and Wine Research, 6,* 131–135.

Crupi, P., Santamaria, M., Vallejo, F., Tomás-Barberán, F. A., Masi, G., Caputo, A. R., et al. (2020). How pre-harvest inactivated yeast treatment may influence the norisoprenoid aroma potential in wine grapes. *Applied Sciences (Switzerland), 10*(10), 3369.

Cynkar, W. U., Cozzolino, D., Dambergs, R. G., Janik, L., & Gishen, M. (2007). Effect of variety, vintage and winery on the prediction by visible and near infrared spectroscopy of the concentration of glycosylated compounds (G-G) in white grape juice. *Australian Journal of Grape and Wine Research, 13*(2), 101–105.

D'Onofrio, C., Cox, A., Davies, C., & Boss, P. K. (2009). Induction of secondary metabolism in grape cell cultures by jasmonates. *Functional Plant Biology: FPB, 36*, 323–338.

Dambergs, R., Gishen, M., & Cozzolino, D. (2015). A review of the state of the art, limitations, and perspectives of infrared spectroscopy for the analysis of wine grapes, must, and grapevine tissue. *Applied Spectroscopy Reviews, 50*(3), 261–278.

Darriet, P., & Pons, A. (2017). Wine. In A. Buettner (Ed.), *Springer Handbook of Odor*. Cham, Switzerland: Springer.

Deluc, L. G., Quilici, D. R., Decendit, A., Grimplet, J., Wheatley, M. D., Schlauch, K. A., et al. (2009). Water deficit alters differentially metabolic pathways affecting important flavor and quality traits in grape berries of Cabernet Sauvignon and Chardonnay. *BMC Genomics, 10*, 212.

Englezos, V., Rantsiou, K., Cravero, F., Torchio, F., Pollon, M., Fracassetti, D., et al. (2018). Volatile profile of white wines fermented with sequential inoculation of *Starmerella bacillaris* and *Saccharomyces cerevisiae*. *Food Chemistry, 257*, 350–360.

Fenoll, J., Manso, A., Hellín, P., Ruiz, L., & Flores, P. (2009). Changes in the aromatic composition of the *Vitis vinifera* grape Muscat Hamburg during ripening. *Food Chemistry, 114*, 420–428.

Ferrandino, A., Carra, A., Rolle, L., Schneider, A., & Schubert, A. (2012). Profiling of hydroxycinnamoyl tartrates and of acylated anthocyanins in the skin of 34 *Vitis vinifera* genotypes. *Journal of Agricultural and Food Chemistry, 60*, 4931–4945.

Ferrandino, A., Pagliarani, C., Carlomagno, A., Novello, V., Schubert, A., & Agati, G. (2017). Improved fluorescence-based evaluation of flavonoid in red and white winegrape cultivars. *Australian Journal of Grape and Wine Research, 23*(2), 207–214.

Ferreira, V., & Lopez, R. (2019). The actual and potential aroma of winemaking grapes. *Biomolecules, 9*, 818.

Ferreira, V., Saenz-Navajas, M. P., Campo, E., Herrero, P., De La Fuente, A., & Fernandez-Zurbano, P. (2016). Sensory interactions between six common aroma vectors explain four main red wine aroma nuances. *Food Chemistry, 199*, 447–456.

Fortea, M. I., López-Miranda, S., Serrano-Martínez, A., Carreño, J., & Núñez-Delicado, E. (2009). Kinetic characterisation and thermal inactivation study of polyphenol oxidase and peroxidase from table grape (Crimson Seedless). *Food Chemistry, 113*(4), 1008–1014.

Fracassetti, D., Gabrielli, M., Encinas, J., Manara, M., Pellegrino, L., & Tirelli, A. (2017). Approaches to prevent the light-struck taste in white wine. *Australian Journal of Grape and Wine Research, 23*, 329–333.

Fracassetti, D., Limbo, S., Pellegrino, L., & Tirelli, A. (2019). Light-induced reactions of methionine and riboflavin in model wine: Effects of hydrolysable tannins and sulfur dioxide. *Food Chemistry, 298*, 124952.

Fracassetti, D., & Vigentini, I. (2018). Occurrence and analysis of sulfur compounds in wine. In A. M. Jordão, & F. Cosme (Eds.), *Grapes and winesadvances in production, processing, analysis and valorization* (1st edn.). Rijeka, Croatia: IntechOpen.

Fuentes, S., Tongson, E. J., De Bei, R., Gonzalez Viejo, C., Ristic, R., Tyerman, S., et al. (2019). Non-invasive tools to detect smoke contamination in grapevine canopies, berries and wine: a remote sensing and machine learning modeling approach. *Sensors, 19*, 3335.

Gambetta, J. M., Bastian, S. E. P., Cozzolino, D., & Jeffery, D. W. (2014). Factors influencing the aroma composition of Chardonnay wines. *Journal of Agricultural and Food Chemistry, 62*, 6512–6534.

Genovese, A., Gambuti, A., Piombino, P., & Moio, L. (2007). Sensory properties and aroma compounds of sweet Fiano wine. *Food Chemistry, 103*, 1228–1236.

Giacosa, S., Giordano, M., Vilanova, M., Cagnasso, E., Río Segade, S., & Rolle, L. (2019a). On-vine withering process of 'Moscato bianco' grapes: Effect of cane-cut system on volatile composition. *Journal of the Science of Food and Agriculture, 99*, 1135–1144.

Giacosa, S., Ossola, C., Botto, R., Río Segade, S., Paissoni, M. A., Pollon, M., et al. (2019b). Impact of specific inactive dry yeast application on grape skin mechanical properties, phenolic compounds extractability, and wine composition. *Food Research Internationa, 116*, 1084–1093.

Giordano, M., Rolle, L., Zeppa, G., & Gerbi, V. (2009). Chemical and volatile composition of three Italian sweet white *passito* wines. *Journal International des Sciences de la Vigne et du Vin, 43*, 159–170.

Giordano, M., Zecca, O., Belviso, S., Reinotti, M., Gerbi, V., & Rolle, L. (2013). Volatile fingerprint and physico-mechanical properties of 'Muscat blanc' grapes grown in mountain area: A first evidence of the influence of water regimes. *Italian Journal of Food Science, 25*, 329–338.

Gishen, M., Dambergs, B. (1998). Some preliminary trials in the application of scanning near infrared spectroscopy (NIRS) for determining the compositional quality of grapes, wine, and spirits. Australian & New Zeland Grape Grower and Winemaker, 414a, 43–45.

Gishen, M., Dambergs, R. G., & Cozzolino, D. (2005). Grape and wine analysis-enhancing the power of spectroscopy with chemometrics. A review of some applications in the Australian wine industry. *Australian Journal of Grape and Wine Research, 11*(3), 296–305.

Godshaw, J., Hjelmeland, A. K., Zweigenbaum, J., & Ebeler, S. E. (2019). Changes in glycosylation patterns of monoterpenes during grape berry maturation in six cultivars of *Vitis vinifera*. *Food Chemistry, 297*, 124921.

González-Barreiro, C., Rial-Otero, R., Cancho-Grande, B., & Simal-Gándara, J. (2015). Wine aroma compounds in grapes: A critical review. *Critical Reviews in Food Science and Nutrition, 55*(2), 202–218.

Granado, J., Felix, G., & Boller, T. (1995). Perception of fungal sterols in plants (subnanomolar concentrations of ergosterol elicit extracellular alkalinization in tomato cells). *Plant Physiology, 107*, 485–490.

Günata, Y. Z., Bayonove, C. L., Baumes, R. L., & Cordonnier, R. E. (1985). The aromas of grapes. I. Extraction and determination of free and glycosidically bound fractions of some grape aroma components. *Journal of Chromatography, 331*, 83–90.

Gutiérrez-Gamboa, G., Garde-Cerdán, T., Martínez-Lapuente, L., Souza da Costa, B., Rubio-Bretón, P., & Pérez-Álvarez, E. P. (2020). Phenolic composition of Tempranillo blanco (*Vitis vinifera* L.) grapes and wines after biostimulation via a foliar seaweed application. *Journal of the Science of Food and Agriculture, 100*(2), 825–835.

Herbst-Johnstone, M., Araujo, L. D., Allen, T. A., Logan, G., Nicolau, L., & Kilmartin, P. A. (2013). Effects of mechanical harvesting on 'Sauvignon blanc' aroma. *Acta Horticulturae, 978*, 179–186.

Iatrou, G., Mourelatos, S., Gewehr, S., Kalaitzopoulou, S., Iatrou, M., & Zartaloudis, Z. (2017). Using multispectral imaging to improve berry harvest for wine making grapes. *Ciência e Técnica Vitivinícola, 32*, 33–41.

Kalua, C. M., & Boss, P. K. (2009). Evolution of volatile compounds during the development of Cabernet Sauvignon grapes (*Vitis vinifera* L.). *Journal of Agricultural and Food Chemistry, 57*, 3818–3830.

Kalua, C. M., & Boss, P. K. (2010). Comparison of major volatile compounds from Riesling and Cabernet Sauvignon grapes (*Vitis vinifera* L.) from fruitset to harvest. *Australian Journal of Grape and Wine Research, 16,* 337−348.

Knauer, U., Matros, A., Petrovic, T., Zanker, T., Scott, E. S., & Seiffert, U. (2017). Improved classification accuracy of powdery mildew infection levels of wine grapes by spatial-spectral analysis of hyperspectral images. *Plant Methods, 13*(1), 47.

Kuhn, N., Guan, L., Dai, Z. W., Wu, B.-H., Lauvergeat, V., Gomès, E., et al. (2014). Berry ripening: Recently heard through the grapevine. *Journal of Experimental Botany, 65*(16), 4543−4559.

Kwasniewski, M. T., Vanden Heuvel, J. E., Pan, B. S., & Sacks, G. L. (2010). Timing of cluster light environment manipulation during grape development affects C13-norisoprenoid and carotenoid concentrations in Riesling. *Journal of Agricultural and Food Chemistry, 58,* 6841−6849.

Laureano, J., Giacosa, S., Río Segade, S., Torchio, F., Cravero, F., Gerbi, V., et al. (2016). Effects of continuous exposure to ozone gas and electrolyzed water on the skin hardness of table and wine grape varieties. *Journal of Texture Studies, 47,* 40−48.

Lee, S., Seo, M., Riu, M., Cotta, J., Block, D., Dokoozlian, N., et al. (2007). Vine microclimate and norisoprenoid concentration in Cabernet sauvignon grapes and wines. *American Journal of Enology and Viticulture, 58,* 291−301.

Li, B., Lecourt, J., & Bishop, G. (2018). Advances in non-destructive early assessment of fruit ripeness towards defining optimal time of harvest and yield prediction—A review. *Plants, 7*(1), 3.

Lohitnavy, N., Bastian, S., & Collins, C. (2010). Berry sensory attributes correlate with compositional changes under different viticultural management of Semillon (*Vitis vinifera* L.). *Food Quality and Preference, 21*(7), 711−719.

López-Miranda, S., Hernández-Sánchez, P., Serrano-Martínez, A., Hellín, P., Fenoll, J., & Núñez-Delicado, E. (2011). Effect of ripening on protein content and enzymatic activity of Crimson Seedless table grape. *Food Chemistry, 127*(2) 481−486.

Marciniak, M., Brown, R., Reynolds, A., & Jollineau, M. (2015). Use of remote sensing to understand the terroir of the Niagara peninsula. Applications in a Riesling vineyard. *Journal International des Sciences de la Vigne et du Vin., 49,* 1−26.

Martin, D. M., Chiang, A., Lund, S. T., & Bohlmann, J. (2012). Biosynthesis of wine aroma: Transcript profiles of hydroxymethylbutenyl diphosphate reductase, geranyl diphosphate synthase, and linalool/nerolidol synthase parallel monoterpenol glycoside accumulation in Gewürztraminer grapes. *Planta, 236,* 919−929.

Matese, A., & Di Gennaro, S. F. (2015). Technology in precision viticulture: A state of the art review. *Int. J. Wine Res., 7,* 69−81.

Mira de Orduña, R. (2010). Climate change associated effects on grape and wine quality and production. *Food Research International, 43,* 1844−1855.

Mpelasoka, B. S., Schachtman, D. P., Treeby, M. T., & Thomas, M. R. (2003). A review of potassium nutrition in grapevines with special emphasis on berry accumulation. *Australian Journal of Grape and Wine Research, 9,* 154−168.

Narayani, M., & Srivastava, S. (2017). Elicitation: A stimulation of stress in in vitro plant cell/tissue cultures for enhancement of secondary metabolite production. *Phytochemistry Reviews, 16*(6), 1227−1252.

Noguerol-Pato, R., Gonzalez-Barreiro, C., Cancho-Grande, B., Martinez, M. C., Santiago, J. L., & Simal-Gandara, J. (2012). Floral, spicy and herbaceous active odorants in Gran Negro grapes from shoulders and tips into the cluster, and comparison with Brancellao and Mouraton varieties. *Food Chemistry, 135,* 2771−2782.

OIV. (2019). *Compendium of international methods of wine and must analysis.* Paris, France: Organisation Internationale de la Vigne et du Vin.

Olarte Mantilla, S. M., Collins, C., Iland, P. G., Johnson, T. E., & Bastian, S. E. P. (2012). Berry sensory assessment: Concepts and practices for assessing winegrapes' sensory attributes. *Australian Journal of Grape and Wine Research, 18*(3), 245−255.

Oliveira, J. M., Faria, M., Sa, F., Barros, F., & Araujo, I. A. (2006). C$_6$-alcohols as varietal markers for assessment of wine origin. *Analytica Chimica Acta, 563,* 300−309.

Ong, P. K. C., & Acree, T. (1999). Similarities in the aroma chemistry of Gewürztraminer variety wines and Lychee (*Litchi chinesis Sonn.*) fruit. *Journal of Agricultural and Food Chemistry, 47,* 665−670.

Pati, S., Crupi, P., Savastano, M. L., Benucci, I., & Esti, M. (2020). Evolution of phenolic and volatile compounds during bottle storage of a white wine without added sulfite. *Journal of the Science of Food and Agriculture, 100*(2), 775−784.

Perestrelo, R., Silva, C., Silva, P., & Câmara, J. S. (2018). Unraveling *Vitis vinifera* L. grape maturity markers based on integration of terpenic pattern and chemometric methods. *Microchemical Journal, Devoted to the Application of Microtechniques in all Branches of Science, 142,* 367−376.

Peyrot Des Gachons, C., Van Leeuwen, C., Tominaga, T., Soyer, J.-P., Gaudillère, J.-P., & Dubourdieu, D. (2005). Influence of water and nitrogen deficit on fruit ripening and aroma potential of *Vitis vinifera* L cv Sauvignon blanc in field conditions. *Journal of the Science of Food and Agriculture, 85*(1), 73−85.

Pinu, F. R., Tumanov, S., Grose, C., Raw, V., Albright, A., Stuart, L., et al. (2019). Juice Index: An integrated Sauvignon blanc grape and wine metabolomics database shows mainly seasonal differences. *Metabolomics: Official Journal of the Metabolomic Society, 15,* 3.

Poni, S., Gatti, M., Palliotti, A., Daic, Z., Duchêned, E., Truongd, T.-T., et al. (2018). Grapevine quality: A multiple choice issue. *Scientia Horticulturae, 234,* 445−462.

Pons, A., Allamy, L., Schüttler, A., Rauhut, D., Thibon, C., & Darriet, P. (2017). What is the expected impact of climate change on wine aroma compounds and their precursors in grape? *OENO One, 51,* 141−146.

Ribéreau-Gayon, P., Glories, Y., Maujean, A., & Dubourdieu, D. (2006). *Handbook of Enology. The chemistry of wine: stabilization and treatments* (2nd Edn.). Chichester, UK: John Wiley & Sons Ltd.

Río Segade, S., Orriols, I., Giacosa, S., & Rolle, L. (2011). Instrumental texture analysis parameters as winegrapes varietal markers and ripeness predictors. *International Journal of Food Properties, 14,* 1318−1329.

Río Segade, S., Paissoni, M. A., Giacosa, S., Bautista-Ortín, A. B., Gómez-Plaza, E., Gerbi, V., et al. (2019). Winegrapes dehydration under ozone-enriched atmosphere: Influence on berry skin phenols release, cell wall composition and mechanical properties. *Food Chemistry, 271,* 673−684.

Ripoll, G., Vazquez, M., & Vilanova, M. (2017). Ultraviolet−visible-near infrared spectroscopy for rapid determination of volatile compounds in white grapes during ripening. *Ciência e Técnica Vitivinícola, 32*(1), 53−61.

Roland, A., Schneider, R., Razungles, A., & Cavelier, F. (2011). Varietal thiols in wine: Discovery, analysis and applications. *Chemical Reviews, 111*(11), 7355−7376.

Rolle, L., Gerbi, V., Schneider, A., Spanna, F., & Río Segade, S. (2011). Varietal relationship between instrumental skin hardness and climate for grapevines (*Vitis vinifera* L.). *Journal of Agricultural and Food Chemistry, 59*, 10624—10634.

Rolle, L., Giordano, M., Giacosa, S., Vincenzi, S., Río Segade, S., Torchio, F., et al. (2012). CIEL*a*b* parameters of white dehydrated grapes as quality markers according to chemical composition, volatile profile and mechanical properties. *Analytica Chimica Acta, 732*, 105—113.

Rolle, L., Siret, R., Río Segade, S., Maury, C., Gerbi, V., & Jourjon, F. (2012). Instrumental texture analysis parameters as markers of table-grapes and winegrape quality: A review. *American Journal of Enology and Viticulture, 63*, 11—28.

Romeyer, F., Macheix, J. J., Goiffon, J. P., Reminiac, C. C., & Sapis, J. C. (1983). The browning capacity of grapes. 3. Changes and importance of hydroxycinnamic acid-tartaric acid esters during development and maturation. *Journal of Agricultural and Food Chemistry, 31*, 346—349.

Rousseau, J. (2001). Quantified description of sensory analysis of berries. *Relationships with wine profiles and consumer tastes. Bulletin OIV, 849—850*, 719—728.

Ruffner, H. P. (1982a). Metabolism of tartaric and malic acids in *Vitis*: A review - Part A. *Vitis, 21*, 247—259.

Ruffner, H. P. (1982b). Metabolism of tartaric and malic acids in *Vitis*: A review - Part B. *Vitis, 21*, 346—358.

Ruiz, J., Kiene, F., Belda, I., Fracassetti, D., Marquina, D., Navascués, E., et al. (2019). Effects on varietal aromas during wine making: A review of the impact of varietal aromas on the flavor of wine. *Applied Microbiology and Biotechnology, 103*(18), 7425—7450.

Ryona, I., Leclerc, S., & Sacks, G. L. (2010). Correlation of 3-isobutyl-2-methoxypyrazine to 3-isobutyl-2-hydroxypyrazine during maturation of bell pepper (*Capsicum annuum*) and wine grapes (*Vitis vinifera*). *Journal of Agricultural and Food Chemistry, 58*, 9723—9730.

Samotica, J., Jara-Palacios, M. J., Hernández-Hierro, J. M., Heredia, F. J., & Wojdylo, A. (2018). Phenolic compounds and antioxidant activity of twelve grape cultivars meausured by chemical and electrochemical methods. *European Food Research and Technology, 244*(11), 1933—1943.

Sarrazin, E., Dubourdieu, D., & Darriet, P. (2007). Characterization of key-aroma compounds of botrytized wines, influence of grape botrytization. *Food Chemistry, 103*, 536—545.

Schievano, E., D'Ambrosi, O. M., Mazzaretto, I., Ferrarini, R., Magno, F., Mammi, S., et al. (2013). Identification of wine aroma precursors in Moscato Giallo grape juice: A nuclear magnetic resonance and liquid chromatography-mass spectrometry tandem study. *Talanta, 116*, 841—851.

Schneider, R., Charrier, F., Moutounet, M., & Baumes, R. (2004). Rapid analysis of grape aroma glycoconjugates using Fourier-transform infrared spectrometry and chemometric techniques. *Analytica Chimica Acta, 513*, 91—96.

Slaghenaufi, D., Guardini, S., Tedeschi, R., & Ugliano, M. (2019). Volatile terpenoids, norisoprenoids and benzenoids as markers of fine scale vineyard segmentation for Corvina grapes and wines. *Food Research International, 125*, 108507.

Sollazzo, M., Baccelloni, S., D'Onofrio, C., & Bellincontro, A. (2018). Combining color chart, colorimetric measurement and chemical compounds for postharvest quality of white wine grapes. *Journal of the Science of Food and Agriculture, 98*, 3532—3541.

Song, M., Fuentes, C., Loos, A., & Tomasino, E. (2018). Free monoterpene isomer profiles of *Vitis vinifera* L. cv. white wines. *Foods, 7*, 27.

Steel, C. C., Blackman, J. W., & Schmidtke, L. M. (2013). Grapevine bunch rots: Impacts on wine composition, quality, and potential procedures for the removal of wine faults. *Journal of Agricultural and Food Chemistry, 61*(22), 5189—5206.

Swiegers, J. H., Bartowsky, E. J., Henschke, P. A., & Pretorius, I. S. (2005). Yeast and bacterial modulation of wine aroma and flavour. *Australian Journal of Grape and Wine Research, 11*, 139—173.

Tahir, H. E., Xiaobo, Z., Jianbo, X., Mahunu, G. K., Jiyong, S., Xu, J. L., et al. (2019). Recent progress in rapid analyses of vitamins, phenolic, and volatile compounds in foods using vibrational spectroscopy combined with chemometrics: A review. *Food Analytical Methods, 12*(10), 2361—2382.

Teixeira, A., Eiras-Dias, J., Castellarin, S. D., & Gerós, H. (2013). Berry phenolics of grapevine under challenging environments. *International Journal of Molecular Sciences, 14*(9), 18711—18739.

Torchio, F., Giacosa, S., Vilanova, M., Río Segade, S., Gerbi, V., Giordano, M., & Rolle, L. (2016). Use of response surface methodology for the assessment of changes in the volatile composition of Moscato Bianco (*Vitis vinifera* L.) grape berries during ripening. *Food Chemistry, 212*, 576—584.

Vilanova, M., Genisheva, Z., Bescansa, L., Masa, A., & Oliveira, J. M. (2012). Changes in free and bound fractions of aroma compounds of four *Vitis vinifera* cultivars at the last ripening stages. *Phytochemistry, 74*, 196—205.

Vincenzi, S., Panighel, A., Gazzola, D., Flamini, R., & Curioni, A. (2015). Study of combined effect of proteins and bentonite fining on the wine aroma loss. *Journal of Agricultural and Food Chemistry, 63*, 2314—2320.

USDA (2020). Grape crush reports listing. United States Department of Agriculture, National Agricultural Statistics Service. https://www.nass.usda.gov/Statistics_by_State/California/Publications/Specialty_and_Other_Releases/Grapes/Crush/Reports/. Accessed 20.08.20.

Walsh, K. B., Blasco, J., Zude-Sasse, M., & Sun, X. (2020). Visible-NIR 'point'spectroscopy in postharvest fruit and vegetable assessment: The science behind three decades of commercial use. *Postharvest Biology and Technology, 168*, 111246.

Wang, J., Abbey, T., Kozak, B., Madilao, L. L., Tindjau, R., Del Nin, J., et al. (2019). Evolution over the growing season of volatile organic compounds in Viognier (*Vitis vinifera* L.) grapes under three irrigation regimes. *Food Research International, 125*, 108512.

Williams, P. J., Cynkar, W., Francis, I. L., Gray, J. D., Iland, P. G., & Coombe, B. G. (1995). Quantification of glycosides in grapes, juices, and wines through a determination of glycosyl glucose. *Journal of Agricultural and Food Chemistry, 43*, 121—128.

Zhang, H., Fan, P., Liu, C., Wu, B., Li, S., & Liang, Z. (2014). Sunlight exclusion from Muscat grape alters volatile profiles during berry development. *Food Chemistry, 164*, 242—250.

Zhao, X., Ju, Y., Wei, X., Dong, S., Sun, X., & Fang, Y. (2019). Significance and transformation of 3-alkyl-2-methoxypyrazines through grapes to wine: Olfactory properties, metabolism, biochemical regulation, and the HP—MP cycle. *Molecules (Basel, Switzerland), 24*, 4598.

Zoecklein, B. W., Fugelsang, K. C., & Gump, B. H. (2010). Practical methods of measuring grape quality. In A. G. Reynolds (Ed.), *Managing wine quality* (pp. 107—133). Cambridge, United Kingdom: Woodhead Publishing.

2

White grape quality monitoring via hyperspectral imaging: from the vineyard to the winery

Gianella Chávez-Segura[1] *and Ricardo Vejarano*[2]

[1]Agroindustrial Engineering Program, Universidad Privada del Norte (UPN), Trujillo, Peru [2]Dirección de Investigación y Desarrollo, Universidad Privada del Norte (UPN), Trujillo, Peru

2.1 Introduction

The implementation of rapid response, automatic, and eco-friendly procedures for the monitoring of technological parameters associated with grape quality poses a critical challenge for the wine industry.

As an alternative, hyperspectral imaging (HSI) is a nondestructive analysis technique applicable in all food production stages (Vejarano, Siche, & Tesfaye, 2017), particularly because it allows producers to quickly, accurately, and simultaneously assess different parameters in different samples, as well as process information in real time (ElMasry, Kamruzzaman, Sun, & Allen, 2012). In addition, HSI requires minimal sample preparation without using solvents and chemical reagents (He & Sun, 2015), which is not always possible in traditional analysis techniques.

2.1.1 Hyperspectral imaging

HSI integrates spectroscopy and image analysis to simultaneously assess sample composition (spectroscopic component) and the sample's external characteristics (image analysis) (Wu & Sun, 2013). As with any spectroscopic technology, all samples reflect, scatter, or absorb energy when subjected to a source of electromagnetic radiation (Mirzaei et al., 2019). Herein, light scattering is related to the physical characteristics of the sample (particle size, cell structure, and tissues), whereas absorption is related to the samples' chemical constituents (Tao, Peng, Gomes, Chao, & Qin, 2015).

Fig. 2.1A shows a typical laboratory experimental setup of the HSI system. The hyperspectral images obtained are three-dimensional data cubes (hypercubes) comprising hundreds of images of the same object at different wavelengths (Fig. 2.1B), wherein the spectrum of each pixel can be used to characterize said specific region (Wu & Sun, 2013).

When interacting with the incident radiation, each sample component presents a "spectral signature" (He, Li, & Deng, 2007) that can be used to characterize, identify, and distinguish different samples. This "spectral signature" is typically represented by the reflectance curve obtained when the sample is subjected to a source of radiation at different wavelengths (Fig. 2.1C), thus generating information on sample constituents at different spectral bands, ranging from ultraviolet–visible (UV–Vis) to near-infrared (NIR) bands (Wu & Sun, 2013).

© 2022 Elsevier Inc. All rights reserved.

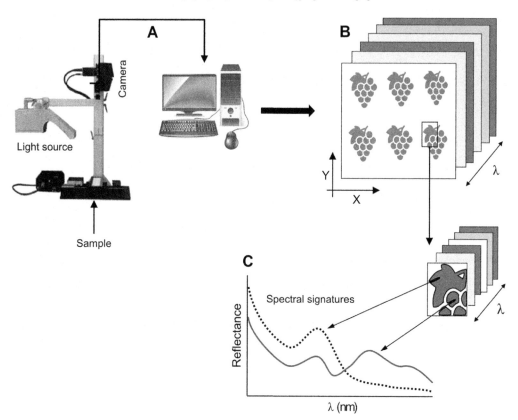

FIGURE 2.1 Schematic of a typical HSI system: (A) Acquisition devices. (B) Hypercube at different wavelengths. (C) Spectral signature for each component of the sample.

2.1.2 HSI processing

The hypercube (Fig. 2.1B) contains gigabytes of spectral information, which must be efficiently processed to reduce the amount of data and exclusively extract information that enables the following:

1. The recognition of common patterns, i.e. information used for discriminating among different varieties, ripeness stages, and phytosanitary states, among others. This can be realized by applying multivariate classification analysis (Fig. 2.2).
2. Establishing relationships between spectral information and sample properties using multivariate regression analysis (Fig. 2.2) for determining parameters such as the sugar content in grape berries.

As a detailed description of the chemometric methods used for these purposes is beyond the scope of this document, a brief overview can be found in ElMasry et al. (2012) and Wu & Sun (2013).

In addition to reducing abundant HSI information, selecting optimal wavelengths enables the construction of multispectral image (MI) models, which are highly suitable for online applications, owing to their higher processing speed (ElMasry et al., 2012; Vejarano et al., 2017). Therefore a majority of studies have focused on identifying optimal or sensitive wavelengths, the obtained information for which is representative of the entire spectrum corresponding to each sample.

In this regard, research aimed at the development of food analysis methodologies using HSI has considerably increased in the last decade; this research includes studies focusing on monitoring grape quality parameters both at the vineyard (hydric status, chlorosis, pathogen presence, and maturity) and the winery (commercial maturity, phenolic maturity, and aromatic profile) levels to guarantee appropriate grape quality and therefore appropriate wine quality.

2.2 Vineyard-level applications

Table 2.1 summarizes some critical vineyard-monitoring HSI applications. Although many previous investigations were conducted at only red grape vineyards, the same methodology may be applied for controlling technological parameters in white grape vineyards.

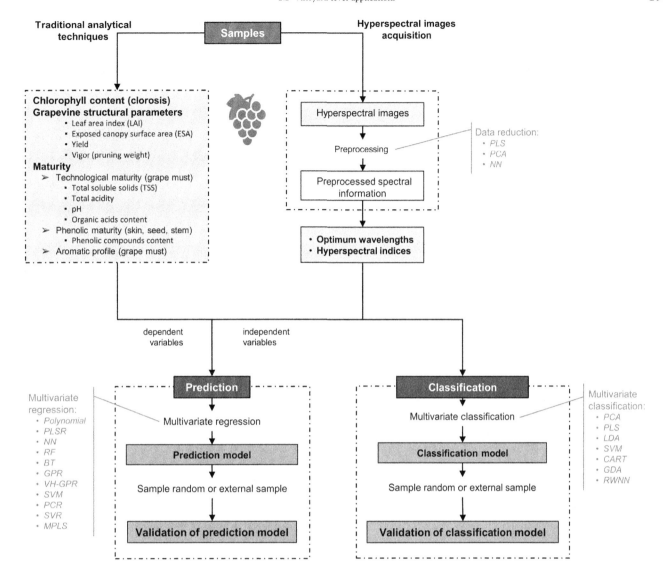

FIGURE 2.2 Flowchart of the main steps for processing hyperspectral images for prediction and classification model development. Abbreviation descriptions: *PLS*: partial least squares; *PLSR*: partial least squares regression; *PCA*: principal component analysis; *PCR*: principal component regression; *NN*: neural networks; *RWNN*: random weight neural networks; *RF*: random forest; *BT*: bagging trees; *GPR*: Gaussian process regression; *VH-GPR*: variational heteroscedastic Gaussian process regression; *SVM*: support vector machine; *SVR*: support vector regression; *MPLS*: modified partial least squares; *LDA*: linear discriminant analysis; *CART*: classification and regression tree; *GDA*: general discriminant analysis.

2.2.1 Grapevine variety recognition

Each wine profile is characterized by grape variety. In fact, production yields for each variety can be estimated by identifying different grape varieties within the same growing area (Galvão, Roberts, Formaggio, Numata, & Breunig, 2009), in addition to certifying the varietal authenticity of the grapes used within a given geographical indication (Álvarez-Cid, García-Díaz, Rodríguez-Araújo, Asensio-Campazas, & de la Torre, 2015).

Traditional variety identification techniques include ampelometry (Fernandes et al., 2018), which is based on the morphological differences among varieties. Its main disadvantage is that it depends on visual identification by trained personnel. As an alternative, ampelographic characterization software have been developed (Soldavini, Schneider, Stefanini, Dallaserra, & Policarpo, 2009) that have attained classification percentages of up to 81%. However, the primary drawback of these software is that they may only be used by experts, particularly during the selection of sample measurement points.

TABLE 2.1 Some hyperspectral imaging (HSI) applications at the vineyard level.

Application	λ (nm)	Accuracy	References
Grapevine variety recognition	350–1028	Rabo de Ovelha: 58.3%, Síria: 64.2%, Fernão Pires: 41.7%, Malvasia Fina: 45.8%, Alvarinho: 55.8%	Fernandes et al. (2018)
Grapevine variety recognition	900–1700	PCA–SVM analysis: ≥81.09%	Cheng et al.(2019)
Grapevine variety recognition	350–2500	ANOVA-PCA–SVM analysis: 89.8%-100%	Mirzaei et al. (2019)
Grapevine water status indicators: at leaf level	400–2000		
- Midday leaf water potential (Ψ_l)		PLSR model: R_{cv}^2 = **0.88**; RMSE$_{CV}$ = **9%**; RMS = **1.36**	Rapaport et al. (2015)
- Stomatal conductance (g_s)		PLSR model: R_{cv}^2 = 0.85; RMSE$_{CV}$ = 12%; RMS = 1.52	Rapaport et al. (2015)
- Nonphotochemical quenching (NPQ)		PLSR model: R_{cv}^2 = 0.81; RMSE$_{CV}$ = 15%; RMS = 1.58	Rapaport et al. (2015)
Prediction of leaf water potential (Ψ_{pd})	325–1075	External validation: GPR model (RMSE: **0.142 MPa**; MAE: **0.111 MPa**; NRMSE: **18.337%**). VH-GPR model (RMSE: **0.142 MPa**; MAE: **0.111 MPa**; NRMSE: **18.370%**). RF model (RMSE: 0.151 MPa; MAE: 0.120 MPa; NRMSE: 19.889%) BT model (RMSE: 0.146 MPa; MAE: 0.119 MPa; NRMSE: 18.876%)	Pôças et al. (2020)
Classification of leaf water potential (Ψ_{pd}):	325–1075		Pôças et al. (2020)
- Low water deficit (0 MPa > Ψ_{pd} > −0.25 MPa)		All models ≈ 100%	
- Moderate water deficit (− 0.25 MPa > Ψ_{pd} > −0.50 MPa)		All models ≥ 81.0%	
- High water deficit (−0.50 MPa > Ψ_{pd})		All models ≥ 78.8%	
Chlorophyll content (C_{ab})	350–1000	R^2 ≥ 0.80	Zarco-Tejada et al. (2005)
GLRaV-3 detection	400–1000	94.1%	MacDonald, Staid, Staid, and Cooper, (2016)

R_{cv}^2: coefficient of determination (cross-validation). RMSE$_{CV}$: root mean squared error of cross-validation. RMS = RMSE$_{CV}$/RMSE$_C$. MAE: mean absolute error. NRMSE: normalized root mean square error. Values in bold indicate the model with the highest coefficient of determination (R^2), and the lowest goodness-of-fit indicator values for the error (RMSE, RMS and MAE).
GLRaV-3: grapevine leafroll-associated virus-3.

Further, analyses based on isozymes (Altube, Cabello, & Ortiz, 1991) and DNA (Pelsy et al., 2010) can be used. While both methods are accurate, they require reagents, high-precision equipment, and trained personnel together with a given period of waiting time before results may be obtained. Thus HSI is a viable alternative.

Fernandes et al. (2018) developed a classification algorithm, Adaptive Boosting-Random Weight Neural Networks (AdaBoost-RWNN), based on HSIs (350–1028 nm) obtained from the stems of 10 grape varieties (5 red grapes and 5 white grapes). The classification rates obtained using AdaBoost-RWNN were 58.3%, 64.2%, 41.7%, 45.8%, and 55.8% for the Rabo de Ovelha, Síria, Fernão Pires, Malvasia Fina, and Alvarinho white grape varieties, respectively (Table 2.1). Here, the Síria and Malvasia Fina varieties reported the lowest false positive rates (FPRs); hence, the algorithm would be more suitable for detecting these varieties. This is crucial because, as the authors reported, the first four varieties exhibit leaf similarities, rendering their ampelometric discrimination difficult. Fernandes et al. (2018) were pioneers in the use of AdaBoost-RWNN to identify grape varieties. AdaBoost-RWNN exhibits better discrimination capacity and processes the same amount of information in less time (∼1 h) when compared to the support vector machine (SVM) analysis ($\sim$45 h). Another innovative contribution of this work is that, unlike most studies, it uses the grape stem instead of grape leaves or berries.

Recently, Cheng, Yang, Liu, Zhang, and Song (2019) classified the Chardonnay, Riesling, Italian Riesling, 8802, 8803, and Ecolly white grape varieties by using HSIs (900–1700 nm) from intact berries. Using principal component analysis (PCA), they selected 26 sensitive wavelengths, which were subsequently processed via SVM, random forest (RF), and an AdaBoost algorithm to develop classification models. In that work, the authors achieved the best results using PCA–SVM, with calibration and validation accuracies of 83.77% and 81.09%, respectively, even when these precisions were lower than those obtained with the full spectrum of 231 wavelengths (90.62% and 85.34%, respectively). Therein, robust classification models were obtained. These models adequately discriminated among varieties, particularly between 8802 and 8803 varieties and between Riesling and Italian Riesling varieties, which have similar structures and chemical composition. In addition, the authors suggest performing field studies; this is because, during vineyard imaging, spectral information may be affected by external factors such as sunlight, water vapor, sample humidity, and air temperature. These effects can be reduced using the indices obtained from spectral information (Prospere, McLaren, & Wilson, 2014).

In a recent study, 2151 wavelengths and 32 spectral indices obtained from leaf and canopy HSIs (350–2500 nm) were used to classify five grape varieties (Mirzaei et al., 2019) and determine the level at which superior classification may be obtained using either wavelengths and spectral indices or leaf and canopy spectra. Sensitive wavelengths and spectral indices were first selected using partial least squares regression (PLSR) and ANOVA-PCA. The wavelengths and indices selected via both methods were subsequently processed using SVM and linear discriminant analysis (LDA). Results revealed that the spectral indices of the canopy provided better discrimination among grape varieties, reporting a classification of up to 100% when using the ANOVA-PCA–SVM combination.

Although PLSR is the most commonly used method for reducing and processing HSI, these results denote that the ANOVA-PCA method can also be used to classify different varieties, with the advantage of directly monitoring the canopy (Clevers, Kooistra, & Schaepman, 2010). HSIs obtained from the canopy, unlike those obtained from individual leaves, include information on structures such as petioles, cordons, and branches in addition to considering factors such as vineyard architecture, leaf orientation, leaf area, and nonphotosynthetic elements (Cho, Sobhan, Skidmore, & De Leeuw, 2008). Therefore HSIs obtained from the canopy enable the development of more complete classification models than those obtained from individual leaves.

2.2.2 Vineyard hydric status

Vine cultivation predominantly occurs in semiarid regions exhibiting high solar radiation, low water availability, and high evaporation. These conditions expose crops to water stress, the improper management of which may affect grape quality (Chaves et al., 2010). In addition, using physiological indicators to monitor water status may be expensive, be intrusive, require specialized personnel, and take time to obtain results (Rapaport, Hochberg, Shoshany, Karnieli, & Rachmilevitch, 2015).

An alternative for assessing water status at the canopy level is using remote sensing through thermal infrared (TIR) spectroscopy (Bellvert, Zarco-Tejada, Girona, & Fereres, 2014); however, the availability of hyperspectral data in said region ($\geq$3000 nm) remains limited.

Another alternative is using leaf HSI in the Vis–NIR (400–2000 nm) range to identify sensitive wavelengths (Rapaport et al., 2015). This spectral information can then be correlated against irrigation management physiological indicators, such as leaf water potential (Ψ_l), stomatal conductance (g_s), and nonphotochemical quenching (NPQ) through methods such as PLSR (Table 2.1).

Rapaport et al. (2015) proposed three new spectral indices sensitive to water deficit (water balance indices, WABI) based on the wavelengths that presented chlorophyll (531, 538, and 550 nm) and water availability (1485, 1490, and 1500 nm) changes. These indices were correlated against the Ψ_l, g_s, and NPQ physiological indicators, obtaining cross-validated coefficient of determination (R_{cv}^2) values of 0.89, 0.80, and 0.86, respectively, at leaf level; these values are higher than those obtained using usual viticulture indices. In fact, these values further increase when validated at the canopy level (R_{cv}^2: 0.94, 0.95, and 0.90, respectively). As they use fewer wavelengths, the new WABIs allow researchers to utilize more affordable multispectral equipment that can easily process information at a fast rate. However, the challenge of transitioning technology toward commercial applications still remains (Rapaport et al., 2015), particularly considering that the new WABIs need to be validated at both the vineyard level and different cultivation stages.

In another recent study, Pôças, Tosin, Gonçalves, and Cunha (2020) estimated the predawn leaf water potential (Ψ_{pd}) from canopy HSIs (325–1075 nm) using two spectral indices ($NRI_{554,561}$ and $WI_{900,970}$) and a novel time-dynamic variable based on Ψ_{pd} (Ψ_{pd_0}) as input variables. In that study, four regression algorithms were applied: RF, bagging trees (BT), Gaussian process regression (GPR), and variational heteroscedastic Gaussian process regression (VH-GPR). Therein, the best results were achieved using GPR and VH-GPR as they reported lower goodness-of-fit indicator values for the prediction error (Table 2.1). Additionally, the Ψ_{pd} values predicted using each algorithm were first grouped into three classes of water deficit (low, moderate, and high), and then compared against the classes corresponding to the observed values of Ψ_{pd} (Table 2.1). In the case of all algorithms, the prediction accuracy for the three water deficit classes ranged between 78.8% and 100%.

2.2.3 Chlorosis diagnosis

Plant stress and deficiencies of nutrients, such as iron and nitrogen, can cause diseases such as chlorosis, which reduces chlorophyll levels and affects the synthesis of sugars, acids, and phenolic compounds (Hall, Lamb, Holzapfel, & Louis, 2002). This subsequently affects vineyard yields and grape quality, thereby affecting the quality of the resulting wine. The common techniques used to assess chlorophyll levels include UV–Vis spectrophotometry (Wellburn, 1994) and high-pressure liquid chromatography (HPLC) (Gutiérrez-Gamboa et al., 2018). Both of these techniques require collecting vineyard leaf samples, preparing laboratory samples, using solvents and reagents, engaging appropriate personnel, and a certain waiting time before obtaining results.

Alternatively, Zarco-Tejada et al. (2005) developed correlation models between the chlorophyll levels (C_{ab}) and 45 spectral indices obtained from HSIs (350–1000 nm) through regression analysis using linear, exponential, and third-order polynomials. In all cases, they obtained $R^2 \geq 0.80$ with the spectral indices calculated from the 700–750 nm wavelengths. These results demonstrate the ability of HSI to remotely identify grapevines affected by chlorosis.

HSIs (430–1650 nm) has also been used to assess vineyards affected by iron chlorosis (Meggio et al., 2010). This was one of the first works that used spectral indices sensitive to carotenoids (C_{ar}) and anthocyanins (A_{nth}) in addition to the traditional normalized difference vegetation index (NDVI) and successful canopy-level chlorophyll indices (TCARI/OSAVI), all of them obtained from HSI. These indices were then correlated via linear regression against foliar parameters (leaf area index [LAI], exposed canopy surface area [ESA], yield, and vigor) and grape berry parameters (including °Brix, total acidity, pH, and tartaric and malic acid content). Finally, $R^2 \geq 0.70$ was obtained for vegetative vigor (using the TCARI/OSAVI index, sensitive to C_{ab}), pH (using the index sensitive to A_{nth}), and malic acid (using indexes sensitive to C_{ar} and A_{nth}). These results suggest that the new C_{ar}- and A_{nth}-sensitive indices produced better relationships than NDVI and C_{ab}-sensitive (TCARI/OSAVI) traditional indices. This can prove extremely useful as one of the symptoms of chlorosis is poor phenolic maturity, which is detected from an increasing level of carotenes and anthocyanins in the leaves, owing to plant stress and micronutrient deficiencies (Chalker-Scott, 1999; Gitelson, Zur, Chivkunova, & Merzlyak, 2002).

2.2.4 Virus detection

Grapevine leafroll disease (GLD) is a recurring disease in vineyards. It is caused by viruses such as grapevine leafroll-associated virus-3 (GLRaV-3). This disease affects grape yields and quality through its action on chlorophyll (Endeshaw, Sabbatini, Romanazzi, Schilder, & Neri, 2014). The alteration can be detected by measuring the spectral differences between infected leaves and uninfected leaves, mainly in the early stages of infection. However, this is not always possible when using techniques such as enzyme-linked immunosorbent assay (ELISA) and polymerase chain reaction (PCR) (Monis & Bestwick, 1997; Rowhani, Uyemoto, & Golino, 1997).

In addition to their destructive nature, some constraints presented by the ELISA and PCR tests are the required use of reagents, high-precision equipment and trained personnel, and long waiting times before obtaining results. These constraints effectively restrict their application at laboratory levels (Zhang & Meng, 2011).

As there is no specific treatment for GLD, its spread can only be constrained through early detection, unless the infected plantations are removed (MacDonald et al., 2016). Then, considerable economic losses in infected vineyards can be prevented using rapid response techniques (Ricketts et al., 2015).

HSI is one of the most promising options. One of the first works to assess GLRaV-3 through vineyard spectral images was developed by Naidu, Perry, Pierce, & Mekuria, 2009). In that study, the authors observed lower absorption of chlorophyll in infected leaves than healthy leaves, particularly around 550 nm and between 600 and 750 nm, without any visible symptoms at the time.

Subsequently, GLRaV-3-infected vineyards have been assessed using HSIs (400—1000 nm) performed at altitudes of 610—1219 m (MacDonald et al., 2016). These HSIs were processed using a classification and regression tree (CART) algorithm, reporting a successful average virus detection of 94.1%. The results were confirmed via PCR analysis.

The above results demonstrate the potential of HSI for detecting GLRaV-3 in white grape vineyards, considering that, in varieties such as Sauvignon Blanc, GLD symptoms are difficult to detect. In addition, other varieties such as the *Vitis labrusca* "Niagara" (white variety) or the rootstocks *Vitis rupestris*, *Vitis riparia*, *Vitis berlandieri*, *Vitis champinii*, and their corresponding hybrid varieties are also asymptomatic to the virus (Naidu, Rowhani, Fuchs, Golino, & Martelli, 2014).

2.3 Grape maturity assessment

Typically, the harvest time is chosen based on technological maturity. However, in conditions where the vineyard is subjected to temperatures higher than usual because of factors such as climate change, there is a gap between the accumulation of sugars and the low levels of organic acids and phenolic compounds (Vejarano, Morata, Loira, González, & Suárez-Lepe, 2013). Therefore phenolic maturity is critical when deciding the optimal harvest times, considering its impact on wine stability and sensory profile.

Traditionally, phenolic content is determined via HPLC (Xu et al., 2011) and UV—Vis spectrophotometry (Espinoza, Gómez, Quispe, Sánchez-González, & León-Vargas, 2017). These are reagent- and solvent-based destructive techniques that involve a long waiting period before obtaining results.

For these purposes, spectroscopy (NIR, Fourier transform NIR [FT-NIR], attenuated total reflection—FT—MIR [ATR-FT-MIR]) has also been used. Although, in some cases, spectroscopy has been applied in discrete growing areas (Musingarabwi, Nieuwoudt, Young, Eyéghè-Bickong, & Vivier, 2016; Urraca, Sanz-Garcia, Tardaguila, & Diago, 2016); most applications are conducted at the laboratory level, under controlled lighting, temperature, and sample placement conditions. However, the control at the vineyard level is important because, in addition to determining the optimum maturity, it can help to divide the vineyard into zones based on grape ripeness, thus facilitating the proper scheduling of harvest time. Besides, at the vineyard level, it is possible to include factors such as environmental light, temperature, and sample placement effects, among others (Gutiérrez, Tardaguila, Fernández-Novales, & Diago, 2018) in order to develop more comprehensive analytical models.

As an alternative, Serrano, González-Flor, and Gorchs (2012) applied regression analysis to estimate yields, total soluble solids (TSS), titratable acidity (TA), and maturation index (IMAD = TSS/TA) in two Chardonnay grape harvests (2007 and 2008) using spectral indices obtained from canopy HSIs (400—1100 nm): NDVI and the water index (WI). For both harvests, the NDVI provided good estimates of yield ($r = 0.56$), and WI also provided good estimates of yield ($r = 0.60$), TA ($r = 0.71$), and IMAD ($r = -0.68$). The obtained results suggest the potential of these spectral indices for the estimation of grape ripeness and yield levels.

Furthermore, the coupling of monitoring systems to on-the-go vehicles has been studied for obtaining information at the vineyard level and in larger sampling areas. One of the pioneering works in this regard was developed by Gutiérrez et al. (2018), wherein the TSS content was predicted from HSIs (400—1000 nm) acquired from a platform moving at 5 km/h (at a distance of 2 m from the vineyard). In that study, after applying epsilon-support vector machines (ε-SVMs), the algorithms obtained values of $R_{cv}^2 \geq 0.91$ (Table 2.2).

Although that study was conducted at Tempranillo grape vineyards, the results demonstrate that HSI sensors, GPS referencing, and automation can all be integrated within a single system capable of real-time monitoring of the ripeness process of any grape variety at the vineyard level.

2.3.1 Phenolic maturity

Nogales-Bueno, Hernández-Hierro, Rodríguez-Pulido, and Heredia (2014) simultaneously evaluated technological and phenolic maturity in Zalema grapes (white variety), applying modified partial least squares (MPLS) regression for correlation between TSS, TA, pH, and total phenols against the spectral information extracted from HSIs (900—1700 nm). They obtained highly robust models for TSS, TA, and pH, with R_{cv}^2 values ranging between 0.93 and 0.95, and a low standard error of prediction (SEP) (Table 2.2). In that study, the authors also recommend

TABLE 2.2 Applications of HSI for estimating grape maturity.

Application	λ (nm)	Accuracy	References
Total soluble solids	400–1000	$R_{cv}^2 = 0.91$; $RMSE_{CV} = 1.358\ °Brix$ $R_p^2 = 0.92$; $RMSE_p = 1.274\ °Brix$	Gutiérrez et al. (2018)
Maturity (Zalema variety)	900–1700	TSS: $R_{cv}^2 = 0.95$; SEP = 1.89 °Brix Titratable acidity: $R_{cv}^2 = 0.93$; SEP = 2.21 g/L pH: $R_{cv}^2 = 0.94$; SEP = 0.18 Total phenols: $R_{cv}^2 = 0.80$; SEP = 2.29 mg/g	Nogales-Bueno et al. (2014)
Maturity (Italia variety)	400–1000	TSS: $R_p^2 = 0.88$, TA: $R_p^2 = 0.78$, pH: $R_p^2 = 0.70$	Piazzolla et al. (2017)
Phenolic maturity in grape seeds (Zalema variety)	897–1752	$R_{cv}^2 \geq 0.91$ Classification for Zalema variety $\geq 91.7\%$	Rodríguez-Pulido et al. (2013)
Flavanols in grape seeds (Zalema variety)	884–1717	Extractability from grounded grape seeds in methanolic solutions: $R_p^2 = 0.82$ Extractability from intact grape seeds in model wine: $R_p^2 = 0.85$	Rodríguez-Pulido et al. (2014)
Phenolic maturity in skins, seeds, and stem (Zalema variety)	884–1717	Flavanols: R_{cv}^2 up to 0.96 Phenolic acids: R_{cv}^2 up to 0.95 Kaempferol: R_{cv}^2 up to 0.98 Quercetin: R_{cv}^2 up to 0.81	Jara-Palacios, Rodríguez-Pulido, Hernanz, Escudero-Gilete, and Heredia, (2016)
Aromatic compounds (Albariño variety)	400–1000	$R^2 > 0.90$	Álvarez-Cid et al. (2015)

TSS: total soluble solids. Total phenols: mg/g skin grape (as gallic acid equivalents). Titratable acidity: g/L (as tartaric acid equivalents). R_{cv}^2: coefficient of determination (cross-validation). R_p^2: coefficient of determination (prediction). $RMSE_{CV}$: root mean squared error of cross-validation. $RMSE_p$: root mean squared error of prediction. SEP: standard error of prediction.

including factors such as different production areas and other grape varieties for developing comprehensive prediction models.

Additionally, Piazzolla, Amodio, and Colelli (2017) estimated technological maturity (TSS, TA, and pH) and total phenols in Italia grapes from HSIs (400–1000 nm). Using PLSR, the best models reported values of $R_p^2 \geq 0.70$ for the technological maturity parameters (Table 2.2).

HSI can also be used to determine grape seed maturation. For example, Rodríguez-Pulido et al. (2013) conducted a study on Zalema grapes, whose state of seed maturation was correlated via PLSR using HSIs (897–1752 nm), and obtained $R_{cv}^2 \geq 0.91$. In addition, using PCA, the authors identified six sensitive wavelengths (928, 940, 1148, 1325, 1620, and 1656 nm), which were subsequently processed through general discriminant analysis (GDA), obtaining classifications exceeding 91%. Then, the same GDA analysis was applied to the full spectrum (240 wavelengths), increasing the classification rate to 100%.

A different study also estimated flavanol contents in Zalema grape seeds (Rodríguez-Pulido et al., 2014). Therein, HSIs (884–1717 nm) were correlated via PLSR with two types of parameters, namely flavonols extracted in methanol:water (crushed seeds) and flavonols extracted in model wines (intact seeds), obtaining R_p^2 values of 0.82 and 0.85, respectively. Finally, the authors recommend including other varieties, from different production areas and different vintages, to obtain comprehensive models.

Thereafter, Jara-Palacios et al. (2016) utilized PLSR to estimate phenolic compound contents in Zalema grape skins, seeds, and stems from HSIs (884–1717 nm), obtaining R_{cv}^2 values of up to 0.98 (Table 2.2). Although the study was conducted using grape residues, the results demonstrate the potential uses of HSI in the estimation of compounds such as flavanols as well as flavonols and phenolic acids in different grape structures, with an important contribution to wine sensory profile.

2.3.2 Aromatic compounds determination

In another study, HSI (400–1000 nm) performed on intact berries was used to estimate the aromatic compound contents in Albariño grapes (Álvarez-Cid et al., 2015). Therein, an SVR model was used to associate PCA-treated spectral information with the contents of different aromatic compounds, obtaining $R^2 \geq 0.90$ values when estimating 1-octen-3-ol, 2-phenylethanol, hexyl acetate, HO-trienol, trans-pyran linalool oxide, vanillin, β-citronellol, and butyrolactone, which are typical of the Albariño variety (Vilanova, Fandiño, Frutos-Puerto, & Cancela, 2019).

The results thereof also prove the potential exhibited by HSI for estimating aromatic profiles of grapes with aromatic typicity, such as the Albariño grapes, in addition to ensuring the varietal authenticity of a specific geographical indication. This, considering the current trend of using aromatic profiles as an additional maturity indicator, along with the traditional technological and phenolic maturities.

2.4 Conclusion and future trends

When making quality wines, HSI has become an interesting alternative for adequately monitoring grape production through the development of rapid response procedures used to validate agricultural and technological parameters. However, some of the challenges that need to be resolved to attain suitable commercial applicability are as follows:

1. Optimization of existing models or creating more robust models that enable adequate monitoring of parameters at the vineyard level, including factors such as the vineyard architecture (rows); the effects of ambient light, shade, and soil; and the effects of nonphotosynthetic materials, temperature, humidity, sample placement, geographical grape origins, and vintage and ripeness levels.
2. Development of systems that quickly and efficiently process the abundant information contained in HSI. Here, the largest disadvantage is the high cost of the hardware and software required to guarantee that valuable information may not be lost.
3. Coupling of the monitoring systems to on-the-go vehicles, which may be used to obtain field information in large sampling areas.

Funding

This work was funded by the Universidad Privada del Norte (UPN), Project UPN-20201003, and National Fund for Scientific, Technological Development and Technological Innovation (FONDECYT, Peru), Project "Agricultura inteligente - Sensores ópticos de espectroscopía del infrarrojo cercano (NIR) e imagen hiperespectral combinados con inteligencia artificial para el control de la calidad interna de agroberries".

Conflicts of interest

No conflicts of interest were reported by the authors.

References

Altube, H., Cabello, F., & Ortiz, J. M. (1991). Caracterización de variedades y portainjertos de vid mediante isoenzimas de los sarmientos. *Vitis*, 30(4), 203−212. Available from https://doi.org/10.5073/vitis.1991.30.203-212.

Álvarez-Cid, M. X., García-Díaz, A., Rodríguez-Araújo, J., Asensio-Campazas, A., & de la Torre, M. V. (2015). Goal-driven phenotyping through spectral imaging for grape aromatic ripeness assessment. *Lecture Notes in Computer Science (Including Subseries Lecture Notes in Artificial Intelligence and Lecture Notes in Bioinformatics)*, 9117, 272−280. Available from https://doi.org/10.1007/978-3-319-19390-8_31.

Bellvert, J., Zarco-Tejada, P. J., Girona, J., & Fereres, E. (2014). Mapping crop water stress index in a 'Pinot-noir' vineyard: Comparing ground measurements with thermal remote sensing imagery from an unmanned aerial vehicle. *Precision Agriculture*, 15(4), 361−376. Available from https://doi.org/10.1007/s11119-013-9334-5.

Chalker-Scott, L. (1999). Environmental significance of anthocyanins in plant stress responses. *Photochemistry and Photobiology*, 70(1). Available from https://doi.org/10.1111/j.1751-1097.1999.tb01944.x.

Chaves, M. M., Zarrouk, O., Francisco, R., Costa, J. M., Santos, T., Regalado, A. P., et al. (2010). Grapevine under deficit irrigation: hints from physiological and molecular data. *Annals of Botany*, 105(5), 661−676. Available from https://doi.org/10.1093/aob/mcq030.

Cheng, Y. L., Yang, S. Q., Liu, X., Zhang, E. Y., & Song, Z. S. (2019). Identification of wine grape varieties based on near-infrared hyperspectral imaging. *Applied Engineering in Agriculture*, 35(6), 959−967. Available from https://doi.org/10.13031/aea.13452959.

Cho, M. A., Sobhan, I., Skidmore, A. K., & De Leeuw, J. (2008). Discriminating species using hyperspectral indices at leaf and canopy scales. *International Archives of the Photogrammetry, Remote Sensing and Spatial Information Sciences - ISPRS Archives*, 37, B7.

Clevers, J. G. P. W., Kooistra, L., & Schaepman, M. E. (2010). Estimating canopy water content using hyperspectral remote sensing data. *International Journal of Applied Earth Observation and Geoinformation*, 12(2), 119−125. Available from https://doi.org/10.1016/j.jag.2010.01.007.

ElMasry, G., Kamruzzaman, M., Sun, D. W., & Allen, P. (2012). Principles and applications of hyperspectral imaging in quality evaluation of agro-food products: A review. *Critical Reviews in Food Science and Nutrition*, *52*(11), 999−1023. Available from https://doi.org/10.1080/10408398.2010.543495.

Endeshaw, S. T., Sabbatini, P., Romanazzi, G., Schilder, A. C., & Neri, D. (2014). Effects of grapevine leafroll associated virus 3 infection on growth, leaf gas exchange, yield and basic fruit chemistry of *Vitis vinifera* L. cv. Cabernet Franc. *Scientia Horticulturae*, *170*, 228−236. Available from https://doi.org/10.1016/j.scienta.2014.03.021.

Espinoza, M., Gómez, E., Quispe, S., Sánchez-González, J. A., & León-Vargas, J. (2017). Physicochemical and nutraceutical characterization of sirimbache fruit (*Gaultheria glomerata* (Cav.) Sleumer). *Scientia Agropecuaria*, *8*(4), 411−417. Available from https://doi.org/10.17268/sci.agropecu.2017.04.12.

Fernandes, A., Utkin, A., Eiras-Dias, J., Silvestre, J., Cunha, J., & Melo-Pinto, P. (2018). Assessment of grapevine variety discrimination using stem hyperspectral data and AdaBoost of random weight neural networks. *Applied Soft Computing Journal*, *72*, 140−155. Available from https://doi.org/10.1016/j.asoc.2018.07.059.

Galvão, L. S., Roberts, D. A., Formaggio, A. R., Numata, I., & Breunig, F. M. (2009). View angle effects on the discrimination of soybean varieties and on the relationships between vegetation indices and yield using off-nadir Hyperion data. *Remote Sensing of Environment*, *113*(4), 846−856. Available from https://doi.org/10.1016/j.rse.2008.12.010.

Gitelson, A. A., Zur, Y., Chivkunova, O. B., & Merzlyak, M. N. (2002). Assessing carotenoid content in plant leaves with reflectance spectroscopy. *Photochemistry and Photobiology*, *75*(3), 272−281. Available from https://doi.org/10.1562/0031-8655(2002)0750272ACCIPL2.0.CO2.

Gutiérrez-Gamboa, G., Marín-San Román, S., Jofré, V., Rubio-Bretón, P., Pérez-Álvarez, E. P., & Garde-Cerdán, T. (2018). Effects on chlorophyll and carotenoid contents in different grape varieties (*Vitis vinifera* L.) after nitrogen and elicitor foliar applications to the vineyard. *Food Chemistry*, *269*, 380−386. Available from https://doi.org/10.1016/j.foodchem.2018.07.019.

Gutiérrez, S., Tardaguila, J., Fernández-Novales, J., & Diago, M. P. (2018). On-the-go hyperspectral imaging for the in-field estimation of grape berry soluble solids and anthocyanin concentration. *Australian Journal of Grape and Wine Research*, *25*(1), 127−133. Available from https://doi.org/10.1111/ajgw.12376.

Hall, A., Lamb, D. W., Holzapfel, B., & Louis, J. (2002). Optical remote sensing applications in viticulture - A review. *Australian Journal of Grape and Wine Research*, *8*(1), 36−47. Available from https://doi.org/10.1111/j.1755-0238.2002.tb00209.x.

He, H. J., & Sun, D. W. (2015). Hyperspectral imaging technology for rapid detection of various microbial contaminants in agricultural and food products. *Trends in Food Science and Technology*, *46*(1), 99−109. Available from https://doi.org/10.1016/j.tifs.2015.08.001.

He, Y., Li, X., & Deng, X. (2007). Discrimination of varieties of tea using near infrared spectroscopy by principal component analysis and BP model. *Journal of Food Engineering*, *79*(4), 1238−1242. Available from https://doi.org/10.1016/j.jfoodeng.2006.04.042.

Jara-Palacios, M. J., Rodríguez-Pulido, F. J., Hernanz, D., Escudero-Gilete, M. L., & Heredia, F. J. (2016). Determination of phenolic substances of seeds, skins and stems from white grape marc by near-infrared hyperspectral imaging. *Australian Journal of Grape and Wine Research*, *22*(1), 11−15. Available from https://doi.org/10.1111/ajgw.12165.

MacDonald, S. L., Staid, M., Staid, M., & Cooper, M. L. (2016). Remote hyperspectral imaging of grapevine leafroll-associated virus 3 in Cabernet sauvignon vineyards. *Computers and Electronics in Agriculture*, *130*, 109−117. Available from https://doi.org/10.1016/j.compag.2016.10.003.

Meggio, F., Zarco-Tejada, P. J., Núñez, L. C., Sepulcre-Cantó, G., González, M. R., & Martín, P. (2010). Grape quality assessment in vineyards affected by iron deficiency chlorosis using narrow-band physiological remote sensing indices. *Remote Sensing of Environment*, *114*(9), 1968−1986. Available from https://doi.org/10.1016/j.rse.2010.04.004.

Mirzaei, M., Marofi, S., Abbasi, M., Solgi, E., Karimi, R., & Verrelst, J. (2019). Scenario-based discrimination of common grapevine varieties using in-field hyperspectral data in the western of Iran. *International Journal of Applied Earth Observation and Geoinformation*, *80*, 26−37. Available from https://doi.org/10.1016/j.jag.2019.04.002.

Monis, J., & Bestwick, R. K. (1997). Serological detection of grapevine associated closteroviruses in infected grapevine cultivars. *Plant Disease*, *81*(7), 802−808. Available from https://doi.org/10.1094/PDIS.1997.81.7.802.

Musingarabwi, D. M., Nieuwoudt, H. H., Young, P. R., Eyéghè-Bickong, H. A., & Vivier, M. A. (2016). A rapid qualitative and quantitative evaluation of grape berries at various stages of development using Fourier-transform infrared spectroscopy and multivariate data analysis. *Food Chemistry*, *190*, 253−262. Available from https://doi.org/10.1016/j.foodchem.2015.05.080.

Naidu, R., Perry, E. M., Pierce, F. J., & Mekuria, T. (2009). The potential of spectral reflectance technique for the detection of Grapevine leafroll-associated virus-3 in two red-berried wine grape cultivars. *Computers and Electronics in Agriculture*, *66*(1), 38−45. Available from https://doi.org/10.1016/j.compag.2008.11.007.

Naidu, R., Rowhani, A., Fuchs, M., Golino, D., & Martelli, G. P. (2014). Grapevine leafroll: A complex viral disease affecting a high-value fruit crop. *Plant Disease*, *98*(9), 1172−1185. Available from https://doi.org/10.1094/PDIS-08-13-0880-FE.

Nogales-Bueno, J., Hernández-Hierro, J. M., Rodríguez-Pulido, F. J., & Heredia, F. J. (2014). Determination of technological maturity of grapes and total phenolic compounds of grape skins in red and white cultivars during ripening by near infrared hyperspectral image: A preliminary approach. *Food Chemistry*, *152*, 586−591. Available from https://doi.org/10.1016/j.foodchem.2013.12.030.

Pelsy, F., Hocquigny, S., Moncada, X., Barbeau, G., Forget, D., Hinrichsen, P., et al. (2010). An extensive study of the genetic diversity within seven French wine grape variety collections. *Theoretical and Applied Genetics*, *120*(6), 1219−1231. Available from https://doi.org/10.1007/s00122-009-1250-8.

Piazzolla, F., Amodio, M. L., & Colelli, G. (2017). Spectra evolution over on-vine holding of Italia table grapes: Prediction of maturity and discrimination for harvest times using a Vis-NIR hyperspectral device. *Journal of Agricultural Engineering*, *48*(2), 109−116. Available from https://doi.org/10.4081/jae.2017.639.

Pôças, I., Tosin, R., Gonçalves, I., & Cunha, M. (2020). Toward a generalized predictive model of grapevine water status in Douro region from hyperspectral data. *Agricultural and Forest Meteorology*, *280*(107793). Available from https://doi.org/10.1016/j.agrformet.2019.107793.

Prospere, K., McLaren, K., & Wilson, B. (2014). Plant species discrimination in a tropical wetland using in situ hyperspectral data. *Remote Sensing*, *6*(9), 8494−8523. Available from https://doi.org/10.3390/rs6098494.

Rapaport, T., Hochberg, U., Shoshany, M., Karnieli, A., & Rachmilevitch, S. (2015). Combining leaf physiology, hyperspectral imaging and partial least squares-regression (PLS-R) for grapevine water status assessment. *ISPRS Journal of Photogrammetry and Remote Sensing, 109*, 88–97. Available from https://doi.org/10.1016/j.isprsjprs.2015.09.003.

Ricketts, K. D., Gomez, M. I., Atallah, S. S., Fuchs, M. F., Martinson, T. E., Battany, M. C., et al. (2015). Reducing the economic impact of grapevine leafroll disease in California: Identifying optimal disease management strategies. *American Journal of Enology and Viticulture, 66*(2), 138–149. Available from https://doi.org/10.5344/ajev.2014.14106.

Rodríguez-Pulido, F. J., Barbin, D. F., Sun, D. W., Gordillo, B., González-Miret, M. L., & Heredia, F. J. (2013). Grape seed characterization by NIR hyperspectral imaging. *Postharvest Biology and Technology, 76*, 74–82. Available from https://doi.org/10.1016/j.postharvbio.2012.09.007.

Rodríguez-Pulido, F. J., Hernández-Hierro, J. M., Nogales-Bueno, J., Gordillo, B., González-Miret, M. L., & Heredia, F. J. (2014). A novel method for evaluating flavanols in grape seeds by near infrared hyperspectral imaging. *Talanta, 122*, 145–150. Available from https://doi.org/10.1016/j.talanta.2014.01.044.

Rowhani, A., Uyemoto, J. K., & Golino, D. A. (1997). A comparison between serological and biological assays in detecting grapevine leafroll associated viruses. *Plant Disease, 81*(7), 799–801. Available from https://doi.org/10.1094/PDIS.1997.81.7.799.

Serrano, L., González-Flor, C., & Gorchs, G. (2012). Assessment of grape yield and composition using the reflectance based water index in Mediterranean rainfed vineyards. *Remote Sensing of Environment, 118*, 249–258. Available from https://doi.org/10.1016/j.rse.2011.11.021.

Soldavini, C., Schneider, A., Stefanini, M., Dallaserra, M., & Policarpo, M. (2009). SuperAmpelo, a software for ampelometric and ampelographic descriptions in *Vitis*. *Acta Horticulturae, 827*, 253–258. Available from https://doi.org/10.17660/ActaHortic.2009.827.43.

Tao, F., Peng, Y., Gomes, C. L., Chao, K., & Qin, J. (2015). A comparative study for improving prediction of total viable count in beef based on hyperspectral scattering characteristics. *Journal of Food Engineering, 162*, 38–47. Available from https://doi.org/10.1016/j.jfoodeng.2015.04.008.

Urraca, R., Sanz-Garcia, A., Tardaguila, J., & Diago, M. P. (2016). Estimation of total soluble solids in grape berries using a hand-held NIR spectrometer under field conditions. *Journal of the Science of Food and Agriculture, 96*(9), 3007–3016. Available from https://doi.org/10.1002/jsfa.7470.

Vejarano, R., Morata, A., Loira, I., González, M. C., & Suárez-Lepe, J. A. (2013). Theoretical considerations about usage of metabolic inhibitors as possible alternative to reduce alcohol content of wines from hot areas. *European Food Research and Technology, 237*(3), 281–290. Available from https://doi.org/10.1007/s00217-013-1992-z.

Vejarano, R., Siche, R., & Tesfaye, W. (2017). Evaluation of biological contaminants in foods by hyperspectral imaging: A review. *International Journal of Food Properties, 20*, 1264–1297. Available from https://doi.org/10.1080/10942912.2017.1338729.

Vilanova, M., Fandiño, M., Frutos-Puerto, S., & Cancela, J. J. (2019). Assessment fertigation effects on chemical composition of *Vitis vinifera* L. cv. Albariño. *Food Chemistry, 278*, 636–643. Available from https://doi.org/10.1016/j.foodchem.2018.11.105.

Wellburn, A. R. (1994). The spectral determination of chlorophylls a and b, as well as total carotenoids, using various solvents with spectrophotometers of different resolution. *Journal of Plant Physiology, 144*(3), 307–313. Available from https://doi.org/10.1016/S0176-1617(11)81192-2.

Wu, D., & Sun, D. W. (2013). Advanced applications of hyperspectral imaging technology for food quality and safety analysis and assessment: A review – Part I: Fundamentals. *Innovative Food Science and Emerging Technologies, 19*, 1–14. Available from https://doi.org/10.1016/j.ifset.2013.04.014.

Xu, Y., Simon, J. E., Welch, C., Wightman, J. D., Ferruzzi, M. G., Ho, L., ... Wu, Q., et al. (2011). Survey of polyphenol constituents in grapes and grape-derived products. *Journal of Agricultural and Food Chemistry, 59*(19), 10586–10593. Available from https://doi.org/10.1021/jf202438d.

Zarco-Tejada, P. J., Berjón, A., López-Lozano, R., Miller, J. R., Martín, P., Cachorro, V., et al. (2005). Assessing vineyard condition with hyperspectral indices: Leaf and canopy reflectance simulation in a row-structured discontinuous canopy. *Remote Sensing of Environment, 99*(3), 271–287. Available from https://doi.org/10.1016/j.rse.2005.09.002.

Zhang, M., & Meng, Q. (2011). Automatic citrus canker detection from leaf images captured in field. *Pattern Recognition Letters, 32*(15), 2036–2046. Available from https://doi.org/10.1016/j.patrec.2011.08.003.

3

Use of glutathione in the winemaking of white grape varieties

Juana Martínez and Lucía González-Arenzana

Institute of Grapevine and Wine Science (Government of La Rioja, University of La Rioja and Superior Council of Scientific Investigations), La Rioja, Logroño, Spain

3.1 Glutathione: Structure and properties

Glutathione (L-glutamyl-L-cysteinyl-glycine) is a tripeptide found in grapes mostly in its reduced form (GSH). Although in low proportion, glutathione might be also in its oxidized form as disulfide (GSSG) (Fig. 3.1). GSH is synthesized in the cytosol and chloroplasts of plant cells in two sequential ATP-dependent reactions. Firstly, γ-glutamylcysteine is synthesized from L-glutamate and L-cysteine by γ- glutamylcysteine synthetase; then, GSH-synthetase catalyzes the addition of glycine to form GSH. In plants, GSH plays several physiological and biochemical roles, the most important ones being redox control, detoxification, and sulfur metabolism (Kritzinger, Bauer, & Du Toit, 2013a). Moreover, it is the most abundant nonprotein thiol in most living organisms, including the wine yeasts *Saccharomyces cerevisiae* (Lavigne, Pons, & Dubourdieu, 2007).

Due to its high antioxidant power, glutathione plays a fundamental role in preventing oxidative processes in musts and wines (Kritzinger et al., 2013a). This compound is characterized by a low redox potential ($E'_0 = 250$ mV a pH 7.0; $E'_0 = -40$ mV a pH 3.0) that provides it an important antioxidant activity, even higher than that of ascorbic acid. Glutathione in its reduced form is capable of neutralizing the ortho-quinones, which are compounds responsible for the enzymatic browning in the must. Through the −SH group, it reacts with caftaric acid, which is one of the most abundant phenolic compounds and is sensitive to oxidation, and generates the so-called GRP (2-S-glutathionylcaftaric acid), a stable and colorless compound (Cheynier, Trousdale, Singleton, Salgues, & Wylde, 1986; Singleton, Salgues, Zaya, & Trousdale, 1985). This mechanism prevents the oxidative processes that lead to browning and deterioration of the quality of musts and wines. Likewise, GSH can compete for quinones with some aromatic thiols (3MH, 3MHA, and 4-MMP) protecting the varietal aromas. Moreover, GSH can also inhibit the decrease of volatile compounds (ethyl esters and terpenes) during the preservation of wine, avoiding the deterioration of sensory quality (Papadopoulou & Roussis, 2008). The use of GSH in wines during bottling reduces the formation of sotolone and 2-aminoacetophenone, compounds responsible for

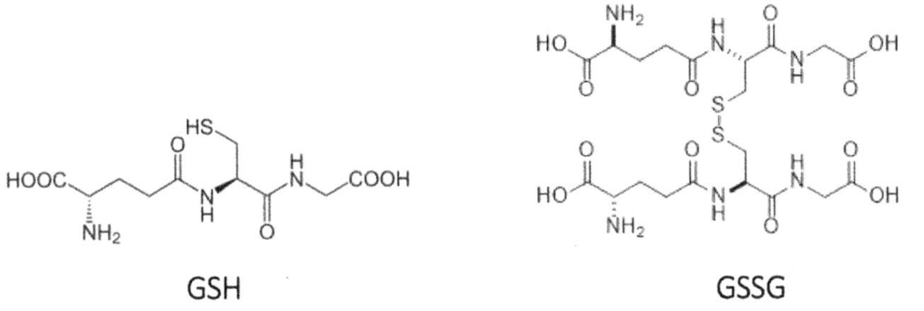

FIGURE 3.1 Molecular structures of reduced glutathione (GHS) and oxidized glutathione (GSSG).

GSH

GSSG

© 2022 Elsevier Inc. All rights reserved.

the aromas of a common aging (El Hosry, Auezova, Sakr, & Hajj-Moussa, 2009; Kritzinger et al., 2013a; Kritzinger, Bauer, & Du Toit, 2013b).

This compound also manifests beneficial effects on human health due to its antioxidant, immune, and detoxifying activity, because it is considered a powerful, versatile, and important self-generating defense molecule (Li, Wei, & Chen, 2004).

In recent years, glutathione has been the subject of numerous studies, although so far many aspects of its behavior and evolution in musts and wines are unknown.

3.2 Glutathione content in white varieties and evolution through winemaking

Glutathione concentration was first quantified in grapes in 1989 by Cheynier, Souquet, and Moutounet (1989). The glutathione levels present in berries are influenced by many factors such as variety, vintage, cultural practices, and nitrogen nutrition. During the maturation of the grape, glutathione is synthesized, and from this point, it increases significantly after veraison (Adams and Liyanage, 1993). Its accumulation in the berry is parallel to that of soluble solids until it reaches 16o Brix, and from this moment, it remains stable. Furthermore, the glutathione present in the grape is closely related to the level of nitrogen nutrition in the vineyard, with lower levels having been observed in the case of vineyards with nitrogen deficiency (Choné et al., 2006).

The levels of glutathione described in grapes are totally linked to the grape variety. Cheynier et al. (1989) conducted a study to analyze the GSH content of both the berries and the respective musts of 28 *Vitis vinifera* grape varieties. The GSH content varied from 56 to 372 μmol/kg (17 − 114 mg/kg) between grape varieties. The Sauvignon blanc variety indicated a higher GSH content than in other varieties with values of up to 114 mg/kg in grapes. They also observed high concentrations in Chardonnay (75 mg/kg) and lower in Macabeo (57.9 mg/kg) and in White Grenache (58 mg/kg). Martínez, García, and Alti (2019) analyzed the concentration of GSH in nine white grape varieties authorized in the Qualified Designation of Origin Rioja (Spain): Chardonnay, White Grenache, Malvasía, White Maturana, White Tempranillo, Turruntés, Sauvignon blanc, Verdejo, and Viura. The GSH content of the grape in the white varieties studied showed significant varietal differences (Fig. 3.2). The GSH values recorded fall within a range between 5.33 and 19.2 μg/g, with the grapes of Verdejo, Sauvignon, and Tempranillo Blanco being those with the highest levels of this compound.

After the grapes are pressed, the polyphenoloxidases (PPO) initiate the oxidation reactions of hydroxycinnamic acids and lead to the formation of the corresponding o-quinones, which can polymerize and condense with many other compounds. In musts, the glutathione plays a fundamental role in limiting enzymatic-type oxidative processes (Fig. 3.3). The glutathione reacts with caftaric acid, generating a colorless compound called GRP (2-S-glutathylcaphatic acid) that is not a substrate for subsequent oxidations by PPO, therefore reducing the enzymatic browning reactions of the mentioned caftaric acid (Singleton et al., 1985). The GRP molecule is so a powerful antioxidant that is the final product in the oxidation process of quinones, although it can be a substrate for the enzyme laccase present in grapes affected by *Botrytis cinerea* (Kritzinger et al., 2013a,b). The presence of high concentrations of GRP in the musts is an indicator of the development of oxidative processes. The

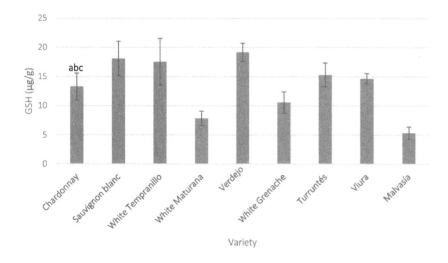

FIGURE 3.2 Glutathione (GSH) concentration (μg/g) of white (W) grape varieties in the Qualified Designation of Origin Rioja (Spain).

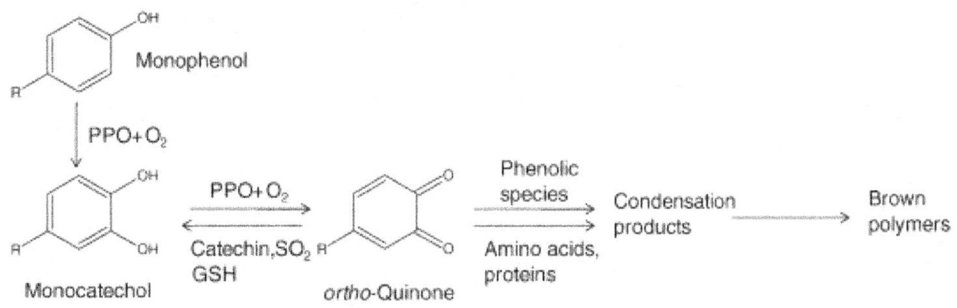

FIGURE 3.3 Enzymatic browning process in grape must (Li et al., 2008).

hydroxycinnamic acids/GSH ratio can be considered a good indicator of the susceptibility of musts to oxidation (Cheynier et al., 1986).

Levels of GSH, ranging from not detectable to 100 mg/L, have been reported in grape must (Kritzinger et al., 2013a). In Sauvignon blanc musts, the described GSH values were between 1.8 and 62 mg/L (Fracassetti et al., 2011; Janeš, Lisjak, & Vanzo, 2010; Pons, Lavigne, Darriet, & Dubourdieu, 2015). The results obtained by Martínez, García, and Alti (2019) in the musts of white varieties of the D.O.Ca. Rioja ranged from 0.84 mg/L in the White Grenache variety to 19.15 mg/L in White Tempranillo. These values corroborated the varietal differences already observed in other studies (Martínez, López, & García-Escudero, 2014). The variability in the GSH concentrations could be related to the influence of the campaign conditions. Pons et al. (2015) observed great variability in the GSH content of Sauvignon blanc musts from different vintages, noting a significant reduction in years with high temperatures and a deficit of water during ripening.

Not only in musts, but also in wines, GSH undergoes modifications due to several factors such as oxygen exposure, tyrosinase activity, and yeast strain (Du Toit, Lisjak, Stander, & Prevoo, 2007; Gabrielli, Aleixandre-Tudo, Kilmartin, Sieczkowski, & Du Toit, 2017). The winemaking technology may exert a great impact on GSH must and wine content. Depending on the quality of the grape, the length of maceration time of the must with the skins can raise the glutathione concentration considerably, since this compound is mainly located in the skins. For instance, Pons et al. (2015) examined the effect of variable maceration times (contact of the must with the grape skins) and different pressing conditions. The authors described higher GSH with an increase of 2%−55% in the first 4 hours of maceration, and they also observed greater GSH protection using inert pressing conditions.

GSH is also involved in the stress response of *S. cerevisiae* being absorbed or excreted during the alcoholic fermentation (AF, Miyake et al., 1998). The content of GSH after AF may depend on the *S. cerevisiae* strain that led the fermentation (Kritzinger et al., 2013b; Lavigne et al., 2007), although the differences observed in some studies were insignificant (Agnolucci et al., 2010). Tomašević, Gracin, Ćurko, & Kovačević Ganić (2017) observed a lower concentration of GSH in white wines fermented with *S. cerevisiae* yeasts than when the fermentation was carried out sequentially by *Torulaspora* and *Saccharomyces* yeasts or by the indigenous flora.

The existing results regarding the evolution of GSH during the AF are contradictory. For example, some authors observed an increase in the GSH concentration (Andújar-Ortiz, Pozo-Bayón, Moreno-Arribas, Martín-Álvarez, & Rodríguez-Bencomo, 2012; Lavigne et al., 2007; Martínez et al., 2014), while other authors described a decrease in GSH (Du Toit, Lisjak, Stander, & Prevoo, 2007). Kritzinger et al. (2013b) observed fluctuations in the concentration of GSH during fermentation depending on the yeast and also on the initial concentration of GSH in the must. These authors suggest the *de novo* synthesis and secretion of GSH by the yeast, as well as the need for future studies with marked GSH to elucidate its evolution in AF.

The GSH concentrations detected in white wines are usually lower than those described for musts, and they are found in the range of non-detectable to 36 mg/L. In Sauvignon blanc wines, the values vary between 1.3 and 35.5 mg/L, in Chardonnay between 4.9 and 35.5 mg/L, and in Chenin Blanc between n.d. and 13.0 mg/L (Fracassetti, Coetzee, Vanzo, Ballabio, & du Toit, 2016; Janeš et al., 2010; Marchand & de Revel, 2010; Martínez et al., 2019). Vallverdú-Queralt, Verbaere, Meudec, Cheynier, & Sommerer (2015) obtained GSH levels that were between 3.94 and 5.80 mg/L in white wines from Viognier, between 5.43 and 9.91 mg/L in rosé wines, and between 1.63 and 2.54 mg/L in red wines from Shiraz. Park, Boulton, & Noble (2000) indicated GSH contents of 5.1 mg/L in Palomino wines. The results obtained by Martínez et al. (2014) showed higher GSH contents in White Tempranillo wines (15.9−26.3 mg/L) than in those made with other white varieties from the area.

Likewise, in Chardonnay and Sauvignon blanc the values were high (21.29 and 17.97 mg/L, respectively) and similar to White Tempranillo in that study (21.23 mg/L).

During the wine preservation process, the glutathione concentration generally decreases continuously over time (Fracassetti & Tirelli, 2015; Kritzinger et al., 2013a,b), being able to be notably affected by exposure to oxygen during aging in barrels (Pons et al., 2015) and bottling (Ugliano et al., 2011).

3.3 Glutathione application during the production and preservation of white wines

In the market, there is a wide range of commercial preparations derived from dry inactive yeasts preparation (DYP) to improve the technological processes of winemaking. The treatment with GSH in must and wines and with specific inactive yeasts with guaranteed glutathione content is authorized by OIV (OIV-OENO 445−2015, OIV-OENO 446−2015, OIV-OENO 532−2017, and OIV-ENO 533−2017) to limit the oxidation of varietal aromatic compounds.

It is recommended to add these products (GSH-DYP) at the beginning of the AF bearing in mind that the level of assimilable nitrogen in the must should be enough to avoid the glutathione yeast metabolism. GSH-DYP can be implemented also during the AF fermentation. Moreover, in wines, it is also advisable to add GSH-DYP at the beginning of aging or during preservation. The maximum dose of added glutathione should not exceed 20 mg/L.

These GSH-DYP products are obtained from *S. cerevisiae* yeasts and classified into four groups depending on the production process: inactive yeasts, yeast autolysates, yeast hulls, and yeast extracts. Their use has become widespread in the wine industry since they have numerous applications depending on their composition and their ability to release some soluble compounds such as peptides, amino acids, and polysaccharides (Pozo-Bayón et al., 2009). Some of these DYP products are specific to preserve the aroma and color in white wines due to their high content of GSH, and hence why they are named GSH-DYP (Andújar-Ortiz, Pozo-Bayón, Moreno-Arribas, Martín-Álvarez, & Rodríguez-Bencomo, 2012; Kritzinger et al., 2013a,b). Studies to determine the natural thiol content in terms of GSH, free amino acids, and protein cysteine in enological dry active yeasts have been done (Tirelli, Fracassetti, & De Noni, 2010) and GSH levels ranging from 39 to 0.92 mmol/100 g were reported. Kritzinger et al. (2013b) and Andújar-Ortiz et al. (2012) tested the amount of total and reduced GSH released into synthetic wines by commercial GSH-DYP. When the preparations were added at the recommended dosage of 0.3 g/L, between 1 and 2.5 mg/L GSH was released into the synthetic wine.

The timing of the addition of GSH-DYP preparations has been the subject of different studies. In one of these studies, Salmon et al. (2012) observed a positive effect on the conservation of varietal thiols when the addition was carried out at the beginning of AF. Kritzinger et al. (2013b) showed that the amount of GSH released by DYP preparations was different depending on the moment of addition during AF, and the highest levels were reached with early additions. In Sauvignon blanc wines, a notable increase in the concentration of GSH was observed, which could not be explained solely by the release of this compound by GSH-DYP preparations. The authors suggested that other components of DYP, such as amino acids and small peptides, might favor the synthesis of GSH by yeasts during fermentation.

In addition to its use in enological processes, yeast extracts can be implemented in the vineyard, since they contain compounds that can act as elicitors, inducing defense mechanisms in plants and favoring the accumulation of secondary metabolites. In red varieties, the foliar application of GSH-DYP preparations increases the concentration of anthocyanins and tannins, depending on the *Vitis vinifera* variety, the vintage conditions, and the time elapsed between the treatment and the harvest (Giacosa et al., 2016; Tellez, Gonzalez, Garcia, Peiro, & Lissarrague, 2015). In Sauvignon blanc, Šuklje et al. (2016) observed an increase in the concentration of amino acids and glutathione in musts and wines. Likewise, they appreciated an increase in different aromatic compounds during the AF and greater stability throughout preservation. Giacosa et al. (2019) observed that its effects on skin thickness depended mainly on the maturity stage. The treatment resulted in the increase of some volatile compounds (acetates and ethyl esters) with great sensory impact on the wines. In the White Tempranillo variety, Martínez et al. (2019) appreciated an increase in the aromatic and nitrogen content of the grapes, as well as that in the fermentative aromas of wines that highlighted the tropical aromatic notes, which are characteristic of this variety.

Reports on GSH in relation to malolactic fermentation (MLF) are scarce. Rauhut et al. (2004) observed that the addition of GSH to low-pH wines promoted the growth of *Oenococcus oeni* and accelerated the speed of MLF. Marchand and de Revel (2010) evaluated the reduced, oxidized, and total glutathione content of five Merlot wines before and after MLF. Their results showed that a tendency could not be deduced from the individual

GSH or GSSG quantifications, although the reduced GSH content generally seemed to either stay constant or decrease during MLF. However, total GSH content decreased significantly during MLF (between 21% and 36% for the five samples). Further investigation regarding the influence of GSH on MLF is, however, required.

Recently, numerous studies have been carried out on the effects of applying glutathione to white wines during bottle storage with the aim of preventing oxidative processes and increasing longevity. Marchante et al. (2020) evaluated the antioxidant potential of different substances added in white and red wine, observing their ability to prevent the formation of acetaldehyde and their ability to consume oxygen. Ascorbic acid treatment showed the greatest antioxidant effect but also produced a greater increase in acetaldehyde. The addition of GSH-DYP was more effective in inhibiting 1-hydroxyethyl radicals in white wine than in red wine and did not show any difference regarding the use of pure GSH at a dose of 2 g/L. Panero et al. (2015) indicated that the addition of pure glutathione increased oxygen consumption during the first month after bottling, although they did not observe any influence on the oxidative processes or on the physical—chemical parameters of the wine after 6 months in the bottle. Tomašević et al. (2017) indicated that the combination of high levels of free sulfur and the addition of glutathione was the most effective treatment in the aromatic preservation of white wines of the Posip variety for 12 months in the bottle, concluding the existence of a synergistic relationship between these two antioxidants.

3.4 Effects of glutathione application on white wine quality

3.4.1 Influence on glutathione content

A large number of factors, including vinification technology, exposure to oxygen, and the strain of yeast used, influence the glutathione content in wines. Kritzinger et al. (2013b) showed that the amount of GSH released by DYP preparations varies depending on the time of addition during AF, reaching the highest levels with early additions. In Sauvignon blanc wines, they saw a marked increase in the concentration of GSH, which cannot be explained only from the release of this compound by GSH-DYP preparations. The authors suggested that other components of GSH-DYP, such as amino acids and small peptides, might promote the synthesis of GSH by yeasts during fermentation. On the other hand, Rodríguez-Bencomo et al. (2016) did not observe any differences in GSH content at the end of fermentation of Verdejo white wines treated with GSH-DYP with respect to control. Moreover, Webber et al. (2014) indicated that the concentration of GSH in sparkling wines was not influenced by the addition of pure GSH to the must. The results obtained by Gonzalo-Diago & Martínez (2018) showed that GSH content increased in DYP-made wines with respect to control in all varieties (Fig. 3.4), although the increase was only significant in White Tempranillo. This compound presented notable varietal differences in the wines, with higher values in White Tempranillo and White Grenache and lower in Malvasía, corroborating those results found in other studies (Martínez et al., 2019).

3.4.2 Influence on the nitrogen composition

Some authors (Kritzinger et al., 2013a,b) have indicated the contribution of the composition of GSH-DYP preparations to the worth of a large amount of nitrogenous compounds. Pozo-Bayón et al. (2009) observed an important release of free amino acids in model wines supplemented with different commercial DYP preparations. Therefore they recommended the use of DYP preparations in varieties with reduced nitrogen content.

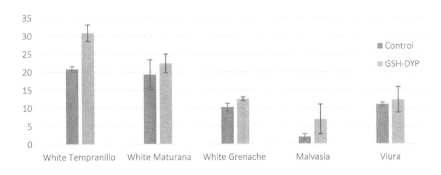

FIGURE 3.4 Content of GSH (mg/L) in different wines of white varieties, control and DYP-treated, with their standard deviation in error flags.

In Rioja wines made with GSH-DYP, the concentration of some amino acids increased significantly compared to control wines, showing important differences in the function of the grape variety. The greatest influence of the GSH-DYP treatment was observed in Malvasía wines, in which the concentration of 13 amino acids, namely asparagine, glutamine, histidine, glycine, alanine, tyrosine, cysteine, valine, tryptophan, phenylalanine, isoleucine, leucine, and proline, was increased relative to control wines. In White Tempranillo, GSH-DYP treatment increased the levels of five amino acids: glutamic acid, glutamine, citrulline, alanine, and tyrosine; in White Grenache, four amino acids: glycine, alanine, ɣ-aminobutyric acid, and cysteine; in Viura, two amino acids: asparagine and glycine; and in White Maturana, only ornithine was increased. The total concentration of amino acids increased by 37% in Malvasía wines made with GSH-DYP with respect to control, 15% in White Tempranillo, 7% in White Maturana, and less than 5% in White Grenache and Viura. Among the amino acids that were increased with the use of GSH-DYP, there were precisely the three compounds, namely glycine, glutamine, and cysteine, that make up the GSH molecule.

3.4.3 Influence on aromatic composition

Numerous authors have observed modifications in the aromatic profile of wines with the use of GSH-DYP in different varieties. Gabrielli et al. (2017) highlighted the increase of numerous volatile compounds, namely isoamyl alcohol, hexanol, 2-phenylethanol, ethyl butyrate, ethyl hexanoate, ethyl octanoate, ethyl decanoate, isoamyl acetate, and 2-phenylethyl acetate, in Sauvignon blanc wines with the addition of GSH-DYP preparations, but not in the case of addition of pure GSH. Therefore they suggested that those responsible for the modifications observed with the addition of DYP preparations are the nitrogen compounds they contained. Webber et al. (2014) found a higher content of superior alcohols (1-propanol, 2-methyl-1-propanol, 2-methyl-1-butanol, 3-methyl-1-butanol, and 2-phenylethanol) in sparkling wines with the addition of GSH in the control. Higher alcohols contribute to the aromatic complexity of wines, and their concentration is determined by the amino acid composition of musts. Cojocaru & Antoce (2019) related the addition of pure GSH in musts of the Feteasca Regala variety with the increase of aromatic complexity in wines, indicating that the effects are proportional to the dose used and are more intense when it is applied in combination with ascorbic acid. Some higher alcohols, such as 2-phenylethanol and 2-methyl-1-butanol, were in lower concentrations in wines treated with reduced glutathione, while the main ethyl fatty acid esters, such as ethyl butanoate, ethyl hexanoate, ethyl octanoate, and ethyl decanoate, were better preserved when higher concentrations of any of the antioxidants were added in must. Gonzalo-Diago & Martínez (2018) observed varietal differences in the influence of the GSH-DYP preparations on the volatile composition of the wines. These differences were more evident in the Malvasía variety, while in Viura, they were not observed. In general, a slight increase in the content of numerous volatile compounds was observed with respect to the control, as well as the different chemical families in which they were grouped; although only some ones showed statistically significant differences. In Malvasía wines, the addition of GSH-DYP increased the concentration of some alcohols (1-propanol and 1-butanol) while others (2-phenylethanol, 3-hexenol, and methionol) decreased compared to the control. The content of methionol, a compound from the nitrogen metabolism of yeasts, with aromatic notes of boiled cabbage and potato, also was decreased in White Tempranillo treated with GSH-DYP; on the contrary, it was increased in White Maturana. The concentration of acetates increased significantly in Malvasía wines made with the addition of GSH-DYP, and the rest of the varieties showed a similar trend (Fig. 3.5). These compounds provide fruity aromas

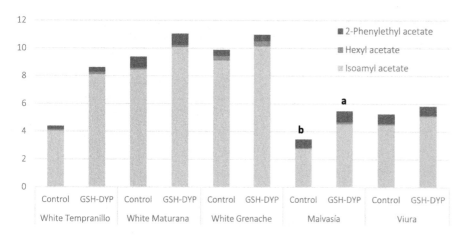

FIGURE 3.5 Acetate concentration (mg/L) in different wines of white varieties, control and DYP-treated. Different letters mean significant differences ($P < .05$).

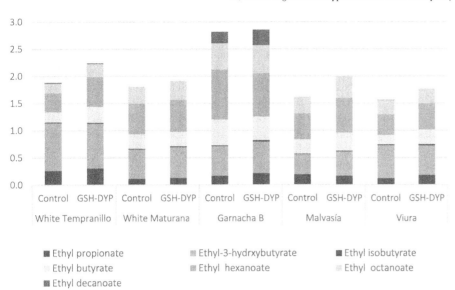

FIGURE 3.6 Ester concentration (mg/L) in different wines of white varieties, control and DYP-treated. Different letters mean significant differences ($P < .05$).

(banana, pear, apple, etc.) and have a great sensory impact on wines since they are found in concentrations that exceed the perception threshold.

Ethyl esters, characterized by the presentation of fruity and floral aromatic notes, were not influenced by the use of GSH-DYP, although the average content was higher than in control wines (Fig. 3.6). The content of ethyl lactate increased compared to control wines in White Tempranillo, White Maturana, and White Grenache varieties. The content of some acids (isobutyric, isovaleric, octanoic, and decanoic acid) and other compounds (acetoin) also underwent modifications in some varieties.

The GSH molecule can compete for quinones with some aromatic thiols (3-mercaptohexanol (3MH), 3-mercaptohexanol acetate (3MHA), and 4-methyl-4-mercaptopentanone (4-MMP)) present in musts and wines, thus protecting the varietal aromas (Kritzinger et al., 2013a). Regarding other stages, the use of GSH in wines during bottling reduces the formation of sotolone and 2-aminoacetophenone, which are the compounds responsible for the aromas of atypical aging (El Hosry et al., 2009; Kritzinger et al., 2013a,b).

As a negative aspect, GSH could also be a potential source of hydrogen sulfide (H_2S) formation from cysteine degradation (Kritzinger et al., 2013a,b). In this context, Ugliano et al. (2011) related the accumulation of SH_2 in bottled wines with the addition of GSH and Wegmann-Herr et al. (2016) observed an increase in sulfur flavors with the addition of glutathione in high doses in Sauvignon blanc musts.

3.4.4 Influence on phenolic composition

The hydroxycinnamic acids are the most abundant phenolic compounds in white wines, and they have a high tendency to oxidative browning, a process that leads to loss of color and aromas, and therefore deterioration of white wine quality. Due to its high antioxidant power, GSH plays a fundamental role in preventing oxidative processes in musts and wines.

One of the main advantages that arises with the use of GSH-DYP in enological processes is the protection against oxidation, although bibliographical results on the impacts of the addition of GSH-DYP on the phenolic composition of wines are scarce. Webber et al. (2014) observed a lower phenolic content (caffeic acid, coumaric acid, and ferulic acid) in sparkling wines made with the addition of GSH-DYP in the musts, while the levels of catechin and epicatechin were not modified in relation to control, and the color index did not vary. Pure GSH directly added at 10 mg/L to Sauvignon blanc wine at bottling significantly decreased the yellow tint of the wine during a 3-year aging period compared to the control wine (Dubourdieu & Lavigne-Cruège, 2004).

The results obtained by our research team showed that the effects of the addition of GSH-DYP preparation on the phenolic composition of the wines were very significant in Malvasía but had little incidence in other varieties. In Malvasía wines treated with GSH-DYP, the content of a large number of phenolic compounds decreased: benzoic acids (gallic, protocatechuic, and syringic acid), cynamic acids (c-caftaric, t-caftaric, t-caffeic, and t-coutaric acid), flavonols (quercetin 3-galactoside and quercetin 3-glucuronide), flavan-3-ols (epicatechin gallate), and stilbenes (t-piceid and t-resveratrol). However, the concentration of GRP increased with the addition of GSH-DYP, which would indicate that the combination of GSH provided with caftaric acid to prevent oxidation. The Malvasía variety was characterized by a higher polyphenolic content, especially hydroxycinnamic acids, so its wines are more prone to browning. The use of GSH-DYP preparations can be an interesting alternative to reduce oxidative processes in this variety.

The oxidation reactions occur primarily during grapes crushing, where the phenolic compounds are enzymatically oxidized by the Polyphenol oxidase (PPO) of the grape, but they may also occur in the wine during the chemistry oxidation (Cheynier et al., 1989). The most susceptible phenolic compounds to chemical oxidation in wine include caffeic acid and its esters, catechin, epicatechin, and gallic acid. It has been demonstrated that sufficient amounts of GSH could inhibit oxidative coloration by delaying the formation of carboxymethine-bridged (+)-catechin dimers formed in the model wine system (Sonni, Clark, Prenzler, Riponi, & Scollary, 2011). Kritzinger et al. (2013a) pointed out that future studies about the role of GSH in real aging conditions are necessary.

3.4.5 Influence on sensorial characteristics

The addition of GSH or DYP can influence the sensory characteristics of the wine, leading to a change in the wine's aroma profile. Gabrielli et al. (2017) indicated that the addition of GSH-DYP modified the sensory profile of Sauvignon blanc wines, increasing the notes of ripe tropical fruit. Andújar-Ortiz et al. (2014) did not observe significant differences in the sensory characteristics of rosé Grenache wines treated with GSH-DYP compared to the control in the early stages of wine preservation, but they did after 9 months in the bottle. The wines made with the DYP preparation enriched in glutathione had more intense fruit aromas, which could be related to the protection of some aromatic compounds in the early stages of winemaking. An evaluation of a panel of consumers did not show differences in the acceptability of both types of wines. The addition of GSH in moderate doses can highlight the fruity character in Riesling and Sauvignon blanc wines but can also cause noticeable sulfurous aromas (Wegmann-Herr et al., 2016). Nikolantonaki (2018) combined metabolomic and sensory analysis to evaluate the effect of the addition of GSH during winemaking, noting that GSH treatment impacted the sensory stability of wines depending on the harvest, which revealed the intrinsic capacity of the matrix of the wine to stop oxidation. Therefore the author suggested that the management of the early stages of the winemaking process is a key factor in estimating the aging potential of wines.

Acknowledgments

This work has been co-financed by the Government of La Rioja and FEDER Funds during 2017.

References

Adams, D. O., & Liyanage, C. (1993). Glutathione increases in grape berries at the onset of ripening. *American Journal of Enology and Viticulture*, 44(3), 333–338. < https://www.ajevonline.org/content/44/3/333 > .

Agnolucci, M., Rea, F., Sbrana, C., Cristani, C., Fracassetti, D., Tirelli, A., & Nuti, M. (2010). Sulphur dioxide affects culturability and volatile phenol production by *Brettanomyces/Dekkera bruxellensis*. *International Journal of Food Microbiology*, 143(1–2), 76–80. Available from https://doi.org/10.1016/j.ijfoodmicro.2010.07.022.

Andújar-Ortiz, I., Chaya, C., Martín-Álvarez, P. J., Moreno-Arribas, M. V., & Pozo-Bayón, M. A. (2014). Impact of using new commercial glutathione enriched inactive dry yeast oenological preparations on the aroma and sensory properties of wines. *International Journal of Food Properties*, 17(5), 987–1001. Available from https://doi.org/10.1080/10942912.2012.685682.

Andújar-Ortiz, I., Pozo-Bayón, M. Á., Moreno-Arribas, M. V., Martín-Álvarez, P. J., & Rodríguez-Bencomo, J. J. (2012). Reversed-phase high-performance liquid chromatography—fluorescence detection for the analysis of glutathione and its precursor γ-glutamyl cysteine in wines and model wines supplemented with oenological inactive dry yeast preparations. *Food Analytical Methods*, 5(1), 154–161. Available from https://doi.org/10.1007/s12161-011-9230-4.

Cheynier, V. F., Trousdale, E. K., Singleton, V. L., Salgues, M. J., & Wylde, R. (1986). Characterization of 2-S-glutathionyl caftaric acid and its hydrolysis in relation to grape wines. *Journal of Agricultural and Food Chemistry*, 34(2), 217–221. Available from https://doi.org/10.1021/jf00068a016.

Cheynier, V., Souquet, J. M., & Moutounet, M. (1989). Glutathione content and glutathione to hydroxycinnamic acid ratio in *Vitis vinifera* grapes and musts. *American Journal of Enology and Viticulture*, 40(4), 320–324. < https://www.ajevonline.org/content/40/4/320 > .

Choné, X., Lavigne-Cruège, V., Tominaga, T., van Leeuwen, C., Castagnède, C., Saucier, C., & Dubourdieu, D. (2006). Effect of vine nitrogen status on grape aromatic potential: flavor precursors (S-cysteine conjugates), glutathione and phenolic content in *Vitis vinifera* L. Cv Sauvignon blanc grape juice. *OENO One*, 40(1), 1–6. Available from https://doi.org/10.20870/oeno-one.2006.40.1.880, SE-Original research articles.

Cojocaru, G. A., & Antoce, A. O. (2019). Influence of glutathione and ascorbic acid treatments during vinification of feteasca regala variety and their antioxidant effect on volatile profile. *Biosensors*, 9(4), 140. Available from https://doi.org/10.3390/bios9040140.

Du Toit, W. J., Lisjak, K., Stander, M., & Prevoo, D. (2007). Using LC-MSMS to assess glutathione levels in South African white grape juices and wines made with different levels of oxygen. *Journal of Agricultural and Food Chemistry*, 55(8), 2765−2769. Available from https://doi.org/10.1021/jf062804p.

Dubourdieu, D., & Lavigne-Cruège, V. (2004). The role of glutathione on the aromatic evolution of dry white wine. *Wine Internet Technical Journal*, 2. < https://www.infowine.com/intranet/libretti/libretto993-01-1.pdf >.

El Hosry, L., Auezova, L., Sakr, A., & Hajj-Moussa, E. (2009). Browning susceptibility of white wine and antioxidant effect of glutathione. *International Journal of Food Science & Technology*, 44(12), 2459−2463. Available from https://doi.org/10.1111/j.1365-2621.2009.02036.x.

Fracassetti, D., Coetzee, C., Vanzo, A., Ballabio, D., & du Toit, W.J. (2016). Oxygen consumption in South African Sauvignon blanc wines: role of glutathione, sulphur dioxide and certain phenolics. *South African Journal of Enology and Viticulture, Vol 34*(No 2). (2013). https://doi.org/10.21548/34-2-1091. < https://www.journals.ac.za/index.php/sajev/article/view/1091 >.

Fracassetti, D., Lawrence, N., Tredoux, A. G. J., Tirelli, A., Nieuwoudt, H. H., & Du Toit, W. J. (2011). Quantification of glutathione, catechin and caffeic acid in grape juice and wine by a novel ultra-performance liquid chromatography method. *Food Chemistry*, 128(4), 1136−1142. Available from https://doi.org/10.1016/j.foodchem.2011.04.001.

Fracassetti, D., & Tirelli, A. (2015). Monitoring of glutathione concentration during winemaking by a reliable high-performance liquid chromatography analytical method. *Australian Journal of Grape and Wine Research*, 21(3), 389−395. Available from https://doi.org/10.1111/ajgw.12139.

Gabrielli, M., Aleixandre-Tudo, L., Kilmartin, P. A., Sieczkowski, N., & Du Toit, W. J. (2017). Additions of glutathione or specific glutathione-rich dry inactivated yeast preparation (DYP) to Sauvignon blanc must: effect on wine chemical and sensory composition. *South African Journal of Enology and Viticulture*, 38(1), 18−28.

Giacosa, S., Ossola, C., Botto, R., Río Segade, S., Paissoni, M. A., Pollon, M., & Rolle, L. (2019). Impact of specific inactive dry yeast application on grape skin mechanical properties, phenolic compounds extractability, and wine composition. *Food Research International*, 116, 1084−1093. Available from https://doi.org/10.1016/j.foodres.2018.09.051.

Giacosa, S., Rio Segade, S., Paissoni, M.A., Ossola, C., Gerbi, V., Caudana, A., ... Rolle, L.G.C. (2016). Foliar spray application of inactive dry yeast at veraison: effect on berry skin thickness, aroma and phenolic quality. In *67th ASEV National Conference: Program and Technical Abstracts* (p. 88). Monterey (CA), USA: American Society of Enology and Viticulture.

Gonzalo-Diago, A., & Martínez, J. (2018). Addition of a commercial preparation enriched in glutathione in white musts: Effect on wines aroma composition. In Macrowine (Ed.), *Abstracts Macrowine* (p. 8).

Janeš, L., Lisjak, K., & Vanzo, A. (2010). Determination of glutathione content in grape juice and wine by high-performance liquid chromatography with fluorescence detection. *Analytica Chimica Acta*, 674(2), 239−242. Available from https://doi.org/10.1016/j.ac.2010.06.040.

Kritzinger, E. C., Bauer, F. F., & Du Toit, W. J. (2013a). Role of glutathione in winemaking: a review. *Journal of Agricultural and Food Chemistry*, 61(2), 269−277. Available from https://doi.org/10.1021/jf303665z.

Kritzinger, E. C., Bauer, F. F., & Du Toit, W. J. (2013b). Influence of yeast strain, extended lees contact and nitrogen supplementation on glutathione concentration in wine. *Australian Journal of Grape and Wine Research*, 19(2), 161−170. Available from https://doi.org/10.1111/ajgw.12025.

Lavigne, V., Pons, A., & Dubourdieu, D. (2007). Assay of glutathione in must and wines using capillary electrophoresis and laser-induced fluorescence detection: Changes in concentration in dry white wines during alcoholic fermentation and aging. *Journal of Chromatography A*, 1139(1), 130−135. Available from https://doi.org/10.1016/j.chroma.2006.10.083.

Li, H., Guo, A., & Wang, H. (2008). Mechanisms of oxidative browning of wine. *Food Chemistry*, 108(1), 1−13. Available from https://doi.org/10.1016/j.foodchem.2007.10.065.

Li, Y., Wei, G., & Chen, J. (2004). Glutathione: a review on biotechnological production. *Applied Microbiology and Biotechnology*, 66(3), 233−242. Available from https://doi.org/10.1007/s00253-004-1751-y.

Marchand, S., & de Revel, G. (2010). A HPLC fluorescence-based method for glutathione derivatives quantification in must and wine. *Analytica Chimica Acta*, 660(1), 158−163. Available from https://doi.org/10.1016/j.ac.2009.09.042.

Marchante, L., Marquez, K., Contreras, D., Izquierdo-Cañas, P. M., García-Romero, E., & Díaz-Maroto, M. C. (2020). Potential of different natural antioxidant substances to inhibit the 1-hydroxyethyl radical in SO2-free wines. *Journal of Agricultural and Food Chemistry*, 68(6), 1707−1713. Available from https://doi.org/10.1021/acs.jafc.9b07024.

Martínez, J., García, S., & Alti, L. (2019). Evaluation of glutathione content in white varieties. *Vitis - Journal of Grapevine Research*, 58, 21−24.

Martínez, J., López, E., & García-Escudero, E. (2014). Efecto varietal sobre el contenido de glutation en mostos y vinos blancos. In *Proceedings XXXVII World Congress of Vine and Wine (OIV)* (pp. 102−121). Mendoza.

Miyake, T., Hazu, T., Yoshida, S., Kanayama, M., Tomochika, K., Shinoda, S., & Ono, B. (1998). Glutathione transport systems of the budding yeast *Saccharomyces cerevisiae*. *Bioscience, Biotechnology, and Biochemistry*, 62(10), 1858−1864. Available from https://doi.org/10.1271/bbb.62.1858.

Nikolantonaki, M., Julien, P., Coelho, C., Roullier-Gall, C., Ballester, J., Schmitt-Kopplin, P., & Gougeon, R. D. (2018). Impact of glutathione on wines oxidative stability: a combined sensory and metabolomic study. *Frontiers in Chemistry*, 6, 182. Available from https://doi.org/10.3389/fchem.2018.00182.

Panero, L., Motta, S., Petrozziello, M., Guaita, M., & Bosso, A. (2015). Effect of SO2, reduced glutathione and ellagitannins on the shelf life of bottled white wines. *European Food Research and Technology*, 240(2), 345−356. Available from https://doi.org/10.1007/s00217-014-2334-5.

Papadopoulou, D., & Roussis, I. G. (2008). Inhibition of the decrease of volatile esters and terpenes during storage of a white wine and a model wine medium by glutathione and N-acetylcysteine. *International Journal of Food Science & Technology*, 43(6), 1053−1057. Available from https://doi.org/10.1111/j.1365-2621.2007.01562.x.

Park, S. K., Boulton, R. B., & Noble, A. C. (2000). Automated HPLC analysis of glutathione and thiol-containing compounds in grape juice and wine using pre-column derivatization with fluorescence detection. *Food Chemistry*, 68(4), 475−480. Available from https://doi.org/10.1016/S0308-8146(99)00227-7.

Pons, A., Lavigne, V., Darriet, P., & Dubourdieu, D. (2015). Glutathione preservation during winemaking with *Vitis vinifera* white varieties: example of sauvignon blanc grapes. *American Journal of Enology and Viticulture*, 66(2), 187−194. Available from https://doi.org/10.5344/ajev.2014.14053.

Pozo-Bayón, M. A. (2009). Characterization of commercial inactive dry yeast preparations for enological use based on their ability to release soluble compounds and their behavior toward aroma compounds in model wines. *Journal of Agricultural and Food Chemistry*, 57(22), 10784−10792. Available from https://doi.org/10.1021/jf900904x.

Pozo-Bayón, M. A., Andújar-Ortiz, I., & Moreno-Arribas, M. V. (2009). Scientific evidences beyond the application of inactive dry yeast preparations in winemaking. *Food Research International*, 42(7), 754−761. Available from https://doi.org/10.1016/j.foodres.2009.03.004.

Rauhut, D., Gawron-Scibek, M., Beisert, B., Kondizior, M., Schwarz, R., Kürbel, H., & Krieger, S. (2004). Der Einfluss von S-haltigen Aminosäuren und Glutathion auf das Wachstum von *Oenococcus oeni* und die malolaktische Gärung. In Lallemand (Ed.), *Weinqualitaät und biologischer Säureabbau: Zusammenfassung der Berichte des XVIes Entretiens Scientifiques 2004* (p. 2123)). Porto, Portugal: Lallemand.

Rodríguez-Bencomo, J. J., Andújar-Ortiz, I., Sánchez-Patán, F., Moreno-Arribas, M. V., & Pozo-Bayón, M. A. (2016). Fate of the glutathione released from inactive dry yeast preparations during the alcoholic fermentation of white musts. *Australian Journal of Grape and Wine Research*, 22(1), 46−51. Available from https://doi.org/10.1111/ajgw.12161.

Salmon, J.-M., Aguera, E., Samson, A., Caille, S., Sieczkowski, N., & Julien-Ortiz, A. (2012). Apport de levures inactivées riches en glutathion en cours de fermentation alcoolique: un nouvel outil pour la protection des vins blancs et rosés contre l'oxydation. *Revue Française d'Oenologie*, 250, 2−11.

Singleton, V. L., Salgues, M., Zaya, J., & Trousdale, E. (1985). Caftaric acid disappearance and conversion to products of ensymic oxidation in grape must and wine. *American Journal of Enology and Viticulture*, 36(1), 50.

Sonni, F., Clark, A. C., Prenzler, P. D., Riponi, C., & Scollary, G. R. (2011). Antioxidant action of glutathione and the ascorbic acid/glutathione pair in a model white wine. *Journal of Agricultural and Food Chemistry*, 59(8), 3940−3949. Available from https://doi.org/10.1021/jf104575w.

Šuklje, K., Antalick, G., Buica, A., Coetzee, Z. A., Brand, J., Schmidtke, L. M., & Vivier, M. A. (2016). Inactive dry yeast application on grapes modify Sauvignon Blanc wine aroma. *Food Chemistry*, 197(Pt B), 1073−1084. Available from https://doi.org/10.1016/j.foodchem.2015.11.105.

Tellez, J., Gonzalez, V., Garcia, E., Peiro, E., & Lissarrague, J.R. (2015). Foliar application of yeast derivatives on grape quality and resulting wines. In *Technical Abstracts 66th ASEV National Conference* (pp. 143−144). Portland, Estados Unidos: ASEV National Conference.

Tirelli, A., Fracassetti, D., & De Noni, I. (2010). Determination of reduced cysteine in oenological cell wall fractions of *Saccharomyces cerevisiae*. *Journal of Agricultural and Food Chemistry*, 58(8), 4565−4570. Available from https://doi.org/10.1021/jf904047u.

Tomašević, M., Gracin, L., Ćurko, N., & Kovačević Ganić, K. (2017). Impact of pre-fermentative maceration and yeast strain along with glutathione and SO2 additions on the aroma of *Vitis vinifera* L. Pošip wine and its evaluation during bottle aging. *LWT - Food Science and Technology*, 81, 67−76. Available from https://doi.org/10.1016/j.lwt.2017.03.035.

Ugliano, M., Kwiatkowski, M., Vidal, S., Capone, D., Siebert, T., Dieval, J.-B., & Waters, E. J. (2011). Evolution of 3-mercaptohexanol, hydrogen sulfide, and methyl mercaptan during bottle storage of Sauvignon blanc wines. effect of glutathione, copper, oxygen exposure, and closure-derived oxygen. *Journal of Agricultural and Food Chemistry*, 59(6), 2564−2572. Available from https://doi.org/10.1021/jf1043585.

Vallverdú-Queralt, A., Verbaere, A., Meudec, E., Cheynier, V., & Sommerer, N. (2015). Straightforward method to quantify GSH, GSSG, GRP, and hydroxycinnamic acids in wines by UPLC-MRM-MS. *Journal of Agricultural and Food Chemistry*, 63(1), 142−149. Available from https://doi.org/10.1021/jf504383g.

Webber, V., Dutra, S. V., Spinelli, F. R., Marcon, Â. R., Carnieli, G. J., & Vanderlinde, R. (2014). Effect of glutathione addition in sparkling wine. *Food Chemistry*, 159, 391−398. Available from https://doi.org/10.1016/j.foodchem.2014.03.031.

Wegmann-Herr, Pascal, Ullrich, Sebastian, Schmarr, Hans-Georg, & Durner, Dominik (2016). Use of glutathione during white wine production - impact on S-off-flavors and sensory production. *BIO Web Conference*, 7, 2031. Available from https://doi.org/10.1051/bioconf/20160702031.

4

White must extraction methods

Juan Manuel Del Fresno[1], Mateo Cardona[2], Pablo Ossorio[3], Iris Loira[1], Carlos Escott[1] and Antonio Morata[1]

[1]enotecUPM, Chemistry and Food Technology Department, Technical University of Madrid, Madrid, Spain [2]Winery Designer and Consultant, Krean S.Coop, Gipuzkoa, Spain [3]Winemaker and Oenology Consultant, Hispano-Suizas Cellar, Valencia, Spain

4.1 Introduction

The quality of the must has special importance in white wine production. A high-quality and low turbidity grape juice determines the final quality of white wine to a large extent. In this way, when the winemaker uses aromatic grape varieties the extraction of varietal aromatic compounds and their precursors has great importance too Ribéreau-Gayon, Dubourdieu, Donèche, & Lonvaud (2006) define the following principles when trying to obtain white must with turbidity of around 200 NTU:

- Low pressing pressure.
- Limited mechanical action capable of triturating grape skins.
- Slow and progressive pressure increases.
- High volume of juice extracted at low pressure.
- Juice extraction at a temperature lower than 20°C.
- Limited crumbling and press-cake breaking during pressing.
- Minimum air contact, rapidly protected from air exposure and sulfited.

Therefore, the main objective of grape juice extraction processes is to obtain clean must with the greatest possible yield. The different processes for the production of white wine as well as the various extraction operations and the machinery currently used thereof are described in this chapter. The winemaker must select the best method according to the type of wine to be produced, taking into account the grape variety, in order to obtain a competitive market price for its product.

4.2 Traditional production process in high-volume wineries

The wineries that produce a large quantity of white wine are clearly conditioned by the time of the operations. The production process is intended to be as continuous as possible; fast extraction of white juice and high yield per kilogram of grape are necessary to obtain a profitable operation. In this process the grapes are obtained from mechanized harvest. The destemming and crushing are not critical points of the process, since these machines are able to work continuously. However, to obtain a high yield of white must, it is necessary to use the combination of dejuicer (screw drainer) and continuous pressing. This juice has high turbidity, high extraction of phenolic compounds, and high pH values. In addition, it is colored and vegetal. These characteristics make it hard to produce high-quality white wines.

The destemming/crushing and the continuous pressing are defined below as fundamental methods of juice extraction in this traditional process.

Fig. 4.1 shows the different steps in the traditional high-volume wineries.

© 2022 Elsevier Inc. All rights reserved.

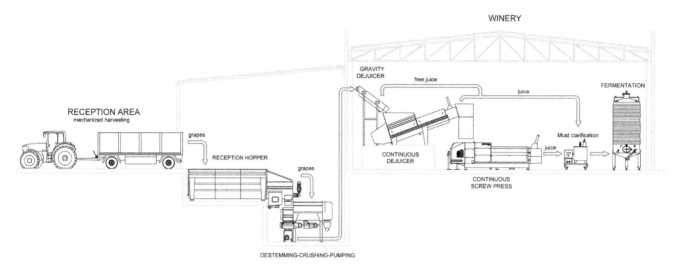

FIGURE 4.1 Scheme of the traditional production process in high-volume wineries.

4.2.1 Destemming and crushing of white grapes

The main objective of destemming is to separate the grape berries from the stalks and other vegetal matter (Catania, De Pasquale, Morello, & Vallone, 2016), and the role of crushing is to break the skin in order to release the pulp and the juice (Ribéreau-Gayon, Dubourdieu et al., 2006). Traditionally, the crushing of grapes is done after the destemming process and both operations are done together using the same machine normally.

These two operations are broadly distributed and their use presents several advantages in white wine production. The destemming before pressing improves the efficiency and the extraction of protein and phenols in the grape juice (Benitez, Castro, Natera, & Garcia-Barroso, 2005). Notably, it minimizes the uptake of phenolics and lipids from vine parts, and the leaf removal before crushing helps to limit the production of C6 aldehydes and alcohols, whose compounds result in herbaceous aromas in wines (Jackson, 2008). In terms of composition the stems have water, scarce sugars, and potassium; therefore their elimination will result in wines with better acidity and alcoholic degree. In addition, the use of these techniques minimizes the size of the press and allows the pumping of the resulting paste and the sulfiting to be done in a more homogeneous way (Ribéreau-Gayon, Dubourdieu et al., 2006).

The winemaker must know that destemming and crushing also have some disadvantages. The contact of the must with the grape skin results in the inoculation of wild yeasts from the pruina on grapes' skins. This is important when the winemaker wants to use a selected yeast strain for the alcoholic fermentation. Furthermore, these mechanical treatments cause an increase in the aeration of the must, which has special importance in partially rotted grapes. In the same way the presence of stems protects the wine's color from oxidasic casse (Ribéreau-Gayon, Dubourdieu, et al., 2006). Finally, the presence of stems improves the extraction of juice during pressing. For this reason, some wineries press in the presence of a part of the total stems. The quantity of the stems is operated from the destemmer.

The machines used to separate the berries from the stems are called destemmers (Coetzee & Lombard, 2013); these machines can have a crusher formed by rollers associated with them. Nearly all commercial destemmers currently in use are horizontally rotating destemmers (Table 4.1).

These machines are formed by two differentiated components called the beater shaft and the rotating drum (Coetzee & Lombard, 2013). The beater shaft is equipped with pins arranged in a spiral around the shaft. The drum has holes through which the grapes pass. However, the stems are retained and discharged at the end of the drum.

Table 4.1 shows the different destemmers and destemmer/crusher machines currently available in the market. All of them are horizontally rotating destemmers with the exception of the model "Delta Oscillys" by Bucher Vaslin.

TABLE 4.1 Destemmers and crushers currently available in the market.

Commercial name	Company	Production (t/h)	Type of machine	Website
DELTA OSCILLYS	Bucher Vaslin	3–20	destemmer	http://www.buchervaslin.com/
DELTA E4/F4	Bucher Vaslin	12–100	destemmer	http://www.buchervaslin.com/
COVIMAN DES15/DES50	Coviman	5–100	destemmer	https://www.coviman.es/
COVIMAN ERC15	Coviman	5–100	crusher	https://www.coviman.es/
VEGA 5–120	Magusa	3.5–120	destemmer	http://www.magusa.es/pdfs/magusa_vinicola_web.pdf
JOLLY 100	Magusa	7–12	crusher	http://www.magusa.es/pdfs/magusa_vinicola_web.pdf
EG	Magusa	4–120	crusher	http://www.magusa.es/pdfs/magusa_vinicola_web.pdf
TOP 120	Agrovin	120	destemmer/crusher	https://www.agrovin.com/
SPECIAL CRUSH	Agrovin	17–25	crusher	https://www.agrovin.com/
SOTIAG E15-E25	Sotiag	14–25	destemmer	https://www.energie-metal.fr/
SOTIAG EO8	Sotiag	6–8	destemmer	https://www.energie-metal.fr/
PERA-PELLENC H-400; H-600; H-800; H-1000	Pellenc pera oenoprocess	5–70	destemmer	https://www.perapellenc.com/
PERA-PELLENC ERAFLOIR H	Pellenc pera oenoprocess	5–70	destemmer	https://www.perapellenc.com/
CMA LUGANA III	CMA	12–20	destemmer	http://www.cmaenologia.com/
SHREDDER 1200	Egretier	50–70	destemmer	https://egretier-viticole.com/
EGRENOIR 450	Egretier	12–15	destemmer	https://egretier-viticole.com/
EGRETIER EGGRENOIR 600	Egretier	18–22	destemmer	https://egretier-viticole.com/
EGRETIER EGGRENOIR 750	Egretier	25–35	destemmer	https://egretier-viticole.com/
EGRETIER EGGRENOIR 900	Egretier	35–45	destemmer	https://egretier-viticole.com/
DELLA TOFFOLA NDC 18	Della Toffola	10–18	destemmer	https://www.dellatoffola.it/
DELLA TOFFOLA NDC 18	Della Toffola	5–10	destemmer	https://www.dellatoffola.it/
PMH VINICOLE OE-9; OE- 16; OE-25; OE-40; OE-70	PMH Vinicole	8–75	destemmer	https://www.pmh-vinicole.fr/
MORI R30; R40; R50;R60; R250	Mori	3–25	destemmer	https://www.moriluigi.com/

4.2.2 Must extraction by dejuicer and continuous pressing

Traditionally, high-production wineries use continuous drainage of the must prior to the continuous pressing of the grapes with a dejuicer (screw drainer). The stemmed and crushed grapes contain a large amount of free juice, and although the pressing cannot be substituted, the drainage process allows an important separation of this must by gravity (Razungles, 2010).

The drainer is a piece of equipment specially designed to separate the free-run juice associated with the skins through some kind of screen (Boulton, Singleton, Bisson, & Kunkee, 2013). The dejuicers are used to increase the efficiency of the drainage and to be able to carry out this process continuously. The continuous dejuicer (screw drainer) consists of a perforated inclined central cylinder that allows the juice to escape and retains the skins at the same time. The crushed grapes are moved up the cylinder by rotation of the central screw. The dejuiced grapes are dumped into a hopper for loading into a press after (Jackson, 2008).

The main advantage of the use of dejuicers is the separation of a significant fraction of the must requiring little energy, thus eliminating the need to use large-capacity presses. In spite of the drainage, the juice generally contains fewer particles and dissolved polysaccharides (Razungles, 2010); the main disadvantage of dejuicers is the additional clarification that may be required from the increased uptake of suspended solids (Jackson, 2008).

4.2.2.1 Continuous presses

The high-production wineries need the process to be as continuous as possible. In this respect, the discontinuous presses cause a great consumption of time due to the filling and emptying cycles. The continuous screw press allows uninterrupted operations (Jackson, 2008). In this type of presses, a fixed helical screw exerts pressure on a perforated wall, allowing the juice to escape. The pressed skins accumulate at the end of the cylinder, where they are periodically discharged through an exit portal.

Traditionally, these presses have been used after a continuous dejuicer, allowing to extract the remaining 30% of the juice contained in the skins (Ribéreau-Gayon, Glories, Maujean, & Dubourdieu, 2006). The screw presses provide continuous operation, which is not possible with batch presses, and most of the equipments have throughputs of between 50 and 100 Tm/h (Jackson, 2008). Therefore the high yield of juice obtained and uninterrupted operation are the main advantages of this machinery.

Ribéreau-Gayon, Glories, Maujean & Dubourdieu (2006) claim that speed is the only advantage of continuous presses. The must obtained this way is bitter, colored, and vegetal. It also has a high concentration of tannins and high pH values. These characteristics are clearly negative when trying to produce quality white wines. The screw causes considerable tearing and grinding of the outer skin tissue, thus leading to the presence of these characteristics in the juice obtained (Boulton et al., 2013). In the same way, the increase in suspended solids requires additional finning or clarification. In addition, the production of fruit esters during fermentation tends to be lower in juices obtained with these presses (Jackson, 2008). All these disadvantages have made the use of continuous screw presses to be less and less popular.

4.3 Current production process in high-volume wineries

Nowadays, there are new alternatives that allow obtaining clean must when processing high amounts of white grapes (Fig. 4.2). For this, once again, it is necessary that the machines work continuously. The use of a decanter centrifuge is a good option as a replacement for the continuous screw presses; with this type of machine, clean musts are obtained while large quantities of grapes are processed. Fig. 4.3 shows the decrease in turbidity of white must after passing through a decanter centrifuge. The use of the decanter centrifuge as a method for extracting white grape juice is defined hereafter.

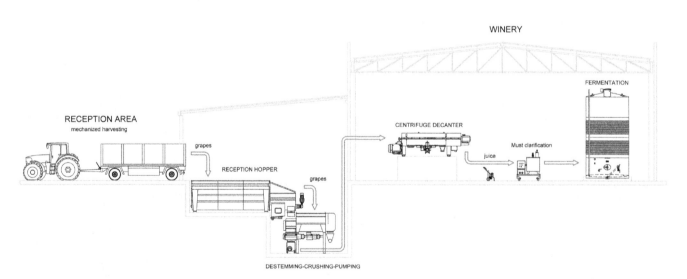

FIGURE 4.2   Scheme of the current production process in high-volume wineries.

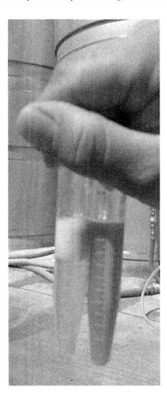

FIGURE 4.3 Treatment with decanter centrifuge; must after the treatment (left vial); must before the treatment (right vial).

4.3.1 Separation of grape must by decanter centrifuge

The method used to separate liquid and solid from crushed grapes has special importance. The collection of clean must will improve other operations in the winery, like the clarification of must. An interesting alternative is the use of decanter centrifuges, which work continuously and can be combined with other centrifuge machines to clarify the white must.

A decanter centrifuge is a versatile machine that is widespread in use in the food industry. It can successfully separate suspended solids from liquids and slurries. It is frequently used in juice processing and in olive oil extraction, especially if large amounts of olives have to be processed in a short time (Beveridge, Harrison, & McKenzie, 1988). The decanter operates mainly by sedimentation, a process causing the separation of suspended solids by virtue of their higher density than the liquid in which they are suspended (Records & Sutherland, 2001). This machine has a cylindrical bowl rotating at high speed. Inside this bowl is a scroll rotating at a slightly different speed. Centrifugal force causes the suspended solids to accumulate at the bowl wall and the clarified liquid flows along the bowl (Records & Sutherland, 2001).

From the review carried out by Beveridge et al. (1988), different advantages provided by the use of decanter centrifuge in the production of white wines can be pointed out:

1. High quality of the product obtained.
2. Higher yields in grape juice extraction. The selection of optimal parameters allows achieving an efficiency greater than 80% (w/w) (Mushtaq, 2018).
3. Productive process automation and compactness in the winery.
4. Decrease in the volume of fermentation lees, thereby lowering processing costs and losses due to lees filtration.
5. Best fermentation control due to the elimination of wild yeast.
6. Reduction of clarifying agents.
7. Reduction of the flavonoid phenol content.

In addition, the decanter provides a clean environment with little or no strange odors (Mushtaq, 2018). The main disadvantage of these machines is the high equipment and maintenance costs. In the same way, the decanter has high energy consumption and is unable to separate out biological solids having small density differences, in particular, cells and viruses (Mushtaq, 2018).

As previously indicated, the decanters are very versatile machines that can be used in a large number of production processes. Table 4.2 shows different specialized decanters for use in the wine industry. However, there are other options that allow the adaptation of the machinery for the extraction of grape juice.

4.4 Juice extraction process in high-quality white wines

The production of high-quality white wines involves new operations in the extraction process of grape juice. Fig. 4.4 shows the scheme of the production process for this type of wine. The harvest is carried out manually and the use of a selection table allows the best grapes to be chosen. These operations involve higher costs and slower grape processing. Therefore all these operations are hardly applicable in high-volume wineries.

The extraction of quality juices requires discontinuous pressing. The most popular presses in white wine elaboration are pneumatic presses. Similarly, static must clarification is used to obtain better quality juices in high-quality winemaking.

4.4.1 Pneumatic presses

Pneumatic presses are widely used in many wineries. The pressure is done by filling a membrane with compressed air. This system allows high control over the pressing and the flexibility of the models adapted to different capacities allows to treat many quantities of vintage (Darias-Martín, Díaz-González, & Díaz-Romero, 2004).

In this type of pressing, the maximum pressure on the grape is 2 bars, and its more important characteristic is the low increase in the concentration of phenolic compounds in the grape must during the pressing cycle (Ribéreau-Gayon, Dubourdieu, et al., 2006). In addition, these presses are easy to use and are standardized so they can be used with other technologies, such as in combination with nitrogen in a closed tank (Catania, Bono,

TABLE 4.2 Decanter centrifuges currently in the market.

Commercial name	Company	Type of machine	Production	Turbidity (NTU)	Website
FOODEC DECANTER	Alfa Laval	Decanter centrifuge	60 t/h	50–300	https://www.alfalaval.com/
GEA WINE DECANTER	Gea	Decanter centrifuge	3–28 m³/h	–	https://www.gea.com/
FLOTTWEG DECANTER Z3E-Z	Flottweg	Decanter centrifuge	1.8–6 m³/h	–	https://www.flottweg.com/
JUBOPRESS	Pieralisi	Decanter centrifuge	–	–	http://www.pieralisi.com/
HILLER DECAFOOD DF31–362S	MSE Hiller	Decanter centrifuge	0–9 m³/h	–	http://www.hiller-us.com/

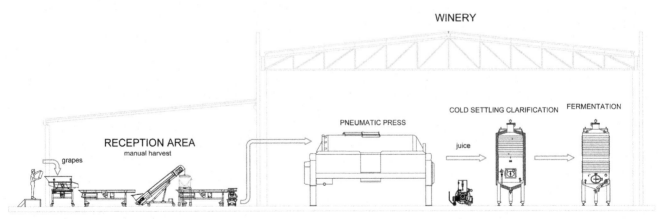

FIGURE 4.4 Scheme of high-quality white wine production.

De Pasquale, & Vallone, 2019), which prevents oxidations in the must. Nowadays, there are many models of pneumatic presses capable of maintaining high yields without reducing the quality of the juice obtained. These new models are fully programmable; the winemaker can operate them with a high number of sequences, choosing parameters such as the pressing time or the number of tank rotations, all of which allow obtaining high-quality white must with low turbidity (Ribéreau-Gayon, Dubourdieu, et al., 2006).

The main drawback of pneumatic presses is the time associated with the filling and emptying cycles that cause a pressing cycle time to be around 1 or 2 h (Jackson, 2008). In the same way, these presses can be considered too gentle (Moreno & Peinado, 2012) when the winemaker is trying to obtain high juice yields.

These presses are fed axially or through an opening at the top. The press rotates to separate the free-run juice through perforated plates. The membrane is filled by using an air compressor, and the grape mass is pressed against the perforated plates inside the press. Lower pressures exerted over a larger surface area liberate juice more quickly and with reduced release of suspended solids and phenolics (Jackson, 2008).

TABLE 4.3 Pneumatic presses currently in the market.

Commercial name	Company	Volume (hL)	Whole grape (t)	Destemmed grape (t)	Type of machine	Website
CF – Pressoir pneumatique à membrane	L.A.S.I.	15–350	–	–	Pneumatic press	https://www.lasi-italia.com/
CO – Pressoir pneumatique à membrane	L.A.S.I.	15–350	–	–	Pneumatic press	https://www.lasi-italia.com/
CHAMAX CF 12000; 4000; 6000;8000 – Pressoir pneumatique Champagne	DIEMME ENOLOGIA	40–260	2–12	–	Pneumatic press	http://www.diemme-enologia.com
CHAMAX CO 2000; 4000; 6000; 8000 – Pressoir pneumatique Champagne	DIEMME ENOLOGIA	40–150	2–8	–	Pneumatic press	http://www.diemme-enologia.com
ENOMET DHEA 32	ENOMET	32	4.8	8	Pneumatic press	https://www.enomet.it/
PE 16; 25; 35 50; 100	DELLA TOFFOLA	16–100	1.1–7	6–40	Pneumatic press	https://www.dellatoffola.it/
PF 160	DELLA TOFFOLA	160	11,12	45	Pneumatic press	https://www.dellatoffola.it/
PPC 45	ENOVENETA	45	3.3	9.5	Pneumatic press	https://www.enoveneta.it/
Pressoir 2000; 4000; 6000; 8000	OENO CONCEPT	40–150	2–8 t	–	Pneumatic press	http://www.oenoconcept.com/
Smart Press SPC 20; 30; 40; 50; 65; 80; 110; 150; 240; 320; 600	PERA-PELLENC	20–600	11.8–55	5–150	Pneumatic press	https://www.perapellenc.com/
Smart Press SPO 40; 50: 65; 80; 110; 150	PERA-PELLENC	40–150	6–40	8–30	Pneumatic press	https://www.perapellenc.com/
TECNOVA CO 15; 23; 34; 40; 50; 65;80; 115	DIEMME Enologia	15–115	0.8–7	2.8–23	Pneumatic press	http://www.diemme-enologia.com/
VELVET CF 34; 40; 50; 65; 80; 115	DIEMME Enologia	34–115	2.1–7	4.7–23	Pneumatic press	http://www.diemme-enologia.com/
VELVET XZ CF 15; 23	DIEMME Enologia	15–23	0.8–5.7	1.8–5.3	Pneumatic press	http://www.diemme-enologia.com/
XPERT 100	BUCHER VASLIN	100	6	16–20	Pneumatic press	http://www.buchervaslin.com/
XPLUS ICS 22; 30; 40; 50; 80	BUCHER VASLIN	22–80	1.3–4.8	3.5–16	Pneumatic press	http://www.buchervaslin.com/
XPRO 5; 8; 15	BUCHER VASLIN	5–15	0.3–0.9	0.8–3	Pneumatic press	http://www.buchervaslin.com/

TABLE 4.4 White juice reference values obtained by different extraction methods.

Machine	Type of extraction	Characteristics of white grape juice				
		Turbidity (NTU)	Suspended solids (%)	Total polyphenol index	Yeast assimilable nitrogen (YAN; mg/L)	Absorbance at 420 nm
Dejuicer	Strong extraction	>800	8–10	high	>200	high; dark or brown must
Decanter centrifuge	Soft extraction	150–400	2–3	medium	150 < YAN < 300	medium; medium color
Pneumatic presses	Gentle extraction	<80	<1	low	100 < YAN < 200	low; pale color

As previously indicated, the pneumatic presses are widely used in white wine wineries. This is the reason why there is a great number of models adapted to different winery sizes. The great advantage of the new pneumatic presses is the replacement of the lateral membrane by the central membrane. This characteristic significantly improves the drainage of the white juice inside the press. Also, the pressing cycle time is cut in half. Table 4.3 shows the different models of pneumatic presses currently in the market. There are presses with capacities from 5 to 600 hL of volume. In this regard, big and small wineries can use these presses to make high-quality white wines.

Regarding the main methods of white must extraction described in this chapter, Table 4.4 shows reference values of the different white grape juice obtained by these three methods. Dejuicer is a machine that allows higher yields. However, a cloudy must is obtained with turbidity values of around 800 NTU and high astringency. In the same way, brown color is characteristic of these types of musts. These parameters are incompatible with the obtention of quality white wines, and a dynamic must clarification is associated with the technique.

The best quality white musts are those obtained by using pneumatic press. These are low-turbidity musts (80 NTU) with a pale color. These juices usually only need a static clarification method that also provides high quality. The main problem with this gentle extraction is the long duration of the pressing and the low yield in must compared to that from the dejuicer machine.

An intermediate option is the use of a decanter centrifuge: turbidity values around 150–400 NTU (Table 4.4) and combination with a quick must clarification method such as flotation clarification allow obtaining must with yeast assimilable nitrogen values between 150 and 300 mg/L. This type of juice is compatible with the requirement for high yield by the big wineries.

4.5 Conclusion

The quality of a white wine depends largely on the juice extraction method. However, wineries that produce large volumes of wine need rapid and continuous production processes. Traditionally, these types of wineries used the combination of dejuicer and continuous pressing, but these machines involve the production of low-quality white must. Nowadays, efficient separation of solids and liquids can be done using a decanter centrifuge; these machines produce better quality musts and work continuously. However, the best quality white must is obtained by discontinuous operations. In this regard, pneumatic presses are the most preferred option by the wineries.

References

Benitez, P., Castro, R., Natera, R., & Garcia-Barroso, C. (2005). Effects of grape destemming on the polyphenolic and volatile content of Fino sherry wine during alcoholic fermentation. *Food Science and Technology International, 11*(4), 233–242.

Beveridge, T., Harrison, J. E., & McKenzie, D. L. (1988). Juice extraction with the decanter centrifuge—A review. *Canadian Institute of Food Science and Technology Journal, 21*(1), 43–49.

Boulton, R. B., Singleton, V. L., Bisson, L. F., & Kunkee, R. E. (2013). *Principles and ractices of winemaking.* Springer Science & Business Media.

Catania, P., Bono, F., De Pasquale, C., & Vallone, M. (2019). Closed tank pneumatic press application to improve Sauvignon blanc wine quality and nutraceutical properties. *Journal of Agricultural Engineering, 50*(4), 159–165.

Catania, P., De Pasquale, C., Morello, G., & Vallone, M. (2016). Influence of grape transport and destemming systems on the quality of Chardonnay wines. *Agricultural Engineering International: CIGR Journal, 18*(2), 260–266.

Coetzee, C. J., & Lombard, S. G. (2013). The destemming of grapes: Experiments and discrete element modelling. *Biosystems Engineering, 114*(3), 232–248.

Darias-Martín, J., Díaz-González, D., & Díaz-Romero, C. (2004). Influence of two pressing processes on the quality of must in white wine production. *Journal of Food Engineering, 63*(3), 335–340.

Moreno, J., & Peinado, R. (2012). *Enological chemistry*. Academic Press.

Jackson, R. S. (2008). *Wine science: Principles and applications*. Academic press.

Mushtaq, M. (2018). *Extraction of fruit juice: An overview. Fruit juices* (pp. 131–159). Academic Press.

Razungles, A. (2010). *Extraction technologies and wine quality. Managing wine quality* (pp. 589–630). Woodhead Publishing.

Records, A., & Sutherland, K. (2001). *Decanter centrifuge handbook*. Elsevier.

Ribéreau-Gayon, P., Dubourdieu, D., Donèche, B., & Lonvaud, A. (Eds.), (2006). *Handbook of enology, volume 1: The microbiology of wine and vinifications* (Vol. 1). John Wiley & Sons.

Ribéreau-Gayon, P., Glories, Y., Maujean, A., & Dubourdieu, D. (Eds.), (2006). *Handbook of enology, volume 2: The chemistry of wine-stabilization and treatments* (Vol. 2). John Wiley & Sons.

5

White must preservation by ultra-high pressure homogenization without SO$_2$

Iris Loira[1], Carlos Escott[1], Juan Manuel Del Fresno[1], María Antonia Bañuelos[2], Carmen González[1], Buenaventura Guamis[3], José Antonio Suárez-Lepe[1] and Antonio Morata[1]

[1]enotecUPM, Chemistry and Food Technology Department, Technical University of Madrid, Madrid, Spain
[2]enotecUPM, Biotechnology-Plant Biology Department, Technical University of Madrid, Madrid, Spain [3]Centre d'Innovació, Recerca i Transferència en Tecnologia dels Aliments (CIRTTA), TECNIO, XaRTA, Departament de Ciència Animal i dels Aliments, Facultat de Veterinària, Autonomous University of Barcelona, Bellaterra, Spain

5.1 Introduction

Phenolic compounds are the main components of grapes responsible for the antioxidant properties, both in musts and in wines. The concentration of these compounds is higher in red wines than in white wines, mainly due to two reasons: on the one hand, red grape varieties contain anthocyanins in the skin (some grape varieties also in the pulp), while white grapes do not have anthocyanins in their composition. So although the content of phenolic compounds other than anthocyanins is quite similar in the skins and seeds of white and red varieties (Rodríguez Montealegre, Romero Peces, Chacón Vozmediano, Martínez Gascueña, & García Romero, 2006), the total content of phenolic compounds is higher in red grape varieties than in white ones; on the other hand, in general, in the production of white wines, there is no maceration stage to favor the diffusion of polyphenols from the solid parts to the must (Ružić, Škerget, Knez, & Runje, 2011). The lower the phenolic compound content, the lower the antioxidant activity of the must (Fuhrman, Volkova, Suraski, & Aviram, 2001). Therefore the handling of white wine in the cellar is much more problematic, from the point of view of protection against oxidation, than that of red wine.

In white vinification, once the processing of the grapes has begun by destemming and crushing, and the must has been obtained through the pressing operation that allows the separation of the must from the solid parts of the grapes, the oxidation processes begin due mainly to the action of the polyphenol oxidase enzymes (e.g., tyrosinase, laccase) and peroxidase (Oliveira, Ferreira, De Freitas, & Silva, 2011). In this enzymatic oxidation process, hydroxycinnamic acid esters (hydroxycinamates such as caffeoyl tartaric acid and *p*-coumaroyl tartaric acid) have been shown to play a crucial role by transforming into *o*-quinones that continue to polymerize and/or react with other compounds in the must, including flavan-3-ols, to provide compounds with a browner color (Oliveira et al., 2011; Recamales, Sayago, González-Miret, & Hernanz, 2006). Therefore the addition of some antioxidant agent such as sulfur dioxide (SO$_2$) is usually desirable in the initial stages of processing. This enological additive provides three major properties to preserve the quality of musts and wines, among other effects, namely antimicrobial activity, antioxidant activity, and antioxidasic activity. This last function represents a great advantage when it comes to protecting white musts from enzymatic browning.

Despite the multifunctional character of SO$_2$ in enology, its use has been highly questioned in recent years due to its capacity to trigger allergic reactions in sensitive consumers and its negative influence on the aroma and color

White Wine Technology.
DOI: https://doi.org/10.1016/B978-0-12-823497-6.00014-4

© 2022 Elsevier Inc. All rights reserved.

of wine when used in excessive concentrations (García Martín & Sun, 2013). Therefore the maximum concentration of SO_2 in wines is delimited by European regulations (EU Regulation No. 606, 2009), and these limits are currently 150 mg/L for dry red wine and 200 mg/L for dry white and rosé wines. In addition, consumers increasingly prefer more natural and chemical-free products. Overall, in the current situation resulting from climate change, where many winemaking processes start with higher pHs in the must, the effectiveness of SO_2 is limited. For this reason, in recent years, much research has been carried out on possible alternatives to the use of SO_2 in wineries, based on the application of both physical and chemical techniques (Giacosa et al., 2018; Lisanti, Blaiotta, Nioi, & Moio, 2019; Morata et al., 2017; Morata, Loira, et al., 2020; Santos, Nunes, Saraiva, & Coimbra, 2012; Yildirim & Darıcı, 2020); but until now, no effective substitute has been found for its three main functions (antioxidant, antioxidasic, and antimicrobial). Among these technological alternatives to reduce the use of SO_2 in the winery are High Hydrostatic Pressures (Bañuelos et al., 2016; Christofi, Malliaris, Katsaros, Panagou, & Kallithraka, 2020; Morata, Loira, et al., 2015), UV radiation and pulsed light (Escott et al., 2017; Falguera, Forns, & Ibarz, 2013; Santamera et al., 2020), ionizing radiation (Błaszak, Nowak, Lachowicz, Migdał, & Ochmian, 2019; Morata, Bañuelos, et al., 2015), and pulsed electric fields (Morata, Bañuelos et al., 2020; Puértolas, López, Condón, Raso, & Álvarez, 2009). Among the chemical methods, the most studied were the use of organic acids with antioxidant or antifungal properties (such as ascorbic and sorbic acids) (Cojocaru & Antoce, 2012), nisin (Fernández-Pérez et al., 2018; Ruiz-Larrea et al., 2007) and lysozyme (Ancín-Azpilicueta, Jiménez-Moreno, Moler, Nieto-Rojo, & Urmeneta, 2016; Gerland, 2002; Sonni, Bastante, Chinnici, Natali, & Riponi, 2009), both with antibacterial activity, and dimethyl dicarbonate (DMDC) (Ancín-Azpilicueta et al., 2016; Costa, Barata, Malfeito-Ferreira, & Loureiro, 2008; Threlfall & Morris, 2002), a synthetic chemical compound with antimicrobial properties. Recently, the use of silver (Garde-Cerdán et al., 2014; Izquierdo-Cañas, García-Romero, Huertas-Nebreda, & Gómez-Alonso, 2012; Moreno-Arribas et al., 2019) both in its colloidal state and in the form of nanoparticles has also been considered to replace totally or partially the use of sulfur, thanks to its high broad-spectrum antimicrobial power. In this line of research, the main efforts have been aimed at replacing the antimicrobial capacity of SO_2 (Lisanti et al., 2019).

Recently, Ultra High Pressure Homogenization (UHPH) must treatment technology has proven to be highly effective in antimicrobial and oxidation protection functions (Bañuelos et al., 2020; Loira, Morata, Bañuelos, et al., 2018). This chapter focuses on showing the benefits and effectiveness of this emerging nonthermal technique for the preservation of white grape must and proving that it a feasible alternative to the use of SO_2 in enology.

5.2 Origin and basis of UHPH technology

Homogenization is a technique traditionally used in the food industry to obtain uniformity in the composition of the food by producing a modification in the size of the particles that form it as a result of pressurization when passing through a narrow valve. The first patent for food homogenization equipment dates from 1899, and the invention was intended for the treatment of milk (Dhankhar, 2014). While conventional homogenization, widely used in the dairy industry, has a range of activity below 100 MPa, in UHPH the range is 200 to 400 MPa, or even higher. However, the design principles are similar in both techniques. The industrial development of this type of equipment has evolved and been perfected over the years, increasing the pressure and decreasing the gap size of the valve. Currently, UHPH equipment is produced for industrial-scale performance with processing capacities of up to 10,000 L/h Table 5.1. The Spanish company Ypsicon Advanced Technologies SL owns the International patent, as well as the exploitation rights, of this UHPH equipment (www.ypsicon.com) Fig. 5.1.

TABLE 5.1 UHPH and microfluidization equipment on the market for industrial-scale processing of liquid foods.

Company	Models	Working pressures (MPa)	Processing capacities (L/h)
Stansted Fluid Power Ltd (Essex, UK)	FPG11300 Currently unavailable -UHPH	400	375
Ypsicon Advanced Technologies SL (Barcelona, Spain)	-UHPH	300–400	10,000
Avestin Inc (Ottawa, Canada)	ErnulsiFlex-C500 & Ernulsiflex-C1000 -Microfluidization	200	500–1000

FIGURE 5.1 The UHPH industrial equipment of the Ypsicon company.

These UHPH equipment are composed of a system of pumps that provide the working pressure to mobilize the food at high speeds and drive it through the homogenizing valve. This homogenizing valve is one of the most important parts of the equipment since the effectiveness of the treatment depends on its design. It is usually made of tungsten carbide or special materials, both capable of withstanding the high pressures and efforts generated. Tungsten carbide is the hardest metal in existence, with a hardness between 8.5 and 9.5 on the Mohs scale. There are also valves made of ceramic materials, highly resistant to ultra-high pressures (Zamora & Guamis, 2015). When passing through this narrow section valve at a high speed, cavitation and turbulent flow phenomena are created, as well as high impact and shearing forces (Desrumaux & Marcand, 2002), resulting in a very rapid rise in temperature. However, this temperature increase is instantaneous since, as the valve passes through, the temperature quickly drops due to the adiabatic expansion phenomenon when depressurization occurs at atmospheric pressure. One of the most immediate consequences of passing through this valve under these conditions is that all the food particles are fragmented to sizes of the order of the nanoscale. Together with the sudden decompression at the valve outlet, the sum of all these effects is the basis on which the UHPH technique achieves the reduction of the present microbial load to sterilization values, i.e., a total elimination of microorganisms is achieved (Morata, Loira, et al., 2020). So far, the only limitation found to its versatility is that only liquids with an average particle size below 500 µm can be processed, in order to avoid homogenization valve clogging. Recently, UHPH technology has been approved by the International Organisation of Vine and Wine (OIV) as an authorized practice in winemaking (Resolution OENOMICRO 16−594B).

5.3 Overview of UHPH employment in different food matrices treated by UHPH

UHPH technology has found a wide field of application in the dairy industry, where its effectiveness as a technique to achieve enzymatic and microbiological inactivation has been verified (Pereda et al., 2008) Table 5.2. With treatments of 250−300 MPa and the temperature increases associated with the UHPH process, complete inactivation of the milk's natural enzymes, plasmin, lactoperoxidase, and alkaline phosphatase is achieved (Hayes, Fox, & Kelly, 2005; Pereda et al., 2008; Picart et al., 2006). Several authors indicate different mechanisms of action of UHPH treatment that justify this enzymatic inactivation, such as temperature increase during processing (Datta, Hayes, Deeth, & Kelly, 2005; Hayes et al., 2005), the mechanical forces generated (Datta et al., 2005; Picart et al., 2006), or the reduction in the size of the milk fat cell, which involves the destruction of its protein membrane, to which these enzymes are attached (Pereda, Ferragut, Quevedo, Guamis, & Trujillo, 2007). In addition, considering other effects on the protein fraction of milk, UHPH decreases the size of the casein micelles and denatures the whey proteins (Kheadr, Vachon, Paquin, & Fliss, 2002; Thiebaud, Dumay, Picart, Guiraud, & Cheftel, 2003), the latter being characterized as thermosensitive.

With regard to microbiological inactivation, reductions in the total bacterial population of 3 logarithmic cycles are achieved when raw cow's milk is treated with 250−300 MPa and a maximum temperature at the valve of 78°C −84°C (Hayes et al., 2005; Thiebaud et al., 2003). However, according to the study published by Pinho et al., (Pinho, Franchi, Tribst, & Cristianinia, 2011), the elimination of bacterial spores is not guaranteed in skimmed milk subjected to treatments of up to 300 MPa, even with milk entry temperatures of 45°C.

At the physicochemical level, the application of UHPH treatment does not significantly affect the color of milk (Hayes et al., 2005), although some authors have reported significant reductions in L* clarity (Pereda et al., 2007),

TABLE 5.2 Examples of effects achieved with the application of UHPH technology on different foods.

Food product	Working pressure (MPa)	Maximum temperature reached (°C)	Effect achieved	Suggested cause	Source
Hondarribi Zuri grape must	300	98	Total elimination of grape microorganisms, including yeast, fungi, and bacteria The UHPH-treated must had a lighter appearance (10%) before fermentation		Loira, Morata, Bañuelos, et al. (2018)
Muscat grape must	300	78	Total elimination of grape microorganisms No fermentation activity in the must for more than 60 days No thermal damage (5-hydroxymethylfurfural was not detected) Preservation of varietal aromas (terpenes)		Bañuelos et al. (2020)
3% solution of soy protein (globulin 115)	>200	86	Improvement of the emulsifying capacity of soya protein	Mechanical forces and high temperatures led the value	Floury, Desrumaux, & Legrand (2002)
Grape and orange juice contaminated with 10^7 cfu/mL of *Listeria monocytogenes* and *Salmonella enterica*	400	74	Partial inactivation of *Listeria monocytogenes* (reduction of approximately 5 log cfu/mL) and total inactivation of *Salmonella enterica*	Mechanical effect and presence of phenolic compounds in the case of grape juice	Velázquez-Estrada et al. (2011)
Whole milk contaminated with 10^5–10^6 cfu/mL of *Listeria innocua*	200		Reduction of casein micelles size (from 200–300 nm in pasteurized milk to 125–150 nm in pressurized milk; increased solubility)	Mechanical effect	Kheadr et al. (2002)
Whole milk contaminated with 10^5–10^6 cfu/mL of *Listeria innocua*	200		*Listeria innocua* population decline between 3–4 log cycles		Kheadr et al. (2002)
Raw cow's milk	300	78	Decrease in the emulsifying capacity of whey proteins	Shearing forces and temperature rise	Thiebaud et al. (2003)
Raw cow's milk	300	78	Inactivation of native microorganisms up to 3 log cycles		Thiebaud et al. (2003)
Raw whole cow's milk	200	>71	Total lipase inactivation	Thermal effect	Datta et al. (2005)
Raw whole cow's milk	200–250	80–84	Reduction of plasmin activity by 90–95%	Mechanical effect	Datta et al. (2005); Hayes et al. (2005)
Raw cow's milk	250	84	Reductions in total mesophilic bacteria count of up to 3 log cycles (milk preheated to 45°C) Total elimination of *Staphylococcus aureus*, coliforms, and psychrotrophic bacteria		Hayes et al. (2005)
Raw whole milk of high microbiological quality (<4 × 10^4 cfu/mL)	200–300	60–75	Partial or total inactivation of alkaline phosphatase	Mechanical forces (shearing, cavitation, and/or impact) on the homogenizing valve	Picart et al. (2006)
Raw whole milk of high microbiological quality (<4 × 10^4 cfu/mL)	200–300	60–75	Decrease in population of native microorganisms between 2 and 3 log cycles	<200 MPa Mechanical effect >250 MPa Mechanical effect and thermal effect	Picart et al. (2006)
Raw cow's milk	300	97–103	Total inactivation of lactoperoxidase	Thermal effect	Pereda et al. (2007)

(*Continued*)

TABLE 5.2 (Continued)

Food product	Working pressure (MPa)	Maximum temperature reached (°C)	Effect achieved	Suggested cause	Source
Raw cow's milk	203–303	78–103	Total alkaline phosphatase inactivation	Thermal effect Shear force Destruction of the membrane by reducing the size of the fat cell (the enzyme is attached to the membrane)	Pereda et al. (2007)
Raw cow's milk	300	103	Reduction of clarity (L* in CIELab coordinates) in milk color	Mechanical effect	Pereda et al. (2007)
Raw cow's milk	300	103	Increased milk viscosity	Mechanical effect	Pereda et al. (2007)
Cow's milk	200–300	54–94	Shorter coagulation time and greater firmness of the curd in cheese making		Zamora, Ferragut, jaramillo, Guamis, & Trujillo, (2007)

without these changes being noticeable to the human eye. The absence of significant thermal damage was confirmed by the lack of thermal indicators after processing, such as lactulose (Pereda, Ferragut, Quevedo, Guamis, & Trujillo, 2009). Thus UHPH is a technique for preserving milk that does not compromise its sensory quality (Pereda et al., 2008).

Its application in the fruit juice–processing industry has also provided important technological outcomes. By treating apple juice with homogenization pressures of 200 and 300 MPa, total inactivation of yeasts, molds, and bacteria is achieved, with initial populations of between 2 and 4 log cfu/mL (Suárez-Jacobo, Gervilla, Guamis, Roig-Sagués, & Saldo, 2010). In a later study, the same authors observed that UHPH has a protective effect on polyphenolic compounds and vitamin C content in apple juice (Suárez-Jacobo et al., 2011). Therefore UHPH can inactivate the polyphenol oxidase enzymes responsible for enzymatic oxidation and preserve a paler color in the juices. In the case of treatment at 400 MPa of orange juice contaminated with a population of 10^7 cfu/mL of *Listeria monocytogenes* and *Salmonella enterica*, microbiological reductions of 5 and 6 log cycles, respectively, were achieved. Applying the same processing protocol on grape juice, reduction of the initial *L. monocytogenes* population to 10 cfu/mL and total elimination of *S. enterica* were achieved, with the antimicrobial contribution of hydroxycinnamic acids from the grape.

UHPH technology has also been tested on other foods such as soy milk (Cruz et al., 2007; Poliseli-Scopel, Hernández-Herrero, Guamis, & Ferragut, 2012; Poliseli-Scopel, Hernández-Herrero, Guamis, & Ferragut, 2013) and the almond milk (Briviba, Gräf, Walz, Guamis, & Butz, 2016; Ferragut, Valencia-Flores, Pérez-González, Gallardo, & Hernández-Herrero, 2015), all of which have advantages associated with nonthermal processing. In the case of the treatment of soybean milk with 200–300 MPa, the bacterial populations were reduced by 2–4 log cycles and the spores by about 2 log cycles (Cruz et al., 2007).

5.4 Must sterilization

In the particular case of the application of this UHPH technique on musts, the must particles are fragmented to up to 100–300 nm, including microorganisms in their vegetative and sporulated forms. This fragmentation of the must particles can be observed under an electron microscope as a cloud of fine dust Fig. 5.2. In addition, breaking up plant cells favors greater extraction of nutrients such as yeast assimilable nitrogen (YAN). YAN is one of the most important nutrients for yeasts as its shortage can cause a deficient development of the fermentation, and it is mainly formed by the ammonium ion NH^{4+} and the nitrogen α-amino from the free amino acids. Thus musts treated with UHPH are more nutritious. Oxidizing enzymes are also inactivated in this process.

One of the great advantages provided by UHPH technology over conventional heat treatments is that it allows sterility to be obtained in foods without significantly increasing the temperature. When a must of the Spanish white grape variety Hondarrabi Zuri (*Vitis vinifera* L.) was treated with UHPH (300 MPa), complete elimination of the yeast and aerobic bacteria present in the must, whose initial population was 1×10^6 and 7×10^3 cfu/mL, respectively,

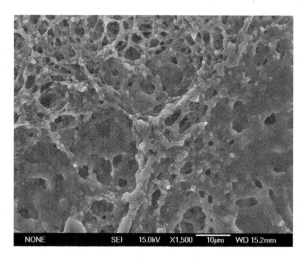

FIGURE 5.2 Scanning electron microscope image of a white must processed by UHPH and filtered through cellulose cartridges with a pore size of 0.22 μm.

was achieved (Loira, Morata, Bañuelos, et al., 2018). Thus in the absence of subsequent external microbiological contamination, there will be no problems of the undesired start of spontaneous fermentations or organoleptic deviations in the wine due to the development of uncontrolled malolactic fermentation or acetic acid bacteria spoilage.

By applying 300 -MPa treatments, it has been observed that *Staphylococcus aureus* is more resistant than *Yersinia enterocolitica* (Diels, Wuytack, & Michiels, 2003). *Streptococcus lactis* has also been shown to be more resistant to high pressure homogenization than *Bacillus subtilis*, indicating that the morphology of the bacteria (coccus or bacillus) is another factor determining resistance to high-pressure treatment (Popper & Knorr, 1990).

UHPH technology does not cause sublethal damage to vegetative and sporulated forms like other processing techniques for microbiological control but is 100% effective in removing yeasts and bacteria present in the must. In addition, it can also destroy bacterial spores when suitable temperatures are reached at the homogenization valve (Morata, Loira, et al., 2020).

In short, the total elimination of the grape's natural microorganisms allows greater control in the development of the vinification with the selected microorganisms (yeast and bacteria) and the chosen fermentation biotechnology, from mixed or sequential inoculations between *Saccharomyces cerevisiae* and non-*Saccharomyces* to yeast/bacteria co-inoculations.

5.5 Oxidation control

The enzyme tyrosinase (the main polyphenol oxidase naturally present in healthy grapes) catalyzes the oxidation reaction of ortho-diphenols to ortho-quinones (catecholase activity), in addition to the conversion of monophenols present in the must to ortho-diphenols, also through an oxidative process (cresolase activity). These ortho-quinones formed have a high redox potential and trigger oxidation reactions of other compounds in the must (including the reducing agents ascorbic acid and SO₂), finally giving rise to the appearance of browner shades, characteristic of oxidized white must (Ribéreau-Gayon, Dubourdieu, Donèche, & Lonvaud, 2006). This enzymatic browning of the must is usually prevented by the use of SO₂ in the prefermentation stages.

Several recent studies have verified the capacity of the UHPH technique to denature and inactivate enzymes as a result of the aggressive physical conditions to which the must is subjected and which cause this strong molecular breakdown. Compared to the polyphenol oxidase activity found in the same SO₂-treated must, the processing of the must by UHPH decreased this enzymatic activity by 90% (Loira, Morata, Bañuelos, et al., 2018). In addition, the antioxidant capacity of UHPH-treated musts was recently found to be 1.5 times that of the same untreated control must (Bañuelos et al., 2020). In laboratory experiments where both the UHPH-treated must and the untreated control were kept under conditions of high oxygen exposure (about 1 cm²/mL) and without the presence of SO₂, it was possible to visually verify the absence of enzymatic oxidation due to the total elimination of polyphenol oxidase enzymes in the UHPH-treated must. So UHPH technology could effectively replace SO₂ in its antioxidant function, thus establishing itself as a suitable methodology for color stabilization in white wine.

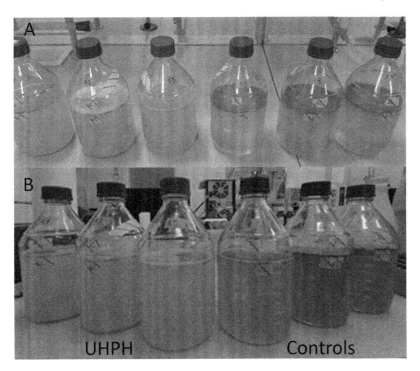

FIGURE 5.3 Visual observation of enzymatic browning in control wines during vinification. (A) comparison between the visual aspect of the control musts (three ISO bottles on the right side) and those treated by UHPH (three ISO bottles on the left side) after the first minute from the start of the fermentation test; (B) comparison between the visual aspect of the control musts (three ISO bottles on the right side) and those treated by UHPH (three bottles on the left side) after the first day from the start of the fermentation test.

This protection against oxidation can be observed during fermentation Fig. 5.3. It is maintained during the month following fermentation, in which control wines without SO_2 addition are oxidized again after clarification (surface browning), while the color of wines from UHPH-treated must remains unchanged.

Therefore UHPH technology offers protection of the must at the sensory level due to enzyme inactivation, which also allows for prolonging its shelf life by keeping the polyphenolic fraction protected against this deterioration pathway.

5.6 Improvement in the implantation of non-*Saccharomyces* yeasts

The UHPH technology, besides providing a high antimicrobial effect, has a very interesting characteristic for its implementation at the industrial level, since it allows the continuous processing of the must to achieve its sterilization in a short time. This is an enormous advantage for microbiological control in the winery, since the implantation of any inoculated microorganism could be guaranteed, including non-*Saccharomyces* yeasts, many of which are characterized by their low resistance to SO_2 and their low tolerance to ethanol (Loira, Morata, Bañuelos, & Suárez-Lepe, 2019). These characteristics make it very difficult to obtain correct implantation in an untreated must, where they have to compete with species much better adapted to the must composition and processing conditions, such as *S. cerevisiae*.

In recent years, the use of this type of yeast has been greatly intensified with the aim of seeking to distinguish and adapt the wines to the new demands of the consumers. In this regard, it is important to make a correct selection of yeasts in order to find those strains that can provide interesting characteristics to wines such as an increase in acidity (*Lachancea thermotolerans*) (Vaquero et al., 2020), an improvement in fresh aroma (*Hanseniaspora spp.*) (Morata, Escott, et al., 2020), greater microbiological stability by consuming malic acid (*Schizosaccharomyces pombe*) (Loira, Morata, Palomero, González, & Suárez-Lepe, 2018), or an increase in glycerol content (*Starmerella bombicolla*, formerly *Candida stellata*) (Ciani, Ferraro, & Fatichenti, 2000), among many others.

5.7 Deseasonalization of wine production

The effectiveness of must sterilization was verified by monitoring the daily weight variation of ultra high pressure homogenized white musts (Bañuelos et al., 2020; Loira, Morata, Bañuelos, et al., 2018), with no change recorded during the months following UHPH treatment. This represents a significant technological and industrial advantage by ensuring the availability of sterile raw material for winemaking at times of the year other than the harvest season.

5.8 Sensory impact of the use of UHPH

As it is a nonthermal technique, it allows the sensory properties of the treated product, in this case must, to be preserved practically intact. Thus by treating the must with UHPH, there are no significant losses of aroma or changes in color caused by oxidation or taste degradation, all of which are common modifications in conventional thermal treatments where heating of the product is essential. The phenolic fractions, which are so important for their antioxidant function and the wine's ability to age, are also protected.

The residence time in the valve is 0.2 s so that exposure to temperatures between 50°C and 100°C, depending on the inlet temperature of the product, is very brief, and therefore the repercussion at a sensorial level of this instantaneous increase in temperature is minimal. Recent studies have shown that the application of UHPH on the must of the Muscat variety does not destroy terpenes or produce 5-hydroxymethylfurfural, a well-known thermal indicator in the food industry (Bañuelos et al., 2020). The latter compound confirms the absence of thermal degradation during UHPH processing.

The nutritional enrichment of the must with UHPH treatment, mainly in nitrogen, means that the aromatic profile of the wines obtained is richer in esters and lower in higher alcohols compared with the untreated and sulfited must (Loira, Morata, Bañuelos, et al., 2018). So the UHPH treatment of the must favors both the preservation of the varietal aroma and the development of the fermentative aroma by increasing the amount of nutrients available in the must for yeast metabolism.

With regard to the effect of high pressure homogenization on the color of white wine, wines obtained from treated musts showed a paler color than wines from untreated musts (Morata, Loira, et al., 2020). These results are in line with the ability of this technique to inhibit the action of polyphenol oxidase enzymes.

In relation to the processes of obtaining cleanliness in the wine, at an experimental level, no major difficulties have been found when clarifying or filtering a wine obtained from must treated with UHPH, despite maintaining a greater quantity of particles in suspension with sizes between 100 and 300 nm (Bañuelos et al., 2020). Immediately after fermentation, these wines from UHPH-treated must remain cloudy, i.e., the natural decanting process promoted by the force of gravity is not started, but these nanoparticles are susceptible to elimination by chemical clarification.

5.9 Conclusion and future prospects

UHPH is a continuous pressurization technique that allows sterility to be obtained in musts, not only without negatively affecting their nutritional and sensory quality but also contributing to increasing the availability of nitrogen for microorganisms by favoring the correct development of fermentation and the greater presence of fermentation esters in the aromatic profile of the wine. In addition, UHPH inactivates the enzymes responsible for enzymatic browning, and pale colors in white wines are better preserved. All these effects give a longer microbiological and physicochemical shelf life to the musts of white grape varieties. The high dependence on SO$_2$ that existed until now in white winemaking is therefore reduced or eliminated.

However, there is still much research to be done regarding its potential application in winemaking. For example, it remains to be verified whether it is useful for breaking down cell structures when processing fluid biomass of yeast and thus accelerating the development of aging on lees. It would also be interesting to evaluate its ability to prevent protein breaks, as well as its ability to protect the varietal aromas typical of other grape varieties (e.g., pyrazines and norisoprenoids).

References

Ancín-Azpilicueta, C., Jiménez-Moreno, N., Moler, J. A., Nieto-Rojo, R., & Urmeneta, H. (2016). Effects of reduced levels of sulfite in wine production using mixtures with lysozyme and dimethyl dicarbonate on levels of volatile and biogenic amines. *Food Additives and Contaminants - Part A Chemistry, Analysis, Control, Exposure and Risk Assessment*, 33(10), 1518–1526. Available from https://doi.org/10.1080/19440049.2016.1232864.

Bañuelos, M. A., Loira, I., Escott, C., Del Fresno, J. M., Morata, A., Sanz, P. D., & Suárez-Lepe, J. A. (2016). Grape processing by high hydrostatic pressure: Effect on use of non-saccharomyces in must fermentation. *Food and Bioprocess Technology, 9*(10), 1769–1778. Available from https://doi.org/10.1007/s11947-016-1760-8.

Bañuelos, M. A., Loira, I., Guamis, B., Escott, C., Del Fresno, J. M., Codina-Torrella, I., & Morata, A. (2020). White wine processing by UHPH without SO_2. Elimination of microbial populations and effect in oxidative enzymes, colloidal stability and sensory quality. *Food Chemistry, 332.* Available from https://doi.org/10.1016/j.foodchem.2020.127417.

Błaszak, M., Nowak, A., Lachowicz, S., Migdał, W., & Ochmian, I. (2019). E-beam irradiation and ozonation as an alternative to the sulphuric method of wine preservation. *Molecules (Basel, Switzerland), 24*(18). Available from https://doi.org/10.3390/molecules24183406.

Briviba, K., Gräf, V., Walz, E., Guamis, B., & Butz, P. (2016). Ultra high pressure homogenization of almond milk: Physico-chemical and physiological effects. *Food Chemistry, 192,* 82–89. Available from https://doi.org/10.1016/j.foodchem.2015.06.063.

Christofi, S., Malliaris, D., Katsaros, G., Panagou, E., & Kallithraka, S. (2020). Limit SO_2 content of wines by applying high hydrostatic pressure. *Innovative Food Science and Emerging Technologies, 62.* Available from https://doi.org/10.1016/j.ifset.2020.102342.

Ciani, M., Ferraro, L., & Fatichenti, F. (2000). Influence of glycerol production on the aerobic and anaerobic growth of the wine yeast *Candida stellata. Enzyme and Microbial Technology, 27*(9), 698–703. Available from https://doi.org/10.1016/S0141-0229(00)00269-6.

Cojocaru, G. A., & Antoce, A. O. (2012). Chemical and biochemical mechanisms of preservatives used in wine: A review. *Horticulture, 56,* 457–466.

Costa, A., Barata, A., Malfeito-Ferreira, M., & Loureiro, V. (2008). Evaluation of the inhibitory effect of dimethyl dicarbonate (DMDC) against wine microorganisms. *Food Microbiology, 25*(2), 422–427. Available from https://doi.org/10.1016/j.fm.2007.10.003.

Cruz, N., Capellas, M., Hernández, M., Trujillo, A. J., Guamis, B., & Ferragut, V. (2007). Ultra high pressure homogenization of soymilk: Microbiological, physicochemical and microstructural characteristics. *Food Research International, 40*(6), 725–732. Available from https://doi.org/10.1016/j.foodres.2007.01.003.

Datta, N., Hayes, M. G., Deeth, H. C., & Kelly, A. L. (2005). Significance of frictional heating for effects of high pressure homogenisation on milk. *Journal of Dairy Research, 72*(4), 393–399. Available from https://doi.org/10.1017/S0022029905001056.

Desrumaux, A., & Marcand, J. (2002). Formation of sunflower oil emulsions stabilized by whey proteins with high-pressure homogenization (up to 350 MPa): Effect of pressure on emulsion characteristics. *International Journal of Food Science and Technology, 37*(3), 263–269. Available from https://doi.org/10.1046/j.1365-2621.2002.00565.x.

Dhankhar, P. (2014). Homogenization fundamentals. *IOSR Journal of Engineering,* 01–08. Available from https://doi.org/10.9790/3021-04540108.

Diels, A. M. J., Wuytack, E. Y., & Michiels, C. W. (2003). Modelling inactivation of *Staphylococcus aureus* and *Yersinia enterocolitica* by high-pressure homogenisation at different temperatures. *International Journal of Food Microbiology, 87*(1–2), 55–62. Available from https://doi.org/10.1016/S0168-1605(03)00050-3.

Escott, C., Vaquero, C., del Fresno, J. M., Bañuelos, M. A., Loira, I., Han, S.-y., & Suárez-Lepe, J. A. (2017). Pulsed light effect in red grape quality and fermentation. *Food and Bioprocess Technology, 10*(8), 1540–1547. Available from https://doi.org/10.1007/s11947-017-1921-4.

EU Regulation No. 606. (2009). *Laying down certain detailed rules for implementing Council Regulation (EC) No. 479/2008 as regards the categories of grapevine products.* COMMISSION REGULATION.

Falguera, V., Forns, M., & Ibarz, A. (2013). UV-vis irradiation: An alternative to reduce SO_2 in white wines? *LWT - Food Science and Technology, 51*(1), 59–64. Available from https://doi.org/10.1016/j.lwt.2012.11.006.

Fernández-Pérez, R., Sáenz, Y., Rojo-Bezares, B., Zarazaga, M., Rodríguez, J. M., Torres, C., & Ruiz-Larrea, F. (2018). Production and antimicrobial activity of nisin under enological conditions. *Frontiers in Microbiology, 9.* Available from https://doi.org/10.3389/fmicb.2018.01918.

Ferragut, V., Valencia-Flores, D., Pérez-González, M., Gallardo, J., & Hernández-Herrero, M. (2015). Quality characteristics and shelf-life of ultra-high pressure homogenized (UHPH) almond beverage. *Foods,* 159–172. Available from https://doi.org/10.3390/foods4020159.

Floury, J., Desrumaux, A., & Legrand, J. (2002). Effect of ultra-high-pressure homogenization on structure and on rheological properties of soy protein-stabilized emulsions. *Journal of Food Science, 67,* 3388–3395. Available from https://doi.org/10.1111/j.1365-2621.2002.tb09595.x.

Fuhrman, B., Volkova, N., Suraski, A., & Aviram, M. (2001). White wine with red wine-like properties: Increased extraction of grape skin polyphenols improves the antioxidant capacity of the derived white wine. *Journal of Agricultural and Food Chemistry, 49*(7), 3164–3168. Available from https://doi.org/10.1021/jf001378j.

García Martín, J. F., & Sun, D. W. (2013). Ultrasound and electric fields as novel techniques for assisting the wine ageing process: The state-of-the-art research. *Trends in Food Science and Technology, 33*(1), 40–53. Available from https://doi.org/10.1016/j.tifs.2013.06.005.

Garde-Cerdán, T., López, R., Garijo, P., González-Arenzana, L., Gutiérrez, A. R., López-Alfaro, I., & Santamaría, P. (2014). Application of colloidal silver versus sulfur dioxide during vinification and storage of Tempranillo red wines. *Australian Journal of Grape and Wine Research, 20*(1), 51–61. Available from https://doi.org/10.1111/ajgw.12050.

Gerland, C. (2002). Applications of lysozyme in enology: Technical report. *Mitt. Klosterneuburg, 52*(3), 116–121.

Giacosa, S., Río Segade, S., Cagnasso, E., Caudana, A., Rolle, L., & Gerbi, V. (2018). *SO_2 in wines: Rational use and possible alternatives. Red wine technology* (pp. 309–321). Italy: Elsevier. Available from https://doi.org/10.1016/B978-0-12-814399-5.00021-9.

Hayes, M. G., Fox, P. F., & Kelly, A. L. (2005). Potential applications of high pressure homogenisation in processing of liquid milk. *Journal of Dairy Research, 72*(1), 25–33. Available from https://doi.org/10.1017/S0022029904000524.

Izquierdo-Cañas, P. M., García-Romero, E., Huertas-Nebreda, B., & Gómez-Alonso, S. (2012). Colloidal silver complex as an alternative to sulphur dioxide in winemaking. *Food Control, 23*(1), 73–81. Available from https://doi.org/10.1016/j.foodcont.2011.06.014.

Kheadr, E. E., Vachon, J. F., Paquin, P., & Fliss, I. (2002). Effect of dynamic high pressure on microbiological, rheological and microstructural quality of Cheddar cheese. *International Dairy Journal, 12*(5), 435–446. Available from https://doi.org/10.1016/S0958-6946(01)00104-2.

Lisanti, M. T., Blaiotta, G., Nioi, C., & Moio, L. (2019). Alternative methods to SO2 for microbiological stabilization of wine. *Comprehensive Reviews in Food Science and Food Safety, 18*(2), 455–479. Available from https://doi.org/10.1111/1541-4337.12422.

Loira, I., Morata, A., Bañuelos, M. A., Puig-Pujol, A., Guamis, B., González, C., & Suárez-Lepe, J. A. (2018). Use of ultra-high pressure homogenization processing in winemaking: Control of microbial populations in grape musts and effects in sensory quality. *Innovative Food Science and Emerging Technologies, 50,* 50–56. Available from https://doi.org/10.1016/j.ifset.2018.10.005.

Loira, I., Morata, A., Bañuelos, M. A., & Suárez-Lepe, J. A. (2019). *Isolation, selection, and identification techniques for non-saccharomyces yeasts of oenological interest. Biotechnological progress and beverage consumption: Volume 19: The science of beverages* (pp. 467–508). Spain: Elsevier. Available from https://doi.org/10.1016/B978-0-12-816678-9.00015-1.

Loira, I., Morata, A., Palomero, F., González, C., & Suárez-Lepe, J. A. (2018). Schizosaccharomyces pombe: A promising biotechnology for modulating wine composition. *Fermentation, 4*(3). Available from https://doi.org/10.3390/fermentation4030070.

Morata, A., Bañuelos, M. A., Loira, I., Raso, J., Álvarez, I., Garcíadeblas, B., & Suárez Lepe, J. A. (2020). Grape must processed by pulsed electric fields: Effect on the inoculation and development of non-saccharomyces yeasts. *Food and Bioprocess Technology, 13*(6), 1087–1094. Available from https://doi.org/10.1007/s11947-020-02458-1.

Morata, A., Bañuelos, M. A., Tesfaye, W., Loira, I., Palomero, F., Benito, S., & Suárez-Lepe, J. A. (2015). Electron beam irradiation of wine grapes: Effect on microbial populations, phenol extraction and wine quality. *Food and Bioprocess Technology, 8*(9), 1845–1853. Available from https://doi.org/10.1007/s11947-015-1540-x.

Morata, A., Escott, C., Bañuelos, M. A., Loira, I., Del Fresno, J. M., González, C., & Suárez-lepe, J. A. (2020). Contribution of non-saccharomyces yeasts to wine freshness. A review. *Biomolecules, 10*(1). Available from https://doi.org/10.3390/biom10010034.

Morata, Antonio, Loira, I., Guamis, B., Raso, J., del Fresno, J. M., Escott, C., & Suárez-Lepe, J. A. (2020). *Emerging technologies to increase extraction, control microorganisms, and reduce SO₂*. IntechOpen. Available from https://doi.org/10.5772/intechopen.92035.

Morata, A., Loira, I., Vejarano, R., Bañuelos, M. A., Sanz, P. D., Otero, L., & Suárez-Lepe, J. A. (2015). Grape processing by high hydrostatic pressure: Effect on microbial populations, phenol extraction and wine quality. *Food and Bioprocess Technology, 8*(2), 277–286. Available from https://doi.org/10.1007/s11947-014-1405-8.

Morata, A., Loira, I., Vejarano, R., González, C., Callejo, M. J., & Suárez-Lepe, J. A. (2017). Emerging preservation technologies in grapes for winemaking. *Trends in Food Science and Technology, 67*, 36–43. Available from https://doi.org/10.1016/j.tifs.2017.06.014.

Moreno-Arribas, M. V., Monge, M., Miralles, B., Armentia, G., Cueva, C., Crespo, J., & González de Llano, D. (2019). Some new findings on the potential use of biocompatible silver nanoparticles in winemaking. *Innovative Food Science and Emerging Technologies, 51*, 64–72. Available from https://doi.org/10.1016/j.ifset.2018.04.01.

Oliveira, C. M., Ferreira, A. C. S., De Freitas, V., & Silva, A. M. S. (2011). Oxidation mechanisms occurring in wines. *Food Research International, 44*(5), 1115–1126. Available from https://doi.org/10.1016/j.foodres.2011.03.050.

Pereda, J., Ferragut, V., Quevedo, J. M., Guamis, B., & Trujillo, A. J. (2007). Effects of ultra-high pressure homogenization on microbial and physicochemical shelf life of milk. *Journal of Dairy Science, 90*(3), 1081–1093. Available from https://doi.org/10.3168/jds.S0022-0302(07)71595-3.

Pereda, J., Ferragut, V., Quevedo, J. M., Guamis, B., & Trujillo, A. J. (2009). Heat damage evaluation in ultra-high pressure homogenized milk. *Food Hydrocolloids, 23*(7), 1974–1979. Available from https://doi.org/10.1016/j.foodhyd.2009.02.010.

Pereda, J., Jaramillo, D. P., Quevedo, J. M., Ferragut, V., Guamis, B., & Trujillo, A. J. (2008). Characterization of volatile compounds in ultra-high-pressure homogenized milk. *International Dairy Journal, 18*(8), 826–834. Available from https://doi.org/10.1016/j.idairyj.2007.12.002.

Picart, L., Thiebaud, M., René, M., Guiraud, J. P., Cheftel, J. C., & Dumay, E. (2006). Effects of high pressure homogenisation of raw bovine milk on alkaline phosphatase and microbial inactivation. A comparison with continuous short-time thermal treatments. *Journal of Dairy Research, 73*(4), 454–463. Available from https://doi.org/10.1017/S0022029906001853.

Pinho, C. R. G., Franchi, M. A., Tribst, A. A. L., & Cristianinia, M. (2011). Effect of high pressure homogenization process on *Bacillus stearothermophilus* and *Clostridium sporogenes* spores in skim milk. *Procedia Food Science, 1*, 869–873. Available from https://doi.org/10.1016/j.profoo.2011.09.131.

Poliseli-Scopel, F. H., Hernández-Herrero, M., Guamis, B., & Ferragut, V. (2012). Comparison of ultra high pressure homogenization and conventional thermal treatments on the microbiological, physical and chemical quality of soymilk. *LWT - Food Science and Technology, 46*(1), 42–48. Available from https://doi.org/10.1016/j.lwt.2011.11.004.

Poliseli-Scopel, F. H., Hernández-Herrero, M., Guamis, B., & Ferragut, V. (2013). Characteristics of soymilk pasteurized by ultra high pressure homogenization (UHPH). *Innovative Food Science and Emerging Technologies, 20*, 73–80. Available from https://doi.org/10.1016/j.ifset.2013.06.001.

Popper, L., & Knorr, D. (1990). Applications of high-pressure homogenization for food preservation: High-pressure homogenization can be used alone or combined with lytic enzyme or chitosan to reduce the microbial population and heat treatment damage in foods. *Food Technology, 44*, 84–89.

Puértolas, E., López, N., Condón, S., Raso, J., & Álvarez, I. (2009). Pulsed electric fields inactivation of wine spoilage yeast and bacteria. *International Journal of Food Microbiology, 130*(1), 49–55. Available from https://doi.org/10.1016/j.ijfoodmicro.2008.12.035.

Recamales, A. F., Sayago, A., González-Miret, M. L., & Hernanz, D. (2006). The effect of time and storage conditions on the phenolic composition and colour of white wine. *Food Research International, 39*(2), 220–229. Available from https://doi.org/10.1016/j.foodres.2005.07.009.

Ribéreau-Gayon, P., Dubourdieu, D., Donèche, B., & Lonvaud, A. (2006). *The microbiology of wine and vinifications* (Vol. 1). John Wiley & Sons, Ltd.

Rodríguez Montealegre, R., Romero Peces, R., Chacón Vozmediano, J. L., Martínez Gascueña, J., & García Romero, E. (2006). Phenolic compounds in skins and seeds of ten grape *Vitis vinifera* varieties grown in a warm climate. *Journal of Food Composition and Analysis, 19*(6–7), 687–693. Available from https://doi.org/10.1016/j.jfca.2005.05.003.

Ruiz-Larrea, F., Rojo-Bezares, B., Sáenz, Y., Navarro, L., Díez, L., Portugal, C. B., & Torres, C. (2007). Bacteriocines for wine microbiological control and reduction of SO₂ levels. *Bulletin de l'OIV, 80*(917), 445–458.

Ružić, I., Škerget, M., Knez, Z., & Runje, M. (2011). Phenolic content and antioxidant potential of macerated white wines. *European Food Research and Technology, 233*(3), 465–472. Available from https://doi.org/10.1007/s00217-011-1535-4.

Santamera, A., Escott, C., Loira, I., del Fresno, J. M., González, C., & Morata, A. (2020). Pulsed light: Challenges of a non-thermal sanitation technology in the winemaking industry. *Beverages, 45*. Available from https://doi.org/10.3390/beverages6030045.

Santos, M. C., Nunes, C., Saraiva, J. A., & Coimbra, M. A. (2012). Chemical and physical methodologies for the replacement/reduction of sulfur dioxide use during winemaking: Review of their potentialities and limitations. *European Food Research and Technology, 234*(1), 1–12. Available from https://doi.org/10.1007/s00217-011-1614-6.

Sonni, F., Bastante, M. J. C., Chinnici, F., Natali, N., & Riponi, C. (2009). Replacement of sulfur dioxide by lysozyme and oenological tannins during fermentation: Influence on volatile composition of white wines. *Journal of the Science of Food and Agriculture, 89*(4), 688−696. Available from https://doi.org/10.1002/jsfa.3503.

Suárez-Jacobo, A., Gervilla, R., Guamis, B., Roig-Sagués, A. X., & Saldo, J. (2010). Effect of UHPH on indigenous microbiota of apple juice. A preliminary study of microbial shelf-life. *International Journal of Food Microbiology, 136*(3), 261−267. Available from https://doi.org/10.1016/j.ijfoodmicro.2009.11.011.

Suárez-Jacobo, A., Rüfer, C. E., Gervilla, R., Guamis, B., Roig-Sagués, A. X., & Saldo, J. (2011). Influence of ultra-high pressure homogenisation on antioxidant capacity, polyphenol and vitamin content of clear apple juice. *Food Chemistry, 127*(2), 447−454. Available from https://doi.org/10.1016/j.foodchem.2010.12.15.

Thiebaud, M., Dumay, E., Picart, L., Guiraud, J. P., & Cheftel, J. C. (2003). High-pressure homogenisation of raw bovine milk. Effects on fat globule size distribution and microbial inactivation. *International Dairy Journal, 13*(6), 427−439. Available from https://doi.org/10.1016/S0958-6946(03)00051-7.

Threlfall, R. T., & Morris, J. R. (2002). Using dimethyldicarbonate to minimize sulfur dioxide for prevention of fermentation from excessive yeast contamination in juice and semi-sweet wine. *Journal of Food Science, 67*(7), 2758−2762. Available from https://doi.org/10.1111/j.1365-2621.2002.tb08811.x.

Vaquero, C., Loira, I., Bañuelos, M. A., Heras, J. M., Cuerda, R., & Morata, A. (2020). Industrial performance of several lachancea thermotolerans strains for pH control in white wines from warm areas. *Microorganisms, 8*(6). Available from https://doi.org/10.3390/microorganisms8060830.

Velázquez-Estrada, R. M., Hernández-Herrero, M. M., López-Pedemonte, T. J., Briñez-Zambrano, W. J., Guamis-López, B., & Roig-Sagués., A. X. (2011). Inactivation of Listeria monocytogenes and Salmonella enterica serovar Senftenberg 775W inoculated into fruit juice by means of ultra high pressure homogenisation. *Food Control, 22,* 313−317. Available from https://doi.org/10.1016/j.foodcont.2010.07.029.

Yildirim, H. K., & Darıcı, B. (2020). Alternative methods of sulfur dioxide used in wine production. *Journal of Microbiology, Biotechnology and Food Sciences,* 675−687. Available from https://doi.org/10.15414/jmbfs.2020.9.4.675-687.

Zamora, A., Ferragut, V., Jaramillo, P. D., Guamis, B., & Trujillo, A. J. (2007). Effects of ultra-high pressure homogenization on the cheese-making properties of milk. *Journal of Dairy Science, 90,* 13−23. Available from https://doi.org/10.3168/jds.S0022-0302(07)72604-8.

Zamora, A., & Guamis, B. (2015). Opportunities for ultra-high-pressure homogenisation (UHPH) for the food industry. *Food Engineering Reviews, 7*(2), 130−142. Available from https://doi.org/10.1007/s12393-014-9097-4.

Use of pulsed electric fields in white grape processing

Carlota Delso[1], *Juan Manuel Martínez*[1], *Diederich Aguilar-Machado*[1], *Marcos Maza*[2], *Antonio Morata*[3], *Ignacio Álvarez*[1] *and Javier Raso*[1]

[1]Department of Animal Production and Food Science, Food Technology, Faculty of Veterinary, Instituto Agroalimentario de Aragón — IA2— (University of Zaragoza-CITA), Zaragoza, Spain [2]National University of Cuyo, Faculty of Agricultural Sciences, Department of Oenological and Food Sciences, Mendoza, Argentina [3]enotecUPM, Chemistry and Food Technology Department, Technical University of Madrid, Madrid, Spain

6.1 Introduction

In recent decades, much research on nonthermal food processing technologies has been carried out to combat spoilage and pathogenic microorganisms, reduce the use of chemical preservatives, improve the functionality of foods, and optimize process efficiency (Sun, 2014). These technologies aim to respond to the demand of different consumer niches and markets for foods with improved nutritional and sensorial properties, but they may also represent eco-innovative approaches to improve the sustainability of food processing.

One of these innovative technologies that aim to achieve these objectives is pulsed electric fields (PEFs). This technique causes the electroporation of cytoplasmic membranes of microorganisms and eukaryote cells of plant and animal tissues using low energy requirements, thus minimizing quality deterioration of the food compounds (Kotnik, Kramar, Pucihar, Miklavcic, & Tarek, 2012). Intense fundamental research efforts have been addressed in recent years to understand the electroporation mechanisms and to identify the critical factors affecting electroporation. In parallel, many studies have been conducted to demonstrate the electroporation efficacy on the inactivation of microbial cells; enhancing mass transfer in different operations of the food industry such as extraction, dehydration, or infusion of compounds; and the modification of food structure (Barba et al., 2015).

The attractive effects deriving from PEF processing have driven the development of pulse power generators that meet the requirements for food processing, and currently, different commercial applications of PEF are available in the market (Toepfl, Kinsella, & Parniakov, 2020).

Wineries are an industrial sector that can take advantage of the attractive effects deriving from PEF. The electroporation of the cells of the grapes before vinification or of the microbial cells involved in winemaking may have a beneficial impact on the different stages involved in winemaking in terms of optimizing process efficiency, reducing the use of chemical preservatives, or saving energy (Saldaña, Luengo, Puértolas, Álvarez, & Raso, 2016). The objective of this chapter is to describe the potential applications of PEF technology in different stages of the white winemaking

6.2 Principles of PEF processing

6.2.1 The electroporation phenomenon

During PEF processing, an external voltage applied between two electrodes generates an electric field whose strength depends on the voltage intensity and the distance between the electrodes. When exposed to a sufficiently

61

© 2022 Elsevier Inc. All rights reserved.

strong electric field, the selective permeability of the cytoplasmic membrane of the cell is modified. This phenomenon, called *electroporation*, is shown by the transport of molecules through the cytoplasmic membrane that are not able to cross it when the structure of the membrane is intact (Saulis, 2010). Therefore the main consequence of membrane electroporation is the influx of compounds into the cell and the outflow of intracellular compounds from the cell. When the intensity of the electric field is not high enough or the exposure to the external electric field is sufficiently short, the membrane may recover its selective permeability, remaining viable (*reversible electroporation*). However, intense electric fields or longer exposures can cause permanent modification in membrane permeability (*irreversible electroporation*). Because both eukaryote and prokaryote cells can be effectively reversibly and irreversibly electroporated, PEF technology can be considered as a universal method that has applications in many different disciplines. Reversible electroporation is a procedure that is typically used in molecular biology and clinical biotechnological applications to gain access to the cytoplasm to introduce or deliver in vivo drugs, oligonucleotides, antibodies, plasmids, etc. (Kotnik et al., 2015; Yarmush, Golberg, Serša, Kotnik, & Miklavčič, 2014). However, the main applications of PEF in the food industry are based on the irreversible electroporation of the cell membranes (Puértolas, Luengo, Álvarez, & Raso, 2012).

It is accepted that for the manifestation of electroporation, an increment in the transmembrane potential of the cells is required. This increment is time- and position-dependent and it is a consequence of the accumulation of oppositely charged ions at both sides of the nonconductive intact membrane (Teissie, Golzio, & Rols, 2005). Transmembrane potential (U_m) generated by an external electric field for a cell can be calculated as a function of the radius of the cell, the external electric field applied, and the angle between the normal to the membrane and the direction of the electric field. As the transmembrane potential depends on an angular parameter, the external electric field induces a position-dependent potential difference linearly related to the intensity of the applied electric field.

A critical value of the external electric field (Ec) that is a function of the cell diameter is required to induce the transmembrane potential (0.2−1.0 V), which induces the structural changes in the membrane that lead to the increment in its permeability. Because the size of eukaryotic cells is around 10−100 times higher than that of microorganisms, the electric field strength required for the electroporation of plant or animal cells (0.1−5 kV/cm) is lower than that required for microbial cells (10−35 kV/cm) (Donsì et al., 2010). The earliest theories proposed to explain beacuse an external electric field causes the electroporation of the cytoplasmic membrane were based on electromechanical models (Zimmermann et al., 1974). According to these models the attraction of charges of opposing sign accumulated on both sides of the membrane that causes an increase in the transmembrane potential produces a viscoelastic deformation of the cell membrane. When this deformation cannot be equilibrated by the surface tension and elasticity of the lipid bilayer, breakdown of the membrane occurs. However, currently, the most accepted theoretical description of the underlying events of electroporation is the aqueous pore formation theory (Weaver & Chizmadzhev, 1996). According to this theory, thermal or mechanical fluctuations may cause the formation and rapid resealing of aqueous pores with radii below a nanometer and lifetime below a nanosecond in the lipid bilayer of a resting cell membrane, permitting water to penetrate at rates much higher than diffusion. The increment in the transmembrane voltage caused by the application of an external electric field induces a certain reduction in the energy required for the spontaneous formation of aqueous pores that are stabilized by the reorientation of the adjacent lipids. Although this mechanism is widely recognized and corroborated by dynamic simulations, some recent studies seem to confirm that PEFs also cause chemical changes to the membrane lipids and modulation of the function of the proteins of the membranes that contribute to the increment in the permeability (Fig. 6.1; Kotnik, Rems, Tarek, & Miklavcic, 2019).

6.2.2 Generation of PEF and PEF treatment system

PEF processing consists of the application of intermittent high-voltage pulses with a duration in the range of microseconds to milliseconds with the product to be processed placed between two electrodes. The voltage applied between the electrodes results in an electric field, the strength of which depends on the distance between the electrodes and the voltage delivered. The main process parameters affecting PEF processing are shown in Fig. 6.2. Generally, PEF treatment efficacy is a function of the applied electric field strength and the duration of the treatment, which depends on the both number of applied pulses and pulse width. Treatment time can be increased either by increasing the number of pulses or by augmenting the pulse length. While for square wave pulses, the pulse width corresponds to the time over which the peak field is maintained, for exponential decay

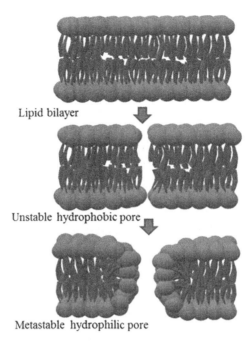

Lipid bilayer

Unstable hydrophobic pore

Metastable hydrophilic pore

FIGURE 6.1   Formation of pores in a lipid bilayer by pulsed electric fields.

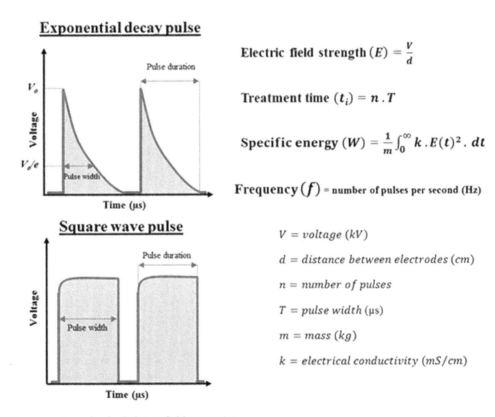

Exponential decay pulse

Electric field strength $(E) = \frac{V}{d}$

Treatment time $(t_i) = n \cdot T$

Specific energy $(W) = \frac{1}{m} \int_0^\infty k \cdot E(t)^2 \cdot dt$

Frequency (f) = number of pulses per second (Hz)

$V = voltage\ (kV)$

$d = distance\ between\ electrodes\ (cm)$

$n = number\ of\ pulses$

$T = pulse\ width\ (\mu s)$

$m = mass\ (kg)$

$k = electrical\ conductivity\ (mS/cm)$

Square wave pulse

FIGURE 6.2   Main parameters of pulsed electric field processing.

pulses, it corresponds to the time until the peak voltage has decayed to 37%. The total specific energy (energy applied per mass unit) of the PEF treatments depends on the applied voltage, pulse width, number of pulses, and resistance of the treatment chamber, which varies according to the treatment chamber's geometry and the electrical conductivity of the treated product.

A PEF processing system consists of a pulse power generator and a treatment chamber. The pulse power supplier transforms the sinusoidal alternating current from the power line to pulses with sufficient peak voltage. The pulse power suppliers currently used in industrial processing for the application of square wave pulses using standard switch devices comprise either pulse transformer setups or semiconductor-based Mark generators (Rebersek, Míklavčič, Bertacchini, & Sack, 2014) (Fig. 6.3). In the first case, an energy storage device such as a capacitor is not required because a low-voltage, high-current switch is used for power switching, with a subsequent transformation to the desired voltage level. In semiconductor-based Mark generators, voltage multiplication occurs by charging a stack of capacitors in parallel that are discharged in series using transistor switches.

A PEF treatment chamber is generally composed of two electrodes, one connected to high voltage and the other to ground, held in position by insulating material that forms an enclosure to contain the food material. Among the different geometrical configurations for the treatment of liquid or pumpable semisolid foods, parallel plates and coaxial or collinear cylinders are the most commonly used (Fig. 6.4). For the treatment of solid products such as tubers or fruits, they are conveyed through an electrode setup immersed in water by a belt.

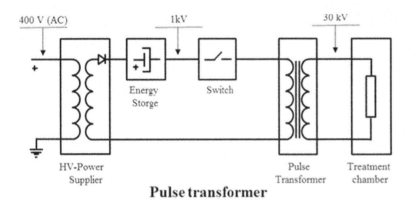

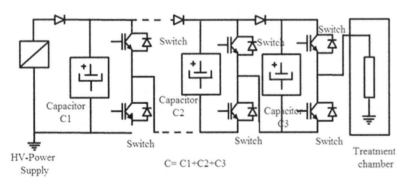

FIGURE 6.3 Electrical circuits for the generation of pulses of high electric field.

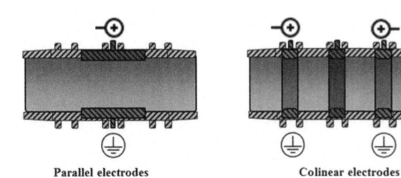

FIGURE 6.4 Treatment chamber configurations to apply Pulsed Electric Field treatment in continuous flow.

6.3 PEF applications in white winemaking

The application of PEF technology on the different stages of white winemaking to improve juice extraction yield and extraction of aroma precursors while preventing microbial spoiling or triggering yeast autolysis has been investigated. Fig. 6.5 shows when the PEF treatment should be applied in the process of elaboration of white wine depending on the objective of the treatment.

6.3.1 Improving juice extraction yield from white grapes by PEF

Different studies have reported that the application of a PEF treatment on various fruits and vegetables such as apples and carrots is an efficient preliminary operation to enhance juice extraction. An increase in juice yield or a reduction in processing times and energy requirements in comparison to enzymatic and thermal treatments of mashes yield has been reported (Grimi, Lebovka, Vorobiev, & Vaxelaire, 2009; Schilling et al., 2007). This improvement, which is highly dependent on the mash structure and dejuicing system, has been related to the electroporation of the cell by PEF that increased membrane permeability and facilitated the loss of the cell content (Jaeger, Schulz, Lu, & Knorr, 2012). Extraction by pressing is a unit operation that is used in wineries at different moments of the winemaking process depending on the type of wine elaboration. In red winemaking the solid parts of the grapes that have been in contact with the fermenting must are pressed at the end of the maceration process to extract the remaining wine. Meanwhile, white and rosé wines are obtained by fermenting the juice, which is obtained by pressing the grapes before the fermentation stage. Juice extraction is a critical step that has a large influence on the final quality of the white wine. Moderate pressures and short pressing times are required to obtain a juice with a low content of solid particles in suspension (turbidity) and a low concentration of pigments and to avoid juice browning due to polyphenol oxidation.

The application of moderate electric fields (0.25−1 kV/cm) to improve the extraction of juice from different grape varieties (Muscadelle, Sauvignon, and Semillon) before pressing or after a given pressing time was

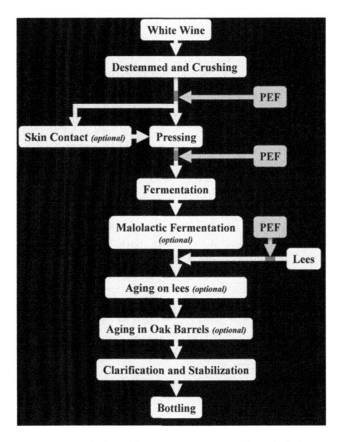

FIGURE 6.5 Stages of the white winemaking in which could result beneficial to apply Pulsed Electric Fields.

investigated for the first time by Praporscic, Lebovka, Vorobiev, and Mietton-Peuchot (2007). The electroporation of the white grapes before pressing (5 bars) was more effective than that after a pressing period. The application of PEF (0.75 kV/cm) increased the juice yield extraction by 24% and the obtained juice had lower turbidity and absorbance at 520 nm. Grimi et al. (2009) investigated the influence of the application of PEF on juice extraction from Chardonnay grapes using two extraction regimes: constant pressure (0.5–1 bar) and progressive pressure increasing up to 1 bar. The constant pressure of 1 bar increased the juice yield by 18% without affecting turbidity and polyphenol content. The most significant effect observed with the progressive pressure application was the rise in the polyphenol content of the juice by 15%. As an increment in the content of polyphenols could result in unbeneficial sensorial properties of white wine, constant pressure seems to be a more suitable extraction regime than progressive pressure increment for pressing white grapes in wineries.

An appreciable increment in juice extraction yield was reported by Comuzzo, Marconi, Zanella, and Querzè (2018) when white grapes of the *Garganega* variety were previously treated with PEF (1.5 kV/cm for 8 and 16 µs). PEF processing affected neither the level of sugars nor the level of yeast assimilable nitrogen in the juice. However, extraction yield improvements decreased from 8.9% to 4.3% when the total specific energy of the treatment was increased from 11 to 22 kJ/kg. Also, the level of suspended solids and the amount of sediments were higher when the grapes were treated with the most intense PEF treatments. These results seem to indicate that more intense modifications in the grape mash structure as a consequence of an increment in the intensity of the PEF treatment could have negative effects on the obtained juice for white winemaking. A combination of PEF and an enzymatic treatment with pectinase was found to be not an efficient approach to improve the juice extraction yield from *Grüner Veltliner* and *Traminer* varieties (Fauster et al., 2020). PEF treatment in combination with the enzymatic treatment did not significantly increase the juice extraction yield when compared with the enzymatic treatment. Meanwhile, the combined treatment increased the turbidity of the juice of the *Traminer* variety by 17% compared with the enzymatic treatment, and no distinct effects of PEF samples were observed on the *Grüner Veltliner* variety.

The variability of the results reported by different authors pertaining to when PEF is applied to white grapes before pressing seems to indicate that an optimization of the PEF treatment conditions is still required depending on the grape variety, pressing characteristics, and levels of applied pressure in order to improve the juice extraction yield without increasing the turbidity.

6.3.2 Improving the extraction of aroma precursor compounds from white grapes by PEF

The effect of the application of a PEF treatment to crushed white grapes of the variety *Garganega* on the extraction of aroma precursors was investigated by Comuzzo et al. (2018). In this study, the grapes mash was immediately pressed after processing in continuous flow (200 L/h) at an electric field of 1.5 kV/cm for different processing times that corresponded to total specific energies of 11 kJ/kg and 22 kJ/kg. PEF did not significantly affect the evolution of the alcoholic fermentation of the juices obtained by pressing or the wine basic composition. However, a higher extraction of varietal aroma precursors without provoking excessive color evolution and extraction of phenolic compounds was observed in the wine obtained with grapes treated at a total specific energy of 22 kJ/kg. However, the concentration of all compounds analyzed that belong to the terpenic and norisoprenoid glycosides families was below the olfactory threshold. The low concentration of these compounds was associated with the fact that *Garganega* is not an aromatic variety.

Fauster et al. (2020) investigated the impact of the combination of the application of a PEF treatment and pectin degradation by enzymes on the fermentation and on the white wine quality obtained from *Traminer* and *Grüner Veltliner* varieties. Crushed grapes were treated in a colinear treatment chamber at a flow of 440 kg/h at 3 kV/cm with total specific inputs of 3 and 10 kJ/kg. Pectolytic enzymes were added just after PEF treatments and pressing of the grape mash was conducted after 4 and 24 h of skin contact at 15°C. While the PEF treatment did not affect the fermentation of the variety *Grüner Veltliner*, the duration of the fermentation was reduced up to 20% for the variety *Traminer*. This effect was related to an increment of about 10% in the total nitrogen released in the grapes of this variety treated with PEF. Unlike total nitrogen release, PEF promoted the release of those polyphenols mainly localized in the pulp in the two varieties. However, a sensory panel did not detect higher astringency and bitterness in the wines obtained by grapes treated with PEF.

In the same study, it was also observed that the PEF treatment did not affect the analyzed aromatic compounds of the variety *Grüner Veltliner*. In general, the concentration of esters analyzed in the wine after 3 months after bottling and of free monoterpenes analyzed in the must before fermentation did not increase when grapes were treated with PEF. However, in the case of *Traminer* variety, an augmentation in the concentration of some esters and a significant increment for all analyzed terpenes in the must were observed. This beneficial effect of PEF observed in the must was neither analytically detectable nor sensorially detectable in the wine after 3 months of bottling.

The benefits derived from the application of PEF on white grapes for the quality of the white wine have not been extensively investigated as compared with the studies conducted to demonstrate the improvements in polyphenol extraction from red grapes. The improvement in the extraction of varietal aroma precursors, without provoking excessive extraction of phenolic compounds, requires further study for optimizing not only the PEF treatment conditions but also the processing parameters of the different white winemaking stages.

6.3.3 Microbial inactivation by PEF in white winemaking

The cytoplasmic membrane is vital for microbial cells because it acts as a semipermeable barrier protecting cells from the surrounding environment. Therefore modification of the integrity and functionality of the microbial membrane by irreversible electroporation caused by PEF leads to the inactivation of vegetative forms of microorganisms. Based on this effect, PEF provides an alternative to conventional thermal treatments for heat-sensitive products.

The thermal process is not a usual technique in wineries because heat may cause undesirable effects on juices and wines. The common practice used in wineries to decrease the risk of microbial spoilage during the winemaking process is the addition of sulfur dioxide (SO_2), which prevents the occurrence of oxidative as well as enzymatic reactions. However, SO_2 has some drawbacks such as the different microbial sensitivity to this compound among different microorganisms and its negative effects on the health of consumers with special sensitivity (Lisanti, Blaiotta, Nioi, & Moio, 2019).

The inactivation or reduction of the indigenous grape yeast microbiota can be produced by the use of PEF; this allows to use lower SO_2 contents to control microbial populations. Additionally, it is possible to better implant non-*Saccharomyces* yeasts by reducing the indigenous yeast population (Morata et al., 2020). The use of non-*Saccharomyces* is a trending biotechnology to improve the sensory profile of neutral varieties by producing fermentative esters or to reveal terpene or thiolic varietal aroma from their grape precursors by beta-glucosidase or carbon-lyase activities of some non-*Saccharomyces* yeasts.

Several studies have demonstrated that PEF is an effective technology to inactivate bacteria and yeast in the juice of white grapes for winemaking. Garde-Cerdán, Marsellés-Fontanet, Arias-Gil, Ancín-Azpilicueta, and Martín-Belloso (2008b) investigated the use of PEF as an alternative to the addition of SO_2 during the elaboration of wine from white grape *Parellada* variety. The applied PEF treatment (35 kV/cm, 1 ms) led to four logarithmic reductions of the microbial population without modifying the content of fatty acids and free amino acids of the juice, which are essential for the development of the yeast during fermentation. After reducing the microbial population by PEF, the juice was inoculated with a starter culture of *Saccharomyces cerevisiae* and fermentations with and without SO_2 addition were compared. The results of this study demonstrated that the PEF treatment of the juice could reduce to safer levels or even eliminate the addition of SO_2. The PEF treatment did not affect the reproducibility of the juice fermentation of must, and the composition of volatile compounds responsible for the typical flavor of wines through alcoholic fermentation and aging of white wine were not significantly modified (Garde-Cerdán, Marsellés-Fontanet, Arias-Gil, Martín-Belloso, & Ancín-Azpilicueta, 2007 and Garde-Cerdán et al., 2008b; Garde-Cerdán, Marsellés-Fontanet, Arias-Gil, Ancín-Azpilicueta, & Martín-Belloso, 2008a; Garde-Cerdán et al. 2007).

Marsellés-Fontanet, Puig, Olmos, Mínguez-Sanz, and Martín-Belloso (2009) investigated the effect of PEF on the inactivation of different types of microorganisms, namely *Kloeckera apiculata, Saccharomyces cerevisiae, Lactobacillus plantarum, Lactobacillus hilgardii*, and *Gluconobacter oxydans*, inoculated on *Parellada* juice. Maximum reductions of 4 log units were obtained for *K. apiculata, S. cerevisiae*, and lactic acid bacteria upon applying electric field strengths in the range of 24.9 to 35 kV/cm, with treatment times between 0.87 and 1 ms. As it has been reported by other authors, bacteria were more PEF resistant than yeast. For achieving the same level of inactivation in lactic acid bacteria than in yeast, the application of more intense PEF treatments was necessary, and only 1.1 log reductions were obtained for *G. oxydans* upon applying the most intense treatment conditions.

The application of PEF as an alternative to SO_2 to stop the fermentation of sweet white wine of the *Semillon* grape variety was investigated by Delsart et al. (2015). The most effective PEF treatment applied (20 kV/cm, 6 ms) reduced the population of *Saccharomyces* and *non-Sacharomyces* yeast up to 3 and 4 log cycles, respectively. Although the yeast inactivation by PEF was less effective than high-voltage discharges (HVED) or the addition of SO_2 (250 mg/L), the wine browning was less pronounced for the samples treated with PEF.

The results on the use of PEF for microbial inactivation for reducing SO$_2$ addition seem to be promising for application in wine winemaking. However, further studies are required for evaluating the potential of PEF for the stabilization of the white wine before bottling and the optimization of the process, because treatment conditions used in the above studies are not applicable at the industrial scale.

6.3.4 Application of PEF for acceleration of yeast autolysis and aging on lees

Yeast autolysis is a lytic event promoted by intracellular yeast enzymes that start after cell death. This self-degradation of yeast cells is essential during the winemaking technique called aging on the lees. This technique, which has been traditionally used in the production of sparkling wines and more recently extended to white and even red wine, involves maintaining the wine in contact with the lees for some time after fermentation (Maza, Delso, Álvarez, Raso, & Martínez, 2020). During this period of time the yeasts, which are the main component of the lees, self-degrade and certain yeast compounds such as proteins, nucleic acids, lipids, and polysaccharides that have positive effects on physicochemical and sensory properties are transferred to the wine (Charpentier, Dos Santos, & Feuillat, 2004).

Wineries are interested in the possibility of accelerating yeast autolysis because prolonged aging on the lees entails elevated labor costs, immobilization of winery stocks, risk of microbial spoilage, and production of unwanted metabolites, such as biogenic amines (González-Marco & Ancín-Azpilicueta, 2006).

The first evidence on the possibility to use PEF for reducing the duration of aging on the lees was reported by Martínez, Cebrián, Álvarez, and Raso (2016). In this study, it was demonstrated that the application of PEF treatments of different intensities (5–25 kV/cm for 30–240 µs) triggered the autolysis of a commercial winemaking strain of *S. cerevisiae*. The autolysis was monitored by the leakage to intracellular materials such as nucleic acids and proteins and the release of mannoproteins into the extracellular environment. The electroporation of the cytoplasmic membrane of the yeast by PEF modified its selective permeability, and a significant amount of intracellular compounds such as nucleic acids and proteins were detected in the extracellular medium after only 24 h of incubation. On the other hand, after 18 days of incubation, the concentration of mannoproteins in the extracellular medium containing PEF-treated yeast was 10-fold higher than in the control. This effect was associated with the release of the enzymes involved in cell-wall degradation (proteases and β-glucanases). Modification of the osmotic pressure of the cytoplasm of the electroporated cells would cause plasmolysis of the organelles that contain these enzymes. On the other hand, the pores created in the cytoplasmic membrane by PEF would facilitate the contact of those released enzymes with the outermost layer of the yeast cell wall where the mannoproteins are located (Fig. 6.6). The influence of the temperature (7°C–43°C), pH (3.5–7.0), and ethanol concentration (6%–25%) on the release of mannoproteins from yeast treated with PEF was also investigated. The highest mannoprotein released agreed with the most favorable condition for activity of the enzymes involved in autolysis (pH = 7, T = 43°C, Et[OH] = 6%).

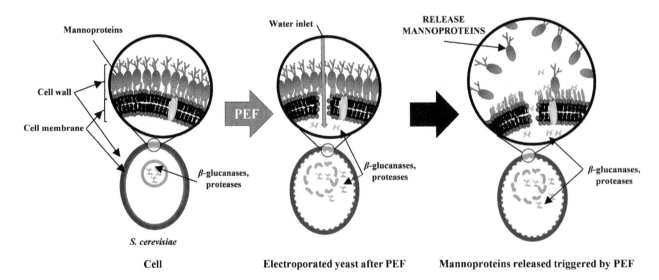

FIGURE 6.6   Mechanism of pulsed electric field triggering yeast autolysis.

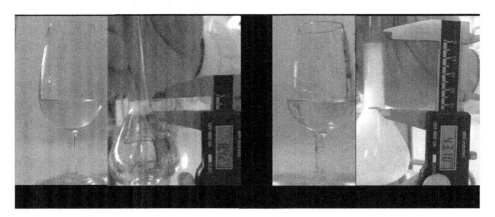

FIGURE 6.7 Effect of mannoproteins released from *S. cerevisiae* treated with Pulsed Electric Field on wine turbidity and foam formation. (A) *Chardonnay* wine containing untreated yeast after 1 month of aging on the lees. (B) *Chardonnay* wine containing Pulsed Electric Field—treated yeast after 1 month of aging on the lees.

The triggering of autolysis by PEF, initially observed in yeast suspended in buffer, was also demonstrated during aging on the lees of white wine of the *Chardonnay* variety (Martínez, Delso, Maza, Álvarez, & Raso, 2019). Mannoprotein release increased drastically in *Chardonnay* wine containing PEF-treated (5 and 10 kV/cm, 75 μs) yeasts. While 180 days of aging on the lees were required to reach the maximum concentration of mannoproteins in the wine containing untreated yeast, this value was achieved after only 30 days in the yeast previously treated with PEF, and 60% of the maximum value was achieved in the first 10 days of aging on the lees.

Analysis of the wine revealed that chromatic characteristics, total polyphenol index, total volatile acidity, pH, ethanol, and CIELAB parameters of the wine aged on the lees with PEF-treated yeast were not affected. On the other hand, mannoproteins released faster from PEF-treated yeast feature similar functional properties in terms of facilitating clarification and foaming properties than mannoproteins released from untreated yeasts (Fig. 6.7).

6.4 Conclusion

The peculiar mechanism of PEF (electroporation) results in a very flexible tool for wineries that can make good use of PEF by applying the technology to impact different processes involved in winemaking. According to the studies presented in this chapter, PEF technology offers an interesting perspective for the wine industry, not only for promoting red winemaking but also for its application in rosé and white wine processing (Puértolas, Saldaña, Álvarez, & Raso, 2011). However, the limited number of studies available on the use of PEF on white grape varieties for different applications means that further research is required for optimizing processing conditions.

The increasing interest of wineries in innovation and emerging technologies to improve its competitiveness in the market and sustainability, the availability of PEF systems capable of responding to the processing capacity demands of wineries, the low energy consumption involved in PEF, and the easy incorporation of PEF technology into existing processing lines all constitute solid arguments for the successful implementation of PEF technology in wineries in the near future for both red and white winemaking.

References

Barba, F. J., Parniakov, O., Pereira, S. A., Wiktor, A., Grimi, N., Boussetta, N., ... Vorobiev, E. (2015). Current applications and new opportunities for the use of pulsed electric fields in food science and industry. *Food Research International*, 77, 773–798. Available from https://doi.org/10.1016/j.foodres.2015.09.015.

Charpentier, C., Dos Santos, A. M., & Feuillat, M. (2004). Release of macromolecules by *Saccharomyces* cerevisiae during ageing of French flor sherry wine *Vin jaune. International Journal of Food Microbiology*, 96(3), 253–262. Available from https://doi.org/10.1016/j.ijfoodmicro.2004.03.019.

Comuzzo, P., Marconi, M., Zanella, G., & Querzè, M. (2018). Pulsed electric field processing of white grapes (cv. *Garganega*): Effects on wine composition and volatile compounds. *Food Chemistry*, 264, 16–23. Available from https://doi.org/10.1016/j.foodchem.2018.04.116.

Delsart, C., Grimi, N., Boussetta, N., Miot Sertier, C., Ghidossi, R., Mietton Peuchot, M., ... Vorobiev, E. (2015). Comparison of the effect of pulsed electric field or high voltage electrical discharge for the control of sweet white must fermentation process with the conventional addition of sulfur dioxide. *Food Research International*, 77, 718–724. Available from https://doi.org/10.1016/j.foodres.2015.04.017.

Donsì, F., Ferrari, G., & Pataro, G. (2010). Applications of pulsed electric field treatments for the enhancement of mass transfer from vegetable tissue. *Food Engineering Reviews*, 2(2), 109–130. Available from https://doi.org/10.1007/s12393-010-9015-3.

Fauster, T., Philipp, C., Hanz, K., Scheibelberger, R., Teufl, T., Nauer, S., ... Jaeger, H. (2020). Impact of a combined pulsed electric field (PEF) and enzymatic mash treatment on yield, fermentation behaviour and composition of white wine. *European Food Research and Technology*, 246(3), 609–620. Available from https://doi.org/10.1007/s00217-020-03427-w.

Garde-Cerdán, T., Marsellés-Fontanet, A. R., Arias-Gil, M., Ancín-Azpilicueta, C., & Martín-Belloso, O. (2008a). Effect of storage conditions on the volatile composition of wines obtained from must stabilized by PEF during ageing without SO2. *Innovative Food Science and Emerging Technologies*, 9(4), 469–476. Available from https://doi.org/10.1016/j.ifset.2008.05.002.

Garde-Cerdán, T., Marsellés-Fontanet, A. R., Arias-Gil, M., Ancín-Azpilicueta, C., & Martín-Belloso, O. (2008b). Influence of SO2 on the evolution of volatile compounds through alcoholic fermentation of must stabilized by pulsed electric fields. *European Food Research and Technology*, 227(2), 401–408. Available from https://doi.org/10.1007/s00217-007-0734-5.

Garde-Cerdán, T., Marsellés-Fontanet, A. R., Arias-Gil, M., Martín-Belloso, O., & Ancín-Azpilicueta, C. (2007). Influence of SO2 on the consumption of nitrogen compounds through alcoholic fermentation of must sterilized by pulsed electric fields. *Food Chemistry*, 103(3), 771–777. Available from https://doi.org/10.1016/j.foodchem.2006.09.018.

González-Marco, A., & Ancín-Azpilicueta, C. (2006). Influence of lees contact on evolution of amines in Chardonnay wine. *Journal of Food Science*, 71(9), C544–C548. Available from https://doi.org/10.1111/j.1750-3841.2006.00182.x.

Grimi, N., Lebovka, N. I., Vorobiev, E., & Vaxelaire, J. (2009). Effect of a pulsed electric field treatment on expression behavior and juice quality of Chardonnay grape. *Food Biophysics*, 4(3), 191–198. Available from https://doi.org/10.1007/s11483-009-9117-8.

Jaeger, H., Schulz, M., Lu, P., & Knorr, D. (2012). Adjustment of milling, mash electroporation and pressing for the development of a PEF assisted juice production in industrial scale. *Innovative Food Science and Emerging Technologies*, 14, 46–60. Available from https://doi.org/10.1016/j.ifset.2011.11.008.

Kotnik, T., Frey, W., Sack, M., Haberl Meglič, S., Peterka, M., & Miklavčič, D. (2015). Electroporation-based applications in biotechnology. *Trends in Biotechnology*, 33(8), 480–488. Available from https://doi.org/10.1016/j.tibtech.2015.06.002.

Kotnik, T., Kramar, P., Pucihar, G., Miklavcic, D., & Tarek, M. (2012). Cell membrane electroporation – Part 1: The phenomenon. *IEEE Electrical Insulation Magazine*, 14–23. Available from https://doi.org/10.1109/MEI.2012.6268438.

Kotnik, T., Rems, L., Tarek, M., & Miklavcic, D. (2019). Membrane electroporation and electropermeabilization: Mechanisms and models. *Annual Review of Biophysics*, 48, 63–91. Available from https://doi.org/10.1146/annurev-biophys-052118-115451.

Lisanti, M. T., Blaiotta, G., Nioi, C., & Moio, L. (2019). Alternative methods to SO2 for microbiological stabilization of wine. *Comprehensive Reviews in Food Science and Food Safety*, 18(2), 455–479. Available from https://doi.org/10.1111/1541-4337.12422.

Marsellés-Fontanet, A. R., Puig, A., Olmos, P., Mínguez-Sanz, S., & Martín-Belloso, O. (2009). Optimising the inactivation of grape juice spoilage organisms by pulse electric fields. *International Journal of Food Microbiology*, 130(3), 159–165. Available from https://doi.org/10.1016/j.ijfoodmicro.2008.12.034.

Martínez, J. M., Cebrián, G., Álvarez, I., & Raso, J. (2016). Release of mannoproteins during *Saccharomyces* cerevisiae autolysis induced by pulsed electric field. *Frontiers in Microbiology*, 7. Available from https://doi.org/10.3389/fmicb.2016.01435.

Martínez, J. M., Delso, C., Maza, M. A., Álvarez, I., & Raso, J. (2019). Pulsed electric fields accelerate release of mannoproteins from *Saccharomyces cerevisiae* during aging on the lees of Chardonnay wine. *Food Research International*, 116, 795–801. Available from https://doi.org/10.1016/j.foodres.2018.09.013.

Maza, M. A., Delso, C., Álvarez, I., Raso, J., & Martínez, J. M. (2020). Effect of pulsed electric fields on mannoproteins release from *Saccharomyces cerevisiae* during the aging on lees of Caladoc red wine. *LWT*, 118. Available from https://doi.org/10.1016/j.lwt.2019.108788.

Morata, A., Bañuelos, M. A., Loira, I., Raso, J., Álvarez, I., Garcíadeblas, B., ... Suárez Lepe, J. A. (2020). Grape must processed by pulsed electric fields: Effect on the inoculation and development of non-*Saccharomyces* yeasts. *Food and Bioprocess Technology*, 13(6), 1087–1094. Available from https://doi.org/10.1007/s11947-020-02458-1.

Praporscic, I., Lebovka, N., Vorobiev, E., & Mietton-Peuchot, M. (2007). Pulsed electric field enhanced expression and juice quality of white grapes. *Separation and Purification Technology*, 52(3), 520–526. Available from https://doi.org/10.1016/j.seppur.2006.06.007.

Puértolas, E., Luengo, E., Álvarez, I., & Raso, J. (2012). Improving mass transfer to soften tissues by pulsed electric fields: Fundamentals and applications. *Annual Review of Food Science and Technology*, 3(1), 263–282. Available from https://doi.org/10.1146/annurev-food-022811-101208.

Puértolas, E., Saldaña, G., Álvarez, I., & Raso, J. (2011). Experimental design approach for the evaluation of anthocyanin content of rosé wines obtained by pulsed electric fields. Influence of temperature and time of maceration. *Food Chemistry*, 126(3), 1482–1487. Available from https://doi.org/10.1016/j.foodchem.2010.11.164.

Rebersek, M., Míklavčič, D., Bertacchini, C., & Sack, M. (2014). Cell Membrane Electroporation—Part 3: The Equipment. *IEEE Electrical Insulation Magazine*, 30(3), 8–18. doi:10.1109/MEI.2014.6804737, In press.

Saldaña, Luengo, E., Puértolas, E., Álvarez, I., & Raso, J. (2016). Pulsed electric fields in wineries: Potential applications. *Handbook of Electroporation*, 1–18.

Saulis, G. (2010). Electroporation of cell membranes: The fundamental effects of pulsed electric fields in food processing. *Food Engineering Reviews*, 2(2), 52–73. Available from https://doi.org/10.1007/s12393-010-9023-3.

Schilling, S., Alber, T., Toepfl, S., Neidhart, S., Knorr, D., Schieber, A., ... Carle, R. (2007). Effects of pulsed electric field treatment of apple mash on juice yield and quality attributes of apple juices. *Innovative Food Science and Emerging Technologies*, 8(1), 127–134. Available from https://doi.org/10.1016/j.ifset.2006.08.005.

Sun, D.-W. (2014). *Emerging technologies for food processing* (2nd ed.). Academic Press. Available from https://doi.org/10.1016/C2012-0-07021-4.

Teissie, J., Golzio, M., & Rols, M. P. (2005). Mechanisms of cell membrane electropermeabilization: A minireview of our present (lack of ?) knowledge. *Biochimica et Biophysica Acta - General Subjects*, 1724(3), 270–280. Available from https://doi.org/10.1016/j.bbagen.2005.05.006.

Toepfl, S., Kinsella, J., & Parniakov, O. (2020). *Industrial scale equipment, patents, and commercial applications* (pp. 269–281). Elsevier BV. Available from https://doi.org/10.1016/b978-0-12-816402-0.00012-4.

Weaver, J. C., & Chizmadzhev, Y. A. (1996). Theory of electroporation: A review. *Bioelectrochemistry and Bioenergetics, 41*(2), 135–160. Available from https://doi.org/10.1016/S0302-4598(96)05062-3.

Yarmush, M. L., Golberg, A., Serša, G., Kotnik, T., & Miklavčič, D. (2014). Electroporation-based technologies for medicine: Principles, applications, and challenges. *Annual Review of Biomedical Engineering, 16*, 295–320. Available from https://doi.org/10.1146/annurev-bioeng-071813-104622.

Zimmermann, U., Pilwat, G., & Riemann, F. (1974). Dielectric breakdown of cell membranes. In *Membrane Transport in Plants*, 146–153.

7

Ultrasound to process white grapes

Encarna Gómez-Plaza, Luis Javier Pérez-Prieto, Paula Pérez-Porras and Ana Belén Bautista-Ortín

Department of Food Science and Technology, University of Murcia, Campus de Espinardo, Murcia, Spain

7.1 What is ultrasound? Definition and basic parameters

Sound waves are a form of vibrational energy with continuous frequencies. Concretely, ultrasound are sound waves whose frequencies exceed the hearing limit of the human ear (>16 kHz).

When ultrasound (US) passes through a liquid medium, the interaction between the ultrasonic waves, liquid, and dissolved gas leads to an exciting phenomenon known as acoustic cavitation, characterized by the generation and evolution (growth and collapse) of microbubbles in the liquid. These bubbles grow to a maximum or critical size in which they become unstable and violently collapse. Bubble collapse is a microscopic implosion that generates high turbulence and the release of heat energy, generating very high temperature (5000K) and pressure (50−1000 atm) conditions (Leong, Ashokkumar, & Kentish, 2011). These hotspots are able to dramatically accelerate the chemical reactivity of the medium (Shirsath, Sonawane, & Gogate, 2012). Also, when acoustic cavitation bubbles collapse near or onto the surface of a solid material, they generate high-speed jets of liquids into the surface and create shockwave damages. Those effects can lead to the fragmentation of materials and localized erosion.

Regarding the chemical reactivity, the high-temperature conditions generated by a collapsing bubble can lead to the formation of radical chemical species. In water, the ultrasonic waves can form hydroxy and hydroxyl radicals by homolytic cleavage (Richards & Loomis, 1927):

$$H_2O \rightarrow H\bullet + OH\bullet$$

These radicals are very reactive and rapidly interact with other radical or chemical species present in solution. These highly oxidizing species can have a significant effect on both biological and chemical species in aqueous solution (Mason & Lorimer, 2002).

In the food industry, ultrasound can be divided into two frequency ranges: high-frequency or low power/intensity ultrasound (100 kHz−1 MHz and <1 W/cm^2) and power ultrasound (16−100 kHz and 10−1000 W/cm^2). High-frequency ultrasound do not produce changes in the molecule structures and are commonly applied as an analytical technique to provide information on the physicochemical properties of food such as ripeness, sugar content, and acidity. On the other hand, power ultrasound, due to their capacity to generate high turbulence and the release of heat energy, have been applied to modify either the physical or chemical properties of food products; they have been used for many years to induce emulsification, disrupt cells, and disperse aggregated materials and for enzyme inactivation, enhancing drying and filtration, and the induction of oxidation reactions (Ferraretto, Cacciola, Ferran Batlló, & Celotti, 2013).

7.2 Basic parameters of US

The effectiveness of US depends on several parameters such as the frequency, ultrasonic energy, amplitude, exposure/contact time, composition and volume of food, and the treatment conditions.

© 2022 Elsevier Inc. All rights reserved.

7.2.1 Frequency

The ultrasonic frequency value determines the cavitation bubble size. Concretely, higher ultrasonic frequencies produce smaller cavitation bubbles than lower frequencies. This is at least partially because cavitation bubbles have more time to grow due to the increased wavelength at a lower frequency and therefore become larger before they implode. So at an ultrasound frequency of 20 kHz, the bubbles generated are relatively large and their collapse results in strong shockwaves, which can be useful for mechanical shearing applications such as emulsification (Leong, Wooster, Kentish, & Ashokkumar, 2009). Between 100 and 1000 kHz, the bubbles generated are much smaller and their collapse induces a higher increase in temperature, which can be more useful for inducing chemical reactions (Suslick & Crum, 1998), while at frequencies over 1 MHz, cavitational effects are much weaker.

7.2.2 Ultrasonic energy

The energy of ultrasound can be expressed as intensity (W/m^2) or energy density (W/m^3). The sonication intensity is increased with an increase in the amplitude. Ultrasonic power (W) influences the number of cavitation bubbles and is increased in proportion to the power level.

7.2.3 Amplitude

The amplitude of ultrasonic vibration determines the cavitation intensity. An increase in this parameter will increase the intensity and a higher US effect will be obtained.

7.2.4 Exposure/contact time

In general, higher effectiveness of US is obtained when the time of exposure is increased, although, in some cases, limited by an increase in temperature or to degradation or chemical reactions.

7.2.5 Treatment conditions

Parameters such as temperature and pressure largely influence the effectiveness of the ultrasound. An increase in the temperature enhances the number of cavitation bubbles (Muthukumaran, Kentish, Stevens, & Ashokkumar, 2006). In the case of the pressure, the cavitation threshold and the intensity of bubble collapse are enhanced when the pressure increases, reducing the processing time (Chemat & Khan, 2011).

7.3 Ultrasound devices for processing grapes and wines

There are different ultrasound systems for food applications in general, depending on the treatment medium and the desired effect. The application of ultrasound through liquid medium is the most used in the food industry.

At the laboratory scale, the ultrasound treatment on grape/wine can be performed in an ultrasonic bath or using an ultrasonic probe. Ultrasonic bath is the most widely used system for the application of ultrasound, in which the transducer (an electromechanical component that generates the ultrasound, made up of one or more coupled piezoelectric ceramics) is fixed at the base of the bath. It generates a vibration in the bottom that is transmitted to the liquid contained in the bath. It works with frequencies between 30 and 50 kHz and a power density of 20−40 W/L. The bath can be used as a reaction vessel, although its normal use involves immersing a glass vessel containing the sample to be treated into the water-filled bath. Its main limitations are the little power it supplies to the environment when compared to other systems, the variation of the acoustic field within the bath, and the difficulty in controlling the temperature (Mulet, Cárcel, Benedito, Rosselló, & Simal, 2003), which can significantly affect the sensory characteristics of the grape, must, or wine. Only lab-scale systems have been described.

In ultrasonic probes, ultrasound is applied directly by means of a vibrating probe. Depending on its geometry, the ultrasonic energy can simply be transmitted or concentrated onto a smaller surface in order to amplify the intensity and therefore its effects (Mason, 1998). In applications with this type of system, the distance between the tip of the probe and the treated sample is an important parameter to be controlled due to the attenuation of the ultrasonic field with distance. Ultrasonic probes use frequencies around 20 kHz and a power density of 20,000 W/L. The main advantage of using this equipment is that the transmitted ultrasonic power can be easily controlled.

One step toward a larger scale can be found in Gambacorta et al. (2017), who reported pilot plant equipment where the grapes were processed using 200-L stainless steel horizontal rotary wine fermenters with a submerged cap, and two of the fermenters were equipped with an ultrasonic delivery system consisting of an ultrasonic generator (25 kHz frequency, 1500 W power output) and a titanium transducer.

Industrial-scale ultrasound equipment are quite new since only now have their application been an approved technique in wineries (Resolution OENO). One of them is the so-called ULTRAWINE equipment (Figs. 7.1 and 7.2), patented by Agrovin S.A. (Alcázar de San Juan, Spain), which are 50 -W equipment that allow working with flow rates of 1000−50,000 kg/h, at low frequency values (15−35 kHz) and with a power density of 0.1−500 W/cm^3.

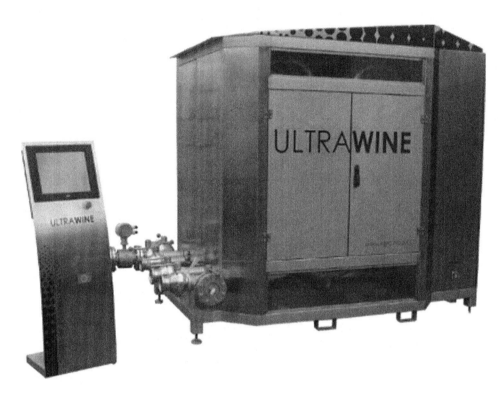

FIGURE 7.1   ULTRAWINE equipment for large wineries.

FIGURE 7.2   Sonoplates in the ULTRAWINE system.

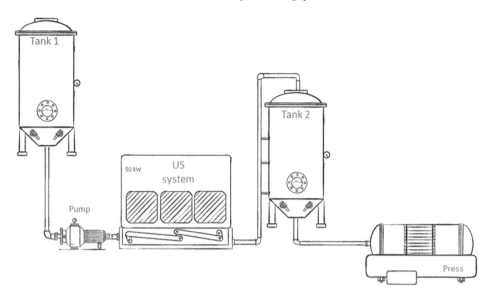

FIGURE 7.3   Scheme of one possible winery flow diagram including the sonication process.

This equipment is composed of several sonoreactors, which are hexagonal pipes in which the sonoplates are coupled. The crushed and destemmed grape must or wine circulates through them, and this is the place where the sonication occurs. The sonoplates or transducers are responsible for transforming an electrical signal into an ultrasonic wave. After receiving the electrical energy, the piezoceramic material inside the sonoplate deforms, producing turbulence in the surrounding environment that will generate the so-called cavitation bubble. The design of the equipment and the process limit the temperature increase to less than 5°C (Fig. 7.3).

7.4 Possible uses of US in white wine technology

The physical (micro-mechanical shocks caused by cavitation effect) and chemical (formation of free radicals) effects of the high-power ultrasound (HPU) can be used to achieve certain advantages during the white wine winemaking process:

7.4.1 Microbiological control and SO₂ reduction

During the production of wine, several types of microorganisms are involved, with some of them being responsible for spoilage problems. These undesired microorganisms have to be controlled, especially at the beginning of the process, to ensure the implantation of the desired organism for the alcoholic and/or malolactic fermentation and for facilitating their rapid completion, avoiding a negative effect on the final composition and organoleptic characteristics of the wine (Sponholz, 1993).

Traditional methods employed to counteract spoilage microorganisms are targeted at removing the undesirable microorganisms and involve the use of chemical preservatives (e.g., sulfur dioxide, dimethyl dicarbonate), thermal pasteurization, or the removal of spoilage microorganisms by filtration. Such interventions are sometimes accompanied by the occurrence of off-flavors or even the loss of aroma, so alternate treatments such as US are always of interest (Gracin et al., 2016).

US below the cavitational threshold have been used to enhance microbial productivity, whereas the cavitation obtained through high-power US is generally associated with the killing of microbial cells. Such killing has been observed for bacteria, yeast, fungi, algae, and protozoa (Jiranek, Grbin, Yap, Barnes, & Bates, 2008).

Ultrasound-assisted microbial inactivation depends on the size, shape, and species of the microorganisms (Hoffmann, Hua, & Höchemer, 1996; Jambrak & Herceg, 2014; Kentish & Ashokkumar, 2011; Raso, Pagán, Condón, & Sala, 1998; Wu, Joyce, & Mason, 2012), with the bigger cells showing more sensitivity to their effects than the smaller ones. The inactivation is due to both the physical and chemical effects generated from the collapse of acoustic cavitation bubbles (Wu et al., 2012). The physical effects involve intense shock waves and shear forces produced by the bubble collapse and acoustic streaming that breakdown biological cell membranes (Cameron, McMaster, & Britz, 2008; Jambrak, Mason, Lelas, Herceg, & Herceg, 2008; Jambrak, Lelas, Mason, Krešić, & Badanjak, 2009; Raso et al., 1998; Wu et al., 2012). The chemical effects are due to the formation of the free radicals (Wu et al. 2012), which can attack cell membranes, leading to lysis of the cell walls (Koda, Miyamoto, Toma, Matsuoka, & Maebayashi, 2009).

Experience has shown that when US is applied alone, intense ultrasound power and long contact times are required to inactivate microorganisms in ambient conditions. Some improvements on the effectivity of ultrasound for microbial inactivation can be obtained by combination with other treatments such as high temperature (thermosonication), pressure (manosonication), or heat and pressure (mano-thermo sonication) (Knorr, Zenker, Heinz, & Lee, 2004; Piyasena, Mohareb, & McKellar, 2003).

The influence of US to inactivate or reduce microbiota on must or wine has been reported in several studies. Luo, Schmid, Grbin, and Jiranek (2012) studied the effect of US on a range of yeasts and bacteria associated with wine production. The viability of these yeasts was more affected than that of the spoilage bacteria investigated in this study because of their higher size and oval shape. In another study, Gracin et al. (2016) employed continuous flow through high-power sonication to reduce spoilage microorganisms. They observed a significant reduction in microbial counts of *Brettanomyces* and lactic acid bacteria in wine. However, the time required for a 90% reduction of bacteria was slightly longer, which can be explained by the greater sensitivity of yeasts and their larger cells when compared to bacteria (Jambrak & Herceg, 2014; Wu et al., 2012). Cui, Lu, Liu, and Wang (2012) investigated different ending fermentation technologies (single SO_2 treatment, combined high-voltage electrostatic field treatment, combined ultrasound/SO_2 treatment, and combined pasteurization/SO_2 treatment) for microbial stability and quality of low-alcohol sweet white wine, observing that combined ultrasound and SO_2 treatment (40 kHz/20 min)/SO_2 (40 mg/L) was the best ending fermentation technology, which had a higher total lethal rate against *Saccharomyces cerevisiae*, and the wines treated in this way were rich in typical Italian Riesling grape variety flavor and attractive aromas, with a pleasant fruity taste and microbial stability.

In a winery, power US could be applied before alcoholic fermentation via a flow-through system during the transfer of musts from tank to tank or by direct treatment of a tank of must or wine. Analogous opportunities for the treatment can be envisaged for the delay or suppression of malolactic fermentation by wine sonication after primary fermentation, which implies that the addition of SO_2 in high amounts or the use of lysozyme in this wine can be avoided. In addition, in places where wines are held in oak barrels, their treatment would also be possible in situ, as the sound emitter (i.e., the sonotrode) can be introduced through the barrel bunghole, thereby obviating the need for transfer of the wine to an open tank (Yap et al., 2007).

All these results of US point to the fact that US may be used as an alternative, or rather as a complement, to SO_2 to control microbes in wine, since this technique alone cannot fully cover the actions of SO_2 (antimicrobial and antioxidant). Concretely, Gracin et al. (2016) found that a concentration of 0.625 mg/L of molecular SO_2 exerted toxic effects on species of the genera *Saccharomyces* and *Brettanomyces* using a high-power ultrasound treatment in a continuous flow system, which might prolong wine stability and allow lower usage of SO_2 when compared to the usual preservation methods. It could be a promising method for the production of quality wines and a low-sulfite content as was reported by Lukić et al. (2020) since no significant differences in the phenolic and volatile composition among sonicated low-sulfite white wine and the one with standard-sulfite content were found by these authors.

Nevertheless, we should not forget that the effect caused by US could also be desirable for wine fermentation in some cases. Low ultrasonic intensities accelerate the growth of *S. cerevisiae* with a resulting reduction in fermentation time. It is clear that although the influence of this technology on the must and wine microbiota has been reported, its effect on the targeted microbiota in real must and winemaking conditions is not clear and more experience is necessary.

7.4.2 Enzymatic control

For the stabilization of must or wine, enzymes such as polyphenoloxidases and peroxidases must be inactivated or their activity must be reduced. Enzyme inactivation can be easily achieved by heat treatment. However, in wine the application of heat to inactivate the enzymes may be a problem as heat can negatively modify some wine properties such as flavor and color, hence the increased interest in an alternative method such as high-power ultrasound in enzyme inactivation.

The physical and chemical effects of cavitation and free radical formation are responsible for enzyme inactivation, although the dominant mechanism depends on enzyme structure. The effects of ultrasonic waves on proteins are very complex. Sensitivity to US also depends on the conditions of the treatment (frequency, ultrasonic intensity, etc.) as well as the nature of the enzyme (McClements, 1995; Terefe et al., 2015).

In the case of white wines, the application of ultrasound on the grape after the crushing and destemming processes, or even on the must after the pressing process, could not only inhibit the enzymes that cause oxidative processes, avoiding or reducing the use of SO_2, but also, the settling process could be favored without the need to apply commercial pectolytic enzymes. Tests carried out in our winery (data not shown) have shown that although after the application of US on a white grape must the amount of suspended solids increases (due to a greater breakdown of the cell walls of the skin and pulp), US breaks the more complex structures of the pectins, thus causing rapid precipitation of suspended solids and obtaining a clearer must.

7.4.3 Extraction of aroma compounds during grape processing

Wine aroma is one of the most important organoleptic characteristics of this product and it can be quite complex, with more than 1000 compounds taken part. The wine aroma can be divided into different classes according to the origin of the compounds: varietal aroma, typical of grape variety; preformentative aroma, originating from the different processes occurring from at harvest to the beginning of the alcoholic fermentation; fermentative aroma, produced by yeast and bacteria during alcoholic and malolactic fermentations; and postfermentative aroma, which is due to all the changes that occurred during conservation and aging of wine, including the apparition of new compounds due to the use of some aging systems such as oak barrels (Zhu et al. 2016).

The importance of wine varietal aroma, especially in white wines and for those elaborated from certain varieties, is paramount. Some studies have shown that the contribution of grape-derived volatile compounds is consistent with the wine quality classification, indicating that the wine aroma quality is clearly associated with the must's free varietal volatile composition (Rocha et al., 2010).

Although grapes are the origin of varietal aroma, we cannot ignore that, since yeasts contribute to wine aroma by utilizing grape juice constituents and biotransforming them into aroma or flavor-impacting components, the aroma profile of wine is also related to the amount of assimilable nitrogen available for the yeast present in the must. It has been shown that the varietal aroma character of certain cultivars could be partially explained by the amino acid composition of the grape must (Montevecchi, Masino, Simone, Cerretti, & Antonelli, 2015). Therefore changes in the must composition and nutrition concentration could result in a significant influence on the growth and metabolism of wine yeasts and alter the formation profiles of aroma compounds (Bell & Henschke, 2005).

Depending on the variety, different molecules are responsible for varietal aromas in white wines. In the case of Muscat-type varieties, monoterpenes are one of the main compounds responsible for their aroma, being present as free volatile compounds or as nonvolatile odorless glycoconjugates (Günata, Bitteur, Brillouet, Bayonove, & Cordonnier, 1988) that represent an important reservoir of glycosidic aroma precursors that can be released under natural conditions and/or by enzymatic or acid hydrolysis during wine elaboration (Tavernini, Ottone, Illanes, & Wilson, 2020).

Another set of varietal aroma compounds released from odorless bound precursors are volatile thiols that give some white wines, such as Sauvignon blanc wines, their characteristic bouquet (Styger, Prior, & Bauer, 2011), the most significant being 4-methyl-4-mercaptopentan-2-one and 3-mercapto-1-hexanol. These compounds are not present in grape juice in their free form but are present in grape must as odorless, nonvolatile, cysteine-bound conjugates, and wine yeast is responsible for the cleaving of the thiol from the precursor during alcoholic fermentation (Styger et al., 2011).

The carotenoids also play a role in varietal aroma. These isoprenoid tetraterpenes originate from the precursor compound mevalonate. Oxidation of these carotenoids produces volatile and strong odor-contributing fragments known as C13-norisoprenoids, which include β-ionone (viola aroma), β-damascenone (exotic fruits), β-damascone (rose), and β-ionol (fruits and flowers), as reported by Iriti and Faoro (2006).

There are several practices commonly used to increase the distinctive aromatic characteristic of white wines, to try and manipulate specific aroma compounds and increase the complexity of wine. In this way, and taking into account that the compounds responsible for varietal aroma are predominant in grape, the preformentative "skin contact" technique can be used to extract these compounds and probably improve wine quality. Low-temperature maceration prior to alcoholic fermentation is a very popular winemaking technology in the production of high-quality wines (Cai et al., 2014). It can extract aroma and polyphenol compounds and, moreover, can also alter the amount of nutrient substances released in grape must, such as nitrogen sources (amino acids), vitamins, and fatty acids, and thus influence cell growth and aromatic compound formation (Luan, Zhang, Duan, & Yan, 2018).

The use of pulsed electric fields (PEF) is another innovative practice for processing white grape varieties, such as in the work of Praporscic, Lebovka, Vorobiev, & Mietton-Peuchot, (2007), working with Muscadelle, Sauvignon, and Semillon varieties; and that of Comuzzo, Marconi, Zanella, and Querzè (2018), who reported that the pretreatment of grapes led to an increase in the extraction of varietal aroma precursors. Fauster et al. (2020) studied the combined application of PEF and an enzymatic treatment on white wine mash and observed an increase in the concentration of selective esters, especially of the variety Traminer.

The use of US could also facilitate the extraction of compounds of interest from grape skins and affect wine aroma, although very few works are available pertaining to white wines in this regard. Regarding red wines, Bautista-Ortín et al. (2017) studied the application of US on crushed grapes, looking for a reduction in the maceration time needed for the extraction of phenolic and volatile compounds from Monastrell grapes, and they found only small differences in the wine volatile composition, although the levels of terpenes and norisoprenoids in wine elaborated with sonicated grapes and 3 days of skin maceration time were similar to those of the control wines made with 8 days of skin maceration. Martínez-Pérez, Bautista-Ortín, Pérez-Porras, Jurado, and Gómez-Plaza (2020) observed that the sum of total alcohols and total esters was not significantly different when compared to the aroma of control and sonicated red wines; only small differences in some esters could be observed.

It seems clear that in varieties without a significant amount of varietal compounds, if US does not change the aminoacidic and nitrogen content of must, the small effect on the aroma compounds is not surprising. The work conducted in our winery, measuring the content of YAN (yeast assimilable nitrogen) in different must from different varieties, did not find significant differences in their content due to the application of US (Fig. 7.4).

Our research group also studied the overall volatile compounds measured from control and sonicated Sauvignon blanc must, before and after being pressed (Fig. 7.5).

The results, expressed as the percentage of increase in aroma compounds compared to control must, showed how all the families increased with the sonication of the crushed grapes, especially terpenes and norisoprenoids as well as higher alcohols in pressed sonicated must.

Also, we compared the effect of the application of US on wine macerated for 8 h to a control wine with no maceration. The results were similar, with a similar increase in alcohols, aldehydes, and esters but with a positive increase in terpenes in sonicated must and a decrease in macerated must (Fig. 7.6).

Sensorial analysis of Sauvignon wines made from these sonicated grapes showed an increase in the structure, acidity, and fruity aroma without an increase in bitterness or astringency (Fig. 7.7).

A very interesting study was conducted by Roman et al. (2020), who applied US to Sauvignon blanc crushed grapes, focusing on volatile thiol composition. When looking at the precursors' concentration, their concentration in grape juice was not positively affected by US, and moreover, that of 3-S-glutathionyl mercaptohexan-1-ol (3MH) showed a negative trend with the US treatment time applied. Looking for an explanation for this effect, ultrasound was applied in a model solution spiked with 3MH and 4-methyl-4-mercaptopentan-2-one (4MMP) precursors, reproducing the conditions of grapes. They found again that US decreased the concentration of precursors but the degradation of precursors was associated with a significant increase in 3MH and 4MMP, which

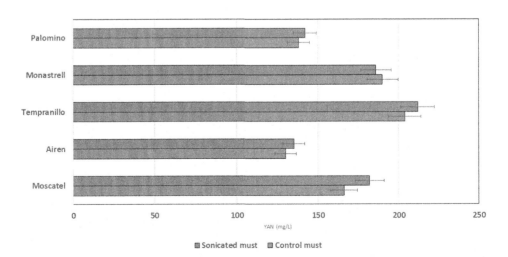

FIGURE 7.4 Content of yeast assimilable nitrogen (YAN) in control and sonicated must from different red and white grape varieties.

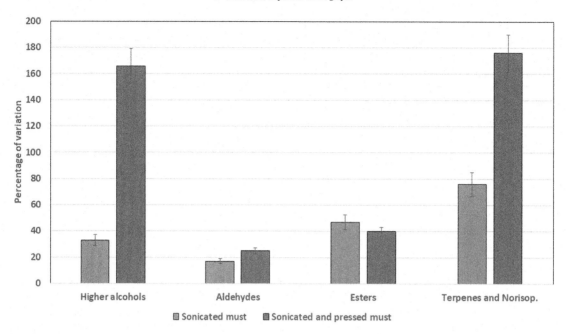

FIGURE 7.5 Percentage of increase in some volatile compounds, with respect to a control must, when crushed grapes were sonicated, before and after crushing.

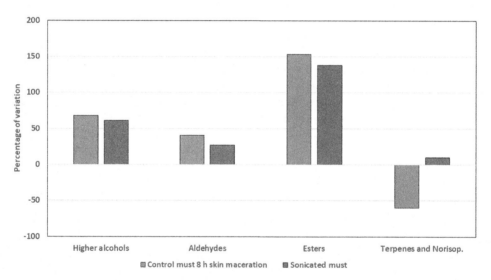

FIGURE 7.6 Percentage of variation of some volatile compounds, with respect to a control must, when the must was cold macerated for 8 h or crushed grapes were sonicated.

was a very interesting result regarding the formation of volatile thiols; following sonication is a finding potentially exploitable in the wine industry through specific vinification protocols.

Another effect that is worth considering is the effect of US on the enzymatic activity of β-D-glucosidase. This enzyme can release aroma precursors to improve the flavor of must and wines, but the hydrolysis efficiency of the enzyme is low. However, the studies of Sun et al. (2019) showed that a low temperature (20°C−45°C), low ultrasonic intensity (<181.53 W/cm^2), and short treatment time (<15 min) led to the activation of β-D-glucosidase. This study suggests that ultrasound combined with β-D-glucosidase can be used in aroma-enhancing.

Some studies also point to the use of US in finished wines. Zhang, Xua, Chen, Zhao, and Xue (2020) evaluated the effect of US on higher alcohol content in wine, reporting a decrease in these compounds, suggesting a modification of wine quality due to the negative effects of excessive content of higher alcohols on wine.

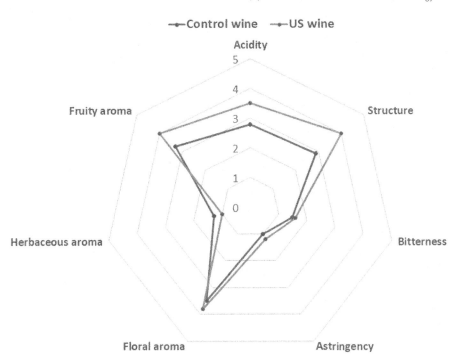

7.4.4 Extraction of oak compounds into wine

The white wine market has been monopolized for many years by young varietal wines, which should be consumed within a short period after bottling to avoid the loss of freshness and fruity character, with different enological practices being used to enhance varietal aromas such as preformentative skin-contact treatment or the use of glycosidic enzymes or more innovative techniques such as PEF and US.

Wood aging in the case of white wines has only been used occasionally, mainly by fermentation in barrels or by aging on lees (Jiménez Moreno & Ancín Azpilicueta, 2007; Liberatone, Pati, Del Nobile, & La Notte, 2010). Aging in wood can increase the variety of white wines, which gives a response to new consumer demands. Usually, for white wines, the objective is to obtain a quality wine with new sensory sensations but without masking the primary and secondary aromas specific to each grape variety (Alañón, Díaz-Maroto, & Pérez-Coello, 2018).

However, not all white wines are suitable for aging in oak barrels, since the aging process can result in a decrease in the wine quality and oak aging is an expensive time-consuming process. Oak wood chips could present a good alternative to the use of barrels for the aging of white wines, leading to a more complex aromatic profile. Additionally, oak wood has an antioxidant capacity, mainly depending on the ellagitannin content that influences the oxidation − reduction potential of wines and therefore its oxidative stability (Alañón et al., 2018). Sanchez-Palomo, Alonso-Villegas, Delgado, and Gonzalez-Viñas (2017) studied the effect that the addition of wood at different stages of the winemaking process has on the volatile composition and sensory characteristics of Verdejo wines, observing an increase in aroma compounds without a decrease in the fresh and fruity characteristics.

The process of aging with oak chips could be accelerated using US. According to Tao, García, and Sun (2014a) and Tao, Wu, Zhang, and Sun (2014b) in a practical application of wine aging with oak chips, the aging period usually lasts for at least a couple of weeks and the acceleration of the extraction of oak-related compounds from oak chips into wine could be really positive. The contact time can be reduced with the use of US. They studied the transference of compounds from chips to a model solution and a significant increase in different compounds in the model wine was observed under ultrasound treatment for 150 min. However, compared to acoustic energy density, the temperature played a more important role in the migration of phenolics from oak chips into wine.

Although no studies regarding its application on white wines could be found, this technology has been applied to accelerate the extraction of wood compounds from oak chips, at the laboratory scale, during the aging of a Sherry Brandy. After 30 days, the proposed method allows obtaining a Brandy with similar analytical and sensorial characteristics to those that have been aged in the traditional way for an average term of between 6 and 18 months (Schwarz, Rodríguez, Sanchez, Guillen, & Garcia-Barroso, 2014). The same research group studied the influence of the movement of the liquor through oak chips, and the results showed that high-power ultrasound, of nearly 40 W/L, in addition to the movement of the spirit, improves the extraction of phenolic compounds up to 33.94% after 7 days of aging. Sensorial analysis of the aged samples showed that the aged spirits had better ratings than the initial distilled wine (Delgado-González et al., 2017).

The application on white wines needs to be studied.

7.4.5 Ultrasound and aging on lees

Aging on lees is a traditional technique for aging white wines. It was originally used in the elaboration of barrel-fermented wines in Burgundy and during bottle aging of sparkling wines. Its importance lies in the peculiar characteristics that it adds to the wines, which originate from the autolysis of yeasts.

Yeast autolysis is a process that occurs after yeast death, consisting of the breakdown and degradation of cell structures by their own enzymes. As a consequence of this breakdown and fragmentation of the cellular material, molecules of different nature are released into the wine, which may affect its colloidal balance, structure, color stability, and aromatic profile, with important organoleptic repercussions. These molecules can come from inside the yeast cell (nucleotides, nucleosides, amino acids, and peptides) or from the cell walls (glycans and mannoproteins).

The impact on the aromatic profile of wine due to volatile molecules from autolysis is remarkable. The nature of these molecules is varied (Lubbers, Charpentier, Feuilliat, & Voilley, 1994): esters, mostly ethyl esters of fatty acids; alcohols, higher and terpenic; aldehydes, highlighting methyl-3-butanal and benzaldehyde; volatile sulfur compounds; and lactones.

On the other hand, the release in the wine of an important colloidal load during the autolysis changes the wine's in-mouth structure, accentuating the sensation of smoothness and density (Fornairon-Bonnefond, Camarasa, Motounet, & Salmon, 2002).

Yeast autolysis can be very rapid in media at pH 4.5−5.0 and at temperatures of 35°C to 40°C, while in real wine conditions, with pH values between 3.0 and 3.8 and temperatures below 15°C, a period of 2−3 months is needed. The autolysis may be intensified if the lees are periodically shaken, as is done with the "bâtonnage" process during the aging of certain white wines in barrels (Caridi, 2006; Doco, Vuchot, Cheynier, & Moutounet, 2003). These long periods of storage time represent an increase in the economic investment for the wineries as well as an increase in the risks of organoleptic and microbiological alteration of the wines, hence the interest in recent years in finding new physical technologies, such as ultrasound treatments, that speed up this process and help minimize the costs and risks associated with aging wines on lees in a traditional way.

Numerous studies have been carried out to verify the effect of ultrasound breakdown on the cellular structures of yeast and the release of the different fractions of their walls to the medium to accelerate the aging process on lees (Celotti & Ferraretto, 2016; García Martín & Sun, 2013; Tao, García et al., 2014; Tao, Wu et al., 2014).

Ferraretto et al. (2013) verified in their tests an increase in soluble colloids from yeast cell walls and that their formation depended on the duration and amplitude of the ultrasound applied. However, an increase in the extraction of soluble colloids was observed for all the studied amplitudes, even within short exposure times, which, from an industrial point of view, may be of great importance when incorporating this technology with the aim of shortening the process times. On the other hand, Cacciola, Ferran Batlló, Ferraretto, Vincenzi, and Celotti (2013) observed that the exposure time for the ultrasound application on lees had a greater impact on the efficacy of the treatment than the increase in the amplitude of the applied waves. The same researchers verified that the application of ultrasound shortened the extraction times of compounds from the lees in comparison with the conventional stirring procedure and/or with the application of enzymes with glucanase activity, obtaining a significant increase in the transfer of total colloids, mainly proteins and polysaccharides, with a reduction in the diameter of the colloidal particles. The same conclusions were drawn by Liu et al. (2016). In this study, it was found that the additional application of the ultrasound treatment on abrasive products significantly improved the efficiency of the process in terms of releasing compounds from the cell walls, but on the other hand, it also significantly decreased the concentration of wine volatile compounds.

Although it seems clear that the use of ultrasound accelerates the aging process on lees and the transfer of compounds from yeast cell walls to wine, it is important to highlight that the optimization of the parameters in the application of the technique is key to obtaining reproducible and stable results over time (Zhang, Fu, & García Martín, 2017). The cavitation phenomenon that occurs with the application of ultrasound can promote the formation of free radicals from the ethanol content of wines, which in turn can initiate various oxidative reactions of phenolic compounds and aromas (Zhang et al., 2015; Zhang, Shen, Fan, & García Martín, 2016).

Another option to avoid degradative and damaging reactions for the general quality of the wines is the application of US on a concentrate of lees from the own vinification. Another option is the generation of a concentrate with the cell walls of a single selected yeast species, produced exogenously, which can even be pasteurized to reduce bacterial contamination, to later add the concentrate to the wines of the winery according to the technological objectives established (Kulkarni et al., 2015).

However, most of the trials and tests with US on wines have been carried out at the laboratory or pilot scale, so it is necessary to carry out more studies to optimize the technique at the industrial level, as well as to establish measurable and reproducible parameters that facilitate the monitoring and management of the process in the most successful way possible.

7.5 Final remarks

Ultrasound technology is a new and promising technology that may have interesting applications during the white wine winemaking process. Ultrasound in winemaking can be applied as a continuous pretreatment of crushed grapes before the pressing, eliminating the prefermentative maceration step and ensuring a better extraction of varietal aromas.

However, many other possible applications could be developed with this technology, and other applications are under study to evaluate their effectiveness in white wine technology, such as the acceleration of aroma extraction from oak chips, the reduction of SO_2 use, and the acceleration of the lees aging process.

Ultrasound can be considered a sustainable, environment-friendly technology that may help optimize the costs of processes, increase must and wine quality, and reduce the processing time. Since it is an interesting technology, already approved by the OIV and with industrial technology available, further studies should be conducted to optimize its use in white wine winemaking processes.

References

Alañón, M. E., Díaz-Maroto, M. C., & Pérez-Coello, M. S. (2018). New strategies to improve sensorial quality of white wines by wood contact. *Beverages*, 4, 91.

Bautista-Ortín, A. B., Jiménez-Martínez, M. D., Jurado, R., Iniesta, J. A., Terrades, S, Andrés, A., & Gómez-Plaza, E. (2017). Application of high-power ultrasounds during red wine vinification. *International Journal of Food Science & Technology*, 52, 1314–1323. Available from https://doi.org/10.1111/ijfs.13411.

Bell, S. J., & Henschke, P. A. (2005). Implications of nitrogen nutrition for grapes, fermentation and wine. *Australian Journal of Grape and Wine Research*, 11, 242–295.

Cacciola, V., Ferran Batlló, I., Ferraretto, P., Vincenzi, S., & Celotti, E. (2013). Study of the ultrasound effects on yeast lees lysis in winemaking. *European Food Research and Technology*, 236, 311–317.

Cai, J., Zhu, B. Q., Wang, Y. H., Lu, L., Lan, Y. B., Reeves, M. J., & Duan, C. (2014). Influence of pre-fermentation cold maceration treatment on aroma compounds of Cabernet Sauvignon wines fermented in different industrial scale fermenters. *Food Chemistry*, 154, 217–229.

Cameron, M., McMaster, L., & Britz, T. (2008). Electron microscopic analysis of dairy microbes inactivated by ultrasound. *Ultrasonic Sonochemistry*, 15, 960–964.

Caridi, A. (2006). Enological functions of parietal yeast mannoproteins. *Antonie van Leeuwenhoek Journal of Microbiology.*, 89, 417–422.

Celotti, E. & Ferraretto, P. (2016). Studies for the ultrasound application in winemaking for a low impact enology. In *Proccedings of the 39th O.I.V. Congress, Brasil.*

Chemat, F., & Khan, M. (2011). Applications of ultrasound in food technology: Processing, preservation and extraction. *Ultrasonics Sonochemistry*, 18, 813–835.

Comuzzo, P., Marconi, M., Zanella, G., & Querzè, M. (2018). Pulsed electric field processing of white grapes (cv. *Garganega*): Effects on wine composition and volatile compounds. *Food Chemistry*, 264, 16–23.

Cui, Y., Lu, W., Liu, J. F., & Wang, B. J. (2012). Effect of different ending fermentation technologies on microbial stability of Italian Riesling low alcohol sweet white wine. *Advances in Material Research*, 393–395, 1165–1168.

Delgado-González, M. J., Sánchez-Guillén, M. M., García-Moreno, M. V., Rodríguez-Dodero, M. C., García-Barroso, C., & Guillén- Sánchez, D. (2017). Study of a laboratory-scaled new method for the accelerated continuous ageing of wine spirits by applying ultrasound energy. *Ultrasonics Sonochemistry*, 36, 226–235.

Doco, T., Vuchot, P., Cheynier, V., & Moutounet, M. (2003). Structural modification of wine arabinogalactans during aging on lees. *American Journal of Enology and Viticulture*, 54, 150–157.

Fauster, T., Philipp, C., Hanz, K., Scheibelberger, R., Teufl, T., Nauer, S., … Jaeger, H. (2020). Impact of a combined pulsed electric field (PEF) and enzymatic mash treatment on yield, fermentation behaviour and composition of white wine. *European Food Research and Technology*, 246, 609–620.

Ferraretto, P., Cacciola, V., Ferran Batlló, I., & Celotti, E. (2013). Ultrasounds application in winemaking: grape maceration and yeast lysis. *Italian Journal of Food Science*, 25, 160–168.

Fornairon-Bonnefond, C., Camarasa, C., Motounet, M., & Salmon, J. M. (2002). New trends on yeast autolysis and wine ageing on lees: A bibliographic review. *Journal International des Sciences d la Vigne et du Vin, 36,* 46−69.

Gambacorta, G., Trania, A., Punzi, R., Fasciano, C., Leo, R., Fracchiolla, G., & Faccia, M. (2017). Impact of ultrasounds on the extraction of polyphenols during winemaking of red grapes cultivars from southern Italy. *Innovative Food Science and Emerging Technologies, 43,* 54−59.

García Martín, J. F., & Sun, D. W. (2013). Ultrasound and electric fields as novel techniques for assisting the wine ageing process: The state-of-the-art research. *Trends in Food Science and Technology, 33,* 40−53.

Gracin, L., Jambrak, A. R., Juretić, H., Dobrović, S., Barukčić, I., Grozdanović, M., & Smoljanić, G. (2016). Influence of high power ultrasound on *Brettanomyces* and lactic acid bacteria in wine in continuous flow treatment. *Applied Acoustics, 103,* 143−147.

Günata, Z., Bitteur, S., Brillouet, J. M., Bayonove, C., & Cordonnier, R. (1988). Sequential enzymic hydrolysis of potentially aromatic glycosides from grape. *Carbohydrate Resesarch, 184,* 139−149.

Hoffmann, M., Hua, I., & Höchemer, R. (1996). Application of ultrasonic irradiation for the degradation of chemical contaminants in water. *Ultrasonic Sonochemistry, 3,* S163−S172.

Iriti, M., & Faoro, F. (2006). Grape phytochemicals: A bouquet of old and new nutraceuticals for human health. *Medical Hypotheses, 67,* 833−838.

Jambrak, A., & Herceg, Z. (2014). Application of ultrasonics in food preservation and processing. In S. Bhattacharya (Ed.), *Conventional and advanced food processing technologies* (pp. 515−536). New Jersey: John Wiley & Sons, Ltd.

Jambrak, A., Lelas, V., Mason, T., Krešić, G., & Badanjak, M. (2009). Physical properties of ultrasound treated soy proteins. *Journal of Food Engineering, 93,* 386−393.

Jambrak, A., Mason, T., Lelas, V., Herceg, Z., & Herceg, I. (2008). Effect of ultrasound treatment on solubility and foaming properties of whey protein suspensions. *Journal of Food Engineering, 86,* 281−287.

Jiménez Moreno, N., & Ancín Azpilicueta, C. (2007). Binding of oak volatile compounds by wine lees during simulation of wine ageing. *LWT-Food Science and Technology, 40,* 619−624.

Jiranek, V., Grbin, P., Yap, A., Barnes, M., & Bates, D. (2008). High power ultrasonics as a novel tool offering new opportunities for managing wine microbiology. *Biotechnology Letters, 30,* 1−6.

Kentish, S., & Ashokkumar, M. (2011). The physical and chemical effects of ultrasound. In H. Feng, G. Barbosa-Canovas, & J. Weiss (Eds.), *Ultrasound technologies for food and bioprocessing* (pp. 1−12). New York: Springer.

Knorr, D., Zenker, M., Heinz, V., & Lee, D. (2004). Application and potential of ultrasonics in food processing. *Trends in Food Science and Technology, 15,* 261−266.

Koda, S., Miyamoto, M., Toma, M., Matsuoka, T., & Maebayashi, M. (2009). Inactivation of Escherichia coli and Streptococcus mutans by ultrasound at 500 kHz. *Ultrasonic Sonochemistry, 16,* 655−659.

Kulkarni, P., Loira, I., Morata, A., Tesfaye, W., González, M. C., & Suárez Lepe, J. A. (2015). Use of non-Saccharomyces yeast strains coupled with ultrasound treatment as a novel technique to accelerate ageing on lees of red wines and its repercussion in sensorial parameters. *LWT-Food Science and Technolgoy, 64,* 1255−1262.

Leong, T., Ashokkumar, M., & Kentish, S. (2011). The fundamentals of power ultrasound. A review. *Acoustic Australia, 39,* 54−68.

Leong, T., Wooster, T., Kentish, S., & Ashokkumar, M. (2009). Minimising oil droplet size using ultrasonic emulsification. *Ultrasonic Sonochemistry, 16,* 721−727.

Liberatone, M. T., Pati, S., Del Nobile, M. A., & La Notte, E. (2010). Aroma quality improvement of Chardonnay White wine by fermentation and ageing in barrique on lees. *Food Research International, 43,* 996−1002.

Liu, L., Loira, I., Morata, A., Suárez-Lepe, J. A., González, M. C., & Rauhut, D. (2016). Shortening the ageing on lees process in wines by using ultrasound and microwave treatments both combined with stirring and abrasion techniques. *European Food Research and Technology, 242,* 559−569.

Luan, Y., Zhang, B., Duan, C., & Yan, G. (2018). Effects of different pre-fermentation cold maceration time on aroma compounds of Saccharomyces cerevisiae co-fermentation with Hanseniaspora opuntiae or Pichia kudriavzevii. *LWT - Food Science and Technology, 92,* 177−186.

Lubbers, S., Charpentier, C., Feulliat, M., & Voilley, A. (1994). Influence of yeast walls on the behavior of aroma compounds in a model wine. *American Journal of Viticulture and Enology, 45,* 29−33.

Lukić, K., Brnčić, M., Ćurko, N., Tomašević, N., Jurinjak, A., Karin, T., & Ganić, K. (2020). Quality characteristics of white wine: The short- and long-term impact of high power ultrasound processing. *Ultrasonics Sonochemistry, 68,* 105194.

Luo, H., Schmid, F., Grbin, P. R., & Jiranek, V. (2012). Viability of common wine spoilage organisms after exposure to high power ultrasonics. *Ultrasonic Sonochemistry, 19,* 415−420.

Martínez-Pérez, P., Bautista-Ortín, A. B., Pérez-Porras, P., Jurado, R., & Gómez-Plaza, E. (2020). A New approach to the reduction of alcohol content in red wines: The use of high-power ultrasounds. *Foods, 9,* 726.

Mason, T. J., & Lorimer, J. P. (2002). *Applied sonochemistry: the uses of power ultrasound in chemistry and processing.* Weinheim: Wiley-VCH.

Mason, T. J. (1998). Power ultrasound in food processing. The way forward. Ultrasound. In M. Povey, & T. Mason (Eds.), *Food Processing.* Londres: Chapman & Hall.

McClements, D. (1995). Advances in the application of ultrasound in food analysis and processing. *Trends in Food Science and Technology, 6,* 293−299.

Montevecchi, G., Masino, F., Simone, G., Cerretti, E., & Antonelli, A. (2015). Aromatic profile of white sweet semi-sparkling wine from Malvasia di Candia Aromatica grapes. *South African Journal for Enology and Viticulture, 36,* 267−276.

Mulet, A., Cárcel, J. A., Benedito, J., Rosselló, C., & Simal, S. (2003). Ultrasonic mass transfer enhancement in food processing. Transport Phenomena. In J. Welti-Chanes, J. Vélez-Ruiz, & G. Barbosa-Canova (Eds.), *Food Processing.* New York: CRC Press.

Muthukumaran, S., Kentish, S., Stevens, G., & Ashokkumar, M. (2006). Application of ultrasound in membrane separation processes: A review. *Reviews in Chemical Engineering, 22,* 155−194.

Piyasena, P., Mohareb, E., & McKellar, R. (2003). Inactivation of microbes using ultrasound: a review. *International Journal of Food Microbiology, 87,* 207−216.

Praporscic, I., Lebovka, N., Vorobiev, E., & Mietton-Peuchot, M. (2007). Pulsed electric field enhanced expression and juice quality of white grapes. *Separation and Purification Technology, 52*, 520–526.

Raso, J., Pagán, R., Condón, S., & Sala, F. (1998). Influence of temperature and pressure on the lethality of ultrasound. *Applied Environmental Microbiology, 64*, 465–471.

Richards, W., & Loomis, A. (1927). The chemical effects of high frequency sound waves I. A preliminary survey. *Journal of the American Chemical Society, 49*, 3086–3100.

Rocha, S., Coutinho, P., Coelho, E., Barros, A., Delgadillo, I., & Coimbra, M. (2010). Relationships between the varietal volatile composition of the musts and white wine aroma quality. A four year feasibility study. *LWT - Food Science and Technology, 43*, 1508–1516.

Roman, T., Tonidandel, L., Nicolini, G., Bellantuono, E., Barp, L., Larcher, R., & Celotti, E. (2020). Evidence of the possible interaction between ultrasound and thiol precursors. *Foods, 9*, 104.

Sanchez-Palomo, E., Alonso-Villegas, R., Delgado, J. A., & Gonzalez-Viñas, J. (2017). Improvement of Verdejo white wines by contact with oak chips at different winemaking stages. *LWT - Food Science and Technology, 79*, 111–118.

Schwarz, M., Rodríguez, M. C., Sanchez, M., Guillen, D., & Garcia-Barroso, C. (2014). Development of an accelerated aging method for Brandy. *LWT - Food Science and Technology, 59*(2014), 108–114.

Shirsath, S., Sonawane, S., & Gogate, P. (2012). Intensification of extraction of natural products using ultrasonic irradiations. A review of current status. *Chemical and Engineering Processes, 53*, 10–23.

Sponholz, W. (1993). Wine spoilage by microorganisms. In G. H. Fleet (Ed.), *Wine microbiology and biotechnology* (pp. 395–420). Reading: Harwood Academic.

Styger, G., Prior, B., & Bauer, F. (2011). Wine flavor and aroma. *Journal of Industrial Microbiology and Biotechnology, 38*, 1145–1159.

Sun, Y., Zeng, L., Xue, Y., Yang, T., Cheng, Z., & Sun, P. (2019). Effects of power ultrasound on the activity and structure of β-D-glucosidase with potentially aroma-enhancing capability. *Food Science and Nutrition, 7*, 1–7.

Suslick, K., & Crum, L. (1998). Sonochemistry and sonoluminescence. In M. J. Crocker (Ed.), *Handbook of Acoustics* (pp. 243–253). New York: Wiley Interscience.

Tao, Y., García, J. F., & Sun, D. W. (2014a). Advances in wine aging technologies for enhancing wine quality and accelerating wine aging process. *Critical Reviews in Food Science and Nutrition, 54*, 817–835.

Tao, Y., Wu, D., Zhang, Q. A., & Sun, D. W. (2014b). Ultrasound assisted extraction of phenolics from wine lees: modeling, optimization and stability of extracts during storage. *Ultrasonic Sonochemistry, 21*, 706–715.

Tavernini, L., Ottone, C., Illanes, A., & Wilson, L. (2020). Entrapment of enzyme aggregates in chitosan beads for aroma release in white wines. *International Journal of Biological Macromolecules, 154*, 1082–1090.

Terefe, N., Buckow, R., & Versteeg, C. (2015). Quality-related enzymes in plant-based products: Effects of novel food-processing technologies part 3: Ultrasonic processing. *Critical Reviews in Food Science and Nutrition, 55*, 147–158.

Wu, X., Joyce, E., & Mason, T. (2012). Evaluation of the mechanisms of the effect of ultrasound on Microcystis aeruginosa at different ultrasonic frequencies. *Water Research, 46*, 2851–2858.

Yap, A., Jiranek, V., Grbin, P., Barnes, M., & Bates, D. (2007). The application of high power ultrasonics to enhance winemaking processes and wine quality. *Australian and New Zealand Wine Industry Journal, 22*, 44–48.

Zhang, Q., Xua, B., Chen, B., Zhao, B., & Xue, C. (2020). Ultrasound as an effective technique to reduce higher alcohols of wines and its influencing mechanism investigation by employing a model wine. *Ultrasonics Sonochemistry, 61*, 104813.

Zhang, Q. A., Fu, X. Z., & García Martín, J. F. (2017). Effect of ultrasound on the interaction between (-)-epicatechin gallate and bovine serum albumin in a model wine. *Ultrasonics Sonochemistry, 37*, 405–413.

Zhang, Q. A., Shen, Y., Fan, X. H., & García Martín, J. F. (2016). Preliminary study of the effect of ultrasound on physicochemical properties of red wine. *CyTA-Journal of Food, 14*, 55–64.

Zhang, Q. A., Shen, Y., Fan, X. H., García Martín, J. F., Wang, W., & Song, Y. (2015). Free radical generation induced by ultrasound in red wine and model wine: An EPR spin trapping study. *Ultrasonic Sonochemistry, 27*, 96–101.

Zhu, F., Du, B., & Li, J. (2016). Aroma compounds in wine. In A. Morata, & I. Loira (Eds.), *Grape and Wine Biotechnology* (pp. 273–283). London: Intechopen.

8

Settling. Clarification of musts

Aude Vernhet

SPO, Univ Montpellier, INRAE, Institut Agro, Montpellier, France

8.1 Introduction

Must cleaning before fermentation is an important step in the production of white wines. Musts most often present high turbidity levels, linked to the presence of suspended solid debris formed during extraction processes. The production of high-quality white wines requires the removal of some of these suspended solids, as the vinification of musts that are too turbid leads to aromatic defects as well as risks of color instability sometimes. However, the level of clarification must be controlled: excessive clarification can lead to sluggish or stuck fermentation. In the usual ranges, the level of clarification also affects the aromatic characteristics and typicity of wines. The positive versus the negative impact of suspended solids is mainly related to their lipid composition and the presence of associated enzymatic activities, which will be detailed first.

Different methods can be used to clarify musts. Their performance is dependent on the characteristics of the musts as well as those of the suspended solids, and their implementation should be carefully considered according to these characteristics, the organization of the work in the cellar, and the targeted wine profiles. These methods will be described and discussed considering the current knowledge of the characteristics of suspended solids.

8.2 Technological and qualitative impact

In white wine winemaking, the degree of must clarification is known to influence the following: (1) yeast fermentation kinetics and cell viability (Alexandre, Nguyen van Long, Feuillat, & Charpentier, 1994; Casalta, Cervi, Salmon, & Sablayrolles, 2013) and (2) the characteristics of finished wines (Karagiannis & Lanaridis, 2002; Singleton, Sieberhagen, Wet, & Wyk, 1975). The impact of suspended solids on fermentation kinetics and yeast viability is mainly attributed to their content in lipids (Deytieux, Mussard, Biron, & Salmon, 2005; Luparia, Soubeyrand, Berges, Julien, & Salmon, 2004). Sterols are essential lipids for maintaining the integrity of the cell membrane, particularly at the end of alcoholic fermentation. They reinforce yeast resistance to ethanol and thus allow better cell viability and decrease the risk of sluggish fermentations. In anaerobic conditions, yeasts can use plant sterols present in musts. Beyond their viability, phytosterol content at the start of fermentation strongly influences the growth rate of the yeasts and therefore fermentation kinetics (Casalta et al., 2013). Although needs are likely strain dependent, a few (around 5) mg/L of phytosterol concentration are needed to reach the maximum growth rates (Casalta, Salmon, Picou, & Sablayrolles, 2019) in the absence of oxygen. In addition to their effect on the lipid status of yeast, must suspended solids, like other particles, may serve as nucleation sites, which favors CO_2 release from the fermentation medium and decreases its concentration and toxicity (Kühbeck, Müller, Back, Kurz, & Krottenthaler, 2007).

Though aroma formation during winemaking is dependent on several other factors (Styger, Prior, & Bauer, 2011), lipids availability in musts also affects the formation of yeast volatiles, i.e., acetate and ethyl esters, higher alcohols, and medium-chain fatty acids (Alexandre et al., 1994; Dubourdieu, Hadjinicolaou, & Bertrand, 1980;

© 2022 Elsevier Inc. All rights reserved.

Ferreira et al., 1995; Rosi & Bertuccioli, 1992; Varela, Torrea, Schmidt, Ancin-Azpilicueta, & Henschke, 2012). This effect is mainly attributed to unsaturated fatty acids. High suspended solid levels enhance the formation of higher alcohols, which have a negative impact, and lead to wines that are less rich in esters (acetates of higher alcohols and ethyl esters of fatty acids) compared to much lower turbidity levels (Dubourdieu et al., 1980). This general trend is also dependent on the management of oxygen during the early phases of fermentation and is likely strain dependent (Varela et al., 2012). High turbidity levels also promote the formation of C_6 alcohols and aldehydes, which are responsible for green and herbaceous flavors in wines. The precursors of these volatile aromas are the unsaturated linoleic and linolenic fatty acids (C18:2 and C18:3) present in must suspended solids, and oxygen. They are mainly formed during the prefermentation steps from a succession of enzymatic reactions, including particularly a lipoxygenase and a hydroperoxide-cleaving enzyme (Crouzet, Flanzy, Günata, Pellerin, & Sapis, 1998). All operations that favor the formation of suspended solids in musts (destemming, crushing, pressing), along with skin maceration, increase the content of C18 unsaturated fatty acids. The latter is then reduced by the clarification operations (Beaumes, Bayonove, Escudier, Cordonnier, & L, 1988; Ferreira et al., 1995). High turbidity may also favor must browning, which is related to the presence of grape polyphenol oxidase activities. These enzymes are partially solubilized during prefermentation operations but most of the activity remains associated with cell debris (Macheix, Fleuriet, Sapis, & Lee, 1991).

The objective in winemaking is therefore to find a balance to benefit from the positive impact associated with suspended solids while avoiding their negative effects. In practice, it is often observed that low turbidity levels favor the formation of fruity aroma whereas wines elaborated from highly turbid musts present heavy, herbaceous aroma with the sometimes formation of off-odors (reductive character). The desired turbidity levels can vary according to the type of wine sought but generally vary between 50 and 200 NTU. As stated before, excessive clarification of musts may lead to sluggish or stuck fermentations. Casalta et al. (2013) pointed out the interactions between phytosterol availability related to grape solids contents and assimilable nitrogen and their effect on fermentation. Excessive clarification of musts, leading to lipid deficiency, is particularly detrimental to the proper course of fermentation in the case of nitrogen-rich musts (see Chapter 13: Nitrogen management during fermentation) (Casalta et al., 2013; Tesnière, Delobel, Pradal, & Blondin, 2013).

8.3 Characteristics of suspended solids in musts

8.3.1 Contents

The levels of suspended solid particles in musts are influenced by the raw material and the conditions under which the grapes are harvested, transported, stored, and pressed (Boulton, Singleton, Bisson, & Kunkee, 1999; Darias-Martín, Díaz-González, & Díaz-Romero, 2004; Dubourdieu, Hadjinicolaou, & Ribéreau-Gayon, 1981; Lafon-Lafourcade, Dubourdieu, Hadjinicolaou, & Ribereau-Gayon, 1980). In winemaking, these levels are most often assessed by turbidity measurements or, less frequently, by the determination of the total wet suspended solids (TWSS, expressed in % w/w). TWSS is obtained by weighing the centrifuge pellets recovered after centrifugation of a given weight of must (around 50 g) under standard laboratory conditions (11,500 g for 10 min). Turbidity is the measurement of must cloudiness, which results from the scattering of light by suspended particles. The mean intensity scattered by the sample is measured under standardized conditions (light intensity and wavelength, observation angle) and the result expressed in NTU (Nephelometric Turbidity Units), by reference to international standards. By contrast to TWSS, turbidity depends not only on the suspended solid fraction but also on particle size, shape, and refractive index (Davies-Colley & Smith, 2001). Thus depending on their nature and size distribution, similar quantities of suspended solid material in musts can result in quite different turbidities (Vernhet, Bes, Bouissou, Carrillo, & Brillouet, 2016). Before clarification, the turbidity of the white musts can vary from 400 to 3000 NTU and their TWSS from 0.7 to 4 g/100 g. The solid content can also be expressed in % volume (volume fraction occupied by solids), this value being determined by a bucket centrifugation test to compact the sediment.

8.3.2 Size distribution, aggregation

There are very little data in the literature on the nature, composition, and size distribution of suspended particles in musts, even though this information would be necessary to control the degree of clarification, its impact on the qualitative characteristics of the wines, and the performances of clarification processes. The size distribution of suspended particles in musts can be evaluated by laser light diffraction after adequate dilution in a 100 -mM NaCl solution (Vernhet et al., 2016). The size distribution is obtained in % volume, which represents the volume fraction (in %) occupied by particles with a given hydrodynamic diameter D_H. The volume-weighted average hydrodynamic diameter (in μm) is given by

$$D_{H,V} = \frac{\sum_i n_i D_{Hi}^4}{\sum_i n_i D_{Hi}^3}.$$

This value is weighted by the larger particles in the medium, which occupy a larger volume fraction. As an example, a 10-μm D_H spherical particle occupies the same volume as 1000 particles of 1-μm D_H. From this measurement, the number-weighted average hydrodynamic diameter, i.e., the arithmetic diameter of particles in the sample (weighted by the number of particles in a size range) can be calculated using the Fraunhofer theory:

$$D_{H,N} = \frac{\sum_i n_i D_{Hi}}{\sum_i n_i}.$$

This theory can be applied when there is no information concerning the refractive index of suspended particles and considering their heterogeneous composition. Experiments performed on white musts and expressed in % volume showed the existence of particles with hydrodynamic diameters ranging from 100 nm to several hundred μm (Vernhet, unpublished data) and underlined the polydispersity of particle size. However, conversion in the number indicated that most of the suspended particles (>80%) had D_H values below 2 μm (Fig. 8.1). These data are in accordance with previous results obtained by Davin & Sahraoui (1993) with a Coulter counter equipped with a 100-μm probe.

Later observation with transmission electron microscopy (TEM) of Sauvignon must lees (Fig. 8.2) confirmed a broad size distribution (from a few ten nm to a few μm) with a majority of small particles. The latter consisted of heterogeneous cell debris (membranes, cell organelles, cell wall fragments, etc.) and aggregates of very different nature and shapes. Thus must particles are essentially small in size and the large ones that form during the prefermentation steps are mostly aggregates of these fines. The latter are redispersed during the growth phase of the alcoholic fermentation under the effect of the agitation created by the release of carbon dioxide (Casalta, Aguera, Nard, & Salmon, 2009). These fines reaggregate progressively during the stationary phase and settle with the yeast lees at the bottom of the tank.

8.3.3 Composition

While very few studies have focused on the overall composition of suspended solids in musts, their impact as a source of lipids influencing yeast fermentation and wine characteristics has long been emphasized (Karagiannis & Lanaridis, 2002; Singleton et al., 1975; Bertrand and Miele, 1984; Valero et al., 1998). The composition of suspended particles (lees) recovered by centrifugation after static settling of a Chardonnay must has been determined first as

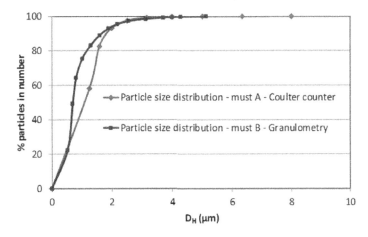

FIGURE 8.1 Size distribution of suspended particles in musts. Cumulative distribution in number (%) of suspended particles in two different white grape musts as a function of their hydrodynamic diameter D_H (μm), as obtained by impedance measurements (Coulter counter, 100-μm aperture probe, adapted from (Davin and Sahraoui, 1993)) and laser light diffraction (Vernhet, unpublished data).

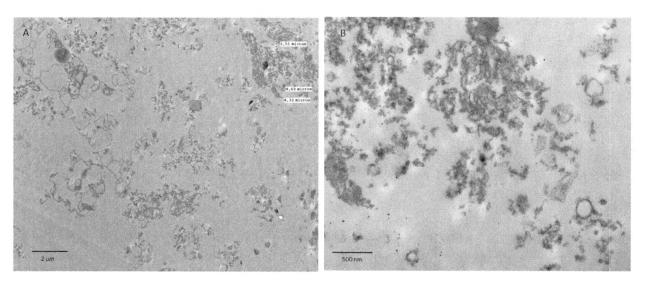

FIGURE 8.2   Observation of suspended solids from a Sauvignon must by Transmission Electron Microscopy. After washing and resuspension in water, particles were fixed as described in Vernhet et al. (2016), dehydrated in successive ethanol baths (from 30% to 100%), and included in Epon. Observations were performed on thin sections (60-nm thickness).

part of a study of their impact as a source of nutrients for yeasts (Alexandre et al., 1994). The dry weight of these particles was 25.6 g/100 g. Their contents in neutral and acidic sugars were determined on dried lees by colorimetric methods after a suitable hydrolysis step. These sugars, related to the presence of cell wall polysaccharides, accounted for 71.9% (neutral sugars in equivalent glucose) and 5.2% (acidic sugars in equivalent galacturonic acid) of the particle dry weight, respectively. Total nitrogen represented 2.6%, total lipids 7.8%, and ash 5.5%. Nitrogen was insoluble and therefore probably associated with proteins or nucleic acids. Lees addition to the must did not influence its total nitrogen content so that they are not a source of assimilable nitrogen for the yeasts. Polysaccharides, proteins, and lipids found in these particles may be associated with the cell organelles and debris (membranes, chloroplasts, cell wall fragments, etc.) as evidenced by TEM. They can also be found in the form of aggregates, formed during juice processing, or be adsorbed on suspended solids. Aggregation phenomena are expected between grape cell proteins due to their extraction in the acidic must (Ferreira, Piçarra-Pereira, Monteiro, Loureiro, & Teixeira, 2002), as well as between soluble pectic polysaccharides and proteins (Beaulieu, Corredig, Turgeon, Wicker, & Doublier, 2005).

When dealing with lipids, special attention has been paid to their content in phytosterols and fatty acids due to their incidence on yeast metabolism (Alexandre et al., 1994; Casalta et al., 2019) and wine aromas. Phytosterols in suspended solids were found to represent from 0.36% to 1.2% w/w dry matter, β-sitosterol representing about 80% of them, as in grapes. Fatty acids, associated with glycerides, phospholipids, and glycolipids account for 1.6−4.6% w/w. Among them, unsaturated fatty acids are in proportions ranging between 48% and 53%, the major ones being oleic and linoleic acids. Major saturated fatty acids are palmitic, stearic, and lauric acids.

Beyond this composition, the important point to keep in mind is that these analyses revealed very variable phytosterol and fatty acid contents (in mg/g dry matter) in the suspended solids from different musts. This can be related to the variety and maturity of the grapes as well as to different pretreatments (such as skin maceration) and extraction processes. These analyses also demonstrated that turbidity does not provide a reliable estimation of the lipid contents in must suspended solids (Casalta et al., 2019).

8.4 Must cleaning—clarification techniques

As stated before, the objective of must cleaning is to remove part of the must suspended solids to avoid their negative impact on wine quality. Another objective can be to achieve the right clarification level for the targeted wine style, which is more or less simple depending on the clarification technique used. In addition, this right clarification level is not so simple to define because of (i) the lack of a direct relationship between turbidity and lipid availability and (ii) the impact of several other factors. The standard procedure is spontaneous settling at low temperatures, with or without the use of commercial enzymes and fining aids. Besides settling, centrifugation and flotation can be used to accelerate the process.

8.4.1 Settling

In static settling, the suspended particles are separated according to their density ρ_p (kg·m^{-3}) with regard to that of the fluid ρ_f. The driving force for the separation is the acceleration due to gravity, g (m·s^{-2}). The two principal forces acting upon a particle suspended in a fluid at rest are the external force causing the motion (here, the gravity) and the drag force resisting motion that arises from the frictional action of the fluid on the particle Fig. 8.3A. This drag force is dependent on the particle settling velocity v_p (m·s^{-1}), its volume and shape, and the fluid viscosity η_f (Pa·s) and density. During sedimentation, it increases with v_p until an equilibrium is reached in which the force of gravity is canceled out by the opposite drag force. The settling velocity then reaches a constant value. According to Stokes' law, the settling velocity of a spherical particle in a diluted suspension (no interference between particles) under laminar flow conditions and a fluid at rest (lack of convection motions) is given by

$$v_p = \frac{D^2}{18\eta_f}\left(\rho_p - \rho_f\right)g,$$

where D (m) in the particle diameter. Though only indicative owing to the hypotheses made, Stokes' law underlines the critical parameters that affect the effectiveness of settling for must clarification. The difference in density between the suspended particles and the fluid, as well as the viscosity, is dependent on the sugar content of the must and the temperature. Musts are usually considered Newtonian liquids, the viscosity of which increases when the temperature decreases. There are few data in the literature concerning the viscosity of musts, and values ranging between 2 and 3 mPa·s depending on the must and the experimental conditions have been reported. Empirical equations have also been proposed to calculate this viscosity at different temperatures from the Brix degree in the case of clarified musts and the absence of enzymatic treatment (Zuritz et al., 2005). Of primary importance is the particle diameter D. Indeed, the settling velocity v_p is decreased by a factor of 100 for a 1-μm particle in comparison to a 10-μm one, and by a factor of 10,000 for a 0.1-μm particle. If we consider, for example, a must with a viscosity of 3 mPa·s and a difference in density between the solids and the must of 0.1 g/mL, a spherical particle with a diameter of 100 μm will take 1.53 h to travel 1 m under the effect of gravity. This time increases to 153 h (6 days) for a 10-μm particle and 15,300 h (almost two years) for a 1-μm one.

Considering the impact of the above-mentioned parameters on particle settling velocity, and the characteristics of must particles, settling will be strongly favored by particle aggregation. Although essentially micronic/submicronic, cellular debris tends to aggregate in musts in large clusters (several tens of μm) that sediment rapidly. This is favored by low temperatures, which decrease the Brownian motion that opposes both aggregation and settling. Pectolytic enzymes, naturally present in grapes, play a determinant part. They degrade the homogalacturonic main smooth chains of the cell wall soluble pectic polysaccharides, allowing the release of rhamnogalacturonan II and neutral sugar side chains (arabinans, arabinogalactans) linked to the rhamnogalacturonan I backbone (hairy regions of pectins) (Saulnier, Brillouet, & Joseleau, 1988; Vidal, Doco, Moutounet, & Pellerin, 2000; Vidal, Williams, O'Neill, & Pellerin, 2001). In musts, pectins help to keep particles in suspension by opposing their aggregation and sedimentation. This effect may be linked to their impact on must viscosity but also to their steric hindrance and the negative charges carried by the homogalacturonic chains, which cause electrostatic repulsions. Although the exact mechanisms are poorly known, the degradation of these homogalacturonic chains promotes the aggregation and sedimentation of

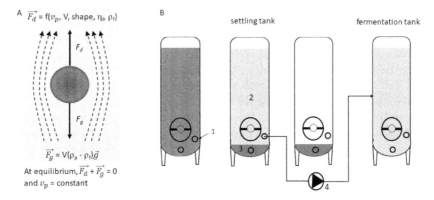

FIGURE 8.3 Must clarification by static settling. (A) Main forces (related to gravity, Fg, and drag force, Fd) acting upon a particle in a liquid at rest and in a diluted suspension. (B) Schematic representation of static settling. 1: racking valve; 2: clarified must; 3: must lees; 4: pump.

suspended solids (Dubourdieu et al., 1981). The use of commercial pectolytic enzymes makes it possible to compensate for the insufficient pectolytic activity in grapes and/or to accelerate this degradation (Canal-Lalaubères, 1993; Canal-Llaubères, 2010).

In addition to pectolytic enzymes, different fining aids can be used to facilitate the aggregation of suspended solids and accelerate their sedimentation. The most commonly used have long been bentonite, potassium caseinate, and polyvinylpolypyrrolidone (PVPP). In addition to their impact on settling velocity, these different fining aids promote the subsequent colloidal stability of wines in the case of bentonite by their interactions with must proteins (Pocock, Salazar, & Waters, 2011) and the color stability (removal of oxidized and/or oxidizable polyphenols) in the case of casein and PVPP (Sims, Eastridge, & Bates, 1995). Vegetable proteins (pea protein, patatin) are now often used as a replacement for casein and PVPP (Gambuti, Rinaldi, Romano, Manzo, & Moio, 2016). Protein fining agents can be used alone or in combination with coaids such as bentonite, enological tannins, or silica sols that facilitate their precipitation in musts. The impact of fining on must clarification is highly dependent on the dose of use and the combinations made.

Temperatures in the range of 8°C−10°C are recommended for static settling to reduce Brownian motion, limit convection, and avoid the start of fermentation. The operation usually lasts between 16 and 24 h. The must is then drawn off above the "solids/clarified juice" interface using a racking arm (Fig. 8.3B). The height of the lees and the thickness of the interface are variable and depend on the extent to which suspended solids compact in the bottom of the tank. Must depectinization increases the packing of lees and then decreases their height. The impact of fining is variable. The desired turbidity levels can vary according to the type of wine sought but generally vary between 100 and 200 NTU. If the turbidity of the clarified must is considered too low, part of the lees can be reincorporated to adjust it to the desired value.

8.4.2 Centrifugation

The time required for must clarification by static settling can be drastically reduced by the use of disk stack centrifuges. In this case, the must is sent continuously into a partitioned bowl that rotates at a high speed around a central axis. Centrifugation differs from settling only because the gravity g is replaced by a centrifugal force γ. The same parameters (particle size distribution and density, must viscosity and density) impact the efficiency of the separation. The centrifugal acceleration at a given radius r in the centrifuge is given by $\gamma = \omega^2 r$, where ω (rad·s^{-1}) is the angular velocity of the bowl. The centrifugal force F_c experienced by a particle in the bowl ($F_c = m\gamma$) thus depends upon the radius and the speed of rotation of the bowl. A given particle will then experience an increasing centrifugal force as it moves away from the center to the periphery of the bowl. The centrifugal force is often expressed as a number of g. The amplification factor Z ($Z = \gamma/g$) is calculated at the maximum radius of the bowl.

In a disk stack centrifuge (Fig. 8.4A), the suspension to be clarified is fed centrally into the bowl through a stationary inlet tube and directed toward the periphery, where the centrifugal acceleration is maximum. The bowl is partitioned by conical plates, stacked around the central axis (disk stack). Both the bowl and the disk stack are rotated at a high speed, producing centrifugal forces on particles more than 10,000 times higher than that of gravity. The heaviest and largest particles directly accumulate at the periphery of the bowl, in the solid holding space. The preclarified must, pushed by the feed, is then distributed between the plates, toward the outlet of the centrifuge. Fine particles are removed as the fluid flows through the spaces between two plates. This step is called polishing (Fig. 8.4B). The particles between the disks are carried along by the fluid flow and are subjected to the centrifugal force that moves them in the radial direction, toward the underside of the upper disk. Separation is achieved when the particle has reached the surface. Separated particles then slide in the form of a layer along the surface and are directed by the centrifugal force toward the periphery where they accumulate and compact in the solid holding space. The low separation space between two disks presents two advantages: it prevents the formation of eddies, which are detrimental to particle separation, even at high fluid feed rates; it strongly decreases the height of fall needed for the separation of fine particles. The clarified fluid that leaves the disk stack is discharged by a turbine. The compacted sediment that accumulates at the periphery of the bowl must be ejected periodically. This operation, called desludging, is performed automatically by opening the bowl for a fraction of a second. The bowl opening is driven by a hydraulic system. The solids are ejected abruptly due to the pressure created by the rotation and the bowl immediately closed again.

A wide range of desludging disk stack centrifuges is available for enology, depending on the flow rates to be treated (from several dozen to several hundred hectoliters per hour). The degree of clarification depends on the characteristics of the particles and the must (as in static settling), the solids content, and the operating parameters. For a given must and equipment (bowl and disk stack geometry, maximum rotational speed), the degree of clarification can be modulated by adjusting the feed rate. Decreasing the feed rate results in an increase in the

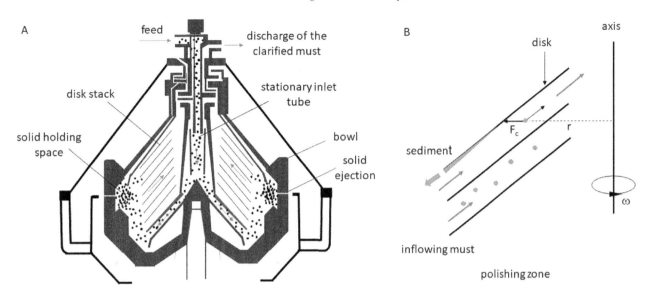

FIGURE 8.4 Must Clarification in a disk stack centrifuge. (A) Schematic representation of a disk stack centrifuge. The charged fluid is directed to the periphery of the centrifuge bowl through a stationary inlet tube and then distributed between the disks (polishing zone). The separated particles accumulate at the periphery of the bowl, in the solid holding space, from where they are periodically ejected. The clarified must is discharged by a turbine. (B) Between the disks, in the polishing zone, the particles conveyed by the fluid are submitted to the centrifugal force F_c that moves them away from the lower disk. The separated particles accumulate on the underside of the disks, where they form a layer that slides under the effect of F_c toward the solid holding space.

residence time of the fluid and suspended particles in the centrifuge bowl, thus improving separation and clarification. In centrifugation, the solid content is most often expressed in % volume. Centrifugation makes it possible to lower the solids content of the musts from 3%−8% vol. to 0.8%−2% vol. in a very short time. With the type of equipment used, however, submicron-sized particles cannot be separated. The turbidity values of the clarified musts are therefore highly variable and can be higher than the targeted values. Considering that must suspended particles are aggregates of fines, this may be related to their redispersion by shear forces during transfer. As in static settling, the efficiency of the operation can be improved by the use of enzymes and fining aids.

8.4.3 Flotation

Another way to speed up must clarification is flotation (Ferrarini, Zironi, & Buiatti, 1995; Wasjfelner, 1989). This process applies to particles whose real (natural flotation) or apparent density (induced flotation) is lower than that of the liquid phase. When dealing with must clarification, the flotation is induced. Induced flotation can be applied where the suspended particles have a good ability to interact with gas bubbles. Their capacity to adhere to the hydrophobic surface of the bubbles must therefore be greater than their wetting by the liquid phase. The particle−gas assembly formed, with lesser density than the liquid phase, rises to the surface (Fig. 8.5). The gas used in winemaking are air and nitrogen. The use of nitrogen is preferred for musts sensitive to oxidation. To allow bubbles to be generated, the gas is injected into the must during transfer, before the must is admitted in a saturator where the gas is dissolved at a pressure of 5−7 bars. The sudden reduction in pressure at the outlet of the saturator induces the formation of numerous small bubbles with diameters between 40 and 80 μm. The mean size of the bubbles and the bubble concentration are important parameters for the success of the flotation. Bubbles that are too large rise very quickly to the surface (Stokes' equation), which is unfavorable to good adhesion of the particles in suspension, and tend to form very aerated foams, which are more difficult to separate. Moreover, it decreases the surface available for particle−bubble interactions. The most commonly used value for the saturation pressure is around 6 bars. This value is adjusted by the operator according to the behavior of the must.

The hydrolysis of pectins in musts is one of the key points for the success of their clarification by flotation. To this end, the use of commercial enzymes is needed. The doses may vary between 0.5 and 4 g/hL, depending on the commercial preparation used and the conditions of their use (early addition in the press or on the must, duration of action, temperature, etc.). The use of β-glucanases may also be required with musts from harvests affected by *Botrytis cinerea*. Considering the variability of raw materials and operating conditions, the best way to ensure the absence of pectins or β-glucans in musts before flotation is using simple and rapid tests. For pectins, this test is based on the addition of two volumes of 96% ethanol acidified with HCl (1% v/v) to one volume of clarified must. After mixing and waiting for 5 min, the formation of a haze indicates that there are still pectins in the must. In this case, the time of action/dose of enzymes must be increased. If the must remains clear (negative test), the flotation process can be started. The same type of test can be applied for harvests altered by *B. cinerea*. The presence of β-glucans can be assessed by modifying the wine to acidified ethanol ratio: two volumes of must are mixed with one volume of acidified ethanol and the formation of a precipitate in the form of filaments is assessed by eye.

The use of fining aids is common for the flotation of white musts. The latter are added just prior to saturation. The objectives are to enhance the formation of large bubble−particle aggregates that quickly rise to the surface and to decrease the volume of foams. Frequently used fining agents are gelatin, whether or not combined with bentonite or silica sol, preparations based on potato (patatin) and pea proteins, or chitin derivatives.

The flotation process can be carried out in batch or continuous mode. In the batch mode (Fig. 8.6), flotation takes place in a simple tank. The must to be clarified is fed into the saturator using a pressurization pump, with gas being injected upstream of the pump with possibly also fining aids. At the outlet of the saturator, the must is either returned to the same tank or sent to another one. The particle−bubble aggregates formed then rise to the surface of the tank to form the flotation foams. When the process is completed (within a few hours), the clarified juice is drawn off from the bottom of the tank and the foams are pumped out to a storage tank. Discontinuous flotation is simple to implement and requires only a limited investment in equipment. Specific systems are

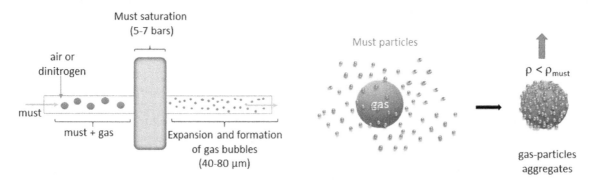

FIGURE 8.5 Principle of induced flotation for must clarification. The must is saturated under pressure by air or nitrogen. Pressure reduction (atmospheric pressure) induces the formation of gas micro-bubbles in the must. Interactions between must particles and gas bubbles lead to the formation of particle−gas assemblies, the density of which is much smaller than that of the liquid phase.

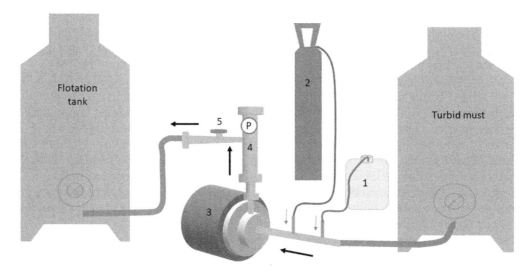

FIGURE 8.6 Must Clarification by discontinuous flotation. 1: Addition of fining aids; 2: gas injection; 3: pressurization pump; 4: saturator; 5: expansion and pressure regulation valve.

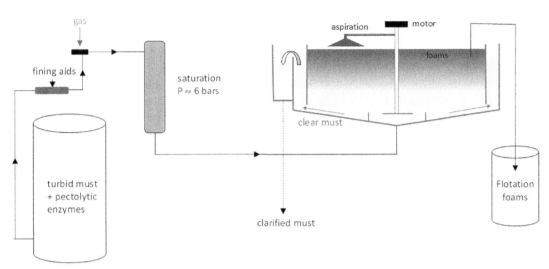

FIGURE 8.7   Schematic representation of the flotation in continuous mode and flotation tank.

available on the market. The topic of interest is the significant reduction in the immobilization time of the settling tanks.

In continuous mode, the must after saturation is sent to a flotation tank (Fig. 8.7). The latter is a small-height and open circular vat, fed from the center of the bottom. A double jacket is connected with the bottom of the tank. The bubble−particle flocs that rise to the surface are continuously removed by rotary aspiration systems and sent to a storage tank. The clear must is itself recovered by the bottom of the tank, in the double jacket. The continuous mode is suitable for units handling large volumes of musts. It allows clarification with flow rates from 100 to 300 hL/h. However, it is more difficult to handle than the batch mode and adjustments are necessary to optimize the process. Flotation also eliminates the need for frigories associated with static settling, as temperatures above 12°C are needed. The process is therefore carried out at room temperature.

8.4.4 Clarification of lees and flotation foams

Settling lees as well as flotation foams represent a non-negligible percentage of the initial volumes of musts (around 2%−15%) and are still very rich in juice (80%−90% of the volume fraction). The valorization of these juices is of obvious economical interest and is very often carried out. Their quality depends on the conditions of production, and their reincorporation into the must should be reasoned considering these conditions and the types of wine to be made. In some cases, this reincorporation may enhance the quality and typicity of wines. This is attributed to the high content of varietal aroma precursors in these musts associated with a prolonged contact with suspended solids.

The clarification of settling lees and flotation foams can first be achieved by earth filtration on rotary drum vacuum or press filters, adapted for the filtration of liquids that present high solid volume fractions. The principle of these operations will not be detailed here and can be found in Gautier (2015). Over the last 10 years, manufacturers have also developed cross-flow microfiltration systems adapted to the treatment of settling lees. Cross-flow microfiltration as proposed for wines is not currently adapted to the clarification of musts due to excessively low flow rates. These low flow rates are no longer prohibitive in the case of lees clarification since the volumes to be treated are small. Cross-flow filtration systems developed for the filtration of lees have, however, specific designs by comparison to those adapted for wines: different geometry and porosity of the membranes used (large diameter capillaries or disks), the use of turbulence sources, etc.

8.5 Conclusion

Must cleaning is then an important step for the elaboration of high-quality white wines. Though different techniques are available, it is not always easy to achieve the targeted clarification level, especially with continuous operations such as centrifugation or flotation. This is related to the variability of the juice composition, especially in cell wall soluble polysaccharides, and the nature and size distribution of suspended solids. Most of them are found within the colloidal size range, and aggregation or adsorption phenomena play a decisive role in the ability to clarify musts.

The target turbidity is most often between 100 and 200 NTU, in order to avoid fermentation problems and to prevent the formation of unpleasant aromas or oxidation reactions. It is now well established that the impact of suspended solids on fermentation kinetics and yeast aromas is related to their contents in phytosterols and fatty acids. However, there is no simple and direct relationship between turbidity, suspended solid weight, and lipid availability. Further characterization, with different varieties and maturity, would be needed to get a better evaluation of the variability of the lipid content as a function of the total suspended solid weight. This would make it possible to better control their impact on the fermentation and aromatic characteristics of the wines, depending on the yeast strains used.

References

Alexandre, H., Nguyen van Long, T., Feuillat, M., & Charpentier, C. (1994). Contribution à l'étude des bourbes: influence sur la fermentescibilité des moûts. *Revue Française d'OEnologie, 146,* 11–20.

Beaulieu, M., Corredig, M., Turgeon, S. L., Wicker, L., & Doublier, J. L. (2005). The formation of heat-induced protein aggregates in whey protein/pectin mixtures studied by size exclusion chromatography coupled with multi-angle laser light scattering detection. *Food Hydrocolloids, 19*(5), 803–812. Available from https://doi.org/10.1016/j.foodhyd.2004.10.025.

Beaumes, R., Bayonove, C. B., Escudier, J. M., Cordonnier, J. L., & L, R. (1988). I. s. l. c., La maceration pelliculaire dans la vinification en blanc. Incidence sur la composante volatile des mouts. *Connaissance de la Vigne et du Vin, 32,* 209–223.

Bertrand, A., & Miele, A. (1984). Influence de la clarification du moût de raisin sur sa teneur en acides gras. *Connaissance de La Vigne et Du Vin, 18*(4), 293–297.

Boulton, R.B., Singleton, V.L., Bisson, L.F., & Kunkee, R.E. (1999). Preparation of must and juices. *Principles and practices of winemaking.*

Canal-Lalaubères, R.-M. (1993). Enzymes in winemaking. *Wine microbiology and biotechnology.*

Canal-Llaubères, R. M. (2010). *Enzymes and wine quality. Managing wine quality: Oenology and wine quality* (pp. 93–132). France: Elsevier Ltd. Available from https://doi.org/10.1016/B978-1-84569-798-3.50004-8.

Casalta, E., Aguera, E., Nard, P. L., & Salmon, J. M. (2009). Physical dynamics of sludges during wine fermentation in liquid phase. *Journal International Des Sciences de La Vigne et Du Vin, 43*(4), 225–230.

Casalta, E., Cervi, M. F., Salmon, J.-M., & Sablayrolles, J.-M. (2013). White wine fermentation: Interaction of assimilable nitrogen and grape solids. *Australian Journal of Grape and Wine Research,* 47–52. Available from https://doi.org/10.1111/j.1755-0238.2012.00205.x.

Casalta, E., Salmon, J.-M., Picou, C., & Sablayrolles, J.-M. (2019). Grape solids: Lipid composition and role during alcoholic fermentation under enological conditions. *American Journal of Enology and Viticulture, 70*(2), 147–154. Available from https://doi.org/10.5344/ajev.2018.18049.

Crouzet, J., Flanzy, C., Günata, Z., Pellerin, P., & Sapis, J.-C. (1998). Les enzymes en oenologieIn C. Flanzy (Ed.), (1st (ed.), pp. 361–411). Londres, Paris, New-York: Lavoisier Tec&Doc, s.

Darias-Martín, J., Díaz-González, D., & Díaz-Romero, C. (2004). Influence of two pressing processes on the quality of must in white wine production. *Journal of Food Engineering, 63*(3), 335–340. Available from https://doi.org/10.1016/j.jfoodeng.2003.08.005.

Davies-Colley, R. J., & Smith, D. G. (2001). Turbidity, suspended sediment, and water clarity: A review. *Journal of the American Water Resources Association, 37*(5), 1085–1101. Available from https://doi.org/10.1111/j.1752-1688.2001.tb03624.x.

Davin, A., & Sahraoui, A. (1993). Débourbage rapide des moûts de raisin par flottation à l'aide de bulles générées au sein du liquide par dépressurisation. *Revue Française d'OEnologie.*

Deytieux, C., Mussard, L., Biron, M. J., & Salmon, J. M. (2005). Fine measurement of ergosterol requirements for growth of *Saccharomyces cerevisiae* during alcoholic fermentation. *Applied Microbiology and Biotechnology, 68*(2), 266–271. Available from https://doi.org/10.1007/s00253-004-1872-3.

Dubourdieu, D., Hadjinicolaou, D., & Bertrand, A. (1980). Observations sur la vinification en blanc sec. Essais de filtration des moûts après débourbage statique. *Connaissance de la Vigne et du Vin, 14,* 247–261.

Dubourdieu, D., Hadjinicolaou, D., & Ribéreau-Gayon, P. (1981). Les polysaccharides solubles du moût: méthode simple d'appréciation; évolution au cours de la maturation. *Journal international des sciences de la vigne et du vin, 15,* 29–40.

Ferrarini, R., Zironi, R., & Buiatti, S. (1995). Recent advances in the process of flotation applied to the clarification of grape musts. *Journal of Wine Research, 6*(1), 19–33. Available from https://doi.org/10.1080/09571269508718014.

Ferreira, B., Hory, C., Bard, M. H., Taisant, C., Olsson, A., & Le Fur, Y. (1995). Effects of skin contact and settling on the level of the C18:2, C18:3 fatty acids and C6 compounds in burgundy chardonnay musts and wines. *Food Quality and Preference, 6*(1), 35–41. Available from https://doi.org/10.1016/0950-3293(94)P4210-W.

Ferreira, R. B., Piçarra-Pereira, M. A., Monteiro, S., Loureiro, V. B., & Teixeira, A. R. (2002). The wine proteins. *Trends in Food Science and Technology, 12,* 230–239.

Gambuti, A., Rinaldi, A., Romano, R., Manzo, N., & Moio, L. (2016). Performance of a protein extracted from potatoes for fining of white musts. *Food Chemistry, 190,* 237–243. Available from https://doi.org/10.1016/j.foodchem.2015.05.067.

Gautier, B. (2015). Practical aspects of wine filtration. Avenir oenologie. oenoplurimedia.

Karagiannis, S., & Lanaridis, P. (2002). Insoluble grape material present in must affects the overall fermentation aroma of dry white wines made from three grape cultivars cultivated in Greece. *Journal of Food Science, 67*(1), 369–374. Available from https://doi.org/10.1111/j.1365-2621.2002.tb11412.x.

Kühbeck, F., Müller, M., Back, W., Kurz, T., & Krottenthaler, M. (2007). Effect of hot trub and particle addition on fermentation performance of *Saccharomyces cerevisiae*. *Enzyme and Microbial Technology*, *41*(6–7), 711–720. Available from https://doi.org/10.1016/j.enzmictec.2007.06.007.

Lafon-Lafourcade, S., Dubourdieu, D., Hadjinicolaou, D., & Ribereau-Gayon, P. (1980). Incidence des conditions de travail des vendanges blanches sur la clarification et la fermentation des moûts. *Connaissance de la Vigne et du Vin*, *14*, 127–138.

Luparia, V., Soubeyrand, V., Berges, T., Julien, A., & Salmon, J. M. (2004). Assimilation of grape phytosterols by *Saccharomyces cerevisiae* and their impact on enological fermentations. *Applied Microbiology and Biotechnology*, *65*(1), 25–32. Available from https://doi.org/10.1007/s00253-003-1549-3.

Macheix, J. J., Fleuriet, A., Sapis, J. C., & Lee, C. Y. (1991). Phenolic compounds and polyphenoloxidase in relation to browning in grapes and wines. *Critical Reviews in Food Science and Nutrition*, *30*(4), 441–486. Available from https://doi.org/10.1080/10408399109527552.

Pocock, K. F., Salazar, F. N., & Waters, E. J. (2011). The effect of bentonite fining at different stages of white winemaking on protein stability. *Australian Journal of Grape and Wine Research*, *17*(2), 280–284. Available from https://doi.org/10.1111/j.1755-0238.2011.00123.x.

Rosi, I., & Bertuccioli, M. (1992). Influences of lipid addition on fatty acid composition of *Saccharomyces cerevisiae* and aroma characteristic of experimental wines. *Journal of the Institute of Brewing*, *98*(4), 305–314. Available from https://doi.org/10.1002/j.2050-0416.1992.tb01113.x.

Saulnier, L., Brillouet, J. M., & Joseleau, J. P. (1988). Structural studies of pectic substances from the pulp of grape berries. *Carbohydrate Research*, *182*(1), 63–78. Available from https://doi.org/10.1016/0008-6215(88)84092-8.

Sims, C. A., Eastridge, J. S., & Bates, R. P. (1995). Changes in phenols, color, and sensory characteristics of muscadine wines by pre-and post-fermentation additions of PVPP, casein, and gelatin. *American Journal of Enology and Viticulture*, *46*, 155–158.

Singleton, V. L., Sieberhagen, H. A., Wet., & Wyk. (1975). Composition and sensory qualities of wines prepared from white grapes by fermentation with and without grape solids. *American Journal of Enology and Viticulture*, *26*, 62–69.

Styger, G., Prior, B., & Bauer, F. F. (2011). Wine flavor and aroma. *Journal of Industrial Microbiology and Biotechnology*, *38*(9), 1145–1159. Available from https://doi.org/10.1007/s10295-011-1018-4.

Tesnière, C., Delobel, P., Pradal, M., & Blondin, B. (2013). Impact of nutrient imbalance on wine alcoholic fermentations: Nitrogen excess enhances yeast cell death in lipid-limited Must. *PLoS One*, *8*(4). Available from https://doi.org/10.1371/journal.pone.0061645.

Valero, E., Millan, M.C., Mauricio, J.C., & Ortega, J.M. (1998). Effect of grape skin maceration on sterol, phospholipid, and fatty acid contents of *Saccharomyces cerevisiae* during alcoholic fermentation. *American Journal of Enology and Viticulture*, *49*(2), 119–124.

Varela, C., Torrea, D., Schmidt, S. A., Ancin-Azpilicueta, C., & Henschke, P. A. (2012). Effect of oxygen and lipid supplementation on the volatile composition of chemically defined medium and Chardonnay wine fermented with *Saccharomyces cerevisiae*. *Food Chemistry*, *135*(4), 2863–2871. Available from https://doi.org/10.1016/j.foodchem.2012.06.127.

Vernhet, A., Bes, M., Bouissou, D., Carrillo, S., & Brillouet, J. M. (2016). Characterization of suspended solids in thermo-treated red musts. *Journal International Des Sciences de La Vigne et Du Vin*, *50*(1), 9–21. Available from https://doi.org/10.20870/oeno-one.2016.50.1.50.

Vidal, S., Doco, T., Moutounet, M., & Pellerin, P. (2000). Soluble polysaccharide content at initial time of experimental must preparation. *American Journal of Enology and Viticulture*, *51*(2), 115–121.

Vidal, S., Williams, P., O'Neill, M. A., & Pellerin, P. (2001). Polysaccharides from grape berry cell walls. Part I: Tissue distribution and structural characterization of the pectic polysaccharides. *Carbohydrate Polymers*, *45*(4), 315–323. Available from https://doi.org/10.1016/S0144-8617(00)00285-X.

Wasjfelner, R. (1989). Application de la flottation à la clarification des moûts. *Journal International Des Sciences de La Vigne et Du Vin*, *23*, 53–57.

Zuritz, C. A., Puntes, E. M., Mathey, H. H., Pérez, E. H., Gascón, A., Rubio, L. A., & Cabeza, M. S. (2005). Density, viscosity and coefficient of thermal expansion of clear grape juice at different soluble solid concentrations and temperatures. *Journal of Food Engineering*, *71*(2), 143–149. Available from https://doi.org/10.1016/j.jfoodeng.2004.10.026.

Application of *Hanseniaspora vineae* to improve white wine quality

Valentina Martin[1], Maria Jose Valera[1], Karina Medina[1], Eduardo Dellacassa[2], Remi Schneider[3], Eduardo Boido[1] and Francisco Carrau[1,4]

[1]Oenology and Fermentation Biotechnology Unit, Food Science and Technology Department, School of Chemistry, Universidad de la Republica, Montevideo, Uruguay [2]Laboratory of Aroma Biotechnology, School of Chemistry, Universidad de la Republica, Montevideo, Uruguay [3]Oenobrands SAS, Montpellier, sur Lez, France [4]Center for Biomedical Research (CEINBIO), School of Medicine, Universidad de la Republica, Montevideo, Uruguay

9.1 Introduction

The ascomycetous yeast genus *Hanseniaspora* and its anamorph *Kloeckera* are morphologically characterized as apiculate yeasts with a bipolar budding shape (Fig. 9.1), while *Saccharomyces cerevisiae* has a well-known multipolar budding process where each daughter always leaves separate scars around the cell surface. In Fig. 9.1A, we can see a scanning electron microscope photograph of a mixed culture of *Hanseniaspora vineae* (bipolar vegetative reproduction) and *S. cerevisiae* (round cells with multipolar vegetative reproduction) obtained during the initial steps of wine fermentation. *H. vineae*, like all apiculate yeasts, is not as well characterized as *Saccharomyces*. With this technique, we can see here the bipolar division process in *H. vineae*, where "multiple scars" are detected (Streiblová, Beran, & Pokorný, 1964) at the two cell poles, showing the individual reproductive capacity of the mother cell. Each rim scar indicates the birth of a daughter cell; so in Fig. 9.1B *H. vineae* has formed four daughters that were born from the upper pole of the bigger cell of the figure, while in the bottom pole the fourth daughter is currently budding (total: eight daughters). Young cells acquire a more typical apiculate form while increasing the number of daughters in the direction of the long axis. Younger *H. vineae* cells are shown at the bottom of the photograph, with just one scar in one of the poles, preparing to start cell division at the opposite pole, and then alternately from each pole during the course of the exponential growth phase.

During grape maturity and the initial steps of alcoholic fermentation, the most frequent genera of non-*Saccharomyces* yeasts are in general *Hanseniaspora/Kloeckera* and *Metschnikowia*, and to a lesser extent *Hansenula*, *Pichia*, and *Rhodotorula* (Fleet & Heard, 1993; Longo, Cansado, Agrelo, & Villa, 1991; Querol, Jiménez, & Huerta, 1990; Schütz & Gafner, 1994). They have a limited fermentation capacity, and some of them contribute to the sensory bouquet due to the production of desired volatile compounds that might increase taste complexity (Gil, Mateo, Jimenez, Pastor, & Huerta, 1996; Herraiz, Reglero, Herraiz, Martin-Alvarez, & Cabezudo, 1990; Mateo, Jimenez, Huerta, & Pastor, 1991; Medina et al., 2013; Moreira, Mendes, Hogg, & Vasconcelos, 2005; Zironi, Romano, Suzzi, Battistutta, & Comi, 1993). The research carried out in recent years concerning the dynamics of yeast populations in wine-related environments (vineyards and wineries) has generated new perspectives on the complex biotransformations that can profoundly improve the character of wine (Varela & Borneman, 2017). Significant variations in yeast diversity in these environments, in different winegrowing areas, are part of the terroir (Alexandre, 2020; Carrau, Boido, & Ramey, 2020; Testa et al., 2020) and have the potential to confer a specific

99

© 2022 Elsevier Inc. All rights reserved.

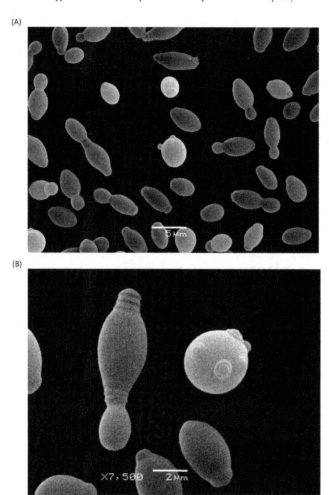

FIGURE 9.1   Scanning electron microscope photograph of *Hanseniaspora vineae*, an apiculate yeast with bipolar vegetative reproduction, obtained during the initial steps of wine fermentation. Bipolar division process is shown in the figure where "multiple scars" at the two cell poles show individual reproductive capacity of mother cells. *H. vineae* has superimposed scars described as concentric funnels with parallel rims in each pole. Each rim scar indicates the birth of a daughter cell; younger cells are shown with just one scar in one of the poles. Photograph: Lic. Magela Rodao, Unidad Microscopía Electrónica Barrido, Facultad de Ciencias, UdelaR, Montevideo, Uruguay. (A) ×3500; (B) ×7500.

regional character on wines (Lappa et al., 2020). This is why studies have been carried out to determine the characteristics of different yeast populations in different wine regions or grape varieties (Beltran et al., 2002; Cadez, Raspor, de Cock, Boekhout, & Smith, 2002; Esteve-Zarzoso, Gostıncar, Bobet, Uruburu, & Querol, 2000; Grangeteau et al., 2015; Jolly, Augustyn, & Pretorius, 2003; Knight, Klaere, Fedrizzi, & Goddard, 2015; Medina et al., 2007; Mendoza, Vega-Lopez, de Ullivarri, & Raya, 2019; Sabate, Cano, Esteve-Zarzoso, & Guillamón, 2002; Torija, Rozes, Poblet, Guillamón, & Mas, 2001; Tristezza et al., 2016; Varela & Borneman, 2017). The diversity and population structure of yeast are affected by several factors, including grape variety, geographical location, climate, and vineyard management (Varela & Borneman, 2017). The way in which these populations of native yeasts act on fermentations that occur spontaneously will give the wines obtained a distinctive authenticity since it strengthens the regional characteristics (Jolly, Varela, & Pretorius, 2014).

Hanseniaspora/Kloeckera species are frequently isolated from various natural sources including vineyards and wineries when starting wine fermentations. These genera have been intensively studied to determine their species' genetic diversity (Cadez et al., 2002). The significant presence of this yeast group in the initial phases of wine fermentation has encouraged the realization of numerous studies. It was concluded that the presence of some of these yeasts contributes to a more complex aroma of wine (Lleixà et al., 2016; Medina et al., 2013; Moreira et al., 2011; Moreira, Mendes, de Pinho, Hogg, & Vasconcelos, 2008; Del Fresno et al., 2020; Martin et al., 2016b; Padilla et al., 2017).

Although for many years apiculate yeasts have been characterized as high volatile acidity producers, with levels between 0.75 and 2.25 g/L of acetic acid and ethyl acetate (Ciani & Piccioti, 1995; Rojas, Gil, Piñaga, & Manzanares, 2003); this behavior was then associated with some species of this genus, such as *H. uvarum*, *H. clermontiae*, and *H. opuntiae*.

H. uvarum and *H. guilliermondii* have been reported to produce high levels of sulfur volatile compounds (Moreira et al., 2005). Testa et al. (2020) found that sequential inoculation of autochthonous *H. guilliermondii* and *S. cerevisiae* yeasts in white grapes of Campania *Vitis vinifera* cv. Fiano allowed the improvement of the sensorial complexity of a traditional wine, thus enhancing its characteristics.

In a study carried out in our laboratory with *H. vineae* native to Uruguay, a significant increase in acetate esters and some interesting phenylpropanoids was observed, as well as a decrease in some of the higher alcohols and carbonyl compounds (β-phenylethyl alcohol, 3-methyl-1-propanol, tyrosol, and 1,3-propanediol) in isovaleric acid and other medium chain fatty acids. It was also observed that the concentration of many of these compounds increases with the mixed inoculation of *Saccharomyces* (cofermentation), obtaining an increase in fruity notes (banana, pear, apple, citrus fruits), cut grass and a decrease in the descriptors related to wet soil (Medina et al., 2013). At the same time, these cofermented wines present more body and greater aroma complexity in the mouth. The presence of increased extracellular C10 medium chain fatty acids, suggests a relatively faster rate of autolysis compared to other yeasts (Ravaglia & Delfini, 1994). For the species *H. vineae*, it has been reported that it presents β-glucosidase activity (López, Mateo, & Maicas, 2016; Pérez et al., 2011), standing out for its high production of desirable aroma compounds that improve the organoleptic quality of wines produced at the industrial scale (Del Fresno et al., 2020; Lleixà et al., 2016; Medina et al., 2013; Viana, Belloch, Vallés, & Manzanares, 2011).

Several studies have been carried out on *H. vineae* vinification of white wines such as Chardonnay (Medina et al., 2007, 2013), and more recently with neutral varieties such as Macabeo and Albillo from Spain (Del Fresno et al., 2020; Lleixà et al., 2016), where it was possible to verify the positive contribution of *H. vineae* strain Hv T02/5F produce to the organoleptic characteristics of the wines obtained, with the aroma profile standing out.

Throughout this chapter, we will make a compilation of the characteristics present within *H. vineae* strains and the contribution of this species to the differentiation of white wines.

9.1.1 Hanseniaspora *in grapes and wine*

This genus is the most frequently found in the world's vineyards and together with *Torulaspora*, *Pichia*, and *Metschnikowia* constitutes the non-*Saccharomyces* yeasts with the highest contribution to wine sensory properties. Within the genus *Hanseniaspora*, the species that is usually found more often is *H. uvarum*. This group of yeasts is currently composed of 10 species associated with the surface of the grapes: *H. meyeri*, *H. clermontiae*, *H. uvarum*, *H. guilliermondii*, *H. opuntiae*, *H. thailandica*, *H. valbyensis*, *H. occidentalis*, *H. vineae*, and *H. osmophila* (Martin, Valera, Medina, Boido, & Carrau, 2018). *Hanseniaspora* is usually one of the dominant genera in the early stages of fermentation (Jolly et al., 2014), perhaps due to its high tolerance to osmotic pressure (over 200 g/L). It is also possible to find them inside the winery, in the harvesting machinery, and in the processing of the grapes (López et al., 2016; Strauss, Jolly, Lambrechts, & Van Rensburg, 2001; Varela & Borneman, 2017). As the fermentation process progresses, their presence decreases due to their low capacity to adapt to increasing levels of ethanol (Díaz-Montaño & Córdova, 2009). In general, this genus shows a medium/low fermentation capacity (in some cases reaching values of up to 10% ethanol) (Martin et al., 2018) but is considered important during winemaking as it produces aroma compounds of interest and modifies the chemical composition of the wines obtained (Herraiz et al., 1990; Mateo et al., 1991; Medina, Boido, Dellacassa, & Carrau, 2018; Moreira et al., 2005; Zironi et al., 1993). However, as there is great variability between the different species of the genus, the same behavior will not be observed in all of them, which makes detailed selection work essential (Ciani & Maccarelli, 1998; Medina et al., 2018; Tristezza et al., 2016). Within the isolations made in our laboratory during different harvests of white grape varieties, we mainly obtained isolates of the species *H. uvarum*. In Chardonnay grapes, we showed that this was the second species present, representing 29% of total yeast diversity (Fig. 9.2).

How grapes are transported to the winery will significantly influence their microbial population profile. In Fig. 9.3, hand-picked grapes in small cases are shown as the way to convey healthy and unbroken berries to the winery, and so the natural flora is sampled in its real population proportions. Other considerations of *Hanseniaspora* spp. metabolism are related to the production of acetoin in must (Romano, Suzzi, Zironi, & Comi, 1993) and the ability to reduce the ochratoxin A content in synthetic must (Angioni et al., 2007).

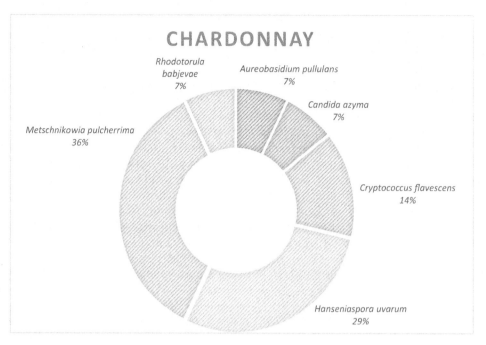

FIGURE 9.2 Yeast diversity in Chardonnay grape musts of Uruguay.

FIGURE 9.3 At harvest time in healthy grapes, the berry skin surface has approximately 10^4 yeast native cells per cm^2. The genus *Hanseniaspora* represents the most abundant species of this population, about 30%−70% of the total yeast diversity depending on the variety and region. Some species with increased fermentation capacity, such as *H. vineae* and *H. osmophila*, usually appear at very low frequency in vines, less than 1%, as happens with *Saccharomyces cerevisiae* strains.

In wine-related environments, it is possible to find *Hanseniaspora* strains in vineyard soils, on grape surfaces, and in the first days of fermentation. Interestingly, our strain collection of *H. vineae* was isolated mainly from red grape varieties compared to white grapes.

9.2 Characterization of species and strain diversity

9.2.1 Strain diversity within Hanseniaspora: *Fermentation capacity*

Medina (2014) conducted a comparative study of the fermentation capacity, cell count, and production of aroma compounds of the following five *Hanseniaspora* yeasts native to Uruguay: *H. vineae* (Hv T02/5F),

H. guilliermondii (Hg T06/09G), *H. opuntiae* (Ho T06/01G), and two *H. clermontiae* (Hc A10/82F and Hc C10/54F). All of them were compared with an *S. cerevisiae* strain native to Uruguay (Sc 882) (Fig. 9.4A and B).

The results obtained showed great diversity in fermentation capacity among the strains, which allowed good discrimination of fermentation performance (Fig. 9.4A). *S. cerevisiae* Sc 882 released the main amount of carbon dioxide during alcoholic fermentation (14 g CO_2/100 mL), followed by *H. vineae* Hv T02/5F (10.59 g CO_2/100 mL). *H. clermontiae* Hc A10/82F (4.94 g CO_2/100 mL) showed an intermediate fermentation capacity, while the strains *H. clermontiae* Hc C10/54F, *H. guilliermondii* Hg T06/09G, and *H. opuntiae* Ho T06/01G presented the lowest fermentation rates, with a weight loss of between 0.81 and 1.28 g CO_2/100 mL. The different fermentation capacity among the non-*Saccharomyces* yeasts was 9.78 g CO_2/100 mL, corresponding to a 92.35% difference in fermentation rates.

At the end of fermentation, total cell count was significantly different in between the yeasts (Fig. 9.4B). *S. cerevisiae* Sc 882 showed the highest cell population count (1.15 ± 0.05 × 10^8 cells/mL), followed by *H. vineae* T02/5F (8.75 ± 0.25 × 10^7 cells/mL).

As shown in Fig. 9.5, when the values for fermentation kinetics (Fig. 9.4A) and total cell count (Fig. 9.4B) are compared, it can be observed that as the fermentation capacity increases, the total cell count increases too, indicating that the yeasts with the highest cell growth are also those with the highest fermentation rates.

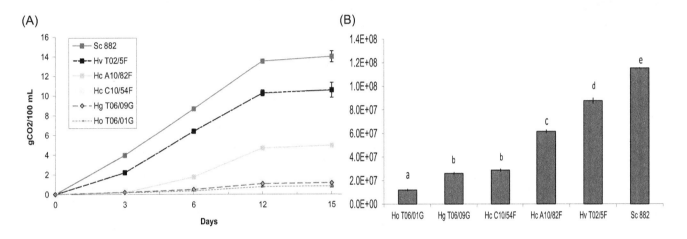

FIGURE 9.4 Fermentation kinetics (A) and cell growth at the end of fermentation (B) at 20°C (cfu colony-forming units). Bars indicate standard deviation of triplicates. Letters indicate significant differences according to the Tukey test.

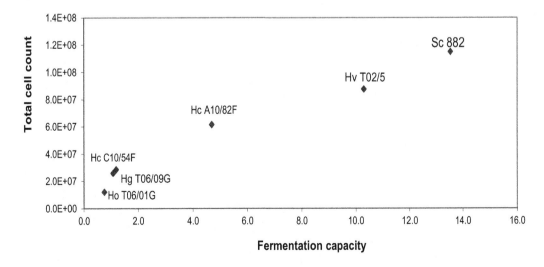

FIGURE 9.5 Correlation between fermentation capacity and total cell count. A direct correlation between the parameters is observed.

9.2.2 Strain diversity within Hanseniaspora: Aroma compounds

As was expected due to the great diversity in fermentation capacity of the different species of this genus, the secondary metabolism that affects flavor compounds is also very diverse. Fig. 9.6 shows the results obtained for each of the most important groups of aroma compounds for some species of *Hanseniaspora* compared with *Saccharomyces*.

The main compounds found within ethyl esters were ethyl succinate, diethyl succinate, and ethyl lactate, except for in *H. vineae*. Ethyl succinate is associated with the descriptors "fruity," "sour," and "bitter" but probably does not contribute to aroma due to its high perception threshold (1200 mg/L) (Meilgaard, 1985). If ethyl succinate is removed from the esters sum, the individual yeasts trends are modified, Sc 882 and Hv T02/5F being the yeasts with the highest production of the rest of the ethyl esters. *H. clermontiae* A10/82F is particularly highlighted for diethyl 2-hydroxyglutarate, associated with the descriptor "cotton candy." *H. opuntiae* also showed the highest concentration of ethyl pyruvate associated with the descriptors "fatty," "oil paint," and "fodder." *H. vineae* is noted for producing the highest concentration of ethyl decanoate, an aroma associated with the descriptors "sweet" and "hazelnut oil." This result is similar to that reported by Medina et al. (2013) with Chardonnay wines. It also produced one of the highest concentrations of ethyl octanoate and diethyl succinate.

Several differences between species were observed for higher alcohol production. *S. cerevisiae* Sc 882 was the highest alcohol producer. The main alcohols produced by the *Hanseniaspora* species were isoamyl alcohol, isobutyl alcohol, and β-phenylethyl alcohol. However, for *H. vineae* the latter alcohol is converted into its acetate ester, so finally, this alcohol is present at a lower concentration in this species. In relation to isoamyl alcohol, described when present in low concentrations as "fruity" and "wine," strain Sc 882 is highlighted as having the highest value, with the rest of the non-*Saccharomyces* yeasts having lower concentrations of this compound. These results are in agreement with a previous report in which higher production of this compound was observed for *Saccharomyces* as compared to *Hanseniaspora* yeasts (Rojas et al., 2003). *H. vineae* showed the highest concentration of 1-butanol, associated with "medicinal" descriptors.

For the ether 3-ethoxy-1-propanol (associated with the descriptor "mature pear"), it was observed that all *Hanseniaspora* yeasts presented a higher concentration than that produced by *S. cerevisiae* Sc 882, particularly *H. clermontiae* Hc A10/82F, which produced the highest concentration (532 μg/L).

Acetate ester production is a clear characteristic of *H. vineae* T02/5F, and it produced the highest concentration of β-phenylethyl acetate compared to the other species. Production of this compound was in the order of five times higher than for the other *Hanseniaspora* species, followed by *H. guilliermondii*. In these experiments, these concentration levels are even higher, up to 20 times that produced by *S. cerevisiae* as was shown previously (Martin et al., 2016b). β-Phenylethyl acetate aroma is described as "rose," "honey," "apple," "sugar," and "tropical fruit" (Lambrechts & Pretorius, 2000; Swiegers & Pretorius, 2005; Swiegers, Bartowsky, Henschke, & Pretorius, 2005; Swiegers, Saerens, & Pretorius, 2016). Although it has also been reported for *H. guilliermondii* (Rojas et al., 2003), it was not found in the strain used for the experiments of Fig. 9.6. It is interesting to note that in the other *Hanseniaspora* species, production of β-phenylethyl acetate was significantly lower, with no significant differences between them. There are two species that are distinguished by their ability to produce high concentrations of β-phenylethyl acetate: *H. osmophila* (Viana, Gil, Vallés, & Manzanares, 2009) and *H. vineae* (Lleixà et al., 2016; Martin et al., 2016b; Medina et al., 2013; Viana et al., 2011). Both species constitute the fermentation clade that was defined recently (Valera, Boido, Dellacassa, & Carrau, 2020a), with increased fermentation capacity when compared to the fruit clade composed by the other species of this genus except *H. occidentalis*.

Medium chain fatty acids showed moderate concentrations in the *Hanseniaspora* species, with *S. cerevisiae* Sc 882 being the yeast with the highest level, followed by *H. vineae* Hv T02/5F. The presence of these fatty acids is often desirable during winemaking since they are ethyl ester precursors and might contribute to well appreciate fermentation aroma compounds (San-Juan, Ferreira, Cacho, & Escudero, 2011).

In reference to lactones, *H. clermontiae* Hc A10/82F, *H. opuntiae* T06/01G, and *H. guilliermondii* T06/09G showed a similar trend, being the yeasts with the highest production of γ-butyrolactone and pantolactone compared to *Saccharomyces* and the other *Hanseniaspora* species.

Volatile phenols production showed a great diversity; *H. clermontiae* Hc A10/82F was the yeast with the highest concentration, mainly due to compounds such as 2,6-dimethoxy-phenol (associated with the descriptors "smoke" and "nut"), 4-vinylguaiacol (associated with "curry" and "clove" descriptors), and guaiacol (associated with the descriptors "hospital," "smoke" and "spicier"). In contrast, *S. cerevisiae* showed the lowest concentrations of the three abovementioned compounds when compared to *Hanseniaspora* strains. The presence of these compounds might be explained by the known β-glycosidase activity present in all the tested yeasts except

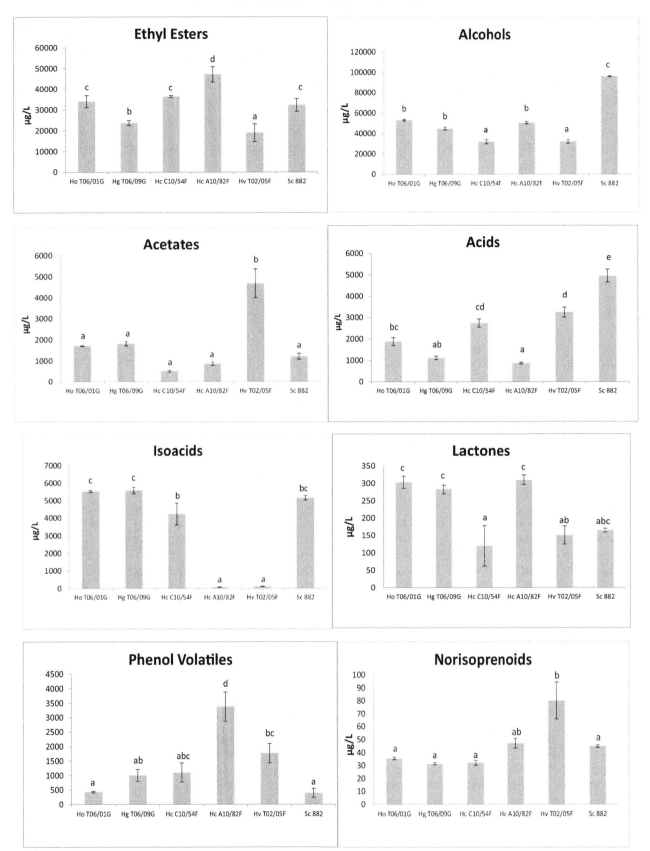

FIGURE 9.6 Performance of different species during similar synthetic white juice microvinifications controlled at the laboratory scale, for each of the different families of aroma compounds. Different letters indicate significant differences according to the Tukey test (95%). Significant differences level for acetates, acids, lactones, isoacids, and volatile phenols were found. Ho, *H. opuntiae*; Hg, *H. guilliermondii*; Hc, *H. clermontiae*; Hv, *H. vineae*; Sc, *S. cerevisiae*.

S. cerevisiae. This activity also might explain the norisoprenoid concentrations. The only compound that could be quantified within this group was 3-oxo-α-ionol, associated with the descriptors "honey" and "apricot." For this compound, the behavior between the yeasts was diverse, with *H. vineae* having a significantly higher concentration (80 μg/L) compared to the rest of the strains, which had concentrations of between 32 and 47 μg/L. This family of compounds has been found in all the grapes in their glycosylated form, together with benzyl and β-phenylethyl alcohols, and terpenes (Williams, Sefton, & Leigh, 1992; Winterhalter & Skouroumounis, 1997). However, these alcohols and the terpenes might be synthesized *de novo* by some yeast as it was shown in synthetic fermentation mediums, particularly by *Hanseniaspora* strains (Carrau et al., 2005; Martin et al., 2016b). Interestingly, while *S. cerevisiae* strains synthesize increasing concentration levels of linalool/α-terpineol, *H. uvarum* also produced three- to fourfold higher levels of geraniol than all *S. cerevisiae* strains tested (Carrau et al., 2005).

9.3 Mixed cultures with *Saccharomyces cerevisiae*

There are currently a few non-*Saccharomyces* strains available for winemakers (Roudil et al., 2020), but in most cases, all these yeasts are not capable of finishing a complete fermentation of mature grapes. Some of the strains can resist up to 10%−11% alcohol, and finally, a *Saccharomyces* strain will finish the processes for obtaining a dry wine. This situation has promoted the application of mixed cultures as a strategy of inoculation to deal with this highly sugary grape must and complete fermentation. The challenge of managing mixed cultures within a commercial application of two different inocula is the usual strong nutrient competition, mainly for some key compounds such as nitrogen and vitamins (Brou, Patricia, Beaufort, & Brandam, 2020; De Koker, 2015; Medina, Boido, Dellacassa, & Carrau, 2012; Roca-Mesa, Sendra, Mas, Beltran, & Torija, 2020; Seguinot et al., 2020), which are present in very low concentrations in natural grape juice.

Coinoculation and sequential inoculation are the two alternatives to manage mixed yeast cultures in grape musts. The first procedure has the advantage of a more balanced competition for nutrients between the two inoculated strains, while the second choice means leaving the non-*Saccharomyces* strain to start the process and so not being excluded faster by the *Saccharomyces* strain. In contrast, *Saccharomyces* strains ferment so rapidly and aggressively in winemaking conditions that in coinoculation, they usually dominate the nonconventional strains, and little contribution will be gained by this strain in the final wine. In our opinion, this fact has promoted the sequential alternative, leaving the non-*Saccharomyces* strains to act for 24 or 72 hours before adding the *Saccharomyces* strain. Using this method, we are simulating a more natural alternative as happens when winemakers do a cold maceration or cold juice treatment at 10°C for 2 or 3 days before inoculating the commercial *Saccharomyces* strain. The natural flora present in the crushed grapes is composed of 99.9% non-*Saccharomyces* strains, and they interact enzymatically and metabolically with the grape juice, contributing potentially to flavors. Recently, we proposed that performing the first inoculation of grape juice to avoid any risk of spontaneous uncontrolled yeast growth should be done with what we call a "friendly" strain. This should be a non-*Saccharomyces* strain with good starter capacity and osmotolerant when it is inoculated in rich sugar juices (Carrau et al., 2020). *H. vineae* has shown these characteristics and increases strain diversity during winemaking, without excluding other strains and letting *Saccharomyces* participate in the second part of the vinification when inoculated after 2 or 3 days.

9.3.1 Nutrient management during vinification

As previously mentioned, there is a concern about the use of mixed cultures with non-*Saccharomyces*, as in the initial steps of growth of some of these cells, they might have an inhibitory effect on the subsequent growth of *S. cerevisiae*. There is always a risk of stuck or sluggish wine fermentations when mixed cultures are inoculated, so the addition of nutrients to improve cell access and availability should be planned to avoid these risks. For example, it is well known that *H. vineae* is not able to synthesize thiamine (Steenwyk et al., 2019; Valera et al., 2020a), a key vitamin for the final steps of alcohol fermentation, so it rapidly removes all the thiamine present in the natural juice (Bataillon, Rico, Sablayrolles, Salmon, & Barre, 1996). Furthermore, pantothenic acid deficiency is also associated with sluggish wine fermentations (Guzzon et al., 2011; Wang, Bohlscheid, & Edwards, 2003). We have obtained excellent results by adding a mixture of vitamins and ammonium salts together with the sequential inoculation of *Saccharomyces* in *H. vineae* in white wine fermentations with 12%−13% potential alcohol.

However, further research is necessary both to improve the drying process in some mixed cultures combinations and for very mature grapes with above 15% of alcohol.

In terms of nitrogen nutrients, today, winemakers can measure the yeast assimilable nitrogen level with simple methods at harvest time in the grape must (Gump, Zoecklein, Fugelsang, & Whiton, 2002; Zoecklein, Fugelsang, & Gump, 2010), allowing the addition of exact quantities of ammonium salts and, at the same time, avoiding the risk of excess residual nitrogen during and after fermentation (Carrau et al., 2020). Such excess nitrogen might increase contamination of dry wines during aging by other yeasts such as *Brettanomyces*, stimulate biogenic amine formation, or result in losses of grape varietal flavor character. An increase of residual nitrogen might occur when yeast assimilable nitrogen levels are above 200 mg N/L during mixed cultures, as recently reported by Seguinot et al. (2020). However, as the determination of vitamin concentrations is a complicated alternative, the addition of vitamins should be prevented. There are interesting nutrients such as yeast autolysates that might contain different vitamins and ergosterol that will help to fit the yeast strains that are fermenting under a balanced competition. Further studies should be carried out to evaluate the effect of nutrient depletion under mixed cultures with *S. cerevisiae*. It has been proposed that competition for nutrients is the main cause of inconsistent results in non-*Saccharomyces*/*Saccharomyces* cofermentations (Medina et al., 2012). The concept that increasing yeast diversity results in superior flavors and sensory complexity has now been clearly proven (Carrau, Gaggero, & Aguilar, 2015). It has also been suggested that understanding nitrogen balance and vitamin requirements for low-nitrogen-demand *Saccharomyces* strains might prevent the production of H₂S aroma notes, improving also the flavor and general positive attributes in the final wine (Carrau et al., 2020; Ciani, Comitini, Mannazzu, & Domizio, 2010; Fleet, 2008; Pretorius, 2020).

9.4 Flavor impact of *Hanseniaspora vineae* in white wines

Regarding the production of white wines at the industrial scale using *H. vineae* yeasts, some examples of studies recently performed were reported (Martin et al., 2018). In the work carried out by Medina et al. (2013), fermentations by coinoculation of a strain of *H. vineae* with a commercial *S. cerevisiae* in Chardonnay must were done in oak barrels. Barrel-fermented Chardonnay wines obtained were characterized by increased fruity characters and intense complexity, as well as full body and a relatively long palate compared to pure fermentations using a commercial *Saccharomyces* strain. In another experiment in 100-L tanks of Macabeo white musts (Lleixà et al., 2016), by inoculating yeasts of the species *H. vineae* the appearance of *S. cerevisiae* occurred naturally at the end of fermentation. Wines were compared to those from the same must inoculated with a commercial *S. cerevisiae* strain. When these wines were evaluated by a trained panel, the judges were able to differentiate between the two wines by triangular tests, with the wines made with *H. vineae* significantly being preferred, standing out for their floral character and greater volume on the palate. In a more recent work carried out by Del Fresno et al. (2020) with the Albillo grape variety from Ribera del Duero, wines with more esters and greater fruit character were obtained using *H. vineae* compared to a commercial *Saccharomyces* strain. They found that this species can be used in aging processes on lees since they are apparently fast in transferring certain desired cellular compounds during the process, as will be discussed.

9.4.1 Aroma complexity and nitrogen metabolism

H. vineae is considered of enological interest mainly due to its positive contribution to wine aroma and palate (Del Fresno et al., 2020) and by its high ethanol tolerance compared with other non-*Saccharomyces* yeast species. Enhancements in positive characteristics by using this apiculate yeast have been reported for white wine flavors (Giorello et al., 2019; Lleixà et al., 2016; Martin et al., 2018; Medina et al., 2007, 2013).

Firstly, *H. vineae* strains are able to release aroma precursors from grapes by their β-glycosidase and protease extracellular activity in grape juices (Martin et al., 2018). Glucosidase enzymes are interesting for the wine industry because they can break the glycosidic bonds of aroma precursors (glycosylated aroma compounds) that remain bound to glucose in grapes or grape must making them nonvolatile (Fernández-González, Di Stefano, & Briones, 2003; Swangkeaw, Vichitphan, Butzke, & Vichitphan, 2011). These enzymes present two subunits codified by *ROT2* and *GTB1* genes. López et al. (2016) detected the increased release of some terpenes, especially linalool, in Muscat wines using two strains of *H. vineae*. Their β-glucosidase activity was tested under different sugar and ethanol concentrations and it was found that with up to 10% (v/v) ethanol and 100 mM sugars, the enzymes present stable hydrolase capacity. This prominent β-glucosidase activity exhibited by *H. vineae* compared with other non-*Saccharomyces* species (Barquet et al., 2012; Pérez et al., 2011) could be related to the high phenylpropanoid and isoprenoid contents of wines vinified with this yeast species (Lleixà et al., 2019).

Secondarily, several routes for aroma production in yeasts are derived from the shikimate pathway (Fig. 9.7). Higher alcohols produced from sugars and amino acids transamination are the precursors of acetate esters. These molecules are liberated in the extracellular medium, enhancing wine aroma. Overall acetate production is increased in *H. vineae* compared with *S. cerevisiae* (Giorello et al., 2019). The argued reason for this difference in production between *H. vineae* and *S. cerevisiae* is the presence of six alcohol acetyltransferase (AATase) genes found in the genome of *H. vineae* compared to the two copies usually appearing in *Saccharomyces* strains (Giorello et al., 2019). In *S. cerevisiae*, acetate ester formation relies on two AATases codified by *ATF1* and *ATF2*. In *H. vineae*, this function is putatively developed by one *ATF2*, four *SLI1*, and one gene that is not homologous to any other described in *S. cerevisiae* (named g4599.t1), attending to the genetic and transcriptional analysis performed at different points of must fermentation. The concentration of acetate ester is increased twofold between days 4 and 10, which agrees with the transcriptomic data obtained for these genes (Giorello et al., 2019).

In *H. vineae* fermentations, acetate esters such as isoamyl acetate or hexyl acetate are detected in similar concentrations to those in fermentations carried out by *S. cerevisiae* (Lleixà et al., 2016; Medina et al., 2013), whereas ethyl acetate and β-phenylethyl acetate have been found in significantly higher concentrations (Del Fresno et al., 2020; Martin et al., 2018; Medina et al., 2013). Among acetate esters, the production of β-phenylethyl acetate is especially increased, up to 50-fold higher than for *S. cerevisiae* (Lleixà et al., 2016; Medina et al., 2013). Chardonnay, Muscat, and Macabeo white wines vinified with *H. vineae* strains using different inoculation strategies present a higher concentration of β-phenylethyl acetate (Table 9.1). Contrariwise, fermentations carried out by *H. vineae* in cofermentation with *S. cerevisiae* performed in Petit Manseng grape must yielded β-phenylethyl acetate at a similar concentration to that in the *S. cerevisiae* control (Martin et al., 2019). Differences in the amino acid profile of this grape variety might explain this low production of the phenylalanine-derived acetate. Although further studies should be done to confirm species characterization, other *Hanseniaspora* species such as *H. guilliermondii* and *H. osmophila* also show enhanced production of β-phenylethyl acetate (Rojas et al., 2003; Viana et al., 2009), which might contribute to the fermentation bouquet of wines.

The biosynthesis of β-phenylethyl acetate uses β-phenylethanol as substrate, which is a higher alcohol produced via the Ehrlich pathway. The production of higher alcohols in yeasts is performed in three steps: transamination, decarboxylation, and reduction (Fig. 9.7). *H. vineae* has increased copies of *ARO* genes, three copies of *ARO8*, and four of *ARO9*, which codify aromatic amino acid transaminases involved in the first step of this route that might explain the high activity of this pathway in this species. In *S. cerevisiae* genomes, just one copy of each of these genes is found. Although no genes with homology to aryl alcohol dehydrogenases have been reported in the genome of *H. vineae* (these enzymes are involved in the reduction step of the Ehrlich pathway), it has been suggested that alcohol dehydrogenases would perform this final step (Giorello et al., 2019). *H. vineae* has eight

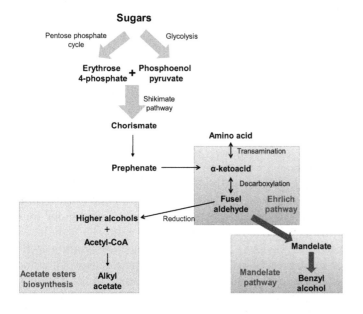

FIGURE 9.7 Metabolic pathways for higher alcohols, acetate esters, and benzenoids biosynthesis in *H. vineae*.

TABLE 9.1 Aroma production in white wines inoculated with *H. vineae* compared with *S. cerevisiae* single inoculated controls under the same conditions.

Grape variety	*H. vineae* strain (reference)	Inoculation strategy	Linalool	β-Phenylethanol	Ethyl acetate	Isoamyl acetate	β-Phenylethyl acetate	Benzyl alcohol
Chardonnay	Hv02/5F (Medina et al., 2013)	Coinoculation *S. cerevisiae* ALG 804	n.r.	↓	↑	≈	↑	n.r.
Macabeu	Hv02/5F (Lleixà et al., 2016)	Single inoculation	≈	↓	n.r.	≈	↑	↑
Petit Manseng	Hv02/5F (Martin et al., 2019)	Coinoculation *S. cerevisiae* ALG 804	n.r.	≈	↑	n.r.	≈	≈
Muscat	CECT11326 CECT11338 (López et al., 2016)	Coinoculation *S. cerevisiae* QA23	↑	↑	n.r.	n.r.	≈	n.r.
Albillo	Hv02/5F (Del Fresno et al., 2020)	Coinoculation *S. cerevisiae* Fermivin 3C	n.r.	≈	↑	≈	↑	n.r.

↓ Significantly lower compared with control fermentation using *S. cerevisiae*.
↑ Significantly higher compared with control fermentation using *S. cerevisiae*.
≈ No significant difference to control fermentation using *S. cerevisiae*.
n.r. Not reported.

ADH genes, as does *S. cerevisiae*, and double the copy numbers of other *Hanseniaspora* species, which might explain the high ethanol resistance of this species (Valera et al., 2020a).

Wines treated with *H. vineae* strains produce more β-phenylethanol in Muscat wines than those inoculated just with *S. cerevisiae* (López et al., 2016). However, in Chardonnay, Trebbiano, Albillo, Semillon, and Macabeo wines, the final concentration of this compound is similar or significantly lower than that produced by *S. cerevisiae* (Del Fresno et al., 2020; Lleixà et al., 2016; Martin et al., 2019; Medina et al., 2013). But it is noteworthy that comparing the sum of β-phenylethanol and β-phenylethyl acetate levels, their production by *H. vineae* is higher than that by *S. cerevisiae* (Giorello et al., 2019; Martin et al., 2016a). Therefore it is shown that *H. vineae* presents enhanced acetylation activity that converts β-phenylethanol faster and more efficiently compared to *Saccharomyces* strains. This fact is interesting for improving wine aroma perception because the odor threshold of β-phenylethyl acetate is 250 μg/L in contrast to the value of 10,000 μg/L for β-phenylethanol (Fariña et al., 2015). Both β-phenylethyl acetate and β-phenylethanol are volatile phenylpropanoids derived from the amino acids phenylalanine or tyrosine. In fact, these molecules and benzenoids have been largely studied in plants (Deng & Lu, 2017), but it has been evidenced that species such as *H. vineae* also contribute greatly to their synthesis under fermentative conditions in synthetic grape must (Martin et al., 2018).

Benzenoids and other volatile phenols found in wine such as 4-vinylguaiacol, tyrosol, tryptophol, and phenyl lactate are also considered phenylpropanoids. Even though they have been commonly detected in white wines inoculated with either *H. vineae* or *S. cerevisiae*; no significant differences were found regarding these two yeast species for 4-vinylguaiacol and phenyl lactate (Lleixà et al., 2016; Medina et al., 2013). Otherwise, tyrosol is found in significantly higher concentrations in *S. cerevisiae* fermentations, but its acetate ester is increased in coinoculations with *H. vineae* (Martin et al., 2019; Medina et al., 2013). *H. vineae* has been reported as a high producer of benzyl alcohol and many derived benzyl and 4-hydroxybenzenoids (Martin et al., 2016a). Recently, this characteristic has allowed to elucidate the biosynthetic pathway of these aromatic compounds in ascomycete yeasts using *H. vineae* as a model microorganism (Valera et al., 2020b). Benzyl alcohol is synthesized from phenylalanine through the so-called mandelate pathway, which has also been validated in *S. cerevisiae* using deletion mutants of key genes such as *ARO10* that codify not only phenylpyruvate decarboxylase but also a benzoylformate decarboxylase function (Valera et al., 2020c). Benzyl alcohol aroma is described as floral, rose, or toasted almonds (Fariña et al., 2015). 4-Hydroxybenzyl alcohol is produced from tyrosine by a parallel route through 4-hydroxymandelate, which has been reported in *S. cerevisiae* and is also active in *H. vineae* (Valera et al., 2020b). This flavor family might contribute to the increased complexity found in neutral white grape wines (Macabeo, Ugni blanc, Albillo, or Trebbiano), described as possessing an increased fruity and full-bodied palate when compared to pure *Saccharomyces* wine processes (Del Fresno et al., 2020; Lleixà et al., 2016; Martin et al., 2018).

The synthesis of phenylpropanoids is clearly related to nitrogen nutrient availability. Feeding experiments have demonstrated that the reduction of diammonium phosphate salts as a nitrogen source for *H. vineae* leads to an increase of these aroma compounds. Yeast assimilable nitrogen of 250 mg N/L reduces the production of benzyl alcohol and β-phenylethyl alcohol (Martin et al., 2016a). These data suggest an inhibitory effect of inorganic nitrogen at some level in the metabolic pathway controlling the biosynthesis of these compounds. Similar behavior has been reported for the synthesis of other higher alcohols in *S. cerevisiae* (Carrau et al., 2008). Therefore the addition of a high concentration of diammonium salts in winemaking, mainly used to increase ester production or avoid H_2S formation, reduces the production of phenylpropanoids, resulting in reduced phenylpropanoid flavor synthesis in the final wines (Martin et al., 2016a). We recently demonstrated that the strategy that should be used to produce more complex white wines under sequential mixed cultures consists of delaying the addition of ammonium salts until the *Saccharomyces* inoculation, which is after 3−6 days of initial fermentation with *H. vineae* (Carrau et al., 2020). During this initial phase, the *H. vineae* strain will produce more phenylpropanoid flavor compounds and the ammonium salts added with *Saccharomyces* will not inhibit their synthesis helping the *Saccharomyces* strain to have assimilable nitrogen available in the medium. Interestingly, the increased presence of phenylalanine in the grape juice will also contribute to the increased formation of phenylpropanoids by *H. vineae* (Valera et al., 2019). During the stationary phase of the vinification process, the level of acetate ester synthesis will also increase due to the redox anaerobic situation (Fariña et al., 2012).

9.4.2 Impact on taste and palate

9.4.2.1 Glycerol

Glycerol is the major yeast metabolite produced during wine fermentation after ethanol and is important in yeast metabolism for regulating redox potential in the cell (Jolly et al., 2014; Prior, Tohl, Jolly, Baccari, & Mortimer, 2000). This compound contributes to smoothness (mouthfeel), sweetness, and complexity in wines (Ciani & Maccarelli, 1998). Maturano et al. (2012) found higher glycerol production by *H. vineae* versus other non-*Saccharomyces* species, which may be explained by the fermentation capacity of *H. vineae* compared to other non-*Saccharomyces* species. However, these authors observed similar production in mixed cultures with *H. vineae* to that in pure *Saccharomyces* cultures.

The gene putatively codifying glycerol-3-phosphate phosphatase in *H. vineae* is homologous to *GPP1* of *S. cerevisiae*. Phosphatase activity is necessary to produce glycerol from glycerol-3-phosphate, a glycolytic intermediate. Under aerobic conditions, *S. cerevisiae* is able to use glycerol as the sole carbon and energy source via gluconeogenesis through phosphorylation of glycerol, mediated by glycerol kinase (Grauslund & Rønnow, 2000). *H. vineae* possesses sequences homologous to *GUT1* and *GUT2* genes that code for glycerol kinase (Valera et al., 2020a). However, *H. vineae*, similarly to other *Hanseniaspora* species, does not present genes for the main gluconeogenic activities: pyruvate carboxylase, phosphoenolpyruvate carboxykinase, and fructose-1,6-bisphosphatase (Seixas et al., 2019). Although further studies will be needed, the weak activity or absence of gluconeogenesis in *H. vineae* might explain the reduced consumption of glycerol that might increase its concentration in the final wine.

9.4.2.2 Polysaccharides, mannoproteins, and cell autolysis

Aging on lees, over yeast biomass, is a process that consists of contact of the wine with lees during aging in bottles, barrels, or tanks. During aging on lees, yeast autolysis occurs (Morata et al., 2020). Yeast cell walls are degraded by the action of enzymes produced by dead yeast cells (Palomero, Morata, Benito, González, & Suárez-Lepe, 2007); this process increases extracellular polysaccharides, mainly mannoproteins, the major cell wall polysaccharides that might increase wine body and palate. Usually, it takes several months under conventional enological conditions for this autolysis process to be completed and mannoproteins to be released (Kulkarni et al., 2015; Palomero, Morata, Benito, Calderón, & Suárez-Lepe, 2009). The different compounds released during yeast autolysis have a direct organoleptic repercussion on the wine palate (Jolly et al., 2014; Morata et al., 2020). Moreover, some non-*Saccharomyces* yeasts increase their polysaccharide content and the speed of cellular autolysis during the aging on lees process (Kulkarni et al., 2015; Morata et al., 2020; Palomero et al., 2009). Degradation of cell membranes releases intracellular compounds into the wine, such as these polysaccharides, medium chain fatty acids, nucleic acids, amino acids, and short peptides. Del Fresno et al. (2020), upon applying *H. vineae* in aging on lees and comparing it with other yeast species, found that samples with *H. vineae* yeast lees showed the highest values for nucleic acids, but no significant differences in protein content were measured after 91 days. The samples with *H. vineae* showed a high content of polysaccharides (around 11 mg/L), but the difference in this concentration was not statistically significant with respect to samples aged on the lees of different *Saccharomyces* yeast strains (Del Fresno et al., 2020). On the other hand, Carrau et al. (2019) studied cell autolysis by flow cytometry analysis, determining the loss of membrane integrity in *H. vineae* at 15 days of the fermentation process, finding a significantly faster rate of degradation compared to *H. uvarum* and *S. cerevisiae* strains. Although shorter periods than 90 days must be studied, the results could explain the strong mouthfeel impact of *H. vineae* white wines compared to *S. cerevisiae* wines after 30 days of wine aging. These results demonstrate that *H. vineae* develops a faster autolysis process and could be an alternative to replace *S. cerevisiae* yeast in the process of aging on lees. The nature of the cell wall of the different yeast species is variable; therefore the use of *H. vineae* for aging on lees techniques might result in wines with increased palate complexity in a shorter time (Morata et al., 2020). As a consequence, studies with this species are also being carried out for the production of base white wines for sparkling processes with a moderate alcohol content that might result in less expensive aging processes than the traditional method of bottle fermentation (Tomas Roman, personal communication).

9.4.2.3 Protease activity

Non-*Saccharomyces* yeasts have been reported to secrete extracellular proteases (Charoenchai, Fleet, Henschke, & Todd, 1997; Dizy & Bisson, 2000). On the other hand, Bilinski, Russell, and Stewart (1987) did not observe protease activity in *Saccharomyces*, while other researchers detected weak activity in this yeast species (Dizy & Bisson, 2000; Lagace & Bisson, 1990).

More recently, extracellular proteases have been observed throughout all fermentations studied using *H. vineae* and *T. delbrueckii* yeasts, whose activity was found to be higher than with pure *S. cerevisiae* cultures on Pedro Ximenez grapes (Maturano et al., 2012). The protease activity of mixed cultures is higher when a higher percentage of non-*Saccharomyces* strains is present in the medium (Maturano et al., 2012). These results can be related to the contribution of non-*Saccharomyces* species to the total activity of this enzyme. However, different factors can modify protease activity. For example, the presence of a readily utilizable nitrogen source was found to repress extracellular proteases (Charoenchai et al., 1997; Strauss et al., 2001).

Hanseniaspora species are reported as good producers of proteases (Charoenchai et al., 1997; López et al., 2016), while Martin (2016) found proteolytic activity in skim milk medium at pH 6 in all *H. vineae* strains studied. However, this activity was strain dependent.

Dizy and Bisson (2000) reported that increased yeast proteolytic activity did not lead to a reduction in haze formation in white wine, but other authors have found a reduction in protein levels by non-*Saccharomyces* proteolytic activity with an increase in protein stability of the end-product (Jolly et al., 2014). Recently, three tests were carried out in our laboratory with two *H. vineae* strains and one *S. cerevisiae* strain in Sauvignon blanc grape must fermented with each of the strains. One of the *H. vineae* strains showed better protein stability, which was confirmed by measuring turbidity in the stability protein test (Fig. 9.8). These results indicate the possibility of improving protein stability using some strains of *H. vineae*.

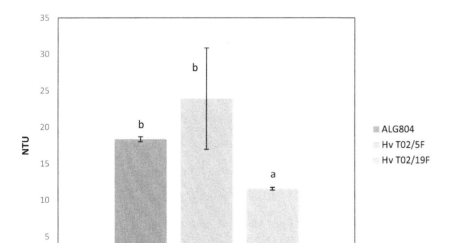

FIGURE 9.8 NTU (nephelometric turbidity unit). Average values for three fermentations in triplicate with different yeasts: *S. cerevisiae* ALG 804 (Oenobrands, France) and two *H. vineae* strains (Hv T02/5F and Hv T02/19F). Error bars indicate standard deviation. Letters indicate significant differences in treatments. The protein stability test was prepared with a heat shock treatment at 80°C for 30 min. Significant protein stability was obtained with strain Hv T02/19F ($P \leq .01$) compared to the other two treatments.

9.5 Conclusions

The sensory characteristics of white wines produced with *H. vineae* have increased flavor and palate complexity, mainly due to the rich metabolic pathways related to the three key aromatic amino acids, namely phenylalanine, tyrosine, and tryptophan, that contribute to the formation of many intense sensory characteristics derived from phenylpropanoid compounds. The capacity of this species to improve the sensory attributes of neutral white varieties is the main attraction in real winemaking conditions evaluated recently in Uruguay, Spain, and Italy. Although further studies should be carried out to improve the pairings under mixed culture fermentation with different *Saccharomyces* strains, *H. vineae*'s capacity for fruity flavor synthesis and increased body should be an interesting application opportunity for the development of young and shorter-aged high-quality wines compared to conventional *Saccharomyces* starters.

Acknowledgments

We thank the following agencies for financial support: CSIC Group Project number 802 (UdelaR, Uruguay), Agencia Nacional de Investigacion e Innovacion (ANII), *Hanseniaspora vineae* FMV 6956 project, and postdoctoral fellowship of Maria Jose Valera (PD_NAC_2016_1_133945). We also thank ANII and Lage y Cia-Lallemand Inc., Uruguay, for supporting our cooperation project on *H. vineae* metabolic characterization (ALI_2_2019_1_155314).

References

Alexandre, H. (2020). Wine yeast terroir: Separating the wheat from the chaff—for an open debate. *Microorganisms, 8*(5), 787.

Angioni, A., Caboni, P., Garau, A., Farris, A., Orro, D., Budroni, M., & Cabras, P. (2007). In vitro interaction between ochratoxin A and different strains of *Saccharomyces cerevisiae* and *Kloeckera apiculata. Journal of Agricultural and Food Chemistry, 55*, 2043–2048.

Barquet, M., Martín, V., Medina, K., Pérez, G., Carrau, F., & Gaggero, C. (2012). Tandem repeat-tRNA (TRtRNA) PCR method for the molecular typing of non-*Saccharomyces* subspecies. *Applied Microbiology and Biotechnology, 93*(2), 807–814.

Bataillon, M., Rico, A., Sablayrolles, J. M., Salmon, J. M., & Barre, P. (1996). Early thiamin assimilation by yeasts under enological conditions: Impact on alcoholic fermentation kinetics. *Journal of Fermentation and Bioengineering, 82*(2), 145–150.

Beltran, G., Torija, M. J., Novo, M., Ferrer, N., Poblet, M., Guillamón, J. M., ... Mas, A. (2002). Analysis of yeast populations during alcoholic fermentation: A six year follow-up study. *Systematic and Applied Microbiology, 25*(2), 287–293.

Bilinski, C. A., Russell, I., & Stewart, G. G. (1987). Applicability of yeast extracellular proteinases in brewing: Physiological and biochemical aspects. *Applied and Environmental Microbiology, 53*(3), 495–499.

Brou, P., Patricia, T., Beaufort, S., & Brandam, C. (2020). Modelling of *S. cerevisiae* and *T. delbrueckii* pure culture fermentation in synthetic media using a compartmental nitrogen model. *OENO One, 54*(2), 299–311.

Cadez, N., Raspor, P., de Cock, A. W., Boekhout, T., & Smith, M. T. (2002). Molecular identification and genetic diversity within species of the genera *Hanseniaspora* and *Kloeckera. FEMS Yeast Research, 1*(4), 279–289.

Carrau, F. M., Medina, K., Boido, E., Farina, L., Gaggero, C., Dellacassa, E., ... Henschke, P. A. (2005). De novo synthesis of monoterpenes by *Saccharomyces cerevisiae* wine yeasts. *FEMS Microbiology Letters, 243*(1), 107–115.

Carrau, F. M., Medina, K., Farina, L., Boido, E., Henschke, P. A., & Dellacassa, E. (2008). Production of fermentation aroma compounds by *Saccharomyces cerevisiae* wine yeasts: Effects of yeast assimilable nitrogen on two model strains. *FEMS Yeast Research, 8*(7), 1196–1207.

Carrau, F., Boido, E., & Ramey, D. (2020). Yeasts for low input winemaking: Microbial terroir and flavor differentiation. In *Advances in applied microbiology* (Vol. 111, pp. 89–121). Massachusetts: Academic Press.

Carrau, F., Gaggero, C., & Aguilar, P. S. (2015). Yeast diversity and native vigor for flavor phenotypes. *Trends in Biotechnology, 33*(3), 148–154.

Carrau, F., Medina, K., Valera, M.J., Dellacassa, E., Martin, V., Boido, E., ... Fariña, L. (2019). Global winemaking application of *Hanseniaspora vineae* under mixed yeast culture conditions. In *Proceeding of ISSY 35, the 35th international specialized symposium on yeasts* (p. 4). Antalya, Turkey.

Charoenchai, C., Fleet, G. H., Henschke, P. A., & Todd, B. E. N. T. (1997). Screening of non-*Saccharomyces* wine yeasts for the presence of extracellular hydrolytic enzymes. *Australian Journal of Grape and Wine Research, 3*(1), 2–8.

Ciani, M., & Maccarelli, F. (1998). Oenological properties of non-*Saccharomyces* yeasts associated with wine-making. *World Journal of Microbiology and Biotechnology, 14*, 199–203.

Ciani, M., & Piccioti, G. (1995). The growth kinetics and fermentation behavior of some non-*Saccharomyces* yeast associated with winemaking. *Biotechnology Letters, 17*, 1247–1250.

Ciani, M., Comitini, F., Mannazzu, I., & Domizio, P. (2010). Controlled mixed culture fermentation: A new perspective on the use of non-*Saccharomyces* yeasts in winemaking. *FEMS Yeast Research, 10*(2), 123–133.

De Koker, S. (2015). *Nitrogen utilisation of selected non-Saccharomyces yeasts and the impact on volatile compound production* (Ph.D. thesis). Stellenbosch University, South Africa. pp. 1–71.

Del Fresno, J. M., Escott, C., Loira, I., Herbert-Pucheta, J. E., Schneider, R., Carrau, F., ... Morata, A. (2020). Impact of *Hanseniaspora vineae* in alcoholic fermentation and ageing on lees of high-quality white wine. *Fermentation, 6*(3), 66.

Deng, Y., & Lu, S. (2017). Biosynthesis and regulation of phenylpropanoids in plants. *Critical Reviews in Plant Sciences, 36*(4), 257–290.

Díaz-Montaño , D. M., & Córdova, J. D. J. R. (2009). The fermentative and aromatic ability of *Kloeckera* and *Hanseniaspora* yeasts. In *Yeast Biotechnology: Diversity and applications* (pp. 281–305). Dordrecht: Springer.

Dizy, M., & Bisson, L. F. (2000). Proteolytic activity of yeast strains during grape juice fermentation. *American Journal of Enology and Viticulture*, *51*(2), 155−167.

Esteve-Zarzoso, B., Gostıncar, A., Bobet, R., Uruburu, F., & Querol, A. (2000). Selection and molecular characterization of wine yeasts isolated from the 'El Penedes' area (Spain). *Food Microbiology*, *17*(5), 553−562.

Fariña, L., Medina, K., Urruty, M., Boido, E., Dellacassa, E., & Carrau, F. (2012). Redox effect on volatile compound formation in wine during fermentation by *Saccharomyces cerevisiae*. *Food Chemistry*, *134*(2), 933−939.

Fariña, L., Villar, V., Ares, G., Carrau, F., Dellacassa, E., & Boido, E. (2015). Volatile composition and aroma profile of Uruguayan Tannat wines. *Food Research International*, *69*, 244−255.

Fernández-González, M., Di Stefano, R., & Briones, A. (2003). Hydrolysis and transformation of terpene glycosides from muscat must by different yeast species. *Food Microbiology*, *20*(1), 35−41.

Fleet, G. H. (2008). Wine yeasts for the future. *FEMS Yeast Research*, *8*(7), 979−995.

Fleet, G. H., & Heard, G. M. (1993). Yeast growth during fermentation. In *Wine Microbiology and Biotechnology* (pp. 27−54). Switzerland: Harwood Academic Publishers.

Gil, J. V., Mateo, J. J., Jimenez, M., Pastor, A., & Huerta, T. (1996). Aroma compounds in wines as influenced by apiculate yeasts. *Journal of Food Science*, *61*, 1247−1266.

Giorello, F., Valera, M. J., Martin, V., Parada, A., Salzman, V., Camesasca, L., ... Carrau, F. (2019). Genomic and transcriptomic basis of *Hanseniaspora vineae*'s impact on flavor diversity and wine quality. *Applied and Environmental Microbiology*, *85*(1), e01959−18.

Grangeteau, C., Gerhards, D., Rousseaux, S., von Wallbrunn, C., Alexandre, H., & Guilloux-Benatier, M. (2015). Diversity of yeast strains of the genus *Hanseniaspora* in the winery environment: What is their involvement in grape must fermentation? *Food Microbiology*, *50*, 70−77.

Grauslund, M., & Rønnow, B. (2000). Carbon source-dependent transcriptional regulation of the mitochondrial glycerol-3-phosphate dehydrogenase gene, *GUT2*, from *Saccharomyces cerevisiae*. *Canadian Journal of Microbiology*, *46*(12), 1096−1100.

Gump, B. H., Zoecklein, B. W., Fugelsang, K. C., & Whiton, R. S. (2002). Comparison of analytical methods for prediction of prefermentation nutritional status of grape juice. *American Journal of Enology and Viticulture*, *53*(4), 325−329.

Guzzon, R., Widmann, G., Settanni, L., Malacarne, M., Francesca, N., & Larcher, R. (2011). Evolution of yeast populations during different biodynamic winemaking processes. *South African Journal of Enology and Viticulture*, *32*(2), 242−250.

Herraiz, T., Reglero, G., Herraiz, M., Martin-Alvarez, P. J., & Cabezudo, M. D. (1990). The influence of the yeast and type of culture on the volatile composition of wines fermented without sulfur dioxide. *American Journal of Enology and Viticulture*, *41*(4), 313−318.

Jolly, N. P., Augustyn, O. P. H., & Pretorius, I. S. (2003). The occurrence of non-*Saccharomyces cerevisiae* yeast species over three vintages in four vineyards and grape musts from four production regions of the Western Cape, South Africa. *South African Journal of Enology and Viticulture*, *24*(2), 35−42.

Jolly, N. P., Varela, C., & Pretorius, I. S. (2014). Not your ordinary yeast: non-Saccharomyces yeasts in wine production uncovered. *FEMS Yeast Research*, *14*(2), 215−237.

Knight, S., Klaere, S., Fedrizzi, B., & Goddard, M. R. (2015). Regional microbial signatures positively correlate with differential wine phenotypes: evidence for a microbial aspect to terroir. *Scientific Reports*, *5*(1), 1−10.

Kulkarni, P., Loira, I., Morata, A., Tesfaye, W., González, M. C., & Suárez-Lepe, J. A. (2015). Use of non-*Saccharomyces* yeast strains coupled with ultrasound treatment as a novel technique to accelerate ageing on lees of red wines and its repercussion in sensorial parameters. *LWT-Food Science and Technology*, *64*(2), 1255−1262.

Lagace, L. S., & Bisson, L. F. (1990). Survey of yeast acid proteases for effectiveness of wine haze reduction. *American Journal of Enology and Viticulture*, *41*(2), 147−155.

Lambrechts, M. G., & Pretorius, I. S. (2000). Yeast and its importance to wine aroma a review. *South African Journal of Enology and Viticulture*, *21*, 97−129.

Lappa, I. K., Kachrimanidou, V., Pateraki, C., Koulougliotis, D., Eriotou, E., & Kopsahelis, N. (2020). Indigenous yeasts: Emerging trends and challenges in winemaking. *Current Opinion in Food Science*, *32*, 133−143.

Lleixà, J., Martín, V., Giorello, F., Portillo, M. C., Carrau, F., Beltran, G., & Mas, A. (2019). Analysis of the NCR mechanisms in *Hanseniaspora vineae* and *Saccharomyces cerevisiae* during winemaking. *Frontiers in Genetics*, *9*, 747.

Lleixà, J., Martín, V., Portillo, M. D. C., Carrau, F., Beltran, G., & Mas, A. (2016). Comparison of fermentation and wines produced by inoculation of *Hanseniaspora vineae* and *Saccharomyces cerevisiae*. *Frontiers in Microbiology*, *7*, 338.

Longo, E., Cansado, J., Agrelo, D., & Villa, T. G. (1991). Effect of climatic conditions on yeast diversity in grape musts from northwest Spain. *American Journal of Enology and Viticulture*, *42*, 141−144.

López, S., Mateo, J. J., & Maicas, S. (2016). Screening of *Hanseniaspora* strains for the production of enzymes with potential interest for winemaking. *Fermentation*, *2*(1), 1.

Martin, M. (2016). Hanseniaspora vineae: caracterización y su *uso en la vinificación* (Ph.D. thesis, pp. 1−211). Universitat Rovira i Virgili, España.

Martin, V., Boido, E., Giorello, F., Mas, A., Dellacassa, E., & Carrau, F. (2016a). Effect of yeast assimilable nitrogen on the synthesis of phenolic aroma compounds by *Hanseniaspora vineae* strains. *Yeast (Chichester, England)*, *33*(7), 323−328.

Martin, V., Fariña, L., Medina, K., Boido, E., Dellacassa, E., Mas, A., & Carrau, F. (2019). Oenological attributes of the yeast *Hanseniaspora vineae* and its application for white and red winemaking. In *BIO web of conferences* (Vol. 12, p. 02010). EDP Sciences.

Martin, V., Giorello, F., Fariña, L., Minteguiaga, M., Salzman, V., Boido, E., ... Carrau, F. (2016b). De novo synthesis of benzenoid compounds by the yeast *Hanseniaspora vineae* increases the flavor diversity of wines. *Journal of Agricultural and Food Chemistry*, *64*(22), 4574−4583.

Martin, V., Valera, M. J., Medina, K., Boido, E., & Carrau, F. (2018). Oenological impact of the *Hanseniaspora/Kloeckera* yeast genus on wines—A review. *Fermentation*, *4*(3), 76.

Mateo, J. J., Jimenez, M., Huerta, T., & Pastor, A. (1991). Contribution of different yeast isolated from musts of Monastrell grapes to the aroma of wine. *International Journal of Food Microbiology*, *14*, 153−160.

Maturano, Y. P., Assaf, L. A. R., Toro, M. E., Nally, M. C., Vallejo, M., de Figueroa, L. I. C., . . . Vazquez, F. (2012). Multi-enzyme production by pure and mixed cultures of *Saccharomyces* and non-*Saccharomyces* yeasts during wine fermentation. *International Journal of Food Microbiology*, 155(1−2), 43−50.

Medina, K. (2014). *Biodiversidad de levaduras no-Saccharomyces: efecto del metabolismo secundario en el color y el aroma de vinos de calidad* (Ph.D. thesis, pp. 1−327). Universidad de la República, Facultad de Química, Montevideo, Uruguay.

Medina, K., Boido, E., Dellacassa, E., & Carrau, F. (2018). Effects of non-*Saccharomyces* yeasts on color, anthocyanin, and anthocyanin-derived pigments of Tannat grapes during fermentation. *American Journal of Enology and Viticulture*, 69(2), 148−156.

Medina, K., Boido, E., Dellacassa, E., & Carrau, F. M. (2012). Growth of non-*Saccharomyces* yeasts affects nutrient availability for *Saccharomyces cerevisiae* during wine fermentation. *International Journal of Food Microbiology*, 157(2), 245−250.

Medina, K., Boido, E., Fariña, L., Gioia, O., Gomez, M. E., Barquet, M., . . . Carrau, F. (2013). Increased flavour diversity of Chardonnay wines by spontaneous fermentation and co-fermentation with *Hanseniaspora vineae*. *Food Chemistry*, 141(3), 2513−2521.

Medina, K., Ferreri, L., Fariña, L., Boido, E., Dellacassa, E., Gaggero, C., & Carrau, F. (2007). *Aplicación de la levadura* Hanseniaspora vineae *en cultivos mixtos con* Saccharomyces cerevisiae *en la vinificación*. Revista Enología, N 2, 4.

Meilgaard, M. C. (1985). Aroma volatiles in beer: Purification, flavour, threshold and interaction. In *Biogeneration of aromas* (pp. 211−254). Washington: American Chemical Society.

Mendoza, L. M., Vega-Lopez, G. A., de Ullivarri, M. F., & Raya, R. R. (2019). Population and oenological characteristics of non-*Saccharomyces* yeasts associated with grapes of Northwestern Argentina. *Archives of Microbiology*, 201(2), 235−244.

Morata, A., Escott, C., Bañuelos, M. A., Loira, I., del Fresno, J. M., González, C., & Suárez-Lepe, J. A. (2020). Contribution of non-*Saccharomyces* yeasts to wine freshness. A review. *Biomolecules*, 10(1), 34.

Moreira, N., Mendes, F., de Pinho, R. G., Hogg, T., & Vasconcelos, I. (2008). Heavy sulphur compounds, higher alcohols and esters production profile of *Hanseniaspora uvarum* and *Hanseniaspora guilliermondii* grown as pure and mixed cultures in grape must. *International Journal of Food Microbiology*, 124, 231−238.

Moreira, N., Mendes, F., Hogg, T., & Vasconcelos, I. (2005). Alcohols, esters and heavy compounds production by pure and mixed cultures of apiculate wine yeasts. *International Journal of Food Microbiology*, 103, 285−294.

Moreira, N., Pina, C., Mendes, F., Couto, J. A., Hogg, T., & Vasconcelos, I. (2011). Volatile compounds contribution of *Hanseniaspora guilliermondii* and *Hanseniaspora uvarum* during red wine vinifications. *Food Control*, 22(5), 662−667.

Padilla, B., Zulian, L., Ferreres, À., Pastor, R., Esteve-Zarzoso, B., Beltran, G., & Mas, A. (2017). Sequential inoculation of native non-*Saccharomyces* and *Saccharomyces cerevisiae* strains for wine making. *Frontiers in Microbiology*, 8(1293), 1−12.

Palomero, F., Morata, A., Benito, S., Calderón, F., & Suárez-Lepe, J. A. (2009). New genera of yeasts for over-lees aging of red wine. *Food Chemistry*, 112(2), 432−441.

Palomero, F., Morata, A., Benito, S., González, M. C., & Suárez-Lepe, J. A. (2007). Conventional and enzyme-assisted autolysis during ageing over lees in red wines: Influence on the release of polysaccharides from yeast cell walls and on wine monomeric anthocyanin content. *Food Chemistry*, 105(2), 838−846.

Pérez, G., Fariña, L., Barquet, M., Boido, E., Gaggero, C., Dellacassa, E., & Carrau, F. (2011). A quick screening method to identify β-glucosidase activity in native wine yeast strains: application of esculin glycerol agar (EGA) medium. *World Journal of Microbiology and Biotechnology*, 27(1), 47−55.

Pretorius, I. S. (2020). Tasting the terroir of wine yeast innovation. *FEMS Yeast Research*, 20(1), foz084.

Prior, B. A., Tohl, T. H., Jolly, N., Baccari, C., & Mortimer, R. K. (2000). Impact of yeast breeding for elevated glycerol production on fermentative activity and metabolite formation in Chardonnay wine. *South African Journal of Enology and Viticulture*, 21(2), 91−92.

Querol, A., Jiménez, M., & Huerta, T. (1990). A study on microbiological and enological parameters during fermentation musts from poor and normal grape-harvest in the region of Alicante (Spain). *Journal of Food Science*, 55, 114−122.

Ravaglia, S., & Delfini, C. (1994). Inhibitory effects of medium chain fatty acids on yeasts cells growing in synthetic nutrient medium and in the sparkling Moscato wine "Asti Spumante". *Wein-Wissenschaft*, 49(1), 40−45.

Roca-Mesa, H., Sendra, S., Mas, A., Beltran, G., & Torija, M. J. (2020). Nitrogen preferences during alcoholic fermentation of different non-*Saccharomyces* yeasts of oenological interest. *Microorganisms*, 8(2), 157.

Rojas, V., Gil, J. V., Piñaga, F., & Manzanares, P. (2003). Acetate ester formation in wine by mixed cultures in laboratory fermentations. *International Journal of Food Microbiology*, 86(1−2), 181−188.

Romano, P., Suzzi, G., Zironi, R., & Comi, G. (1993). Biometric study of acetoin production in *Hanseniaspora guilliermondii* and *Kloeckera apiculata*. *Applied and Environmental Microbiology*, 59, 1838−1841.

Roudil, L., Russo, P., Berbegal, C., Albertin, W., Spano, G., & Capozzi, V. (2020). Non-*Saccharomyces* commercial starter cultures: Scientific trends, recent patents and innovation in the wine sector. *Recent Patents on Food, Nutrition and Agriculture*, 11(1), 27−39.

Sabate, J., Cano, J., Esteve-Zarzoso, B., & Guillamón, J. M. (2002). Isolation and identification of yeasts associated with vineyard and winery by RFLP analysis of ribosomal genes and mitochondrial DNA. *Microbiological Research*, 157(4), 267−274.

San-Juan, F., Ferreira, V., Cacho, J., & Escudero, A. (2011). Quality and aromatic sensory descriptors (mainly fresh and dry fruit character) of Spanish red wines can be predicted from their aroma-active chemical composition. *Journal of Agricultural and Food Chemistry*, 59, 7916−7924.

Schütz, M., & Gafner, J. (1994). Dynamics of the yeast strain population during spontaneous alcoholic fermentation determined by CHEF gel electrophoresis. *Letters in Applied Microbiology*, 19, 253−257.

Seguinot, P., Bloem, A., Brial, P., Meudec, E., Ortiz-Julien, A., & Camarasa, C. (2020). Analysing the impact of the nature of the nitrogen source on the formation of volatile compounds to unravel the aroma metabolism of two non-*Saccharomyces* strains. *International Journal of Food Microbiology*, 316, 108441.

Seixas, I., Barbosa, C., Mendes-Faia, A., Güldener, U., Tenreiro, R., Mendes-Ferreira, A., & Mira, N. P. (2019). Genome sequence of the non-conventional wine yeast *Hanseniaspora guilliermondii* UTAD222 unveils relevant traits of this species and of the *Hanseniaspora* genus in the context of wine fermentation. *DNA Research*, 26, 67−83.

Steenwyk, J. L., Opulente, D. A., Kominek, J., Shen, X. X., Zhou, X., Labella, A. L., ... DeVirgilio, J. (2019). Extensive loss of cell-cycle and DNA repair genes in an ancient lineage of bipolar budding yeasts. *PLoS Biology, 17*(5), e3000255.

Strauss, M. L. A., Jolly, N. P., Lambrechts, M. G., & Van Rensburg, P. (2001). Screening for the production of extracellular hydrolytic enzymes by non-*Saccharomyces* wine yeasts. *Journal of Applied Microbiology, 91*(1), 182–190.

Streiblová, E., Beran, K., & Pokorný, V. (1964). Multiple scars, a new type of yeast scar in apiculate yeasts. *Journal of Bacteriology, 88*(4), 1104–1111.

Swangkeaw, J., Vichitphan, S., Butzke, C. E., & Vichitphan, K. (2011). Characterization of β-glucosidases from *Hanseniaspora* sp. and *Pichia anomala* with potentially aroma-enhancing capabilities in juice and wine. *World Journal of Microbiology and Biotechnology, 27*(2), 423–430.

Swiegers, J. H., Bartowsky, E. J., Henschke, P. A., & Pretorius, I. (2005). Yeast and bacterial modulation of wine aroma and flavour. *Australian Journal of Grape and Wine Research, 11*(2), 139–173.

Swiegers, J. H., Saerens, S. M., & Pretorius, I. S. (2016). Novel yeast strains as tools for adjusting the flavor of fermented beverages to market specifications. *Biotechnology Flavor Production, 62*–132.

Swiegers, J. H., & Pretorius, I. S. (2005). Yeast modulation of wine flavour. *Advances in Applied Microbiology, 57*, 131–175.

Testa, B., Lombardi, S. J., Iorizzo, M., Letizia, F., Di Martino, C., Di Renzo, M., ... Sorrentino, E. (2020). Use of strain *Hanseniaspora guilliermondii* BF1 for winemaking process of white grapes *Vitis vinifera* cv Fiano. *European Food Research and Technology, 246*(3), 549–561.

Torija, M. J., Rozes, N., Poblet, M., Guillamón, J. M., & Mas, A. (2001). Yeast population dynamics in spontaneous fermentations: Comparison between two different wine-producing areas over a period of three years. *Antonie Van Leeuwenhoek, 79*(3–4), 345–352.

Tristezza, M., Tufariello, M., Capozzi, V., Spano, G., Mita, G., & Grieco, F. (2016). The oenological potential of *Hanseniaspora uvarum* in simultaneous and sequential co-fermentation with *Saccharomyces cerevisiae* for industrial wine production. *Frontiers in Microbiology, 7*, 670.

Valera, M. J., Boido, E., Dellacassa, E., & Carrau, F. (2020a). Comparison of the Glycolytic and Alcoholic Fermentation Pathways of *Hanseniaspora vineae* with *Saccharomyces cerevisiae* Wine Yeasts. *Fermentation, 6*(3), 78.

Valera, M. J., Boido, E., Ramos, J. C., Manta, E., Radi, R., Dellacassa, E., & Carrau, F. (2020b). The mandelate pathway, an alternative to the PAL pathway for the synthesis of benzenoids in yeast. *Applied and Environmental Microbiology, 86*(17), e00701–e00720.

Valera, M. J., Zeida, A., Boido, E., Beltran, G., Torija, M. J., Mas, A., ... Carrau, F. (2020c). Genetic and transcriptomic evidences suggest *ARO10* genes are involved in benzenoid biosynthesis by yeast. *Yeast (Chichester, England)*. Available from https://doi.org/10.1002/yea.3508.

Valera, M.J., Olivera, V., Perez, G., Boido, E., Dellacassa, E. & Carrau, F. (2019). Effect of phenylalanine addition during vinification of Vitis vinifera cv Chardonnay with the yeast Hanseniaspora vineae. In *Proceeding of XVI congreso latinoamericano de viticultura y enología* (pp. 391–339). Ica, Perú.

Varela, C., & Borneman, A. R. (2017). Yeasts found in vineyards and wineries. *Yeast (Chichester, England), 34*(3), 111–128.

Viana, F., Belloch, C., Vallés, S., & Manzanares, P. (2011). Monitoring a mixed starter of *Hanseniaspora vineae-Saccharomyces* cerevisiae in natural must: Impact on 2-phenylethyl acetate production. *International Journal of Food Microbiology, 151*, 235–240.

Viana, F., Gil, J. V., Vallés, S., & Manzanares, P. (2009). Increasing the levels of 2-phenylethyl acetate in wine through the use of a mixed culture of *Hanseniaspora osmophila* and *Saccharomyces cerevisiae*. *International Journal of Food Microbiology, 135*(1), 68–74.

Wang, X. D., Bohlscheid, J. C., & Edwards, C. G. (2003). Fermentative activity and production of volatile compounds by *Saccharomyces* grown in synthetic grape juice media deficient in assimilable nitrogen and/or pantothenic acid. *Journal of Applied Microbiology, 94*(3), 349–359.

Williams, P.J., Sefton, M.A. & Leigh, F. (1992). Glycosidic precursors of varietal grape and wine flavor. Flavor precursors: thermal and enzymatic conversions. In *American chemical society symposium* (pp. 74–86). Washington, United States.

Winterhalter, P., & Skouroumounis, G. K. (1997). Glycoconjugated aroma compounds: occurrence, role and biotechnological transformation. in *Advances in biochemical engineering biotechnology* (Vol. 55, pp. 73–105). Berlin, Heidelberg: Springer.

Zironi, R., Romano, P., Suzzi, G., Battistutta, F., & Comi, G. (1993). Volatile metabolites produced in wine by mixed and sequential cultures of *Hanseniaspora guilliermondii* or *Kloeckera apiculata* and *Saccharomyces cerevisiae*. *Biotechnology Letters, 15*, 235–238.

Zoecklein, B. W., Fugelsang, K. C., & Gump, B. H. (2010). Practical methods of measuring grape quality. in *Managing wine quality* (pp. 107–133). Cambridge: Woodhead Publishing.

10

Improving white wine aroma and structure by non-*Saccharomyces* yeasts

Maurizio Ciani, Laura Canonico and Francesca Comitini

Department of Life and Environmental Sciences, Polytechnic University of Marche, Ancona, Italy

10.1 Introduction

Wine is an acidic hydroalcoholic solution produced from the alcoholic fermentation of grape by yeasts and/or bacteria, characterized by peculiar sensory properties that have attracted consumers for centuries (Waterhouse, Sacks, & Jeffery, 2016). The chemical composition of wine is a complex result determined by many factors, including grape variety, geographical origin, viticultural condition of grape cultivation, microbial ecology of the grape, fermentation process, and winemaking practices (Cole & Noble, 1995). Higher alcohols, acids, and esters are quantitatively dominant in aroma (Zhu, Du, & Li, 2016), especially in the quality criteria of white wine (Moreira, Lopes, Ferreira, Cabral, & de Pinho, 2018). Nevertheless, it is during the aging in bottle or during wood storage that wine composition gradually changes to reach the final sensory quality (Liu, Lu, Duan, & Yan, 2016). During the whole process, the microbiological component plays a relevant role in the analytical characterization and sensorial profile formulation in the final product. Indeed, the complex interactions among yeasts, bacteria, and other fungi influence wine composition in all winemaking steps, from the vineyard to the winery. Even if *Saccharomyces cerevisiae* is recognized as mainly responsible for the grape juice fermentation process (Pretorius, 2000), other yeasts species classified as non-*Saccharomyces* (NS) can affect both the analytical composition and the aroma profile of final wine (Jolly, Varela, & Pretorius, 2014). These yeast species are predominant on the surface of ripe grapes and colonize the winery environment (Agarbati et al. 2019; Albergaria & Arneborg, 2016; Barata, Malfeito-Ferreira, & Loureiro, 2012; Cocolin, Pepe, Comitini, Comi, & Ciani, 2004). In recent years, NS yeasts with different fermentation behavior are being studied in applied research, evidencing their important role in wine fermentation under certain controlled conditions such as mixed fermentations (Ciani, Comitini, Mannazzu, & Domizio, 2010). In this regard, these microbial cofermentations led to the production of volatile compounds and secondary metabolites, favoring the complexity of the wine aroma or modifying the structure of the final wine and undoubtedly representing new biotechnological opportunities to exploit in wine production.

The choice of yeast, as a microbiological strategy, defines the uniqueness of fermented products, including color, mouthfeel, and aroma complexity, due to the variations in the metabolic characteristics among yeast–yeast interactions.

In mixed fermentation the interactions between selected strains of *S. cerevisiae* and NS yeasts may give positive effects on wine composition (Ciani et al., 2016; Sadoudi et al., 2012). Indeed, studies have reported that NS species show advantages that can improve specific parameters of wine quality (Comitini et al., 2011; Padilla et al., 2017), depending on the specific yeast species and strains used; on the other hand, the microbial interactions could be profitably used to control undesired microorganisms in winemaking (Albergaria, Francisco, Gori, Arneborg, & Gírio, 2010; Branco et al., 2014; Mehlomakulu, Setati, & Divol, 2014; Oro, Ciani, & Comitini, 2014; Villalba, Sáez, del Monaco, Lopes, & Sangorrín, 2016). White wine technology generally favors the reduction of "wild" yeasts widely present on the grape surface, favoring the control of fermentation. In this regard, the practice of the inoculum of selected *S. cerevisiae* starter culture in combination with the use of sulfur dioxide (SO_2) is a consolidated procedure to control the winemaking process. In this way, the management of selected NS yeasts used in cofermentation or sequential fermentations for white wine production could be more useful. On the other hand, in view of a reduction in the use of SO$_2$ in winemaking, selected NS yeasts may have a role of a biocontroller during the fermentation process and, specifically, in the production of white wines, during the prefermentative clarification and/or cold maceration phases.

10.2 Non-*Saccharomyces* mixed fermentation

The use of *S. cerevisiae* strains as a starter culture undoubtedly determines sure control of the fermentation process but, at the same time, can lead to standardization of wines, limiting the potential positive contribution of other yeasts (Vigentini et al., 2014).

The reevaluation of the role of NS yeasts and their use in mixed inoculated fermentation is an innovative biotechnological strategy suitable for emphasizing the analytical and aroma complexity of wine.

The practice of mixed fermentation aims to simulate spontaneous fermentations under controlled conditions through time, levels, and modalities of inoculation of selected yeast species to avoid stuck fermentation (Benito, Calderón, & Benito, 2019b; Ciani et al., 2016). The application of mixed fermentation practice in winemaking is an approach already proposed in the 1960s when *T. delbrueckii/S. cerevisiae* sequential fermentation was used to reduce volatile acidity in wine. To date, this fermentation practice is used for several other objectives such as to enhance the aroma complexity and to modulate the production of some other compounds (glycerol, ethanol, and lactic or malic acids), which can affect the structure of wines (Padilla et al., 2017; Varela, 2016). Mixed fermentation can be carried out in the following two ways: (i) *S. cerevisiae* strain inoculated at the same time with NS yeast strain; (ii) *S. cerevisiae* strain inoculated sequentially after 48−72 hours with NS strain to allow the less competitive species to affect the final quality of wine via their metabolic pathways. The ability of NS yeasts to contribute to the aroma complexity of wine is strictly related to different features. Firstly, some of the NS yeasts belonging to different genera have specific enzymatic activities such as protease, β-glucosidase, and esterase activity that could enhance the aromatic compounds. Another aspect is related to the capacity to produce secondary metabolites linked to carbon metabolism that contribute to the sensorial profile (Benito, Calderón, & Benito, 2019a, 2019b; Dutraive et al., 2019).

Obviously, in controlled mixed fermentation, there are physiological and biochemical interactions between NS and *S. cerevisiae* metabolic pathways (Ciani & Comitini, 2015; Ciani et al., 2010). These features are strictly related to extrinsic environmental factors. For these reasons, the management of mixed fermentation like the inoculation level and timing and temperature management is crucial. In this way, it is possible to obtain a synergistic interaction between NS and *S. cerevisiae* yeasts (Andorrà, Berradre, Mas, Esteve-Zarzoso, & Guillamón, 2012; Canonico, Comitini, & Ciani, 2015; Sadoudi et al., 2012) (Tables 10.1 and 10.2).

10.3 Yeast interaction in mixed fermentation for white wine production

During multistarter wine fermentations, it is important to determine the impacts of different NS strains in the wine phenotype and particularly the aroma outputs in different inoculation schemes and under fermentation conditions. In this regard, the interactions between NS yeasts and *Saccharomyces* (Ciani & Comitini, 2015; Ciani et al., 2010) play an important role in the final composition of wines. These complex interactions depend on several factors, which can be classified as abiotic and biotic factors. Among the abiotic factors, the temperature of fermentation plays an important role in yeast interactions, influencing the dominance and/or the persistence of different inoculated starter yeasts. In white wine production the temperature during the prefermentative stages may also influence the dynamics of grape juice microbiota before the fermentation. High temperature in synergy with increasing ethanol concentration affects membrane permeability and integrity. In this contest, some works indicated that ethanol does not provide a clear advantage to *S. cerevisiae* at a low temperature (<15°C) since a higher persistence of NS yeast at a low temperature in mixed fermentation has been recognized (Ciani, Beco, & Comitini, 2006; Gao & Fleet, 1988). The antagonistic effect between *S. cerevisiae* and *Lachancea thermotolerans* is temperature dependent. *L. thermotolerans* showed higher inhibitory activity on *S. cerevisiae* at 20°C than at 30°C (Gobbi, Comitini, D'Ignazi, & Ciani, 2013). Oxygen availability, particularly during the first steps of fermentation, can affect the growth and fermentation performance of wine yeasts, having a selective action among the various yeast species (Brandam, Lai, Julien-Ortiz, & Taillandier, 2013; Ciani et al., 2006; Jolly et al., 2014; Taillandier, Lai,

TABLE 10.1 Influence of NS yeast on aroma compounds in different white wines.

Species	White wine	Aroma compounds	References
Torulaspora delbueckii	Verdejo, Sauvignon blanc	↑ Thiols	Belda et al. (2017), Renault et al. (2015)
	Soave, Chardonnay and Vino Santo	↑ Higher alcohols and esters	Azzolini et al. (2015)
	Sauvignon blanc	↓ Higher alcohols and esters	Renault et al. (2015)
	Verdicchio	↑ β-phenylethanol, acetaldehyde	Canonico et al. (2015)
	Riesling and Garnacha	↑ Isoamyl alcohol ↓ Aromatic acetates fatty acids (FA) and ethyl esters (EE)	Oliveira and Ferreira (2019)
	Verdicchio (sparkling wine)	↑ Ethyl hexanoate, ethyl octanoate ↓ n-propanol, isobutanol, isoamyl alcohol, acetaldehyde	Canonico et al. (2018)
Lachancea thermotolerans	Chardonnay and Airen	↓ Higher alcohols	Benito, Calderón, et al. (2016), Shekhawat et al. (2018)
	Riesling	↑ Ethyl lactate and terpenes	Benito et al. (2015)
	Riesling and Garnacha	↓ FA and EE	Oliveira and Ferreira (2019)
Metschnikowia pulcherrima	Verdejo	↑ Terpenes or thiols, ethyl octanoate	Ruiz et al. (2018)
	Riesling	↑ Fruity esters	Benito et al. (2015)
Pichia kluyveri	Riesling	↑ β-phenyl ethyl acetate, ethyl octanoate, ethyl acetate isoamyl acetate, amyl acetate, and ethyl phenylacetate	Benito et al. (2015)
Schizosaccharomyces pombe	[a]	↓ Higher alcohols and esters ↓ Ethyl acetate	Benito (2019)
Hanseniaspora vineae	Chardonnay	↑ β-phenyl ethyl acetate and ethyl acetate	Medina et al. (2013)
	Macabeo	↓ Higher alcohols	Lleixà et al. (2016)
H. guilliermondii	Muscat	↑ β-phenyl ethanol, terpenes, and norisoprenoids	López et al. (2014)
H. osmophila	Muscat	↑ β-phenyl ethanol, terpenes, and norisoprenoids	López et al. (2014)
H. uvarum	Muscat	↑ β-phenyl ethanol, terpenes, and norisoprenoids	López et al. (2014)
Zygosaccharomyces bailii	Chardonnay	↑ Ethyl esters	Garavaglia et al. (2015)
Starmerella bacillaris	Sauvignon blanc	↑ Thiols	Englezos et al. (2018)

[a]*Grape variety not specified.*

Julien-Ortiz, & Brandam, 2014). The interactions between *S. cerevisiae* and NS yeasts may also be influenced by the competition for nutrients (nitrogen and vitamins).

Among biotic factors, the production of antimicrobial compounds and cell-to-cell contact mechanism plays a main role in yeast—yeast interactions of controlled multistarter fermentations.

Killer toxins are certainly involved in yeast—yeast interactions in mixed fermentations. Several works have focused on the study of NS strains able to counteract the development of *Brettanomyces* spp., a relevant and dangerous yeast in the cellar (Boynton, 2019; Mannazzu et al., 2019; Pinto, Baruzzi, Cocolin, & Malfeito-Ferreira, 2020). In this regard, the use of *Wickerhamomyces anomalus* and *Kluyveromyces wickerhamii* killer yeasts in sequential fermentation with *S. cerevisiae* starter strain was evaluated to control *Dekkera/Brettanomyces* spoilage yeasts (Comitini, Ingeniis De, Pepe, Mannazzu, & Ciani, 2004). Others killer toxins produced by other NS yeasts such as *Torulaspora delbrueckii*, and *W. anomalus* were studied to be effective natural agents involved in the biocontrol activity against the spoilage wine yeasts (de Ullivarri, Mendoza, & Raya, 2018; Villalba et al., 2016).

The interaction among yeast strains may also be influenced by cell-to-cell contact. Nissen and Arneborg (2003) demonstrated this phenomenon by carrying out single- and mixed-culture fermentations using both *L. thermotolerans* and *T. delbrueckii* with *S. cerevisiae* in a new double-compartment fermenter. More recently, Renault, Albertin, & Bely

TABLE 10.2   Influence of NS yeast on main analytical compounds of white wines.

Species	White wine	Enological characters	References
Torulaspora delbueckii	Chardonnay, Palomino, Sémillon	↓ Acetic acid, ethanol ↑ Glycerol	Bely et al. (2008), Canonico et al. (2019), Puertas et al. (2017)
	Macabeo (base wine for sparkling wine)	↑ Glycerol, foam, protein ↓ Volatile acidity	González-Royo et al. (2015), Medina-Trujillo et al. (2017)
Lachancea thermotolerans	Macabeo	Variable production in acetic acid production	Escribano et al. (2018)
	Chardonnay	↑ Glycerol ↓ Ethanol	Shekhawat et al. (2018)
	Pinot gris	↑ Volatile acidity ↓ Total acidity, lactic acid	Binati et al. (2020)
Metschnikowia pulcherrima	Charonnay, Verdicchio	↓ Ethanol	Canonico et al. (2019, 2019b)
	Riesling	↑ Glycerol	Benito et al. (2015), Dutraive et al. (2019), Hranilovic et al. (2020)
Pichia kluyveri	Riesling wine	↓ Ethanol ↑ Glycerol	Röcker et al. (2016)
Schizosaccharomyces pombe	[a]	↑ Acetic acid	Lambrechts and Pretorius (2000)
Schizosaccharomyces japonicus	Synthetic grape juice	↑ Polysaccharides	Domizio et al. (2018)
	Commercial white grape juice	↑ Malic and acetic acids	Romani et al. (2018)
Zygosaccharomyces bailii	[a]	↓ Acetic acid	Romano and Suzzi (1993)
Zygosaccharomyces fermentati	[a]	↓ Acetic acid	Romano and Suzzi (1993)
Zygotorulaspora florentina	Vino Santo	↓ Volatile acidity	Lencioni et al. (2018)
Starmerella bombicola/ bacillaris	Trebbiano, Toscano	↑ Glycerol	Ciani and Ferraro (1996, 1998), Ferraro et al. (2000)
	Verdicchio	↓ Ethanol	Canonico et al. (2016)
	Treixadura	↓ Ethanol	Castrillo et al. (2019)

[a]*Grape variety not identified.*

(2013) found that physical contact between *S. cerevisiae* and *T. delbrueckii* induced rapid death of the NS yeast. On the other hand, the early death of *L. thermotolerans* during anaerobic mixed-culture fermentations with *S. cerevisiae* was caused by a combination of cell-to-cell contact and antimicrobial peptides (Kemsawasd et al., 2015).

In summary, the abiotic and biotic factors strongly influence the specific interactions between the couple of starter yeasts in mixed fermentations. From this, it follows that the management of mixed fermentations, inoculation level, inoculation modalities (coculture or sequential fermentation), and timing of inoculation is fundamental for obtaining the desired results in terms of the final composition and aromatic profile of wine.

10.4 Biocontrol activity of non-*Saccharomyces*

The application of emerging technologies based on bioactive microorganisms for food and beverage control and preservation is a relevant topic also in the wine industry. Indeed, these biotechnological processes are very promising to reduce the growth of spoilage microorganisms as well as the production of undesirable microbial metabolites during processing.

It has been widely demonstrated that some NS strains can control spoilage yeasts or filamentous fungi both in the grapes and during the early stages of fermentation (de Ullivarri et al., 2018; Kuchen et al., 2019; Velázquez,

Zamora, Álvarez, & Ramírez, 2019). Certainly, the enhancement of the aromatic trait of the final wine is a common prerogative that each NS yeast must possess to be selected in winemaking. Indeed, after the general revaluation of the role of NS yeasts, several studies have focused attention on multiple advantages (Ciani et al., 2020).

Under Section 10.3 (Yeast interaction in mixed fermentation for white wine production), we have already mentioned the effect of micocins production by NS yeasts in the interactions during mixed fermentations and their use against *Brettanomyces* spp. In this regard, other species off-flavors or films in bulk wine producer such as *Zygosaccharomyces rouxii, Kloeckera apiculata, Pichia* spp., and *Candida* spp. may be easily prevented by the use of NS yeasts producers of active antimicrobial extracellular molecules (Escott, Del Fresno, Loira, Morata, & Suárez-Lepe, 2018). Regarding this, natural control strategies may involve the use of antimicrobial peptides such as Lactoferricin B (Lfcin B) and the use of killer toxins produced by NS yeasts as biological tools to counteract contamination by spoilage yeast in white wine (Alonso, Belda, Santos, Navascués, & Marquina, 2015; Escott, Loira, Morata, Bañuelos, & Suárez-Lepe, 2017). Biological control agents are potential alternatives for the chemical fungicides presently used in agriculture to counteract plant diseases, limit the spoilage of industrial fermentations, and control the deterioration of foods and beverages. In winemaking, there is a growing interest to reduce and rationalize the use of SO_2. Two main efforts are paced: (i) limit compounds that permanently bind SO_2; (ii) search for new preservative substances of natural origin. In this regard, the use of NS yeasts that produce zymocins was proposed as an alternative to the massive use of sulfur dioxide (Comitini and Ciani, 2010; Oro, Ciani, Bizzaro, & Comitini, 2016). Some of these biotechnical strategies are already in use in the winemaking industry, and others may be explored by different processes of stages from vine to bottled wine, to preserve varietal aromas, improving the overall quality of final product.

10.5 NS aroma and structure of white wine

In order to promote the impact of NS yeast on the final wine aroma, several factors need to be considered. Obviously, the NS choice is the starting point to manage the process. During the years, several studies have highlighted that NS yeasts, affected the analytical structure and aroma complexity of wine through their metabolic pathways (Domizio et al.,2007; Jolly, Augustyn, & Pretorius, 2006; Varela & Borneman, 2017). In this regard, NS yeasts belonging to *Torulaspora, Metschnikowia, Pichia, Starmerella, Zygosaccharomyces, Schizosaccharomyces*, and *Lachancea* genera positively contribute to enhancing wine aroma and structure (Anfang, Brajkovich, & Goddard, 2009; Hranilovic et al., 2018; Lin et al., 2020; Rollero, Bloem, Ortiz-Julien, Camarasa, & Divol, 2018; Ruiz et al., 2018). The contributions to the wine aroma by NS yeasts involve the direct biosynthesis of norisoprenoids, terpenoids, esters, higher alcohols, acetaldehyde, organic acids, volatile fatty acids (FA), and carbonyl and sulfur compounds (Capozzi, Garofalo, Chiriatti, Grieco, & Spano, 2015).

Likewise, NS yeasts can strongly influence the wine structure by varying the quantitative composition of some key wine compounds. Some NS yeasts can enhance or reduce the total acidity by relevant production of lactic acid (*L. thermotolerans*) or by complete degradation of malic acid (*S. pombe*). Several other NS yeasts species (*S. bombicola, S. bacillaris, M. pulcherrima L. thermotolerans*) are instead characterized by a high level of glycerol that influences the sweetness and softness of the wine, and it is associated with flavor persistence and oiliness (Laguna, Bartolomé, & Moreno-Arribas, 2017). Due to the specific fermentation metabolic pathways and respirofermentative regulation, the use of NS yeasts is a promising way to reduce ethanol content in wine.

Ethanol reduction is one of the most important topics of investigation in winemaking. The increase in ethanol content in wine in the last two decades, due to several factors such as new consumer requests and climate change, impact the wine flavor negatively and can have harmful effects on human health. In this regard, several works addressed the interspecies and/or intraspecies variability in the ethanol yield of NS wine yeasts (Contreras et al., 2014; Contreras et al., 2015; Gobbi et al., 2014; Magyar & Tóth, 2011). The results indicated that ethanol yield is a species-related trait but a wide intraspecies variability was also evident (Ciani & Maccarelli, 1998; Comitini et al., 2011; Domizio et al., 2011). Another additional feature of NS yeasts to reduce the alcohol content in wine is the development of respiration-based methods in NS yeast strains showing no or weak Crabtree effect.

The results of several investigations showed that selected NS yeast strains in controlled aeration conditions with consequent sequential inoculation with *S. cerevisiae* are suitable to reduce the ethanol content in wine. According to the results obtained by controlled aeration fermentations, the ethanol reduction was 1.6% v/v for *M. pulcherrima*, 0.9%—1.5% v/v for *T. delbrueckii*, and 1.0%—2.0% v/v for *Z. bailii* (Canonico, Solomon, Comitini, Ciani, & Varela, 2019; Contreras et al., 2015).

The reduction of the volatile acidity is another important feature that involves the use of NS yeasts in mixed fermentation (*T. delbrueckii, S. bacillaris, Z. florentina*). In this regard, NS yeasts are characterized by a general enhancement of polysaccharides production in comparison with *S. cerevisiae* starter culture, but some species as *Schizosaccharomyces* spp. may produce very high amounts, determining a substantial increase in the body and structure of wine (Domizio, Liu, Bisson, & Barile, 2017).

An increase in the body of the wine can be made from organic acids produced during fermentation. In addition to lactate, some NS yeasts can produce high levels of succinic acid (*S. bombicola*) (Ciani & Ferraro, 1998) or fumaric acid (*M. pulcherrima*) (Hranilovic, Gambetta, Jeffery, Grbin, & Jiranek, 2020). The aroma compounds' production behavior differs among different NS species. The high production of some aroma compounds is a peculiarity at the species level even if strain variation within the species occurred. Generally, *H. uvarum* increases isoamyl acetate (banana aroma), while *H. guilliermondii* and *H. vineae* leads to an increase in 2-phenyl ethyl acetate; *M. pulcherrima* is characterized by isoamyl acetate, 2-phenyl ethyl ethanol (rose aroma), and volatile thiols (exotic fruit), while *T. delbrueckii* in sequential fermentation favors the production of β-phenyl ethyl ethanol.

In the following paragraphs the main distinctive characteristics of NS yeasts used to positively influence the aroma and structure of white wines are reported.

10.5.1 *Torulaspora sp*

One of the most investigated NS yeasts in mixed fermentation used to improve quality parameters of white wines are *T. delbrueckii* strains, (Benito, 2018a, 2018b). However, some distinctive features such as good fermentative power, tolerance to high ethanol concentrations, ability to reduce the volatile acidity and ethanol content are also relevant for the use of this species in white wine production (Bely, Stoeckle, Masneuf-Pomarède, & Dubourdieu, 2008; Canonico et al., 2019; Liu, Laaksonen, Kortesniemi, Kalpio, & Yang, 2018). The contribution of *T. delbrueckii* to white wine quality is mainly focused on analytical and aroma profiles. Investigations on this NS in Chardonnay, Palomino, and Sémillon wine highlighted a reduction in acetic acid production, ethanol content, and increase of glycerol in final wine (Bely et al., 2008; Canonico et al., 2019; Puertas, Jiménez, Cantos-Villar, Cantoral, & Rodríguez, 2017). The use of *T. delbrueckii* in sequential fermentation in aeration conditions led to a decrease in ester compounds and an increase in higher alcohols (Canonico et al., 2019).

Investigations on the use of *T. delbrueckii* in Verdejo grapes and Sauvignon blanc showed an improvement of the intensity and quality of wine aroma, increasing the overall impression and the varietal and fruity characters through the release of higher concentrations of thiols (Belda et al., 2017; Renault et al., 2015). Regarding the production of higher alcohols and ester formation, the studies showed a different trend that has been explained by the high strain variability within the species. Indeed, while an increase in higher alcohols and esters production has been reported in Soave, Chardonnay, and Vino Santo (sweet white wine) (Azzolini, Tosi, Lorenzini, Finato, & Zapparoli, 2015), in Sauvignon blanc the concentration of these compounds was decreased (Renault et al., 2015). Canonico et al. (2015) evaluated the effect of *T. delbrueckii* selected strain in sequential fermentation with two different *S. cerevisiae* starter strains on Verdicchio wine (produced by the homonym white grape variety) during three vintages. The results indicated that the contribution of *T. delbrueckii* was influenced by not only the *S. cerevisiae* starter strain used but also the vintage (grape juice variation). However, a general trend exhibited by the use of this yeast combination was an increase in β-phenylethanol and acetaldehyde related to the presence of *T. delbrueckii*. These sequential fermentations would thus be particularly suitable to enhance the complexity of a specific well-structured Verdicchio wine style.

Oliveira and Ferreira (2019) used the sequential fermentation with different NS yeasts, including *T. delbrueckii*, to modulate wine fermentation, inducing aromatic changes of industrial interest. After fermentation, the fermentative aroma profiles of Riesling and Garnacha were consistently changed due to a reduced level of isoamyl alcohol, high levels of aromatic acetates, and high levels of FA and their ethyl esters (EE). In particular, *T. delbrueckii* produced large levels of branched acids. The contribution of *T. delbrueckii* was investigated also in Riesling wines, where Dutraive et al. (2019) showed that the use of *T. delbrueckii* in sequential fermentation affects the final wine aroma through esters, acetates, higher alcohols, FA, and low volatile sulfur increasing.

The use of *T. delbrueckii* was proposed also for the fermentation of base wine for sparkling wine production (Macabeo) (González-Royo et al., 2015; Medina-Trujillo et al., 2017). The use of sequential inoculation of *T. delbrueckii* increased glycerol concentration, reduced volatile acidity, and employed a positive effect on foam (González-Royo et al., 2015). Medina-Trujillo et al. (2017) showed that the use of selected *T. delbrueckii* in sequential fermentation increased the protein concentration and improved the foaming properties of a base wine. The use of *T. delbrueckii* for secondary fermentation of the base wine (Verdicchio) in pure and mixed fermentation with *S. cerevisiae* was also evaluated through evaluation of its impact on the analytical composition and aromatic profile of the sparkling wine produced. Pure and mixed fermentation with *T. delbrueckii* leads to an increase in ethyl hexanoate (fruity, green apple, brandy, wine-like) and ethyl octanoate (fruity, strawberry, sweet); reduction in n-propanol, isobutanol, isoamyl alcohol (alcohol, ripe fruit); and significant reductions in acetaldehyde content. Moreover, *T. delbrueckii* pure cultures and their mixed fermentations showed higher scores for the aromatic descriptors as white flowers, citrus, honey, odor intensity, and softness.

Thus a plethora of investigations carried out in the last few years particularly in white wine, confirmed the relevant influence of *T. delbrueckii* in mixed fermentation on enhancing the aroma complexity and peculiarity of several white wines.

10.5.2 Lachancea sp

L. thermotolerans (previously classified as *Kluyveromyces thermotolerans*) is an NS species characterized by a peculiar fermentation metabolism with the ability to produce significant amounts of lactate that determine an increase of total acidity in wine (Ciani and Comitini, 2011; Gobbi et al., 2013). A recent investigation has shown that the pathway of lactate formation is from pyruvate through the enzyme lactate dehydrogenase enzyme (Hranilovic et al., 2018). The amount of lactic acid production by *L. thermotolerans* depends on intraspecific variability and the fermentation temperature (Gobbi et al., 2013) and can vary from 0.2−0.3 to several grams per liter.

This NS yeast is also characterized by other important fermentation features. Regarding the production of acetic acid, during the years, several studies have revealed a different behavior related to a great intraspecific variability (Hranilovic et al., 2018; Kapsopoulou, Mourtzini, Anthoulas, & Nerantzis, 2007). Recent studies reported *L. thermotolerans* strains to show an intraspecific variability of up to 49% in acetic acid production in Macabeo wine (Escribano et al., 2018). The use of *L. thermotolerans* in sequential fermentation in Chardonnay wine highlighted an increase in glycerol content and a decrease in ethanol content in the particular fermentation condition of fermentation supplied by oxygen (Shekhawat, Porter, Bauer, & Setati, 2018). Regarding the production of higher alcohols, some studies have reported that *L. thermotolerans* sequential fermentations produce lower concentrations of these compounds in Chardonnay and Airen wines (Benito, Calderón, Palomero, & Benito, 2016; Shekhawat et al., 2018), while in another investigation, *L. thermotolerans* showed an opposite trend (Comitini et al., 2011). In Riesling wine, this NS yeast exhibited an increase in ethyl lactate and terpenes (Benito et al., 2015). In Riesling and Garnacha wines, *L. thermotolerans* produced reduced levels of FA and their EE (Oliveira & Ferreira, 2019). Binati et al. (2020), analyzing the effect of *L. thermotolerans* in Pinot gris (frence), found a reduced volatile acidity and an enhancement in total acidity due to lactic acid production.

Despite some different behaviors of *L. thermotolerans* in the various investigations, with the use of appropriately selected strains, it is possible to obtain wines with higher acidity and an aromatic profile characterized by the metabolic pathway of this NS yeast.

10.5.3 Metschnikowia sp

The influence on the white wine aromatic profile related to the use of *M. pulcherrima* could be mainly due to the ability to exhibit the cystathionine-β-lyase activity involved in varietal thiols released from their S-conjugated nonvolatile precursors.

These aromatic compounds are the most important quality indicators in wine varieties such as Sauvignon blanc or other regional varieties such as Verdejo. Using NS yeasts is a novel approach to enhancing the thiol profiles in wine. In this regard, *M. pulcherrima* is a promising species to be used to enhance the formation of volatile thiols as the presence of β-lyase activity was found (Zott et al., 2011).

A recent investigation indicated that the use of *M. pulcherrima* not only affected the varietal aroma compounds of Verdejo (thiols and terpenes) but also showed an increase in ethyl octanoate and fruity aromas related to pineapple, which are usually considered pleasant and a very positive feature (Ruiz et al., 2018).

Other studies have reported that *M. pulcherrima* is a higher producer of fruity esters in Riesling wine (Benito et al., 2015) and in other white varieties (Rodriguez et al., 2010; Sadoudi et al., 2012) The use of *M. pulcherrima* in sequential fermentation led to the reduction of higher alcohols compared to *S. cerevisiae* (Benito et al., 2015). Higher alcohols show the most irregular results of all the reported enological parameters for *M. pulcherrima*. Some studies using sequential fermentation with *M. pulcherrima* showed an increment in the total concentration of higher alcohols, while other studies (Escribano et al., 2018; Rodriguez et al., 2010) found a contrary effect, with a reduction of these compounds (Benito et al., 2015; Comitini et al., 2011; Ruiz et al., 2018; Varela et al., 2016). On the other hand, 2-phenylethanol, an important aromatic compound, showed a constant increase in *M. pulcherrima* sequential fermentations (Dutraive et al., 2019; Rodriguez et al., 2010; Sadoudi et al., 2012).

The use of *Metschnikowia* strains increases the final concentration of glycerol in wines. Recent works have revealed that the increase in glycerol during sequential inoculation ranges from 4% to 20% (Benito et al., 2015; Dutraive et al., 2019), or even up to more than 40% (Hranilovic et al., 2020).

M. pulcherrima was also investigated for ethanol reduction in wine. In this regard, free or immobilized cells were inoculated under different aeration conditions in sequential fermentations with *S. cerevisiae*. In immobilized form and under aeration conditions, *M. pulcherrima*, besides leading to a significant reduction in ethanol, increased ethyl butyrate and isoamyl acetate content (Canonico et al., 2019). A negative aspect of the use of *M. pulcherrima* in free form and under aeration conditions is excessive ethyl acetate content. Ethyl acetate is an ester that in low concentrations imparts a fruity aroma; concentrations above 150–160 mg/L are associated with undesirable "nail polish remover" and "solvent" sensory descriptors (Ribéreau-Gayon et al., 2006).

After fermentation, a real volatile profile enhancement and an evident reduction of alcohol were demonstrated in Chardonnay wines.

10.5.4 Pichia sp

Pichia kluyveri is one of the most investigated species within the *Pichia* genus for its high production of volatile compounds. Some studies have reported that this NS species in sequential fermentation produced higher levels of esters such as β-phenyl ethyl acetate by ethyl octanoate (Benito et al., 2015). The total terpene concentration was also enhanced by about 20%, contributing to the increase in the typicality of the grape variety.

Pichia guilliermondii and *P. kluyveri* in Riesling wine in sequential fermentation led to an ethanol reduction in wines obtained by *S. cerevisiae* and also affected the aroma and analytical profiles (Röcker, Strub, Ebert, & Grossmann, 2016). Indeed, *P. kluyveri* increased the glycerol content and formed high amounts of ethyl acetate isoamyl acetate, amyl acetate, and ethyl phenylacetate. On the contrary, *P. guilliermondi* formed high amounts of ethyl propionate, ethyl isobutyrate, and ethyl phenylacetate and high amounts of isobutyric and isovaleric acids.

10.5.5 Starmerella spp

Starmerella is a yeast genus that was quite recently established (Rosa & Lachance, 1998), and some characterized strains previously classified as *Candida stellata* have passed into this genus (Šipiczki, Ciani, & Csoma, 2005). In this regard, older works, carried out on *C. stellata* (now reclassified as *Starmerella bombicola* or *Starmerella bacillaris*) and related to the enological environment, highlighted the expression of some interesting features. Indeed, *C. stellata* in cofermentation with *S. cerevisiae* showed a positive influence on the aroma and composition of Chardonnay wine (Soden, Francis, Oakey, & Henschke, 2000). Other works highlighted specific fermentation behavior such as the wide production of glycerol (Ciani & Ferraro, 1996; Ciani Ferraro, 1998) that positively influenced the mouthfeel sensation and flavor of wine (Jolly et al., 2014). This behavior was then confirmed by trials at the pilot-scale fermentation using Trebbiano Toscano white grape juice (Ferraro, Fatichenti, & Ciani, 2000). More recently, *Starmerella bacillaris* (formerly *C. stellata*) was evaluated and proposed from an application perspective for the reduction in ethanol content (Englezos et al., 2016) and for the reduction of the acetic acid produced from high sugar musts (Lencioni, Taccari, Ciani, & Domizio, 2018; Rantsiou et al., 2012).

Starmerella bombicola (formerly *C. stellata*) was also investigated for the reduction of the final ethanol content in wine using sequential fermentation with immobilized cells (Canonico, Comitini, Oro, & Ciani, 2016) More recently (Castrillo, Rabuñal, Neira, & Blanco, 2019) the potential application of selected NS wine yeasts to mitigate the ethanol increase by sequential fermentation in Treixadura white grape must was proposed. In this case, it was found that mixed fermentations with *S. bacillaris* effectively reduced the alcohol content (1.1% vol). Regarding the influence of *S. bacillaris* on the aromatic profile of white wines, a recent work conducted on various musts of white grape varieties indicated that there were no negative influences on the aromatic profile of the wines, while the aromatic profiles of Sauvignon blanc wines fermented by mixed cultures contained significantly higher levels of thiols (Englezos et al., 2018).

Summarizing the characteristics of the species belonging to the genus *Starmerella*, with particular reference to *S. bacillaris* and *S. bombicola*, we can conclude that these species have various positive characteristics such as the reduction of volatile acidity and ethanol, the increase of glycerol, as well as some volatile compounds. These features make *Starmerella* a promising NS yeast to be used in the production of white wines.

10.5.6 Zygosaccharomyces spp

Within the genus *Zygosaccaromyces*, *Zygosaccharomyces bailii* is a notorious food spoilage yeast in the food and beverage industry due to its ability to resist antiseptic compounds such as SO_2 and benzoic and sorbic acids. However, its potential use in winemaking was evaluated as a costarter for the improvement of EE production (Garavaglia et al., 2015). White wine vinifications conducted with mixed cultures containing different proportions of *Z. bailii* BCV 08 and *S. cerevisiae* showed an enhancement in EE in comparison to the vinification control with an increase in the aromatic complexity of wine. On the other hand, the potential use of *Zygosaccharomyces* species has been previously proposed. *Z. bailii* and *Z. fermentati* showed considerable differences in some enological characteristics such as a high fermentative vigor and the production of lower amounts of acetic acid (Romano & Suzzi, 1993).

Another species, previously classified as *Zygosaccaromyces florentinus* and now reclassified as *Zygotorulaspora florentina*, showed some promising application in white wine production. In particular, multistarter fermentation *Z. florentina/S. cerevisiae* was evaluated in the production of Vino Santo wine, a typical dessert wine produced in Tuscany, in particular, and in other areas of central and northern Italy. In this regard, the inoculation of *Z. florentina* determined a reduction of volatile acidity in these specific high sugar musts (Lencioni et al., 2018).

10.5.7 Hanseniaspora spp

Species belonging to *Hanseniaspora* genus are characterized by low ethanol tolerance (c. 4% ethanol). For this reason, the presence of these yeasts is limited to the first phase of the wine fermentation process (Pina, Santos, Couto, & Hogg, 2004). In pure fermentation, these yeasts have the tendency to produce high levels of volatile acidity, and for this reason, they are limited through the use of SO_2 and the inoculation of selected cultures of *S. cerevisiae*. Despite this, several studies in white wine production have highlighted that the use of selected strains of *Hanseniaspora* genus, in mixed fermentation with *S. cerevisiae*, can positively affect the quality of the final wine due to their enzymatic activities like glucosidase, xylosidase, glycosidase, and protease (Hu, Jin, Xu, & Tao, 2018; Lleixà et al., 2016). An investigation on a strain of *Hanseniaspora vineae* in sequential fermentation showed a relevant influence on the aroma profile of Chardonnay wine, showing an increase in 2-phenylethyl acetate and ethyl acetate (Medina et al., 2013) and a decrease in higher alcohols content. Moreover, *H. vineae* in Macabeo wine fermentation produced low levels of higher alcohols and a fivefold greater concentration of the acetates (Lleixà et al., 2016). López, Mateo, and Maicas (2014) investigated different strains of *H. guilliermondii*, *H. osmophila*, *H. uvarum*, and *H. vineae* in Muscat wine. The results reported showed increases in the levels of 2-phenylethanol, terpenes, and norisoprenoids in the wine.

10.5.8 Schizosaccharomyces spp

The main technological use in the winemaking of yeast strains belonging to the genus *Schizosaccharomyces* is the reduction of acidity of grape juices. Indeed, for a long time the use of *Schizosaccharomyces pombe* was suggested to reduce malic acid in grape juice and/or wine (Munyon & Nagel, 1977; Rankine, 1966). Several studies have been carried out with the aim of deacidifying the grape must or wine through malic acid degradation using mixed fermentations of *S. pombe* and *S. cerevisiae* with, in some cases, immobilized *S. pombe* cells to obtain a more controlled biological deacidification (Ciani, 1995; Magyar & Panyik, 1989; Snow & Gallander, 1979; Yokotsuka, Otaki, Naitoh, & Tanaka, 1993). In these last conditions, the undesirable effects of *S. pombe* on the wine quality were limited or eliminated.

The ability to degrade malic acid by *Schizosaccharomyces* yeasts is a very promising resource for wet regions, where undesirable fungal attacks are common. In addition to this important enological feature, *Schizosaccharomyces* yeasts showed other favorable enological characteristics. Selected *Schizosaccharomyces* yeasts can reduce gluconic acid, avoiding negative effects of this compound in wine (Peinado, Moreno, Maestre, & Mauricio, 2007), and they may have also a positive effect on the presence of biogenic amines and Ochratoxin A in wine. On the one hand, wines fermented by *Schizosaccharomyces* no longer contain malic acid that could be used by spontaneous lactic bacteria to generate biogenic amines, and on the other hand, *S. pombe* was found to be a very active species in the removal of Ochratoxin A (70% of the initial concentration) (Cecchini, Morassut, Moruno, & Di Stefano, 2006).

In contrast, *S. pombe* showed the tendency to produce acetic acid, a compound that negatively affects the analytical composition of wine determining over the threshold of 0.8 g/L, a vinegar character that is one of the most serious faults in wine sensory perception (Lambrechts and Pretorius, 2000). In this regard, the selection of strain (Benito, Palomero, Calderón, Palmero, & Suárez-Lepe, 2014; Benito, Palomero, et al., 2014; Benito, Jeffares, et al., 2016) and the use of mixed and sequential inoculations with selected commercial *S. cerevisiae* strains are the possible solutions for that problem (Benito et al., 2013). Regarding the other main compounds, several studies indicated *Schizosaccharomyces* to be a high glycerol producer (Domizio et al., 2017), while it usually produces higher concentrations of acetaldehyde than *S. cerevisiae* Benito et al. (2013). Most studies report that *S. pombe* strains influenced the aroma profile of the wine, producing lower amounts of higher alcohols and esters with the exception of ethyl acetate (Benito, 2019).

Another important feature of the genus *Schizosaccharomyces* is the release of high amounts of cell wall polysaccharides starting from the onset of alcoholic fermentation (Domizio, Liu, Bisson, & Barile, 2014). These polysaccharides have been characterized, and the results obtained are in agreement with the composition of the cell wall of yeasts related to the genus *Schizosaccharomyces*, the only one having galacto-mannoproteins located in the outer layer of the cell wall (Domizio et al., 2017). Yeast polysaccharides have many positive effects on wine quality: reducing protein and tartrate instability, increasing the "fullness" sensation, and retaining aromatic compounds (Romani et al., 2018). Regarding the production of mannoproteins by *Schizosaccharomyces* yeasts, Benito (2019) reported that approximately 2.5 times more hydrolyzed mannose was released from polysaccharides in wines fermented by *S. pombe* in comparison with *S. cerevisiae* control wines, while Domizio, Lencioni, Calamai, Portaro, and Bisson (2018) reported that *Schizosaccharomyces japonicus* produced five times more polysaccharides than *S. cerevisiae* control trial. Based on these results, it can be deduced that *S. japonicus* releases significantly higher amounts of polysaccharides as compared to *S. pombe*. In addition to the increase in polysaccharide release, *S. japonicus* showed a protection effect of polysaccharides against protein haze (Domizio et al., 2018). Beyond that, the results of the analytical profiles of the wines produced by mixed starter cultures *S. japonicus/S. cerevisiae* indicated that this NS yeast modulates the concentration of malic and acetic acids and some of the most important volatile compounds, in an inoculum-ratio-dependent fashion (Romani et al., 2018).

10.6 Conclusions

The positive contribution of NS yeasts in wine has become a reality of the 21st century. The studies carried out during these years pointed out that the use of NS yeasts is a strategy to enhance sensory quality, including wine aroma, color, and structure. In white wines, these features are particularly important to emphasize the specific characteristics of a cultivar or a characteristic geographical area of a wine. Moreover, the production of organic acids and bioprotective molecules helps to get safer and more stable wines with a consequent reduction of the content of wine preservatives such as sulfites.

Obviously, to obtain the best expression of the fermentation potential of each NS species and to emphasize the varietal characteristics of different grapes, the management of the fermentation process is crucial. For this purpose, now, there are different commercial starter cultures containing ready-to-use NS species/strains that facilitate the management of multistarter fermentations.

References

Agarbati, A., Canonico, L., Mancabelli, L., Milani, C., Ventura, M., Ciani, M., & Comitini, F. (2019). The influence of fungicide treatments on mycobiota of grapes and its evolution during fermentation evaluated by metagenomic and culture-dependent methods. *Microorganisms, 7*(5), 114.

Albergaria, H., & Arneborg, N. (2016). Dominance of *Saccharomyces cerevisiae* in alcoholic fermentation processes: Role of physiological fitness and microbial interactions. *Applied Microbiology and Biotechnology, 100*(5), 2035−2046.

Albergaria, H., Francisco, D., Gori, K., Arneborg, N., & Gírio, F. (2010). *Saccharomyces cerevisiae* CCMI 885 secretes peptides that inhibit the growth of some non-*Saccharomyces* wine-related strains. *Applied Microbiology and Biotechnology, 86*(3), 965−972.

Alonso, A., Belda, I., Santos, A., Navascués, E., & Marquina, D. (2015). Advances in the control of the spoilage caused by *Zygosaccharomyces* species on sweet wines and concentrated grape musts. *Food Control, 51*, 129−134.

Andorrà, I., Berradre, M., Mas, A., Esteve-Zarzoso, B., & Guillamón, J. M. (2012). Effect of mixed culture fermentations on yeast populations and aroma profile. *LWT, 49*(1), 8−13.

Anfang, N., Brajkovich, M., & Goddard, M. R. (2009). Co-fermentation with *Pichia kluyveri* increases varietal thiol concentrations in Sauvignon blanc. *Australian Journal of Grape and Wine Research, 15*(1), 1–8.

Azzolini, M., Tosi, E., Lorenzini, M., Finato, F., & Zapparoli, G. (2015). Contribution to the aroma of white wines by controlled *Torulaspora delbrueckii* cultures in association with *Saccharomyces cerevisiae*. *World Journal of Microbiology and Biotechnology, 31*(2), 277–293.

Barata, A., Malfeito-Ferreira, M., & Loureiro, V. (2012). The microbial ecology of wine grape berries. *International Journal of Food Microbiology, 153*, 243–259.

Belda, I., Ruiz, J., Beisert, B., Navascués, E., Marquina, D., Calderón, F., … Santos, A. (2017). Influence of *Torulaspora delbrueckii* in varietal thiol (3-SH and 4-MSP) release in wine sequential fermentations. *International Journal of Food Microbiology, 257*, 183–191.

Bely, M., Stoeckle, P., Masneuf-Pomarède, I., & Dubourdieu, D. (2008). Impact of mixed *Torulaspora delbrueckii-Saccharomyces cerevisiae* culture on high-sugar fermentation. *International Journal of Food Microbiology, 122*, 312–320.

Benito, Á., Calderón, F., & Benito, S. (2019a). Mixed alcoholic fermentation of *Schizosaccharomyces pombe* and *Lachancea thermotolerans* and its influence on mannose-containing polysaccharides wine composition. *AMB Express, 9*(1), 17. Available from https://doi.org/10.1186/s13568-019-0738-0.

Benito, Á., Calderón, F., & Benito, S. (2019b). The influence of non-*Saccharomyces* species on wine fermentation quality parameters. *Fermentation, 5*(3), 54.

Benito, Á., Calderón, F., Palomero, F., & Benito, S. (2016). Quality and composition of airén wines fermented by sequential inoculation of *Lachancea thermotolerans* and *Saccharomyces cerevisiae*. *Food Technol. Biotechnol., 54*, 135–144.

Benito, Á., Jeffares, D., Palomero, F., Calderón, F., Bai, F., Bähler, J., & Benito, S. (2016). Selected *Schizosaccharomyces pombe* strains have characteristics that are beneficial for winemaking. *PLoS One, 11*, e0151102. Available from https://doi.org/10.1371/journal.pone.0151102.

Benito, S. (2018a). Analytical impact of *Metschnikowia pulcherrima* in the volatile profile of Verdejo white wines. *Applied Microbiology and Biotechnology, 102*, 8501–8509.

Benito, S. (2018b). The impact of *Torulaspora delbrueckii* yeast in winemaking. *Applied Microbiology and Biotechnology, 102*, 3081–3094.

Benito, S. (2019). The impacts of *Schizosaccharomyces* on winemaking. *Applied Microbiology and Biotechnology, 103*, 4291–4312.

Benito, S., Hofmann, T., Laier, M., Lochbühler, B., Schüttler, A., Ebert, K., … Rauhut, D. (2015). Effect on quality and composition of Riesling wines fermented by sequential inoculation with non-*Saccharomyces* and *Saccharomyces cerevisiae*. *Eur. Food Res. Technol., 241*, 707–717.

Benito, S., Hofmann, T., Laier, M., Lochbühler, B., Schüttler, A., Ebert, K., & Rauhut, D. (2015). Effect on quality and composition of Riesling wines fermented by sequential inoculation with non-Saccharomyces and Saccharomyces cerevisiae. *European Food Research and Technology, 241*(5), 707–717.

Benito, S., Palomero, F., Calderón, F., Palmero, D., & Suárez-Lepe, J. A. (2014). Selection of appropriate *Schizosaccharomyces* strains for winemaking. *Food Microbiology, 42*(218), 224. Available from https://doi.org/10.1016/j.fm.2014.03.014.

Benito, S., Palomero, F., Gálvez, L., Morata, A., Calderón, F., Palmero, D., & Suárez-Lepe, J. A. (2014). Quality and composition of red wine fermented with *Schizosaccharomyces pombe* as sole fermentative yeast, and in mixed and sequential fermentations with *Saccharomyces cerevisiae*. *Food Technology and Biotechnology, 52*, 376–382.

Benito, S., Palomero, F., Morata, A., Calderon, F., Palmero, D., & Suárez-Lepe, J. A. (2013). Physiological features of *Schizosaccharomyces pombe* of interest in making of white wines. *European Food Research and Technology, 236*(1), 29–36. Available from https://doi.org/10.1007/s00217-012-1836-2.

Binati, R. L., Junior, W. J. L., Luzzini, G., Slaghenaufi, D., Ugliano, M., & Torriani, S. (2020). Contribution of non-*Saccharomyces* yeasts to wine volatile and sensory diversity: A study on *Lachancea thermotolerans*, *Metschnikowia* spp. and *Starmerella bacillaris* strains isolated in Italy. *International Journal of Food Microbiology, 318*, 108470.

Boynton, P. J. (2019). The ecology of killer yeasts: Interference competition in natural habitats. *Yeast (Chichester, England), 36*(8), 473–485.

Branco, P., Francisco, D., Chambon, C., Hébraud, M., Arneborg, N., Almeida, M. G., … Albergaria, H. (2014). Identification of novel GAPDH-derived antimicrobial peptides secreted by *Saccharomyces cerevisiae* and involved in wine microbial interactions. *Applied Microbiology and Biotechnology, 98*(2), 843–853.

Brandam, C., Lai, Q. P., Julien-Ortiz, A., & Taillandier, P. (2013). Influence of oxygen on alcoholic fermentation by a wine strain of *Torulaspora delbrueckii*: Kinetics and carbon mass balance. *Bioscience, Biotechnology, and Biochemistry, 77*(9), 1848–1853.

Canonico, L., Comitini, F., & Ciani, M. (2015). Influence of vintage and selected starter on *Torulaspora delbrueckii/Saccharomyces cerevisiae* sequential fermentation. *European Food Research and Technology, 241*(6), 827–833.

Canonico, L., Comitini, F., & Ciani, M. (2018). *Torulaspora delbrueckii* for secondary fermentation in sparkling wine production. *Food microbiology, 74*, 100–106.

Canonico, L., Comitini, F., & Ciani, M. (2019b). *Metschnikowia pulcherrima* selected strain for ethanol reduction in wine: Influence of cell immobilization and aeration condition. *Foods, 8*(9), 378.

Canonico, L., Comitini, F., Oro, L., & Ciani, M. (2016). Sequential fermentation with selected immobilized non-*Saccharomyces* yeast for reduction of ethanol content in wine. *Frontiers in Microbiology, 7*, 278.

Canonico, L., Solomon, M., Comitini, F., Ciani, M., & Varela, C. (2019). Volatile profile of reduced alcohol wines fermented with selected non-*Saccharomyces* yeasts under different aeration conditions. *Food Microbiology, 84*, 103247.

Capozzi, V., Garofalo, C., Chiriatti, M. A., Grieco, F., & Spano, G. (2015). Microbial terroir and food innovation: the case of yeast biodiversity in wine. *Microbiological Research, 181*, 75–83.

Castrillo, D., Rabuñal, E., Neira, N., & Blanco, P. (2019). Oenological potential of non-*Saccharomyces* yeasts to mitigate effects of climate change in winemaking: Impact on aroma and sensory profiles of Treixadura wines. *FEMS Yeast Research, 19*(7), foz065.

Cecchini, F., Morassut, M., Moruno, E. G., & Di Stefano, R. (2006). Influence of yeast strain on ochratoxin A content during fermentation of white and red must. *Food Microbiology, 23*(5), 411–417. Available from https://doi.org/10.1016/j.fm.2005.08.003.

Ciani, M. (1995). Continuous deacidification of wine by immobilized *Schizosaccharomyces pombe* cells: evaluation of malic acid degradation rate and analytical profiles. *Journal of Applied Bacteriology, 79*(6), 631–634.

Ciani, M., Beco, L., & Comitini, F. (2006). Fermentation behaviour and metabolic interactions of multistarter wine yeast fermentations. *International Journal of Food Microbiology, 108*(2), 239–245.

Ciani, M., Canonico, L., Oro, L., & Comitini, F. (2020). Footprint of non-conventional yeasts and their contribution in alcoholic fermentations. In *Biotechnological progress and beverage consumption*, (pp. 435–465). Academic Press.

Ciani, M., Capece, A., Comitini, F., Canonico, L., Siesto, G., & Romano, P. (2016). Yeast interactions in inoculated wine fermentation. *Frontiers in Microbiology*, 7, 555.

Ciani, M., & Comitini, F. (2011). Non-Saccharomyces wine yeasts have a promising role in biotechnological approaches to winemaking. *Annals of microbiology*, 61(1), 25–32.

Ciani, M., & Comitini, F. (2015). Yeast interactions in multi-starter wine fermentation. *Current Opinion in Food Science*, 1, 1–6.

Ciani, M., Comitini, F., Mannazzu, I., & Domizio, P. (2010). Controlled mixed culture fermentation: A new perspective on the use of non-*Saccharomyces* yeasts in winemaking. *FEMS Yeast Research*, 10, 123–133.

Ciani, M., & Ferraro, L. (1996). Enhanced glycerol content in wines made with immobilized *Candida stellata* cells. *Applied and Environmental Microbiology*, 62(1), 128–132.

Ciani, M., & Ferraro, L. (1996). Enhanced glycerol content in wines made with immobilized Candida stellata cells. *Applied and environmental microbiology*, 62(1), 128.

Ciani, M., & Ferraro, L. (1998). Combined use of immobilized *Candida stellata* cells *and Saccharomyces cerevisiae* to improve the quality of wines. *Journal of Applied Microbiology*, 85, 247–254.

Ciani, M., & Maccarelli, F. (1998). Oenological properties of non-*Saccharomyces* yeasts associated with wine-making. *World Journal of Microbiology and Biotechnology*, 14(2), 199–203.

Cocolin, L., Pepe, V., Comitini, F., Comi, G., & Ciani, M. (2004). Enological and genetic traits of *Saccharomyces cerevisiae* isolated from former and modern wineries. *FEMS Yeast Research*, 5(3), 237–245.

Cole, V. C., & Noble, A. C. (1995). Flavour chemistry and assessment. In *Fermented beverage production*. Boston, MA: Springer.

Comitini, F., & Ciani, M. (2010). The zymocidial activity of Tetrapisispora phaffii in the control of Hanseniaspora uvarum during the early stages of winemaking. *Letters in applied microbiology*, 50(1), 50–56.

Comitini, F., Gobbi, M., Domizio, P., Romani, C., Lencioni, L., Mannazzu, I., & Ciani, M. (2011). Selected non-*Saccharomyces* wine yeasts in controlled multistarter fermentations with *Saccharomyces cerevisiae*. *Food Microbiology*, 28(5), 873–882.

Comitini, F., Ingeniis De, J., Pepe, L., Mannazzu, I., & Ciani, M. (2004). Pichia anomala and *Kluyveromyces wickerhamii* killer toxins as new tools against *Dekkera/Brettanomyces* spoilage yeasts. *FEMS Microbiology Letters*, 238(1), 235–240.

Contreras, A., Hidalgo, C., Henschke, P. A., Chambers, P. J., Curtin, C., & Varela, C. (2014). Evaluation of non-*Saccharomyces* yeasts for the reduction of alcohol content in wine. *Applied and Environmental Microbiology*, 80(5), 1670–1678.

Contreras, A., Hidalgo, C., Schmidt, S., Henschke, P. A., Curtin, C., & Varela, C. (2015). The application of non-*Saccharomyces* yeast in fermentations with limited aeration as a strategy for the production of wine with reduced alcohol content. *International Journal of Food Microbiology*, 205, 7–15.

de Ullivarri, M. F., Mendoza, L. M., & Raya, R. R. (2018). Characterization of the killer toxin KTCf20 from *Wickerhamomyces anomalus*, a potential biocontrol agent against wine spoilage yeasts. *Biological Control*, 121, 223–228.

Domizio, P., Lencioni, L., Calamai, L., Portaro, L., & Bisson, L. F. (2018). Evaluation of the yeast Schizosaccharomyces japonicus for use in wine production. *American Journal of Enology and Viticulture*, 69, 266–277. Available from https://doi.org/10.5344/ajev.2018.18004.

Domizio, P., Lencioni, L., Ciani, M., Di Blasi, S., Pontremolesi, C. D., & Sabatelli, M. P. (2007). Spontaneous and inoculated yeast populations dynamics and their effect on organoleptic characters of Vinsanto wine under different process conditions. *International Journal of Food Microbiology*, 115, 281–289.

Domizio, P., Liu, Y., Bisson, L. F., & Barile, D. (2014). Use of non-*Saccharomyces* wine yeasts as novel sources of mannoproteins in wine. *Food Microbiology*, 43, 5–15. Available from https://doi.org/10.1016/j.fm.2014.04.005.

Domizio, P., Liu, Y., Bisson, L. F., & Barile, D. (2017). Cell wall polysaccharides released during the alcoholic fermentation by *Schizosaccharomyces pombe* and *S. japonicus*: Quantification and characterization. *Food Microbiology*, 61, 136–149.

Domizio, P., Romani, C., Lencioni, L., Comitini, F., Gobbi, M., Mannazzu, I., & Ciani, M. (2011). Outlining a future for non-*Saccharomyces* yeasts: selection of putative spoilage wine strains to be used in association with *Saccharomyces cerevisiae* for grape juice fermentation. *International Journal of Food Microbiology*, 147(3), 170–180.

Dutraive, O., Benito, S., Fritsch, S., Beisert, B., Patz, C. D., & Rauhut, D. (2019). Effect of sequential inoculation with non-*Saccharomyces* and *Saccharomyces* yeasts on Riesling wine chemical composition. *Fermentation*, 5(3), 79.

Englezos, V., Rantsiou, K., Cravero, F., Torchio, F., Ortiz-Julien, A., Gerbi, V., . . . Cocolin, L. (2016). *Starmerella bacillaris* and *Saccharomyces cerevisiae* mixed fermentations to reduce ethanol content in wine. *Applied Microbiology and Biotechnology*, 100(12), 5515–5526.

Englezos, V., Rantsiou, K., Cravero, F., Torchio, F., Pollon, M., Fracassetti, D., . . . Cocolin, L. (2018). Volatile profile of white wines fermented with sequential inoculation of *Starmerella bacillaris* and *Saccharomyces cerevisiae*. *Food Chemistry*, 257, 350–360.

Escott, C., Del Fresno, J. M., Loira, I., Morata, A., & Suárez-Lepe, J. A. (2018). Zygosaccharomyces rouxii: control strategies and applications in food and winemaking. *Fermentation*, 4(3), 69.

Escott, C., Loira, I., Morata, A., Bañuelos, M. A., & Suárez-Lepe, J. A. (2017). Wine spoilage yeasts: Control strategy. *Yeast-Industrial Applications*, 89–116.

Escribano, R., González-Arenzana, L., Portu, J., Garijo, P., López-Alfaro, I., López, R., . . . Gutiérrez, A. R. (2018). Wine aromatic compound production and fermentative behaviour within different non-*Saccharomyces* species and clones. *Journal of Applied Microbiology*, 124(6), 1521–1531.

Ferraro, L., Fatichenti, F., & Ciani, M. (2000). Pilot scale vinification process using immobilized *Candida stellata* cells and *Saccharomyces cerevisiae*. *Process Biochemistry*, 35(10), 1125–1129.

Gao, C., & Fleet, G. H. (1988). The effects of temperature and pH on the ethanol tolerance of the wine yeasts, *Saccharomyces cerevisiae, Candida stellata* and *Kloeckera apiculata*. *Journal of Applied Bacteriology*, 65(5), 405–409.

Garavaglia, J., de Souza Schneider, R. D. C., Mendes, S. D. C., Welke, J. E., Zini, C. A., Caramão, E. B., & Valente, P. (2015). Evaluation of Zygosaccharomyces bailii BCV 08 as a co-starter in wine fermentation for the improvement of ethyl esters production. *Microbiological Research*, 173, 59–65.

Gobbi, M., Comitini, F., D'Ignazi, G., & Ciani, M. (2013). Effects of nutrient supplementation on fermentation kinetics, H 2 S evolution, and aroma profile in Verdicchio DOC wine production. *European Food Research and Technology, 236*(1), 145–154.

Gobbi, M., De Vero, L., Solieri, L., Comitini, F., Oro, L., Giudici, P., & Ciani, M. (2014). Fermentative aptitude of non-*Saccharomyces* wine yeast for reduction in the ethanol content in wine. *European Food Research and Technology, 239*(1), 41–48.

González-Royo, E., Pascual, O., Kontoudakis, N., Esteruelas, M., Esteve-Zarzoso, B., Mas, A., ... Zamora, F. (2015). Oenological consequences of sequential inoculation with non-*Saccharomyces* yeasts (*Torulaspora delbrueckii* or *Metschnikowia pulcherrima*) and *Saccharomyces cerevisiae* in base wine for sparkling wine production. *European Food Research and Technology, 240*(5), 999–1012.

Hranilovic, A., Gambetta, J. M., Jeffery, D. W., Grbin, P. R., & Jiranek, V. (2020). Lower-alcohol wines produced by *Metschnikowia pulcherrima* and *Saccharomyces cerevisiae* co-fermentations: The effect of sequential inoculation timing. *International Journal of Food Microbiology*, 108651.

Hranilovic, A., Gambetta, J. M., Schmidtke, L., Boss, P. K., Grbin, P. R., Masneuf-Pomarede, I., ... Jiranek, V. (2018). Oenological traits of *Lachancea thermotolerans* show signs of domestication and allopatric differentiation. *Scientific Reports, 8*, 14812–14825.

Hu, K., Jin, G. J., Xu, Y. H., & Tao, Y. S. (2018). Wine aroma response to different participation of selected *Hanseniaspora uvarum* in mixed fermentation with *Saccharomyces cerevisiae*. *Food Research International, 108*, 119–127.

Jolly, N.P., Augustyn, O.P.H., & Pretorius, I.S. (2006). The role and use of non-*Saccharomyces* yeasts in wine production. *South African Journal for Enology and Viticulture, 27*:15–39.

Jolly, N. P., Varela, C., & Pretorius, I. S. (2014). Not your ordinary yeast: Non-*Saccharomyces* yeasts in wine production uncovered. *FEMS Yeast Research, 14*(2), 215–237.

Kapsopoulou, K., Mourtzini, A., Anthoulas, M., & Nerantzis, E. (2007). Biological acidification during grape must fermentation using mixed cultures of *Kluyveromyces thermotolerans* and *Saccharomyces cerevisiae*. *World Journal of Microbiology and Biotechnology, 23*, 735–739.

Kemsawasd, V., Viana, T., Ardö, Y., & Arneborg, N. (2015). Influence of nitrogen sources on growth and fermentation performance of different wine yeast species during alcoholic fermentation. *Applied Microbiology and Biotechnology, 99*(23), 10191–10207.

Kuchen, B., Maturano, Y. P., Mestre, M. V., Combina, M., Toro, M. E., & Vazquez, F. (2019). Selection of native non-*Saccharomyces* yeasts with biocontrol activity against spoilage yeasts in order to produce healthy regional wines. *Fermentation, 5*(3), 60.

Laguna, L., Bartolomé, B., & Moreno-Arribas, M. V. (2017). Mouthfeel perception of wine: Oral physiology, components and instrumental characterization. *Trends in Food Science & Technology, 59*, 49–59.

Lambrechts, M. G., & Pretorius, I. S. (2000). Yeast and its importance to wine aroma-a review. *South African Journal of Enology and Viticulture, 21*(1), 97–129.

Lencioni, L., Taccari, M., Ciani, M., & Domizio, P. (2018). *Zygotorulaspora florentina* and *Starmerella bacillaris* in multistarter fermentation with *Saccharomyces cerevisiae* to reduce volatile acidity of high sugar musts. *Australian Journal of Grape and Wine Research, 24*(3), 368–372.

Lin, M. M. H., Boss, P. K., Walker, M. E., Sumby, K. M., Grbin, P. R., & Jiranek, V. (2020). Evaluation of indigenous non-*Saccharomyces* yeasts isolated from a South Australian vineyard for their potential as wine starter cultures. *International Journal of Food Microbiology, 312*, 108373.

Liu, P. T., Lu, L., Duan, C. Q., & Yan, G. L. (2016). The contribution of indigenous non-*Saccharomyces* wine yeast to improved aromatic quality of Cabernet Sauvignon wines by spontaneous fermentation. *LWT-Food Science and Technology, 71*, 356–363.

Liu, S., Laaksonen, O., Kortesniemi, M., Kalpio, M., & Yang, B. (2018). Chemical composition of bilberry wine fermented with non-*Saccharomyces* yeasts (*Torulaspora delbrueckii* and *Schizosaccharomyces pombe*) and *Saccharomyces cerevisiae* in pure, sequential and mixed fermentations. *Food Chemistry, 266*, 262–274.

Lleixà, J., Martín, V., Portillo, M. D. C., Carrau, F., Beltran, G., & Mas, A. (2016). Comparison of fermentation and wines produced by inoculation of *Hanseniaspora vineae* and *Saccharomyces cerevisiae*. *Frontiers in Microbiology, 7*, 338.

López, S., Mateo, J. J., & Maicas, S. (2014). Characterisation of *Hanseniaspora* isolates with potential aroma-enhancing properties in Muscat wines. *South African Journal of Enology and Viticulture, 35*(2), 292–303.

Magyar, I., & Panyik, I. (1989). Biological deacidification of wine with *Schizosaccharomyces pombe* entrapped in Ca-alginate gel. *American Journal of Enology and Viticulture, 40*(4), 233–240.

Magyar, I., & Tóth, T. (2011). Comparative evaluation of some oenological properties in wine strains of *Candida stellata, Candida zemplinina, Saccharomyces uvarum* and *Saccharomyces cerevisiae*. *Food Microbiology, 28*(1), 94–100.

Mannazzu, I., Domizio, P., Carboni, G., Zara, S., Zara, G., Comitini, F., ... Ciani, M. (2019). Yeast killer toxins: from ecological significance to application. *Critical Reviews in Biotechnology, 39*(5), 603–617.

Medina, K., Boido, E., Fariña, L., Gioia, O., Gomez, M. E., Barquet, M., ... Carrau, F. (2013). Increased flavour diversity of Chardonnay wines by spontaneous fermentation and co-fermentation with *Hanseniaspora vineae*. *Food Chemistry, 141*, 2513–2521.

Medina-Trujillo, L., González-Royo, E., Sieczkowski, N., Heras, J., Canals, J. M., & Zamora, F. (2017). Effect of sequential inoculation (*Torulaspora delbrueckii/Saccharomyces cerevisiae*) in the first fermentation on the foaming properties of sparkling wine. *European Food Research and Technology, 243*(4), 681–688.

Mehlomakulu, N. N., Setati, M. E., & Divol, B. (2014). Characterization of novel killer toxins secreted by wine-related non-*Saccharomyces* yeasts and their action on *Brettanomyces* spp. *International Journal of Food Microbiology, 188*, 83–91.

Moreira, N., Lopes, P., Ferreira, H., Cabral, M., & de Pinho, P. G. (2018). Sensory attributes and volatile composition of a dry white wine under different packing configurations. *Journal of Food Science and Technology, 55*(1), 424–430.

Munyon, J. R., & Nagel, C. W. (1977). Comparison of methods of deacidification of musts and wines. *American Journal of Enology and Viticulture, 28*, 79–87.

Nissen, P., & Arneborg, N. (2003). Characterization of early deaths of non-*Saccharomyces* yeasts in mixed cultures with *Saccharomyces cerevisiae*. *Archives of Microbiology, 180*(4), 257–263.

Oliveira, I., & Ferreira, V. (2019). Modulating fermentative, varietal and aging aromas of wine using non-*Saccharomyces* yeasts in a sequential inoculation approach. *Microorganisms, 7*(6), 164.

Oro, L., Ciani, M., Bizzaro, D., & Comitini, F. (2016). Evaluation of damage induced by Kwkt and Pikt zymocins against *Brettanomyces/Dekkera* spoilage yeast, as compared to sulphur dioxide. *Journal of Applied Microbiology, 121*(1), 207–214.

Oro, L., Ciani, M., & Comitini, F. (2014). Antimicrobial activity of *Metschnikowia pulcherrima* on wine yeasts. *Journal of Applied Microbiology, 116*(5), 1209–1217.

Padilla, B., Zulian, L., Ferreres, À., Pastor, R., Esteve-Zarzoso, B., Beltran, G., & Mas, A. (2017). Sequential inoculation of native non-*Saccharomyces* and *Saccharomyces cerevisiae* strains for wine making. *Frontiers in Microbiology, 8*, 1293–1305.

Peinado, R. A., Moreno, J. J., Maestre, O., & Mauricio, J. C. (2007). Removing gluconic acid by using different treatments with a *Schizosaccharomyces pombe* mutant: Effect on fermentation byproducts. *Food Chemistry, 104*(2), 457–465.

Pina, C., Santos, C., Couto, J. A., & Hogg, T. (2004). Ethanol tolerance of five non-*Saccharomyces* wine yeasts in comparison with a strain of *Saccharomyces cerevisiae*—Influence of different culture conditions. *Food Microbiology, 21*, 439–447.

Pinto, L., Baruzzi, F., Cocolin, L., & Malfeito-Ferreira, M. (2020). Emerging technologies to control *Brettanomyces* spp. in wine: Recent advances and future trends. *Trends in Food Science & Technology, 99*, 88–100.

Pretorius, I. S. (2000). Tailoring wine yeast for the new millennium: novel approaches to the ancient art of winemaking. *Yeast (Chichester, England), 16*(8), 675–729.

Puertas, B., Jiménez, M. J., Cantos-Villar, E., Cantoral, J. M., & Rodríguez, M. E. (2017). Use of *Torulaspora delbrueckii* and *Saccharomyces cerevisiae* in semi-industrial sequential inoculation to improve quality of Palomino and Chardonnay wines in warm climates. *Journal of Applied Microbiology, 122*, 733–746.

Rankine, B. C. (1966). Decomposition of L-malic acid by wine yeasts. *Journal of the Science of Food and Agriculture, 17*, 312–316.

Rantsiou, K., Dolci, P., Giacosa, S., Torchio, F., Tofalo, R., Torriani, S., . . . Cocolin, L. (2012). *Candida zemplinina* can reduce acetic acid produced by *Saccharomyces cerevisiae* in sweet wine fermentations. *Applied and Environmental Microbiology, 78*, 1987–1994.

Renault, P., Coulon, J., de Revel, G., Barbe, J. C., & Bely, M. (2015). Increase of fruity aroma during mixed *T. delbrueckii/S. cerevisiae* wine fermentation is linked to specific esters enhancement. *International Journal of Food Microbiology, 207*, 40–48.

Renault, P. E., Albertin, W., & Bely, M. (2013). An innovative tool reveals interaction mechanisms among yeast populations under oenological conditions. *Applied microbiology and biotechnology, 97*(9), 4105–4119.

Ribéreau-Gayon, P., Dubourdieu, D., Donèche, B., & Lonvaud, A. (Eds.), (2006). Handbook of enology. In: *The microbiology of wine and vinifications* (1(Vol. 1).). John Wiley & Sons.

Röcker, J., Strub, S., Ebert, K., & Grossmann, M. (2016). Usage of different aerobic non-*Saccharomyces* yeasts and experimental conditions as a tool for reducing the potential ethanol content in wines. *European Food Research and Technology, 242*(12), 2051–2070.

Rodríguez, M. E., Lopes, C. A., Barbagelata, R. J., Barda, N. B., & Caballero, A. C. (2010). Influence of Candida pulcherrima Patagonian strain on alcoholic fermentation behaviour and wine aroma. *International journal of food microbiology, 138*(1–2), 19–25.

Rollero, S., Bloem, A., Ortiz-Julien, A., Camarasa, C., & Divol, B. (2018). Altered fermentation performances, growth, and metabolic footprints reveal competition for nutrients between yeast species inoculated in synthetic grape juice-like medium. *Frontiers in Microbiology, 9*, 196.

Romani, C., Lencioni, L., Gobbi, M., Mannazzu, I., Ciani, M., & Domizio, P. (2018). *Schizosaccharomyces japonicus*: A polysaccharide-overproducing yeast to be used in winemaking. *Fermentation, 4*(1), 14.

Romano, P., & Suzzi, G. (1993). Potential use for *Zygosaccharomyces* species in winemaking. *Journal of Wine Research, 4*(2), 87–94.

Rosa, C. A., & Lachance, M. A. (1998). The yeast genus *Starmerella* gen. nov. and *Starmerella bombicola* sp. nov., the teleomorph of *Candida bombicola* (Spencer, Gorin & Tullock) Meyer & Yarrow. *International Journal of Systematic and Evolutionary Microbiology, 48*(4), 1413–1417.

Ruiz, J., Belda, I., Beisert, B., Navascués, E., Marquina, D., Calderón, F., . . . Benito, S. (2018). Analytical impact of *Metschnikowia pulcherrima* in the volatile profile of Verdejo white wines. *Applied Microbiology and Biotechnology, 102*(19), 8501–8509.

Sadoudi, M., Tourdot-Maréchal, R., Rousseaux, S., Steyer, D., Gallardo-Chacón, J. J., Ballester, J., . . . Alexandre, H. (2012). Yeast–yeast interactions revealed by aromatic profile analysis of Sauvignon Blanc wine fermented by single or co-culture of non-*Saccharomyces* and *Saccharomyces* yeasts. *Food Microbiology, 32*(2), 243–253.

Shekhawat, K., Porter, T. J., Bauer, F. F., & Setati, M. E. (2018). Employing oxygen pulses to modulate *Lachancea thermotolerans*–*Saccharomyces cerevisiae* Chardonnay fermentations. *Annals Of Microbiology, 68*, 93–102.

Šipiczki, M., Ciani, M., & Csoma, H. (2005). Taxonomic reclassification of *Candida stellata* DBVPG 3827. *Folia Microbiologica, 50*(6), 494–498.

Snow, P. G., & Gallander, J. F. (1979). Deacidification of white table wines through partial fermentation with *Schizosaccharomyces pombe*. *American Journal of Enology and Viticulture, 30*(1), 45–48.

Soden, A., Francis, I. L., Oakey, H., & Henschke, P. A. (2000). Effects of co-fermentation with *Candida stellata* and *Saccharomyces cerevisiae* on the aroma and composition of Chardonnay wine. *Australian Journal of Grape and Wine Research, 6*(1), 21–30.

Taillandier, P., Lai, Q. P., Julien-Ortiz, A., & Brandam, C. (2014). Interactions between *Torulaspora delbrueckii* and *Saccharomyces cerevisiae* in wine fermentation: Influence of inoculation and nitrogen content. *World Journal of Microbiology and Biotechnology, 30*(7), 1959–1967.

Varela, C. (2016). The impact of non-*Saccharomyces* yeasts in the production of alcoholic beverages. *Applied Microbiology and Biotechnology, 100*, 9861–9874.

Varela, C., & Borneman, A. R. (2017). Yeasts found in vineyards and wineries. *Yeast (Chichester, England), 34*, 111–128.

Varela, C., Sengler, F., Solomon, M., & Curtin, C. (2016). Volatile flavour profile of reduced alcohol wines fermented with the non-conventional yeast species Metschnikowia pulcherrima and Saccharomyces uvarum. *Food Chemistry, 209*, 57–64.

Velázquez, R., Zamora, E., Álvarez, M. L., & Ramírez, M. (2019). Using *Torulaspora delbrueckii* killer yeasts in the elaboration of base wine and traditional sparkling wine. *International Journal of Food Microbiology, 289*, 134–144.

Vigentini, I., Fabrizio, V., Faccincani, M., Picozzi, C., Comasio, A., & Foschino, R. (2014). Dynamics of *Saccharomyces cerevisiae* populations in controlled and spontaneous fermentations for Franciacorta, D.O.C.G. base wine production. *Annals of Microbiology, 64*, 639–651.

Villalba, M. L., Sáez, J. S., del Monaco, S., Lopes, C. A., & Sangorrín, M. P. (2016). TdKT, a new killer toxin produced by *Torulaspora delbrueckii* effective against wine spoilage yeasts. *International Journal of Food Microbiology, 217*, 94–100.

Waterhouse, A. L., Sacks, G. L., & Jeffery, D. W. (2016). *Understanding wine chemistry*. John Wiley & Sons.

Yokotsuka, K., Otaki, A., Naitoh, A., & Tanaka, H. (1993). Controlled simultaneous deacidification and alcohol fermentation of a high-acid grape must using two immobilized yeasts, *Schizosaccharomyces pombe* and *Saccharomyces cerevisiae*. *American Journal of Enology and Viticulture, 44*(4), 371–377.

Zhu, F., Du, B., & Li, J. (2016). Aroma compounds in wine. *Grape and Wine Biotechnology, 273*.

Zott, K., Thibon, C., Bely, M., Lonvaud-Funel, A., Dubourdieu, D., & Masneuf-Pomarede, I. (2011). The grape must non-*Saccharomyces* microbial community: Impact on volatile thiol release. *International Journal of Food Microbiology, 151*(2), 210.

11

Biological acidification by *Lachancea thermotolerans*

Antonio Morata[1], Iris Loira[1], Carmen González[1], María Antonia Bañuelos[2], Rafael Cuerda[3], José María Heras[4], Cristian Vaquero[1] and José Antonio Suárez-Lepe[1]

[1]enotecUPM, Chemistry and Food Technology Department, Universidad Politécnica de Madrid, Campus Ciudad Universitaria, Madrid, Spain [2]enotecUPM, Biotechnology-Plant Biology Department, Universidad Politécnica de Madrid, Campus Ciudad Universitaria, Madrid, Spain [3]Bodegas Comenge, Valladolid, Spain [4]Lallemand Ibérica SL., Madrid, Spain

11.1 Introduction

Global warming produces an intense impact on viticulture and enology. The grape maturation is often affected, producing higher sugar contents, which subsequently increases the alcohol degree of wines, and low acidity. The pH is a key parameter in wine technology and in fermentation and aging microbiology since high values facilitate the development of spoilage microorganisms, which can affect wine quality and favor the formation of toxic molecules such as biogenic amines. The effectiveness of sulfites is also reduced at high pH, which not only increases microbial problems but also affects chemical evolution and stability. *Lachancea thermotolerans* (Lt), formerly *Kluyveromyces thermotolerans* (Kurtzman, 2003), is a ubiquitous yeast that can be easily found in grapes but also in soils, insects, and plant tissues, and distributed around the globe. Lt has shown a high prevalence in soils (Chandra, Mota, Silva, & Malfeito-Ferreira, 2020) and sour-rotten grapes (Barata, González, Malfeito-Ferreira, Querol, & Loureiro, 2008). Several recent review papers have reported significant information concerning its ecology, physiology, metabolisms, and impact on wine production (Hranilovic et al., 2018; Hranilovic, Bely, Masneuf-Pomarede, Jiranek, & Albertin, 2017; Morata et al., 2018; Porter, Divol, & Setati, 2019). Lt has a spherical or ellipsoidal morphology, which makes it impossible to distinguish from *Saccharomyces cerevisiae* (Sc) by microscopy. Asexual reproduction is by multipolar budding. It is a teleomorph yeast forming 1−4 spherical spores. Lt can be found frequently in grapes as part of the indigenous yeast population (Fig. 11.1), together with other typical apiculate yeasts genera such as *Kloeckera* spp. or *Hanseniaspora* spp.

The current enology uses the acids of the grape, mainly tartaric acid, to correct the acidity, and therefore the pH. Tartaric acid is quite strong but unstable in wine conditions and usually precipitates as tartrates, reducing its effect on pH. Ion exchangers are a highly effective technology, but with a strong impact on wine quality, as they also produce large amounts of highly polluting saline effluents, and mineral acids must be used to regenerate the resins. An ecological and natural alternative is the use of acidifying yeasts able to decrease wine pH through the formation of organic acids.

The main application of Lt in enology is the reduction of pH by the formation of lactic acid from sugars. So far, the maximum lactic acid reported is 16.6 g/L (Banilas et al., 2016); however, it is common to find strains that produce between 1 and 5 g/L (Morata et al., 2018). Fermentative formation of succinic acid has also been reported (Aponte & Blaiotta, 2016). Its fermentative power is moderate, and most of the strains can ferment

© 2022 Elsevier Inc. All rights reserved.

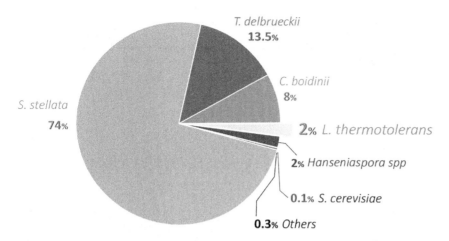

FIGURE 11.1 Indigenous yeast population in grapes of *Vitis vinifera* L. cv. Muscat of Alexandria by next-generation sequencing.

musts, reaching alcohol contents of 7%−10% v/v (Aponte & Blaiotta, 2016; Comitini et al., 2011; Hranilovic et al., 2018; Morata et al., 2018; Vaquero et al., 2020). The suitable fermentation temperature range is 25°C−30°C, with slower growth and fermentations below 20°C (Schnierda, Bauer, Divol, van Rensburg, & Görgens, 2014; Vaquero et al., 2020).

It shows medium production of H_2S (Aponte & Blaiotta, 2016; Comitini et al., 2011) and good resistance to SO_2, easily fermenting in the presence of 40−60 mg/L SO_2 (Morata et al., 2018) (Table 11.1), which is a usual dose in white wine fermentation. Glycerol formation during fermentation is moderate; however, some strains able to reach 8 g/L have been described (Hranilovic et al., 2018). The production of volatile acidity during fermentation by Lt is moderate and usually below or around 0.4 g/L (Comitini et al., 2011; Hranilovic et al., 2018; Morata et al., 2019). Moreover, it has been described that it is able to decrease the volatile acidity of wines, but under aerobic conditions, that can be incompatible with conventional wine fermentations (Vilela, 2018). Many Lt strains can slightly degrade malate; however, some of them have shown a certain capacity to produce small amounts of malic acid, reaching 0.3 g/L (Hranilovic et al., 2018). This yeast species has also been described as an interesting wine aroma modulator, producing floral esters such as 2-phenylethyl acetate (Comitini et al., 2011; Morata et al., 2019) and releasing thiols such as 4-methyl-4-mercaptopentan-2-one (4MMP, box tree aroma) and 3-mercaptohexan-1-ol (3MH, grape and passion fruit aroma) (Zott et al., 2011).

11.2 Microbiological and molecular identification

11.2.1 Selective and differential media

Lt can be identified by selective/differential media and molecular techniques (Loira, Morata, Bañuelos, & Suárez-Lepe, 2019). Using conventional synthetic media, Lt grows on lysine agar selective media like many non-*Saccharomyces* yeasts, but it does not form colonies on either yeast extract peptone dextrose (YPD) media at a high temperature (37ₒC) or nitrate agar media (Loira et al., 2019). Differential media can be used for the isolation and presumptive identification of Lt. In CHROMagar Candida media, Lt shows a creamy appearance and a red-orange color, while *S. cerevisiae* displays a soft purple color (Fig. 11.2).

11.2.2 Molecular techniques

Several molecular techniques have been used to identify Lt, such as PCR-denaturing gradient gel electrophoresis (PCR-DGGE); microsatellite markers and multilocus SSR analysis; sequencing of the ribosomal rDNA genes (internal transcribed spacer [ITS] rDNA, 26S rDNA) and RAPD fingerprinting; restriction patterns of amplified regions of ribosomal rDNA; and PCR of the intron 2 of the mitochondrial COX1 gene. The performance, and especially the sensitivity, is variable depending on the molecular technique. (Table 11.2).

In recent years, the use of massive sequencing techniques is increasing, not only as a powerful metagenomic tool for the study of microbial populations that can be found in complex ecosystems with dense microbial communities but also

TABLE 11.1 Fermentation of *Lachancea thermotolerans* compared with *Saccharomyces cerevisiae*.

Yeast	pH reduction (pH units)	Lactic acid production (g/L)	Alcohol reduction (% vol.)	Volatile acidity (g/L)	Total SO$_2$ resistance (mg/L)	Nitrogen requirements YAN (mg/L)	2-Phenylethyl acetate (mg/L)
Lt	0.1–0.5	1–7	0.2–0.5	<0.4	40–60	150–200	8–50
Sc	–	<0.2	–	0.2–0.4	>80	≃150	4–6

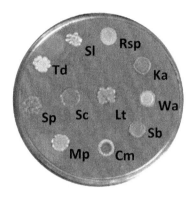

FIGURE 11.2 Color and appearance in CHROMagar Candida Chromogenic differential isolation medium of the colonies of Cm, *Candida magnoliae*; Ka, *Kloeckera apiculata*; Lt, *Lachancea thermotolerans*; Mp, *Metschnikowia pulcherrima*; Rsp, *Rhodotorula* sp.; Sc, *Saccharomyces cerevisiae*; Sb, *Starmerella bacillaris*; Sl, *Saccharomycodes ludwigii*; Sp, *Schizosaccharomyces pombe*; Td, *Torulaspora delbrueckii*; Wa, *Wickerhamomyces anomalus*.

TABLE 11.2 Molecular techniques to detect and identify *Lachancea thermotolerans*.

Technique	Sample preparation	Specificity	Sensitivity	References
PCR-DGGE	Culture-independent	Identification of non-*Saccharomyces* species	Low. Need 2-log CFU/mL	(Prakitchaiwattana, Fleet, & Heard, 2004)
Microsatellite markers	Culture-dependent	Strain diversity of *Lachancea thermotolerans*	–	Banilas et al. (2016)
26S rDNA	Culture-dependent	Identification of non-*Saccharomyces* species	–	Kurtzman & Robnett (1998), Lopandic et al. (2008)
rDNA PCR-RFLP	Culture-dependent	Cluster at species level	–	Esteve-Zarzoso, Belloch, Uruburu, & Quero (1999), Baleiras Couto, Reizinho, & Duarte (2005)
PCR of the intron 2 for the COX1	Culture-independent	Specific primers for *Lachancea thermotolerans*. Able to discriminate 1000/1 Sc/Lt in wine	Low. Need 4-log CFU/mL	Zara et al. (2014)
Next-generation sequencing (NGS)	High-throughput techniques. Culture-independent	Cluster at species level. Needs bio-computational analysis	Low	Kioroglou, LLeixá, Mas, and del Carmen Portillo (2018)

for monitoring the evolution of these fermentative microorganisms during the various stages of the alcoholic fermentation (Kioroglou et al., 2018). Massive sequencing techniques, also known as next-generation sequencing (NGS), are high-throughput methods that allow analyzing complex communities within a short time at a moderate cost and with high resolution, clustering the microorganisms at the species level. To process a large amount of data, NGS analysis requires the use of powerful bioinformatics pipelines (a set of bioinformatics algorithms, executed according to a predefined sequence) to manage millions of short DNA sequences used to classify complex microbial ecosystems.

11.3 Commercially available strains of *Lachancea thermotolerans*

There are several commercial strains of Lt available for wine fermentation within the wide catalog of non-*Saccharomyces* wine yeasts, which has been increasing a lot in recent years (Morata et al., 2020; Morata &

Suárez-Lepe, 2016). Most commercial Lt strains are able to ferment to 7%–10% ethanol. They are usually described as being able to increase freshness, with low production of volatile acidity and acetaldehyde. Some of them can increase the production of ethyl acetate under experimental conditions. Depending on the strain, the effect on the aroma is quite variable. Most of the strains are described as having higher nitrogen nutritional requirements than *S. cerevisiae*.

There is not much information about the performance when used in low-temperature fermentation, or about the SO_2 levels they can tolerate. Recently, Vaquero et al. (2020) observed that several selected and commercial Lt strains were able to ferment with 75 mg/L of total SO_2 but showed delayed fermentation kinetics compared to the *S. cerevisiae* control. The maximum differences among the strains with that amount of SO_2 were found on day 3 of fermentation at 27°C, with a total difference of 3.77% v/v ethanol in alcoholic strength between the highest and lowest contents. At a lower temperature (17°C) and with the same SO_2 content, the maximum differences were observed slightly delayed, on day 5 of fermentation, with a value of 3.44% v/v ethanol.

These data show that it is possible to use Lt over a wide range of fermentation temperatures (17°C–27°C) and different amounts of SO_2 but in the typical range used for white wine production (Vaquero et al., 2020). A suitable performance at a low temperature (17°C) was also observed, which is close to the usual fermentation temperatures to preserve the varietal aroma.

11.4 Lactic acid production and pH control during fermentation

11.4.1 pH control

L. thermotolerans has the rare ability to increase acidity during fermentation with the production of lactic acid from sugars (Comitini et al., 2011; Jolly, Varela, & Pretorius, 2014; Morata et al., 2018). Several applications have been reported in wine production, with it being used sequentially or in mixed cultures with other *Saccharomyces* and

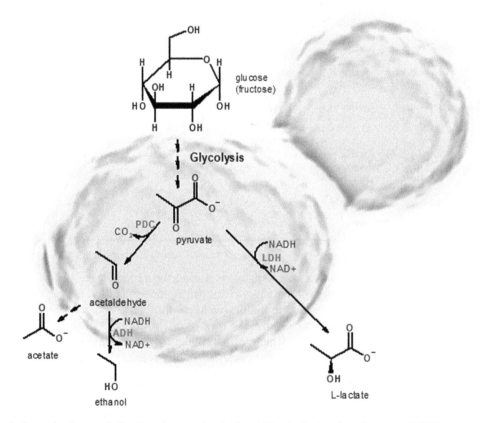

FIGURE 11.3 Pathway for the metabolization of sugars into lactic acid by the lactate dehydrogenase (LDH) enzyme of *Lachancea thermotolerans*, in direct competition with the formation of ethanol from pyruvate and therefore from sugars. *ADH*, alcohol dehydrogenase; *PDC*, pyruvate decarboxylase.

non-*Saccharomyces* yeasts (Comitini et al., 2011; Escott et al., 2018; Gobbi et al., 2013; Morata et al., 2019). The production of lactic acid by Lt is driven by the enzyme lactate dehydrogenase (LDH) to recover the NAD^+ cofactor (Fig. 11.3). The DNA sequence encoding this enzyme has been sequenced for several strains. It has been reported that the Lt genome includes three LDH, namely LDH1, LDH2, and LDH3, and two alcohol dehydrogenase (ADH), namely ADH1 and ADH2, paralogous genes (Sgouros, Mallouchos, Filippousi, Banilas, & Nisiotou, 2020).

The acidification pattern also depends on the fermentation kinetics that is affected by temperature and SO_2 content. When evaluating acidification at a low fermentation temperature (17°C) and a high temperature (27°C), it is observed that most of the selected and commercial strains produce a stronger pH reduction when they ferment at warmer temperatures (Vaquero et al., 2020). As for the effect of SO_2, in some of them, a lower pH can be observed at low SO_2 concentrations.

Most of the acidification occurs during the early stages of fermentation (Morata et al., 2018), usually during days 2−5 (Fig. 11.4). This helps to ensure effective acidification under industrial conditions, even when stronger fermentative species such as *S. cerevisiae* are developed in a mixed fermentation.

Acidification starts at a significant level when the yeast population is greater than 6-log CFU/mL (Vaquero et al., 2020), in addition to being conditioned by nutritional requirements. When Lt is used to ferment a diluted must from concentrate, usually with low nutrients, it is difficult to reach Lt populations higher than 5-log. So, in this case, the lactic acid production, even for strong acidifying strains, is usually between 1 and 2 g/L. However, the same strain fermenting a fresh must can reach populations higher than 6-log and the acidification can exceed 5 g/L of lactic acid, depending on the strain.

There is wide strain-dependent variability in lactic acid production, ranging from <1 g/L to more than 16 g/L (Banilas et al., 2016). Therefore suitable strains can be used for each level of acidification according to the type of wine to be produced. In addition, the industrial application of the acidification by Lt can be controlled by the sequential inoculation of the complementary Sc. The sooner the Sc is inoculated, the earlier it will begin to compete with Lt, thus reducing its metabolic impact on the formation of lactic acid from sugars.

11.4.2 Reduction of ethanol content

Pyruvate can be metabolized alternatively to lactate by LDH or to acetaldehyde and ethanol (Fig. 11.3) by the sequence pyruvate decarboxylase (PDC) and ADH. In this regard, an effect on the degree of alcohol

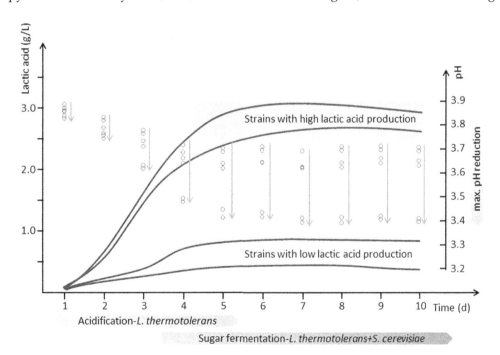

FIGURE 11.4 Acidification pattern by six strains of *Lachancea thermotolerans* and the subsequent pH reduction during fermentation. The open circles are the pH values for each strain and the continuous lines are the lactic acid contents. Continuous lines show the acidification performance of strains with high and low lactic acid production.

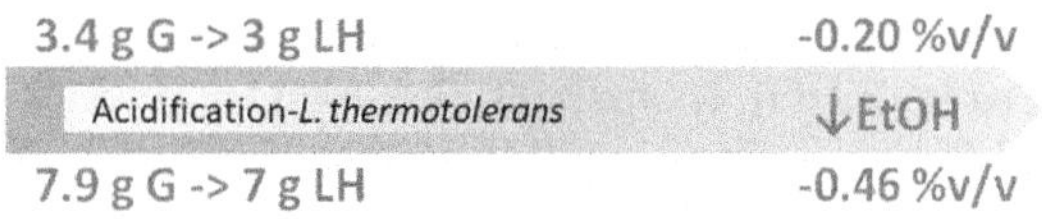

3.4 g G -> 3 g LH	-0.20 %v/v
Acidification-*L. thermotolerans*	↓EtOH
7.9 g G -> 7 g LH	-0.46 %v/v

FIGURE 11.5 Potential ethanol reduction by acidification with *Lachancea thermotolerans* depending on the amount of glucose metabolized into lactic acid. Glucose (G), lactic acid (LH), and ethanol (EtOH).

reduction could occur when fermenting with Lt. Lactic acid is produced from sugars; therefore with strains able to produce higher lactic acid contents during fermentation, a significant reduction in alcohol content is expected. Among other non-*Saccharomyces* species (Lt, *Metschnikowia* spp., and *Starmerella bacillaris*), Lt has been described as the most effective in lowering the ethanol content under fermentative conditions (Binati et al., 2020). *S. bacillaris* metabolism has the peculiarity of producing higher concentrations of glycerol from sugars, probably because of the low ADH activity and the high glycerol-3-phosphate dehydrogenase activity (García, Esteve-Zarzoso, Cabellos, & Arroyo, 2018). Alternatively, ethanol decrease in Lt fermentations can be correlated with the lactic acid formed from the sugars during fermentation: 0.2%–0.5% v/v ethanol lowered for 3–7 g/L of lactic acid formed (Fig. 11.5). The ethanol yield in g/g of Lt has been described as low (average 0.34 ± 0.04) with a maximum value of 0.4 for 58 studied strains (Hranilovic et al., 2018). Therefore another minor side effect of using Lt is the slight reduction in ethanol content that can be useful in wines from warm areas.

11.4.3 Lachancea thermotolerans *and malolactic fermentation*

A concomitant positive effect on wine acidity and freshness is the inhibition of malolactic fermentation (MLF) due to the high content of lactic acid. Concentrations higher than 4 g/L really hinder the development of MLF (Morata, Bañuelos, et al., 2020), thus making it possible to keep red wines with the initial malic acid content with greater stability and without the risk of developing further bottle fermentations. White wines are less likely to develop spontaneous undesired MLF because of their composition as well as the higher levels of sulfites used to control oxidation and the more intense membrane filtrations, usually with smaller pore size. However, it can sometimes happen in warm areas, especially when molecular SO_2 levels can decrease. Therefore acidification with Lt in which more than 4 g/L of lactic acid is produced strongly inhibits MLF, preserving malic acidity and protecting freshness.

Alternatively, if MLF is desired, simultaneous coinoculation with *Oenococcus oeni* (Morata et al., 2019) or *Lactobacillus plantarum* can be used to facilitate malic acid metabolization at the beginning of fermentation with low concentrations of lactic acid (Krieger-Weber, Heras, & Suarez, 2020) (Fig. 11.6). It was observed that the simultaneous coinoculation of Lt and *O. oeni* also resulted in higher acidification and lower pH values (Morata et al., 2019), maybe by the formation of lactic acid by *O. oeni* or alternatively by synergistic interactions between Lt and *O. oeni*.

11.4.4 *Alternative fermentation biotechnologies with* Lachancea thermotolerans

Therefore Lt can be used either in coinoculation or sequentially with *S. cerevisiae* to control the pH during grape must fermentation by increasing the acidity (Fig. 11.6). It is only possible to control the reached acidification by optimizing the addition time of the complementary Sc or alternatively, in sequential fermentations, by selecting the appropriate ratio of populations (Lt:Sc), which can modulate the pH 0.05–0.3 units when ranging from 7log-Lt:7log-Sc to 7log-Lt:3log-Sc (Comitini et al., 2011). Another potential tool to manage wine pH can be the blending of wines after promoting strong acidification in a tank with Lt and later balancing the total acidity of the wine by mixing it with another conventional high-pH wine.

The use of ternary yeast cultures with Lt and other non-*Saccharomyces* in sequential or mixed fermentation with Sc to completely deplete the sugars can be a powerful tool to increase complexity and freshness in neutral white varieties of warm areas (Fig. 11.6). The role of Lt would be acidification and aroma improvement, which can be enhanced with the complementary enzymatic activities or fermentative aroma production by several non-*Saccharomyces* species; among them, *Hanseniaspora vineae*, *Metschnikowia pulcherrima*, or *Torulaspora delbrueckii* can be successful combinations and should be explored (Vaquero et al., 2021).

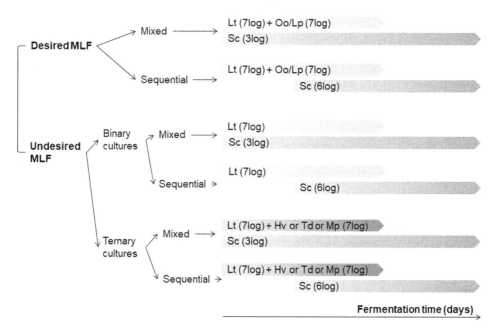

FIGURE 11.6 Alternative fermentation biotechnologies with *Lachancea thermotolerans* to control the pH of the wine and increase its freshness through mixed or sequential cultures with lactic acid bacteria, *Saccharomyces cerevisiae*, or non-Saccharomyces yeasts. *Lachancea thermotolerans* (Lt), *Saccharomyces cerevisiae* (Sc), *Hanseniaspora vineae* (Hv), *Torulaspora delbrueckii* (Td), *Metschnikowia pulcherrima* (Mp), *Oenococcus oeni* (Oo), and *Lactobacillus plantarum* (Lp). Inoculation ratios in log CFU/mL.

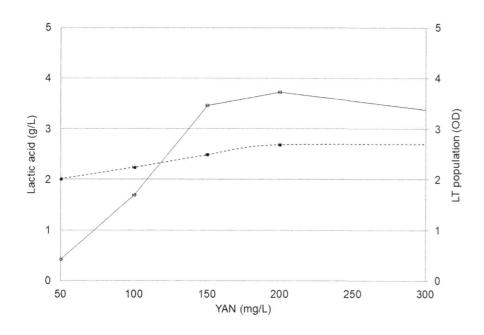

FIGURE 11.7 Lactic acid produced (continuous line) and *Lachancea thermotolerans* counts (dashed line) measured by optical density according to the yeast assimilable nitrogen of the media.

11.5 Nutritional requirements

Even when Lt can ferment a conventional must (220–240 g/L of sugars) with a yeast assimilable nitrogen (YAN) content of 150 mg/L, especially since it is not able to completely ferment the sugars and its main fermentative work is done at the beginning of fermentation, we have observed that higher YAN concentrations, around 200 mg/L, favor higher cell populations and, consequently, a better acidification performance (Fig. 11.7) (Hernández, 2018; Morata et al., 2018).

Lt has been reported to be highly sensitive to low levels of oxygen, which affect its viability more than ethanol content (Lachance & Kurtzman, 2011). Additionally, when a synthetic growth media, suitable for *S. cerevisiae* but with specific nitrogen and vitamin composition, is used instead of general YPD media with peptone and yeast extract, the growth of Lt is weaker, probably because of the lack of specific nutrients that are unnecessary for Sc, but affects the development of Lt. We have also observed a better acidification yield when it is used in the fermentation of crushed red grapes instead of some white grape juices, probably due to the presence of certain nutrients that are richer or better extracted from the skins during the fermentation of red grapes.

11.6 Effect of pH on SO_2 levels and wine stability

In warm areas and vineyards affected by climatic change, the grapes show frequently high pH values, which has an impact on the subsequent microbiological and physicochemical wine stability, if not corrected. Another key problem of high-pH wines is the low effectiveness of sulfur dioxide. The main active chemical species of this antimicrobial is molecular SO_2; however, its concentration in wines strongly depends on pH values. The typical protective concentration of 30 mg/L of free SO_2 can mean 0.8 mg/L of molecular SO_2 at pH 3.5, but less than 0.5 mg/L at pH higher than 3.8, which is quite usual in wines from these areas. Biological acidification by Lt is a natural way of keeping the pH at suitable values (Table 11.3), promoting the effectiveness of SO_2 during the winemaking process and later during storage.

11.7 Impact on wine sensory profile

11.7.1 Acidity and freshness

The main impact of Lt on wine flavor is related to acidity. White wines, in general, are preferred fresher than red wines. Freshness is strongly connected to acidity, so wines with higher acidity levels or lower pH, in general, are perceived as more refreshing, and this attribute is very valuable in most white wines. There is a wrong perception that lactic acidity can be perceived as dairy taste, perhaps because of the slight yogurt or cheese taints

TABLE 11.3 Effect of acidification by *Lachancea thermotolerans* on molecular SO_2 levels.

Initial pH	Free SO_2 (mg/L)	Molecular SO_2 before acidification (mg/L)	Lactic acid production by Lt (g/L)	pH after acidification	Molecular SO_2 after acidification (mg/L)	Protective effect
3.8	30	<0.5				Low
			2	3.6	0.65	Middle
			3	3.4	1.0	High

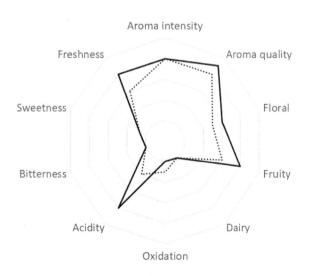

FIGURE 11.8 Typical sensory profile of sequential fermentation of *Lachancea thermotolerans* with *Saccharomyces cerevisiae* (continuous black line) compared with a control of single *S. cerevisiae* fermentation (dotted line).

that accompany MLF. However, this dairy profile is more correlated with some of the side metabolites of MLF such as diacetyl or acetoin. When water ethanolic solutions (12%−13% v/v) are added with lactic acid and tasted, the perceived acidity is quite citric with a marked refreshing profile. Therefore the production of lactic acid by Lt is a very useful biotool to improve the wine sensory profile by increasing freshness, especially in a global warming context, in which the winey and flat sensations have a high prevalence in many white wines. The sensory profile of a must from a neutral variety fermented by Sc or Lt shows typical differences, especially in the mouthfeel, due to the higher acidity, and also in the aromatic profile, depending on the strain (Fig. 11.8).

11.7.2 Aroma

Lt improves the freshness and fruitiness of wines by producing fruity and floral esters at higher levels than Sc. The contents of 10 fruity esters have been affected depending on the Lt strain including ethyl esters (ethyl propanoate, ethyl octanoate, ethyl decanoate, ethyl 9-decenoate, diethyl succinate), acetates (ethyl acetate, isobutyl acetate, isoamyl acetate, and 2-phenylethyl acetate), and amyl lactate (Hranilovic et al., 2018). The same authors, using Lt in the fermentation of Chardonnay musts, also found one strain (1/58) able to produce significant differences in the concentration of the norisoprenoid ß-damascenone and another influencing the amount of the terpenol β-citronellol. Lt has been described to increase ethyl isobutyrate (strawberry aroma) in sequential fermentations with Sc (Sgouros et al., 2020). In addition, sequential and mixed fermentations of Lt can improve the content of 2-phenylethyl acetate (rose petals) (Morata et al., 2019).

A recent publication analyzing the fermentative volatiles produced by six Lt strains fermenting at different SO_2 concentrations and at different temperatures has shown that, in general, the content of higher alcohols was slightly higher at a low temperature (17°C), regardless of the SO_2 content, and that the effect of the SO_2 content was not clear in terms of higher alcohol formation (Vaquero et al., 2020). Three of the six Lt produced more than 350 mg/L of higher alcohols, which may influence the sensory profile with a more vinous and flattened aroma. The majority of compounds were isoamyl alcohols ranging from 148 to 214 mg/L, below the perception threshold and with sweet fruit and berry aromas (Balikci, Tanguler, Jolly, & Erten, 2016; Saberi, Cliff, & van Vuuren, 2012).

As for the carbonylic compounds, including acetoin and diacetyl, they were generally higher at high temperatures. Their predominant compound was acetoin, although well below the perception threshold of 150 mg/L, which can confer buttery flavors (Peinado, Moreno, Medina, & Mauricio, 2004). No clear differences were observed in these volatiles with respect to the effect of SO_2 at low temperatures (Vaquero et al., 2020). However, at high temperatures, there was a clear significant trend toward the formation of larger amounts of these compounds at high SO_2 concentrations. It was observed that diacetyl, with a perception threshold between 4 and 12 mg/L, in fermentations at 17°C, was lower than in fermentations at 27°C.

The aromatic esters, which include isobutyl acetate, ethyl butyrate, ethyl lactate, isoamyl acetate, and 2-phenylethyl acetate, were also analyzed. In general, there was a clear influence of temperature at low SO_2 levels on the formation of these esters; they were found in higher concentrations in fermentations at 27°C (Vaquero et al., 2020). Ethyl butyrate and ethyl lactate appear in higher concentrations in the Lt fermentations (compared to the control fermentation of *S. cerevisiae*); these esters correspond to the descriptors of pineapple and strawberry/coffee, respectively. The higher content of ethyl lactate is a direct consequence of the strong production of lactic acid by most Lt strains. However, it should be noted that the sensory threshold of ethyl lactate is 150 mg/L, which shows sweet, lactic, and fruit aromas, and this value was not reached in the fermentations. It should be added that isoamyl acetate (sweet, banana, fruity with a hint of ripe aroma) was not detected and isobutyl acetate (sweet aroma of ethereal banana and tropical fruit) was below 1 mg/L (Gobbi et al., 2013).

The following extracellular enzymatic activities have been observed in strains of Lt with application in the formation or revelation of wine aroma or phenol extraction: Esterase, Esterase-Lipase, ß-glucosidase, Pectinase, Cellulase, Xylanase, and β-xylosidase (Escribano et al., 2017; Porter et al., 2019). The enzymes were expressed with variable intensity and with different substrate specificities. The release of the thiols 4MMP and 3MH has been reported as a result of its enzymatic activities (Zott et al., 2011).

11.7.3 Color

Acidity in white wines also has some impact in terms of color. Low pH wines usually have higher molecular SO_2 levels with the same total content. This helps to preserve phenols from oxidation, thus promoting the protection of paleness in white wines.

11.8 Biocontrol by *Lachancea thermotolerans*

Biocontrol is the antimicrobial effect produced by some yeast species against mainly fungal pathogens, but also against yeasts or bacteria. The main protective mechanisms in plants are competition, enzyme secretion, toxin production, volatiles, mycoparasitism, and induction of resistance (Freimoser, Rueda-Mejia, Tilocca, & Migheli, 2019; Suzzi, Romano, Ponti, & Montuschi, 1995). Lt has been described as a biocontrol agent against *Aspergillus*, inhibiting its development during grape fermentations without affecting the growth of Sc and wine quality (Nally et al., 2018). The use of Lt and its derivatives (cells or volatile organic compounds) when applied to pathogenic fungus (*A. parasiticus*, *P. verrucosum*, and *F. graminearum*) have shown effects not only on fungal growth but also on sporulation and toxins production (Zeidan, Ul-Hassan, Al-Thani, Balmas, & Jaoua, 2018). These studies open the possibilities for yeast-based biofungicides with protective activities in the vineyard or in grape fermentations.

The strong acidification produced by some Lt strains can be a protective mechanism to exclude some fermentative microorganisms during fermentation. Actually, as previously mentioned, MLF is strongly inhibited at a lactic acid of 4 g/L (Morata, Bañuelos, et al., 2020).

11.9 Special wines

11.9.1 Sweet wines

In sweet wines from concentrated musts, such as late harvest or ice wines, the use of Lt is an interesting way to balance palatability because excess sugar can be improved at a sensory level with suitable acidity. As previously stated, the citric profile produced by high levels of lactic acid helps to reduce the perception of sweetness and to obtain a natural freshness during fermentation (Morata et al., 2018). As the natural alcohol level obtained with Lt usually ranges from 8% to 10% vol., it can help to obtain a typically low ethanol content in these wines, thus preserving grape sugars during fermentation. Additionally, reducing the pH can help to lower the typical excessive SO_2 contents that are frequent in sweet wines to ensure their stability.

11.9.2 Sparkling wines

In warm areas, natural sparkling wines, with second fermentation and aging on lees in bottle, show reduced freshness by the low acidity contents of the base wines. Moreover, the development of unwanted MLF during the long aging on lees in the bottle is currently very likely. Lt is a powerful biotool for improving the quality of base wines for sparkling production, not only by increasing acidity but also by inhibiting subsequent unwanted MLF when lactic acid is at high levels (Morata, Bañuelos et al., 2020).

11.10 Conclusions

Lt is a powerful tool to modulate the flavor while improving the microbiological and physicochemical stability of white wines in warm areas. By controlling the pH, the proportion of molecular SO_2 is increased, producing more stable and healthier wines by reducing total SO_2 levels. Several available strains from biotechnological wine companies ensure a wide range of possibilities to cover several winemaking situations and allow the use of an appropriate strain for each white wine style. Early acidification and suitable adaptation to low fermentation temperatures and SO_2 contents provide good implantation in industrial fermentations.

References

Aponte, M., & Blaiotta, G. (2016). Potential role of yeast strains isolated from grapes in the production of taurasi DOCG. *Frontiers in Microbiology*, 7. Available from https://doi.org/10.3389/fmicb.2016.00809. (MAY).

Baleiras Couto, M. M., Reizinho, R. G., & Duarte, F. L. (2005). Partial 26S rDNA restriction analysis as a tool to characterise non-Saccharomyces yeasts present during red wine fermentations. *International Journal of Food Microbiology*, 102(1), 49−56. Available from https://doi.org/10.1016/j.ijfoodmicro.2005.01.005.

Balikci, E. K., Tanguler, H., Jolly, N. P., & Erten, H. (2016). Influence of *Lachancea thermotolerans* on cv. Emir wine fermentation. *Yeast (Chichester, England)*, 33(7), 313−321. Available from https://doi.org/10.1002/yea.3166.

Banilas, G., Sgouros, G., & Nisiotou, A. (2016). Development of microsatellite markers for *Lachancea thermotolerans* typing and population structure of wine-associated isolates. *Microbiological Research, 193*, 1−10. Available from https://doi.org/10.1016/j.micres.2016.08.010.

Barata, A., González, S., Malfeito-Ferreira, M., Querol, A., & Loureiro, V. (2008). Sour rot-damaged grapes are sources of wine spoilage yeasts. *FEMS Yeast Research*. Portugal: Blackwell Publishing Ltd. Available from https://doi.org/10.1111/j.1567−1364.2008.00399.x.

Binati, R. L., Lemos Junior, W. J. F., Luzzini, G., Slaghenaufi, D., Ugliano, M., & Torriani, S. (2020). Contribution of non-*Saccharomyces* yeasts to wine volatile and sensory diversity: A study on *Lachancea thermotolerans, Metschnikowia* spp. and *Starmerella bacillaris* strains isolated in Italy. *International Journal of Food Microbiology, 318*. Available from https://doi.org/10.1016/j.ijfoodmicro.2019.108470.

Chandra, M., Mota, M., Silva, A. C., & Malfeito-Ferreira, M. (2020). Forest oak woodlands and fruit tree soils are reservoirs of wine-related yeast species. *American Journal of Enology and Viticulture*. Available from https://doi.org/10.5344/ajev.2020.19067, ajev.2020.19067.

Comitini, F., Gobbi, M., Domizio, P., Romani, C., Lencioni, L., Mannazzu, I., & Ciani, M. (2011). Selected non-*Saccharomyces* wine yeasts in controlled multistarter fermentations with *Saccharomyces cerevisiae*. *Food Microbiology, 28*(5), 873−882. Available from https://doi.org/10.1016/j.fm.2010.12.001.

Escott, C., Morata, A., Ricardo-Da-Silva, J. M., Callejo, M. J., Del Carmen González, M., & Suarez-Lepe, J. A. (2018). Effect of *Lachancea thermotolerans* on the formation of polymeric pigments during sequential fermentation with *schizosaccharosmyces pombe* and *Saccharomyces cerevisiae*. *Molecules (Basel, Switzerland), 23*(9). Available from https://doi.org/10.3390/molecules23092353.

Escribano, R., González-Arenzana, L., Garijo, P., Berlanas, C., López-Alfaro, I., López, R., ... Santamaría, P. (2017). Screening of enzymatic activities within different enological non-*Saccharomyces* yeasts. *Journal of Food Science and Technology, 54*(6), 1555−1564. Available from https://doi.org/10.1007/s13197-017-2587-7.

Esteve-Zarzoso, B., Belloch, C., Uruburu, F., & Quero, A. (1999). Identification of yeasts by RFLP analysis of the 5.8S rRNA gene and the two ribosomal internal transcribed spacers. *International Journal of Systematic Bacteriology, 49*, 329−337. Available from https://doi.org/10.1099/00207713-49-1-329.

Freimoser, F. M., Rueda-Mejia, M. P., Tilocca, B., & Migheli, Q. (2019). Biocontrol yeasts: mechanisms and applications. *World Journal of Microbiology and Biotechnology, 35*(10). Available from https://doi.org/10.1007/s11274-019-2728-4.

García, M., Esteve-Zarzoso, B., Cabellos, J. M., & Arroyo, T. (2018). Advances in the study of Candida stellata. *Fermentation, 4*(3). Available from https://doi.org/10.3390/fermentation4030074.

Gobbi, M., Comitini, F., Domizio, P., Romani, C., Lencioni, L., Mannazzu, I., & Ciani, M. (2013). *Lachancea thermotolerans* and *Saccharomyces cerevisiae* in simultaneous and sequential co-fermentation: A strategy to enhance acidity and improve the overall quality of wine. *Food Microbiology, 33*(2), 271−281. Available from https://doi.org/10.1016/j.fm.2012.10.004.

Hernández, P. (2018). *Use of Lachancea thermotolerans to improve pH in red wines. Effect of the coinoculation with Oenococcus oeni* (presented at the Ms thesis).

Hranilovic, A., Bely, M., Masneuf-Pomarede, I., Jiranek, V., & Albertin, W. (2017). The evolution of *Lachancea thermotolerans* is driven by geographical determination, anthropisation and flux between different ecosystems. *PLoS One, 12*(9). Available from https://doi.org/10.1371/journal.pone.0184652.

Hranilovic, A., Gambetta, J. M., Schmidtke, L., Boss, P. K., Grbin, P. R., Masneuf-Pomarede, I., ... Jiranek, V. (2018). Oenological traits of *Lachancea thermotolerans* show signs of domestication and allopatric differentiation. *Scientific Reports, 8*(1). Available from https://doi.org/10.1038/s41598-018-33105-7.

Jolly, N. P., Varela, C., & Pretorius, I. S. (2014). Not your ordinary yeast: Non-*Saccharomyces* yeasts in wine production uncovered. *FEMS Yeast Research, 14*(2), 215−237. Available from https://doi.org/10.1111/1567-1364.12111.

Kioroglou, D., LLeixá, J., Mas, A., & del Carmen Portillo, M. (2018). Massive sequencing: A new tool for the control of alcoholic fermentation in wine? *Fermentation, 4*(1). Available from https://doi.org/10.3390/fermentation4010007.

Krieger-Weber, S., Heras, J. M., & Suarez, C. (2020). *Lactobacillus plantarum*, a new biological tool to control malolactic fermentation: A review and an outlook. *Beverages, 23*. Available from https://doi.org/10.3390/beverages6020023.

Kurtzman, C. P., & Robnett, C. J. (1998). Identification and phylogeny of ascomycetous yeasts from analysis of nuclear large subunit (26S) ribosomal DNA partial sequences. *Antonie Van Leeuwenhoek*. Available from https://doi.org/10.1023/A:1001761008817.

Kurtzman, C. P. (2003). Phylogenetic circumscription of *Saccharomyces, Kluyveromyces* and other members of the *Saccharomycetaceae*, and the proposal of the new genera Lachancea, Nakaseomyces, Naumovia, Vanderwaltozyma and Zygotorulaspora. *FEMS Yeast Research, 4*(3), 233−245. Available from https://doi.org/10.1016/S1567-1356(03)00175-2.

Lachance, M. A., & Kurtzman, C. P. (2011). LachanceaKurtzman, *The yeasts* (Vol. 2, pp. 511−519). Canada: Elsevier (2003). Available from https://doi.org/10.1016/B978-0-444-52149-1.00041-0.

Loira, I., Morata, A., Bañuelos, M. A., & Suárez-Lepe, J. A. (2019). Isolation, selection, and identification techniques for non-Saccharomyces yeasts of oenological interest. *Biotechnological progress and beverage consumption: volume 19: The science of beverages* (pp. 467−508). Spain: Elsevier. Available from https://doi.org/10.1016/B978-0-12-816678-9.00015-1.

Lopandic, Ksenija, Tiefenbrunner, Wolfgang, Gangl, Helmut, Mandl, Karin, Berger, Susanne, Leitner, Gerhard, ... Prillinger, Hansjörg (2008). Molecular profiling of yeasts isolated during spontaneous fermentations of Austrian wines. *FEMS Yeast Research, 8*(7), 1063−1075. Available from https://doi.org/10.1111/j.1567-1364.2008.00385.x.

Morata, A., Bañuelos, M. A., López, C., Song, C., Vejarano, R., Loira, I., ... Suarez Lepe, J. A. (2020). Use of fumaric acid to control pH and inhibit malolactic fermentation in wines. *Food Additives and Contaminants - Part A Chemistry, Analysis, Control, Exposure and Risk Assessment, 37*(2), 228−238. Available from https://doi.org/10.1080/19440049.2019.1684574.

Morata, A., Bañuelos, M. A., Vaquero, C., Loira, I., Cuerda, R., Palomero, F., ... Bi, Y. (2019). *Lachancea thermotolerans* as a tool to improve pH in red wines from warm regions. *European Food Research and Technology, 245*(4), 885−894. Available from https://doi.org/10.1007/s00217-019-03229-9.

Morata, A., Escott, C., Bañuelos, M. A., Loira, I., Del Fresno, J. M., González, C., & Suárez-lepe, J. A. (2020). Contribution of non-*Saccharomyces* yeasts to wine freshness. A review. *Biomolecules, 10*(1). Available from https://doi.org/10.3390/biom10010034.

Morata, A., Loira, I., Tesfaye, W., Bañuelos, M. A., González, C., & Suárez Lepe, J. A. (2018). *Lachancea thermotolerans* applications in wine technology. *Fermentation, 4*(3). Available from https://doi.org/10.3390/fermentation4030053.

Morata, A., & Suárez-Lepe, J. A. (2016). New Biotechnologies for wine fermentation and ageing. *Advances in Food Biotechnology*. Available from https://doi.org/10.1002/9781118864463.ch17.

Nally, M. C., Ponsone, M. L., Pesce, V. M., Toro, M. E., Vazquez, F., & Chulze, S. (2018). Evaluation of behaviour of *Lachancea thermotolerans* biocontrol agents on grape fermentations. *Letters in Applied Microbiology*, *67*(1), 89−96. Available from https://doi.org/10.1111/lam.13001.

Peinado, R. A., Moreno, J., Medina, M., & Mauricio, J. C. (2004). Changes in volatile compounds and aromatic series in sherry wine with high gluconic acid levels subjected to aging by submerged flor yeast cultures. *Biotechnology Letters*, *26*(9), 757−762. Available from https://doi.org/10.1023/B:BILE.0000024102.58987.de.

Porter, T. J., Divol, B., & Setati, M. E. (2019). *Lachancea* yeast species: Origin, biochemical characteristics and oenological significance. *Food Research International*, *119*, 378−389. Available from https://doi.org/10.1016/j.foodres.2019.02.003.

Prakitchaiwattana, C. J., Fleet, G. H., & Heard, G. M. (2004). Application and evaluation of denaturing gradient gel electrophoresis to analyse the yeast ecology of wine grapes. *FEMS Yeast Research*, *4*(8). Available from https://doi.org/10.1016/j.femsyr.2004.05.004.

Saberi, S., Cliff, M. A., & van Vuuren, H. J. J. (2012). Impact of mixed *S. cerevisiae* strains on the production of volatiles and estimated sensory profiles of Chardonnay wines. *Food Research International*, *48*(2), 725−735. Available from https://doi.org/10.1016/j.foodres.2012.06.012.

Schnierda, T., Bauer, F. F., Divol, B., van Rensburg, E., & Görgens, J. F. (2014). Optimization of carbon and nitrogen medium components for biomass production using non-*Saccharomyces* wine yeasts. *Letters in Applied Microbiology*, *58*(5), 478−485. Available from https://doi.org/10.1111/lam.12217.

Sgouros, G., Mallouchos, A., Filippousi, M. E., Banilas, G., & Nisiotou, A. (2020). Molecular characterization and enological potential of a high lactic acid-producing *Lachancea thermotolerans* vineyard strain. *Foods*, *9*(5). Available from https://doi.org/10.3390/foods9050595.

Suzzi, G., Romano, P., Ponti, I., & Montuschi, C. (1995). Natural wine yeasts as biocontrol agents. *Journal of Applied Bacteriology*, *78*(3), 304−308. Available from https://doi.org/10.1111/j.1365-2672.1995.tb05030.x.

Vaquero, C., Loira, I., Bañuelos, M. A., Heras, J. M., Cuerda, R., & Morata, A. (2020). Industrial performance of several *Lachancea thermotolerans* strains for pH control in white wines from warm areas. *Microorganisms*, *8*(6). Available from https://doi.org/10.3390/microorganisms8060830.

Vaquero, C., Loira, I., Heras, J. M., Carrau, F., González, C., & Morata, A. (2021). Biocompatibility in ternary fermentations with *Lachancea thermotolerans*, other non-*Saccharomyces* and *Saccharomyces cerevisiae* to control pH and improve the sensory profile of wines from warm areas. *Front. Microbiol.*, *12*656262. Available from https://doi.org/10.3389/fmicb.2021.656262.

Vilela, A. (2018). *Lachancea thermotolerans*, the non-*Saccharomyces* yeast that reduces the volatile acidity of wines. *Fermentation*, *4*(3). Available from https://doi.org/10.3390/fermentation4030056.

Zara, G., Ciani, M., Domizio, P., Zara, S., Budroni, M., Carboni, A., & Mannazzu, I. (2014). A culture-independent PCR-based method for the detection of *Lachancea thermotolerans* in wine. *Annals of Microbiology*, *64*. Available from https://doi.org/10.1007/s13213-013-0647-4.

Zeidan, R., Ul-Hassan, Z., Al-Thani, R., Balmas, V., & Jaoua, S. (2018). Application of low-fermenting yeast *Lachancea thermotolerans* for the control of toxigenic fungi *Aspergillus parasiticus*, *Penicillium verrucosum* and *Fusarium graminearum* and their mycotoxins. *Toxins*, *10*(6). Available from https://doi.org/10.3390/toxins10060242.

Zott, K., Thibon, C., Bely, M., Lonvaud-Funel, A., Dubourdieu, D., & Masneuf-Pomarede, I. (2011). The grape must non-*Saccharomyces* microbial community: Impact on volatile thiol release. *International Journal of Food Microbiology*, *151*(2), 210−215. Available from https://doi.org/10.1016/j.ijfoodmicro.2011.08.026.

12

Nitrogen management during fermentation

Jean-Marie Sablayrolles and Jean-Roch Mouret

SPO-Sciences For Enology, Univ Montpellier, Montpellier SupAgro, Montpellier, France

12.1 Introduction

During alcoholic fermentation, the key role of yeast assimilable nitrogen (YAN) is now widely recognized, and most winemakers consider it as one of the major parameters to control during fermentation. However, optimized management still remains difficult. Indeed, especially in white winemaking, the objective is to improve fermentation kinetics but also more and more to act on wine aromatic characteristics. In addition, in recent years, (1) many new nutrients based on organic nitrogen have been developed, and their effect is sometimes different from the usual ammoniacal salts; and (2) more and more different yeasts are used, including non-*Saccharomyces* yeasts, whose nitrogen metabolism is still not completely understood, and such yeasts are sometimes present in the mixture. The objective of this chapter is to synthesize current knowledge relating to YAN and its management during fermentation, which has evolved considerably in recent years. After describing the nitrogen metabolism and its variability according to strains and yeast species, we will tackle the question of nitrogen management during fermentation. This question is challenging because it must take into account the characteristics of the raw material and those of the targeted product and must also consider possible interactions with other nutrients.

12.2 Nitrogen metabolism in yeasts

The major sources of assimilable nitrogen in grape must are amino acids and ammonium (Bell & Henschke, 2005; Bely, Sablayrolles, & Barre, 1990; Torrea et al., 2011). These nutrients are used by yeasts not only to support growth but also to synthesize volatile compounds involved in wine flavor (Gobert, Tourdot-Maréchal, Sparrow, Morge, & Alexandre, 2019).

12.2.1 Saccharomyces cerevisiae

Within the species *S. cerevisiae*, the consumption kinetics of total nitrogen and nitrogen requirements are strain dependent (Brice, Sanchez, Tesnière, & Blondin, 2014; Crépin, Nidelet, Sanchez, Dequin, & Camarasa, 2012). Nevertheless, the order of consumption of nitrogen sources is very similar for the different studied strains (Crépin et al., 2012), indicating that the molecular mechanisms regulating nitrogen consumption are relatively conserved in this yeast species.

12.2.1.1 Assimilation of nitrogen sources

Amino acids and ammonium are transported into the cell by various transporters. Ammonium is transported by three permeases: Mep1p, Mep2p, and Mep3p. These three proteins consist of uniport systems and have different affinities for this substrate (Gobert et al., 2019). Amino acids are assimilated by various, more or less, selective transporters. Amino-acid permeases (AAPs) are active symport systems (Gobert et al., 2019). Among them, the

© 2022 Elsevier Inc. All rights reserved.

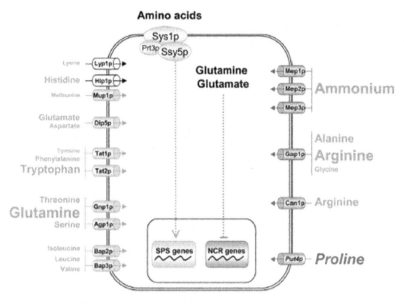

FIGURE 12.1 The transporters represented include carriers of one or a few specific amino acid (Lyp1p, Hip1p, Mup1, Dip5p, Tat2p, Gnp1, Agp1p, Bap3p, and Can1p) and the general amino acid carriers Gap1p, the arginine transporter Can1p, and the ammonium permeases (Mep1p, Mep2p, and Mep3). The regulatory mechanisms controlling the expression of nitrogen permease genes are indicated in orange (SPS, on the left side) or in green (NCR, on the right-hand side). The order of nitrogen source consumption during wing fermentation is indicated at the bottom.

general amino-acid permease (Gap1p) allows the transport of all amino acids. Other AAPs (presented in Fig. 12.1) are more selective and transport only one or a group of amino acids.

These different nitrogen transporters are regulated by two mechanisms (Crépin et al., 2012; Gobert et al., 2019; Zhang, Du, Zhou, & Chen, 2018). The major regulation mechanism, called nitrogen catabolite repression (NCR), prevents the uptake of the poorest nitrogen sources when better sources are available. The second mechanism involving the SPS sensor regulates the expression of genes encoding for transporters specific to certain amino acids. These two regulation systems make it possible for the yeast to selectively use preferred sources of nitrogen when they are available in the grape must. Then, when these substrates are exhausted, the other nitrogen compounds are consumed. So it is possible to classify the nitrogenous nutrients into "preferred" and "non-preferred" nitrogen sources. Different methods have been used to make this classification (Gobert et al., 2019); the most recent one is based on both NCR and SPS mechanisms (Crépin et al., 2012).

From this ranking (presented in Fig. 12.1), nitrogen compounds can be separated into three groups, according to their order of use: prematurely consumed (Lys), early consumed (Asp, Thr, Glu, Leu, His, Met, Ile, Ser, Gln, and Phe), and late consumed (ammonium, Val, Arg, Ala, Trp, and Tyr).

Early consumed amino acids are transported by specific permeases under SPS-mediated control, which are expressed at the beginning of yeast growth. Most nitrogen compounds consumed late are transported by permeases under NCR control; others (Val, Trp, and Tyr) are transported by SPS-regulated low-affinity permeases (Crépin et al., 2012). This order of assimilation depends neither on the concentration of nitrogen compounds available in the medium nor on the strain used (14 strains tested).

12.2.1.2 Role of nitrogen in volatile compound metabolism

The production of volatile compounds involved in wine flavor is highly impacted by nitrogen availability. At first, some fermentative aromas (higher alcohols and acetate esters, involved in wine fruity flavor) are directly produced from nitrogen compounds (Fig. 12.2), via the Ehrlich pathway (Hazelwood, Daran, Van Maris, Pronk, & Dickinson, 2008; Hirst & Richter, 2016; Styger, Jacobson, Prior, & Bauer, 2013).

For each amino acid, this metabolic pathway consists of three steps: (1) transamination of amino acid into keto acid, (2) decarboxylation into aldehyde, and (3) conversion in either acid or higher alcohol (Hazelwood et al., 2008). Finally, higher alcohol is converted into acetate ester by alcohol acetyltransferases, encoded by the genes AT1 and ATF2 (Verstrapen et al., 2003). It is important to note that keto acids precursor of higher alcohols can be produced both from the catabolism of amino acids and from central carbon metabolism (as shown in Fig. 12.2).

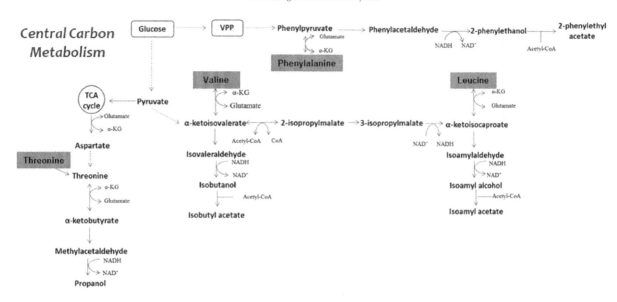

FIGURE 12.2 Scheme of the biosynthesis of higher alcohols and acetate esters.

Therefore even if several studies have shown a significant correlation between initial nitrogen content and the final concentration of fermentative aromas in wine (Carrau et al., 2008; Fairbairn, McKinnon, Musarurwa, Ferreira, & Bauer, 2017; Hernández-Orte, Ibarz, Cacho, & Ferreira, 2005; Mouret et al., 2014; Rollero et al., 2015), recent studies suggest that this link may be much weaker than previously thought (Crépin et al., 2017; Rollero et al., 2017). Indeed, the studies performed, using labeled ^{13}C compounds, have shown that a very small share (less than 10%) of higher alcohols (and corresponding acetate esters) are produced using the carbon skeletons of amino acids (Crépin et al., 2017; Rollero et al., 2017). This last result is coherent with two additional observations: (1) consumed amino acids are mainly used for *de novo* synthesis of proteinogenic amino acids (Crépin et al., 2017) and (2) with the exception of propanol, higher alcohols are mainly synthesized during the stationary phase, i.e., when nitrogen is exhausted (Mouret et al., 2014).

Secondly, the production of varietal thiols, involved in fruity flavor, is also impacted by nitrogen (Gobert et al., 2019). 4-Methyl-4-sulfanylpentan-2-one (4MSP) and 3-sulfanyl-hexan-1-ol (3SH) are produced from inodorous nonvolatile precursors by *S. cerevisiae* during alcoholic fermentation and involve a carbon—sulfur β-lyase activity (Thibon et al., 2008; Tominaga, Peyrot des gachons, & Dubourdieu, 1998). Irc7p, a putative cystathionine β-lyase, is the main protein catalyzing the release of 4MSP and 3SH (Thibon et al., 2008). It has been shown that the transcriptional regulation of IRC7 is associated with the NCR system (Thibon et al., 2008). The NCR system is also involved at another step of the process of thiol genesis: the uptake of the thiol precursors into the cell. Therefore in presence of ammonium, the activity of Gap1p is inhibited; it results in the limitation of the transport of the precursors into the cell and in a lower final concentration of 4MSP and 3SH in wine (Subileau, Schneider, Salmon, & Degryse, 2008).

12.2.2 Non-Saccharomyces yeasts

The nitrogen metabolism of non-*Saccharomyces* (NS) yeasts during winemaking fermentation has been studied only recently. It is not possible to present a general overview of nitrogen metabolism of NS yeasts because the nitrogen requirement and the ability to consume a particular source of nitrogen depend on the NS species as well as the strain (Englezos et al., 2018; Gobert et al., 2017; Harle et al., 2020; Kemsawasd, Viana, Ardö, & Arneborg, 2015; Su et al., 2020). For example, for three strains of *Torulaspora delbrueckii*, *Metschnikowia pulcherrima*, and *Metschnikowia fructicola*, it has been shown that the assimilation sequence of nitrogen sources is generally comparable to the one obtained for *S. cerevisiae* even if the time to consume the nitrogen resources is very different (Su et al., 2020). On the opposite side, for a strain of *Starmerella bacillaris* (Englezos et al., 2018), the assimilation sequence changes completely: amino acids are poorly assimilated whereas ammonium is entirely consumed. Moreover, this specific strain synthesizes several amino acids during its growth (such as alanine, glutamic acid, glycine, and valine), whereas such a phenomenon has never been reported before.

Whatever the tested NS strain, it appears that the *S. cerevisiae* and the NS yeast strains compete for nitrogen resources and that a significant proportion of YAN can be assimilated by the NS yeast strain before the predominance of *S. cerevisiae* in the fermentation (Harle et al., 2020). As a consequence, the growth and fermentation kinetics of *S. cerevisiae* may be negatively affected (Gobert et al., 2017; Kemsawasd et al., 2015; Medina, Boido, Dellacassa, & Carrau, 2012; Rollero, Bloem, Ortiz-Julien, Camarasa, & Divol, 2018). In particular, it has been shown that NS yeasts can consume between 66 and 215 mg/L of YAN, depending on the species (Gobert et al., 2017). In some cases, this competition for resources can lead to sluggish or stuck fermentation (Medina et al., 2012). Therefore when using an NS yeast strain and an *S. cerevisiae* strain in sequential inoculation, it appears essential to add nitrogen to the grape must to cover the needs of *S. cerevisiae* and get an optimal alcoholic fermentation.

12.3 Assimilable nitrogen in musts

As said previously, YAN is primarily constituted by α-amino nitrogen and ammonium ions (Bell & Henschke, 2005; Bely et al., 1990; Torrea et al., 2011) but peptides can also contribute (Becerra-Rodríguez, Marsit, & Galeote, 2020).

12.3.1 YAN measurement

YAN measurement is not obvious and the simplest methods are not fully satisfactory. Formol titration does not distinguish between the nitrogen from amino acids (i.e., free amino nitrogen, FAN) and ammonium ions (i.e., inorganic nitrogen). Furthermore, it tends to underestimate YAN (Filipe-Ribeiro & Mendes-Faia, 2007). A method such as NOPA (nitrogen by o-phthaldialdehyde) (Gump, Zoecklein, Fugelsang, & Whiton, 2002) paired with enzymatic ammonia may be preferred, as it enables the determination of not only the total amount of YAN but also the proportion of FAN to inorganic nitrogen. Unfortunately, YAN is also usually underestimated.

Different measurement methods of YAN in grape musts have been compared, using actual assimilated nitrogen as a reference method (Casalta, Sablayrolles, & Salmon, 2013). The authors concluded that using an amino acids analyzer to quantify individual amino acids and an enzymatic assay for determining ammoniacal nitrogen was more reliable. Individual amino acid determination combines high-performance liquid chromatography (HPLC) methods (Dukes & Butzke, 1998) with different detection methods. The more widely used derivatizing agents are ninhydrin and OPA, but different others (Callejón, Troncoso, & Morales, 2010) are available. It is noteworthy that two main pitfalls in the determination of assimilable nitrogen in musts are imino acids and arginine, with 4N but only 3 of them being assimilated (Casalta, Sablayrolles, et al., 2013). Such analyses are time consuming and remain expensive. Furthermore, they are not suitable for big data collection (Petrovic, Aleixandre-Tudo, & Buica, 2019b). Infrared (IR) spectroscopy, coupled with chemometrics, is undoubtedly an excellent alternative for routine analysis (Petrovic, Aleixandre-Tudo, & Buica, 2020) with the best compromise between convenience and reliability. It is now the most used method in wine laboratories but it remains a correlative method whose accuracy is fully dependent on the quality of the database used in method development.

12.3.2 YAN concentrations in musts

YAN concentrations can vary considerably depending on the geographic location (Rapp & Versini, 1995), the climate (Ribéreau-Gayon, Glories, Maujean, & Dubourdieu, 2006), the cultivar (Paul Schreiner, Osborne, & Skinkis, 2018), and viticulture techniques (Spayd et al., 1994). The assimilable nitrogen content of 600 musts from different Italian regions ranged from a few tens of mg/L to almost 400 mg/L, with an average concentration of 136 mg/L (Nicolini, Larcher, & Versini, 2004). This average value is much lower than that (213 mg/L) reported (Butzke, 1998) for 1500 musts from Oregon, California, and Washington. In the musts from the main French wine regions (30 cultivars, 3 harvests), the reported values varied from 53 to 444 mg/L, with an average value of 175 mg/L (Bely et al., 1990). Recently, 805 commercial juices from South Africa were analyzed (Petrovic et al., 2019b). YAN concentrations ranged from 45 to 484 mg/L (mean: 191 mg/L).

The percentages of ammoniacal nitrogen are generally between 25% and 30%. Usually, a decrease in YAN during maturation caused mainly by a decrease in ammonia is observed (Dubois, Manginot, Roustan, Sablayrolles, & Barre, 1996; Nisbet, Martinson, & Mansfield, 2014). The amino acid composition varies widely, between cultivars, arginine and proline being generally the two main ones. However, it is very difficult to statistically discriminate between grape varieties according to their amino acid content (Petrovic, Aleixandre-Tudo, & Buica, 2019a). All these data confirm the great variability in the nitrogen composition of musts, whatever the country or even the region, and therefore the importance of nitrogen management during fermentation.

12.3.3 Effect of YAN on the kinetics of fermentation

A direct relationship between the concentrations of assimilable nitrogen and the maximum fermentation rates has been described (Bely et al., 1990), indicating that assimilable nitrogen is the main limiting nutrient in musts and mainly responsible for "slow fermentations" characterized by a low fermentation rate throughout the process. These fermentations must be distinguished from sluggish or stuck fermentations (Blateyron & Sablayrolles, 2001), which are characterized by a very slow fermentation rate only at the end due to the very low viability of yeasts, whose probability increased when the initial sugar concentration was high or when the limiting nutrient was not nitrogen. This last phenomenon was recently explained (Duc et al., 2017; Duc, Noble, Tesnière, & Blondin, 2019; Tesnière, Delobel, Pradal, & Blondin, 2013). These authors demonstrated that nitrogen availability plays a key role in yeast cell death in interaction with micronutrient limitations such as lipids or vitamins and concluded that the risks of cell mortality are lower in the case of nitrogen-limited musts.

12.3.4 Nitrogen is not always limiting

In most cases, assimilable nitrogen is the limiting nutrient, but there are exceptions, especially in white winemaking. Two main phenomena can be responsible for such situations:

- very high concentrations of assimilable nitrogen in the musts. Above about 400 mg/L, nitrogen has no positive effect on the kinetics of fermentation (Malherbe, Fromion, Hilgert, & Sablayrolles, 2004) and can therefore be considered as non-limiting from a metabolic point of view. This value must be distinguished from the classic value of 140 mg/L, which represents a technological nitrogen deficiency threshold (Beltran, Esteve-Zarzoso, Rozès, Mas, & Guillamón, 2005; Bely et al., 1990; Bisson, 1999; Kemsawasd et al., 2015);
- low concentrations of other nutrients. Several nutrients can be limiting, such as vitamins, especially thiamine in the case of contaminated musts (Bataillon, Rico, Sablayrolles, Salmon, & Barre, 1996). However, the case of highly clarified musts represents, by far, the situation with the main technological consequences in white winemaking. Indeed, lipids, particularly in the forms of phytosterols and fatty acids, are a major source of nutrients for fermenting yeasts and, due to their strong hydrophobic status, lipids are mostly provided by the solid particles. An interaction between grape solids and assimilable nitrogen content has been highlighted (Casalta, Cervi, Salmon, & Sablayrolles, 2013): adding solids in a solid-free must had a very positive effect on the fermentation kinetics, especially in the case of nitrogen-rich musts. The authors explained that the addition of solids resulted in increased consumption of assimilable nitrogen and consequently in a higher fermentation rate and cell population. This phenomenon was confirmed (Ochando, Mouret, Humbert-Goffard, Sablayrolles, & Farines, 2017) by adding phytosterols instead of solids. To exhaust nitrogen in musts with 50 mg/L and 250 mg/L YAN, 3 and 8 mg/L of phytosterols, respectively, were necessary. In practice, such conditions are not always met, and musts excessively clarified and lipid limited are quite common in white winemaking.

12.4 Nitrogen management

Adding nitrogen has become common practice. Ammonium salts are still the main sources, although organic sources are increasingly used. These additions are very effective on the fermentation parameters and can also act on the fermentation aromas, which are of primary interest in white winemaking. In all cases, these additions must be thoughtful and controlled.

12.4.1 Control of fermentation kinetics

Many studies describe the consequences of nitrogen additions on the fermentation parameters, and some of them are contradictory (Gobert et al., 2019). To go further into understanding this, it is necessary to distinguish between different situations.

12.4.1.1 Case of "slow fermentations"

Slow fermentations are common. As described above, they are characterized by nitrogen as the only limiting nutrient. In such fermentation, the addition of ammonium salts (diammonium phosphate [DAP] or ammonium sulfate) is a highly efficient way to increase fermentation rate and lower fermentation duration (Carrau et al., 2008; Malherbe et al., 2004; Miller, Wolff, Bisson, & Ebeler, 2007; Ugliano, Siebert, Mercurio, Capone, & Henschke, 2008), but the timing of this addition is crucial: (i) if nitrogen is added at the time of inoculation, it is metabolized and used for additional yeast growth. Initial supplementations are justified mainly in the case of musts with very low YAN; (ii) if added at the start of the stationary phase, it is mostly used to reactivate the existing yeasts (Sablayrolles, 2019). Compared to the initial addition, its efficiency may be equivalent (Bely et al., 1990) or higher, particularly in the case of sugar-rich musts (Seguinot et al., 2018), as indicated in Fig. 12.3.

Ammonium salts are sometimes replaced by organic sources (Garde-Cerdán & Ancín-Azpilicueta, 2008; Hernández-Orte et al., 2005; Martínez-Moreno, Morales, Gonzalez, Mas, & Beltran, 2012). By comparing with the same amount of assimilable nitrogen added, it was shown that the efficiency of the additions of ammoniacal and amino nitrogen was identical (Seguinot et al., 2018).

12.4.1.2 Case of sluggish and stuck fermentations

Despite improvements in fermentation management, such as using selected strains and controlling temperature, risks of sluggish (without residual sugar) or stuck (with remaining residual sugar) fermentations persist, mostly because of an increase in the average sugar content of the musts.

Adding nitrogen and oxygen makes it possible to combine the positive effect of nitrogen on the activity of yeasts with that of oxygen on their survival. However, this addition must be correctly timed, the best moment being at the start of the stationary phase, when about 5% ethanol has been produced (Blateyron, Julien, & Sablayrolles, 2000). The effectiveness of such combined supplementation was evaluated on 72 musts leading to sluggish or stuck fermentations (Blateyron & Sablayrolles, 2001). In all cases, such additions led to (i) a dramatic decrease (by 44% on average) in the duration of the fermentation (for sluggish fermentations) or (ii) sugar exhaustion (for stuck fermentations). This effect was independent of the variety and origin of the must and the

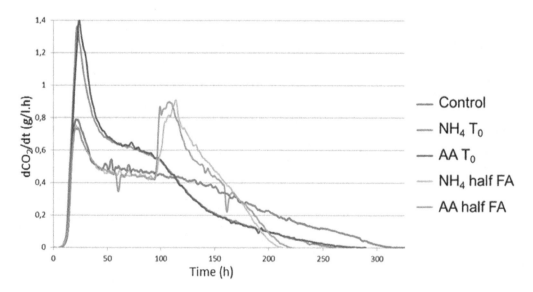

FIGURE 12.3 Changes in the fermentation kinetics depending on the addition of nitrogen: control without addition (red curve); addition at the beginning of fermentation of DAP (dark green curve) or amino acids (dark blue curve); addition during the stationary phase of DAP (light green curve) or amino acids (light blue curve).

yeast strain, and the authors did not observe a negative impact on the quality of the product. Therefore they considered that this supplementation could be carried out even in the absence of a precise diagnosis of the type of fermentation problem.

12.4.1.3 Case of very clarified musts

As previously described, the addition of solid particles to highly clarified musts dramatically increases the fermentation rate (Fig. 12.4). This effect is due to greater consumption of assimilable nitrogen with a switch from a lipid deficiency to a nitrogen deficiency (Fig. 12.5) Therefore it can be considered that, in the case of nitrogen-rich musts, the best strategy is to combine (i) sufficient turbidity, i.e., 50−150 NTU, to increase nitrogen assimilation and therefore increase the fermentation rate and cell population and move toward a nitrogen-limited situation with (ii) oxygen addition, which is the best way to improve yeast viability (Sablayrolles, 2019). It is also possible to use commercial products that are combinations of different nutrients, including at least ammoniacal nitrogen, thiamine, and inactivated yeasts. Some activators containing inactivated dry yeasts (IDY) are usable during the rehydration phase because they facilitate the rehydration of active dry yeasts (ADY) due to the transfer of sterols from IDY to ADY (Soubeyrand et al., 2005).

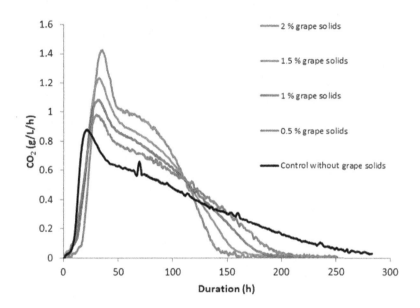

FIGURE 12.4 Changes in the fermentation kinetics depending on the addition of grape solids in a highly clarified N-rich must (425 mg/L): control without grape soldis (dark); addition of 0.5% (blue), 1% (red), 1.5% (green) or 2% grape solids.

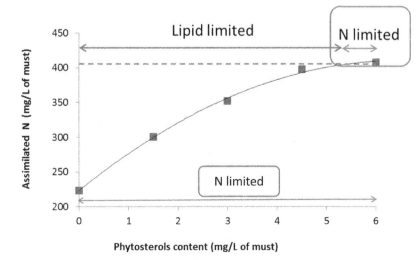

FIGURE 12.5 Effect of adding phytosterols on nitrogen assimilation. Must with low nitrogen content (<200 mg/L) (green): always nitrogen limitation. Switch from lipid limitation to nitrogen limitation when 5 mg/L phytosterols were added.

12.4.2 Impact on production of fermentative aromas by Saccharomyces cerevisiae

In this part, we are going to present the impact of the addition of nitrogen on the synthesis of fermentative aromas when nitrogen is the limiting nutrient of the fermentation process, i.e., when the must has not been highly clarified. To our knowledge, the impact of nitrogen addition when lipids are the limiting nutrient has never been evaluated.

Many studies have been performed within the scientific community concerning the impact of nitrogen on aroma production (Beltran et al., 2005; Carrau et al., 2008; Garde-Cerdán & Ancín-Azpilicueta, 2008; Hernández-Orte et al., 2005; Jiménez-Martí, Aranda, Mendes-Ferreira, Mendes-Faia, & li del Olmo, 2007; Mouret et al., 2014; Rollero et al., 2015; Seguinot et al., 2018; Torrea et al., 2011; Ugliano et al., 2008). The obtained results are very dispersed and sometimes contradictory (Gobert et al., 2019), mainly because the strain, grape must, and operating conditions are very different depending on the studies. Nevertheless, some trends can be highlighted: (1) the impact of nitrogen addition differs depending on the chemical family of the considered aroma (higher alcohols, acetate esters, ethyl esters); (2) the impact of nitrogen addition depends on both the timing of addition (at the beginning of the process or during the stationary phase) and the quality (ammonium or amino acids) of added nitrogen (Seguinot et al., 2018).

For an initial addition of nitrogen, it has been shown that adding nitrogen to must with low nitrogen content increases the production of higher alcohols (Barbosa, Falco, Mendes-Faia, & Mendes-Ferreira, 2009; Mouret et al., 2014; Rollero et al., 2015). However, when the assimilable nitrogen concentration exceeds $200-300$ mg/L, the production of higher alcohols decreases (Mouret et al., 2014; Rollero et al., 2015). The quality of added nitrogen (ammonium or amino acids) does not impact the synthesis of higher alcohols (Seguinot et al., 2018). Propanol does not have the same behavior as the other higher alcohols. Its production ends when the assimilable nitrogen is exhausted and is proportional to the initial amount of nitrogen (Mouret et al., 2014; Rollero et al., 2015). It should be noted that propanol production also depends on the nature of added nitrogen; its synthesis is higher following an ammonium addition compared to an amino acid addition (Seguinot et al., 2018). These data indicate that propanol can be considered as a marker of both the quantity and quality of available nitrogen (mineral or organic). Acetate esters and ethyl esters show a simpler relationship with nitrogen concentration: an increase in initial nitrogen content is associated with an increase in ester production (Hernández-Orte et al., 2005; Mouret et al., 2014; Rollero et al., 2015; Torrea et al., 2011). For acetate esters, adding organic nitrogen slightly increases their production compared to the supply of ammonium. For ethyl ester, the effect of the nature of nitrogen differs depending on the studied molecule: for ethyl hexanoate, synthesis is slightly higher when amino acids are added compared to ammonium; for ethyl octanoate, an opposite finding is observed (Seguinot et al., 2018).

Following an addition during the stationary phase (Fig. 12.6), the production of ethyl esters is slightly higher (Seguinot et al., 2018); the synthesis of higher alcohols remains unchanged, whereas acetate esters are overproduced (Hernández-Orte et al., 2005; Seguinot et al., 2018).

12.5 Prospect

Optimizing nitrogen management by taking into account its effect on the fermentation kinetics and on the aromatic quality of the wine as well as interactions with other parameters such as temperature is very complex. To tackle these challenges, an innovative strategy has already been developed (Aceves Lara et al., 2018) as a proof of concept. It is based on (1) online monitoring of the kinetics of the alcoholic fermentation and the major fermentative aromas; (2) the integration of different models predicting the fermentation kinetics, with a key role of nitrogen (Malherbe et al., 2004), the power required to cool the tank (Colombié, Malherbe, & Sablayrolles, 2007), the gas—liquid ratio (Morakul et al., 2011), and the production kinetics of five fermentative aromas (Mouret, Farines, Sablayrolles, & Trelea, 2015); and (3) laws of control optimization looking for compromises between the main reaction (fermentation duration, tank, and energy use) and aroma production (maximization of production, minimization of losses).

In addition, it will also be necessary to consider the variability between strains (S. cerevisiae or non-Saccharomyces) and to take into account mixed cultures or natural ecosystems. Fortunately, the effects of nitrogen are quite comparable depending on the strains of S. cerevisiae (reference). On the other hand, knowledge is still far too partial with regard to non-Saccharomyces strains and, even more, mixtures of strains.

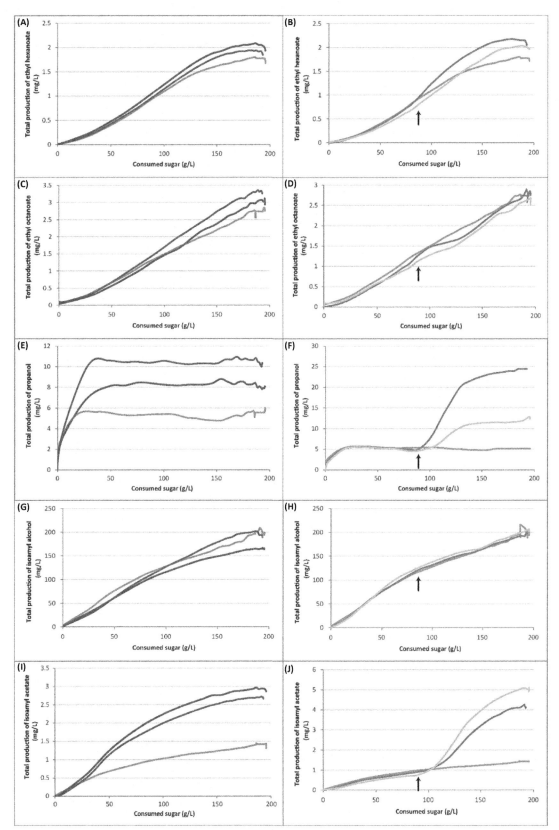

FIGURE 12.6 Changes in the production kinetics of ethyl hexanoate, ethyl octanoate, propanol, isoamyl alcohol, and isoamyl acetate depending on the addition of nitrogen: control without addition (green curve); addition at the beginning of fermentation of DAP (dark blue curve) or amino acids (red curve); addition during the stationary phase of DAP (light blue curve) or amino acids (orange curve). (*For interpretation of the references to color in this figure legend, see the color version*).

References

Aceves Lara, C., Athes, V., Buche, P., Della Valle, G., Farines, V., Fonseca, F., ... Thomopoulos, R. (2018). The virtual food system: Innovative models and experience feedback in the technologies of winemaking, cereals chain, packaging and eco-design of starter production. *Innovative Food Science and Emerging Technologies, 46*, 54−64.

Barbosa, C., Falco, V., Mendes-Faia, A., & Mendes-Ferreira, A. (2009). Nitrogen addition influences formation of aroma compounds, volatile acidity and ethanol in nitrogen deficient media fermented by *Saccharomyces cerevisiae* wine strains. *Journal of Bioscience and Bioengineering, 108*, 99−104.

Bataillon, M., Rico, A., Sablayrolles, J. M., Salmon, J. M., & Barre, P. (1996). Early thiamin assimilation by yeasts under enological conditions: Impact on alcoholic fermentation kinetics. *Journal of Fermentation and Bioengineering, 82*(2), 145−150. Available from https://doi.org/10.1016/0922-338X(96)85037-9.

Becerra-Rodríguez, C., Marsit, S., & Galeote, V. (2020). Diversity of oligopeptide transport in yeast and its impact on adaptation to winemaking conditions. *Frontiers in Genetics.* Available from https://doi.org/10.3389/fgene.2020.00602.

Bell, S. J., & Henschke, P. A. (2005). Implications of nitrogen nutrition for grapes, fermentation and wine. *Australian Journal of Grape and Wine Research, 11*(3), 242−295. Available from https://doi.org/10.1111/j.1755-0238.2005.tb00028.x.

Beltran, G., Esteve-Zarzoso, B., Rozès, N., Mas, A., & Guillamón, J. M. (2005). Influence of the timing of nitrogen additions during synthetic grape must fermentations on fermentation kinetics and nitrogen consumption. *Journal of Agricultural and Food Chemistry, 53*(4), 996−1002. Available from https://doi.org/10.1021/jf0487001.

Bely, M., Sablayrolles, J. M., & Barre, P. (1990). Automatic detection of assimilable nitrogen deficiencies during alcoholic fermentation in oenological conditions. *Journal of Fermentation and Bioengineering, 70*(4), 246−252. Available from https://doi.org/10.1016/0922-338X(90)90057-4.

Bisson, L. F. (1999). Stuck and sluggish fermentations. *American Journal of Enology and Viticulture, 50*(1), 107−119.

Blateyron, L., Julien, & Sablayrolles, J. (2000). Stuck fermentations − O2 and nitrogen requirements − importance of optimizing their addition. *Lallemand Research Meeting.*

Blateyron, L., & Sablayrolles, J. M. (2001). Stuck and slow fermentations in enology: Statistical study of causes and effectiveness of combined additions of oxygen and diammonium phosphate. *Journal of Bioscience and Bioengineering, 91*(2), 184−189. Available from https://doi.org/10.1016/S1389-1723(01)80063-3.

Brice, C., Sanchez, I., Tesnière, C., & Blondin, B. (2014). Assessing the mechanisms responsible for differences between nitrogen requirements of *Saccharomyces cerevisiae* wine yeasts in alcoholic fermentation. *Applied and Environmental Microbiology, 80*(4), 1330−1339.

Butzke, C. E. (1998). Survey of yeast assimilable nitrogen status in musts from California, Oregon, and Washington. *American Journal of Enology and Viticulture, 49*(2), 220−224.

Callejón, R. M., Troncoso, A. M., & Morales, M. L. (2010). Determination of amino acids in grape-derived products: A review. *Talanta, 81*(4−5), 1143−1152. Available from https://doi.org/10.1016/j.talanta.2010.02.040.

Carrau, F. M., Medina, K., Farina, L., Boido, E., Henschke, P. A., & Dellacassa, E. (2008). Production of fermentation aroma compounds by *Saccharomyces cerevisiae* wine yeasts: Effects of yeast assimilable nitrogen on two model strains. *FEMS Yeast Research, 8*(7), 1196−1207.

Casalta, E., Cervi, M. F., Salmon, J. M., & Sablayrolles, J. M. (2013). White wine fermentation: Interaction of assimilable nitrogen and grape solids. *Australian Journal of Grape and Wine Research,* 47−52.

Casalta, E., Sablayrolles, J. M., & Salmon, J. M. (2013). Comparison of different methods for the determination of assimilable nitrogen in grape musts. *LWT − Food Science and Technology, 54*(1), 271−277. Available from https://doi.org/10.1016/j.lwt.2013.05.009.

Colombié, S., Malherbe, S., & Sablayrolles, J. M. (2007). Modeling of heat transfer in tanks during wine-making fermentation. *Food Control, 18*(8), 953−960. Available from https://doi.org/10.1016/j.foodcont.2006.05.016.

Crépin, L., Nidelet, T., Sanchez, I., Dequin, S., & Camarasa, C. (2012). Sequential use of nitrogen compounds by *Saccharomyces cerevisiae* during wine fermentation: A model based on kinetic and regulation characteristics of nitrogen permeases. *Applied and Environmental Microbiology, 78*(22), 8102−8111.

Crépin, L., Truong, N. M., Bloem, A., Sanchez, I., Dequin, S., & Camarasa, C. (2017). Management of multiple nitrogen sources during wine fermentation by *S. cerevisiae. Applied and Environmental Microbiology, 83.* Available from https://doi.org/10.1128/AEM.02617-16, e02617−16.

Dubois, C., Manginot, C., Roustan, J. L., Sablayrolles, J. M., & Barre, P. (1996). Effect of variety, year, and grape maturity on the kinetics of alcoholic fermentation. *American Journal of Enology and Viticulture, 47*(4), 363−368.

Duc, C., Noble, J., Tesnière, C., & Blondin, B. (2019). Occurrence of yeast cell death associated with micronutrient starvation during wine fermentation varies with nitrogen sources. *Oeno One, 53*(3), 445−456. Available from https://doi.org/10.20870/oeno-one.2019.53.3.2408.

Duc, C., Pradal, M., Sanchez, I., Noble, J., Tesnière, C., & Blondin, B. (2017). A set of nutrient limitations trigger yeast cell death in a nitrogen-dependent manner during wine alcoholic fermentation. *PLoS One, 12*(9). Available from https://doi.org/10.1371/journal.pone.0184838.

Dukes, B. C., & Butzke, C. E. (1998). Rapid determination of primary amino acids in grape juice using an o-phthaldialdehyde/N-acetyl-L-cysteine spectrophotometric assay. *American Journal of Enology and Viticulture, 49*(2), 125−134.

Englezos, V., Cocolin, L., Rantsiou, K., Ortiz-Julien, A., Bloem, A., Dequin, S., & Camarasa, C. (2018). Specific phenotypic traits of *Starmerella bacillaris* related to nitrogen source consumption and central carbon metabolite production during wine fermentation. *Applied and Environmental Microbiology, 4.* Available from https://doi.org/10.1128/AEM.00797-18, e00797−18.

Fairbairn, S., McKinnon, A., Musarurwa, H. T., Ferreira, A. C., & Bauer, F. F. (2017). The impact of single amino acids on growth and volatile aroma production by *Saccharomyces cerevisiae* strains. *Frontiers in Microbiology, 8*, 1−12.

Filipe-Ribeiro, L., & Mendes-Faia, A. (2007). Validation and comparison of analytical methods used to evaluate the nitrogen status of grape juice. *Food Chemistry, 100*(3), 1272−1277. Available from https://doi.org/10.1016/j.foodchem.2005.10.006.

Garde-Cerdán, T., & Ancín-Azpilicueta, C. (2008). Effect of the addition of different quantities of amino acids to nitrogen-deficient must on the formation of esters, alcohols, and acids during wine alcoholic fermentation. *LWT − Food Science and Technology, 41*(3), 501−510. Available from https://doi.org/10.1016/j.lwt.2007.03.018.

Gobert, A., Tourdot-Maréchal, R., Morge, C., Sparrow, C., Liu, Y., Quintanilla-Casas, B., ... Alexandre, H. (2017). Non-*Saccharomyces* yeasts nitrogen source preferences: Impact on sequential fermentation and wine volatile compounds profile. *Frontiers in Microbiology, 8*, 2175. Available from https://doi.org/10.3389/fmicb.2017.02175.

Gobert, A., Tourdot-Maréchal, R., Sparrow, C., Morge, C., & Alexandre, H. (2019). Influence of nitrogen status in wine alcoholic fermentation. *Food Microbiology*, *83*, 71−85.

Gump, B. H., Zoecklein, B. W., Fugelsang, K. C., & Whiton, R. S. (2002). Comparison of analytical methods for prediction of prefermentation nutritional status of grape juice. *American Journal of Enology and Viticulture*, *53*(4), 325−329.

Harle, O., Legrand, J., Tesniere, C., Pradal, M., Mouret, J. R., & Nidelet, T. (2020). Investigations of the mechanisms of interactions between four non-conventional species with *Saccharomyces cerevisiae* in oenological conditions. *PloS One*, *15*(5), e0233285. Available from https://doi.org/10.1371/journal.pone.0233285.

Hazelwood, L. A., Daran, J. M., Van Maris, A. J. A., Pronk, J. T., & Dickinson, J. R. (2008). The Ehrlich pathway for fusel alcohol production: A century of research on *Saccharomyces cerevisiae* metabolism. *Applied and Environmental Microbiology*, *74*(8), 2259−2266.

Hernández-Orte, P., Ibarz, M. J., Cacho, J., & Ferreira, V. (2005). Effect of the addition of ammonium and amino acids to musts of Airen variety on aromatic composition and sensory properties of the obtained wine. *Food Chemistry*, *89*(2), 163−174. Available from https://doi.org/10.1016/j.foodchem.2004.02.021.

Hirst, M. B., & Richter, C. L. (2016). Review of aroma formation through metabolic pathways of *Saccharomyces cerevisiae* in beverage fermentations. *American Journal of Enology and Viticulture*, *67*, 361−370.

Jiménez-Martí, E., Aranda, A., Mendes-Ferreira, A., Mendes-Faia, A., & li del Olmo, M. (2007). The nature of the nitrogen source added to nitrogen depleted vinifications conducted by a *Saccharomyces cerevisiae* strain in synthetic must affects gene expression and the levels of several volatile compounds. *Antonie Van Leeuwenhoek*, *92*, 61−75.

Kemsawasd, V., Viana, T., Ardö, Y., & Arneborg, N. (2015). Influence of nitrogen sources on growth and fermentation performance of different wine yeast species during alcoholic fermentation. *Applied Microbiology and Biotechnology*, *99*(23), 10191−10207. Available from https://doi.org/10.1007/s00253-015-6835-3.

Malherbe, S., Fromion, V., Hilgert, N., & Sablayrolles, J. M. (2004). Modeling the effects of assimilable nitrogen and temperature on fermentation kinetics in enological conditions. *Biotechnology and Bioengineering*, *86*(3), 261−272. Available from https://doi.org/10.1002/bit.20075.

Martínez-Moreno, R., Morales, P., Gonzalez, R., Mas, A., & Beltran, G. (2012). Biomass production and alcoholic fermentation performance of *Saccharomyces cerevisiae* as a function of nitrogen source. *FEMS Yeast Research*, *12*(4), 477−485.

Medina, K., Boido, E., Dellacassa, E., & Carrau, F. (2012). Growth of non-Saccharomyces yeasts affects nutrient availability for *Saccharomyces cerevisiae* during wine fermentation. *International Journal of Food Microbiology*, *157*, 245−250.

Miller, A. C., Wolff, S. R., Bisson, L. F., & Ebeler, S. E. (2007). Yeast strain and nitrogen supplementation: Dynamics of volatile ester production in chardonnay juice fermentations. *American Journal of Enology and Viticulture*, *58*(4), 470−483.

Morakul, S., Mouret, J. R., Nicolle, P., Trelea, I. C., Sablayrolles, J. M., & Athes, V. (2011). Modelling of the gas−liquid partitioning of aroma compounds during wine alcoholic fermentation and prediction of aroma losses. *Process Biochemistry*, *46*(5), 1125−1131. Available from https://doi.org/10.1016/j.procbio.2011.01.034.

Mouret, J. R., Camarasa, C., Angenieux, M., Aguera, E., Perez, M., Farines, V., & Sablayrolles, J. M. (2014). Kinetic analysis and gas−liquid balances of the production of fermentative aromas during winemaking fermentations: Effect of assimilable nitrogen and temperature. *Food Research International*, *62*, 1−10. Available from https://doi.org/10.1016/j.foodres.2014.02.044.

Mouret, J. R., Farines, V., Sablayrolles, J. M., & Trelea, I. C. (2015). Prediction of the production kinetics of the main fermentative aromas in winemaking fermentations. *Biochemical Engineering Journal*, *103*, 211−218. Available from https://doi.org/10.1016/j.bej.2015.07.017.

Nicolini, G., Larcher, R., & Versini, G. (2004). Status of yeast assimilable nitrogen in Italian grape musts and effects of variety, ripening and vintage. *Vitis − Journal of Grapevine Research*, *43*(2), 89−96.

Nisbet, M. A., Martinson, T. E., & Mansfield, A. K. (2014). Accumulation and prediction of yeast assimilable nitrogen in New York winegrape cultivars. *American Journal of Enology and Viticulture*, *65*(3), 325−332. Available from https://doi.org/10.5344/ajev.2014.13130.

Ochando, T., Mouret, J. R., Humbert-Goffard, A., Sablayrolles, J. M., & Farines, V. (2017). Impact of initial lipid content and oxygen supply on alcoholic fermentation in champagne-like musts. *Food Research International*, *98*, 87−94. Available from https://doi.org/10.1016/j.foodres.2016.11.010.

Paul Schreiner, R., Osborne, J., & Skinkis, P. A. (2018). Nitrogen requirements of pinot noir based on growth parameters, must composition, and fermentation behavior. *American Journal of Enology and Viticulture*, *69*(1), 45−58. Available from https://doi.org/10.5344/ajev.2017.17043.

Petrovic, G., Aleixandre-Tudo, J. L., & Buica, A. (2019a). Grape must profiling and cultivar discrimination based on amino acid composition and general discriminant analysis with best subset. *South African Journal of Enology and Viticulture*, *40*(2), 1−13. Available from https://doi.org/10.21548/40-2-3373.

Petrovic, G., Aleixandre-Tudo, J. L., & Buica, A. (2019b). Unravelling the complexities of wine: A big data approach to yeast assimilable nitrogen using infrared spectroscopy and chemometrics. *Oeno One*, *53*(2), 107−127. Available from https://doi.org/10.20870/oeno-one.2019.53.2.2371.

Petrovic, G., Aleixandre-Tudo, J. L., & Buica, A. (2020). Viability of IR spectroscopy for the accurate measurement of yeast assimilable nitrogen content of grape juice. *Talanta*, *206*. Available from https://doi.org/10.1016/j.talanta.2019.120241.

Rapp, A., & Versini, G. (1995). Influence of nitrogen compounds in grapes on aroma compounds of wines. *Developments in Food Science*, *37*(C), 1659−1694. Available from https://doi.org/10.1016/S0167-4501(06)80257-8.

Ribéreau-Gayon, P., Glories, Maujean, A., & Dubourdieu, D. (2006). *Handbook of enology: The microbiology of wine and vinifications*. John Wiley & Sons.

Rollero, S., Bloem, A., Camarasa, C., Sanchez, I., Ortiz-Julien, A., Sablayrolles, J. M., . . . Mouret, J. R. (2015). Combined effects of nutrients and temperature on the production of fermentative aromas by *Saccharomyces cerevisiae* during wine fermentation. *Applied Microbiology and Biotechnology*, *99*, 2291−2304. Available from https://doi.org/10.1007/s00253-014-6210-9.

Rollero, S., Bloem, A., Ortiz-Julien, A., Camarasa, C., & Divol, B. (2018). Altered fermentation performances, growth, and metabolic footprints reveal competition for nutrients between yeast species inoculated in synthetic grape juice-like medium. *Frontiers in Microbiology*, *9*, 1−12. Available from https://doi.org/10.3389/fmicb.2018.00196.

Rollero, S., Mouret, J. R., Bloem, A., Sanchez, I., Ortiz-Julien, A., Sablayrolles, J. M., . . . Camarasa, C. (2017). Quantitative 13C-isotope labelling-based analysis to elucidate the influence of environmental parameters on the production of fermentative aromas during wine fermentation. *Microbial Biotechnology*, *10*, 1649−1662.

Sablayrolles, J.M. (2019). Wine yeast nutrients in alcoholic fermentation: Role and management to improve alcoholic fermentation kinetics. 2nd International Wine Nutrition School.

Seguinot, P., Rollero, S., Sanchez, I., Sablayrolles, J. M., Ortiz-Julien, A., Camarasa, C., & Mouret, J. R. (2018). Impact of the timing and the nature of nitrogen additions on the production kinetics of fermentative aromas by *Saccharomyces cerevisiae* during winemaking fermentation in synthetic media. *Food Microbiology, 76*, 29−39.

Soubeyrand, V., Luparia, V., Williams, P., Doco, T., Vernhet, A., Ortiz-Julien, A., & Salmon, J. M. (2005). Formation of micella containing solubilized sterols during rehydration of active dry yeasts improves their fermenting capacity. *Journal of Agricultural and Food Chemistry, 53*(20), 8025−8032. Available from https://doi.org/10.1021/jf050907m.

Spayd, S. E., Wample, R. L., Evans, R. G., Stevens, R. G., Seymour, B. J., & Nagel, C. W. (1994). Nitrogen fertilization of White Riesling grapes in Washington. Must and wine composition. *American Journal of Enology and Viticulture, 45*(1), 34−42.

Styger, G., Jacobson, D., Prior, B. A., & Bauer, F. F. (2013). Genetic analysis of the metabolic pathways responsible for aroma metabolite production by *Saccharomyces cerevisiae*. *Applied Microbiology and Biotechnology, 97*, 4429−4442.

Su, Y., Seguinot, P., Sanchez, I., Ortiz-Julien, A., Heras, J. M., Querol, A., ... Guillamon, J. M. (2020). Nitrogen sources preferences of non-*Saccharomyces* yeasts to sustain growth and fermentation under winemaking conditions. *Food Microbiology, 85*. Available from https://doi.org/10.1016/j.fm.2019.103287.

Subileau, M., Schneider, R., Salmon, J. M., & Degryse, E. (2008). Nitrogen catabolite repression modulates the production of aromatic thiols characteristic of Sauvignon blanc at the level of precursor transport. *FEMS Yeast Research, 8*, 771−780. Available from https://doi.org/10.1111/j.1567-1364.2008.00400.x.

Tesnière, C., Delobel, P., Pradal, M., & Blondin, B. (2013). Impact of nutrient imbalance on wine alcoholic fermentations: Nitrogen excess enhances yeast cell death in lipid-limited must. *PLoS One* (4).

Thibon, C., Marullo, P., Claisse, O., Cullin, C., Dubourdieu, D., & Tominaga, T. (2008). Nitrogen catabolic repression controls the release of volatile thiols by *Saccharomyces cerevisiae* during wine fermentation. *FEMS Yeast Research, 8*, 1076−1086.

Tominaga, T., Peyrot des gachons, C., & Dubourdieu, D. (1998). A new type of flavor precursors in *Vitis vinifera* L. cv Sauvignon blanc: S-cysteine conjugates. *Journal of Agriculture and Food Chemistry, 46*, 5215−5219.

Torrea, D., Varela, C., Ugliano, M., Ancin-Azpilicueta, C., Leigh Francis, I., & Henschke, P. A. (2011). Comparison of inorganic and organic nitrogen supplementation of grape juice—Effect on volatile composition and aroma profile of a Chardonnay wine fermented with *Saccharomyces cerevisiae* yeast. *Food Chemistry, 127*(3), 1072−1083.

Ugliano, M., Siebert, T., Mercurio, M., Capone, D., & Henschke, P. A. (2008). Volatile and color composition of young and model-aged shiraz wines as affected by diammonium phosphate supplementation before alcoholic fermentation. *Journal of Agricultural and Food Chemistry, 56* (19), 9175−9182. Available from https://doi.org/10.1021/jf801273k.

Verstrapen, K., Derdelinckx, G., Dufour, J., Winderickx, J., Pretorius, I., Thevelein, J., & Delvaux, F. (2003). The Saccharomyces cerevisiae alcohol acetyl transferase gene is a target of the cAMP/PKA and FGM nutrient-signalling pathways. *FEMS Yeast Research, 4*, 285−296. Available from https://doi.org/10.1016/S1567-1356(03)00166-1.

Zhang, W., Du, G., Zhou, J., & Chen, J. (2018). Regulation of sensing, transportation, and catabolism of nitrogen sources in *Saccharomyces cerevisiae*. *Microbiology and Molecular Biology Reviews, 82*(1). Available from https://doi.org/10.1128/MMBR.00040-17.

13

Tasting the *terroir* of wine yeast innovation

Isak S. Pretorius

ARC Centre of Excellence in Synthetic Biology, Macquarie University, Sydney, NSW, Australia

13.1 The past and future of wine is ever present in its tradition of innovation

Wine is an archetypal traditional fermented beverage with strong territorial and socio-cultural connotations (Pretorius & Bauer, 2002; Pretorius & Høj, 2005). Its 7000-year history is patterned by a *tradition* of *innovation* (McGovern et al., 2004). Every value-adding innovation, whether in the vineyard, winery, supply chain, or marketplace, that led to the *invention* of a new *tradition* spurred progress and created a brighter future from past developments. In a way, wine *traditions* can be defined as *remembered innovations* from the distant past inherited knowledge and wisdom that withstood the test of time (Jagtap, Jadhav, Bapat, & Pretorius, 2017; Pretorius, 2020). Therefore it should not be assumed *a priori* that tradition and innovation are polar opposites. The relations between the forces driven by the *anchors of tradition* and the *wings of innovation* do not necessarily involve displacement, conflict, or exclusiveness. Innovation can strengthen wine tradition, and the reinvention of a tradition-bound practice, approach, or concept can foster innovation. In cases where a paradigm-shifting innovation disrupts a tradition, the process of such an innovation transitioning into a radically new tradition can become protracted while proponents of divergent opinions duke it out (Pretorius, 2020). Sometimes, these conflicting opinions are based on fact, and sometimes not. The imperfections of such a debate between the "ancients" and the "moderns" can, from time to time, obscure the line between myth and reality. Therefore finding the right balance between *traditions worth keeping* and *innovations worth implementing* can be complex. The intent here is to harness the creative tension between *science fiction* and *science fact* when innovation's first principles challenge the *status quo* by reexamining the foundational principles about a core traditional concept, such as *terroir* (Robinson, 1999). Poignant questions are raised about the importance of the *terroir* (the biogeography, environmental and spatial constructs of microbial communities inhabiting particular vineyards) of yeasts and the value of the microbiome of grapes to wine quality. This chapter imagines a metaphorical *terroir* free from cognitive biases where diverse perspectives can converge to uncork the effervescent power of territorial yeast populations as well as "nomadic" yeast starter cultures. At the same time, this chapter also engages in *mental time-travel*. A future scenario is imagined, explored, tested, and debated where *terroir*-less *yeast avatars* are equipped with designer genomes to safely and consistently produce, individually or in combination with region-specific wild yeasts and or other starter cultures, high-quality wine according to the preferences of consumers in a range of markets. The purpose of this review is to look beyond the horizon and to synthesize a link between *what we know now* and *what could be.*

Steering toward the future, it is likely that vintners will continue to be inspired by a custom to innovate through tradition. In this context, the value proposition of any future innovation at any point across the entire *from-grapes-to-glass* value chain (Fig. 13.1) will be critically assessed by producers and consumers through a lens tinted by a deep *respect for the past*, an indomitable spirit to *lead the present*, and a quest to *secure the future* of wine (Pretorius, 2020). This chapter recognizes that these three core drivers of progress in the ancient art of winemaking shape the *mental terroir* that will also determine the successful implementation of current and future wine yeast innovations. Today's wine stakeholders must be wise enough to learn from the past, smart enough to utilize the present, and imaginative enough to anticipate the future with realism and optimism.

© 2022 Elsevier Inc. All rights reserved.

FIGURE 13.1 The *from-grapes-to-glass value chain* in wine production. The traditional *production-driven* view of the *supply chain* has been largely replaced by a more innovative *market-driven* approach. This new mindset among contemporary vintners has placed their products in the high-tension field between the forces of *market-pull* and *technology-push*, where tradition and innovation must coexist.

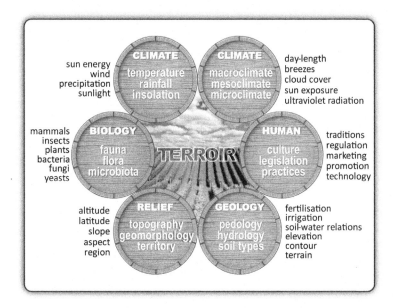

FIGURE 13.2 The concept of *terroir* in wine production. *Terroir* entails the total natural environment for specific viticultural sites, including climate as measured by temperature and rainfall; sunlight energy (or insolation) received per unit of land surface area; relief (topography or geomorphology) comprising altitude, slope, and aspect; geology and pedology, determining the basic physical and chemical characteristics of various soil types; and hydrology or soil–water relations.

13.2 Stamping yeast signatures onto the terroir and vintage labels of wine

Bottled poetry is a delicious phrase that often rolls off the tongue of wine connoisseurs when they refer to *fine* wine. In the context of this metaphor, grapes and yeasts are the *ink* and *pen* with which the *poets* viticulturists and enologists cowrite *fine wine poetry* on the *terroir sheets* of their vineyards and the vintages of their wine. When these two natural *ink-and-pen* companions combine, a complex mixture of grape- and yeast-derived (and in some cases bacterial and oak-derived) compounds emerge, which largely define a wine's appearance, aroma, flavor, and mouthfeel properties (Swiegers et al., 2005; Swiegers 2005; Swiegers et al., 2007; Swiegers et al., 2007). The grape-derived compounds provide varietal distinction in addition to giving wine its basic structure, while yeast fermentation gives wine its vinous character (Pretorius, 2000). Wine attributes are the result of an almost infinite number of variations in production, whether in the vineyard or the winery (Fig. 13.2).

In *spontaneous* fermentations (uninoculated ferments) there is a progressive growth pattern of indigenous yeasts (also referred to as *autochthonous*, *natural*, *wild*, or *feral* yeasts) with the final stages invariably being dominated by the Crabtree-positive, alcohol-tolerant strains of *S. cerevisiae* (universally known as *the wine yeast*). The primary role of wine yeast is to catalyze the rapid, complete, and efficient conversion of grape sugars (glucose and fructose) to ethanol, carbon dioxide, and other minor, but important, flavor-active metabolites (e.g., acids, alcohols, carbonyls, esters, terpenes, thiols), without the development of off-flavors (e.g., hydrogen sulfide) (Cordente, Cordero-Bueso, Pretorius, & Curtin, 2013; Cordente, Curtin, Varela, & Pretorius, 2012; Cordente, Heinrich, Pretorius, & Swiegers, 2009; Lilly et al., 2006; Lilly, Styger, Bauer, Lambrechts, & Pretorius, 2006; Swiegers et al., 2005). To achieve this outcome, a Crabtree-positive carbon metabolism is the most efficient strategy for grape sugar utilization (with a preference for glucose over fructose) that maximizes ethanol production. This adaptation in *Saccharomyces* enables energy generation under fermentative or anaerobic conditions and limits the growth of competing microbes including non-*Saccharomyces* yeasts by producing toxic metabolites, such as ethanol and carbon dioxide. Therefore *S. cerevisiae*'s effective *make-accumulate-tolerate-consume* alcohol strategy makes it the preferred yeast species for initiating wine fermentations in *inoculated* (or *guided*) ferments (Goold et al., 2017; Varela et al., 2018).

Grape must is not sterile and naturally contains, among other microbes, a mixture of *Saccharomyces* and non-*Saccharomyces* yeast species; therefore wine fermentation is not a *single-species* fermentation process. The indigenous non-*Saccharomyces* yeasts, often already present in the must at much greater numbers than *S. cerevisiae*, are adapted to the specific environment and in an active growth state, which gives them a competitive edge. However, the eventual dominance of *S. cerevisiae* in both spontaneous and inoculated ferments is essential to ferment wine to dryness (1−2 g/L of residual sugar). The length of time during which the non-*Saccharomyces* and non-*cerevisiae Saccharomyces* yeasts are allowed to participate in these *multispecies* ferments is the choice of the winemaker. More than 40 of the 1500 known yeast species have been isolated from grape musts (Jolly, Varela, & Pretorius, 2014; Van Wyk, Pretorius, & Von Wallbrunn, 2020). Some of these wine-related non-*Saccharomyces* yeasts (e.g., *Brettanomyces bruxellensis*) cause off-flavors (e.g., volatile acidity and phenolic odors) while others contribute positively to a wine's aromatic *complexity* and *textural roundness*. For example, if a winemaker deems that the risk of high concentrations of *vinegary* volatile acidity or *leathery* ethylphenols outweighs the beneficial metabolites of non-*Saccharomyces* yeasts in a particular grape must, the pH of the fermenting grape juice can be lowered and a higher dosage of sulfite can be added to restrict the feral spoilage yeasts, thereby allowing the starter culture strain of *S. cerevisiae* with which the ferment was inoculated to gain dominance faster (Curtin, Borneman, Chambers, & Pretorius, 2012; Curtin & Pretorius, 2014). It is generally accepted that risk management with *spontaneous* (*multispecies*) ferments is more complex than with *guided* (mostly inoculated with a single species) ferments (Van Wyk et al., 2020). To gain the benefits of both practices without risking the negative effects of spoilage yeasts, some winemakers prefer to inoculate their grape musts with selected non-*cerevisiae* and/or non-*Saccharomyces* yeast strains along with one of more of the 250 different commercially available wine strains of *S. cerevisiae* (Fig. 13.3).

Over the millennia, thousands of clonal varieties of *Vitis vinifera* vines originating from Transcaucasia have evolved and spread across the world while vintners matched their traits to the *terroir* of each newly planted vineyard site and the likings of their targeted consumers (Gambino, Dal Molin, & Boccacci, 2017; Roach et al., 2018; Vivier & Pretorius, 2000; Vivier & Pretorius, 2002). At the same time, the commensal microbial flora that coexisted with the vines in those new vineyards similarly evolved. However, to date, the *microbiome* of wine-producing regions and the potential influence of *microbial terroirs* on wine quality have not received the same level of scientific scrutiny (Gilbert, van der Lelie, & Zarraonaindia, 2014). Investigation of the importance of yeast communities associated with the grapes from a particular vineyard or region is complex because some ambient yeasts might originate from a neighboring vineyard or be imported via oak barrels and commercial yeast starter cultures used in a nearby winery (Knight, Klaere, Fedrizzi, & Goddard, 2015). Conditions and practices applied in both the vineyard and winery can dramatically alter the composition of grape and wine-related *yeastomes* in a particular setting (Bokulich et al., 2016; Bokulich et al., 2013; Bokulich et al., 2013; Setati, Jacobson, Andong, & Bauer, 2012).

For clarity, the following nomenclature is used here to differentiate flavor-active yeast types that might constitute a specific *yeastome* relevant to wine quality. The indigenous *Saccharomyces* and non-*Saccharomyces* yeasts inhabiting a niche vineyard are referred to as *natives* and yeasts that migrated and accumulated in a winery are referred to as *settlers*. Commercial active-dried yeast starter strains used in many wineries across the world are referred to as *nomads*. Laboratory-bred strains of *S. cerevisiae* are termed supermodel *mannequins* while genetically engineered and semisynthetic yeasts with reinvented or edited genomes are called *avatars*. The following sections explore the importance of residential *natives* (indigenous yeasts from *somewhere*), colonizing *settlers* (migrant yeasts from *everywhere*), imported *nomads* (commercial starter yeasts from *anywhere*), prototypical *mannequins* (quintessential model yeasts from *elsewhere*), and alien *avatars* (genetically modified yeasts from *nowhere*) in terms of our understanding of their role in current wine ferments or potential role in future winemaking practices (Fig. 13.3).

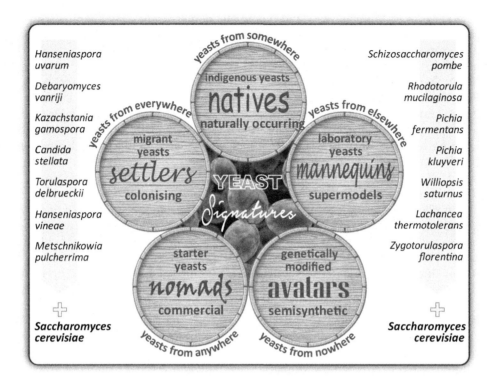

FIGURE 13.3   **Yeast imprints on the sensory profile of wine.** A diverse range of flavor-active yeasts can contribute to the overall quality of a wine. Flavor-active yeast types that might constitute a specific *yeastome* relevant to wine quality can include the following kinds of yeasts: the indigenous *Saccharomyces* and non-*Saccharomyces* yeasts inhabiting a niche vineyard are referred to as *natives* and yeasts that migrated and accumulated in a winery are referred to as *settlers*. Commercial active-dried yeast starter strains used in many wineries across the world are referred to as *nomads*. Laboratory-bred strains of *S. cerevisiae* are termed supermodel *mannequins*, while genetically engineered and semisynthetic yeasts with reinvented or edited genomes are called *avatars*. The following sections explore the importance of residential *natives* (indigenous yeasts from *somewhere*), colonizing *settlers* (migrant yeasts from *everywhere*), imported *nomads* (commercial starter yeasts from *anywhere*), prototypical *mannequins* (quintessential model yeasts from *elsewhere*), and alien *avatars* (genetically modified yeasts from *nowhere*) in terms of our understanding of their role in current wine ferments or potential role in future winemaking practices.

13.2.1 *Yeasts from somewhere*

Native yeasts residing in a niche site represent an important component of the *microbiome* of a vineyard. Across viticultural zones and over time, the microbiota (e.g., acetic acid bacteria, botrytis fungi, and yeasts) that inhabited a vineyard and colonized the phyllosphere of *V. vinifera* can substantially affect grapevine health, fruit development, and ripening, as well as the quality of grapes and wine (Gilbert, van der Lelie, & Zarraonaindia, 2014). There is mounting evidence that the nonrandom biogeographical distribution patterns of microbial assemblages of grape surface microbiota in vineyards can be modulated by a combination of several factors. These factors include geographical location, farming system, soil, cultivar, vintage, and climate to varying degrees. However, not all members of a vineyard's microbiome can complete the *from-vine-to-wine* journey because many of them cannot withstand the low-pH, high-ethanolic, and anaerobic conditions of wine fermentations (Bokulich et al., 2016; Bokulich et al., 2013; Bokulich et al., 2013). Nevertheless, there is renewed interest to establish an indisputable link between differential *geographic phenotypes* and *sensorial signatures* as encapsulated by the concept of *terroir*. The key hypothesis of such investigations is that nonrandom yeast inhabitants with specific *vineyard postcodes* form part of the *microbial terroir* of grapes harvested from well-established vineyards and can influence the chemical and sensorial profile and the so-called *typicity* of the wine in a unique, reproducible, and recognizable manner (Belda, et al., 2017).

The microbiome of a grapevine plant has direct and indirect relationships with its host (Gilbert, van der Lelie, & Zarraonaindia, 2014). For instance, these relationships are affected by the availability of organic matter and essential nutrients in the soil (including nitrogen fixation) and environmental stresses (e.g., water stress caused by drought or stresses caused by the presence of phytotoxic contaminants). Other factors that play a role include the degree of phytopathogens activity in terms of competition for space and nutrients, antibiosis, production of inhibitory enzymes (e.g., hydrolytic enzymes), and systemic induction of plant defense mechanisms. Soil endophytes in the rhizosphere of the vine and those that migrate through the plant to colonize aerial tissues internally or externally (epiphytes) can, for example, have several metabolic activities that support vine health by either promoting the physiology or suppressing disease-causing pathogens, which in turn can alter the microbial composition that ends up in grape must (Gilbert, van der Lelie, & Zarraonaindia, 2014).

It would seem logical to assume that the same physical and chemical criteria that determine which vines grow well in conditions prevailing in certain sites (e.g., soil nutrients levels, solar radiation, temperature, humidity, and precipitation) would also impact the biography of microorganisms in a vineyard's ecosystem (Gilbert, van der Lelie, & Zarraonaindia, 2014). To understand the biogeographical regionalization of microbial communities of site-specific vineyards and regions, it would also be necessary to determine if the grapevine plants themselves select for different microbiota based on their physiological response to different environmental conditions and viticultural practices.

This century has seen a marked increase in the sophistication of technologies with which the microbiome of a vineyard can be investigated. Recent studies are starting to shed some light on how farming practices (e.g., soil cultivation, fertilization, irrigation, and the application of herbicides, pesticides, and fungicides) and some enological practices in wineries (repeated use of selective yeast starter cultures) are shaping the composition of the microbiome of vineyards (Belda et al., 2017; Canfora, Vendramin, Felici, & Luo, 2018; Cordero-Bueso, Arroyo, & Serrano, 2011; Cordero-Bueso, Arroyo, & Serrano, 2011; De Celis et al., 2019; Morrison-Whittle, Lee, & Goddard, 2017). For example, by using next-generation sequencing of 16 S rRNA and internal transcribed spacer (ITS) sequences of ribosomal DNA to determine the relative abundances of bacteria, mycelial fungi, and yeasts, it has become clear that vineyard under-vine floor management alters the microbial composition of soil but does not seem to affect any shifts in the fruit-associated microbiome (Chou, Vanden Heuvel, Bell, Panke-Buisse, & Kao-Kniffin, 2018). However, other vineyard management practices and environmental factors are more influential in shaping not only the grape-associated microbiome but also its later behavior in wine fermentation (Grangeteau et al., 2016).

By using a high-throughput, short-amplicon sequencing approach, researchers were able to demonstrate that regional and site-specific factors along with grape variety-specific factors shape the fungal and bacterial consortia inhabiting the surfaces of grape berries (Gilbert, van der Lelie, & Zarraonaindia, 2014). These communities were shown to be correlated to specific climatic features, thereby demonstrating a link between environmental conditions and microbial inhabitation patterns in vineyards. It was shown that the degree of differentiation among these grape-surface microbial communities from different regions was substantially increased when the biogeography was investigated within a grape variety of a particular vintage (Bokulich et al., 2016; Bokulich et al., 2013; Bokulich et al., 2013). It was found that the host genotype, and therefore the phenotype of the grape cultivar, along with local and interannual (seasonal) climate variation (vintage), play a significant role in determining the nature of the microbial assemblages on the surface of wine grapes. The authors of this study suggested that these factors appear to shape the unique inputs to regional wine fermentations. They further proposed the nonrandom existence of *microbial terroir* as a determining factor among grapes harvested from different grape varieties and from regions with different climatic conditions.

13.2.2 Yeasts from everywhere

Yeast *settlers* can colonize a wide variety of natural and man-made habitats, including vineyards and wineries, to form microbial communities associated with specialized niches, such as vineyard soils, grapevine plants, grape skins, and the surfaces of equipment used within wineries. Gaining deeper insights into the composition, population dynamics, dispersal, and maintenance of these yeast communities along their journey from the vineyard to the winery can potentially clarify the relations between the microbiomes associated with vine health, grape yield, grape and must quality, and the metabolome of wine impacting the sensorial profile of the end-product.

The composition of microbial communities associated with precrushed grapes including the variety and quantity of yeast species depend on factors, such as the method of harvest (hand-picked or mechanical), grape temperature (day or night harvest), grape condition (biotic and abiotic damage, degree of ripeness, grape variety), sulfite addition, and the time between harvesting and crushing of the grapes (distance and duration of transport from the vineyard to the winery, ambient temperature and initial grape temperature) (Gilbert, van der Lelie, & Zarraonaindia, 2014). The population profile of yeasts present in grape must can also be significantly influenced by the method and intensity of grape destemming and crushing (grape-stomping or mechanical crushing with various types of wine presses), cellar hygiene (sanitation protocols and disinfectants used), must pretreatment (aeration, sulfite addition, enzyme treatment, clarification protocol, temperature), and inoculation with starter yeast cultures (Jolly, Varela, & Pretorius, 2014).

Untreated grape must provides a rich nutritive niche for yeasts to grow. However, cellar hygiene practices, low pH conditions, high osmotic pressure, sulfite concentrations, and temperature can make the grape pulp and winery environment much harsher for the less robust species (Jolly, Varela, & Pretorius, 2014). These factors and the anaerobic conditions that set in when fermentation commences are bound to stack the odds against bacteria and fungi with oxidative metabolisms. This is also true for yeast species and any other microbes that are more sensitive to high sugar levels (varying from 150 to 250 g/L in ripe grapes) and sulfite concentrations (free SO_2 levels ranging between 25 and 30 mg/L) in conjunction with low pH levels (varying between pH 3.0 and 3.6), suboptimal fermentation temperatures (ranging from 12°C to 18°C for white wines, and from 20°C to 30°C for red wines), and high alcohol levels (12%−15%) toward the end of fermentation (Jolly, Varela, & Pretorius, 2014).

It is reasonable to expect that microbes able to survive these harsh conditions could accumulate on the surfaces of large specialized equipment, oak barrels, and other winery tools and surfaces. Wineries could therefore serve as reservoirs of resident microbial communities, which might shape the microbiota in wine fermentations and perhaps even vector wine spoilage organisms (Bokulich et al., 2013). Robust data on the degree that winery milieus influence the microbial profile in fermenting grape juice are relatively scant and not well understood at a systems level. However, there are some reports that indicate that winery surfaces harbor seasonally fluctuating microbial populations with site-specific dependencies shaped by technological practices, processing stage, and season. During each vintage, grape- and fermentation-associated microbes populate most winery surfaces, serving as potential reservoirs for microbial transfer between fermentations (Bokulich et al., 2013). Winery surfaces usually house a fair amount of alcohol-tolerant *S. cerevisiae* strains and other yeasts, which could potentially act as an important vector of these yeasts in wine fermentations. However, there is mounting evidence that resident microbial assemblages on winery surfaces, before and after harvest, comprise microorganisms with no known link to wine fermentations with almost no spoilage microbes, suggesting that winery surfaces do not overtly vector wine spoilage organisms under normal cleaning and operating conditions (Bokulich et al., 2013).

Regardless of the umpteen variables in the grape harvest and winery operating conditions, the yeast species generally found on grapes and in wines are similar everywhere in the world. However, the proportion or yeast population profile in various wine-producing regions projects distinct differences.

13.2.3 Yeasts from anywhere

Yeast *nomads* can be imported from anywhere as commercial starter culture strains to take control of the fermentation process by outcompeting bacteria and fungi, as well as *native* and *settler* yeasts. The concept of inoculating grape must with a starter strain of *S. cerevisiae* originated in the late 1800s but was based on discoveries that date back to the late 1600s. It all started in the 1670s when Antonie van Leeuwenhoek observed and described microscopically tiny creatures for the first time and Louis Pasteur, in the 1850s, proved that these subvisible entities were living yeast cells responsible for fermentation (Chambers & Pretorius, 2010; Pretorius, 2000). When Emil Christian Hansen succeeded in isolating the first pure yeast culture, Julius Wortmann and Herman Müller-Thurgau were quick to introduce the concept of inoculating wine ferments with pure yeast cultures in the 1890s. But it was not until 1965 that this innovation was taken a step further when the first pure culture of an *S. cerevisiae* wine yeast strain from Red Star became commercially available to winemakers in California (Jagtap, Jadhav, Bapat, & Pretorius, 2017). Since then, the commercial production of yeast used in the food and fermented beverage industries exceeds 1.8 million tons per year (Joseph & Bachhawat, 2014). It is estimated that there are approximately 250 commercial strains available to the global wine industry as *active dry yeast* (ADY) cultures.

Despite the availability and abundance of *easy-to-use* commercial starter culture strains with proven desirable enological characteristics, there is still an ongoing debate as to whether wine ferments should be allowed to occur by the action of grapevine-associated *native yeasts* and winery-residential *settler* yeasts or be driven by inoculated *nomadic* strains. However, this debate has now matured and transitioned from an *either-or* paradigm into a new paradigm with multiple options, i.e., spontaneous ferments, single-species ferments inoculated with one or more strains of *S. cerevisiae*, and mixed-species ferments inoculated with one or more non-*Saccharomyces* yeasts alongside at least one robust wine strain of *S. cerevisiae* (Belda et al., 2017; Jolly, Varela & Pretorius, 2014; Van Wyk et al., 2020). These kinds of fermentation options are referred to as *multistarter*, *mixed-culture*, and *coculture* ferments, in which strains can either be *sequentially* or *simultaneously* inoculated. These decision options enable modern-day winemakers to choose whether to leave their uninoculated grape musts until fermentation commences spontaneously or to guide their wine fermentations by using *single-strain* or *multistrain S. cerevisiae* inoculation strategies along with a mixture of selected non-*Saccharomyces* strains.

Every winemaker knows that when crushed grapes or must remain in a vat, fermentation will commence spontaneously after a while. They also know that, during the initial phases of fermentation, non-*Saccharomyces* yeasts are both present and active in all ferments until an alcohol level of 3%−4% is reached and before *S. cerevisiae* starts to dominate, whether inoculated or not (Fig. 13.4). The most sulfite- and alcohol-sensitive non-*Saccharomyces* yeast species will die off at this point, but some of the more resilient species could remain metabolically active in later phases of the fermentation. There are three groups of these non-*Saccharomyces* yeasts (Jolly, Varela, & Pretorius, 2014). The first group includes yeasts that are mostly aerobic, such as species of *Candida*, *Cryptococcus*, *Debaryomyces*, *Pichia*, and *Rhodotorula*. The second group comprises apiculate yeasts with low fermentative activity, such as *Hanseniaspora uvarum* (*Kloeckera apiculata*), *Hanseniaspora guilliermondii* (*Kloeckera apis*), and *Hanseniaspora occidentalis* (*Kloeckera javanica*). The third group consists of yeasts with a fermentative metabolism like *Kluyveromyces marxianus* (*Candida kefyr*), *Metschnikowia pulcherrima* (*Candida pulcherrima*), *Torulaspora delbrueckii* (*Candida colliculosa*), and *Zygosaccharomyces bailii*.

The only non-*Saccharomyces* yeast that is able to remain metabolically active at the end of fermentation when the alcohol concentration reaches 12%−15% is the spoilage yeast *Brettanomyces* (*Dekkera*). It is important to note that viable *Brettanomyces* cells are often not "culturable" in the laboratory (Capozzi et al., 2016). In practice, this means that, even in the presence of relatively high concentrations of SO_2, *Brettanomyces* cells can remain dormant as opposed to being eliminated. Winemakers should therefore always be on guard to maintain appropriate SO_2 levels during fermentation to avoid the "resuscitation" of this spoilage yeast during wine aging. The latter poses an ongoing challenge to the global wine industry in that strains of *B. bruxellensis*, which are more tolerant to higher levels of ethanol and sulfite than other non-*Saccharomyces* yeasts, can contaminate winery equipment, such as oak barrels, and when they remain metabolically active postfermentation, they could produce off-flavors. During wine maturation, particularly in oak barrels, *B. bruxellensis* can survive for prolonged periods of time and decarboxylate

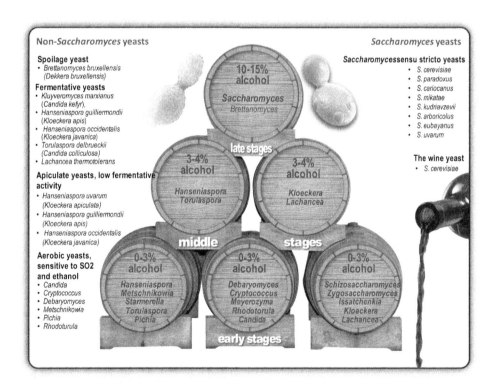

FIGURE 13.4 **The sequential participation of various yeast species during wine fermentation.** The selective pressures prevailing in fermenting grape must (e.g., sugar-induced osmotic pressure), fermentation conditions (e.g., temperature and pH), and winemaking practices (e.g., sulfite additions) inhibit and/or eliminate non-*Saccharomyces* yeast species during the course of fermentation. The selective nature of grape must becomes more pronounced once anaerobic conditions are established, nutrients become depleted, and rising alcohol levels start to constrain the survival of ethanol-sensitive yeasts.

hydroxycinnamic acids to *p*-coumaric acid, ferulic acid, and caffeic acid to their corresponding vinyl derivatives, which in turn can subsequently be reduced to 4-ethylphenol, 4-ethylguaiacol, and 4-ethylcatechol. These compounds impart an undesirable *leathery, medicinal*, or *metallic* aroma (Curtin & Pretorius, 2014; Dashko et al., 2015). The most practical way to keep *B. bruxellensis* at bay is for winemakers to apply a strict hygiene regime in their wineries (especially oak barrels), to maintain appropriate concentrations of free SO_2, and to inoculate the grape must with a fast-fermenting wine strain of *S. cerevisiae*. In other words, an effective way to safeguard wine against *Brettanomyces* spoilage is to conduct fermentation in clean wineries and under low-pH, high-sulfite, high-ethanol, and anaerobic conditions that restrict *Brettanomyces* but still allow *S. cerevisiae* to prevail and dominate (Dashko et al., 2015). However, this frontline in the battle against "Brett" might change if *B. bruxellensis* strains evolve and become more resistant to standard sulfite dosages used in winemaking. That is why genomic insights into the evolution of *B. bruxellensis* have gained so much interest in recent times.

13.2.4 Yeasts from elsewhere

S. cerevisiae S288c is a laboratory-bred *mannequin* yeast with superstar status in the world of supermodels and breakthrough scientific discoveries. This *out-of-this-world* single-cell model eukaryote tops the league table of A-listed supermodel organisms. *S. cerevisiae* stands tall in Life Sciences' *hall-of-fame* among other supermodel organisms, such as the bacterium *Escherichia coli*, the fruit fly *Drosophila melanogaster*, the nematode *Caenorhabditis elegans*, the plant *Arabidopsis thaliana*, and the zebrafish, *Danio rerio*. Some of the most profound fundamental principles of biology were uncovered by using these supermodel organisms. These landmark discoveries include the definition of subcellular structures and the function of organelles; the basic principles of heredity, including the structure of genomes, chromosomes and genes, the genetic code, the rules of DNA replication and transcription, mRNA translation and protein synthesis; the regulation of genes and metabolic pathways; and the development of recombinant DNA technologies (e.g., gene cloning and transformation) and in vitro amplification, sequencing, editing, and synthesis of DNA (Alfred & Baldwin, 2015).

Just as the role of the fashion world's supermodels is to herald what the future wears, rather than accurately representing the ordinary version of *Homo sapiens* in the street, model organisms, such as *S. cerevisiae* S288c, are fashioned for experimental use in laboratories and not necessarily to be a mirror-image of other members of their own species or their nearest relatives living at the wild frontier of the real world. For example, *S. cerevisiae* S288c does not have the robustness and effervescent power to rise bread dough, brew beer, or sparkle wine. However, building on the foundational milestone discoveries by Antonie van Leeuwenhoek (1676), Antoine-Laurent Lavoisier (1789), Joseph Gay-Lussac (1815), Friedrich Erxleben (1818), Charles Cagniard de la Tour, Friedrich Kützing, Theodor Schwann (1825), Julius Meyen (1837), Louis Pasteur (1857), Emil Hansen (1888), Eduard Buchner (1897), Øjvind Winge (1935), and Carl Lindegren (1943), strain S288c was specifically developed to control and manipulate its life cycle and genetics for research purposes (Pretorius, 2000). By several rounds of crossing selected parental strains and sporulation inductions, S288c was isolated as a strain that can be maintained as a stable haploid. Very early on, yeast researchers adopted S288c as the reference strain for studying the life and cell cycles of *S. cerevisiae*, genetic hybridization, genetic engineering, genome sequencing, and genome engineering. Today, S288c and its derivate haploid strains of both mating-types (a and α) are the mainstay yeast strains for genetic, genomic, transcriptomic, proteomic, metabolomic, fluxomic, and interactomic research, to name but a few (Liti, 2015).

The superstar status of *S. cerevisiae* is boosted by its proven safety track record as a long-time domesticated, food-grade yeast, its compartmentalized subcellular structure (including an encapsulated nucleus) typical of a eukaryote, and the uncomplicated, inexpensive way of culturing it rapidly in the laboratory (Pretorius, 2000, Pretorius, 2017). Under optimal culturing and nutritional conditions, all three basic cell types (a, α, and a/α) can double their mass every 90 min through an asexual mitotic budding process. Heterothallic a/α diploids lacking the *HO* mating-type switch gene, can be induced to undergo meiosis and sporulate, thereby generating stable haploids of both mating-types (*MAT*a and *MAT*α). In turn, *MAT*a and *MAT*α haploids can mate and give rise to *MAT*a/*MAT*α diploids capable of sporulating and generating four new ascospores per ascus (tetrad), two of each mating-type. Homothallic haploids are less useful for classical genetic analyses because they can switch their mating-types from *MAT*a to *MAT*α and vice versa, and then self-mate. So haploid ascospores derived from homothallic diploids can establish diploid lines (i) by mating with their own mitotic daughter cells after a mating-type switching event (*haploselfing*); (ii) by mating with another sibling ascospore stemming from the same meiotic event (intra-tetrad mating); or, more rarely, (iii) by mating with an unrelated individual (*outcrossing*) (Liti, 2015). In homothallic haploids, the *HO* gene can utilize the information from the silent *HML* and *HMR* loci located on either side of the *MAT* locus on Chromosome 3 and dictate the switching between *MAT*a and *MAT*α. The *haplontic* phase of homothallic strains is therefore much shorter than the *diplontic* phase of their sexual life cycle. Both heterothallic and homothallic strains can asexually reproduce in the *MAT*a and *MAT*α haploid state or the state of higher ploidy (from diploids to heptaploids) and aneuploidy (abnormal number of chromosomes per cell, i.e., not 16, 32, 48, etc.). Most laboratory strains of *S. cerevisiae* are heterothallic haploid or diploid strains, whereas industrial wine strains can be either heterothallic or homothallic, and they are mostly diploid or aneuploid, and occasionally polyploid (Pretorius, 2000, 2017, 2017, 2020). The ability to control the life cycle of *S. cerevisiae* and to switch it between mitotic and meiotic reproduction, and to develop strains with different ploidies surpass the experimental flexibility of any other model organism.

To fully understand what makes wine-related yeasts tick, it is important that we bridge the faultline between *cellular-molecular-developmental* research into the decontextualized *S. cerevisiae* S288c model strain and *ecological-evolutionary* research into the natural histories of both *Saccharomyces* and non-*Saccharomyces* yeasts in the wild (Liti, 2015). For example, recent population genomics have illuminated the evolutionary history and natural genetic variations within subpopulations of *S. cerevisiae* (Almeida et al., 2015; Belda, Ruiz, Santos, Van Wyk, & Pretorius, 2019; Goddard & Greig, 2015; Hyma, Saerens, Verstrepen, & Fay, 2011; Langdon et al., 2019; Legras et al., 2018; Marsit & Dequin, 2015; Steensels, Gallone, Voordeckers, & Verstrepen, 2019). By studying its life cycle in natural settings with fluctuating environmental conditions, yeast ecologists and molecular biologists are closer to understanding the origin of *S. cerevisiae*. One of the natural habitats of *S. cerevisiae* is oak bark, which is subject to seasonal changes and cycles of sap flow in *Quercus* oak trees. Cells of *S. cerevisiae* are therefore likely to spend most of their lifetime in a nondividing (*quiescence*) state. Population genomic studies also indicated that budding yeast mostly reproduce asexually and that outcrossing (outbreeding) is rare in the wild but is not restricted to mating within a species (Liti, 2015). Introgressed genomic regions and interspecies hybrids between *S. cerevisiae* and other members of the *Saccharomyces sensu stricto* complex can generate viable hybrids when interbred. For example, hybridization between *S. cerevisiae* and *S. eubayanus* generated the hybrid species *S. pastorianus*, which is now widely used in the brewing industry (Langdon et al., 2019; Liti, 2015). Deliberate genetic breeding (crossing, spheroplast fusion, and rare-mating) has also been successfully applied to generate superior bread, beer, and wine strains of *S. cerevisiae*.

In addition to human-related environments, such as baking, brewing, and winemaking, population genomics has also uncovered *S. cerevisiae* strains in primary forests in China that are remote from human activity, thereby indicating that this yeast species has a distribution more widespread than what was previously been postulated (Wang, Liu, Liti, Wang, & Bai, 2012). This brings into question a long-held notion that *S. cerevisiae* is a *man-made organism* and queries whether it is only a coincidence that winemakers have favored oak (*Quercus alba*, *Quercus petraea*, and *Quercus robur*) as cooperage material for fermentation or maturation vessels.

Genetic variants within some lineages of *S. cerevisiae* have been shown to be nearly unique to subpopulations, such as European wine strains; Malaysian bertram palm-associated strains; strains from North American woodlands; Japanese saké strains; and West African strains associated with food and beverage fermentations (Borneman et al., 2011, 2012; Borneman, Forgan, Pretorius, & Chambers, 2008; Borneman et al., 2013; Borneman et al., 2013). In some cases, phenotypic variation tends to follow population structure and some of these lineages are characteristic of domesticated breeds linked to distinct fermentation processes. In one sense, domesticated strains transcend geographic boundaries, share recent ancestry, and reflect human migration history, including the transport of wine in oak barrels. In other cases, it is clear that some lineages are not linked to human activity; rather they are characteristic of specific geographic areas. For instance, the ancient and surprisingly divergent lineages within the well-structured population of Chinese isolates from primeval forests showed to be a remarkable reservoir of natural genetic variants for future investigations and potential applications.

Distinctions between model and nonmodel yeasts will become increasingly clear as yeast ecologists, organismal biologists, and molecular geneticists apply metagenomic tools to their *broadband* mega field surveys of nonmodel yeast strains and their *laser-focused* laboratory-based genomic analyses of *S. cerevisiae* model strains (Alfred & Baldwin, 2015; Liti, 2015). This does not imply that the in-depth molecular probing of a laboratory-bred strain like S288c should be slowed down. Its well-understood sexual and asexual reproduction cycles, along with the accessibility of its genetic system and the ease of gene transformation procedures make *S. cerevisiae* irreplaceable as a trailblazer for yeast research. This tractable supermodel yeast of GRAS status can rise to almost any challenge that contemporary science can pose. *S. cerevisiae* is amenable to nearly all types of genetic modification (breeding, mutagenesis, and cloning) in the pursuit of probing the fundamental intricacies regarding molecular and cellular aspects of biology (Pretorius, Curtin, & Chambers, 2012).

The versatility of *S. cerevisiae* as an unequaled eukaryotic supermodel organism in research laboratories and the most used microbial workhorse in many fermentation industries is evidenced by several *world-firsts* (Pretorius, 2017b). In ancient times, unknowingly and serendipitously, this yeast was the first microbe to be domesticated for the production of wine and other fermented products. It was also the first microorganism to be observed under a microscope and described as a living biochemical agent responsible for the transformation of sugar into alcohol and carbon dioxide. In modern times, *S. cerevisiae* became the first host organism for the production of a recombinant vaccine (against hepatitis B) and a recombinant food enzyme (the milk-coagulating enzyme, chymosin, for cheese making). Nowadays, *S. cerevisiae* is the most popular microbial cell factory for a range of products.

These practical applications are backed up by an extensive, online searchable database (www.yeastgenome.org) and research tools. The S288c strain became the first eukaryote to have its whole genome sequenced, and from there a full set of libraries of gene deletions, overexpression mutants, and genes tagged by reporter genes were developed (Belda, Ruiz, Santos, Van Wyk, & Pretorius, 2019; Goffeau, Barrell, & Bussey, 1996; Oliver, 1996). The availability of such powerful genomic toolkits enabled researchers to investigate its 12 Mb (nonredundant) genome, distributed over 16 linear chromosomes (varying in size between 200 kb and 2000 kb), inside and out. The total genome size of 14 Mb includes 12.07 Mb of chromosomal DNA, 85 kb of mitochondrial DNA, and 6.3 kb of episomal plasmids (2μ). The genome contains 6604 open reading frames (ORFs) with 79% of the ORFs verified, 11% uncharacterized, and 10% regarded as dubious. A total of 1786 ORFs are still assigned to unknown functions. The S288c genome carries 428 RNA genes (299 tRNA, 77 snoRNA, 27 rRNA, 18 ncRNA, 6 snRNA), one telomerase RNA, and 295 introns in 280 genes with 9 genes containing more than one intron (Engel et al., 2014; Peter et al., 2018). By comparing the S288c genome to the exponentially growing genome sequences of other *S. cerevisiae* strains, at least 55 genes of the best-studied strains were found to be absent in S288c. There are more than 500 sets of paralogs. As more *S. cerevisiae* genomes are being sequenced, the Saccharomyces Genome Database (SGD) is constantly being updated and equipped with online search and analysis software. Remarkable discoveries facilitated by these genomic toolkits include the unraveling of the genetic and protein interaction networks, a prelude to full comprehension of the yeast interactome.

These highly useful resources and supporting frameworks developed for *S. cerevisiae* will increasingly open up opportunities to accelerate research into nonmodel *S. cerevisiae* strains, other members of the *Saccharomyces sensu stricto* group, and non-*Saccharomyces* yeasts. For example, these metagenomics, transcriptomics, and proteomics tools spearheaded by S288c could shed a brighter light on the natural history of *S. cerevisiae*: how the life cycle of native strains progress in the wild; how they interact with other microbes in natural habitats; and what the extent of strain variation is in those natural habitats. Such information could help population biologists to conduct *reverse ecology* and to establish more definitive lineages within domesticated and wild yeast populations (Liti, 2015). The efforts of the yeast research community will be well supported by applying these multiomics and synthetic genomic technologies to non-*Saccharomyces* wine yeasts, such as *Torulaspora delbrueckii*, *Pichia kluyveri*, *Lachancea thermotolerans*, *Candida/Metschnikowia pulcherrima*, *H. guilliermondii*, and *H. uvarum* (Liu, Redden, & Alper, 2013; Seixas et al., 2016).

In summary, *S. cerevisiae* celebrity strains are the trendsetting *mannequins* modeling the future world of yeast-related discoveries, applications, and innovations. Research findings with these supermodel strains could assist in extending foundational knowledge on and novel technologies of nonmodel *S. cerevisiae* strains and non-*Saccharomyces* yeasts.

13.2.5 Yeasts from nowhere

Yeast avatars are prototypic creations of Synthetic Biology and, in extreme instances, might even be viewed by some as *yeasts with no ancestry*: computer-designed yeasts from *nowhere*. With the confluence of modern-day biomolecular sciences, information technology, and engineering, the DNA of yeasts can now be redesigned, reinvented, rewritten, and edited with astounding precision (Duan et al., 2010; Gibson & Venter, 2014; Lajoie et al., 2013; Mercy et al., 2017; Pretorius & Boeke, 2018). Engineering the biology of model and nonmodel yeast strains (including clonal variants of natural isolates, mutants, hybrids, and genetically engineered genetically modified [GM] strains) with laser-sharp accuracy can stretch the realms of possibility in yeast research and wine yeast innovation. By applying basic engineering principles (involving the classic rational *design*, *build*, *test*, and *learn* cycle) in high-throughput, automated biofoundries with robotic workflows and technology platforms (Chao, Mishra, Si, & Zhao, 2017; Dixon, Curach, & Pretorius, 2020; Hillson et al., 2019; Walker & Pretorius, 2018), the speed with which synthetic and semisynthetic prototypes can be developed is accelerating at a break-neck pace (Fig. 13.5). These biofoundry-based workflows encompass the computational design of DNA genetic parts, physical assembly of

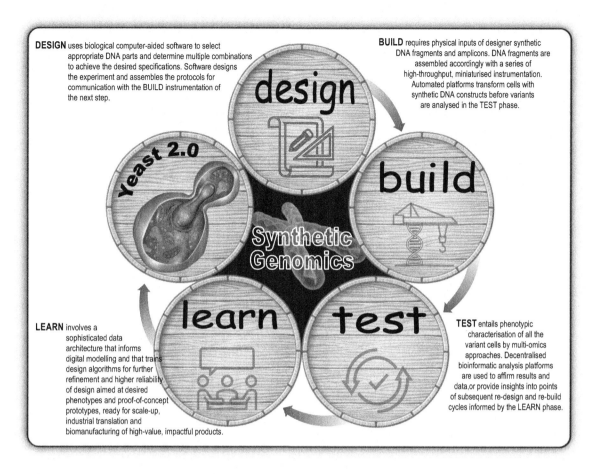

FIGURE 13.5   The *design-build-test-learn* (DBTL) biological engineering cycle. Application of the DBTL cycle can be accelerated in high-throughput, automated biofoundries with robotic workflows and technology platforms in Synthetic Biology. Recent rapid advances in high-throughput DNA sequencing (*reading*) and DNA synthesis (*writing* and *editing*) techniques are enabling the design and construction of new biological parts (genes), devices (gene networks), and modules (biosynthetic pathways), and the redesign of biological systems (cells and organisms) for useful purposes.

designed DNA parts, and prototyping and testing the performance of designs in living cells, followed by applying modeling and computational learning tools to inform the design process. Iterations of the *design-build-test-learn* cycle in biofoundries result in genetic designs that aim to fulfill the design specifications. For the time being, most of the Engineering Biology research into eukaryotes is focused on *S. cerevisiae*, primarily to unearth more secrets of the inner workings of this supermodel yeast as a prelude to expanding that knowledge to other *Saccharomyces* and non-*Saccharomyces* yeasts and to push the performance of industrial strains to greater heights.

Numerous game-changing technologies and milestone breakthroughs that followed the first yeast transformation in 1978 and the release of the first whole-genome sequence of *S. cerevisiae* in 1996 have hastened the capacity to analyze and manipulate the yeast genome with remarkable success. The wealth of genomic data generated in this way ushered yeast research into an age of Synthetic Biology where genetic variation is rationally designed and evolution is harnessed and fast-tracked. At the core of this new era of Engineering Biology lies hundreds of publicly available yeast genome sequences and the capacity to *de novo* synthesize genes, chromosomes, and genomes (Annaluru et al., 2014; Mitchell et al., 2017; Pennisi, 2014; Pretorius & Boeke, 2018; Richardson et al., 2017; Shen et al., 2017; Wu et al., 2017; Xie et al., 2017; Zhang et al., 2017) to edit DNA with *Clustered Regularly Interspaced Short Palindromic Repeats* (CRISPR-Cas9, CRISPR-CPF1, and other variations) technologies and to generate large-scale genetic diversity with *Synthetic Chromosome Rearrangement and Modification by LoxP-mediated Evolution* (SCRaMbLE) via *Cre*-recombinase induction (Blount et al., 2018; Hillson et al., 2019; Jia et al., 2018; Liu et al., 2018; Luo, Wang, & Wang, 2018; Shen, Wu, & Yang, 2018; Shen et al., 2016; Wu et al., 2018).

In much the same way as a genome sits at the heart of a yeast cell, the synthetic Sc2.0 genome occupies the center of the application of the CRISPR and SCRaMbLE technologies to karyotype engineering, genetic variant generation, and protype strain development. The Sc2.0 genome was designed to probe some vexing questions about the fundamental properties of chromosomes, genome organization, chromosome number, gene content, and annotation; function of RNA splicing; the extent to which small RNAs play a role in yeast biology; the distinction between prokaryotes and eukaryotes; and questions relating to the genome structure and evolution while recognizing that the eventual "synthetic yeast" being designed and refined could ultimately play an important practical role (see www.syntheticyeast.org).

The first draft of the 16 synthetic chromosomes is now complete, but there remains work to be undertaken to ensure that the growth of a strain carrying these redesigned chromosomes in a single cell would be on par with that of the original strain (Pretorius & Boeke, 2018; Sliva, Yang, Boeke, & Mathews, 2015). The Sc2.0 genome was designed to contain specific base substitutions within some of the ORFs to accommodate desirable enzyme recognition sites or deletions of undesirable enzyme recognition sites. This designer genome also includes recognizable PCRTags (short recoded sequences within certain ORFs to enable a polymerase chain reaction [PCR]-based assay) so that the synthetic DNA can be differentiated from native DNA. Other important variations between the Sc2.0 genome and that of the native strain are the addition of multiple loxPsym sites for future genome reshuffling purposes; all TAG stop codons being recoded to TAA to free up one codon for future inclusion of unusual amino acids into *new-to-nature* proteins and enzymes; all repetitive and dispensable sequences like the five families Ty retrotransposons (a total of 50 copies, each flanked by long terminal repeat [LTR] sequences), pre-tRNA and pre-mRNA introns, subtelomeric regions, and silent *HML* and *HMR* mating-type loci being omitted from the design; and all tRNA genes being relocated to a novel neochromosome (Richardson et al., 2017). The decision to delete the retrotransposons and their LTRs from the design was to remove as much dispersed repetitive DNA as possible from the genome, thereby potentially delivering a more stable synthetic genome free of mobile elements. The pre-mRNA introns were accurately deleted from the design, excepting (for now) those genes with evidence of fitness defects caused by intron omission (Richardson et al., 2017). For example, the *HAC1* intron, which uses separate splicing mechanisms and is known to play a critical role in the unfolded protein response, was retained in the design. The rationale for the relocation of all tRNA genes to a specialized neochromosome encoding only tRNA species was based on the fact that tRNA genes lead to genome instability by replication fork collapse (Richardson et al., 2017). The latter might be caused by a collision with tRNA polymerase PolIII and/or the formation of R-loops at actively transcribed tRNA genes, which impede the replication fork in a polar manner, ultimately causing replication fork stalling and subsequent repair through recombination.

The draft set of 16 synthetic Sc2.0 chromosomes has already been put to good use to answer some profound fundamental biological questions, thereby helping researchers to understand what basic genomic features and genetic combinations are essential for cell viability. For example, one of the most fundamental characteristics of the *S. cerevisiae*'s genome queried was the number of chromosomes in each cell and the industrial relevance thereof (Gorter de Vries, Pronk, & Daran, 2017; Luo et al., 2018). Chromosome number varies wildly across eukaryotes. Thus two basic questions to ask would be why does a haploid *S. cerevisiae* strain distribute its genomic DNA along 16 chromosomes, and how well would it tolerate a change in its chromosome number without substantial changes to its genome content. Two independent studies were conducted to answer these intriguing questions. In one study, CRISPR-Cas9-mediated genome editing was used to fuse *S. cerevisiae*'s chromosomes and generate a near-isogenic series of strains with progressively fewer chromosomes until the whole genome was compacted into two chromosomes (Luo et al., 2018). These researchers found that as the number of chromosomes dropped below 16, spore viability decreased dramatically. However, homotypic crosses between pairs of strains with 8, 4, and 2 chromosomes produced good spore viability, demonstrating that eight chromosome fusion events suffice to isolate strains reproductively. In another study, the DNA of *S. cerevisiae*'s 16 native linear chromosomes was squeezed into a single chromosome by successive end-to-end chromosome fusions and centromere deletions (Shao et al., 2018). Although the strain carrying the giant chromosome supported viability, it did show reduced growth and less competitiveness across different culturing conditions. Nevertheless, these karyotype engineering experiments uncovered the surprising insight that *S. cerevisiae* copes remarkably well with one or two mega chromosomes instead of sixteen. These two studies indicated that chromosome number seems to reflect 'accidents of genome history', such as telomer−telomer fusions and genome duplication events (Luo et al., 2018; Shao et al., 2018).

The learnings from the Sc2.0 project are also advancing the frontiers of researchers' understanding of how they can expand the genetic range of yeast prototypes and reshuffle the genome for industrial applications. To improve the complex rearrangements achieved by SCRaMbLE, the original method has been adapted to accelerate and expand the diversity of prototypical strains and to rapidly identify them. Two such adaptations are MuSIC (*Multiplex SCRaMbLE Iterative Cycling*) and ReSCuES (*Reporter of SCRaMbLEd Cells using Efficient Selection*)

(Jia et al., 2018; Luo et al., 2018). In one example, MuSIC was used to put yeast strains through five cycles of SCRaMbLE under selective pressure, thereby driving increases in catenoid titers. In another example, ReSCuES was utilized to quickly screen for successfully SCRaMbLEd cells by integrating two selective markers, one of which was functional before SCRaMbLE and one only after the *Cre* recombinase inverted their orientation (Luo et al., 2018). To enhance recombination control, red light was employed to control *Cre* (Hochrein, Mitchell, Schulz, Messerschmidt, & Mueller-Roeber, 2018). In addition, the principles of SCRaMbLE have been extended from an in vivo to an in vitro technique. It also appears that mating *S. cerevisiae* carrying synthetic Sc2.0 chromosomes with wild-type *S. cerevisiae* strains or closely related *Saccharomyces* species can overcome the potential for highly desirable recombination events even if they impose severe growth defects.

These improvisations boosted the SCRaMbLE toolkit to rapidly generate an enormous amount of genetic variation for both basic research and applied industrial purposes. For example, by applying SCRaMbLE in a test tube rather than a cell, a pathway or a set of genes can now be removed from the complexity of the genome and used to quickly generate huge prototype variations in a single pathway (Liu et al., 2018; Wu et al., 2018). In other instances, when SCRaMbLE was applied to strains containing the synthetic version of Chromosome V, strains were generated with marked increases in their capacity to produce violacein and penicillin, or in their capacity to utilize xylose as a carbon source (Blount et al., 2018).

SCRaMbLE is now also being applied to hundreds of different *S. cerevisiae* strains with distinctive phenotypes that provide an advantage in a specific environmental niche or industry, such as baking, brewing, and winemaking. These phenotypic differences are the direct result of specific genetic variation among strains and this can range from single nucleotide polymorphisms (SNPs) to the presence of strain-specific genes or gene clusters (Borneman et al., 2011; Borneman et al., 2013; Borneman et al., 2013). The presence or absence of these genes among strains can have remarkable phenotypic consequences, including providing strains with the ability to synthesize vitamins or to endure specific types of stress or inhibitory compounds. To provide greater insight into the role of these strain-specific genes, researchers identified over 200 kb of nonrepetitive DNA encoding 75 ORFs, which exist across the breadth of strain-specific ORF diversity of the *S. cerevisiae* pan-genome, but which are absent from the laboratory strain used for Sc2.0 (Borneman et al., 2011; Borneman et al., 2013; Borneman et al., 2013). These sequences have been synthesized and assembled into a neochromosome. This pan-genome neochromosome is now being analyzed to determine the phenotypes that it can impart in a laboratory strain background, while also providing a resource for introducing additional variation into the Sc2.0 genome through processes such as SCRaMbLE.

These developments bode well for future applications of SCRaMbLE for the improvement of a diverse range of unrelated metabolic pathways that are of commercial interest, including pathways that could enhance the aroma of wine or lead to a reduction in the level of alcohol concentration in light-bodied wines (Goold et al., 2017; Kutyna & Borneman, 2018; Timmins, Kroukamp, Paulsen, & Pretorius, 2020; Van Wyk, Kroukamp, & Pretorius, 2018). The in vivo and in vitro prototyping of rearranged pathways involving pan-genomic neochromosomes or *avatar* strains of *S. cerevisiae* with altered karyotypes will undoubtedly accelerate the uncovering and harnessing of the full potential of diversity found in the yeast genome.

Another frontier that is currently igniting the imagination of some adventurous synthetic biologists is the potential to generate synthetic yeast organelles. Compartmentalization is a fundamental mechanism for eukaryotic cells to segregate and sequester valuable biomolecules (Pretorius, 2000). Yeast organelles compartmentalize important cellular processes, such as energy production in mitochondria and the decoration of proteins with sugars and their packaging into membrane-bound vesicles for intracellular sorting and secretion by the Golgi apparatus. The ability to concentrate substrates together for particular reaction pathways, and separate them from competing reactions, enables highly efficient biosynthesis of valuable compounds (e.g., desirable flavor-active compounds). The separation of toxic chemicals or production of chemical gradients allows unique reactions to occur that would be impossible in a single compartment. However, despite the benefits of compartmentalization, natural organelles are limited by numerous demands imposed by yeast cells or the convoluted evolutionary trajectories that created them. In the medium to long future, the untapped potential of compartmentalization in yeast cells is set to be unleashed by the re-engineering of existing organelles and building completely new organelles from scratch (Pretorius, 2000).

These transformative synthetic genomic tools and concomitant prototypical yeast *avatars* will only translate into practical applications in the winery if they can tread lightly over the genetically modified organism (GMO) quicksand where consumer perceptions are their reality regardless of numerous scientific reassurances of safety. In this quagmire that has trapped so many GM food products, it has been shown repeatedly that it is irrelevant whether a product is safe if people refuse to consume it for psychosocial or cultural reasons. Although CRISPR editing technologies provide researchers with the ability to redesign with surgical precision specific genes and gene networks without having to introduce foreign DNA from other organisms, it remains to be seen whether consumers will take advantage of what *avatar* wine yeasts are able to offer them. At present, producing commercial wine with *avatar* strains seems distant. So, what will it take to persuade the public that the benefits of, for example, an *avatar* yeast strain capable of producing low-alcohol wines with superior favor profiles outweigh any safety risks and fears for the erosion of the sacrosanct concept of *terroir*? Understanding what drives public sensitivities and consumer perceptions, preferences, and purchase decisions is key to answering this question in the context of an archetypal traditional product such as wine with such strong territorial and socio-cultural connotations (Pretorius, 2020).

13.3 Linking yeast with wine quality and consumer gratification

We humans like the good things in life. Our dogged search for pleasure inspires and motivates us and drives industries such as the wine sector to meet demand. But that demand is changing. At any time on any given day, someone somewhere is enjoying a good glass of wine. But what do we mean by good? Is it the appearance, aroma, taste, or texture? Professional wine tasters would say all of the above. The reality is more complex. Last century, it was acceptable for wines to just be *good*. Today, wines have to be *great* in order to stand out in overcrowded consumer markets. *Good* is not enough when your competition is 30 billion liters of wine and an annual global surplus of 15% (Pretorius, 2016). *Great* requires a total commitment to innovate. Today's fickle consumer expects their wine to provide a sensory experience; to be safe; to be produced in an environmentally sustainable way; and to measure up to an indefinable mystique (Pretorius & Høj, 2005). Inventive winemakers recognize that everything worthwhile in today's consumer market is uphill and, to successfully meet the ever-changing demands of wine drinkers, ingenuity and innovation are required every step of the way, every day, all the way. Innovative winemakers realize that the success of their carefully crafted wines is often at the mercy of increasingly tech-savvy consumers who communicate their likes and dislikes globally and instantaneously via websites, blogs, and social media (Jagtap, Jadhav, Bapat, & Pretorius, 2017). In this *click-like/dislike-share* era, consumers expect more than products and services in return for their money. They expect a pleasurable multisensory experience that embraces a *value chain* from where a grape is grown to how the wine is produced and packaged. For winemakers, innovation means understanding their consumers and exceeding their expectations. They must anticipate what wine drinkers of today and tomorrow want to see, smell, touch, taste, feel, and hear (Swiegers et al., 2005). To succeed in doing so, winemakers increasingly rely on rigorous science and technical know-how to create great experiences with wines. Truly great wines are born of great marriages between grape variety and *terroir* on the one hand, and technology, innovation, and craftsmanship on the other. In this century, winemaking has become an industry where art meets science including the science that underpins yeast biology and the fermentation process. Backed by robust research and evidence-based data, winemakers can put their own individual stamp on the production chain by making well-informed and insightful decisions as to what yeast strain or combination of yeast strains will best guide their products to market success (Cordente, Cordero-Bueso, Pretorius, & Curtin, 2013; Cordente, Curtin, Varela, & Pretorius, 2012; Cordente, Heinrich, Pretorius, & Swiegers, 2009; Dzialo, Park, Steensels, Lievens, & Verstrepen, 2017; Gallone et al., 2016; Hyma, Saerens, Verstrepen, & Fay, 2011; Swiegers et al., 2005; Swiegers et al., 2005; Van Wyk, Grossmann, Wendland, Von Wallbrunn, & Pretorius, 2019; Van Wyk et al., 2020).

The expectations and preferences of wine drinkers often use the terms *quality* and *value* when they compare different wines for purchase. In this context, the term *quality* relates to the "intrinsic" quality of the wine, meaning how the wine gratifies on appearance, the nose, and the palate, as well as the perceived value. When consumers use the term *value*, they usually refer to both the *intrinsic value* and the *image* of a wine in relation to the price (Pretorius & Høj, 2005). The image of a product depends on how a wine is marketed, the origin, regionality, and *terroir* of a wine, how environmentally sound the winery's practices are, how many medals have been awarded at wine shows, how high sommeliers and other market influencers rate a wine, and price. Consumers will consider a wine high in value if the product is sensorially pleasing and recognizable, and perceived as high in the image at a competitive price.

In terms of how much *value* consumers attach to a product, the single most important factor that defines a consumer's perception of wine is its organoleptic quality, which involves four senses: sight, smell, taste, and touch (Fig. 13.6). Professional wine tasters would normally describe what they sense by referring to a wine's appearance, aroma, flavor, and texture (Pretorius, 2016; Timmins, Kroukamp, Paulsen, & Pretorius, 2020). *Appearance* (sight, e.g., cloudy, hazy, deposit in the bottom of the glass, depth of color, hue, mousse), *nose* (smell, e.g., aroma and bouquet), and *palate* (taste and touch—flavor and mouthfeel). The term *aroma* is typically used to describe the fragrant smell

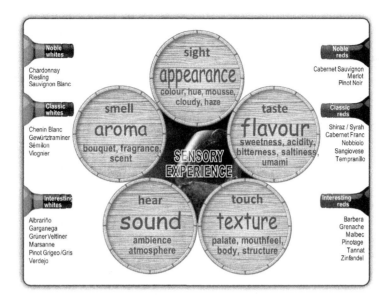

FIGURE 13.6 **The sensorial quality of wine involves all five senses: sight, smell, taste, touch, and sound.** The *appearance* of wine can be affected by cloudiness, haziness, a deposit in the bottom of the glass, and the depth of color, hue, and mousse. The *smell* of wine refers to both the *aroma* and the *bouquet*. A wine's *palate*, *taste*, and *touch* refer to its *flavor* and *mouthfeel*.

of a young, fresh wine whose *primary* aromas and upfront fresh-fruit notes originate during yeast fermentation. *Bouquet* is the term for more mature wines that are less fresh but more complex thanks to *secondary* aromas (e.g., stewed fruit notes) stemming from oak maturation and *tertiary* aromas (e.g., dried fruit and honey notes) forming during bottle aging. The term *flavor* refers to sweetness, acidity, bitterness, saltiness, and the taste of umami. The descriptor of *mouthfeel* relates to the texture and body of wine as affected by factors such as alcohol strength (sensation of warmth) and tannins (drying sensation). The *structure* of a wine describes its acidity, sweetness, bitterness (occasionally), tannin (in red wine), alcohol, palate weight and length, mouthfeel, mousse (in sparkling wine), as well as the intensity of fruit aroma and flavor, and complexity (diversity and layers of flavor).

A wine's sensory quality is largely determined by the presence of desirable flavor compounds and metabolites in a well-balanced ratio, and the absence of undesirable ones. The *absolute* and *relative* concentrations of these flavor-active compounds are important in shaping a wine's smell and taste (Swiegers et al., 2005). The choice of an *S. cerevisiae* starter strain in single-strain fermentations, or a combination of *Saccharomyces* and/or non-*Saccharomyces* yeasts in multistrain and multispecies fermentations, is by far the most cost-effective, time-efficient, and flexible way to shape the aroma and flavor according to changing consumer preferences than to replant a vineyard, manage climatic conditions, or adapt viticultural practices (Jolly, Varela, & Pretorius, 2014). This is the primary driver behind the never-ending search for naturally occurring yeast variants inhabiting different *terroirs*, the genetic improvement of well-characterized yeast strains, and the exploration of various combinations of compatible *Saccharomyces* and non-*Saccharomyces* starter yeast cultures aimed at meeting the shifting consumer preferences in specific segments of the market.

13.3.1 Flavor-active non-Saccharomyces wine yeasts

To meet the preferences of consumer markets where the appeal of *terroir, typicity*, and wine *complexity* dominates purchasing decisions, a growing number of winemakers are experimenting with non-*Saccharomyces* yeast starter cultures. Some winemakers pursue this goal by conducting spontaneous fermentation, while others, who wish to avoid the risks of sluggish/stuck ferments and spoilage often associated with uninoculated fermentations, search for coculturing candidate yeasts in the biodiversity treasure trove of their vineyards and culture collections. There is a growing list of these so-called *unconventional* yeasts as potential candidates for coculturing with stalwart *S. cerevisiae* wine strains. So far, the non-*Saccharomyces* species that have received the most attention in this regard include *Candida stellata, Candida railenensis, Debaryomyces hansenii, Debaryomyces vanriji, H. uvarum, Hanseniaspora vineae, Kazachstania gamospora, L. thermotolerans, Metschnikowia pulcherrima, Nakazawaea ishiwadae, Pichia fermentans, P. kluyveri, Rhodotorula mucilaginosa, Schizosaccharomyces pombe, T. delbrueckii, Williopsis saturnus*, and *Zygotorulaspora florentina* (Jolly, Varela, & Pretorius, 2014; Van Wyk et al., 2020; Van Wyk et al., 2020; Van

Wyk et al., 2020). These multispecies fermentation strategies seem to be an effective way to mimic flavor-diversity and product complexity outcomes obtained by successful spontaneous fermentation without forfeiting fermentation performance, reliability, and overall wine quality.

In addition to the use of these non-*Saccharomyces* yeasts in cocultured ferments, winemakers also apply a range of different inoculation regimes (Pretorius, 2000, 2016, 2017; Van Wyk et al., 2020). Some inoculate their ferments sequentially first with the non-*Saccharomyces* species and allow those yeasts to participate in the fermentation process before they inoculate the fermenting must with a selected *S. cerevisiae* starter culture strain(s). In this way, the non-*Saccharomyces* species have sufficient time to contribute to the sensory profile of the wine before they are constrained by high alcohol concentrations stemming from fast-fermenting alcohol-tolerant *S. cerevisiae*. The potential downside of such a sequential inoculation regime is that the non-*Saccharomyces* could use up some nutrients that are essential to *S. cerevisiae* during the later stages of the fermentation process. It is for that reason that some winemakers choose to coinoculate non-*Saccharomyces* and *S. cerevisiae* starter cultures at the same time. However, there are also variations within this strategy. Winemakers could opt for simultaneous inoculation of non-*Saccharomyces* and *S. cerevisiae* starter cultures but start off by using a much lower dosage (cell count) of *S. cerevisiae*, thereby giving some "first-mover" advantage to the slower non-*Saccharomyces* fermenters in the early phase of fermentation. The initial ratio of non-*Saccharomyces* versus *S. cerevisiae* yeasts in the starter culture is therefore another important decision point in conducting controlled cocultured wine fermentations. These choices will depend on the specific species and strains of non-*Saccharomyces* and *S. cerevisiae* (and their state of metabolic activity) in the cocultured "yeast starter cocktails" (Pretorius, 2016; Van Wyk et al., 2020). It is also important that the selected non-*Saccharomyces* species be compatible (nonantagonistic) with the companion *S. cerevisiae* strain, whether sequentially or simultaneously inoculated. Another point to consider is whether the non-*Saccharomyces* species can be produced as active dried cultures. It is well known that yeast manufacturers often struggle to optimize the growth conditions and/or drying conditions in their factories for some of these non-*Saccharomyces* yeasts. Despite these challenges, there are now commercialized yeast starter culture products available to winemakers, and some of these multispecies and multistrain 'yeast blends' are prepackaged in optimal ratios for fermentation performance and sensory outcomes.

Varying degrees of success with multispecies ferments have been reported. Generally, a decrease in volatile acidity (responsible for vinegary aromas) is observed with such coculturing because several non-*Saccharomyces* species (e.g., *T. delbrueckii*) produce low concentrations of acetic acid (Bely, Stoeckle, Masneuf-Pomarède, & Dubourdieu, 2008). Multispecies ferments also tend to produce an increase in ester production, especially with *Hanseniaspora* and *Metschnikowia* species (Martin, Valera, Medina, Boido, & Carrau, 2018). Some non-*Saccharomyces* species secrete flavor-enhancing enzymes, such as β-glucosidase and β-lyase (Belda, 2016; Belda, Ruiz, Navascués, Marquina, & Santos, 2016). It is important to note that there are significant strain variations within the same non-*Saccharomyces* species in terms of their capacity to influence the sensory profile of the final product.

Several studies have reported positive sensory outcomes when certain non-*Saccharomyces* species were paired up with *S. cerevisiae* in multispecies ferments. For example, *D. vanriji* (Garcia et al., 2002) was reported to increase the concentration of geraniol in Muscat wines while *K. gamospora* (Dashko et al., 2015) was found to increase the levels of phenylethyl alcohol and phenylethyl acetate in Ribolla Gialla. In another study, *H. vineae* also produced increased levels of phenylethyl acetate in Tempranillo (Viana, Belloch, Vallés, & Manzanares, 2011). *P. kluyveri* is known to increase the concentration of volatile thiols in Sauvignon blanc (Anfang, Brajkovich, & Goddard, 2009). Increased ester concentrations were observed in Cabernet Sauvignon, Riesling, Merlot, and Sangiovese when *L. thermotolerans* (Gobbi et al., 2013), *M. pulcherrima* (Röcker, Strub, Kristin, & Grossmann, 2016), *T. delbrueckii* (Renault, Coulon, de Revel, Barbe, & Bely, 2015), and *Z. florentina* (Lencioni et al., 2016), respectively, were used in cocultured fermentations. *W. saturnus* was reported to increase isoamyl acetate in Emir wines (Tanguler, 2013). *S. pombe* is used to remove the sharp-tasting malic acid and decrease the "tartness" in Ezerfürtű wine (Yokotsuka, Otaki, Naitoh, & Tanaka, 1993). *R. mucilaginosa* was reported to improve the aroma of Ecolly wine (Wang et al., 2017). These are only a few examples of the benefits of including non-*Saccharomyces* species in conjunction with *S. cerevisiae* wine strains in multispecies ferments. There is no doubt that the hunting for beneficial non-*Saccharomyces* companions for *S. cerevisiae* wine yeasts will continue and that biodiversity prospecting in vineyards will unearth more interesting yeasts in the years to come.

13.3.2 Flavor-active *Saccharomyces wine yeasts*

Since cracking the genetic code of the first wine yeast strain (AWRI1631) in 2008, the genomes of several other widely used commercial wine yeast strains including AWRI1796, EC1118, QA23, VIN7, VIN13, and VL3 were sequenced and compared with the genomes of laboratory strains of *S. cerevisiae* (S288c and Sigma1278b) as well as genomes of commercial *Saccharomyces* strains used in the baking, brewing, biofuel, ragi, and saké industries (Borneman, Forgan, Pretorius, & Chambers, 2008; Galeote et al., 2010). An additional stretch of 200 kb of DNA present in some of these industrial strains, but found to be lacking in the laboratory strains, include strain-specific loci. These loci reside in the hypervariable subtelomeric regions and, in some cases, they distinguish specific classes of industrial strains. For example, a member of a subtelomeric three-gene cluster, the *RTM1* locus, is chiefly present in ale and distilling strains and to some extent in strains that carry a set of genes specific to wine yeasts (Borneman, Forgan, Pretorius, & Chambers, 2008). This wine-specific suite of genes comprises a second industry-defining locus, which consists of a cluster of five genes, displaying strain differences in copy number, genomic location, and gene order most likely due to mobilization into, and throughout, wine-strain genomes as a circular intermediate via an unknown process (Borneman, Forgan, Pretorius, & Chambers, 2008;). The third industry-specific locus entails the evolutionary differences in biotin prototrophy among certain industrial strains. More specifically, while the majority of wine and beer strains are biotin auxotrophs, saké strains acquired the capacity to synthesize biotin *de novo* over time, presumably because of evolutionary pressures in the low-biotin fermentations of saké mash (Borneman, Forgan, Pretorius, & Chambers, 2008).

Three wine-strain-specific genes were also identified: (1) the *FSY1* gene encoding an H^+/fructose symporter; (2) two paralogues, *MPR1* and *MPR2*, conferring resistance to L-azetidine-2-carboxylic acid; (3) and the -lyase encoding gene, *IRC7* (Borneman, Forgan, Pretorius, & Chambers, 2008). It is reasonable to argue that an effective H^+/fructose symporter could provide a selective advantage to strains expressing *FSY1* in highly concentrated mixtures of glucose and fructose during the fermentation of grape must. Wine strains that are capable of consuming fructose efficiently are less likely to produce sluggish or stuck fermentations during vintages in which heat waves distort the usual equilibrium between glucose and fructose in grape juice. Strains that carry the *MPR*-family paralogues cope better with stressful fermentation conditions because they are able to decrease the toxic effects of reactive oxygen. It is believed that the fermentation performance of *MPR*-carrying strains is more robust. Whereas, while the *FSY1* and *MPR1/2* genes are thought to convey fermentation robustness and performance, the *IRC7* gene might be associated with aroma enhancement in wine. *IRC7*-expressing strains seem to release more volatile thiols during fermentation, thereby increasing the fruitiness of wine. Paradoxically, the functional version of *IRC7* is rarely found in wine strains (Belda, Ruiz, Navascués, Marquina, & Santos, 2016).

Insightful observations were also made by analyzing the genomes of some members of the *Saccharomyces sensu stricto* clade. This clade consists of the *S. cerevisiae* complex (comprising well-defined "pure" lineages based strictly around geographic or industrial parameters, and mosaic strains that appear to be the result of outcrossing between multiple pure lineages) and seven distinct *Saccharomyces* species: *S. arboricolus, S. cariocanus, S. eubayanus, S. kudriavzevii, S. mikatae, S. paradoxus,* and *S. uvarum.* For instance, comparative genomic analyses revealed that the thiol-releasing wine yeast, VIN7, has an allotriploid hybrid genome with *S. cerevisiae* and *S. kudriavzevii* origins (Borneman et al., 2012). That explained the genetic basis of this VIN7's unique capacity to produce wines with a distinctive guava-like aroma. However, additional analyses of more natural hybrids like VIN7 are required to uncover more genes responsible for distinct flavors before bioengineers would be able to construct complex aroma-enhancing metabolic pathways in yeast strains customized for specific wine styles.

In this regard, with the construction of the first semisynthetic wine strain, it was demonstrated that customized flavor-activity falls with the realm of possibility (Kutyna & Borneman, 2018; Lee, Lloyd, Pretorius, & Borneman, 2016). A haploid wine strain (AWRI1631) of *S. cerevisiae* was equipped with a biosynthetic pathway, which consists of four separate enzymatic activities required for the production of the raspberry ketone, 4-[4-hydroxyphenyl]butan-2-one. This phenylpropanoid is the principal aroma compound found in raspberries, and it is also present, to a lesser extent, in blackberries, grapes, and rhubarb. The phenylpropanoid pathway begins with the conversion of phenylalanine to *p*-coumaric acid via cinnamate or directly from tyrosine to *p*-coumaric acid. Conversion of *p*-coumaric acid to raspberry ketone requires three additional enzymatic steps including a condensation reaction between coumaroyl-CoA and malonyl-CoA. To construct a raspberry ketone biosynthetic pathway in a haploid wine strain (AWRI1631), the following codon-optimized genes were synthesized and integrated into its *HO* locus: the phenylalanine ammonia lyase from an oleaginous yeast, *Rhodosporidium toruloides*; the cinnamate-4-hydroxylase from *A. thaliana*; and the coumarate CoA ligase 2 gene from parsley, *Petroselinum crispum*, fused by a rigid linker to the benzalacetone synthase from rhubarb, *Rheum palmatum*. This semisynthetic wine yeast was able to synthesize raspberry ketone at concentrations nearly two orders of magnitude above its predicted sensory threshold in Chardonnay grape juice under standard wine fermentation conditions while retaining the ability to ferment the must to dryness (Lee, Lloyd, Pretorius, & Borneman, 2016).

For now, though, the practical implications of the Sc2.0 lab yeast and this raspberry flavor-active wine yeast all lie in the future. These breakthroughs can best be thought of as a *Sputnik moment* in yeast research by itself; Sputnik did nothing except for orbiting the Earth while beeping, but it proved a concept and grabbed the attention of the world's futurists. These yeast *avatars* send a strong signal to synbio skeptics that the promise of synthetic biology can be realized in theory as well as in practice. It is a red-hot research field but it is also a field prone to *hype-horror-hope* oscillations; a great deal of research remains to be done before yeast *avatars* will be used for real-world tasks in commercial wineries.

13.4 Inventing a brighter future for wine with yeast

In an era of transformative technologies, unprecedented breakthroughs, and award-winning research in the converging fields of Synthetic Biology, Artificial Intelligence, and Quantum Computing, we could do worse than remember that science prizes tend to be bestowed upon those researchers who have made esoteric, if profound, scientific advances in the laboratory rather than practical ones in, for example, the winery. We need to separate the justified excitement in the laboratory from the opportunistic hyperbole in the winery and marketplace (Dixon, Curach, & Pretorius, 2020).

Beyond the scholarly contributions to an academic research field, there are often "soft" and "hard" impacts along the nonlinear *inputs-activities-outputs-outcomes-impacts* innovation pathway. These two kinds of impacts can have intended and unintended consequences. In this context, *hard impacts* refer to those tangible benefits that can directly be attributed to a research initiative (e.g., the *Yeast 2.0* project) or invention (e.g., the raspberry flavor-active wine yeast) whereas *soft impacts* tend to occur via uptake and use of new ideas, knowledge, or innovations by independent parties under indirect (or no) influence from the original researchers. The *agri-industrial* era is evolving into an unpredictable *bio-informational* future (Dixon, Curach, & Pretorius, 2020). It is vital that all wine industry stakeholders, namely the researchers, industry practitioners, policymakers, regulators, commentators, and consumers, stay attuned to developments in these future-shaping technologies. This includes the rollout of algorithm-enabled automation in biofoundries focused on accelerating and prototyping biological designs for engineering-biology applications. Stakeholders need to embrace the creative spark of uncertainty, as we all live in a world where we have to expect the unexpected. In doing so, we can ignite the ingenuity of researchers and practitioners to think anew about, for example, *microbiomes* of grapes from different viticultural sites, *signature microbial terroirs*, and the coculturing of region-specific *Saccharomyces* and non-*Saccharomyces* strains in multistarter ferments.

As described in previous sections, vineyard- and winery-related yeasts live as members of complex multispecies microbial communities that provide robustness to environmental perturbations (*terroirs*) and extended metabolic capacities. Such naturally occurring microbial assemblages are shaped by environmental gradients and resource availability and are spatially and functionally organized in a way that optimizes organism fitness and overall community productivity (Pretorius, 2020). The inherent division of "labor" among microbial populations with specialized subfunctions allows communities to reduce the metabolic burden imposed on each population and to carry out tasks that no single organism could undertake. Cooperative interactions among community members in the vineyard overcome the physiological and metabolic constraints of individual microbes and allow mixed microbial populations to execute otherwise incompatible functions simultaneously. Blending the synergistic modular design of naturally occurring microbial communities in grape must with the engineering capacity of Synthetic Biology opens up new vistas for studying, modeling, and optimizing spontaneous wine fermentation scenarios.

In the same way that computers are built using different specialized hardware, rationally connected to increase overall system performance, *synthetic microbial communities* can be "biologically wired" to collectively perform complex higher-order tasks unattainable by using individual populations from the vineyard or grape must. By employing such a "plug-and-play" approach, specialized permutations of compatible *Saccharomyces* and non-*Saccharomyces* yeast populations can be differentially combined in multistarter ferments depending on the desired outcome from a wine quality perspective. This will allow researchers to harness synergistic wine yeast consortia properties and tap into the unparalleled functional versatility of *terroir*-specific *Saccharomyces* and non-*Saccharomyces* yeasts. Building such *synthetic yeast starter-culture consortia* for modeling purposes will require an acceleration of the current *biodiversity prospecting* in vineyards with recognizable *microbial terroir* effects and the purposeful hunting for non-*Saccharomyces* yeasts in those vineyards and associated wineries (Pretorius, 2020).

It will not be a surprise if concepts such as synthetic microbial consortia that include yeasts with recoded single-chromosome genomes, carrying synthetic organelles, expressing reengineered biosynthetic pathways for highly reactive and orthogonal cofactors, and producing "new-to-nature" compounds sound like geeky "sci-fi" abracadabra to most winemakers. However, for many researchers involved in the engineering of biology, this is *science fiction* in the process of transitioning to *science fact*.

A few doubt that a wine yeast *avatar* completely powered by a computer-designed and synthetically made genome is possible in principle (Pretorius & Boeke, 2018); however, the consensus is that the use of such *avatars* in commercial wine production is not imminent. Anyone counting on the widespread application of wine yeast *avatars* to tantalize consumers' taste buds and open their wallets for repeat purchases in the near future will do well by remembering this cautionary note. At the same time, it is prudent to not lose sight of the wine industry's *tradition of innovation*, a metaphorical *terroir* free from cognitive prejudices where the newest ideas in wine yeast innovation can be powered by the oldest traditions in winemaking.

Acknowledgments

The *Yeast 2.0* research is financially supported by Macquarie University, Bioplatforms Australia, the New South Wales (NSW) Chief Scientist and Engineer, and the NSW Government's Department of Primary Industries. Australian Government funding through its investment agency, the *Australian Research Council*, toward the Macquarie University—led *ARC Centre of Excellence in Synthetic Biology* is gratefully acknowledged. This chapter is a shortened and slightly reworked version of an Open Access (CC BY license) article published in 2020 in *FEMS Yeast Research* (doi: 10.1093/femsyr/foz084). The information and diagrams are reproduced here with the kind permission of the Editor-in-Chief of *FEMS Yeast Research*, Dr. John Morrissey, and the publisher of the journal, Oxford Press. Bronte Turner from Serpentine Studio in Adelaide, Australia was commissioned to create the artwork for the six Figures.

References

Alfred, J., & Baldwin, I. T. (2015). New opportunities at the wild frontier. *eLife*, 4, 06956.

Almeida, P., Barbosa, R., Zalar, P., Imanishi, Y., Shimizu, K., Turchetti, B., . . . Sampaio, J. P. (2015). A population genomics insight into Mediterranean origins of wine yeast domestication. *Molecular Ecology*, 24, 5412–5427.

Anfang, N., Brajkovich, M., & Goddard, M. R. (2009). Co-fermentation with *Pichia kluyveri* increases varietal thiol concentrations in Sauvignon Blanc. *Australian Journal of Grape Wine Research*, 15, 1–8.

Annaluru, N., Muller, H., Mitchell, L. A., Ramalingam, S., Stracquadanio, G., Richardson, S. M., . . . Chandrasegaran, S. (2014). Total synthesis of a functional designer eukaryotic chromosome. *Science (New York, N.Y.)*, 344, 55–58.

Belda, I., Ruiz, J., Alastruey-Izquierdo, A., Navascués, E., Marquina, D., & Santos, A. (2016). Unraveling the enzymatic basis of wine "flavor-ome": A phylo-functional study of wine related yeast species. *Frontier Microbiology*, 7, 1–13.

Belda, I., et al. (2017). Microbial contribution to wine aroma and its intended use for wine quality improvement. *Molecules (Basel, Switzerland)*, 22, 1–29.

Belda, I., Ruiz, J., Navascués, E., Marquina, D., & Santos, A. (2016). Improvement of aromatic thiol release through the selection of yeasts with increased β-lyase activity. *International Journal of Food Microbiology*, 225, 1–8.

Belda, I., Ruiz, J., Santos, A., Van Wyk, N., & Pretorius, I. S. (2019). Saccharomyces cerevisiae. *Trends Genetics*, 35, 956–957.

Belda, I., et al. (2017). From vineyard soil to wine fermentation: Microbiome approximations to explain the *terroir* concept. *Frontiers Microbiology*, 8, 821.

Bely, M., Stoeckle, P., Masneuf-Pomarède, I., & Dubourdieu, D. (2008). Impact of mixed *Torulaspora delbrueckii-Saccharomyces cerevisiae* culture on high-sugar fermentation. *International Journal of Food Microbiology*, 122, 312–320.

Blount, B. A., Gowers, G. O. F., Ho, J. C. H., Ledesma-Amaro, R., Jovicevic, D., McKiernan, R. M., . . . Ellis, T. (2018). Rapid host strain improvement by in vivo rearrangement of a synthetic yeast chromosome. *Nature Communication*, 9, 1932.

Bokulich, N. A., Collins, T. S., Masarweh, C., Allen, G., Heymann, H., Ebeler, S. E., & Mills, D. A. (2016). Associations among wine grape microbiome, metabolome, and fermentation behaviour suggest contribution to regional wine characteristics. *mBio ASM*, 7, 1–12.

Bokulich, N. A., et al. (2013). Monitoring season changes in winery-resident microbiota. *PLoS One*, 8, e66437.

Bokulich, N. A., et al. (2013). Microbial biogeography of wine grapes is conditioned by cultivar, vintage, and climate. *Proceeding of the National Academy Sciences of the USA*, 111, 139–148.

Borneman, A. R., Desany, B. A., Riches, D., Affourtit, J. P., Forgan, A. H., Pretorius, I. S., . . . Chambers, P. J. (2011). Whole genome comparison reveals novel genetic elements that characterize the genome of industrial strains of *Saccharomyces cerevisiae*. *PLoS Genetics*, 7, e1001287.

Borneman, A. R., Desany, B. A., Riches, D., Affourtit, J. P., Forgan, A. H., Pretorius, I. S., . . . Chambers, P. J. (2012). The genome sequence of the wine yeast VIN7 reveals an allotriploid hybrid genome with *Saccharomyces cerevisiae* and *Saccharomyces kudriavzevii*. *FEMS Yeast Research*, 12, 88–96.

Borneman, A. R., Forgan, A. H., Pretorius, I. S., & Chambers, P. J. (2008). Comparative genome analysis of a *Saccharomyces cerevisiae* wine strain. *FEMS Yeast Research*, 8, 1185–1195.

Borneman, A. R., Pretorius, I. S., & Chambers, P. J. (2013). Comparative genomics: a revolutionary tool for wine yeast strain development. *Current Opinion in Biotechnology*, 24, 192–199.

Borneman, A. R., et al. (2013). At the cutting-edge of grape and wine biotechnology. *Trends in Genetics: TIG*, 29, 263–271.

Canfora, L., Vendramin, E., Felici, B., & Luo, L. (2018). Vineyard variations during different fertilisation practices revealed by 16S rRNA gene sequencing. *Applied Soil Ecology*, 125, 71–80.

Capozzi, V., Rosario Di Toro, M., Grieco, F., Michelotti, V., Salma, M., Lamontanara, A., . . . Spano, G. (2016). Viable but not culturable (VBNC) state of *Brettanomyces bruxellensis* in wine: New insights on molecular basis of VBNC behaviour using a transcriptomic approach. *Food Microbiology*, 59, 196–204.

Chambers, P. J., & Pretorius, I. S. (2010). Fermenting knowledge: The history of winemaking, science and yeast research. *EMBO Reports, 11*, 1–7.

Chao, R., Mishra, S., Si, T., & Zhao, H. (2017). Engineering biological systems using automated biofoundries. *Metabolic Engineering, 42*, 98–108.

Chou, M. Y., Vanden Heuvel, J., Bell, T. H., Panke-Buisse, K., & Kao-Kniffin, J. (2018). Vineyard under-vine floor management alters soil microbial composition, while the fruit microbiome shows no corresponding shifts. *Scientific Reports, 8*, 11039.

Cordente, A. G., Cordero-Bueso, G., Pretorius, I. S., & Curtin, C. D. (2013). Novel wine yeast with mutations in *YAP1* that produce less acetate during fermentation. *FEMS Yeast Research, 13*, 62–73.

Cordente, A. G., Curtin, C. D., Varela, C., & Pretorius, I. S. (2012). Flavour-active wine yeast. *Applied Microbiology and Biotechnology, 2012*(96), 601–618.

Cordente, A. G., Heinrich, A. J., Pretorius, I. S., & Swiegers, J. H. (2009). Isolation of sulfite reductase variants of a commercial wine yeast with significantly reduced hydrogen sulfide production. *FEMS Yeast Research, 9*, 446–459.

Cordero-Bueso, G., Arroyo, T., Serrano, A., et al. (2011). Influence of the farming system and vine variety on yeast communities associated with grape berries. *International Journal of Food Microbiology, 145*, 132–139.

Cordero-Bueso, G., Arroyo, T., & Serrano, A. (2011). Influence of different floor management strategies of the vineyard on the natural yeast population associated with grape berries. *International Journal of Food Microbiology, 148*, 23–29.

Curtin, C. D., Borneman, A. R., Chambers, P. J., & Pretorius, I. S. (2012). *De-novo* assembly and analysis of the triploid genome of the wine spoilage yeast *Dekkera bruxellensis*. *PLoS One, 7*, 1–9.

Curtin, C. D., & Pretorius, I. S. (2014). Genomic insights into the evolution of industrial yeast species *Brettanomyces bruxellensis*. *FEMS Yeast Research, 14*, 997–1005.

Dashko, S., Zhou, N., Tinta, T., Sivilotti, P., Lemut, M. S., Trost, K., ... Piskur, J. (2015). Use of non-conventional yeast improves the wine aroma profile of Ribolla Gialla. *Journal of Industrial Microbiology & Biotechnology, 42*, 997–1010.

De Celis, M., Ruiz, J., Martin-Santamaría, M., Rizk, Z., Bianco, A., Zara, G., ... Budroni, M. (2019). Diversity of *Saccharomyces cerevisiae* yeasts associated to spontaneous and inoculated fermenting grapes from Spanish vineyards. *Letters in Applied Microbiology, 68*, 580–588.

Dixon, T., Curach, N., & Pretorius, I. S. (2020). Bio-informational futures: The convergence of artificial intelligence and synthetic biology. *EMBO Reports, 21*, 1–5, e50036.

Duan, Z., Andronescu, M., Schutz, K., McIlwain, S., Kim, Y. J., Lee, C., ... Noble, W. S. (2010). A three-dimensional model of the yeast genome. *Nature, 465*, 363–367.

Dzialo, M. C., Park, R., Steensels, J., Lievens, B., & Verstrepen, K. J. (2017). Physiology, ecology and industrial applications of aroma formation in yeast. *FEMS Microbiology Reviews, 41*, S95–S128.

Engel, S. R., Dietrich, F. S., Fisk, D. G., Binkley, G., Balakrishnan, R., Costanzo, M. C., ... Cherry, J. M. (2014). The reference genome sequence of *Saccharomyces cerevisiae*: then and now. *Genes Genomes Genetics, 4*, 389–398.

Galeote, V., Novo, M., Salema-Oom, M., Brion, C., Valério, E., Gonçalves, P., & Dequin, S. (2010). *FSY1*, a horizontally transferred gene in the *Saccharomyces cerevisiae* EC1118 wine yeast strain, encodes a high-affinity fructose/H + symporter. *Microbiology (Reading, England), 56*, 3754–3761.

Gallone, B., Steensels, J., Prahl, T., Soriaga, L., Saels, V., Herrera-Malaver, B., ... Verstrepen, K. J. (2016). Domestication and divergence of *Saccharomyces cerevisiae* beer yeasts. *Cell, 166*, 1397–1410.

Gambino, G., Dal Molin, A., Boccacci, P., et al. (2017). Whole-genome sequencing and SNV genotyping of "Nebbiolo" (*Vitis vinifera* L.) clones. *Scientific Reports, 7*, 17294.

Garcia, A., Carcel, C., Dulau, L., Samson, A., Aguera, E., Agosin, E., & Günata, Z. (2002). Influence of a mixed culture with *Debaryomyces vanriji* and *Saccharomyces cerevisiae* on the volatiles of a Muscat wine. *Journal of Food Science, 67*, 1138–1143.

Gibson, D. G., & Venter, J. G. (2014). Construction of a yeast chromosome. *Nature, 509*, 168–169.

Gilbert, J. A., van der Lelie, D., & Zarraonaindia, I. (2014). Microbial *terroir* for wine grapes. *Proceedings of the National Academy of the Sciences USA, 111*, 5–6.

Gobbi, M., Comitini, F., Domizio, P., Romani, C., Lencioni, L., Mannazzu, I., & Ciani, M. (2013). *Lachancea thermotolerans* and *Saccharomyces cerevisiae* in simultaneous and sequential co-fermentation: A strategy to enhance acidity and improve the overall quality of wine. *Food Microbiology, 33*, 271–281.

Goddard, M. R., & Greig, D. (2015). *Saccharomyces cerevisiae*: A nomadic yeast with no niche? *FEMS Yeast Research, 15*, 1–6.

Goffeau, A., Barrell, B. G., & Bussey, H. (1996). Life with 6000 genes. *Science (New York, N.Y.), 274*, 546–567.

Goold, H. D., Kroukamp, H., Williams, T. C., Paulsen, I. T., Varela, C., & Pretorius, I. S. (2017). Yeast's balancing act between ethanol and glycerol production in low-alcohol wines. *Microbiology and Biotechnology, 10*, 264–278.

Gorter de Vries, A. R., Pronk, J. T., & Daran, J. M. G. (2017). Industrial relevance of chromosomal copy number variation in *Saccharomyces* yeasts. *Applied and Environmental Microbiology, 83*, e03206-16.

Grangeteau, C., Roullier-Gall, C., Rousseaux, S., Gougeon, R. D., Schmitt-Kopplin, P., Alexandre, H., & Guilloux-Benatier, M. (2016). Wine microbiology is driven by vineyard and winery anthropogenic factors. *Microbiology and Biotechnology*. Available from https://doi.org/10.1111/1751-7915.12428.

Hillson, N., Caddick, M., Cai, Y., Carrasco, J. A., Chang, M. W., Curach, N. C., ... Freemont, P. S. (2019). Building a global alliance of biofoundries. *Nature Communication, 10*, 2040.

Hochrein, L., Mitchell, L. A., Schulz, K., Messerschmidt, K., & Mueller-Roeber, B. (2018). L-SCRaMbLE as a tool for light-controlled Cre-mediated recombination in yeast. *Nature Communication, 9*, 1931.

Hyma, K. E., Saerens, S. M., Verstrepen, K. J., & Fay, J. C. (2011). Divergence in wine characteristics produced by wild and domesticated strains of *Saccharomyces cerevisiae*. *FEMS Yeast Research, 11*, 540–551.

Jagtap, U. B., Jadhav, J. P., Bapat, V. A., & Pretorius, I. S. (2017). Synthetic biology stretching the realms of possibility in wine yeast research. *International Journal of Food Microbiology, 252*, 24–34.

Jia, B., Wu, Y., Li, B. Z., Mitchell, L. A., Liu, H., Pan, S., ... Yuan, Y. J. (2018). Precise control of SCRaMbLE in synthetic haploid and diploid yeast. *Nature Communication, 9*, 1933.

Jolly, N. P., Varela, C., & Pretorius, I. S. (2014). Not your ordinary yeast: Non-*Saccharomyces* yeasts in wine production uncovered. *FEMS Yeast Research, 14*, 15−237.

Joseph, R., & Bachhawat, A. K. (2014). Yeasts: Production and commercial uses. *Encyclopedia of food microbiology*, 823−830. (2nd edn).

Knight, S., Klaere, S., Fedrizzi, B., & Goddard, M. R. (2015). Regional microbial signatures positively correlate with differential wine phenotypes: Evidence for a microbial aspect to *terroir. Scientific Reports, 5*, 14233.

Kutyna, D. R., & Borneman, A. R. (2018). Heterologous production of flavour and aroma compounds in *Saccharomyces cerevisiae. Genes, 9*, 326.

Lajoie, M. J., Rovner, A. J., Goodman, D. B., Aerni, H. R., Haimovich, A. D., Kuznetsov, G., ... Isaacs, F. J. (2013). Genomically recoded organisms expand biological functions. *Science (New York, N.Y.), 342*, 357−360.

Langdon, Q. K., Peris, D., Baker, E. C. P., Opulente, D. A., Nguyen, H. V., Bond, U., ... Hittinger, C. T. (2019). *Nature Ecology and Evolution, 3*, 1576−1586.

Lee, D., Lloyd, N., Pretorius, I. S., & Borneman, A. R. (2016). Heterologous production of raspberry ketone in the wine yeast *Saccharomyces cerevisiae* via pathway engineering and synthetic enzyme fusion. *Microbiology Cell Factory, 15*, 49−55.

Legras, J. L., Galeote, V., Bigey, F., Camarasa, C., Marsit, S., Nidelet, T., ... Dequin, S. (2018). Adaptation of *S. cerevisiae* to fermented food environments reveals remarkable genome plasticity and the footprints of domestication. *Molecular Biology and Evolution, 35*, 1712−1727.

Lencioni, L., Romani, C., Gobbi, M., Comitini, F., Ciani, M., & Domizio, P. (2016). Controlled mixed fermentation at winery scale using *Zygotorulaspora florentina* and *Saccharomyces cerevisiae. International Journal of Food Microbiology, 234*, 36−44.

Lilly, M., Bauer, F. F., Lambrechts, M. G., Swiegers, J. H., Cozzolino, D., & Pretorius, I. S. (2006). The effect of increased alcohol acetyl transferase and esterase activity on flavour profiles of wine and distillates. *Yeast (Chichester, England), 23*, 641−659.

Lilly, M., Styger, G., Bauer, F. F., Lambrechts, M. G., & Pretorius, I. S. (2006). The effect of increased yeast branched-chain amino acid transaminase activity and the production of higher alcohols on the flavour profiles of wine and distillates. *FEMS Yeast Research, 6*, 726−743.

Liti, G. (2015). The fascinating and secret wild life of the budding yeast *S. cerevisiae. eLife, 4*, 05835.

Liu, L., Redden, H., & Alper, H. S. (2013). Frontiers of yeast metabolic engineering: diversifying beyond ethanol and ethanol and *Saccharomyces. Current Opinion. Biotechnology (Reading, Mass.), 24*, 1023−1030.

Liu, W., Luo, Z., Yun Wang, Y., Pham, N. T., Tuck, L., Pérez-Pi, I., ... Cai, Y. (2018). Rapid pathway prototyping and engineering using *in vitro* and *in vivo* synthetic SCRaMbLE-in methods. *Nature Communication, 9*, 1936.

Luo, J., et al. (2018). Karyotype engineering by chromosome fusion leads to reproductive isolation in yeast. *Nature, 560*, 392−396.

Luo, Z., et al. (2018). Identifying and characterizing SCRaMbLEd synthetic yeast using ReSCuES. *Nature Communication, 9*, 1930.

Marsit, S., & Dequin, S. (2015). Diversity and adaptive evolution of *Saccharomyces* wine yeast: A review. *FEMS Yeast Research, 15*, 1−12.

Martin, V., Valera, M., Medina, K., Boido, E., & Carrau, F. (2018). Oenological impact of the *Hanseniaspora/Kloeckera* yeast genus on wines—A review. *Fermentation, 4*, 76.

McGovern, P., Zhang, J., Tang, J., Zhang, Z., Hall, G. R., Moreau, R. A., ... Wang, C. (2004). Fermented beverages of pre- and proto-historic China. *Proceedings of the National Academy of the Sciences USA, 101*, 17593−17598.

Mercy, G., Mozziconacci, J., Scolari, V. F., Yang, K., Zhao, G., Thierry, A., ... Koszul, R. (2017). 3D organization of synthetic and scrambled chromosomes. *Science (New York, N.Y.), 355*, 1050-eaaf4597.

Mitchell, L. A., Wang, A., Stracquadanio, G., Kuang, Z., Wang, X., Yang, K., ... Boeke, J. D. (2017). Synthesis, debugging, and effects of synthetic chromosome consolidation: SynVI and beyond. *Science (New York, N.Y.), 355*, 1045-eaaf4831.

Morrison-Whittle, P., Lee, S. A., & Goddard, M. R. (2017). Fungal communities are differentially affected by conventional and biodynamic agricultural management approaches in vineyard ecosystems. *Agriculture Ecosystems Environment, 246*, 306−313.

Oliver, S. G. (1996). From DNA sequence to biological function. *Nature, 379*, 597−600.

Pennisi, E. (2014). Building the ultimate yeast genome. *Science (New York, N.Y.), 343*, 1426−1429.

Peter, J., De Chiara, M., Friedrich, A., Yue, J. X., Pflieger, D., Bergström, A., ... Wincker, P. (2018). Genome evolution across 1,011 *Saccharomyces cerevisiae* isolates. *Nature, 556*, 339−344.

Pretorius, I. S. (2016). Conducting wine symphonics with the aid of yeast genomics. *Beverages, 2*, 36. Available from https://doi.org/10.3390/beverages2040036.

Pretorius, I. S. (2017). Synthetic genome engineering forging new frontiers for wine yeast. *Critical Reviews in Biotechnology, 37*, 112−136.

Pretorius, I. S. (2017). Solving yeast jigsaw puzzles over a glass of wine: Synthetic genome engineering pioneers new possibilities for wine research. *EMBO Reports, 18*, 1875−1884.

Pretorius, I. S. (2000). Tailoring wine yeast for the new millennium: Novel approaches to the ancient art of winemaking. *Yeast (Chichester, England), 16*, 675−729.

Pretorius, I. S. (2020). Tasting the terroir of wine yeast innovation. *FEMS Yeast Research, 20*. Available from https://doi.org/10.1093/femsyr/foz084.

Pretorius, I. S., & Bauer, F. F. (2002). Meeting the consumer challenge through genetically customised wine yeast strains. *Trends in Biotechnology, 20*, 426−432.

Pretorius, I. S., & Boeke, J. D. (2018). Yeast 2.0 Connecting the dots in the construction of the world's first functional synthetic eukaryotic genome. *FEMS Yeast Research, 18*, 1018, 18:1−15.

Pretorius, I. S., Curtin, C. D., & Chambers, P. J. (2012). The winemaker's bug: From ancient wisdom to opening new vistas with frontier yeast science. *Bioengineered, 3*, 47−156.

Pretorius, I. S., & Høj, P. B. (2005). Grape and wine biotechnology: Challenges, opportunities and potential benefits. *Australian Journalof Grape Wine Research, 11*, 83−108.

Renault, P., Coulon, J., de Revel, G., Barbe, J. C., & Bely, M. (2015). Increase of fruity aroma during mixed *T. delbrueckii/S. cerevisiae* wine fermentation is linked to specific esters enhancement. *International Journal of Food Microbiology, 207*, 40−48.

Richardson, S. M., Mitchell, L. A., Stracquadanio, G., Yang, K., Dymond, J. S., DiCarlo, J. E., ... Bader, J. S. (2017). Design of a synthetic yeast genome. *Science (New York, N.Y.), 355*, 1040−1044.

Roach, M. J., Johnson, J. L., Bohlman, J., Van Vuuren, H. J. J., Jones, S. J. M., Pretorius, I. S., ... Borneman, A. R. (2018). Population sequencing reveals clonal diversity and ancestral inbreeding in the grapevine cultivar Chardonnay. *PLoS Genetics, 20*, 1−24.

Robinson, J. (1999). *The Oxford companion to wine*. New York: Oxford Press.

Röcker, J., Strub, S., Kristin, E., & Grossmann, M. (2016). Usage of different aerobic non-*Saccharomyces* yeasts and experimental conditions as a tool for reducing the potential ethanol content in wines. *European Food Research and Technology, 242*, 2051–2070.

Seixas, I., Barbosa, C., Salazar, S. B., Mendes-Faia, A., Wang, Y., Güldener, U., ... Mira, N. P. (2016). Genome sequence of the unconventional wine yeast *Hanseniaspora guilliermondii* UTAD222. *Genome Announcements, 5*, e01515–e01516.

Setati, M. E., Jacobson, D., Andong, U. C., & Bauer, F. (2012). The vineyard yeast microbiome, a mixed model microbial map. *PLoS One, 7*, e52609.

Shao, Y., Lu, N., Wu, Z., Cai, C., Wang, S., Zhang, L. L., ... Qin, Z. (2018). Creating a functional single chromosome yeast. *Nature, 560*, 331–335.

Shen, M. J., Wu, Y., Yang, K., Li, Y., Xu, H., Zhang, H., ... Boeke, J. D. (2018). Heterozygous diploid and interspecies SCRaMbLEing. *Nature Communication, 9*, 1934.

Shen, Y., Stracquadanio, G., Wang, Y., Li, Y., Xu, H., Zhang, H., ... Boeke, J. D. (2016). SCRaMbLE generates designed combinatorial stochastic diversity in synthetic chromosomes. *Genome Research, 26*, 36–49.

Shen, Y., Wang, Y., Chen, T., Yang, K., Mitchell, L. A., Xue, Y., ... Bader, J. S. (2017). Deep functional analysis of synII, a 770-kilobase synthetic yeast chromosome. *Science (New York, N.Y.), 355*, 1047-eaaf4791.

Sliva, A., Yang, H., Boeke, J. D., & Mathews, D. J. (2015). Freedom and responsibility in synthetic genomics: The synthetic yeast project. *Genetics, 200*, 1021–1028.

Steensels, J., Gallone, B., Voordeckers, K., & Verstrepen, K. J. (2019). Domestication of industrial microbes. *Current Biology, 29*, 381–393.

Swiegers, J. H., et al. (2005). Yeast and bacterial modulation of wine aroma and flavour. *Australian Journalof Grape Wine Research, 11*, 139–173.

Swiegers, J. H., et al. (2007). Engineering volatile thiol release in *Saccharomyces cerevisiae* for improved wine aroma. *Yeast (Chichester, England), 24*, 561–574.

Swiegers, J. H., et al. (2005). The genetics of olfaction and taste. *Australian Journalof Grape Wine Research, 11*, 109–113.

Swiegers, J. H., et al. (2007). Modulation of volatile sulfur compounds by wine yeast. *Applied Biochemistry and Biotechnology, 74*, 954–960.

Tanguler, H. (2013). Influence of temperatures and fermentation behaviour of mixed cultures of *Williopsis saturnus* var. *saturnus* and *Saccharomyces cerevisiae* associated with winemaking. *Food Sciences and Technology Research, 19*, 781–793.

Timmins, J. J. B., Kroukamp, H., Paulsen, I. T., & Pretorius, I. S. (2020). The sensory significance of apocarotenoids in wine: Importance of carotenoid cleavage dioxygenase 1 (CCD1) in the production of β-ionone. *Molecules (Basel, Switzerland), 25*, 2779. Available from https://doi.org/10.3390/molecules2522779.

Van Wyk, N., Grossmann, M., Wendland, J., Von Wallbrunn, C., & Pretorius, I. S. (2019). The whiff of wine yeast innovation: strategies for enhancing aroma production by yeast during wine fermentation. *Journal of Agricultural and Food Chemistry, 67*, 13496–13505.

Van Wyk, N., Kroukamp, H., Espinosa, M. I., Von Wallbrunn, C., Wendland, J., & Pretorius, I. S. (2020). Blending wine yeast phenotypes with the aid of CRISPR DNA editing technologies. *International Journal of Food Microbiology, 324*. Available from https://doi.org/10.1016/j.ijfoodmicro.2020.108615.

Van Wyk, N., Kroukamp, H., & Pretorius, I. S. (2018). The smell of synthetic biology: Engineering strategies for aroma compound production in yeast. *Fermentation, 4*, 1–18.

Van Wyk, N., Michling, F., Bergamo, D., Brezina, S., Pretorius, I. S., Von Wallbrunn, C., & Wendland, J. (2020). Effect of isomixing on grape must fermentations of *ATF1*-overexpressing wine yeast strains. *Foods, 9*. Available from https://doi.org/10.3390/foods9060717.

Van Wyk, N., Pretorius, I. S., & Von Wallbrunn, C. (2020). Assessing the oenological potential of *Nakazawaea ishiwadae, Candida railenensis* and *Debaryomyces hansenii* strains in a mixed-culture grape must fermentation with *Saccharomyces cerevisiae*. *Fermentation, 6*, 49. Available from https://doi.org/10.3390/fermentation6020049.

Varela, C., Schmidt, S. A., Borneman, A. R., Nam, C., Pang, I., Krömerx, J. O., ... Chambers, P. J. (2018). Systems-based approaches enable identification of gene targets which improve the flavour profile of low-ethanol wine yeast strains. *Metabolic Engineering, 49*, 178–191.

Viana, F., Belloch, C., Vallés, S., & Manzanares, P. (2011). Monitoring a mixed starter of *Hanseniaspora vineae-Saccharomyces cerevisiae* in natural must: Impact on 2-phenylethyl acetate production. *International Journal of Food Microbiology, 151*, 235–240.

Vivier, M. A., & Pretorius, I. S. (2000). Genetic improvement of grapevine: Tailoring grape varieties for the third millennium − A review. *South African Journal for Enology Viticulture, 21*, 5–26.

Vivier, M. A., & Pretorius, I. S. (2002). Genetically tailored grapevines for the wine industry. *Trends in Biotechnology, 20*, 472–478.

Walker, R. S., & Pretorius, I. S. (2018). Applications of synthetic biology geared towards the production of biopharmaceuticals. *Genes, 9*, 1–23.

Wang, Q. M., Liu, W. Q., Liti, G., Wang, S. A., & Bai, F. Y. (2012). Surprisingly diverged populations of *Saccharomyces cerevisiae* in natural environments remote from human activity. *Molecular Ecology, 21*, 5404–5417.

Wang, X. C., Li, A. H., Dizy, M., Ullah, N., Sun, W. X., & Tao, Y. S. (2017). Evaluation of aroma enhancement for "Ecolly" dry white wines by mixed inoculation of selected *Rhodotorula mucilaginosa* and *Saccharomyces cerevisiae*. *Food Chemistry, 228*, 550–559.

Wu, Y., Li, B. Z., Zhao, M., Xie, Z. X., Lin, Q. H., Wang, X., ... Yuan, Y. J. (2017). Bug mapping and fitness testing of chemically synthesized chromosome X. *Science (New York, N.Y.), 355*, 1048–eaaf4706.

Wu, Y., Zhu, R. Y., Mitchell, L. A., Ma, L., Liu, R., Zhao, M., ... Boeke, J. D. (2018). *In vitro* DNA SCRaMbLE. *Nature Communication, 9*, 1935.

Xie, Z. X., Li, B. Z., Mitchell, L. A., Wu, Y., Qi, X., Jin, Z., ... Yuan, Y. J. (2017). Perfect designer chromosome V and behavior of a ring derivative. *Science (New York, N.Y.), 355*, 1046-eaaf4704.

Yokotsuka, K., Otaki, A., Naitoh, A., & Tanaka, H. (1993). Controlled simultaneous deacidification and alcohol fermentation of a high-acid grape must using two immobilized yeasts, *Schizosaccharomyces pombe* and *Saccharomyces cerevisiae*. *Amercian Journal of Enology Viticulture, 44*, 371–377.

Zhang, W., Zhao, G., Luo, Z., Lin, Y., Wang, L., Guo, Y., ... Dai, J. (2017). Engineering the ribosomal DNA in a megabase synthetic chromosome. *Science (New York, N.Y.), 355*, 1049-eaaf3981.

14

Malolactic fermentation in white wines

Isabel Pardo and Sergi Ferrer

Institute of Biotechnology and Biomedicine (BioTecMed) and Department of Microbiology and Ecology, University of Valencia, Valencia, Spain

Malolactic fermentation (MLF) is an important process in wine production. Although in fact, from a biochemical point of view, it is actually not a "fermentation" process, like alcoholic fermentation (AF), it does involve gas emissions, so winemakers called it "malolactic fermentation." The term "malolactic fermentation" has been used since the early 20th century when it was related to the activity of lactic acid bacteria (LAB) (Lonvaud-Funel, 2010). As we all know, MLF involves the bioconversion of malic acid (a dicarboxylic acid) into lactic acid (monocarboxylic acid), thereby producing CO_2 molecules (Gil-Sánchez, Bartolomé Suáldea, & Victoria Moreno-Arribas, 2019). Besides, many other biochemical reactions occur simultaneously, and several other grape must and wine compounds are also modified besides malic acid. Therefore in addition to lactic acid, many other molecules accumulate in wine (Gil-Sánchez et al., 2019). Some can improve quality, while others may be detrimental to sensory quality or health (Lonvaud-Funel, 2010). Among them, there are the synthesis of carbonyl compounds (e.g., diacetyl, pyruvate, lactate, acetoin, 2,3-butanediol, acetate, etc.), esters (ethyl fatty acid esters as ethyl lactate, ethyl acetate, ethyl hexanoate, and ethyl octanoate), or monoterpenes (Bartowsky, Francis, Bellon, & Henschke, 2002; Cappello, Zapparoli, Logrieco, & Bartowsky, 2017; Hernández-Orte et al., 2009; Lerm, Engelbrecht, & du Toit, 2010). Off-flavors, such as volatile phenols, can also be produced, with lactobacilli and pediococci having a higher capacity for producing these compounds than *Oenococcus oeni* (Cavin, Andioc, Etievant, & Diviès, 1993). Besides, other detrimental metabolites that LAB can produce in wine are acetamide or mousy odor (heterocyclic compounds such as 2-ethyltetrahydropyridine, 2-acetyltetrahydopyridine, and 2-acetylpyrroline) (Costello, Francis, & Bartowsky, 2012; Costello & Henschke, 2002). Other metabolites such as methanethiol, dimethyl disulfide, methionol, and 3-(methylsulfanyl) propionic acid can positively contribute or decrease the wine quality depending on their concentrations (Landaud, Helinck, & Bonnarme, 2008; Swiegers, Bartowsky, Henschke, & Pretorius, 2005).

MLF usually occurs after AF and affects the overall quality of wine. The role of MLF in winemaking is triple: reducing the acidity of the wine, stabilizing wine by removing potential nutrients for other microorganisms, and producing aroma and flavor changes (Bartowsky, 2014; Virdis, Sumby, Bartowsky, & Jiranek, 2021). Throughout this chapter, the term "MLF" will refer to all these complex reactions, metabolites, and activities and not only to the simple decarboxylation of malic acid, unless specified.

In the past, MLF was mainly used to produce red wine, and only from time to time very acidic white wine. In the production of white wines, MLF is usually used to reduce the acidity of wines produced in cool climates or to impart a certain degree of sensory complexity to white wines with a lighter taste such as Chardonnay (Morenzoni, 2015). Recent studies have explored the impact of MLF on white wines from different viticultural regions.

In recent years, there has been a tendency to induce MLF also in white wines. Formerly, apart from some particular cases, white wines were not considered to conduct MLF, as they were not typically aged in barrels and used to be consumed early as "young" and fresh wines. However, in the last years, MLF has received more attention for white wine, and every year more and new white wines are coming to the market after performing this fermentation.

Reasons for this increasing interest are diverse and largely depend on the variety, characteristics of the wine, and intention of the winemaker. For some producers, malic acid contributes to a green taste, which is considered

© 2022 Elsevier Inc. All rights reserved.

detrimental for the quality; the growth of LAB-performing MLF can therefore fix this problem by converting it into lactic acid: smoother and with a different particular taste. For others, FML can bring a bioprotection to the wines, avoiding on one side spoilage or decreases in wine quality (e.g., preventing synthesis of biogenic amines, or growth of spoilage bacteria or yeasts such as *Brettanomyces*) (Berbegal, 2014; Berbegal et al., 2014; Berbegal, Benavent-Gil, Navascués, et al., 2017). Another aspect of the bioprotection of induced MLF in white wines, especially interesting in sparkling ones with a second AF in bottles, is the prevention of hazing in bottles when a late and undesired MLF is produced (Virdis et al., 2021). This cloudiness is extremely difficult to remove, and this provokes a big economic loss whenever it happens. Finally, the third reason is the change in organoleptic characteristics that an MLF induces in wine, and whites are no exception. Greater complexity in wine profile is also obtained when MLF is conducted in white wines, which can be used to obtain new and different wines in the market, more complex and with a different personality. The growth necessities of LAB in white wine are similar as in red wine, but when they interact with the components of the white wine matrix, the products they form are different because of the different composition between these types of wine (Morenzoni, 2015). This is very interesting and useful because it can help the winemaking community to seek to make new styles of wines, looking for products with unique, pleasant, and distinguishing sensorial characteristics.

14.1 Effect of MLF on pH changes

When MLF occurs, dicarboxylic acid L-malic acid is converted to monocarboxylic acid L-lactic acid. As an acidic carboxyl group of malic acid is converted into carbon dioxide, the pH value will increase accordingly. This tends to reduce the acidity of the wine, making it softer with a better taste. On average, after the MLF is completed, the pH will increase by approximately 0.2 units (Morenzoni, 2015). As white wines typically are of low pH value (lower than red ones) (Darias-Martín, Socas-Hernández, Díaz-Romero, & Díaz-Díaz, 2003), it is not so problematic, in some cases, to suffer this concomitant increase in pH and a slight decrease in total acidity due to MLF. For other wines, however, to lose acidity is a drama, and MLF is frowned upon.

To counterbalance this increase, applying a biological strategy is possible: to use microorganisms (yeasts or bacteria) for producing organic acids from sugars. Yeasts belonging to the *Saccharomyces* genus (Dequin, Baptista, & Barre, 1999; Yeramian, 2003), *Lachancea thermotolerans* (syn. *Kluyveromyces thermotolerans*) (Kapsopoulou, Mourtzini, Anthoulas, & Nerantzis, 2007), and bacteria *Lactiplantibacillus plantarum* (former *Lactobacillus plantarum*) (Lucio, Pardo, Krieger-Weber, Heras, & Ferrer, 2016; Onetto & Bordeu, 2015) can be used to increase acids during winemaking. Yeast can synthesize acids from sugars, mainly malic acid, lactic acid, or succinic acid (Dequin et al., 1999; Lucio, 2014; Morata et al., 2018; Morata et al., 2019; Su, Wang, Wang, Li, & Li, 2014; Thoukis, Ueda, & Wright, 1965; Vilela, 2019; Yeramian, 2003; Yéramian, Chaya, & Suárez Lepe, 2007). When used in coinoculation with *S. cerevisiae* in grape musts, LAB find available sugars enough to allow their growth and low ethanol content (see detailed explanation below). Under these circumstances, *L. plantarum* is able to increase the titratable acidity of Pedro Ximénez by 0.7–1.6 g/L (Val, Ferrer, & Pardo, 2002). Besides, up to 11 g/L of lactic acid (but most typically 2.5–6 g/L) can be produced from sugars in low-acidity grape musts when these two microorganisms are inoculated simultaneously at the beginning of the vinification (Lucio, 2014; Lucio, Pardo, Heras, Krieger, & Ferrer, 2014; Onetto & Bordeu, 2015). This acidification process can thus be simultaneous to the MLF, and the unpleasant and "green" taste of malic acid is removed from the wine. Besides, the synthesis of lactic acid through sugar fermentation results in a relative decrease in pH, counterbalancing the pH rise provoked by MLF, or even surpassing it and decreasing the final net pH and increasing the acidity (Lucio et al., 2016).

To achieve this acidification effect, it is preferable to inoculate homofermentative or facultative heterofermentative bacteria such as *L. plantarum* to avoid the production of acetic acid, although sometimes heterofermenters such as *O. oeni* have been employed for this purpose (Lucio et al., 2016). Besides, a good coexistence with yeasts during the first days of fermentation is necessary, avoiding competition among inoculated microorganisms (Lucio et al., 2014). In part, this is possible given the temporary lack of competition that yeast exerts on the growth of bacteria in the first moments (Henick-Kling & Park, 1994). But it is very important a right choice of "the proper companion" of yeasts and bacteria to be used overlapped in the winemaking, as sometimes antagonistic responses are observed and AF and/or MLF is jeopardized (Capucho & San Romão, 1994; Edwards & Beelman, 1987; Lonvaud-Funel, Joyeux, & Desens, 1988; Osborne & Edwards, 2007). Caridi and Corte (1997) observed that white wines produced by certain cryotolerant strains of *Saccharomyces cerevisiae* are more resistant to MLF than white wines produced by normal strains. This effect is probably related to the high production of metabolites, such as succinic acid and 2-phenylethanol, which inhibit the growth of

LAB (Caridi & Corte, 1997). Besides, winemaking conditions can also alter the MLF, as described by Mota et al. (2018) when they found that AF temperature lower than 12°C reduces the metabolism of LAB, and when it is associated with a pH below 3.2 and free SO_2 above 15 mg/L, it may impair their growth and activity after the lysis of yeast cells.

14.2 MLF as a bioprotection agent

One of the advantages of provoking MLF in white wines is the bioprotection that it confers against detrimental or spoilage microorganisms. Besides the "positive" yeasts and bacteria that contribute to AF and MLF, there are other microorganisms present in wine, with some of them causing unnecessary spoilage. The most relevant is the yeast species *Brettanomyces/Dekkera bruxellensis*, which produces off-flavors in wine. *Saccharomycodes ludwigii* and *Zygosaccharomyces bailii* are responsible for refermentation and turbidity in sweet white wines. Regarding bacteria, the uncontrolled activity of acetic acid bacteria in the wine eventually turns it into vinegar, while LAB may cause undesirable changes in wine taste and flavor (Malfeito-Ferreira, 2014). Especially interesting in sparkling wines with a second alcoholic fermentation in bottles, is the early induction of MLF in base wines, preventing thus of hazing when a late and undesired MLF is produced, avoiding a big economic loss whenever it happens. Another spoilage risk, with putative health implications, is the synthesis by LAB of biogenic amines or citrulline, a precursor of carcinogenic ethyl carbamate (urethane). Indeed, the overall levels of biogenic amines in white wines are generally lower than in reds (Ganic, Gracin, Komes, Curko, & Lovric, 2009), mainly because white wines are not so frequently aged in barrels and are consumed as "young" wines. However, aging of white wines for several weeks before bottling is more and more frequent, and this could increase the risk of biogenic amines' formation. During this process, many white wines also are aged on lees, and residual yeast lees in wine have been identified as an important risk factor for increased citrulline or biogenic amines' formation (Smit & du Toit, 2013; Smit, du Toit, Stander, & du Toit, 2013; Terrade & Orduña, 2006), among other undesirable compounds.

As MLF can produce an increase in the pH or undesired changes in organoleptic properties, many times, this process is annoying in white wines. This is a real and risky possibility increasing undoubtedly in the last years (because of climate change among other causes), and it is especially undesired in sparkling wines, where a second AF occurs in bottles from which the wine cannot be removed. Besides an increase in the pH or undesired changes in organoleptic properties, an undesired and uncontrolled MLF can produce some other alterations such as the appearance of turbidity and ropiness. To prevent this undesired MLF in white wines or base wines for sparkling wines, some treatments are applicable to wines such as filtration before bottling or adding chemicals such as DMDC, lysozyme, or chitosan. For lysozyme treatment, this compound has even been immobilized on chitosan beads to treat white wine continuously in a fluidized-bed reactor with success (Cappannella et al., 2016; Liburdi, Benucci, & Esti, 2014).

Another form of bioprotection is to avoid the increases in pH due to MLF. Interestingly, LAB can not only counterbalance the intrinsic rise in pH from MLF but also even decrease the final pH of the wines when growing in grape musts. This acidification effect is only possible when bacteria can grow. In some strategies, when high cell numbers on non-proliferating cells are employed to develop MLF, malic acid can be metabolized because of the enzymatic machinery of these cells and converted into lactic acid, but as sugar catabolism is not driven, other organic acids are not synthesized. An advantage of this strategy of early inoculation allowing growth in grape must is that sugars are converted into lactic acid, contributing to the net acidification of the final wine. Lactic acid, besides its organoleptic characteristics, does not precipitate as tartaric acid does and is much softer than malic acid, giving wine more complexity.

However, among other drawbacks, these chemical treatments cannot guarantee an ulterior growth of undesired LAB, even if only a few bacteria bypass the control measures. If some bacteria can survive the control measures and there are enough nutrients present, they can grow and spoil the wine in a relatively short time. An alternative to these chemical treatments (non-exclusive and able to be used jointly with them) is using LAB as bioprotectors and conducting a controlled MLF in wine or even in grape musts. If selected safe starters of LAB are employed, they can conduct MLF under controlled conditions, displacing the natural microbiota and removing metabolic resources that are no longer available for the refermentations, i.e., malic acid as a possible energy source (Henick-Kling, 1986; Henick-Kling, 1995; Henick-Kling, Cox, & Olsen, 1991; Henick-Kling, Cox, & Olsen, 1992; Lerm et al., 2010). Among these biological alternatives, the use of custom-selected yeasts and bacterial strains inoculated to promote AF and MLF as biological control agents have been advocated to avoid the "Brett" character in wine (Berbegal, Spano, Fragasso, et al., 2017).

14.3 Impact of MLF on wine organoleptic properties

MLF can be beneficial in white wines when replacing the strong "green" taste of L-malic acid with the less aggressive taste of lactic acid (Volschenk et al., 2006). Talking about aromas, there exist also contradictory tendencies about the benefits or drawbacks of MLF in white wines, as the sensory changes caused during MLF are not always satisfactory in all wine styles and varieties. In highly aromatic wines such as Riesling, Muscat, Sauvignon blanc, or Gewürztraminer, MLF is undesired and hence avoided and proscribed. This is because LAB degrade many terpenes and other flavor molecules that reduce varietal fruity aromas (Volschenk et al., 2006). It is considered that for these cases, the respective wines lose aroma character after MLF, making wines flatter. However, for some other varieties such as Chardonnay or Viognier MLF is considered often beneficial in Burgundy and Champagne regions (Gambetta, Bastian, Cozzolino, & Jeffery, 2014). For Chardonnay, significant differences in aroma composition and taste have been found according to the MLF timing and inoculation conditions and temperature (Sereni, Phan, Osborne, & Tomasino, 2020). The temperature also has a greater effect on the aroma components of sequential inoculation in this grape variety, while the temperature difference of simultaneous fermentation is small (Sereni et al., 2020). Regarding other white varieties, *L. plantarum* enhanced floral notes in white Falanghina wines (Cinquanta, De Stefano, Formato, Niro, & Panfili, 2018). These authors, among others, also found that the change in the aroma profile of white wines is not so evident when *O. oeni* is inoculated as a malolactic starter than when MLF is conducted by lactobacilli (Cinquanta et al., 2018). There is an increase in aromas because of the enzymatic activities of the LAB (glucosidases….), with lactobacilli being much more active enzymatically than oenococci, resulting in this activity in more complex and aromatic wines (Esteban-Torres et al., 2015; Lucio et al., 2016). Otherwise, pediococci have also been employed as MLF starters for white wines, as in the Caíño grape variety (Juega et al., 2014). In this case, autochthonous *Pediococcus damnosus* strains were able to predominate over the commercial *O. oeni* inoculated starter and perform MLF in Caíño wine. Juega et al. (2014) also demonstrated that neither the commercial strain nor indigenous *Pediococcus* carried out MLF in Albariño wine. However, MLF was achieved when autochthonous strains that predominated in Caíño were inoculated into Albariño. Finally, Albariño increased its body and softness, while Caíño results in a more mature wine.

For Sauvignon white wine, when MLF was developed in barrels, Bertrand et al. described that carbonyl substances were formed in connection with a more or less fast bacterial growth and the degradation of citric acid (Bertrand, De Revel, & Pripis Nicolau, 2000). Besides, they observed that the concentration of the compounds resulting from wood was higher in the wines after MLF in barrels compared to wines not having undergone bacterial growth were observed. The greater complexity was attributed to buttered, spiced, roasted, vanilla, and smoked notes (Bertrand et al., 2000); differences according to starter cultures were noted, and the chemical analyses confirmed these differences for some compounds such as vanillin, oak lactone, and eugenol (de Revel, Marin, Pripis-Nicolau, Lonvaud-Funel, & Bertand, 1999).

Not always is the impact of MLF in white wines beneficial, and many times, some spoilage metabolites are synthesized by some LAB species. As an example, the synthesis of diacetyl usually masks the characteristics of white wine with a strong butter or cheese flavor; it can be considered an unpopular flavor (Martineau and Acree 1995; Martineau and Henick-Kling 1995a, 1995b; Martineau et al., 1995). Certain LAB strains, especially some *Lactobacillus*, have the ability to degrade L-tartaric acid may cause serious defects in wine acidity, which is defined as *tourne* (Ribéreau-Gayon, Dubourdieu, Donèche, & Lonvaud, 2000a; Ribéreau-Gayon, Dubordieu, Donèche, & Lonvaud, 2000b). Another smell related to *Lactobacillus* is caused by dehydration of glycerol when converted to 3-hydroxypropionaldehyde and further reduced to 1,3-propanediol (Schütz & Radler 1984a, 1984b). One more problem can arise by the combination of acrolein derived from 3-hydroxypropionaldehyde with the phenolics groups of anthocyanins, producing bitterness (Bartowsky, 2009). *Lactobacillus hilgardii* can also produce a mousy off-flavor in wine by the synthesis of the N-heterocyclic bases 2-ethyltetrahydropyridine, 2-acetyl-1-pyrroline, and 2-acetyltetrahydropyridine, intermediates derived from the metabolism of L-ornithine and L-lysine (Costello & Henschke, 2002).

An interesting finding was achieved by Knoll (2011), who demonstrated that the change of volatile aroma components is not necessarily related to the complete MLF, and partial MLF still produced a significant impact on the aroma characteristics of white wine. It is also well known that the wine matrix and pH and alcohol concentration affect the MLF and final volatile aroma (Knoll et al., 2011). Altogether, this reflects that the final effect of the microbiota on the sensorial characteristics of wine is not always easy to foresee, and it is not necessary to get a 100% implantation of a starter to obtain noticeable changes, positive or negative, on the organoleptic properties in a wine.

Knoll (2011) and Knoll et al. (2011) studied the effects of pH and ethanol on aroma compounds related to MLF in white wine varieties. They reported only slight changes in higher alcohol content but pointed out that rose and floral aromas (2-phenylethanol) increased at lower ethanol concentrations, while the concentrations of the ethyl esters of hexanoic, octanoic, and decanoic acids increased at higher alcohol concentrations. They found that pH and ethanol are the main variables that affect the distribution of volatile components in Riesling wine after completing MLF.

14.4 New findings, new ways to employ MLF, and new applications

For white wines, as in other types, the possibility to conduct an early MLF in grape musts can bring an extra advantage if correctly driven. When bacteria grow in wine, few sugars remain available for them, and several other drawbacks are present: high ethanol, higher sulfite content, etc. This makes the growth in wine more diffi- cult for LAB than in grape must, and MLF is more difficult and risky in wines, especially for lactobacilli. An alternative is early MLF in grape musts, before resources are scarce and the environment becomes harsher (Guzzon et al., 2016). Then, LAB can be inoculated in grape musts just before yeasts, at the same time, or just after them, with a high concentration of sugars and low ethanol available (Pardo & Ferrer, 2019). This environ- ment allows very rapid growth of bacteria and, besides a rapid MLF within a few days, the synthesis of lactic acid from the metabolized sugars. Therefore the greatest dynamics and efficiency in deacidification were observed when AF and MLF occurred simultaneously (Lasik-Kurdyś, Gumienna, & Nowaket al., 2017). Besides, the fermentation time was significantly reduced in coinoculation strategies, without an increase in volatile acidity or effect on the metabolism of citric acid or the final ethanol and glycerol concentration, and the lowest residual sugar concentrations were obtained.

The coinoculation of yeast and bacteria may be a very useful technique for wine production in cool climates, especially for grape must with extremely high acidity (Lasik-Kurdyś et al., 2017). Guzzon et al. (2016) described that simultaneous inoculation of two *O. oeni* strains ensured a regular evolution of MLF in a high-acidity Incrocio Manzoni grape must. In general, carrying out AF and MLF at the same time can greatly reduce the fermentation time, but the grape must pH is still an important factor for the success of the fermentation and determines the final concentration of various wine metabolites. The inoculation time affects the formation and degradation kinetics of organic acids and acetaldehyde (Jackowetz & Mira de Orduña, 2012; Osborne, Dubé Morneau, & Mira de Orduña, 2006; Osborne, Mira de Orduña, Pilone, & Liu, 2000; Pan, Jussier, Terrade, Yada, & Mira de Orduña, 2011).

LAB species chosen for this simultaneous inoculation with yeasts can be lactobacilli (better homofermentative species), or even heterofermenters such as *O. oeni*. Many studies have demonstrated that the final production of acetic acid is not so high even with heterofermenters when correctly employed. It is questionable whether selected bacterial species and strains are specific or not for white and red wines. Many starter producers differentiate spe- cific strains for both types of wines, and even for low-acidity or high-acidity and low-ethanol or high ethanol wines. On the contrary, many winemakers use these strains indistinctly for reds and whites. Is it such a difference to justify employing different LAB strains for white and red wines? It seems so. Campbell-Sills et al. (2017) propose the existence of two groups of *O. oeni* strains, one associated with red wines and the other with white wines. A common ancestor should have colonized two different substrates, namely red and white wine—associated environ- ments, diverging over time and disseminating to various regions. The same researchers describe that the biogeogra- phy of *O. oeni* reveals distinctive but non-specific populations in wine-producing regions (El Khoury et al., 2017). When used to ferment a Chardonnay wine, the final volatile composition depended on the group of origin of these strains (Campbell-Sills et al., 2017). Following this tendency, some researchers focus their election programs on specific strains for white wines (Wang et al., 2016).

The genome sequence of *O. oeni* OE37, an autochthonous strain isolated from an Italian white wine, has been described by Costantini et al. (2020). This strain was isolated from a Chardonnay wine during spontaneous MLF. The observation of the genes contained in the OE37 genome confirmed the presence of genes to conduct MLF, as well as the diacetyl reductase gene related to citrate metabolism. Other genes concerning the organoleptic charac- teristics are present in this strain, such as β-glucosidase and other glucosidase enzyme genes. Genes coding for exopolysaccharide synthesis are also present.

Cofermentation is a winemaking practice based on combining two or more different grape cultivars at crushing such that the musts would "coferment" during the maceration period (Casassa, Mawdsley, Stoffel, Williams, & Dodson Peterson, 2020). This is a traditional practice in some Old World regions, and probably the chemical and sen- sory benefits of this practice are relatively small. When comparing cofermentation with post-MLF blending of Syrah with selected Rhône white cultivars, Casassa et al. (2020) found that post-MLF blending increases aromatics and color, and when compared to cofermentation, it was less detrimental for anthocyanin concentration, resulted in higher pigment concentration, and yielded higher values of chroma, red color component, and saturation. Post-MLF blended wines were also higher in black fruit and cherry aromas than cofermented wines, which suggests that it increased the diversity of the aromatic profile of the resulting wines, and these effects were perceivable from a sensory viewpoint (Casassa et al., 2020).

An interesting approach to develop MLF is the use of immobilized cells (Agouridis, Bekatorou, Nigam, & Kanellaki, 2005; Agouridis, Kopsahelis, Plessas, Koutinas, & Kanellaki, 2008; Berbegal, Spano, Tristezza, Grieco, & Capozzi, 2017; Claus, 2017; Genisheva, Mussatto, Oliveira, & Teixeira, 2013; Janssen, Maddox, & Mawson, 1993; Liu, 2002; Maicas, Pardo, & Ferrer, 2001; Naouri, Chagnaud, Arnaud, & Galzy, 1990; Pardo & Ferrer, 2019; Rodríguez-Nogales et al., 2020; Servetas et al., 2013; Shieh & Tsay, 1990; Simó, Fernández-Fernández, Vila-Crespo, Ruipérez, & Rodríguez-Nogales, 2017; Trueck & Hammes, 1989; Vila-Crespo, Rodriguez-Nogales, Fernández-Fernández, & Hernanz-Moral, 2010; Virdis et al., 2021). Specifically for white wines, *Lactobacillus casei* cells encapsulated in calcium pectate gel has been carried out (Kosseva & Kennedy, 2004). In this case, the degree of conversion of malic acid in wine by the encapsulated cells was twice as high as that obtained by the free cells. MLF can take place at high ethanol concentrations, and the operational stability of calcium pectate gel capsules was 6 months (Kosseva & Kennedy, 2004).

Acknowledgments

The present work has been partially financed by Project AGL2015−71227-R (ERDF and the Spanish Ministry of Economy and Competitiveness).

References

Agouridis, N., Bekatorou, A., Nigam, P., & Kanellaki, M. (2005). Malolactic fermentation in wine with *Lactobacillus casei* cells immobilized on delignified cellulosic material. *Journal of Agricultural and Food Chemistry, 53*(7), 2546−2551. Available from http://pubs.acs.org/cgi-bin/article.cgi/jafcau/2005/53/i07/pdf/jf048736t.pdf, Retrieved from.

Agouridis, N., Kopsahelis, N., Plessas, S., Koutinas, A. A., & Kanellaki, M. (2008). *Oenococcus oeni* cells immobilized on delignified cellulosic material for malolactic fermentation of wine. *Bioresource Technology, 99*, 9017−9020. Available from http://www.sciencedirect.com/science/article/B6V24-4SK0BYY-3/1/c83352e27dc4a7ce4c0e47324960efc9, Retrieved from.

Bartowsky, E. J. (2009). Bacterial spoilage of wine and approaches to minimize it. *Letters in Applied Microbiology, 48*(2), 149−156. Available from https://doi.org/10.1111/j.1472-765X.2008.02505.x, Retrieved from.

Bartowsky, E. J. (2014). WINES | malolactic fermentation A2 − Batt, Carl A. In M. L. Tortorello (Ed.), *Encyclopedia of food microbiology* (Second Edition, pp. 800−804). Oxford: Academic Press.

Bartowsky, E. J., Francis, I. L., Bellon, J. R., & Henschke, P. A. (2002). Is buttery aroma perception in wines predictable from the diacetyl concentration. *Australian Journal of Grape and Wine Research, 8*(3), 180−185. Available from https://doi.org/10.1111/j.1755-0238.2002.tb00254.x.

Berbegal, C. (2014). Novel liquid starter cultures for malolactic fermentation in wine (Tesis Doctoral Ph.D. Thesis). Universidad de Valencia.

Berbegal, C., Benavent-Gil, Y., Navascués, E., Calvo, A., Albors, C., Pardo, I., & Ferrer, S. (2017). Lowering histamine formation in a red Ribera del Duero wine (Spain) by using an indigenous *O. oeni* strain as a malolactic starter. *International Journal of Food Microbiology, 244*, 11−18. Available from https://doi.org/10.1016/j.ijfoodmicro.2016.12.013.

Berbegal, C., Benavent-Gil, Y., Pardo, I., Izcara, E., Navascués, E., & Ferrer, S. (2014). Indigenous *O. oeni* strain selection as a malolactic fermentation starter culture to avoid the histamine production in wine. *Infowine.* Available from http://www.infowine.com/default.asp?scheda = 13744, Retrieved from.

Berbegal, C., Spano, G., Fragasso, M., Grieco, F., Russo, P., & Capozzi, V. (2017). Starter cultures as biocontrol strategy to prevent *Brettanomyces bruxellensis* proliferation in wine. *Applied Microbiology and Biotechnology, 102*(2), 569−576. Available from https://doi.org/10.1007/s00253-017-8666-x.

Berbegal, C., Spano, G., Tristezza, M., Grieco, F., & Capozzi, V. (2017). Microbial resources and innovation in the wine production sector. *South African Journal of Enology and Viticulture, 38*(2), 156−166.

Bertrand, A., De Revel, G., & Pripis Nicolau, L. (2000). Sensory evaluation of consequences of malolactic fermentation for white wine in barrels. *Bulletin OIV, 831*, 313−321, *832*.

Campbell-Sills, H., El Khoury, M., Gammacurta, M., Miot-Sertier, C., Dutilh, L., Vestner, J., & Lucas, P. (2017). Two different *Oenococcus oeni* lineages are associated to either red or white wines in Burgundy: Genomics and metabolomics insights. *OENO One, 51*(3), 309−322. Available from https://doi.org/10.20870/oeno-one.2017.51.4.1861.

Cappannella, E., Benucci, I., Lombardelli, C., Liburdi, K., Bavaro, T., & Esti, M. (2016). Immobilized lysozyme for the continuous lysis of lactic bacteria in wine: Bench-scale fluidized-bed reactor study. *Food Chemistry, 210*, 49−55. Available from https://doi.org/10.1016/j.foodchem.2016.04.089.

Cappello, M. S., Zapparoli, G., Logrieco, A., & Bartowsky, E. J. (2017). Linking wine lactic acid bacteria diversity with wine aroma and flavour. *International Journal of Food Microbiology, 243*, 16−27. Available from https://doi.org/10.1016/j.ijfoodmicro.2016.11.025.

Capucho, I., & San Romão, M. V. (1994). Effect of ethanol and fatty acids on malolactic activity of *Leuconostoc oenos. Applied Microbiology and Biotechnology, 42*, 391−395.

Caridi, A., & Corte, V. (1997). Inhibition of malolactic fermentation by cryotolerant yeasts. *Biotechnology Letters, 19*(8), 723−726. Available from https://doi.org/10.1023/a:1018319705617.

Casassa, L. F., Mawdsley, P. F. W., Stoffel, E., Williams, P., & Dodson Peterson, J. C. (2020). Chemical and sensory effects of cofermentation and post-malolactic fermentation blending of Syrah with selected Rhône white cultivars. *Australian Journal of Grape and Wine Research*, 26(1), 41−52. Available from https://doi.org/10.1111/ajgw.12413.

Cavin, J. F., Andioc, V., Etievant, P. X., & Diviès, C. (1993). Ability of wine lactic acid bacteria to metabolize phenol carboxylic acids. *American Journal of Enology and Viticulture*, 44, 76−80.

Cinquanta, L., De Stefano, G., Formato, D., Niro, S., & Panfili, G. (2018). Effect of pH on malolactic fermentation in southern Italian wines. *European Food Research and Technology*, 244(7), 1261−1268. Available from https://doi.org/10.1007/s00217-018-3041-4.

Claus, H. (2017). Microbial enzymes: Relevance for winemaking. In H. König, G. Unden, & J. Fröhlich (Eds.), *Biology of microorganisms on grapes, in must and in wine* (pp. 315−338). Cham: Springer International Publishing.

Costantini, A., Blazquez, A. B., Cerutti, F., Vaudano, E., Peletto, S., Saiz, J.-C., & Garcia-Moruno, E. (2020). Genome sequence of *Oenococcus oeni* OE37, an autochthonous strain isolated from an Italian white wine. *Microbiology Resource Announcements*, 9(39), e00582−00520. Available from https://doi.org/10.1128/mra.00582-20.

Costello, P. J., Francis, I. L., & Bartowsky, E. J. (2012). Variations in the effect of malolactic fermentation on the chemical and sensory properties of Cabernet Sauvignon wine: interactive influences of *Oenococcus oeni* strain and wine matrix composition. *Australian Journal of Grape and Wine Research*, 18(3), 287−301. Available from https://doi.org/10.1111/j.1755-0238.2012.00196.x.

Costello, P. J., & Henschke, P. A. (2002). Mousy off-flavor of wine: Precursors and biosynthesis of the causative N-heterocycles 2-ethyltetrahydropyridine, 2-acetyltetrahydropyridine, and 2-acetyl-1-pyrroline by *Lactobacillus hilgardii* DSM 20176. *Journal of Agricultural and Food Chemistry*, 50(24), 7079−7087. Available from https://doi.org/10.1021/jf020341r.

Darias-Martín, J., Socas-Hernández, A., Díaz-Romero, C., & Díaz-Díaz, E. (2003) Comparative study of methods for determination of titrable acidity in wine. Journal of Food Composition and Analysis, 16, 555−562.

de Revel, G., Marin, N., Pripis-Nicolau, L., Lonvaud-Funel, A., & Bertand, A. (1999). Contribution to the knowledge of malolactic fermentation influence of wine aroma. *Journal of Agricultural and Food Chemistry*, 47, 4003−4008.

Dequin, S., Baptista, E., & Barre, P. (1999). Acidification of grape musts by *Saccharomyces cerevisiae* wine yeast strains genetically engineered to produce lactic acid. *American Journal of Enology and Viticulture*, 50(1), 45−50.

Edwards, C. G., & Beelman, R. B. (1987). Inhibition of the malolactic bacterium *Leuconostoc oenos* (PSU-1), by decanoic acid and subsequent removal of the inhibition by yeast ghosts. *American Journal of Enology and Viticulture*, 38, 239−242.

El Khoury, M., Campbell-Sills, H., Salin, F., Guichoux, E., Claisse, O., & Lucas, P. M. (2017). Biogeography of *Oenococcus oeni* reveals distinctive but non-specific populations in wine-producing regions. *Applied and Environmental Microbiology*, 83(3), e02322−02316. Available from https://doi.org/10.1128/aem.02322-16.

Esteban-Torres, M., Landete, J. M., Reveron, I., Santamaría, L., de las Rivas, B., & Muñoz, R. (2015). A *Lactobacillus plantarum* esterase active on a broad range of phenolic esters. *Applied and Environmental Microbiology*, 81(9), 3235−3242. Available from https://doi.org/10.1128/aem.00323-15.

Gambetta, J. M., Bastian, S. E. P., Cozzolino, D., & Jeffery, D. W. (2014). Factors influencing the aroma composition of Chardonnay wines. *Journal of Agricultural and Food Chemistry*, 62(28), 6512−6534. Available from https://doi.org/10.1021/jf501945s.

Ganic, K. K., Gracin, L., Komes, D., Curko, N., & Lovric, T. (2009). Changes of the content of biogenic amines during winemaking of Sauvignon wines. *Croatian Journal of Food Science and Technology*, 1(2), 21−27. < http://hrcak.srce.hr/index.php?show = clanak&id_clanak_jezik = 81239 > , Retrieved from.

Genisheva, Z., Mussatto, S. I., Oliveira, J. M., & Teixeira, J. A. (2013). Malolactic fermentation of wines with immobilised lactic acid bacteria-influence of concentration, type of support material and storage conditions. *Food Chemistry*, 138(2−3), 1510−1514.

Gil-Sánchez, I., Bartolomé Suáldea, B., & Victoria Moreno-Arribas, M. (2019). Malolactic fermentation. In A. Morata (Ed.), *Red Wine Technoloogy* (pp. 85−98). Elsevier.

Guzzon, R., Moser, S., Davide, S., Villegas, T. R., Malacarne, M., Larcher, R., & Krieger-Weber, S. (2016). Exploitation of simultaneous alcoholic and malolactic fermentation of Incrocio Manzoni, a traditional Italian white wine. *South African Journal of Enology and Viticulture*, 37(2), 124−131. < http://www.journals.ac.za/index.php/sajev/article/view/828 > , Retrieved from.

Henick-Kling, T. (1986). *Growth and metabolism of Leuconostoc oenos and Lactobacillus plantarum in wine. Thesis (Ph.D.)--Dept. of Agricultural Biochemistry*. Waite Agricultural Research Institute, University of Adelaide.

Henick-Kling, T. (1995). Control of malolactic fermentation in wine: energetics, flavour modification and methods of starter culture preparation. *Journal of Applied Bacteriology*, 79, 29−S37.

Henick-Kling, T., Cox, D. J., & Olsen, E. B. (1991). Production de l'énergie durant la fermentation malolactique. *Revue Française d'OEnologie*, 132, 63−66.

Henick-Kling, T., Cox, D. J., & Olsen, E. B. (1992). Energy from malolactic fermentation. *Biologia Oggi*, 6, 9−14.

Henick-Kling, T., & Park, Y. H. (1994). Considerations for the use of yeast and bacterial starter cultures: SO_2 and timing of inoculation. *American Journal of Enology and Viticulture*, 45, 464−469.

Hernández-Orte, P., Cersosimo, M., Loscos, N., Cacho, J., García-Moruno, E., & Ferreira, V. (2009). Aroma development from non-floral grape precursors by wine lactic acid bacteria. *Food Research International*, 42(7), 773−781. < http://www.sciencedirect.com/science/article/B6T6V-4VR9FFN-4/2/fe0c751a663dc08576534b44c700fd6e > , Retrieved from.

Jackowetz, J. N., & Mira de Orduña, R. (2012). Metabolism of SO_2 binding compounds by *Oenococcus oeni* during and after malolactic fermentation in white wine. *International Journal of Food Microbiology*, 155(3), 153-157. < http://www.sciencedirect.com/science/article/pii/S016816051200058X > , Retrieved from.

Janssen, D. E., Maddox, I. S., & Mawson, A. J. (1993). An immobilized cell bioreactor for the malolactic fermentation of wine. *Australian and New Zealand Wine Industry Journal*, 8, 161−165.

Juega, M., Costantini, A., Bonello, F., Cravero, M.-C., Martinez-Rodriguez, A. J., Carrascosa, A. V., & Garcia-Moruno, E. (2014). Effect of malolactic fermentation by *Pediococcus damnosus* on the composition and sensory profile of Albariño and Caiño white wines. *Journal of Applied Microbiology*, 116, 586−595. Available from https://doi.org/10.1111/jam.12392.

Kapsopoulou, K., Mourtzini, A., Anthoulas, M., & Nerantzis, E. (2007). Biological acidification during grape must fermentation using mixed cultures of *Kluyveromyces thermotolerans* and *Saccharomyces cerevisiae*. *World Journal of Microbiology and Biotechnology*, 23(5), 735−739. Available from https://doi.org/10.1007/s11274-006-9283-5.

Knoll, C. (2011). *Evaluating the influence of stress parameters on Oenococcus oeni and the subsequent volatile aroma composition of white wine.* (Doctor agriculturae). Justus-Liebig-University Gießen, Germany.

Knoll, C., Fritsch, S., Schnell, S., Grossmann, M., Rauhut, D., & du Toit, M. (2011). Influence of pH and ethanol on malolactic fermentation and volatile aroma compound composition in white wines. *LWT - Food Science and Technology, 44*(10), 2077–2086. Available from https://doi.org/10.1016/j.lwt.2011.05.009.

Kosseva, M. R., & Kennedy, J. F. (2004). Encapsulated lactic acid bacteria for control of malolactic fermentation in wine. *Artificial Cells, Blood Substitutes, and Biotechnology, 32*(1), 55–65.

Landaud, S., Helinck, S., & Bonnarme, P. (2008). Formation of volatile sulfur compounds and metabolism of methionine and other sulfur compounds in fermented food. *Applied Microbiology and Biotechnology, 77*(6), 1191–1205. Available from https://doi.org/10.1007/s00253-007-1288-y.

Lasik-Kurdyś, M., Gumienna, M., & Nowak, J. (2017). Influence of malolactic bacteria inoculation scenarios on the efficiency of the vinification process and the quality of grape wine from the Central European region. *European Food Research and Technology.* Available from https://doi.org/10.1007/s00217-017-2919-x.

Lerm, E., Engelbrecht, L., & du Toit, M. (2010). Malolactic termentation: The ABC's of MLF. *South African Journal of Enology and Viticulture, 31*(2), 186–212.

Liburdi, K., Benucci, I., & Esti, M. (2014). Lysozyme in wine: An overview of current and future applications. *Comprehensive Reviews in Food Science and Food Safety, 13*(5), 1062–1073. Available from https://doi.org/10.1111/1541-4337.12102.

Liu, S.-Q. (2002). Malolactic fermentation in wine-beyond deacidification. *Journal of Applied Microbiology, 92*(4), 589–601. Available from http://www.blackwell-synergy.com/loi/jam, Retrieved from.

Lonvaud-Funel, A. (2010). 3-Effects of malolactic fermentation on wine quality. In A. G. Reynolds (Ed.), *Managing wine quality* (pp. 60–92). Woodhead Publishing.

Lonvaud-Funel, A., Joyeux, A., & Desens, C. (1988). Inhibition of malolactic fermentation of wines by products of yeast metabolism. *Journal of the Science of Food and Agriculture, 44*, 183–191.

Lucio, O. (2014). *Acidificación biológica de vinos de pH elevado mediante la utilización de bacterias lácticas.* (Tesis Doctoral Ph.D. Thesis). Universidad de Valencia.

Lucio, O., Pardo, I., Heras, J.M., Krieger, S., & Ferrer, S. (2014). Effect of yeasts/bacteria co-inoculation on malolactic fermentation of Tempranillo wines. Infowine. <http://www.infowine.com/default.asp?scheda = 13630 >, Retrieved from.

Lucio, O., Pardo, I., Krieger-Weber, S., Heras, J. M., & Ferrer, S. (2016). Selection of *Lactobacillus* strains to induce biological acidification in low acidity wines. *LWT – Food Science and Technology, 73*, 334–341. Available from https://doi.org/10.1016/j.lwt.2016.06.031.

Maicas, S., Pardo, I., & Ferrer, S. (2001). The potential of positively-charged cellulose sponge for malolactic fermentation of wine, using *Oenococcus oeni. Enzyme and Microbial Technology, 28*(4–5), 415–419. Available from http://www.sciencedirect.com/science/article/B6TG1-42G6M8J-P/2/c851db03213d40da39505245e522e5ce, Retrieved from.

Malfeito-Ferreira, M. (2014). WINES | Wine spoilage yeasts and bacteria. In M. L. T. Carl, & A. Batt (Eds.), *Encyclopedia of food microbiology* (Second Edition, pp. 805–810). Oxford: Academic Press.

Martineau, B., Acree, T. E., & Henick-Kling, T. (1995). Effect of wine type on the detection threshold for diacetyl. *Food Research International, 28*, 139–143.

Martineau, B., & Henick-Kling, T. (1995a). Formation and degradation of diacetyl in wine during alcoholic fermentation th *Saccharomyces cerevisiae* strain EC11 and malolactic fermentation with *Leuconostoc oenos* strain MCW. *American Journal of Enology and Viticulture, 46*, 442–448.

Martineau, B., & Henick-Kling, T. (1995b). Performance and diacetyl production of commercial strains of malolactic bacteria in wine. *Journal of Applied Bacteriology, 78*, 526–536.

Martineau, B. H.-K., & Acree, T. T. (1995). Reassessment of the influence of malolactic fermentation on the concentration of diacetyl in wines. *American Journal of Enology and Viticulture, 46*, 385–388.

Morata, A., Bañuelos, M. A., Vaquero, C., Loira, I., Cuerda, R., Palomero, F., & Bi, Y. (2019). *Lachancea thermotolerans* as a tool to improve pH in red wines from warm regions. *European Food Research and Technology, 245*(4), 885–894. Available from https://doi.org/10.1007/s00217-019-03229-9.

Morata, A., Loira, I., Tesfaye, W., Bañuelos, M., González, C., & Suárez Lepe, J. (2018). *Lachancea thermotolerans* applications in wine technology. *Fermentation, 4*(3), 53. Available from https://doi.org/10.3390/fermentation4030053.

Morenzoni, R. (2015). Wine sensory components and malolactic fermentation. In K. S. Specht (Ed.), *Malolactic fermentation- importance of wine lactic acid bacteria in winemaking* (pp. 49–52). Montréal, Canada: Lallemand Inc.

Mota, R. V. d, Ramos, C. L., Peregrino, I., Hassimotto, N. M. A., Purgatto, E., Souza, C. R. d, & Regina, M. d A. (2018). Identification of the potential inhibitors of malolactic fermentation in wines. *Food Science and Technology, 38*, 174–179. < http://www.scielo.br/scielo.php?script = sci_arttext&pid = S0101-20612017005026114&nrm = iso >, Retrieved from.

Naouri, P., Chagnaud, P., Arnaud, A., & Galzy, P. (1990). Essais de bioconversion malolactique des vins au moyen d'un réacteur à lit fluidisé et à cellules immobilisées. *Revue Suisse de Viticulture. Arboriculture et Horticulture, 22*, 207–213.

Onetto, C., & Bordeu, E. (2015). Pre-alcoholic fermentation acidification of red grape must using *Lactobacillus plantarum. Antonie Van Leeuwenhoek, 108*(6), 1469–1475. Available from https://doi.org/10.1007/s10482-015-0602-4.

Osborne, J. P., Dubé Morneau, A., & Mira de Orduña, R. (2006). Degradation of free and sulfur-dioxide-bound acetaldehyde by malolactic lactic acid bacteria in white wine. *Journal of Applied Microbiology, 101*(2), 474–479. Available from http://www.blackwell-synergy.com/doi/abs/10.1111/j.1365-2672.2006.02947.x, Retrieved from.

Osborne, J. P., & Edwards, C. G. (2007). Inhibition of malolactic fermentation by a peptide produced by *Saccharomyces cerevisiae* during alcoholic fermentation. *International Journal of Food Microbiology, 118*(1), 27–34. Available from https://doi.org/10.1016/j.ijfoodmicro.2007.05.007.

Osborne, J. P., Mira de Orduña, R., Pilone, G. J., & Liu, S. Q. (2000). Acetaldehyde metabolism by wine lactic acid bacteria. *FEMS Microbiology Letters, 191*(1), 51–55. < http://www.ncbi.nlm.nih.gov/pubmed/11004399 >, Retrieved from.

Pan, W., Jussier, D., Terrade, N., Yada, R. Y., & Mira de Orduña, R. (2011). Kinetics of sugars, organic acids and acetaldehyde during simultaneous yeast-bacterial fermentations of white wine at different pH values. *Food Research International, 44*(3), 660–666. Available from https://doi.org/10.1016/j.foodres.2010.09.041.

Pardo, I., & Ferrer, S. (2019). Chapter 7 — Yeast-bacteria coinoculation. In A. Morata (Ed.), *Red Wine technology* (pp. 99–114). Elsevier.

Ribéreau-Gayon, J., Dubourdieu, D., Donèche, B., & Lonvaud, A. (2000a). *Metabolism of lactic acid bacteria. Handbook of enology. The microbiology of wine and vinifications* (pp. 129–148). (2nd ed.), Vol. 1, pp.

Ribéreau-Gayon, P., Dubordieu, D., Donèche, B., & Lonvaud, A. (2000b). Lactic acid bacteria development in wine. *Handbook of enology. The microbiology of wine and vinifications*, 149–177, Vol. 1, pp.

Rodríguez-Nogales, J. M., Simó, G., Pérez-Magariño, S., Cano-Mozo, E., Fernández-Fernández, E., Ruipérez, V., & Vila-Crespo, J. (2020). Evaluating the influence of simultaneous inoculation of SiO₂-alginate encapsulated bacteria and yeasts on volatiles, amino acids, biogenic amines and sensory profile of red wine with lysozyme addition. *Food Chemistry*, 327, 126920. Available from https://doi.org/10.1016/j.foodchem.2020.126920.

Schütz, H., & Radler, F. (1984a). Anaerobic reduction of glycerol to propanediol-1.3 by *Lactobacillus brevis* and *Lactobacillus buchnerii*. *Systematic and Applied Microbiology*, 5, 169–178.

Schütz, H., & Radler, F. (1984b). Propanediol-1,2-dehydratase and metabolism of glycerol of *Lactobacillus brevis*. *Archives of Microbiology*, 139, 366–370.

Sereni, A., Phan, Q., Osborne, J., & Tomasino, E. (2020). Impact of the timing and temperature of malolactic fermentation on the aroma composition and mouthfeel properties of Chardonnay wine. *Foods*, 9(6), 802. < https://www.mdpi.com/2304-8158/9/6/802 >, Retrieved from.

Servetas, I., Berbegal, C., Camacho, N., Bekatorou, A., Ferrer, S., Nigam, P., & Koutinas, A. A. (2013). *Saccharomyces cerevisiae* and *Oenococcus oeni* immobilized in different layers of a cellulose/starch gel composite for simultaneous alcoholic and malolactic wine fermentations. *Process Biochemistry*, 48(0), 1279–1284. Available from https://doi.org/10.1016/j.procbio.2013.06.020.

Shieh, Y. M., & Tsay, S. S. (1990). Malolactic fermentation by immobilized *Leuconostoc* sp. M-1. *Journal of the Chinese Agricultural Chemical Society*, 28, 246–257.

Simó, G., Fernández-Fernández, E., Vila-Crespo, J., Ruipérez, V., & Rodríguez-Nogales, J. M. (2017). Silica—alginate-encapsulated bacteria to enhance malolactic fermentation performance in a stressful environment. *Australian Journal of Grape and Wine Research*, 23(3), 342–349. Available from https://doi.org/10.1111/ajgw.12302.

Smit, A. Y., & du Toit, M. (2013). Evaluating the influence of malolactic fermentation inoculation practices and ageing on lees on biogenic amine production in wine. *Food and Bioprocess Technology*, 6(1), 198–206. Available from https://doi.org/10.1007/s11947-011-0702-8.

Smit, A. Y., du Toit, W. J., Stander, M., & du Toit, M. (2013). Evaluating the influence of maceration practices on biogenic amine formation in wine. *LWT — Food Science and Technology*, 53(1), 297–307. Available from https://doi.org/10.1016/j.lwt.2013.01.006.

Su, J., Wang, T., Wang, Y., Li, Y.-Y., & Li, H. (2014). The use of lactic acid-producing, malic acid-producing, or malic acid-degrading yeast strains for acidity adjustment in the wine industry. *Applied Microbiology and Biotechnology*, 98(6), 2395–2413. Available from https://doi.org/10.1007/s00253-014-5508-y.

Swiegers, J. H., Bartowsky, E. J., Henschke, P. A., & Pretorius, I. S. (2005). Yeast and bacterial modulation of wine aroma and flavour. *Australian Journal of Grape and Wine Research*, 11(2), 139–173. Available from https://doi.org/10.1111/j.1755-0238.2005.tb00285.x.

Terrade, N., & Orduña, R. M. D. (2006). Impact of winemaking practices or arginine and citrulline metabolism during and after malolactic fermentation. *Journal of Applied Microbiology*, 101, 406–411. http://doi.org/10.11117j.1365-2672.2006.02978.x.

Thoukis, G., Ueda, M., & Wright, D. (1965). The formation of succinic acid during alcoholic fermentation. *American Journal of Enology and Viticulture*, 16(1), 1–8.

Trueck, H. U., & Hammes, W. P. (1989). The application of a fluidized-bed reactor for malolactic fermentation in wine with immobilized cells. *Chemie Mikrobiologie Technologie der Lebensmittel*, 12, 119–126.

Val, P., Ferrer, S., & Pardo, I. (2002). Acidificación biológica de vinos con bajo contenido en ácidos. *Enólogos*, 20, 35–42.

Vila-Crespo, J., Rodriguez-Nogales, J., Fernández-Fernández, E., & Hernanz-Moral, M. (2010). Strategies for the enhancement of malolactic fermentation in the new climate conditions. *Current Research, Technology and Education Topics in Applied Microbiology and Microbial Biotechnology, Microbiology Series*, 2, 920–929.

Vilela, A. (2019). Use of nonconventional yeasts for modulating wine acidity. *Fermentation*, 5(1), 27.

Virdis, C., Sumby, K., Bartowsky, E., & Jiranek, V. (2021). Lactic acid bacteria in wine: Technological advances and evaluation of their functional role. *Frontiers in Microbiology*, 11, 1–16. Available from https://doi.org/10.3389/fmicb.2020.612118.

Volschenk, H., van Vuuren, H. J. J., & Viljoen-Bloom, M. (2006). Malic acid in wine: Origin, function and metabolism during vinification. *South African Journal of Enology and Viticulture*, 27, 123–136.

Wang, P., Li, A., Sun, H., Dong, M., Wei, X., & Fan, M. (2016). Selection and characterization of *Oenococcus oeni* strains for use as new malolactic fermentation starter cultures. *Annals of Microbiology*, 1–8. Available from https://doi.org/10.1007/s13213-016-1217-3.

Yeramian, N. (2003). *Acidificación biológica de mostos en zonas cálidas (Ph.D. Thesis)*. Universidad Politécnica de Madrid.

Yéramian, N., Chaya, C., & Suárez Lepe, J. A. (2007). L-(−)-Malic acid production by *Saccharomyces* spp. during the alcoholic fermentation of wine. *Journal of Agricultural and Food Chemistry*, 55(3), 912–919. Available from https://doi.org/10.1021/jf061990w.

Pinking

Luís Filipe-Ribeiro[1], Jenny Andrea-Silva[1,2], Fernanda Cosme[1,3] and Fernando M. Nunes[1,4]

[1]Chemistry Research Centre-Vila Real (CQ-VR), Food and Wine Chemistry Laboratory, University of Trás-os-Montes and Alto Douro, Vila Real, Portugal [2]Adega Cooperativa de Figueira de Castelo Rodrigo, Portugal [3]Biology and Environment Department, School of Life Sciences and Environment, University of Trás-os-Montes and Alto Douro, Vila Real, Portugal [4]Chemistry Department, School of Life Sciences and Environment, University of Trás-os-Montes and Alto Douro, Vila Real, Portugal

15.1 Introduction

At present, reductive or hyperreductive vinification of white wines is one of the most used winemaking processes for its production, being particularly advantageous for varieties rich in varietal aromas sensitive to oxidation like Sauvignon Blanc, Colombard, Petit Manseng, Chenin, and Gewürztraminer (Baiano, Scrocco, Sepielli, & Del Nobile, 2016). In this process, the vinification is performed by limiting the contact with oxygen throughout the winemaking process and ensuring good protection against enzymatic oxidation in the prefermentation phase with an early addition of SO_2 combined with ascorbic acid (Baiano et al., 2012). The hyperreductive winemaking process uses inert gases such as CO_2 and N_2 to guarantee protection against oxidation also during the crushing phase. This also ensures a satisfactory extraction of desirable sensory compounds, which are mainly located in the grape skin. Reductive winemaking can be performed in combination with a prefermentative cryomaceration step (Naviglio et al., 2018). It has been shown that besides preserving or increasing the content of varietal aromas (Antonelli, Arfelli, Masino, & Sartini, 2010), also increases the content of phenolic compounds in the final wine due to the SO_2 solubilizing effect (Gomez-Plaza, Gil-Munoz, López-Roca, Martínez-Cutillas, & Fernandez-Fernandez, 2002), and also as they are better preserved during the winemaking process, especially after grape crushing. Nevertheless, this type of winemaking also seems to be responsible for the appearance of one visual defect in white wines named pinking. This defect is characterized by the appearance of a salmon-red color in white wines produced exclusively from white grapes after exposure to oxygen (Andrea-Silva et al., 2014; Du Toit, Marais, Pretorius, & Du Toit, 2006; Jones, 1989; Simpson, 1977; Singleton, 1972). Pinking is considered an undesirable visual defect in white wines, both for wine consumers and for the wine industry (Andrea-Silva et al., 2014; Du Toit et al., 2006; Jones, 1989; Simpson, 1977). This chapter reviews the existing knowledge about the pinking phenomena of white wines, analytical techniques for its prediction, and techniques for the prevention of its appearance and removal after appearance.

15.2 Origin of the pinking phenomena of white wines

Pinking occurrence in white wines has been described worldwide with predominance in some white wines produced from *Vitis vinifera* L. white grape varieties such as Chardonnay, Chenin Blanc, Crouchen, Muscat Gordo Blanco, Palomino, Riesling, Sauvignon Blanc, Semillon, Sultana, Pinot Grigio, Albariño, Verdejo, Garnatxa Blanca,

White Wine Technology.
DOI: https://doi.org/10.1016/B978-0-12-823497-6.00024-7

© 2022 Elsevier Inc. All rights reserved.

Síria, and Thompson Seedless, however, with seasonal and regional variations (Andrea-Silva et al., 2014; Du Toit et al., 2006; Lamuela-Raventós, Huix-Blanquera, & Waterhouse, 2001; Simpson, 1977; Singleton, 1972). This phenomenon is generally observed in white wines after bottling and storage or after alcoholic fermentation, although occasionally it occurs as soon as grape must is extracted (Simpson, Bennett, & Miller, 1983; Simpson, Miller, & Orr, 1982; Singleton, Trousdale, & Zaya, 1979). Although most authors agree that pinking is mainly observed when white wines are produced under reducing conditions (Du Toit et al., 2006, Jones, 1989, Simpson, 1977, Van Wyk, Louw, & Rabie, 1976) and that white wines developed a pink coloration after exposure to air (Du Toit et al., 2006), the chemical explanation for this visual defect is not consensual. It was described that the appearance of the pink color in white wines is derived from at least 10 compounds and polymeric material (Jones, 1989). Several hypotheses have been raised in the past and at present for explaining this phenomena, although in some of them, there is a lack of evidence. One of the first hypotheses is that the pinking phenomenon may occur due to the slow dehydration of leucoanthocyanidins (flavan-3,4-diols) to their corresponding flavenes (flav-3-en-3-ol) under a highly reductive medium and then quick oxidation to their corresponding colored flavylium cations (cyanidin) upon the exposure to oxygen (Fig. 15.1A) (Singleton, 1972; Zoecklein, Fugelsang, Gump, & Nury, 1995). Another alternative hypothesis for the appearance of pink color in white wines is attributed to the slow acid catalysis cleavage of interflavan bonds of certain proanthocyanidins present in grape skins to their corresponding carbocation intermediate, which, following an oxygen exposure, turn into flavylium cations (Fig. 15.1B) (Simpson, 1977). Also, the pink color developed in white wines after exposure to oxygen has been attributed to the formation of unknown red-colored compounds after oxidation of the 2-S-glutathionilcaftaric acid (Fig. 15.1C) (Van Wyk et al., 1976). The more recent hypothesis is that the pinking of white wines is due to the presence of small amounts of anthocyanins in white grapes that are extracted and after reversible reaction with the hydrogen sulfite ion resulting from the use of sulfur dioxide at the crushing of grapes renders the flavene-4-sulfonate that is colorless. These compounds, the anthocyanins, are also preserved from oxidation due to the reductive conditions used during the wine-making process, end up in the white wine most of the time unnoticed. When the amount of sulfur dioxide in the wines lowers, for example during storage or upon exposure to oxygen, there is observed an increase in the flavylium form of the anthocyanins in the wine due to the dissociation of flavene-4-sulfonate. When the concentration of the colored form of anthocyanins reaches a certain concentration the pink color can be visually detected in the white wine (Fig. 15.1D) (Andrea-Silva et al., 2014).

Although it is generally believed that white grapes do not contain anthocyanins, their occurrence has been recently described in several grape varieties. The presence of anthocyanins in the skins and pulp of the Síria variety, a Portuguese white grape variety with a high tendency for pinking, was clearly shown (Further discussed below Andrea-Silva et al., 2014). Arapitsas, Oliveira, and Mattivi (2015) also found that the white grape varieties Chardonnay, Sauvignon Blanc, and Riesling were able to synthesize and accumulate anthocyanins although in lower concentrations than red and pink grape varieties.

Low levels of anthocyanins, mainly malvidin-3-O-glucoside, were also detected in white wines produced from Malvasia Fina, Loureiro, Sauvignon Blanc, and Albariño (Cosme et al., 2019). In fact, it has been suggested that the progenitors of all white grape varieties had "red grape varieties" and that white varieties have derived from these colored varieties (Slinkard & Singleton, 1984). White grapes contain the genes responsible for the synthesis of all enzymes in the anthocyanin pathway; some of them shared in the formation of other polyphenols present in white grapes (Holton & Cornish, 1995). The absence of red color in the white varieties (i.e., the absence of anthocyanins) results from multiallelic mutations of the VvMYBA1 and VvMYBA2 genes, (Kobayashi, Goto-Yamamoto, & Hirochika, 2004; This, Lacombe, Cadle-Davidson, & Owens, 2007; Walker et al., 2007; Walker, Lee, & Robinson, 2006) which controls the last biosynthetic step of anthocyanin synthesis, a glycosylation reaction mediated by the UDP-glucose flavonoid 3-O-glucosyltransferase enzyme. Therefore external environmental conditions and vineyard practices can switch these genes on to start the anthocyanin metabolic pathways (Boss, Davies, & Robinson, 1996).

The second argument against the flavylium nature of the pigments causing the pinking phenomena derives from the claimed insensitivity of the pinking pigment toward pH changes. Simpson (1977) showed that the influence of pH on the absorbance of the pinking materials in the range of 2.75−4.00 was slight and acidification to pH < 1 of the same wines after treatment with hydrogen peroxide (75 mg/L, 24 hours at 25°C) did not enhance the pink color. On the other hand, in the same work, it was shown that the formation of pinking compounds after oxidation with hydrogen peroxide was dependent on the initial concentration of free SO_2, with levels of free sulfur dioxide higher than 40 mg/L, inhibiting their formation after treatment with hydrogen peroxide.

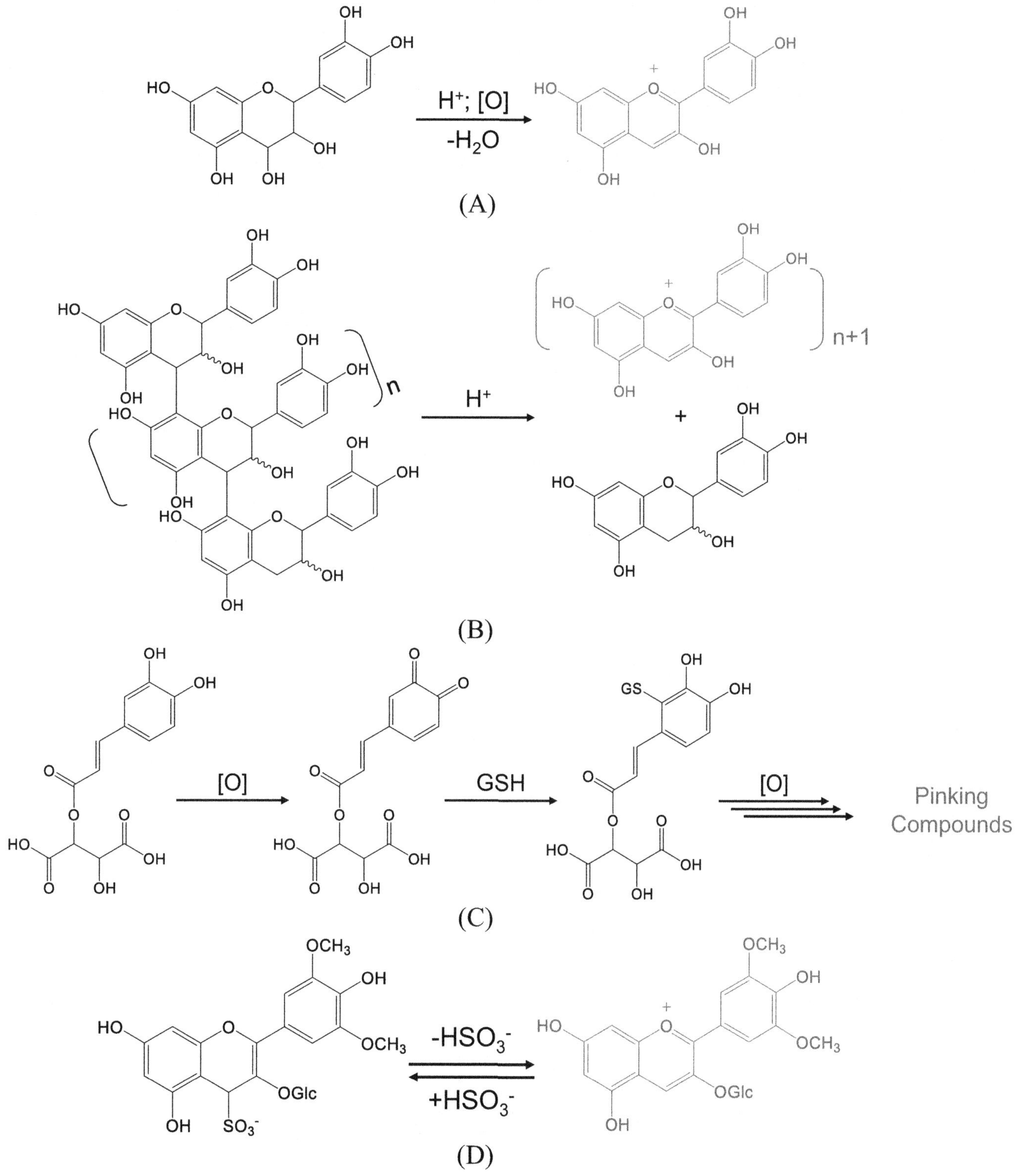

FIGURE 15.1 Hypothesis raised for the chemical explanation of the pinking phenomena of white wines: A) Dehydration of leucoanthocyanidins (flavan-3,4-diols) to their corresponding flavenes (flav-3-en-3-ol) under a highly reductive medium and then quick oxidation to their corresponding colored flavylium cations (cyanidin) upon the exposure to oxygen (Singleton, 1972; Zoecklein, Fugelsang, Gump, & Nury, 1995). (B) Acid catalysed cleavage of interflavan bonds of certain proanthocyanidins present in grape skins to their corresponding carbocation intermediate, which upon oxygen exposure turn into flavylium cations (Simpson, 1977). (C) Formation of unknown red-colored compounds after oxidation of the 2-S-glutathionilcaftaric acid (Van Wyk et al., 1976). (D) Reversible reaction of the red coloured flavylium form of anthocyanins present in small amounts in white grapes with the hydrogen sulfite ion resulting in the formation of the colorless flavene-4-sulfonate (Andrea-Silva et al., 2014).

15.3 Pinking of Síria white wines

In Figueira de Castelo Rodrigo, the designation of origin Beira Interior, the subregion of Castelo Rodrigo (Portugal), white wines are produced mainly from the Síria grape variety (Fig. 15.2). Síria variety is clearly a white grape cultivar, and at technological maturity, no pigmentation of the skins and pulp are observed by the naked eye.

Local producers state that pinking has appeared at least for the past 18 years (Fig. 15.3), although not with the same severity each year. In this period, the white wines have been produced under reducing conditions.

In some years, the must already contains a pink color (Fig. 15.4). The Síria grape must color changed with the pH variation, being more intense at pH 1.39, and disappeared entirely at pH 4.01 (Fig. 15.4). At pH 4.01, no pink color could be detected, which supports the notion that the anthocyanins were the only compounds with the pink color present in the grape must.

The pink-colored compounds were isolated from wines where the pinking phenomena occurred naturally in the bottle and were removed after treatment with polyvinylpolypyrrolidone (PVPP) (further discussed below) or from non-pinked Síria white wines by solid-phase extraction after acidification to pH 1. The isolated compounds were analyzed by reversed-phase high-performance liquid chromatography (RP-HPLC) (Fig. 15.5A) and anthocyanins were identified by comparison of their retention times and UV-Vis spectra (Fig. 15.5B) with those of authentic standards as well by tandem mass spectrometry after electrospray ionization (Fig. 15.5C). The results clearly show that Síria white wines contain small amounts of anthocyanins, mainly malvidin-3-O-glucoside, and that the pink color recovered by treatment with PVPP from a naturally occurring pinking wine were also anthocyanins. The amount of total monomeric anthocyanins required in order for the wine to have a visible pink color

FIGURE 15.2   Síria white grapes variety.

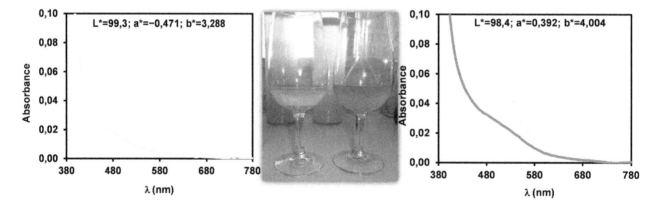

FIGURE 15.3   The visible absorption spectra of white wine (left, Síria 2010) and the same white wine where pinking was induced by standing in a glass for 24 h (right). After 24 h, the color of the wine changed from a clear pale yellow ($L^* = 99.3$; $a^* = -0.471$; $b^* = 3.288$) to a pale salmon ($L^* = 98.4$; $a^* = 0.392$; $b^* = 4.004$), with an evident increase in the absorbance in the visible region ($\Delta E^* 51.39$).

FIGURE 15.4 Effect of pH (1.39; 3.02, and 4.01) on the color of the grape must of the Síria white grape variety from Figueira de Castelo Rodrigo vineyards.

pH=1.39 pH=3.02 pH=4.01

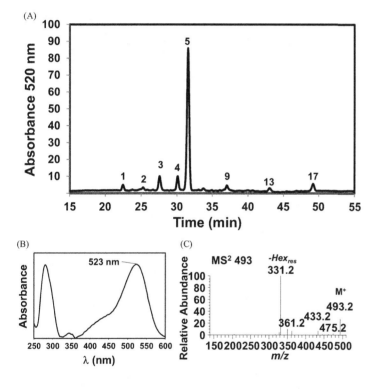

FIGURE 15.5 (A) RP-HPLC chromatograms of the extracts obtained by elution of the PVPP obtained from the treatment of a pinking white wine from the Síria grape variety, vintage 2010; (B) Online UV-Vis spectrum of peak 5 identified as malvidin-3-O-glucoside detected by HPLC–diode-array detector (DAD) analysis of the PVPP-retained and PVPP-recovered pinking compounds of Síria 2010 white wine; (C) Ms2 spectra of 17—malvidin-3-O-(6-O-coumaroyl)-glucoside obtained by ESI-MS[n] analysis of the PVPP-retained and PVPP-recovered pinking compounds of Síria 2010 white wine. Peaks 1—delphinidin-3-O- glucoside; 2—cyanidin-3-O-glucoside; 3—petunidin-3-O-glucoside; 4—peonidin-3-O-glucoside; 5—malvidin-3-O-glucoside; 13—delphinidin-3-O-(6-O-coumaroyl)-glucoside; 17—malvidin-3-O-(6-O-coumaroyl)-glucoside.

in the case of Síria variety was close to 0.3 mg/L. In the Síria variety, anthocyanins were shown to occur both in the skin and pulp of the grape (Andrea-Silva et al., 2014).

Although it cannot be excluded that the pinking of different grape varieties of white grapes can have a different origin from that observed for the Síria grape variety, all the technical descriptions of the pinking phenomena fit with the known chemical properties and reactivity of anthocyanins. The presence of anthocyanins in some white grape varieties that are sensitive to the pinking phenomena is without a doubt firmly shown. On the other hand, the insensitivity of the pinking compounds to the pH changes is one of the strongest points against the involvement of anthocyanins in the pinking phenomena. One explanation to this point is that anthocyanins also in the white wine matrix were shown to polymerize. In Fig. 15.6, it is shown the anthocyanins recovered after 1 year of bottling of the Síria 2010 white wine. It can be clearly observed that the polymerization of anthocyanins has occurred. It is widely known that polymerized pigments are insensible to pH changes and the addition of sulfur dioxide. Therefore anthocyanin polymerization can also occur when the white wines are forced to be oxidized by the addition of the excess of hydrogen peroxide as used in the forced oxidation process employed

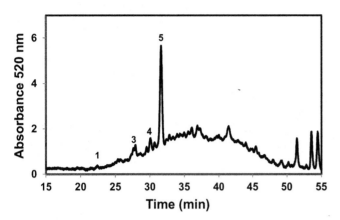

FIGURE 15.6 RP–solid-phase extraction (SPE) extracts of white wine from the Síria grape variety, vintage 2010 analyzed 1 year after the first analysis.

by Simpson (1977). As the color of the polymeric anthocyanins is resistant to sulfur dioxide bleaching (Somers, 1971), the polymerization of anthocyanins can also contribute to the appearance of pinking during bottling.

15.4 Methods to predict the pinking susceptibility of white wine

The development of pinking in white wines can cause several problems for wine producers. When formed prior to bottling, although the options available at present are limited, it can be handled by fining (see next section); nevertheless, when formed after bottling, there can be serious economic losses as for the treatment of these wines, they have to be removed from the bottles. Also when exposed in the market the development of the pink color can reduce the attractiveness of the wines for the consumers by, depending on the initial white wine color, changing its color to pink or conferring tones that can be confused with oxidized wines. Therefore it is important to know if a given white wine will have the tendency to pink after bottling. There are several simple methods that can be used for predicting the tendency for a white wine to pink: for example, by exposing the wine to oxygen in a glass (Fig. 15.3) or by adding a few drops of hydrogen peroxide. On the other hand, Simpson (1977) developed a method in which the pink color is forced to be formed by the application of hydrogen peroxide, a strong oxidant, and the pinking tendency of the wine is quantitatively measured. There are several variations of the original method and the values above which the wines are at risk of pinking. This method is based on the measurement of the difference in absorbance at 500 nm (although other wavelengths have been used, for example, 520 nm) of the test wine sample treated (12 mL with a free sulfur dioxide level below 80 mg/L) with 3 mL of a 0.3% (w/v) solution of hydrogen peroxide and incubated 24 hours at 25°C, in the dark, against a control wine sample without treatment. For convenience, the difference in the absorbance at 500 nm before and after the treatment is multiplied by 1000, although the use of the value of 100 has also been described. Nevertheless, as the application of hydrogen peroxide usually also increases browning, easily observed by the increase of the absorbance at 420 nm, and as the presence of an intense wine browning can cause an increase in the baseline also at 500 nm, a hypothetical baseline after the hydrogen peroxide treatment can be estimated by using the absorbance's at 400, 410, 420, 625, and 650 nm and fitting and exponential equation, or another equation presenting a good fit, to these values in order to estimate the hypothetical baseline value at 500 nm. Then, the absorbance difference between the real value and hypothetical baseline value is calculated and multiplied by 1000 (Fig. 15.7). The values used to indicate that the pinking is serious enough to be worried about are also dependent on the method used but a value of 5 or higher indicates that the wine is susceptible to pinking (Simpson, 1977).

15.5 Preventive and curative treatments for pinking

As previously discussed the pink color appears in white wines produced under reductive conditions by exposure to oxygen; therefore the exclusion of oxygen during bottling, the use of adequate levels of free sulfur dioxide, and the addition of ascorbic acid are preventive measures that can be used in order to avoid the appearance of the pink color after bottling during storage. It has been described that ascorbic acid at levels of 45 mg/L can

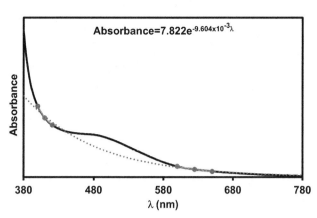

Absorbance=7.822e$^{-9.604\times10^{-3}\lambda}$

Absorbance

λ (nm)

FIGURE 15.7 Visible spectra of a wine forced to pink by application of hydrogen peroxide and extrapolation of the baseline by the exponential fitting of the values at 400, 410, 420, 625, and 650 nm.

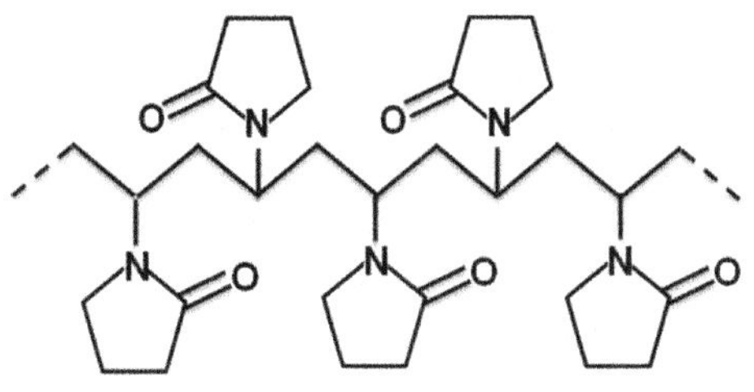

FIGURE 15.8 Polyvinylpolypyrrolidone (PVPP) structure.

completely prevent the appearance of pinking (Lamuela-Raventós et al., 2001). The use of high levels of ascorbic acid for preventing pinking can have some drawbacks. High levels of ascorbic acid have been shown to increase the browning potential of white wines (Bradshaw, Cheynier, Scollary, & Prenzler, 2003). It is therefore important that the free SO_2 concentration of the wine should always be kept at 35–45 mg/L if ascorbic acid is used. Therefore if a white wine shows a great tendency to pinking, as measured by the method developed by Simpson (1977), then the removal of the precursors is at present the best solution. There are some enological treatments that can be used, with the most frequently used being the application of PVPP. PVPP associated with bentonite can also be used (Lamuela-Raventós et al., 2001).

PVPP is a water-insoluble synthetic polymer produced by the cross-linking of polyvinylpyrrolidone (PVP), Fig. 15.8 (Haaf, Sanner, & Straub, 1985). Due to its excellent adsorbent properties, selectivity, stability, inertness, and nonallergenicity, (Folch-Cano, Olea-Azar, & Speisky, 2013) it is widely used in wine (up to a maximum dose of 80 g/hL) as a fining agent for removing low-molecular-weight phenolic compounds ((EU) No. 1169/2011; Haaf et al., 1985). However, the use of these treatments increases wine production costs and can change wine sensory properties (Sanborn, Edwards, & Ross, 2010), therefore the treatments should be controlled by measuring the pinking susceptibility of the wine before treatment.

Simpson (1980) suggested that when wine pinks in the bottle, it can be exposed to direct sunlight or UV light for about 10 min to reduce the pink discoloration. UV exposure might lead to reduced pinking but can have other negative effects. In a recent work, where darkness and UV radiation were tested to treat pinking in wines from the grape varieties Sauvignon blanc and Chardonnay, it was shown that pinking can be reduced by about 25% and 17% with UV light and darkness treatments, respectively, and for the Chardonnay wine, the pinking was only slightly reduced by UV light treatment (8%) or darkness (20%) (Cojocaru & Antocea, 2019). These treatments help, but they do not sufficiently decrease the pinking in the wines. The authors concluded that UV light treatment shows more potential to reverse pinking than darkness, and it may be even better on bottles with lower UV light filtering power (Antique green glass bottles, with 70%–80% UV reduction effect, were used).

15.6 Conclusions

Pinking is a white wine visual defect resulting in the appearance of a salmon-red color in wines produced under reductive or hyperreductive winemaking conditions and exposed to oxygen, especially during bottling.

As its appearance can impact the acceptability of the wine by the consumers, the evaluation of the susceptibility of white wine for pinking is important for the wine producers in order to decide if and what treatments should be performed in order to avoid its appearance. The nature of the chemical compounds responsible for the pinking phenomena in white wine is still a matter of debate; nevertheless, for the Síria white wine, its appearance is without a doubt due to the presence of small amounts of anthocyanins, mainly malvidin-3-O-glucoside, present in the pulp and skin that are extracted and conserved due to protective conditions of the reductive winemaking process. Nevertheless, small amounts of anthocyanins have also been detected in the skin and white wines produced from other white grape varieties widely described to be pinking-sensitive grape varieties. Although preventive measures are at present available, such as the use of ascorbic acid for avoiding the appearance of pinking after bottling or the use of PVPP for removing the pinking precursors before the appearance of the salmon color or for the removal of the pink color already present in the white wine.

References

Andrea-Silva, J., Cosme, F., Filipe-Ribeiro, L., Moreira, A. S. P., Malheiro, A. C., Coimbra, M. A., . . . Nunes, F. M. (2014). Origin of the pinking phenomenon of white wines. *Journal of Agricultural and Food Chemistry, 62*, 5651–5659.

Antonelli, A., Arfelli, G., Masino, F., & Sartini, E. (2010). Comparison of traditional and reductive winemaking: Influence on some fixed components and sensorial characteristics. *European Food Research and Technology, 231*, 85–91.

Arapitsas, P., Oliveira, J., & Mattivi, F. (2015). Do white grapes really exist? *Food Research International, 69*, 21–25.

Baiano, A., Scrocco, C., Sepielli, G., & Del Nobile, A. M. (2016). Wine processing: A critical review on physical, chemical, and sensory implications of innovative vinification procedures. *Critical Reviews in Food Science and Nutrition, 56*, 2391–2407.

Baiano, A., Terracone, C., Longobardi, F., Ventrella, A., Agostiano, A., & Del Nobile, M. A. (2012). Effects of different vinification technologies on physical and chemical characteristics of Sauvignon blanc wines. *Food Chemistry, 135*, 2694–2701.

Boss, P. K., Davies, C., & Robinson, S. P. (1996). Anthocyanin composition and anthocyanin pathway gene expression in grapevine sports differing in berry skin colour. *Australian Journal of Grape and Wine Research, 2*(3), 163–170.

Bradshaw, M., Cheynier, V., Scollary, G., & Prenzler, P. (2003). Defining the ascorbic acid crossover from anti-oxidant to pro-oxidant in a model wine matrix containing (+)-catechin. *Journal of Agricultural and Food Chemistry, 51*(14), 4126–4132.

Cojocaru, G. A., & Antocea, A. O. (2019). Effect of certain treatments to prevent or partially reverse the pinking phenomenon in susceptible white wines. *BIO Web of Conferences, 15*, 02003.

Commission Regulation (EU) No. 1169/2011 (2011). European Parliament and of the Council, of 25 October 2011. *Official Journal of the European Union*, L 304.

Cosme, F., Andrea-Silva, J., Filipe-Ribeiro, L., Moreira, A. S. P., Malheiro, A. C., Coimbra, M. A., . . . Nunes, F. M. (2019). The origin of pinking phenomena in white wines: An update. *Bio Web of Conferences, 12*. Available from https://doi.org/10.1051/bioconf/20191202013.

Du Toit, W., Marais, J., Pretorius, I., & Du Toit, M. (2006). Oxygen in must and wine: A review. *South African Journal for Enology and Viticulture, 27*, 76–94.

Folch-Cano, C., Olea-Azar, C., & Speisky, H. (2013). Structural and thermodynamic factors on the adsorption process of phenolic compounds onto polyvinylpolypyrrolidone. *Colloids and Surfaces A, 418*, 105–111.

Gomez-Plaza, E., Gil-Munoz, R., López-Roca, J. M., Martínez-Cutillas, A., & Fernandez-Fernandez, J. I. (2002). Maintenance of color composition of a red wine during storage. Influence of prefermentative practices, maceration time and storage. *Lebensmittel-Wissenschaft und-Technologie, 35*, 46–53.

Haaf, F., Sanner, A., & Straub, F. (1985). Polymers of N-vinylpyrrolidone: Synthesis, characterization and uses. *Polymer Journal, 17*, 143–152.

Holton, T. A., & Cornish, E. C. (1995). Genetics and biochemistry of anthocyanin biosynthesis. *Plant Cell, 7*, 1071–1083.

Jones, T. (1989). *Pinking of white table wines further studies* (Master of Science thesis). University of California, Davis, CA, USA, 125 pp.

Kobayashi, S., Goto-Yamamoto, N., & Hirochika, H. (2004). Retrotransposon-induced mutations in grape skin color. *Science, 304*, 982.

Lamuela-Raventós, R., Huix-Blanquera, M., & Waterhouse, A. (2001). Treatments for pinking alteration in white wines. *American Journal of Enology and Viticulture, 52*, 156–158.

Naviglio, D., Formato, A., Scaglione, G., Montesano, D., Pellegrino, A., Villecco, F., & Gallo, M. (2018). Study of the grape cryo-maceration process at different temperatures. *Foods, 7*(7), 107.

Sanborn, M., Edwards, C., & Ross, C. (2010). Impact of fining on chemical and sensory properties of Washington State Chardonnay and Gewürztraminer wines. *American Journal of Enology and Viticulture, 61*(1), 31–41.

Simpson, R. (1977). Oxidative pinking in white wines. *Vitis, 16*, 286–294.

Simpson, R., Bennett, S., & Miller, G. (1983). Oxidative pinking of whites wines: A note on the influence of sulfur dioxide and ascorbic acid. *Food Technology in Australia, 35*, 34–36.

Simpson, R., Miller, G., & Orr, L. (1982). Oxidative pinking of whites wines: Recent observations. *Food Technology in Australia, 34*, 46–47.

Simpson, R. F. (1980). Some aspects of oxidation and oxidative browning in white table wines. *Australian Grapegrower and Winemaker, 193*, 20–21.

Singleton, V. (1972). Common Plant Phenols other than anthocyanins, contribution to coloration and discoloration. In O. Chichester (Ed.), *The chemistry of plant pigments. Advances in Food Research, supplement* (3, pp. 143–191). New York: Academic Press.

Singleton, V. L., Trousdale, E., & Zaya, J. (1979). Oxidation of wines. I. Young white wines periodically exposed to air. *American Journal of Enology and Viticulture, 30*, 49–54.

Slinkard, K., & Singleton, V. L. (1984). Phenol content of grape skins and the loss of ability to make anthocyanins by mutation. *Vitis, 23*, 175–178.

Somers, T. (1971). The polymeric nature of wine pigments. *Phytochemistry, 10*, 2175–2186.

This, P., Lacombe, T., Cadle-Davidson, M., & Owens, L. (2007). Wine grape (*Vitis vinifera* L.) color associates with allelic variation in the domestication gene VvmybA1. *Theoretical and Applied Genetics, 114,* 723–730.

Van Wyk, C., Louw, A., & Rabie, M. (1976). The effect of reductive wine making conditions on wine quality and composition. In *Proceedings of the 11th international oenological symposium, June 3–5, Sopron, Budapest, Hungary* (pp. 180–200). International Association for Winery Technology and Management: Breisach, Germany.

Walker, A., Lee, E., Bogs, J., McDavid, D., Thomas, M., & Robinson, S. (2007). White grapes arose through the mutation of two similar and adjacent regulatory genes. *Plant Journal, 49,* 772–785.

Walker, A., Lee, E., & Robinson, S. (2006). Two new grape cultivars, bud sports of Cabernet Sauvignon bearing pale-coloured berries, are the result of deletion of two regulatory genes of the berry colour locus. *Plant Molecular Biology, 62,* 623–635.

Zoecklein, B., Fugelsang, K., Gump, B., & Nury, F. (1995). *Phenolic compounds and wine color. Wine analysis and production* (pp. 115–151). New York: Chapman and Hall.

16

Origin, prevention, and mitigation of light-struck taste in white wine

Luís Filipe-Ribeiro[1], Fernanda Cosme[1,2] and Fernando M. Nunes[1,3]

[1]Chemistry Research Centre-Vila Real (CQ-VR), Food and Wine Chemistry Laboratory, University of Trás-os-Montes and Alto Douro, Vila Real, Portugal [2]Biology and Environment Department, School of Life Sciences and Environment, University of Trás-os-Montes and Alto Douro, Vila Real, Portugal [3]Chemistry Department, School of Life Sciences and Environment, University of Trás-os-Montes and Alto Douro, Vila Real, Portugal

16.1 Introduction

When white wines bottled in transparent bottles are exposed to light, several harmful changes to the wine quality can occur, namely the production of off-flavors (Dozon & Noble, 1989), color changes (Benítez, Castro, Natera, & García Barroso, 2006; Dias, Smith, Ghiggino, & Scollary, 2012), and the depletion of sulfur dioxide (Blake, Kotseridis, Brindle, Inglis, & Pickering, 2010), and therefore light exposure can lower the quality and reduce the shelf life of bottled white wine. One of the most frequent white wine quality changes, when exposed to light, is the development of an off-flavor named light-struck, which produces unpleasantly tasting and smelling compounds. This defect is associated with the formation of volatile sulfur compounds with unpleasant aroma notes, formed by the methionine degradation catalyzed by the photochemically activated riboflavin. Methanethiol and dimethyl disulfide are the main compounds responsible for the light-struck taste in white wine termed as "cooked cabbage" (Maujean & Seguin, 1983b; Maujean, Haye, & Feuillat, 1978). There are other wine defects that can occur when white wine is exposed to light, such as the increase of acetaldehyde odor (Dias, Clark, Smith, Ghiggino, & Scollary, 2013), furfural that has been positively correlated to the "cooked vegetable" aroma of white wines stored under oxidative conditions (Escudero, Asensio, Cacho, & Ferreira, 2002; Jung et al., 2007). Exposure of wine to light at wavelengths close to 370 or 442 nm, corresponding to the highest visible light absorption of riboflavin, is particularly effective in inducing the light-struck taste (Maujean & Seguin, 1983b; Satter & Demand, 1977), manly when clear glass bottles are used (Dias et al., 2012). The changes depend on the irradiation conditions (i.e., the spectrum of the light source, the light intensity, the optical properties of the bottle) and the duration of light exposure. Bottles may be exposed to light in different conditions and for different periods of time, and this may contribute to variation in sensory attributes and shelf life between bottles.

The tendency of white wines to degrade due to light to present the light-struck taste defect is not the same in all wines, although not all the factors responsible for the light-struck taste defect are well understood (Fracasseti et al., 2017).

16.2 Origin of the light-struck taste

Hydrogen sulfide, methanethiol, and dimethyl disulfide are among the main volatile sulfur compounds responsible for "reduced" off-flavor in bottled wine (Ugliano et al., 2013), while methional is an important aroma compound in wines spoiled by oxidation (Escudero, Hernández-Orte, Cacho, & Ferreira, 2000); these compounds

© 2022 Elsevier Inc. All rights reserved.

have also been implicated in the light-induced off-flavor of bottled white wine. The development of the light-struck taste in white wines bottled in clear bottles exposed to light is directly correlated with the levels of riboflavin and methionine in wines. Riboflavin (vitamin B2) is a water-soluble vitamin highly sensitive to light (Fig. 16.1) (Powers, 2003). Riboflavin absorbs light to generate the lowest excited singlet state, which may undergo intersystem crossing to form the lowest excited triplet state (Maujean & Seguin, 1983b). Excited triplet riboflavin may interact with a quencher to become ground-state riboflavin (Type I pathway) or may interact with triplet oxygen ($^3O_2$) to produce singlet oxygen ($^1O_2$) (Type II pathway). Singlet oxygen may naturally decay, react

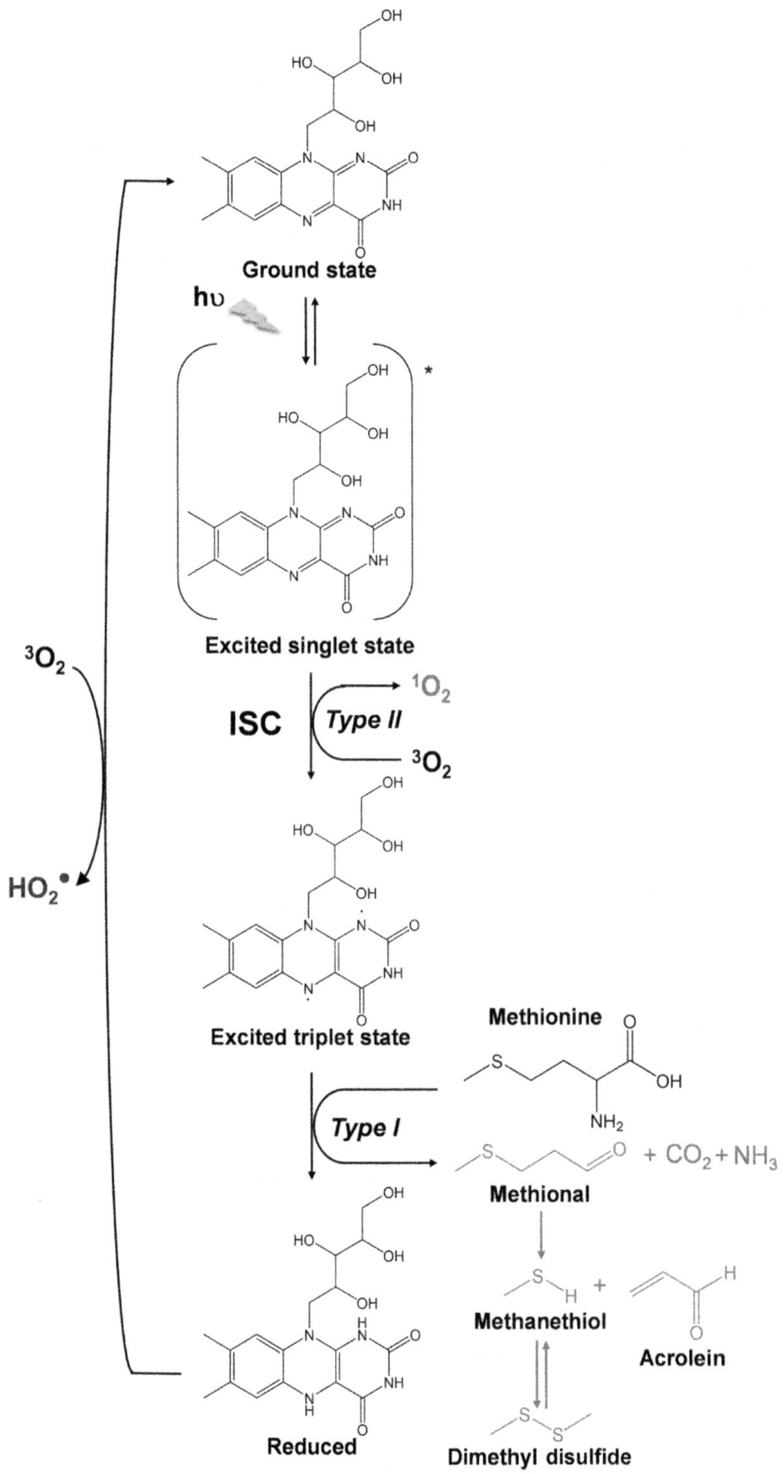

FIGURE 16.1 Riboflavin-sensitized oxidation of methionine in bottled white wine exposed to light.

with singlet riboflavin to form oxidized riboflavin, or be quenched physically or chemically by quenchers (Min & Boff, 2002). For flavins, the quantum yield (number of moles of product formed per mole of photons [Einstein] absorbed) of the triplet state in neutral aqueous solutions exposed to 347 nm light is around 0.7, indicating that this process is highly efficient (Grodowski, Veyret, & Weiss, 1977). The efficiency decreases as pH decreases, primarily due to the quenching of the excited states by hydrogen ions (Grodowski et al., 1977). For riboflavin in aqueous solutions at pH 2.2, the quantum yield of the triplet state is 0.4 (Grodowski et al., 1977). Hence it is expected that at wine pH, the quantum yield for this process would be between 0.4 and 0.7.

In the absence of oxygen, only type I processes occur, whereas, in the presence of oxygen, both type I and type II processes occur to some extent. Triplet riboflavin is reactive toward triplet oxygen and a range of different compounds including ascorbic acid, phenolic compounds, and amino acids (Cardoso, Libardi, & Skibsted, 2012; Choe, Huang, & Min, 2005). The photosensitized production of singlet oxygen may contribute to the increased rate of oxygen consumption in wine exposed to light (Singleton, 1987). Maujean and Seguin (1983b) proposed that in white wine exposed to light, triplet riboflavin oxidizes methionine to methional. Methional, the product of the reaction, is in turn photodegraded through retro-Michael reactions to form dimethylsulfide, dimethyl disulfide, methanethiol, and hydrogen sulfide (Maujean & Seguin, 1983a; Pripis-Nicolau, de Revel, Bertrand, & Maujean, 2000) (Fig. 16.1). The presence of these compounds in wine, with negative sensory notes described as cooked cabbage, corn nuts, wet dog/wet wool, and soy/marmite aromas (Table 16.1), reduces the wine fruit and varietal sensory notes (Dozon & Noble, 1989).

In grapes, the riboflavin concentration is usually low: 3−60 µg/L in grape juice (Ribéreau-Gayon, Glories, Maujean, & Dubourdieu, 2006); however, the riboflavin concentration can increase significantly during alcoholic fermentation by the metabolic activity of *Saccharomyces cerevisiae* yeast (Ournac, 1968; Santos, García-Ramírez, & Revuelta, 1995) and by the addition of nutrients containing riboflavin commonly used to stimulate yeast growth (Fracasseti et al., 2017). Wines can have 150 µg/L of riboflavin (Mattivi, Monetti, Vrhovšek, Tonon, & Andrés-Lacueva, 2000) or even higher depending on the yeast strain (Fracasseti et al., 2017) (Table 16.2). Mattivi et al. (2000) reported that in 85 commercial white wines the average concentration of riboflavin was about 100 µg/L, while the levels of flavin mononucleotide and flavin adenine dinucleotide were negligible. Similar values have been reported in other studies (Andrés-Lacueva, Mattivi, & Tonon, 1998; Cataldi, Nardiello, Scrano, & Scopa, 2002). On the other hand, the wine methionine concentration is 3 mg/L on average (Amerine & Ough, 1980; Grant-Preece, Barril, Schmidtke, & Clark, 2018), i.e., a molar concentration of about 40 times higher than that of riboflavin, and its oxidation leads to the formation of methional (Table 16.1).

More recently, it was observed that the addition of ascorbic acid to a model wine solution containing methionine, iron, and other wine compounds increased the production of methional under conditions that favored oxidation (Grant-Preece, Fang, Schmidtke, & Clark, 2013). In wine conditions, ascorbic acid is degraded to form α-dicarbonyl compounds including dehydroascorbic acid and xylosone (Bradshaw, Barril, Clark, Prenzler, & Scollary, 2011), and this may account for the greater degradation of methionine to methional in the presence of ascorbic acid (Schonberg & Moubacher, 1952).

Maujean and Seguin (1983a) assessed the use of chemical means to inhibit the riboflavin-sensitized production of volatile sulfur off-flavor compounds in white wine. The addition of copper (II) to Champagne prevented the production of methanethiol and dimethyl disulfide during exposure to light; however, dimethyl disulfide was detected

TABLE 16.1 Odour detection threshold of volatile sulfur compounds formed by riboflavin-sensitized oxidation of methionine in bottled white wine exposed to light.

Compound	Odor description	Odour detection threshold (ODT)	References
Hydrogen sulfide	Rotten egg	1.6 µg/L in white wine	Maujean et al. (1978), Mestres, Busto, and Guasch (2000), Ugliano et al. (2013), Siebert, Bramley, and Solomon (2009)
Methional	Boiled potatoes, cooked vegetables	0.5 µg/L in 11% ethanol model wine	Escudero et al. (2000), Silva Ferreira, Guedes de Pinho, Rodrigues, and Hogg (2002)
Methanethiol	Cooked cabbage	0.3 µg/L in model wine, 2−10 µg/L in wine	Mestres et al. (2000), Pripis-Nicolau et al. (2004), Solomon, Geue, Osidacz, and Siebert (2010)
Dimethyl disulfide (DMDS)	Cabbage, cooked cabbage, onion notes	20−45 µg/L	Mestres et al. (2000), Goniak and Noble (1987)

TABLE 16.2 Levels of riboflavin and methionine in white grape musts and wines.

Riboflavin (µg/L)		Methionine (mg/L)		Origin	References
Grape must	Wine	Grape must	Wine		
3–60	8–133			France	Ribéreau-Gayon, Peynaud, and Ribéreau-Gayon (1975)
			7	France, Champagne (Chardonnay)	Desportes, Charpentier, Duteurtre, Maujean, and Duchiron (2000)
5.2	50–170	1–8		Italy (Chardonnay)	Fracasseti et al. (2017)
			3	United States	Amerine and Ough (1980)
			1.02–2.63	Albariño, Godello, Treixadura	Mirás-Avalos, Bouzas-Cid, Trigo-Córdoba, Orriols, and Falqué (2020)
55	110–170				Ournac (1968)
	11–79			United States (Chenin Blanc, Sauvignon Blanc, Rhineskeller, Rhine Garten, Rhine Castle)	Voigt, Eitenmiller, Powersand, and Ware (1978)
	100–110				Hall, Brinner, Amerine, and Morgan(1956)
	90				Leake and Silverman (1966)
	8.2–200.3			Italy (Chardonnay,Pinot Gris), Slovenia (Chardonnay, Pinot Gris), Spain (Penedes [still], Cava [sparkling])	Mattivi et al. (2000)
	69–151			Italian wines	Cataldi et al. (2002)
	97–281			Italian wines	Bonamore et al. (2016)

in the irradiated samples the following day(Maujean & Seguin, 1983a). It is possible that copper (II) formed complexes with methanethiol, which eventually degraded to form dimethyl disulfide (Maujean & Seguin, 1983a). However, the addition of a mixture of (+)-catechin and (−)-epicatechin monomers and dimers to a model wine solution containing riboflavin and methionine slowed the production of the volatile sulfur compounds during exposure to light (Maujean & Seguin, 1983a). Therefore it appears that in addition to the concentration of riboflavin, the concentrations of sulfur-containing amino acids and other triplet riboflavin quenchers, including phenolic compounds, ascorbic acid, and oxygen, would also influence the development of light-struck flavor in white wine.

16.3 Preventive treatments and strategies to avoid light-struck taste in white wines

The changes that occur in bottled white wine exposed to light are dependent on wine composition, irradiation conditions, light exposure time, bottle glass quality, and color. The light exposure induces changes in the volatile composition, color, and concentrations of oxygen and sulfur dioxide in bottled white wine. As mentioned before, this taste defect is strongly associated with high riboflavin concentrations (Mattivi et al., 2000; Pichler, 1996); nevertheless, for riboflavin concentrations lower than 80–100 µg/L, the occurrence of the light-struck taste defect is reduced. Therefore the majority of the preventive treatments aim to reduce the riboflavin levels in grape juice and wine. There are some enological practices or strategies that can be used to reduce or prevent the light-struck taste defect in white wines (Table 16.3). There are several classic authorized fining agents, such as bentonite and activated carbon, that can be used to reduce riboflavin levels in white wine (Pichler, 1996). Bentonite can reduce riboflavin levels by as much as 40% (Fracasseti et al., 2017; Pichler, 1996). The efficiency of sodium and calcium bentonites was similar (Fracasseti et al., 2017). Also, the concentration of riboflavin in grape must be reduced by 50% during winemaking with the application of bentonite for the grape must clarification. Also, sulfite treatment of the grape must before fermentation results in similar decreases in riboflavin concentration (Morgan, Nob, Wiens, Margb, & Winkler, 1939). Another strategy that can be used to reduce the levels of riboflavin in the final wine is the selection of low riboflavin-producing yeasts for the alcoholic fermentation (Fracasseti et al., 2017; Pichler, 1996).

TABLE 16.3 Treatments or preventive strategies to avoid light-struck taste on white wine.

Preventive strategies and treatments		References
Authorized		
Selected yeasts	Low riboflavin production	Ournac and Décor (1967), Ournac (1968), Fracasseti et al. (2019)
Avoid contact with yeast lees	Increased riboflavin concentration	Ournac and Décor (1967)
Avoid yeast-based nutrients	These nutrients could present riboflavin in its composition	Fracasseti et al. (2019)
SO_2	Sulfur dioxide enhanced the methionine degradation although the light-struck taste was not perceived when sulfur dioxide concentration was higher than 50 mg/L (model wine)	Fracasseti et al. (2019)
Ascorbic acid	Ascorbic acid is effective in preventing "sunlight flavor" due to its photosensitive, reducing properties. Photochemical interaction between the two vitamins reduces riboflavin's interaction with methionine. The reaction is reversible and therefore when wines are no longer exposed to light, the ascorbic acid is totally recovered	Ribéreau-Gayon et al. (2006)
Activated charcoal	Riboflavin reduction of 70%	Fracasseti et al. (2017)
Bentonite	Riboflavin reduction of 40%	Fracasseti et al. (2017)
PVPP	Not efficient	Fracasseti et al. (2017)
Silica	Not efficient	Fracasseti et al. (2017)
Avoid clear bottles	Bottles with low wavelengths transmission, below 450 nm	Grant-Preece et al. (2015) Dozon and Noble (1989) Ribéreau-Gayon et al. (2006)
Experimental		
Zeolite	Reduction of 50% of riboflavin	Fracasseti et al. (2017)
Cupric cations, sodium dithionite, tannins		Maujean and Seguin (1983a,b)
Hydrolyzable tannins	Hydrolyzable tannins, in particular, nutgall tannins, can effectively hamper methionine degradation and volatile sulfur compounds formation. Quinones can readily react with the methanethiol formed from methionine and thus lead to a lower amount of DMDS (model wine)	Fracasseti et al. (2019)
Catechin tannins (condensed tannins)	Due to their high absorption capacity of UV light, especially that absorbed by riboflavin at 370 nm, which prevents it from reacting with methionine. This explains why red wines, with their high procyanidin content, are much less light sensitive	Ribéreau-Gayon et al. (2006)

Fracasseti et al. (2017), using 15 commercial *Saccharomyces* strains, found riboflavin concentrations ranging from 30 to 170 µg/L in white wine. The level of riboflavin in wine can be higher (160–318 µg/L) if the wine is left in contact with yeast for 4 to 6 days after fermentation is completed (Ournac & Décor, 1967; Ournac, 1968).

Besides riboflavin, the levels of methionine in white wine need to be considered for preventing the light-struck taste. Methionine concentration itself is detrimental, even more than riboflavin, as an increase in methionine levels causes strong increases in both the volatile sulfur compounds amount and the ratio of degraded methionine/degraded riboflavin. Sulfur dioxide enhances methionine degradation, even at low concentrations, but its addition does not result in the sensory perception of light-struck taste. Hydrolyzable tannins have a positive influence on photo-degradative mechanisms preventing the light-struck taste in model wine (Fracasseti, Limbo, Pellegrino, & Tirelli, 2019). Hydrolyzable tannins, in particular gallnut tannins, can effectively hamper methionine degradation and the formation of volatile sulfur compounds under oxidative conditions.

The exposure of white wine to light (sunlight or fluorescent light) in clear bottles is also an important factor that stimulates the light-struck taste formation (Dozon & Noble, 1989). According to Maujean (1984), possible reasons for the increase in the light-struck taste defect is the decrease in the bottle glass quality, resulting in lower wine protection against the light. However, in recent years, prompted by a desire to make wine bottling practices more sustainable, bottle producers have extended their range of lightweight bottles, and these bottles have been widely adopted by winemakers. Although bottle color is known to have an important role in the preservation of bottled white wine that may be exposed to light, the contribution of bottle weight and other factors such as the degree of UV protection of clear bottles is not well understood (Grant-Preece, Barrila, Schmidtke, Scollary, & Clark, 2015). Dark-colored bottles such as green and amber bottles provide greater protection from light than clear and light-colored bottles; however, the latter are sometimes preferred for marketing reasons (Grant-Preece et al., 2015). Carbon costs can also influence the selection of both the bottle color and the weight, (WRAP, 2012). However, through studies performed by Dozon and Noble (1989) simulating the conditions in retail outlets, the impact of fluorescent lighting on the white wine aroma profiles stored in green and clear glass bottles was observed, and it was concluded that green glass provided more protection against negative light effects, requiring longer exposure time before detectable sensory differences were observed. Bottling the wine in clear and/or lighter, thinner bottles has implications for its quality as the bottle will offer less protection to the detrimental effects of light, and this could be perceived as a barrier to the adoption of lighter or clear bottles for use in the wine trade. Therefore bottling wine with light protection and changing the glass compositions to modify optical properties, avoiding exposure of wine to the sun, are important practices to reduce or avoid completely the light-struck taste sensory defect.

The application of SO_2 and ascorbic acid after bottling can also be a preventive strategy to limit the light-struck taste mechanism. Heelis, Parsons, Phillips, and McKellar (1981) suggested that ascorbic acid can be used because of the formation of a triplet flavin-ascorbic acid complex. Some preventive additives like cupric cations, sodium dithionite, and tannins were investigated (Maujean & Seguin, 1983a). The addition of copper (II) to Champagne prevented the production of methanethiol and dimethyl disulfide during exposure to light; however, dimethyl disulfide was detected in the irradiated samples the following day (Maujean & Seguin, 1983a). It is possible that copper (II) formed complexes with methanethiol, which eventually degraded to form dimethyl disulfide (Maujean & Seguin, 1983a; Smith, Reed, & Hill, 1994). The results for sodium dithionite and tannins were more promising, but their efficiency and effect on sensory characteristics need to be examined. Also, zeolite application resulted in a decrease of 50% in the riboflavin concentration (Fracasseti et al., 2017).

16.4 Conclusions

UV—visible light exposure can induce negative changes in the sensory attributes and reduce the shelf life of bottled white wine through the formation of negative volatile compounds such as methanethiol and dimethyl disulfide with a cooked cabbage aroma. Riboflavin and methionine present in white wines have an important role in the formation of these sulfur compounds during white wine light exposure. Critical wavelengths have been identified at 340, 380, and 440 nm. Also, the duration of light exposure, temperature, oxygen levels or oxidative or reductive conditions, and level of sulfur dioxide and ascorbic acid can significantly influence the levels of these volatile sulfur compounds. Recent works have shown the positive effect of hydrolyzable tannins in the prevention of this defect through inhibition of the light-driven reactions due to the reaction of the quinones with methanethiol, reducing the formation of dimethyl disulfide. The use of yeast with low riboflavin production and wine fining with bentonite and activated carbon can also reduce the riboflavin levels in wine. Although many of the factors influencing the light-struck appearance in wines are already known, further studies are needed on the influence of the wine composition on the levels of the different negative volatile sulfur compounds responsible for the negative sensory notes in order to be able to predict their formation. Also, further studies are needed to develop more efficient fining treatments and technological solutions for preventing and controlling their appearance after wine bottling.

References

Amerine, M. A., & Ough, C. S. (1980). Alcohols. In M. A. Amerine, & C. S. Ough (Eds.), *Methods for wine and must analysis*. New York: John Wiley and Sons.

Andrés-Lacueva, C., Mattivi, F., & Tonon, D. (1998). Determination of riboflavin, flavin mononucleotide and flavin-adenine dinucleotide in wine and other beverages by high performance liquid chromatography with fluorescence detection. *Journal of Chromatography A, 823*, 355–363.

Benítez, P., Castro, R., Natera, R., & García Barroso, C. (2006). Changes in the polyphenolic and volatile content of "Fino" sherry wine exposed to high temperature and ultraviolet and visible radiation. *European Food Research and Technology, 222*, 302–309.

Blake, A., Kotseridis, Y., Brindle, I. D., Inglis, D., & Pickering, G. J. (2010). Effect of light and temperature on 3-alkyl-2-methoxypyrazine concentration and other impact odourants of Riesling and Cabernet Franc wine during bottle ageing. *Food Chemistry, 119*, 935–944.

Bonamore, A., Gargano, M., Calistil, L., Francioso, A., Mosca, L., Boffi, A., & Federico, R. (2016). A novel direct method for determination of riboflavin in alcoholic fermented beverages. *Food Analytical Methods, 9*, 840–844.

Bradshaw, M. P., Barril, C., Clark, A. C., Prenzler, P. D., & Scollary, G. R. (2011). Ascorbic acid: A review of its chemistry and reactivity in relation to a wine environment. *Critical Reviews in Food Science and Nutrition, 51*, 479–498.

Cardoso, D. R., Libardi, S. H., & Skibsted, L. H. (2012). Riboflavin as a photosensitizer. Effects on human health and food quality. *Food & Function, 3*, 487–502.

Cataldi, T. R. I., Nardiello, D., Scrano, L., & Scopa, A. (2002). Assay of riboflavin in sample wines by capillary zone electrophoresis and laser-induced fluorescence detection. *Journal of Agricultural and Food Chemistry, 50*, 6643–6647.

Choe, E., Huang, R., & Min, D. B. (2005). Chemical reactions and stability of riboflavin in foods. *Journal of Food Science, 70*, R28–R36.

Desportes, C., Charpentier, M., Duteurtre, B., Maujean, A., & Duchiron, F. (2000). Liquid chromatographic fractionation of small peptides from wine. *Journal of Chromatography A, 9893*, 281.

Dias, D. A., Clark, A. C., Smith, T. A., Ghiggino, K. P., & Scollary, G. R. (2013). Wine bottle color and oxidative spoilage: whole bottle light exposure experiments under controlled and uncontrolled temperature conditions. *Food Chemistry, 138*, 2451–2459.

Dias, D. A., Smith, T. A., Ghiggino, K. P., & Scollary, G. R. (2012). The role of light, temperature, and wine bottle colour on pigment enhancement in white wine. *Food Chemistry, 135*, 2934–2941.

Dozon, N. M., & Noble, A. C. (1989). Sensory study of the effect of fluorescent light on a sparkling wine and its base wine. *American Journal of Enology and Viticulture, 40*, 265–271.

Escudero, A., Asensio, E., Cacho, J., & Ferreira, V. (2002). Sensory and chemical changes of young white wines stored under oxygen. An assessment of the role played by aldehydes and some other important odorants. *Food Chemistry, 77*, 325–331.

Escudero, A., Hernández-Orte, P., Cacho, J., & Ferreira, V. (2000). Clues about the role of methional as character impact odorant of some oxidized wines. *Journal of Agricultural and Food Chemistry, 48*, 4268–4272.

Fracasseti, D., Gabrielli, M., Encinas, J., Manara, M., Pellegrino, L., & Tirelli, A. (2017). Approaches to prevent the light-struck taste in white wine. *Australian Journal of Grape and Wine Research, 12295*, 1–5.

Fracasseti, D., Limbo, S., Pellegrino, L., & Tirelli, A. (2019). Light-induced reactions of methionine and riboflavin in model wine: Effects of hydrolysable tannins and sulfur dioxide. *Food Chemistry, 298*, 124952.

Goniak, O. J., & Noble, A. C. (1987). Sensory study of selected volatile sulfur compounds in white wine. *American Journal of Enology and Viticulture, 38*, 223–227.

Grant-Preece, P., Barril, C., Schmidtke, L. M., & Clark, A. C. (2018). Impact of fluorescent lighting on the browning potential of model wine solutions containing organic acids and iron. *Food Chemistry, 243*, 239–248.

Grant-Preece, P., Barrila, C., Schmidtke, L. M., Scollary, G. R., & Clark, A. C. (2015). Light-induced changes in bottled white wine and underlying photochemical mechanisms. *Critical Reviews in Food Science and Nutrition, 57*(4).

Grant-Preece, P., Fang, H., Schmidtke, L. M., & Clark, A. C. (2013). Sensorially important aldehyde production from amino acids in model wine systems: Impact of ascorbic acid, erythorbic acid, glutathione and sulphur dioxide. *Food Chemistry, 141*, 304–312.

Grodowski, M. S., Veyret, B., & Weiss, K. (1977). Photochemistry of flavins II. Photophysicalproperties of alloxazines and isoalloxazines. *Photochemistry and Photobiology, 26*, 341–352.

Hall, A. P., Brinner, L., Amerine, M. A., & Morgan, A. F. (1956). The B-vitamin content of grapes, musts and wine. *Food Research, 21*, 362.

Heelis, P. F., Parsons, G. J., Phillips, G. O., & McKellar, J. F. (1981). The flavin sensitised photooxidation of ascorbic acid: A continuous and flash photolysis study. *Photochemistry and Photobiology, 33*, 7–13.

Jung, R., Hey, M., Hoffmann, D., Leiner, T., Patz, C. D., Rauhut, D., ... Wirsching, M. (2007). Lichteinfluss bei der Lagerrung von Wein. *Mitteilungen Klosterneuburg, 57*, 224–231.

Leake, C. D., & Silverman, M. (1966). *Alcoholic Beverages in Clinical Medicine*. Cleveland: The World Publishing Co.

Mattivi, F., Monetti, A., Vrhovšek, U., Tonon, D., & Andrés-Lacueva, C. (2000). High-performance liquid chromatographic determination of the riboflavin concentration in white wines for predicting their resistance to light. *Journal of Chromatography A, 888*, 121–127.

Maujean, A., Haye, M., & Feuillat, M. (1978). Contribution a l'étude des "goûts de lumière" dans le vin de Champagne. II. Influence de la lumière sur le potentiel d'oxydoreduction. Correlation avec la teneur en thiols du vin. *Connaissance de la Vigne et du Vin, 12*, 277–290.

Maujean, A., & Seguin, N. (1983a). Contribution à l'étude des goûts de lumière dans les vins de Champagne. 4. Approaches a une solution oenologique des moyens de prévention des goûts de lumière. *Science des Aliments, 3*, 603–613.

Maujean, A., & Seguin, N. (1983b). Contribution à l'étude des goûts de lumière dans les vins de Champagne. 3. Les réactions photochimiques responsables des goûts de lumière dans le vin de Champagne. *Sciences des Aliments, 3*, 589–601.

Maujean, M. A. (1984). L'influence de la lumière sur les vins. *Revue Française d'Oenologie, 96*, 3–10.

Mestres, M., Busto, O., & Guasch, J. (2000). Analysis of organic sulfur compounds in wine aroma. *Journal of Chromatography A, 881*, 569–581.

Min, D. B., & Boff, J. M. (2002). Chemistry and reaction of singlet oxygen in foods. *Comprehensive Reviews in Food Science and Food Safety, 1*, 58–72.

Mirás-Avalos, J. M., Bouzas-Cid, Y., Trigo-Córdoba, E., Orriols, I., & Falqué, E. (2020). Amino acid profiles to differentiate white wines from three autochthonous galician varieties. *Foods, 9*, 114.

Morgan, A. F., Nob, A. L., Wiens, A., Margb, G. L., & Winkler, A. J. (1939). The B-vitamins of California grape juices and wines. *Food Research, 4*, 217.

Ournac, A. (1968). Riboflavine pendant la fermentation du jus de raisin et la conservation du vin sur lies. *Annales de Technologie Agricole, 17*, 67–75.

Ournac, A., & Décor, M. (1967). Riboflavin in the grape during the course of its development. *Annals of Agricultural Sciences, 16*, 309.

Pichler, U. (1996). Analisi della riboflavina nei vini bianchi e influenza della sua concentrazione. *L'Enotecnico, 32*, 57–62.

Powers, H. J. (2003). Riboflavin (vitamin B-2) and health. *The American Journal of Clinical Nutrition, 77*(6), 1352–1360.

Pripis-Nicolau, L., de Revel, G., Bertrand, A., & Lonvaud-Funel, A. (2004). Methionine catabolism and production of volatile sulphur compounds by Œnococcus œni. *Journal of Applied Microbiology, 96*, 1176–1184.

Pripis-Nicolau, L., de Revel, G., Bertrand, A., & Maujean, A. (2000). Formation of flavor components by thefigure reaction of amino acid and carbonyl compounds in mild conditions. *Journal of Agricultural and Food Chemistry, 48*, 3761–3766.

Ribéreau-Gayon, P., Glories, Y., Maujean, A., & Dubourdieu, D. (2006). *Handbook of Enology. Volume 2: The chemistry of wine stabilization and treatments*. Hoboken, NJ, USA: John Wiley and Sons Inc.

Ribéreau-Gayon, J., Peynaud, E., & Ribéreau-Gayon, P. (1975). Sciences et Techniques Du Vin Dunod Publishing: Paris, France.

Santos, M. A., García-Ramírez, J. J., & Revuelta, J. L. (1995). Riboflavin biosynthesis in Saccharomyces cerevisiae. *Journal of Biological Chemistry, 270*, 437–444.

Satter, A., & Demand, J. (1977). Light-induced degradation of vitamins. I. Kinetic studies on riboflavin decomposition in solution. *Canadian Institute of Food Science and Technology, 10*(1), 61–64.

Schonberg, A., & Moubacher, R. (1952). The Strecker degradation of α-amino acids. *Chemical Reviews, 50*, 261–277.

Siebert, T., Bramley, B., & Solomon, M. (2009). Hydrogen sulfide: Aroma detection threshold study in white and red wine. *The Australian Wine Research Institute Technical Review, 183*, 14–16.

Silva Ferreira, A. C., Guedes de Pinho, P., Rodrigues, P., & Hogg, T. (2002). Kinetics of oxidative degradation of white wines and how they are affected by selected technological parameters. *Journal of Agricultural and Food Chemistry, 50*, 5919–5924.

Smith, R. C., Reed, V. D., & Hill, W. E. (1994). Oxidation of thiols by copper(II). *Phosphorus, Sulfur and Silicon and the Related Elements, 90*, 147–154.

Singleton, V. L. (1987). Oxygen with phenols and related reactions in musts, wines, and model systems: Observations and practical implications. *American Journal of Enology and Viticulture, 38*, 69–77.

Solomon, M., Geue, J., Osidacz, P., & Siebert, T. (2010). Aroma detection threshold study of methanethiol in white and red wine. *The Australian Wine Research Institute Technical Review, 186*, 8–10.

Ugliano, M., Dieval, J.-B., Dimkou, E., Wirth, J., Cheynier, V., Jung, R., & Vidal, S. (2013). Controlling oxygen at bottling to optimize post bottling development of wine. *Practical Winery and Vineyard, 34*, 44–50.

Voigt, M. N., Eitenmiller, R. R., Powersand, J., & Ware, G. (1978). Water-soluble vitamin content of some California wines. *Journal of Food Science, 43*, 1071–1073.

WRAP (2012). Waste and Resources Action Program (WRAP). < http://www.wrap.org.uk >.

17

White wine polyphenols and health

Celestino Santos-Buelga[1,2], *Susana González-Manzano*[1,2] *and Ana M. González-Paramás*[1,2]

[1]Polyphenols Research Group (GIP-USAL), University of Salamanca, Campus Miguel de Unamuno, Salamanca, Spain
[2]Unit of Excellence Agricultural Production and Environment (AGRIENVIRONMENT), Scientific Park, University of Salamanca, Salamanca, Spain

17.1 Introduction

A great deal of epidemiological evidence has accumulated suggesting the existence of a relationship between light to moderate alcohol consumption and a lower incidence of and mortality by cardiovascular disease (CVD) (e.g., Corrao, Rubbiati, Bagnardi, Zambon, & Poikolainen, 2000; Di Castelnuovo et al., 2006; Larsson, Wallin, & Wolk, 2018; Reynolds et al., 2003) as well as other chronic conditions like type 2 diabetes (Carlsson, Hammar, & Grill, 2005; Djoussé, Biggs, Mukamal, & Siscovick, 2007; Huang, Wang, & Zhang, 2017; Koppes, Dekker, Hendriks, Bouter, & Heine, 2005) or dementia and cognitive decline in the old age (Letenneur, 2004; Peters, Peters, Warner, Beckett, & Bulpitt, 2008; Pinder & Sandler, 2004). The relationship between alcohol intake and cardiovascular mortality has been described as a J-shaped curve with a minimum situated at a level of consumption around 10 and 30 g alcohol/day (Fig. 17.1), with some gender differences, with greater putative beneficial effects situated at lower alcohol intakes in women than in men, owing to the distinct body composition and alcohol pharmacokinetics (Mumenthaler, Taylor, O'Hara, & Yesavage, 1999).

Such observational studies are not free from controversy, as they are attributed to suffering from a series of methodological limitations, which may have led to misinterpretations or biased conclusions (Fillmore, Kerr, Stockwell, Chikritzhs, & Bostrom, 2006; Fillmore, Stockwell, Chikritzhs, Bostrom, & Kerr, 2007). A common criticism is that former and occasional drinkers who have given up alcohol for health reasons are usually included in the studies as abstainers, which would artifactually raise the CVD risk in the nondrinking reference group (Fillmore et al., 2007). Actually, the association between moderate alcohol consumption and diminished CVD risk seems not to apply in ex-drinkers, in whom alcohol always leads to a greater risk of heart failure compared with nondrinkers or occasional drinkers (Larsson et al., 2018). The existence of different behavioral or dietary habits among drinkers and nondrinkers that might affect coronary heart disease (CHD) risk has also been indicated as another possible confounding factor. Naimi et al. (2005) reported that 27 of 30 factors associated with cardiovascular risk that they assessed were significantly more prevalent among nondrinkers than in light to moderate drinkers. Naimi et al. (2017) highlighted the importance of considering former and occasional drinkers separately from abstainers as those groups displayed significantly elevated CHD mortality risk. The same authors also observed that the CHD protection effect for moderate drinkers was higher in people older than 55 years, which could be explained by the existence of an increased CHD risk at the baseline group including former and occasional drinkers. They concluded that in older cohorts the apparent cardio-protection would rather reflect the selection biases accumulated over lifetime, resulting in the selected continuous drinkers being compared with increasingly unhealthy current abstainers (Naimi et al., 2017; Zhao, Stockwell, Roemer, Naimi, & Chikritzhs, 2017). On the contrary, binge drinking episodes can occur among people included as moderate drinkers according to their average alcohol intake, which would increase the cardiovascular risk in this group (Roerecke & Rehm, 2014). Actually,

© 2022 Elsevier Inc. All rights reserved.

standardization of alcohol consumption is tricky considering the different drinking patterns and, in the end, its measurement in epidemiological studies normally relies on self-reported reports, raising doubts about the accurate recall of alcohol intakes (Chikritzhs, Fillmore, & Stockwell, 2009). Another point to take into account is that chronic diseases take years to develop, so that their incidence would not relate with "current" alcohol consumption, but rather respond to lifestyle and drinking patterns throughout their lifetime. This "time-lag" effect was argued by Law and Wald (Law & Wald, 1999) to criticize the studies that led to postulate the "French Paradox." According to them, if animal fat consumption and serum cholesterol levels in France up till 30 years ago were considered, no inverse correlation would have been observed between CHD mortality and moderate wine intake.

Despite these and other possible biases, many authors agree that when confounding factors are specifically adjusted, observational trials still continue showing remarkably consistent data and do not usually result in a great reduction of benefits found for low to moderate alcohol intake on CVD morbidity and mortality (Hansel et al., 2010; Naimi et al., 2005; Rehm & Roerecke, 2017; Smyth et al., 2015; Zhao et al., 2017).

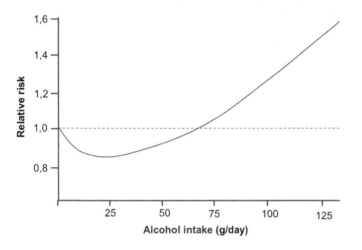

FIGURE 17.1   J-shaped relationship between average alcohol intake and cardiovascular mortality (freely adapted from different sources).

17.2 Wine and polyphenols

An issue of discussion is whether the intended beneficial effects of alcoholic beverages are due to ethanol or other components also have a role. In many studies, associations between alcohol intake and CVD or other chronic diseases are established for overall alcohol consumption without differentiating drinks. Indeed, ethanol itself may exert positive effects on the cardiovascular system by increasing high-density lipoprotein cholesterol, preventing platelet aggregation and enhancing fibrinolysis, which favorably affect thrombolytic processes (Covas, Gambert, Fitó, & de la Torre, 2010). However, when differentiation among drinks is made, it is usually concluded that wine provides higher health benefits than other alcoholic drinks like beer or spirits, either in the protection against CVD (Gronbaek et al., 1995; Haseeb, Alexander, & Baranchuk, 2017; Liberale, Bonaventura, Montecucco, Dallegri, & Carbone, 2019; Snopek et al., 2018; Truelsen, Grønbæk, Schnohr, & Boysen, 1998), type 2 diabetes (Huang et al., 2017), or dementia (Pinder & Sandler, 2004; Truelsen et al., 1998). This perception has also been supported by the results obtained in some intervention trials (Chiva-Blanch, Arranz, Lamuela-Raventos, & Estruch, 2013; Droste et al., 2013; Estruch et al., 2011; Toth et al., 2014). Schutte, Papageorgiou, Najlah, Huisman, Ricci, Zhang, and Schutte (2020) have even reported that among alcoholic drinks, only wine is associated with a decreased cardiovascular risk, whereas beer and spirits would be detrimental at any level of consumption.

The intended superior effects of wine have been largely associated with polyphenols, compounds that are not present (or are present to less extent) in other alcoholic drinks, and for which a range of biological activities and beneficial effects on human health have been described (Fernandes et al., 2017).

Wine polyphenols are a variable mixture of flavonoid and nonflavonoid compounds, extracted from the grape during winemaking. Since white wines are not usually submitted to maceration with grape solids, they contain different phenolic composition and lower phenolic amounts than red wines. Total polyphenol content in red wines is usually well above 1 g/L, while in white wines, it does not commonly exceed a few hundred mg/L (Waterhouse, 2002).

Flavan-3-ols (i.e., catechins and procyanidins), anthocyanins, and, to a less extent, flavonols are the main classes of flavonoids in grape skins and consequently in red wines. Hydroxycinnamic acids and especially their tartaric esters are the most important phenolic compounds in grape pulp and white wines. Other phenolic compounds, like hydroxybenzoic acids, stilbenes (e.g., resveratrol), or dihydroflavonols (Fig. 17.2) can also be found in grapes and wines, although they are usually present in low concentrations (Waterhouse, 2002).

Procyanidins (i.e., catechin oligo/polymers) are considered the main vasoactive components among grape and wine polyphenols. In general, these compounds are recognized to possess a range of biological activities and are thought to contribute to the disease-preventive properties of plant-derived foods (Rasmussen, Frederiksen, Krogholm, & Poulsen, 2005; Santos-Buelga & Scalbert, 2000). Among others, they have been shown to be able to suppress endothelin-1 synthesis, a key factor in the development of vascular disease and atherosclerosis (Corder et al., 2001). Red wine is actually one of the richest dietary sources of this class of compounds, while they are hardly present in white wines (Manach, Scalbert, Morand, Rémésy, & Jiménez, 2004).

Resveratrol (3,5,4'-trihydroxystilbene) is another wine component for which a variety of potential health benefits, ranging from chemoprevention to cardiovascular protection, have been claimed (Das, Mukherjee, & Ray, 2010). Nevertheless, it seems unlikely that such effects can be produced from moderate consumption of wine, owing to its very small contents and low bioavailability. Actually, in order to get the amounts required to obtain some of the beneficial effects reported in in vitro and animal assays, several dozens of wine bottles would be required (Santos-Buelga & González-Manzano, 2011).

Other phenolic compounds present in wines like anthocyanins, flavonols, phenolic acids, or lignans are also known to possess biological activities, such as antioxidant, antiinflammatory, antiproliferative, estrogenic, antimicrobial, or gene modulation abilities, as established in different in vitro, ex vivo, and animal studies (Ferreira, Martins, & Barros, 2017; Santos-Buelga, González-Paramás, Oludemi, Ayuda-Durán, & González-Manzano, 2019). Despite these compounds being considered, at least in part, responsible for the health-protective effects of fruit- and vegetable-rich diets, they have been less frequently associated with the positive health effects of wines.

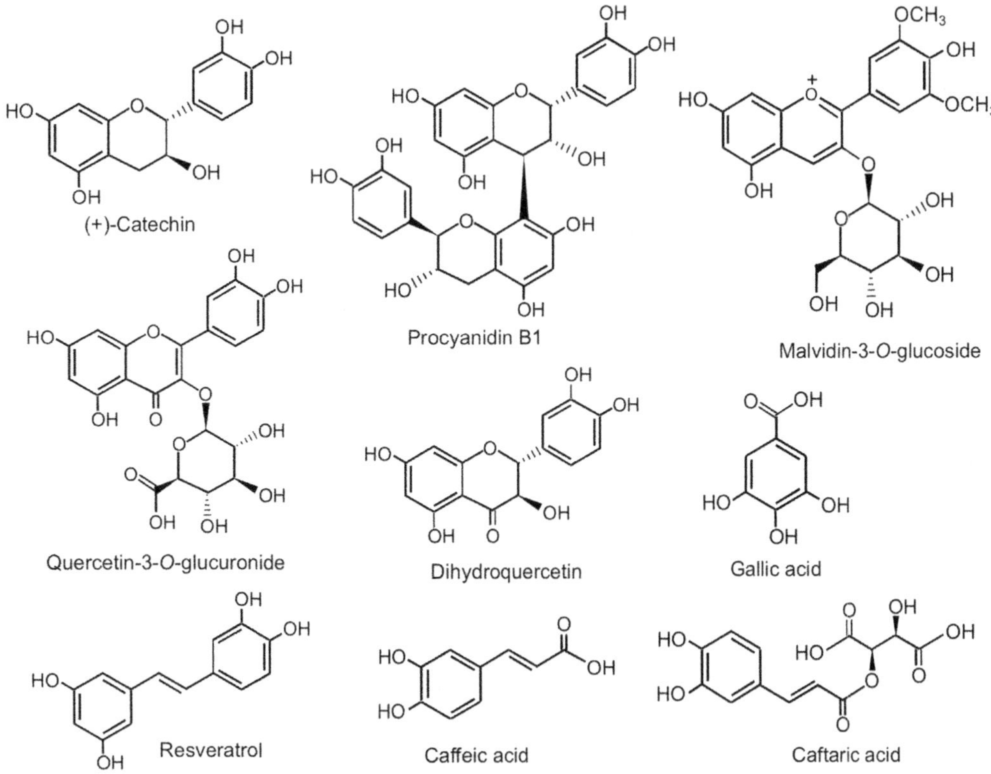

FIGURE 17.2   Structures of some characteristic grape and wine polyphenols.

17.3 Bioactive components in white wine

17.3.1 Grape polyphenols

Total polyphenol content in a range from around 100 to more than 800 mg/L has been reported in white wines from different origins (Alvarez-Casas, Pajaro, Lores, & Garcia-Jares, 2016; Makris, Psarra, Kallithraka, & Kefalas, 2003; Ragusa et al., 2017; Rastija, Srečnik, & Marica-Medić-Šarić, 2009), being highly variable depending on the grape variety, edaphoclimatic factors, winemaking process, aging, and storage. An untargeted UHPLC-ESI/QTOF-Ms comprehensive metabolomic study of the phenolic composition in Chardonnay wines was performed by Rocchetti, Gatti, Bavaresco, and Lucini (2018), allowing the detection of up to 258 compounds, including hydroxycinnamic and hydroxybenzoic acid derivatives, flavonoids, simple phenolics, stilbenes, coumarins, and lignans.

Hydroxycinnamates are the major phenolic compounds in white wine, with concentrations that could be as high as those present in red wines (Waterhouse, 2002). Total hydroxycinnamate contents ranging from 90 to more than 750 mg/L (mean values of 324.7 mg/L) were determined by Makris et al. (2003) in a large screening on Greek white wines, and concentrations reported by other authors in wines from different countries usually fall within this range. The predominant compound is always caftaric acid, normally accounting for 50% or more of this fraction. Other commonly reported hydroxycinnamates are caffeic acid; ethyl caffeoate; 2-S-glutathionyl-caftaric acid; campesteryl ferulate; and coutaric, fertaric, coumaric, and ferulic acids (Alvarez-Casas et al., 2016; Makris et al., 2003; Recamales, Sayago, González-Miret, & Hernanz, 2006; Rocchetti et al., 2018; Rodríguez-Bernaldo, López-Hernández & Lage-Yusty, 2009).

Concentrations of total nonhydroxycinnamate phenolics in white wines from 1.9 to 370.8 mg/L (mean value 52.2 mg/L) were determined by Makris et al. (2003). Catechin, epicatechin, gallic acid, and tyrosol are typically the main nonhydroxycinnamates, although other compounds are also frequently reported, such as protocatechuic acid, p-hydroxybenzoic acid, ethyl gallate, procyanidin dimers, quercetin glycosides, or resveratrol (Alvarez-Casas et al., 2016; Makris et al., 2003; Recamales et al., 2006; Rocchetti et al., 2018; Rodríguez-Bernaldo et al., 2009).

Catechin and epicatechin are generally the most important nonhydroxycinnamates, with contents that oscillate within a wide range, from practical absence to around 150 mg/L, depending on wine origin and winemaking characteristics (Hernanz et al., 2007; Makris et al., 2003; Recamales et al., 2006; Rocchetti et al., 2018). Some amounts of procyanidin oligomers are also usually found; for instance, the dimer B1 was reported by Rodríguez-Bernaldo et al. (2009) to be the predominant flavan-3-ol in white wines from the Albariño variety, with mean values of 17.8 mg/L, representing about 30% of total flavan-3-ols. These compounds are highly susceptible to oxidation and polymerization and, together with hydroxycinnamic-tartaric esters, are the compounds most frequently associated with browning in white wines; hence winemaking practices are frequently applied to limit their levels so as to prevent oxidation (Du Toit, Marais, Pretorius, & du Toit, 2006; Singleton, 1987).

Benzoic acids are minor components in wines, mostly deriving from the hydrolysis of hydrolyzable tannins released from the barrel during wood aging; so their concentrations are usually very low in nonaged wines, including most white wines. Average concentrations of gallic acid, as the main benzoic acid, of 13.8 mg/L were determined by Makris et al. (2003) in white wines.

Total stilbene contents in white wines are usually below 1 mg/L (Waterhouse, 2002), although concentrations of up to 4 mg/L have been determined for resveratrol in some wines (Alvarez-Casas et al. 2016; Makris et al. 2003; Ragusa et al., 2017; Rastija et al. 2009). As for lignans, mean values around 5 mg/L, with medioresinol and isomeric forms of lariciresinol as the most abundant ones, were reported by Rocchetti et al. (2018) in Chardonnay wines.

17.3.2 Other bioactive compounds

Polyphenols are not the only bioactive compounds that can be present in wine, but other phenolic and nonphenolic components are also present that might contribute to the putative health benefits associated with wine consumption and, in some cases, to explaining possible differential effects between red and white wines.

During the fermentation process, yeasts catabolize aromatic amino acids, such as tyrosine, tryptophan, and phenylalanine, to their respective aromatic alcohols, tyrosol, tryptophol, and phenyl ethanol (Fig. 17.3; Mas et al., 2014). These compounds may participate in wine flavor (for instance, 2-phenyl ethanol contributes to the typical rose scent of some white wines), but they also possess bioactive properties, which have been associated with some of the beneficial effects of moderate wine consumption. Tyrosol has been reported to be the second most abundant nonhydroxycinnamate phenolic in many white wines, with concentrations that may reach up to

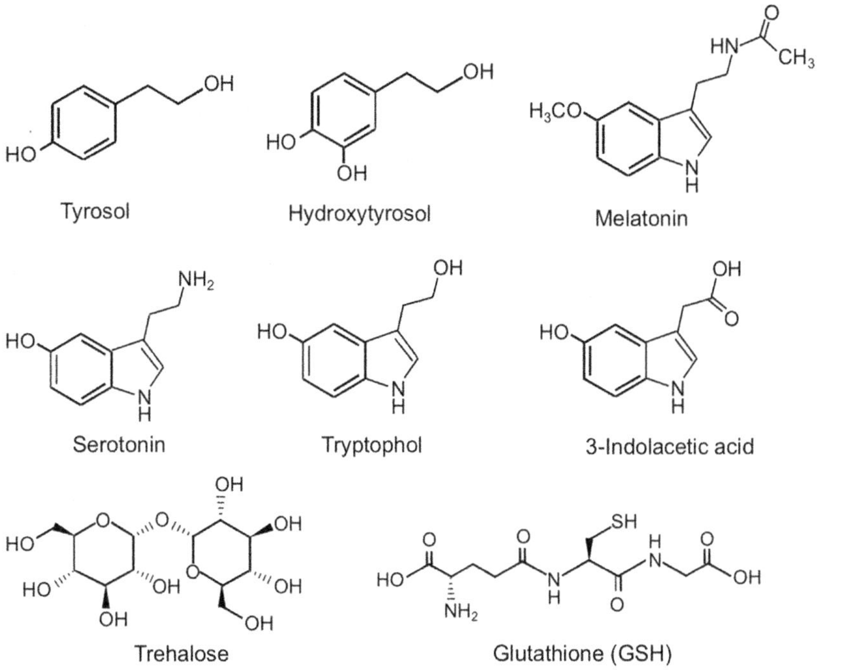

FIGURE 17.3 Structures of some wine bioactive compounds from microbial metabolism.

95 mg/L (Hernanz et al., 2007; Recamales et al., 2006; Rocchetti et al., 2018), similar or even higher than those found in red wines (Bordiga et al., 2016; Gris et al., 2010; Piñeiro et al., 2011). The presence of hydroxytyrosol has also been reported in wine, resulting from tyrosol hydroxylation (Mas et al., 2014), in levels that usually range from 1 to 10 mg/L (Bordiga et al., 2016; Piñeiro et al., 2011; Rocchetti et al., 2018). Concentrations of tryptophol from 0.4 to 5.4 mg/L have been quantified in different wines; in this case, greater contents, above 3 mg/L, are found in white and rosé wines than in red ones (Arapitsas, Guella, & Mattivi, 2018; Bordiga et al., 2016).

Other bioactive products from tryptophan metabolism found in wine are 3-indoleacetic acid (IAA), melatonin (*N*-acetyl-5-methoxytryptamine), and serotonin (Fig. 17.3). Although some amounts of these indolic compounds can be present in grapes, their content in wine is mostly influenced by the fermentation process, with the used yeast strain and the fermentation time being the most influential factors (Fernández-Cruz, Cerezo, Cantos-Villar, Troncoso, & García-Parrilla, 2019). Up to nine melatonin isomers have been detected in wine, fluctuating within a very wide range from a few μg/L to more than 150 μg/L (Kocadağli, Yilmaz, & Gökmen, 2014; Mercolini et al., 2008; Rodríguez-Naranjo, Torija, Mas, Cantos-Villar, & García-Parrilla, 2012; Stege, Sombra, Messina, Martinez, & Silva, 2010). Indeed, some wines can be considered a significant source of dietary melatonin, as they possess concentrations higher than those found in most fruits and vegetables, usually situated in the low ng/g level, with only a few commodities, such as mushrooms, coffee beans, or some berries showing contents in the μg/g range (Meng et al., 2017).

Serotonin contents from 2 to 23 mg/L have been reported in wine, being mainly formed during the malolactic fermentation, especially by *Lactobacillus plantarum* (Mas et al., 2014). Since malolactic fermentation is not usually performed in white wines, serotonin concentrations are much lower in them than in red wines. In addition to serotonin, other biogenic amines are also produced from the decarboxylation of amino acids by bacteria involved in the malolactic fermentation. Among them, tyramine and histamine are relevant, as they might be harmful to human health. For instance, histamine can cause headaches and allergies, while tyramine is a strong vasoconstrictor. Red wines can show concentrations of these amines greater than 10 mg/L, while in white and rosé, wines they are usually below 2 mg/L (Galgano, Caruso, & Favati, 2009; Landete, Ferrer, & Pardo, 2007). In view of the risks that histamine poses for human health, several countries (e.g., Australia, Netherlands, Germany, Belgium, France, and Switzerland) have placed limits on its concentrations in a range from 2 to 10 mg/L (Guo, Yang, Peng, & Han, 2015).

Other bioactive compounds that might have some relevance in wine are glutathione and disaccharide trehalose. Reduced glutathione (GSH) plays an important role in wine quality, especially in white wines, as it contributes to preventing browning and also has an influence on wine aroma. After grape crushing, polyphenol oxidases (PPO) produce rapid oxidation of phenolic compounds, and especially hydroxycinnamic acids, leading to the formation of quinones that polymerize to give rise to brown pigments. GSH reacts with quinones, preventing their polymerization and producing the so-called grape reaction product (e.g., 2-S-glutathionylcaftaric acid; GRP) that cannot be oxidized by PPO (Kritzinger, Bauer, & Du Toit, 2013). The levels of GSH largely depend on the winemaking process and yeast metabolism, although they are usually higher in white than in red wines. GSH concentrations in the range of 1.7−7.1 mg/L in white wines and 0.3−0.7 mg/L in red wines were determined by Marchand and de Revel (2010). Trehalose (α-D-glucopyranosyl-(1 → 1)-α-D-glucopyranoside) is not naturally present in mammals but widespread in plants and microorganisms, including wine yeasts, with a function of protecting cells against desiccation. It has been shown that to improve resistance to dehydration, yeasts grown in nutrient-poor media accumulate large amounts of trehalose, while they tend to accumulate glutathione in nutrient-rich medium (Camara et al., 2005). Concentrations of trehalose in white wines below 1 mg/L and up to 26.1 mg/L in rosé and red wines were determined by Ruiz-Matute, Sanz, Moreno-Arribas, and Martínez-Castro (2009).

All the previous compounds are present in both white and red wines, although the very low contents of flavonoids in white wine make nonflavonoid compounds and, in particular, those resulting from yeast activity more relevant to explaining the positive health effects that might be associated with this type of wines.

17.4 Putative positive health effects of white wine and components and mechanisms involved

In most observational studies, attention is only paid to overall wine intake without differentiating between white and red wine, thus making it impossible to ascribe effects to one or another. Nevertheless, health benefits have been more commonly related to red wine consumption, owing to its much higher concentration of polyphenols. Indeed, being components of the grape skin, white wines contain limited amounts of flavonoids, with which the health properties of wine have been usually associated. There are, however, studies comparing both types of wines where no relevant differences between the protective effects of red and white wine were found, as well as studies that show that moderate white wine consumption is specifically associated with some health benefits.

In a long-life, large epidemiological survey on adults in Northern California (United States), Klatsky, Friedman, Armstrong, and Kipp (2003) found that light drinkers of either red or white wine, or combinations of both, had a lower risk of mortality than did beer or liquor drinkers, which was especially evident for deaths attributed to coronary disease. Similarly, Whelan et al. (2004) did not detect significant differences pertaining to acute ingestion of moderate amounts of red or white wine in the improvement of endothelial function, with both increasing endothelium-dependent vasodilation by two- to threefold in individuals with coronary artery disease. The same group also reported that ingestion of moderate amounts of both red and white wine with a light meal increased plasma interleukin-6 concentration and flow-mediated dilation of the brachial artery, without appreciably altering plasma concentrations of VCAM-1 and ICAM-1 cell adhesion molecules related to the development of atherosclerotic lesions (Williams, Sutherland, Whelan, McCormick, & De Jong, 2004). The authors postulated that the moderate intake of either wine would have an acute beneficial influence on the vasculature that might contribute to the decreased risk of CHD. Mansvelt et al. (2002) compared the effect of the consumption of red and white wine (intake of 32 g of alcohol/day in men, and 23 g/day in women, for 28 days) on platelet aggregation and fibrinolytic factors in healthy humans. They found that, although only red wine was able to decrease platelet aggregation, similar favorable effects on the fibrinolytic factors were induced by both types of wine. Kikura et al. (Kikura, Levy, Safon, Lee, & Szlam, 2004) reported that acute intake of 300−350 mL of either red or white wine during dinner does not induce detrimental effects on platelets and viscoelastic properties of blood and does not compromise the coagulation function in healthy volunteers. Gepner et al. (2015), in a clinical trial on patients with controlled type 2 diabetes, found that, in slow ethanol metabolizers, both red and white wine were equally efficient in improving glycemic control, although red wine seemed superior in the control of the lipoprotein profile.

In studies carried out in rats, Cui et al. (2002) showed that consumption of ethanol-free extracts of white wine (100 mg/100 g of body weight) for three weeks provided a cardioprotective effect, improving the postischemic contractile recovery and reducing myocardial infarct size. This effect was attributed to the antioxidant ability of the phenolic components of the extracts, mostly consisting of catechin, tyrosol, and caftaric acid. Thirunavukkarasu et al. (2008) also reported protection against myocardial infarction in rats gavaged with white wine, containing caffeic acid, tyrosol, and shikimic acid as the main active components, at a daily dosage of 6.5 mL/kg for a month. Decreased myocardial infarct size and cardiomyocyte apoptosis, and increased myocardial functions were found in the rats treated with white wine. The same authors (Samuel, Thirunavukkarasu, Penumathsa, Paul, & Maulik, 2008) further explored the mechanism for the cardioprotective effect of tyrosol in the same rat model, concluding that it was related to the ability of this phenylethanoid to activate the Akt/FOXO3a/eNOS stress response pathway, as well as to enhance expression of SIRT1 protein, resulting in decreased cell death response and increased stress resistance and longevity. Vauzour et al. (2007) showed that tyrosol, caffeic acid, and gallic acid from extracts of Champagne wine protected primary cortical neurons against oxidative injury in vitro at concentrations between 0.1 and 10 μM, suggesting a neuroprotective potential.

The possibility that the antioxidant and antiinflammatory properties of tyrosol and caffeic acid could be, at least in part, responsible for the beneficial effects attributed to moderate consumption of white wine has been proposed by different authors. Despite these components being also present in red wine, their relevance would be higher in white wines, owing to their lower contents of other phenolic compounds. Bertelli, Migliori, Bertelli et al. (2002) and Bertelli, Migliori, Panichi et al. (2002) explored the ability of both compounds to regulate inflammatory responses using a cell model of human peripheral blood mononuclear cells (PBMCs). They found that these phenolics were able to inhibit the lipopolysaccharide (LPS)-induced production of the proinflammatory cytokines tumor necrosis factor-alpha (TNF-α) and interleukins IL-1β and IL-6, with a synergistic effect when both compounds were tested together at nanomolar concentrations. Pignatelli et al. (2004) found that the plasma levels of caffeic acid, catechin, and resveratrol significantly increased following an intake of 300 mL/day of red or white wine by healthy subjects for 15 days. Although greater plasmatic levels of catechin (0.056 vs. 0.013 μM) and resveratrol (1.33 vs. 1.72 μM) were reached in the individuals taking red instead of white wine, similar concentrations of caffeic acid, around 0.19 μM, were found in subjects drinking either wine. Those authors also observed that whereas concentrations of a single polyphenol below 1 μM failed to inhibit low-density lipoprotein (LDL) oxidation, the mixture of the three compounds in a total concentration under 1 μM significantly inhibited LDL oxidation, indicating the existence of a synergistic effect among polyphenols. In assays in healthy men, Caccetta, Croft, Beilin, and Puddey (2000) also found that caffeic acid was absorbed after ingestion of wine and dealcoholized wine (5 mL of wine containing 11 mg caffeic acid/L per kg body weight), resulting in an acute increase in its plasmatic levels, although the concentrations reached (below 100 nmol/L) would not be sufficient to produce an effect on LDL or serum oxidative status. In a previous work, the same authors estimated that concentrations of caffeic acid, around 0.6 μM (IC$_{50}$), were required to significantly inhibit the copper-initiated oxidation of human LDL (Abu-Amsha, Croft, Puddey, Proudfoot, & Beilin, 1996). Plasma concentrations of 0.09 μM, 0.05 μM, and 0.03 μM were measured by Nardini et al. (2009) for caffeic, p-coumaric, and ferulic acids, respectively, at peak time, after consumption by individuals of a single dose of 250 mL of white wine, containing about 30 mg of total hydroxycinnamic acids. These authors noticed, however, that these compounds were mostly circulating in the blood as glucuronide and sulfate conjugates, which promotes their excretion in urine and bile and may modify their biological activity. They also highlighted that wine would be always a minor contributor to the intake of hydroxycinnamates, mostly provided in the diet by fruits and vegetables, with an average intake that could be above 200 mg/day (Clifford, 2000). Therefore possible beneficial effects of these compounds would be rather expected to be provided from food sources different to wine, in which putative health benefits should be explained by other compounds, such as tyrosol and hydroxytyrosol.

Hydroxytyrosol has been shown to possess antiatherogenic and antioxidant properties, related to its strong free radical scavenging and metal-chelating abilities (Granados-Principal, Quiles, Ramirez-Tortosa, Sanchez-Rovira, & Ramirez-Tortosa, 2010). In 2011, the European Food Safety Authority (EFSA) released a health claim on the benefits of the regular daily ingestion of 5 mg of hydroxytyrosol and related compounds (i.e., tyrosol and oleuropein complex) from olive oil on the protection of LDL particles from oxidative damage (EFSA Panel on Dietetic Products, Nutrition, and Allergies NDA, 2011). It is recognized that extra virgin olive oil is the richest dietary source of these compounds, with concentrations that may range from a few tens to several hundred mg/kg, depending on the olive variety and cultivation and oil processing practices, half of which could correspond to hydroxytyrosol (Ramírez-Tortosa et al., 2006). As indicated above, the presence of hydroxytyrosol has been reported in wine in levels ranging from 1 to 10 mg/L, while concentrations of tyrosol in white wines usually range from 10 to 95 mg/L. However, although hydroxytyrosol contents in wine are (much) lower than those in virgin olive oil, it could contribute to the dietary intake of this compound, taking into account that tyrosol can be converted into hydroxytyrosol in the human body. In a clinical trial, De La Torre, Covas, Pujadas, Fitó, & Farré (2006) found that the consumption of moderate doses of wine or olive oil by healthy subjects led to a higher increase in the urinary concentrations of hydroxytyrosol in the wine group, despite the amount of hydroxytyrosol administered being fivefold greater in the olive oil group (1.7 mg vs. 0.35 mg). As the recovery of hydroxytyrosol in urine could not be explained by the dispensed dose, they proposed that wine may promote endogenous hydroxytyrosol generation. Those authors concluded that a single glass of wine is at least equivalent to 25 mL (22 g) of virgin olive oil in its capacity to increase hydroxytyrosol concentrations in the body, thus leading to similar beneficial effects. Actually, the same group found that there was a direct association between wine consumption and the urinary concentrations of tyrosol and hydroxytyrosol determined in individuals at cardiovascular risk included in the PREDIMED study (Pérez-Mañá et al., 2015). Taking into account data obtained from human and animal studies, they concluded that hydroxytyrosol was mostly produced from the endogenous biotransformation of tyrosol as a result of a shift in dopamine and tyramine oxidative metabolism to a reduced pathway favored by ethanol; besides, the alcohol would help increase the bioavailability of tyrosol present in wine. According to them, the endogenous formation of hydroxytyrosol might contribute to explain part of the cardiovascular benefits associated with light-to-moderate wine consumption.

Another compound that has been related to the benefits of moderate wine consumption is melatonin. This is a neurohormone secreted from the pineal gland, with well-characterized antioxidant, antiinflammatory, and immune-modulating properties, that contributes to the regulation of the circadian rhythms and has been attributed to tumor inhibitory activities and positive effects on the cardiovascular system, lipid and glucose metabolism, and neuroprotection (Mas et al., 2014; Meng et al., 2017). Circulating plasma levels of melatonin oscillate between around 10 pg/mL during the day and 200 pg/mL at the maximal night level (Bonnefont-Rousselot & Collin, 2010), although its dietary intake has been shown to increase the melatonin blood levels and total antioxidant capacity in plasma (Aguilera et al., 2016; González-Flores et al., 2012; Reiter & Tan, 2003). Some wines possess relevant contents of melatonin, higher than those present in most fruits and vegetables (Meng et al., 2017), so that it is not unlikely that melatonin could be a significant contributor to the purported beneficial effects of wine consumption (Vitalini et al., 2011).

Other amino acid—derived compounds, such as IAA, serotonin, or tryptophol, are also acknowledged to possess bioactive properties. Serotonin is a neurotransmitter that is known to modulate cognition and learning, gastrointestinal motility, vasoconstriction, blood pressure, and platelet function (Berger, Gray, & Roth, 2009); furthermore, it is a precursor of melatonin synthesis (Mas et al., 2014). IAA has been shown to enhance cell proliferation and to possess antioxidant properties (Vilela, 2019), while tryptophol acts as a functional analog or precursor of serotonin and melatonin (Mas et al., 2014), and it is also a precursor in the synthesis of indoramin, an alpha-adrenoreceptor-blocking drug used to treat hypertension and in the treatment of benign prostatic hyperplasia (Vilela, 2019). However, despite their biological properties, the contribution of these compounds to the possible beneficial effects of wine intake seems unlikely, owing to their low contents (Fernández-Cruz et al., 2019).

17.5 Wine, lifestyle, and sociological factors

Wine components are not the only factor that has been related to the putative health benefits of wine consumption; contextual aspects, such as food choices, lifestyle and drinking patterns, or socioeconomic determinants, have also been shown to have similar or greater influence. Many authors have pointed to a tendency of moderate wine drinkers to have healthier dietary and lifestyle behaviors than people that preferentially choose other alcoholic beverages, like beer or liquors, which has been proposed as a further explanation for the beneficial effects of wine consumption. For instance, healthier dietary habits are usually associated with some Mediterranean countries, where wine is the traditionally consumed alcoholic drink. The so-called Mediterranean diet has been linked to a range of health benefits, such as prevention of CVD and cognitive decline, lower incidence of type 2 diabetes (Ditano-Vázquez et al., 2007), reduced rates of obesity (Castro-Barquero, Lamuela-Raventós, Doménech, & Estruch, 2018), or longer life expectancy (Trichopoulou & Critselis, 2004). This diet is characterized by including a variety of fruits and vegetables, nuts, cereals, fish, olive oil as the main fat, low intake of red meat, and moderate consumption of wine. These food choices contribute adequate amounts of components like dietary fiber, antioxidants (e.g., vitamins E and C, carotenoids, and polyphenols), and long-chain n-3 fatty acids, as well as a balanced n-6/n-3 ratio of essential fatty acids, which are traits that are all considered to provide health benefits (Román et a., 1996).

Indeed, when interpreting the findings on the relationships between alcohol consumption and health, the underlying lifestyle and dietary patterns can be as influential on the health outcome as the type of drink. In a study on different populations in the United States and Europe, Sluik, Bezemer, Sierksma, and Feskens (2016) showed that a preference for wine was strongly associated with healthier dietary patterns, while persons that preferred beer displayed, in general, less favorable dietary behavior. Burke, Puddey, and Beilin (1995), in a health screening on working men aged 25—51 years in Australia, found that a preference for wine was related to greater consumption of fruit and vegetables and bread, while meat, fried foods, eggs, and salt were associated with a preference for beer. In a survey on people aged 25—64 years in Finland, Mannistö, Uusitalo, Roos, Fogelholm, and Pietinen (1997) observed that wine drinkers had significantly higher intakes of antioxidants (carotenoids and vitamin C) in their diet, indicating a greater consumption of fruit and vegetables, and that they also tended to have higher educational level than groups with other drink preferences. A higher intake of fruits, salads, cooked vegetables, fish, and olive oil was also found by Tjønneland, Grønbœk, Stripp, and Overvad (1999) in those that preferred wine, as compared with other alcoholic drinks, in a cross-sectional study conducted in Denmark among men and women aged 50–64 years. Similar outcomes were obtained by Barefoot et al. (2002) in a screening on drinking and health behaviors among the Alumni of the University of North Carolina, where individuals that preferred wine consumed more fruit and vegetables and less red or fried meats than those who preferred beer or spirits. A survey on Danish supermarket food choices revealed that people that bought wine were more likely to include in their basket healthy food items, such as olives, fruit and vegetables, poultry, and low-fat cheese, milk, and meat, while those that chose beer showed a greater tendency to purchase ready-to-cook dishes, sugar, margarine, chips, pork, sausages, lamb, or soft drinks (Johansen, Friis, Skovenborg, & Grønbæk, 2006). The authors also reported that wine drinkers also tended to have a higher socioeconomic status and better psychological functioning than beer consumers. In the same way, Sluik, Van Lee, Geelen, and Feskens (2014), in a representative sample of people from the Netherlands, detected large differences in dietary habits between individuals with a preference for wine. In general, wine drinkers consumed less energy and more vegetables and fruit juices, while beer choice was associated with a higher intake of meat, soft drinks, margarine, and snacks. Interestingly, this type of associations has not been found in studies performed in some Mediterranean countries, as, in the case of some Italian (Chatenoud, Negri, La Vecchia, Volpato, & Franceschi, 2000) or Spanish cohorts (Alcácera et al., 2008; Carmona-Torre et al., 2008), no significant correlation between wine consumption and healthier dietary habits was observed in relation to nondrinkers or consumers of other alcoholic beverages.

Besides the diet, the preference for wine has also been associated with more favorable behavioral and socioeconomic factors and health status, especially in countries where wine is not a traditional alcoholic drink. In a large health screening in Stockholm, Sweden, Rosell, de Faire, and Hellénius (2003) observed that, in addition to a healthier diet, moderate wine drinkers exhibited better lifestyle patterns and metabolic parameters than beer and spirits drinkers. As for the metabolic profile, the wine group showed lower fasting insulin and fibrinogen concentrations, and higher HDL-cholesterol levels. In a survey about health determinants on people included in the National Longitudinal Study of Adolescent Health in the United States, Paschall & Lipton (2005) found that wine drinkers generally showed better dietary and exercise habits, formal education, and health indicators and were also less likely to report smoking or problem drinking, compared to beer or liquor drinkers. These findings led the authors to propose that wine preference in young adulthood could help to explain the association between light-moderate wine consumption and morbidity, and mortality risk in later adulthood. McCann et al. (2003), analyzing a series of case-control studies on alcohol use and disease incidence in people from the western area of New York State, concluded that, besides better diets, wine drinkers had a lower prevalence of smoking and higher household incomes and educational level than beer and spirits drinkers. Mortensen, Jensen, Sanders, and Reinisch (2001), in a cross-sectional study of characteristics associated with beverage choice in a Danish cohort, observed that wine drinking was significantly associated with higher intelligence quotient, parental educational level, and socioeconomic status, whereas beer drinking was significantly associated with lower scores on the same variables. They concluded that wine drinking could be a general indicator of optimal social, cognitive, and personality development also associated with better health, which might explain the apparent health benefits of wine.

Another aspect that may also have an influence on the distinct health outcomes usually obtained for wine and other alcoholic beverages is the drinking pattern. Wine tends to be more usually associated with meals and often sipped slowly than other alcoholic drinks, a factor that may provide metabolic advantages. Among others, the concomitant presence of food in the stomach slows down gastric emptying and subsequent delivery of ethanol to the duodenum and liver, which favors hepatic metabolism and clearance, lowering the peak of alcohol concentration in blood (Jones, Jönsson, & Kechagias, 1997). The simultaneous presence of food might also reduce the amount of alcohol available to the oral microbiota, which has the capacity to metabolize ethanol to acetaldehyde, a compound associated with the tumorigenic effects of the ethanol in the upper gastrointestinal tract (Homann, Jousimies-Somer, Jokelainen, Heine, & Salaspuro, 1997). Furthermore, when wine is consumed with meals, the onset of the plasma uric acid elevation coincides with the period of postprandial oxidative stress produced after a meal, which may contribute to the wine's protective effects (Boban et al., 2016). Finally, irregular (binge) drinking, a behavior that is related with higher mortality rates than steady drinking, seems more frequent among beer and spirits drinkers than in wine consumers (Grønbæk, 2003; Grønbæk, Jensen, Johansen, Sørensen, & Becker, 2004; Rehm, Sempos, & Trevisan, 2003).

On their whole, these observations pointed to that, at least in some countries, moderate wine consumption might represent a marker of higher social level and superior health status, with a subsequent decrease in the risk of disease. Furthermore, they raise the question of if the lower cardiovascular incidence and mortality associated with moderate wine consumption in epidemiological studies are rather a surrogate of the influence of other healthy behaviors. It should be indicated that most of the previously discussed studies just compare wine with other alcoholic drinks without differentiating between white and red wine, which, regarding lifestyle and sociological determinants, could have a similar influence. Interestingly, in a cross-cultural study on wine preference and consumption behavior among Korean and Australian people, Yoo, Saliba, MacDonald, Prenzler, and Ryan (2013) found that whereas Koreans drank significantly more red than white wine, with no differences between genders, similar consumption of both types of wine was made by Australians, with significantly higher consumption of white wine by women. According to the authors, the preference for white wine would be typical for Western women, who consume much more white wine than red.

17.6 Alcohol risks

There is no doubt that excessive consumption of alcohol is detrimental to human health, and, in fact, alcohol drinking is one of the main causes of mortality, morbidity, and social dysfunction worldwide. Indeed, alcohol intake is directly related to the incidence of a variety of noncommunicable diseases, like liver cirrhosis, chronic pancreatitis, or different types of cancer, as well as a higher risk of infectious diseases, like HIV, tuberculosis, or pneumonia, and traffic accidents (Anderson & Baumberg, 2006). It is pertinent to highlight that the consumption of wine or any type of alcoholic drink is not free from risks, even at light or moderate alcohol intakes. The International Agency for Research on Cancer (2010) classifies alcohol as a Group 1 carcinogen, causally associated with the development of cancers of the oral cavity, pharynx, larynx, esophagus, and liver, and suggested to be positively associated with rectal and breast cancer, without substantial differences among the type of drink. In women, the regular consumption of 18 g alcohol/day has been estimated to increase breast cancer risk by 1.13-fold, and an intake of 50 g alcohol/day by 1.5-fold, compared to nondrinkers. Since there is no sufficient information about the level of alcohol intake that is completely safe during pregnancy, its consumption must be avoided by pregnant women, as well as during the period of breastfeeding (Larroque & Kaminski, 1998; Olson et al., 1997). Similarly, for some groups of individuals, there is no level of alcohol that could be considered safe. It is the case of young people, like adolescents, for which clear positive relationships between alcohol consumption and total mortality have been observed at all levels of consumption, as well as with higher risk for development of chronic alcohol use disorders in adulthood (Anderson & Baumberg, 2006; Emberson & Bennett, 2006). Furthermore, alcohol consumption favors the adoption of dangerous behaviors, leading to an increased risk of accidents, violent episodes, injuries, suicide, or unwanted pregnancy (Nordqvist, Holmqvist, Nilsen, Bendtsen, & Lindqvist, 2006). The consumption of alcohol must also be avoided by people that have to drive, operate machinery, or develop activities that require attention, skill, or coordination, as well as by individuals under treatments with drugs that may interact with alcohol or with conditions that entail an increased risk, such as mental illnesses, familiar history of alcohol abuse or dependence, liver and pancreatic pathologies, or reduced activity of the enzyme alcohol dehydrogenase (e.g., some ethnic groups like Aboriginal Australians, Native Americans, and some East Asians). Particular caution must be taken by elder people, as they become progressively more sensitive to the adverse effects of alcohol, due to a decrease in lean body mass and water proportion, which modify alcohol metabolization and distribution. Besides, they are usually subjected to more medication, and their nutritional and health conditions usually differ from those in earlier age (Mukamal et al., 2005).

The association between alcohol and obesity is another point to consider. Alcohol provides 7 kcal/g, and its intake is usually added to the daily energy intake and may further act as an appetite stimulant, increasing the risk of developing obesity and subsequently associated diseases such as type 2 diabetes, stroke, or coronary heart disease. Interestingly, regular consumption of small amounts of alcohol has been associated with smaller waist and hip circumferences and body mass indices by some authors, which might represent an additional cardiovascular-protective factor, given the relationship between abdominal fat and cardiovascular risks (Tolstrup et al., 2005). Another health issue on which there is no consensus regarding alcohol's effects is hypertension. Whereas some authors maintain that all levels of alcohol intake are harmful (Corrao, Bagnardi, Zambon, & La Vecchia, 2004; Marmot et al., 1994), others suggest that there is a threshold, below which alcohol would not only not increase hypertension risk, but it might even be beneficial. Thus, Sesso et al. (2008) reported that an intake of up to 1 drink/day (about 10 g ethanol/day) is related to a decreased risk of hypertension in wine drinkers, while the same increases if the intake exceeds 1 drink/day. Halanych et al. (2010), analyzing the results of the CARDIA study, a 20-year longitudinal survey on coronary artery disease risk factors in Americans, found distinct results for men and women, as light to moderate alcohol consumption was associated with lower risk of hypertension in European-American women but with an increased risk in men and African Americans. Different results were also observed among white and black American men by Fuchs et al. (1995), who found that light to moderate alcohol intake increased the risk of hypertension in black men but not in white men.

Given the addictive effect of alcohol, and hence the difficulty of keeping consumption moderate in some individuals, there is, by no means, a consensus on whether moderate wine drinking should be recommended or not (Perez-Vizcaino & Fraga, 2018). Even in healthy adults, it is hazardous to establish a limit where the putative benefits of wine consumption would be surpassed by its risks. Current estimates tend to situate this limit in the range of 20–30 g alcohol/day (i.e., 2 drinks/day) for men and half this amount for women, which is the level corresponding to a moderate alcohol intake that could be considered of "low risk" (Chiva-Blanch & Badimon, 2020).

17.7 Conclusions

The putative positive health effects associated with moderate wine consumption have been mostly related to red wine based on its greater content and diversity of polyphenols compared with white wine. Nevertheless, despite having lower concentrations of phenolic compounds and the practical absence of flavonoids, white wines still possess a variety of phenolics, particularly hydroxycinnamic acids and derivatives, as well as a range of bioactive products formed during winemaking as a result of the microbial metabolism, which might also play a role in the purported beneficial effects of wine. Actually, most of the observational studies do not differentiate between white and red wines in their relationship with health outcomes, raising questions about the possible influence of one or another. On the other hand, there are sociocultural patterns that would apply to both red and white wines and might contribute to indirectly explaining the superior beneficial effects of moderate wine consumption compared with other alcoholic drinks. Some studies have concluded that moderate wine drinkers normally have healthier dietary and lifestyle habits and, at least in some countries, tend to possess better health, educational, and/or social status than people that show a preference for other alcoholic beverages. Nonetheless, we shall never forget that alcohol, either in the form of wine or any other drink, is toxic and that its consumption may have undesirable consequences for health, especially when it is heavy or irresponsible. It is therefore necessary to be very cautious about spreading any positive message about wine intake, even in light or moderate amounts, as it could be misinterpreted or used for spurious purposes. As highlighted by Boban et al. (2016), it is very important that all scientific and medical information regarding the effects of wine on health is communicated in a competent, credible, and disinterested manner. The risks of alcohol must never be minimized and always be taken in the first place. Wine has to be regarded as a fruitive food, whose moderate and responsible consumption in the context of a suitable diet, and provided that there are no reasons that do not advise their intake, can be an additional element of a healthy lifestyle, but it must never be consumed by its supposed therapeutic properties and in view to gain health benefits.

References

Abu-Amsha, R., Croft, K. D., Puddey, I. B., Proudfoot, J. M., & Beilin, L. J. (1996). Phenolic content of various beverages determines the extent of inhibition of human serum and low-density lipoprotein oxidation in vitro: Identification and mechanism of action of some cinnamic acid derivatives from red wine. *Clinical Science, 91*(4), 449–458. Available from https://doi.org/10.1042/cs0910449.

Aguilera, Y., Rebollo-Hernanz, M., Herrera, T., Cayuelas, L. T., Rodríguez-Rodríguez, P., De Pablo, Á. L. L., & Martin-Cabrejas, M. A. (2016). Intake of bean sprouts influences melatonin and antioxidant capacity biomarker levels in rats. *Food and Function, 7*(3), 1438–1445. Available from https://doi.org/10.1039/c5fo01538c.

Alcácera, M. A., Marques-Lopes, I., Fajó-Pascual, M., Foncillas, J. P., Carmona-Torre, F., & Martínez-González, M. A. (2008). Alcoholic beverage preference and dietary pattern in Spanish university graduates: The SUN cohort study. *European Journal of Clinical Nutrition, 62*(10), 1178–1186. Available from https://doi.org/10.1038/sj.ejcn.1602833.

Alvarez-Casas, M., Pajaro, M., Lores, M., & Garcia-Jares, C. (2016). Polyphenolic composition and antioxidant activity of galician monovarietal wines from native and experimental non-native white grape varieties. *International Journal of Food Properties, 19*(10), 2307–2321. Available from https://doi.org/10.1080/10942912.2015.1126723.

Anderson, P., & Baumberg, B. (2006). *Alcohol in Europe*. A report for the European Commission. Institute of Alcohol Studies.

Arapitsas, P., Guella, G., & Mattivi, F. (2018). The impact of SO2 on wine flavanols and indoles in relation to wine style and age. *Scientific Reports, 8*, 858. Available from https://doi.org/10.1038/s41598-018-19185-5.

Barefoot, J. C., Grønbæk, M., Feaganes, J. R., McPherson, R. S., Williams, R. B., & Siegler, I. C. (2002). Alcoholic beverage preference, diet, and health habits in the UNC Alumni Heart Study. *American Journal of Clinical Nutrition, 76*(2), 466–472.

Berger, M., Gray, J. A., & Roth, B. L. (2009). The expanded biology of serotonin. *Annual Review of Medicine, 60*, 355–366. Available from https://doi.org/10.1146/annurev.med.60.042307.110802.

Bertelli, A. A. E., Migliori, M., Panichi, V., Longoni, B., Origlia, N., Ferretti, A., & Giovannini, L. (2002). *Oxidative stress and inflammatory reaction modulation by white wine. Annals of the New York Academy of Sciences.* Italy: New York Academy of Sciences. Available from https://doi.org/10.1111/j.1749-6632.2002.tb02929.x.

Bertelli, A., Migliori, M., Bertelli, A. A. E., Origlia, N., Filippi, C., Panichi, V., & Giovannini, L. (2002). Effect of some white wine phenols in preventing inflammatory cytokine release. *Drugs under Experimental and Clinical Research, 28*(1), 11–15.

Boban, M., Stockley, C., Teissedre, P. L., Restani, P., Fradera, U., Stein-Hammer, C., & Ruf, J. C. (2016). Drinking pattern of wine and effects on human health: Why should we drink moderately and with meals? *Food and Function, 7*(7), 2937–2942. Available from https://doi.org/10.1039/c6fo00218h.

Bonnefont-Rousselot, D., & Collin, F. (2010). Melatonin: Action as antioxidant and potential applications in human disease and aging. *Toxicology, 278*(1), 55–67. Available from https://doi.org/10.1016/j.tox.2010.04.008.

Bordiga, M., Lorenzo, C., Pardo, F., Salinas, M. R., Travaglia, F., Arlorio, M., & Garde-Cerdán, T. (2016). Factors influencing the formation of histaminol, hydroxytyrosol, tyrosol, and tryptophol in wine: Temperature, alcoholic degree, and amino acids concentration. *Food Chemistry, 197*, 1038–1045. Available from https://doi.org/10.1016/j.foodchem.2015.11.112.

Burke, V., Puddey, I. B., & Beilin, L. J. (1995). Mortality associated with wines, beers, and spirits. *BMJ, 311*(7013), 1166. Available from https://doi.org/10.1136/bmj.311.7013.1166a.

Caccetta, R. A. A., Croft, K. D., Beilin, L. J., & Puddey, I. B. (2000). Ingestion of red wine significantly increases plasma phenolic acid concentrations but does not acutely affect ex vivo lipoprotein oxidizability. *American Journal of Clinical Nutrition, 71*(1), 67–74.

Camara, A. A., Nguyen, T. D., Jossier, A., Endrizzi, A., Saurel, R., Simonin, H., & Husson, F. (2005). Improving total glutathione and trehalose contents in *Saccharomyces* cerevisiae cells to enhance their resistance to fluidized bed drying. *Process Biochemistry, 69*, 1051–1054. Available from https://doi.org/10.1016/j.procbio.2018.03.013.

Carlsson, S., Hammar, N., & Grill, V. (2005). Alcohol consumption and type 2 diabetes: *Meta*-analysis of epidemiological studies indicates a U-shaped relationship. *Diabetologia, 48*(6), 1051–1054. Available from https://doi.org/10.1007/s00125-005-1768-5.

Carmona-Torre, F. A., García-Arellano, A., Marques-Lopes, I., Basora, J., Corella, D., Gómez-Gracia, E., & Martinez-Gonzalez, M. A. (2008). Relationship of alcoholic beverage consumption to food habits in a mediterranean population. *American Journal of Health Promotion, 23*(1), 27–30. Available from https://doi.org/10.4278/ajhp.07050143.

Castro-Barquero, S., Lamuela-Raventós, R. M., Doménech, M., & Estruch, R. (2018). Relationship between mediterranean dietary polyphenol intake and obesity. *Nutrients, 10*(10). Available from https://doi.org/10.3390/nu10101523.

Chatenoud, L., Negri, E., La Vecchia, C., Volpato, O., & Franceschi, S. (2000). Wine drinking and diet in Italy. *European Journal of Clinical Nutrition, 54*, 177–179. Available from https://doi.org/10.1038/sj.ejcn.1600913.

Chikritzhs, T., Fillmore, K., & Stockwell, T. (2009). A healthy dose of scepticism: Four good reasons to think again about protective effects of alcohol on coronary heart disease. *Drug and Alcohol Review, 28*(4), 441–444. Available from https://doi.org/10.1111/j.1465-3362.2009.00052.x.

Chiva-Blanch, G., Arranz, S., Lamuela-Raventos, R. M., & Estruch, R. (2013). Effects of wine, alcohol and polyphenols on cardiovascular disease risk factors: Evidences from human studies. *Alcohol and Alcoholism, 48*(3), 270–277. Available from https://doi.org/10.1093/alcalc/agt007.

Chiva-blanch, G., & Badimon, L. (2020). Benefits and risks of moderate alcohol consumption on cardiovascular disease: Current findings and controversies. *Nutrients, 12*(1), 108. Available from https://doi.org/10.3390/nu12010108.

Clifford, M. N. (2000). Chlorogenic acids and other cinnamates – Nature, occurrence, dietary burden, absorption and metabolism. *Journal of the Science of Food and Agriculture, 80*(7), 1033–1043, https://doi.org/10.1002/(SICI)1097-0010(20000515)80:7 < 1033::AID-JSFA595 > 3.0. CO;2-T.

Corder, R., Douthwaite, J. A., Lees, D. M., Khan, N. Q., Viseu dos Santos, A. C., Wood, E. G., & Carrier, M. J. (2001). Endothelin-1 synthesis reduced by red wine. *Nature*, 863–864. Available from https://doi.org/10.1038/414863a.

Corrao, G., Bagnardi, V., Zambon, A., & La Vecchia, C. (2004). A *meta*-analysis of alcohol consumption and the risk of 15 diseases. *Preventive Medicine, 38*(5), 613–619. Available from https://doi.org/10.1016/j.ypmed.2003.11.027.

Corrao, G., Rubbiati, L., Bagnardi, V., Zambon, A., & Poikolainen, K. (2000). Alcohol and coronary heart disease: A *meta*-analysis. *Addiction, 95* (10), 1505–1523. Available from https://doi.org/10.1046/j.1360-0443.2000.951015056.x.

Covas, M. I., Gambert, P., Fitó, M., & de la Torre, R. (2010). Wine and oxidative stress: Up-to-date evidence of the effects of moderate wine consumption on oxidative damage in humans. *Atherosclerosis, 208*(2), 297–304. Available from https://doi.org/10.1016/j.atherosclerosis.2009.06.031.

Cui, J., Tosaki, A., Cordis, G. A., Bertelli, A. A. E., Bertelli, A., Maulik, N., & Das, D. K. (2002). Cardioprotective abilities of white wine. *Annals of the New York Academy of Sciences, 957*, 308–316. Available from https://doi.org/10.1111/j.1749-6632.2002.tb02931.x.

Das, D. K., Mukherjee, S., & Ray, D. (2010). Resveratrol and red wine, healthy heart and longevity. *Heart Failure Reviews, 15*(5), 467–477. Available from https://doi.org/10.1007/s10741-010-9163-9.

De La Torre, R., Covas, M. I., Pujadas, M. A., Fitó, M., & Farré, M. (2006). Is dopamine behind the health benefits of red wine? *European Journal of Nutrition, 45*(5), 307–310. Available from https://doi.org/10.1007/s00394-006-0596-9.

Di Castelnuovo, A., Costanzo, S., Bagnardi, V., Donati, M. B., Iacoviello, L., & De Gaetano, G. (2006). Alcohol dosing and total mortality in men and women: An updated *meta*-analysis of 34 prospective studies. *Archives of Internal Medicine, 166*(22), 2437–2445. Available from https://doi.org/10.1001/archinte.166.22.2437.

Ditano-Vázquez, P., Torres-Peña, J. D., Galeano-Valle, F., Pérez-Caballero, A. I., Demelo-Rodríguez, P., Lopez-Miranda, J., & Alvarez-Sala-Walther, L. A. (2007). The fluid aspect of the mediterranean diet in the prevention and management of cardiovascular disease and diabetes: The role of polyphenol content in moderate consumption of wine and olive oil. *Nutrients, 11*, 1758–1765.

Djoussé, L., Biggs, M. L., Mukamal, K. J., & Siscovick, D. S. (2007). Alcohol consumption and type 2 diabetes among older adults: The cardiovascular health study. *Obesity, 15*(7), 1758–1765. Available from https://doi.org/10.1038/oby.2007.209.

Droste, D. W., Iliescu, C., Vaillant, M., Gantenbein, M., Bremaeker, N. D., Lieunard, C., & Gilson, G. (2013). A daily glass of red wine associated with lifestyle changes independently improves blood lipids in patients with carotid arteriosclerosis: Results from a randomized controlled trial. *Nutrition Journal, 12*, 147. Available from https://doi.org/10.1186/1475-2891-12-147.

Du Toit, W. J., Marais, J., Pretorius, I. S., & Du Toit, M. (2006). Oxygen in must and wine: A review. *South African Journal of Enology and Viticulture, 27*(1), 76−94. Available from https://doi.org/10.21548/27-1-1610.

EFSA Panel on Dietetic Products, Nutrition, and Allergies (NDA). (2011). Scientific opinion on the substantiation of health claims related to polyphenols in olive oil and protection of LDL particles from oxidative damage. *EFSA Journal, 9*, 2033. https://doi.org/10.2903/j.efsa.2011.2033.

Emberson, J. R., & Bennett, D. A. (2006). Effect of alcohol on risk of coronary heart disease and stroke: Causality, bias, or a bit of both? *Vascular Health and Risk Management, 2*(3), 239−249. Available from https://doi.org/10.2147/vhrm.2006.2.3.239.

Estruch, R., Sacanella, E., Mota, F., Chiva-Blanch, G., Antúnez, E., Casals, E., & Urbano-Marquez, A. (2011). Moderate consumption of red wine, but not gin, decreases erythrocyte superoxide dismutase activity: A randomised cross-over trial. *Nutrition, Metabolism and Cardiovascular Diseases, 21*(1), 46−53. Available from https://doi.org/10.1016/j.numecd.2009.07.006.

Fernandes, I., Pérez-Gregorio, R., Soares, S., Mateus, N., De Freitas, V., Santos-Buelga, C., & Feliciano, A. S. (2017). Wine flavonoids in health and disease prevention. *Molecules, 22*(2), 292. Available from https://doi.org/10.3390/molecules22020292.

Fernández-Cruz, E., Cerezo, A. B., Cantos-Villar, E., Troncoso, A. M., & García-Parrilla, M. C. (2019). Time course of l-tryptophan metabolites when fermenting natural grape musts: Effect of inoculation treatments and cultivar on the occurrence of melatonin and related indolic compounds. *Australian Journal of Grape and Wine Research, 25*, 92−100. Available from https://doi.org/10.1111/ajgw.12369.

Ferreira, I. C. F. R., Martins, N., & Barros, L. (2017). *Phenolic compounds and its bioavailability: In vitro bioactive compounds or health promoters?* Advances in food and nutrition research (Vol. 82, pp. 1−44). Academic Press Inc. Available from https://doi.org/10.1016/bs.afnr.2016.12.004.

Fillmore, K. M., Kerr, W. C., Stockwell, T., Chikritzhs, T., & Bostrom, A. (2006). Moderate alcohol use and reduced mortality risk: Systematic error in prospective studies. *Addiction Research and Theory, 14*(2), 101−132. Available from https://doi.org/10.1080/16066350500497983.

Fillmore, K., Stockwell, T., Chikritzhs, T., Bostrom, A., & Kerr, W. (2007). Moderate alcohol use and reduced mortality risk: systematic error studies and new hypotheses. *Annals of Epidemiology, 17*, 16−23. Available from https://doi.org/10.1016/j.annepidem.2007.01.005.

Fuchs, C. S., Stampfer, M. J., Colditz, G. A., Giovannucci, E. L., Manson, J. E., Kawachi, I., & Rosner, B. (1995). Alcohol consumption and mortality among women. *New England Journal of Medicine, 332*(19), 1245−1250. Available from https://doi.org/10.1056/NEJM199505113321901.

Galgano, F., Caruso, M., & Favati, F. (2009). *Biogenic amines in wines: A review.* In Red wine and health (pp. 173−203). Italy: Nova Science Publishers, Inc Retrieved from. Available from https://www.novapublishers.com/catalog/product_info.php?products_id = 9092.

Gepner, Y., Golan, R., Harman-Boehm, I., Henkin, Y., Schwarzfuchs, D., Shelef, I., et al. (2015). Effects of initiating moderate alcohol intake on cardiometabolic risk in adults with type 2 diabetes: A 2-year randomized, controlled trial. *Annals of Internal Medicine, 163*(8), 569−579. Available from https://doi.org/10.7326/M14-1650.

González-Flores, D., Gamero, E., Garrido, M., Ramírez, R., Moreno, D., Delgado, J., & Paredes, S. D. (2012). Urinary 6-sulfatoxymelatonin and total antioxidant capacity increase after the intake of a grape juice cv. Tempranillo stabilized with HHP. *Food and Function, 3*(1), 34−39. Available from https://doi.org/10.1039/c1fo10146c.

Granados-Principal, S., Quiles, J. L., Ramirez-Tortosa, C. L., Sanchez-Rovira, P., & Ramirez-Tortosa, M. C. (2010). Hydroxytyrosol: From laboratory investigations to future clinical trials. *Nutrition Reviews, 68*(4), 191−206. Available from https://doi.org/10.1111/j.1753-4887.2010.00278.x.

Gris, E. F., Mattivi, F., Ferreira, E. A., Vrhovsek, U., Filho, D. W., Pedrosa, R. C., & Bordignon-Luiz, M. T. (2010). Stilbenes and tyrosol as target compounds in the assessment of antioxidant and hypolipidemic activity of Vitis vinifera red wines from southern Brazil. *Journal of Agriculture and Food Chemistry, 68*, 7954−7961. Available from https://doi.org/10.1021/jf2008056.

Grønbæk, M. (2003). Type of alcoholic beverage and cardiovascular disease − Does it matter? *Journal of Cardiovascular Risk, 10*(1), 5−10. Available from https://doi.org/10.1097/00043798-200302000-00002.

Grønbæk, M., Deis, A., Sorensen, T. I. A., Becker, U., Schnohr, P., & Jensen, G. (1995). Mortality associated with moderate intakes of wine, beer, or spirits. *BMJ, 310*, 1165−1169. Available from https://doi.org/10.1136/bmj.310.6988.1165.

Grønbæk, M., Jensen, M. K., Johansen, D., Sørensen, T. I. A., & Becker, U. (2004). Intake of beer, wine and spirits and risk of heavy drinking and alcoholic cirrhosis. *Biological Research, 37*(2), 195−200.

Guo, Y. Y., Yang, Y. P., Peng, Q., & Han, Y. (2015). Biogenic amines in wine: A review. *International Journal of Food Science and Technology, 50*(7), 1523−1532. Available from https://doi.org/10.1111/ijfs.12833.

Halanych, J. H., Safford, M. M., Kertesz, S. G., Pletcher, M. J., Kim, Y. I., Person, S. D., & Kiefe, C. I. (2010). Alcohol consumption in young adults and incident hypertension: 20-year follow-up from the coronary artery risk development in young adults study. *American Journal of Epidemiology, 171*(5), 532−539. Available from https://doi.org/10.1093/aje/kwp417.

Hansel, B., Thomas, F., Pannier, B., Bean, K., Kontush, A., Chapman, M. J., & Bruckert, E. (2010). Relationship between alcohol intake, health and social status and cardiovascular risk factors in the urban Paris-Ile-De-France Cohort: Is the cardioprotective action of alcohol a myth. *European Journal of Clinical Nutrition, 64*(6), 561−568. Available from https://doi.org/10.1038/ejcn.2010.61.

Haseeb, S., Alexander, B., & Baranchuk, A. (2017). Wine and Cardiovascular Health. *Circulation, 136*(15), 1434−1448. Available from https://doi.org/10.1161/CIRCULATIONAHA.117.030387.

Hernanz, D., Recamales, A. F., González-Miret, M. L., Gómez-Míguez, M. J., Vicario, I. M., & Heredia, F. J. (2007). Phenolic composition of white wines with a prefermentative maceration at experimental and industrial scale. *Journal of Food Engineering, 80*(1), 327−335. Available from https://doi.org/10.1016/j.jfoodeng.2006.06.006.

Homann, N., Jousimies-Somer, H., Jokelainen, K., Heine, R., & Salaspuro, M. (1997). High acetaldehyde levels in saliva after ethanol consumption: Methodological aspects and pathogenetic implications. *Carcinogenesis, 18*(9), 1739−1743. Available from https://doi.org/10.1093/carcin/18.9.1739.

Huang, J., Wang, X., & Zhang, Y. (2017). Specific types of alcoholic beverage consumption and risk of type 2 diabetes: A systematic review and *meta*-analysis. *Journal of Diabetes Investigation, 8*(1), 56−68. Available from https://doi.org/10.1111/jdi.12537.

International Agency for Research on Cancer. (2010). IARC monographs on the evaluation of carcinogenic risks to humans. *Volume 96 − Alcohol consumption and ethyl carbamate.* Lyon: International Agency for Research on Cancer.

Johansen, D., Friis, K., Skovenborg, E., & Grønbæk, M. (2006). Food buying habits of people who buy wine or beer: Cross sectional study. *British Medical Journal*, *332*(7540), 519−521. Available from https://doi.org/10.1136/bmj.38694.568981.80.

Jones, A. W., Jönsson, K. A., & Kechagias, S. (1997). Effect of high-fat, high-protein, and high-carbohydrate meals on the pharmacokinetics of a small dose of ethanol. *British Journal of Clinical Pharmacology*, *44*(6), 521−526. Available from https://doi.org/10.1046/j.1365-2125.1997.t01-1-00620.x.

Kikura, M., Levy, J. H., Safon, R. A., Lee, M. K., & Szlam, F. (2004). The influence of red wine or white wine intake on platelet function and viscoelastic property of blood in volunteers. *Platelets*, *15*(1), 37−41. Available from https://doi.org/10.1080/0953710032000158772.

Klatsky, A. L., Friedman, G. D., Armstrong, M. A., & Kipp, H. (2003). Wine, liquor, beer, and mortality. *American Journal of Epidemiology*, *158*(6), 585−595. Available from https://doi.org/10.1093/aje/kwg184.

Kocadağli, T., Yilmaz, C., & Gökmen, V. (2014). Determination of melatonin and its isomer in foods by liquid chromatography tandem mass spectrometry. *Food Chemistry*, *153*, 151−156. Available from https://doi.org/10.1016/j.foodchem.2013.12.036.

Koppes, L. L. J., Dekker, J. M., Hendriks, H. F. J., Bouter, L. M., & Heine, R. J. (2005). Moderate alcohol consumption lowers the risk of type 2 diabetes: A *meta*-analysis of prospective observational studies. *Diabetes Care*, *28*(3), 719−725. Available from https://doi.org/10.2337/diacare.28.3.719.

Kritzinger, E. C., Bauer, F. F., & Du Toit, W. J. (2013). Role of glutathione in winemaking: A review. *Journal of Agricultural and Food Chemistry*, *61*(2), 269−277. Available from https://doi.org/10.1021/jf303665z.

Landete, J. M., Ferrer, S., & Pardo, I. (2007). Biogenic amine production by lactic acid bacteria, acetic bacteria and yeast isolated from wine. *Food Control*, *18*(12), 1569−1574. Available from https://doi.org/10.1016/j.foodcont.2006.12.008.

Larroque, B., & Kaminski, M. (1998). Prenatal alcohol exposure and development at preschool age: Main results of a french study. *Alcoholism: Clinical and Experimental Research*, *22*(2), 295−303. Available from https://doi.org/10.1111/j.1530-0277.1998.tb03652.x.

Larsson, S. C., Wallin, A., & Wolk, A. (2018). Alcohol consumption and risk of heart failure: *Meta*-analysis of 13 prospective studies. *Clinical Nutrition*, *37*(4), 1247−1251. Available from https://doi.org/10.1016/j.clnu.2017.05.007.

Law, M., & Wald, N. (1999). Why heart disease mortality is low in France: The time lag explanation. *British Medical Journal*, *318*(7196), 1471−1480.

Letenneur, L. (2004). Risk of dementia and alcohol and wine consumption: A review of recent results. *Biological Research*, *37*(2), 189−193. Available from https://doi.org/10.4067/S0716-97602004000200003.

Liberale, L., Bonaventura, A., Montecucco, F., Dallegri, F., & Carbone, F. (2019). Impact of red wine consumption on cardiovascular health. *Current Medicinal Chemistry*, *26*(19), 3542−3566. Available from https://doi.org/10.2174/0929867324666170518100606.

Makris, D. P., Psarra, E., Kallithraka, S., & Kefalas, P. (2003). The effect of polyphenolic composition as related to antioxidant capacity in white wines. *Food Research International*, *36*(8), 805−814. Available from https://doi.org/10.1016/S0963-9969(03)00075-9.

Manach, C., Scalbert, A., Morand, C., Rémésy, C., & Jiménez, L. (2004). Polyphenols: Food sources and bioavailability. *American Journal of Clinical Nutrition*, *79*(5), 727−747. Available from https://doi.org/10.1093/ajcn/79.5.727.

Mannistö, S., Uusitalo, K., Roos, E., Fogelholm, M., & Pietinen, P. (1997). Alcohol beverage drinking, diet and body mass index in a cross-sectional survey. *European Journal of Clinical Nutrition*, *51*(5), 326−332. Available from https://doi.org/10.1038/sj.ejcn.1600406.

Mansvelt, E. P. G., Van Velden, D. P., Fourie, E., Rossouw, M., Van Rensburg, S. J., & Marius Smuts, C. (2002). The in vivo antithrombotic effect of wine consumption on human blood platelets and hemostatic factors. *Annals of the New York Academy of Sciences*, *957*, 329−332. Available from https://doi.org/10.1111/j.1749-6632.2002.tb02935.x.

Marchand, S., & de Revel, G. (2010). A HPLC fluorescence-based method for glutathione derivatives quantification in must and wine. *Analytica Chimica Acta*, *660*(1−2), 158−163. Available from https://doi.org/10.1016/j.ac.2009.09.042.

Marmot, M. G., Elliott, P., Shipley, M. J., Dyer, A. R., Ueshima, H. U., Beevers, D. G., & Stamler, J. (1994). Alcohol and blood pressure: The INTERSALT study. *BMJ*, *308*, 1263−1267. Available from https://doi.org/10.1136/bmj.308.6939.1263.

Mas, A., Guillamon, J. M., Torija, M. J., Beltran, G., Cerezo, A. B., Troncoso, A. M., & Garcia-Parrilla, M. C. (2014). Bioactive compounds derived from the yeast metabolism of aromatic amino acids during alcoholic fermentation. *BioMed Research International*, *2014*, 898045. Available from https://doi.org/10.1155/2014/898045.

McCann, S. E., Sempos, C., Freudenheim, J. L., Muti, P., Russell, M., Nochajski, T. H., & Trevisan, M. (2003). Alcoholic beverage preference and characteristics of drinkers and nondrinkers in western New York (United States). *Nutrition, Metabolism and Cardiovascular Diseases*, *13*(1), 2−11. Available from https://doi.org/10.1016/S0939-4753(03)80162-X.

Meng, X., Li, Y., Li, S., Zhou, Y., Gan, R. Y., Xu, D. P., & Li, H. B. (2017). Dietary sources and bioactivities of melatonin. *Nutrients*, *9*(4), 367. Available from https://doi.org/10.3390/nu9040367.

Mercolini, L., Saracino, M. A., Bugamelli, F., Ferranti, A., Malaguti, M., Hrelia, S., & Raggi, M. A. (2008). HPLC-F analysis of melatonin and resveratrol isomers in wine using an SPE procedure. *Journal of Separation Science*, *31*(6−7), 1007−1014. Available from https://doi.org/10.1002/jssc.200700458.

Mortensen, E. L., Jensen, H. H., Sanders, S. A., & Reinisch, J. M. (2001). Better psychological functioning and higher social status may largely explain the apparent health benefits of wine: A study of wine and beer drinking in young Danish adults. *Archives of Internal Medicine*, *161*(15), 1844−1848. Available from https://doi.org/10.1001/archinte.161.15.1844.

Mukamal, K. J., Chung, H., Jenny, N. S., Kuller, L. H., Longstreth, W. T., Mittleman, M. A., & Siscovick, D. S. (2005). Alcohol use and risk of ischemic stroke among older adults: The cardiovascular health study. *Stroke*, *36*(9), 1830−1834. Available from https://doi.org/10.1161/01.STR.0000177587.76846.89.

Mumenthaler, M. S., Taylor, J. L., O'Hara, R., & Yesavage, J. A. (1999). Gender differences in moderate drinking effects. *Alcohol Research and Health*, *23*(1), 55−64.

Naimi, T. S., Brown, D. W., Brewer, R. D., Giles, W. H., Mensah, G., Serdula, M. K., & Stroup, D. F. (2005). Cardiovascular risk factors and confounders among nondrinking and moderate-drinking United States adults. *American Journal of Preventive Medicine*, *28*(4), 369−373. Available from https://doi.org/10.1016/j.amepre.2005.01.011.

Naimi, T. S., Stockwell, T., Zhao, J., Xuan, Z., Dangardt, F., Saitz, R., & Chikritzhs, T. (2017). Selection biases in observational studies affect associations between 'moderate' alcohol consumption and mortality. *Addiction*, *112*(2), 207−214. Available from https://doi.org/10.1111/add.13451.

Nardini, M., Forte, M., Vrhovsek, U., Mattivi, F., Viola, R., & Scaccini, C. (2009). White wine phenolics are absorbed and extensively metabolized in humans. *Journal of Agricultural and Food Chemistry*, *57*(7), 2711−2718. Available from https://doi.org/10.1021/jf8034463.

Nordqvist, C., Holmqvist, M., Nilsen, P., Bendtsen, P., & Lindqvist, K. (2006). Usual drinking patterns and non-fatal injury among patients seeking emergency care. *Public Health, 120*(11), 1064–1073. Available from https://doi.org/10.1016/j.puhe.2006.06.007.

Olson, H. C., Streissguth, A. P., Sampson, P. D., Barr, H. M., Bookstein, F. L., & Thiede, K. (1997). Association of prenatal alcohol exposure with behavioral and learning problems in early adolescence. *Journal of the American Academy of Child and Adolescent Psychiatry, 36*(9), 1187–1194. Available from https://doi.org/10.1097/00004583-199709000-00010.

Paschall, M., & Lipton, R. I. (2005). Wine preference and related health determinants in a United States national sample of young adults. *Drug and Alcohol Dependence, 78*(3), 339–344. Available from https://doi.org/10.1016/j.drugalcdep.2004.12.004.

Pérez-Mañá, C., Farré, M., Rodríguez-Morató, J., Papaseit, E., Pujadas, M., Fitó, M., & de la Torre, R. (2015). Moderate consumption of wine, through both its phenolic compounds and alcohol content, promotes hydroxytyrosol endogenous generation in humans. A randomized controlled trial. *Molecular Nutrition and Food Research, 59*(6), 1213–1216. Available from https://doi.org/10.1002/mnfr.201400842.

Perez-Vizcaino, F., & Fraga, C. G. (2018). Research trends in flavonoids and health. *Archives of Biochemistry and Biophysics, 646*, 107–112. Available from https://doi.org/10.1016/j.abb.2018.03.022.

Peters, R., Peters, J., Warner, J., Beckett, N., & Bulpitt, C. (2008). Alcohol, dementia and cognitive decline in the elderly: A systematic review. *Age and Ageing, 37*(5), 505–512. Available from https://doi.org/10.1093/ageing/afn095.

Pignatelli, P., Ghiselli, A., Buchetti, B., Carnevale, R., Natella, F., Germano, G., & Violi, F. (2004). Polyphenols synergistically inhibit oxidative stress in subjects given red and white wine. *Atherosclerosis, 188*, 449–456. Available from https://doi.org/10.1016/j.atherosclerosis.2005.10.025.

Pinder, R. M., & Sandler, M. (2004). Alcohol, wine and mental health: Focus on dementia and stroke. *Journal of Psychopharmacology, 18*(4), 449–456. Available from https://doi.org/10.1177/0269881104047272.

Piñeiro, Z., Cantos-Villar, E., Palma, M., & Puertas, B. (2011). Direct liquid chromatography method for the simultaneous quantification of hydroxytyrosol and tyrosol in red wines. *Journal of Agricultural and Food Chemistry, 59*(21), 11683–11689. Available from https://doi.org/10.1021/jf202254t.

Ragusa, A., Centonze, C., Grasso, M. E., Latronico, M. F., Mastrangelo, P. F., Sparascio, F., & Maffia, M. (2017). A comparative study of phenols in Apulian Italian Wines. *Foods, 6*, 24.

Ramírez-Tortosa, M. C., Granados, S., & Quiles, J. L. (2006). *Chemical composition types and characteristics of olive oil. In Olive oil and health* (pp. 45–62). Spain: CABI Publishing Retrieved from. Available from http://bookshop.cabi.org/.

Rasmussen, S. E., Frederiksen, H., Krogholm, K. S., & Poulsen, L. (2005). Dietary proanthocyanidins: Occurrence, dietary intake, bioavailability, and protection against cardiovascular disease. *Molecular Nutrition and Food Research, 49*(2), 159–174. Available from https://doi.org/10.1002/mnfr.200400082.

Rastija, V., Srečnik, G., & Marica-Medić-Šarić. (2009). Polyphenolic composition of Croatian wines with different geographical origins. *Food Chemistry, 115*(1), 54–60. Available from https://doi.org/10.1016/j.foodchem.2008.11.071.

Recamales, A. F., Sayago, A., González-Miret, M. L., & Hernanz, D. (2006). The effect of time and storage conditions on the phenolic composition and colour of white wine. *Food Research International, 39*(2), 220–229. Available from https://doi.org/10.1016/j.foodres.2005.07.009.

Rehm, J., & Roerecke, M. (2017). Cardiovascular effects of alcohol consumption. *Trends in Cardiovascular Medicine, 27*(8), 534–538. Available from https://doi.org/10.1016/j.tcm.2017.06.002.

Rehm, J., Sempos, C. T., & Trevisan, M. (2003). Average volume of alcohol consumption, patterns of drinking and risk of coronary heart disease − A review. *Journal of Cardiovascular Risk, 10*, 15–20. Available from https://doi.org/10.1177/174182670301000104.

Reiter, Manchester, & Tan, D.-X. (2003). Melatonin in walnuts: Influence on levels of melatonin and total antioxidant capacity of blood. *Nutrition, 21*, 579–588. Available from https://doi.org/10.1016/j.nut.2005.02.005.

Reynolds, K., Lewis, L. B., Nolen, J. D. L., Kinney, G. L., Sathya, B., & He, J. (2003). Alcohol consumption and risk of stroke: A Meta-analysis. *Journal of the American Medical Association, 289*(5), 579–588. Available from https://doi.org/10.1001/jama.289.5.579.

Rocchetti, G., Gatti, M., Bavaresco, L., & Lucini, L. (2018). Untargeted metabolomics to investigate the phenolic composition of Chardonnay wines from different origins. *Journal of Food Composition and Analysis, 71*, 87–93. Available from https://doi.org/10.1016/j.jfc.2018.05.010.

Rodríguez-Bernaldo, Q. A., López-Hernández, J., & Lage-Yusty, M. A. (2009). HPLC-analysis of polyphenolic compounds in Spanish white wines and determination of their antioxidant activity by radical scavenging assay. *Food Research International, 42*, 1018–1022. Available from https://doi.org/10.1016/j.foodres.2009.04.009.

Rodríguez-Naranjo, M. I., Torija, M. J., Mas, A., Cantos-Villar, E., & García-Parrilla, M. C. (2012). Production of melatonin by *Saccharomyces* strains undergrowth and fermentation conditions. *J. Pineal Res, 53*, 219–224. Available from https://doi.org/10.1111/j.1600-079X.2012.00990.x.

Roerecke, M., & Rehm, J. (2014). Alcohol consumption, drinking patterns, and ischemic heart disease: A narrative review of *meta*-analyses and a systematic review and *meta*-analysis of the impact of heavy drinking occasions on risk for moderate drinkers. *BMC Medicine, 12*(1), 182. Available from https://doi.org/10.1186/s12916-014-0182-6.

Román, G. C., Jackson, R. E., Gadhia, R., Román, A. N., & Reis, J. (1996). Mediterranean diet: The role of long-chain v-3 fatty acids in fish; polyphenols in fruits, vegetables, cereals, coffee, tea, cacao and wine; probiotics and vitamins in prevention of stroke, age-related cognitive decline, and Alzheimer disease. *Revue Neurologique, 175*, 2124–2128.

Rosell, M., de Faire, U., & Hellénius, M. L. (2003). Low prevalence of the metabolic syndrome in wine drinkers—Is it the alcohol beverage or the lifestyle? *European Journal of Clinical Nutrition, 57*, 227–234. Available from https://doi.org/10.1038/sj.ejcn.1601548.

Ruiz-Matute, A. I., Sanz, M. L., Moreno-Arribas, M. V., & Martínez-Castro, I. (2009). Identification of free disaccharides and other glycosides in wine. *Journal of Chromatography A, 1216*(43), 7296–7300. Available from https://doi.org/10.1016/j.chroma.2009.08.086.

Samuel, S. M., Thirunavukkarasu, M., Penumathsa, S. V., Paul, D., & Maulik, N. (2008). Akt/FOXO3a/SIRT1-mediated cardioprotection by n-tyrosol against ischemic stress in rat in vivo model of myocardial infarction: Switching gears toward survival and longevity. *Journal of Agricultural and Food Chemistry, 56*(20), 9692–9698. Available from https://doi.org/10.1021/jf802050h.

Santos-Buelga, C., & González-Manzano, S. (2011). Wine and health relationships. A question of moderation. *Ciencia e Tecnica Vitivinicola, 26* (1), 33–44, Retrieved from. Available from http://www.scielo.gpeari.mctes.pt/pdf/ctv/v26n1/v26n1a04.pdf.

Santos-Buelga, C., González-Paramás, A. M., Oludemi, T., Ayuda-Durán, B., & González-Manzano, S. (2019). *Plant phenolics as functional food ingredients, Advances in food and nutrition research* (Vol. 90, pp. 183–257). Academic Press Inc. Available from https://doi.org/10.1016/bs.afnr.2019.02.012.

Santos-Buelga, C., & Scalbert, A. (2000). Proanthocyanidins and tannin-like compounds - Nature, occurrence, dietary intake and effects on nutrition and health. *Journal of the Science of Food and Agriculture, 80*(7), 1094–1117. Available from https://doi.org/10.1002/(SICI)1097-0010 (20000515)80:7 < 1094::AID-JSFA569 > 3.0.CO;2-1.

Schutte, R., Papageorgiou, M., Najlah, M., Huisman, H. W., Ricci, C., Zhang, J., ... Schutte, A. E. (2020). Drink types unmask the health risks associated with alcohol intake-prospective evidence from the general population. *Clinical Nutrition, 39*(10), 3168–3174. Available from https://doi.org/10.1016/j.clnu.2020.02.009.

Sesso, H. D., Cook, N. R., Buring, J. E., Manson, J. A. E., & Gaziano, J. M. (2008). Alcohol consumption and the risk of hypertension in women and men. *Hypertension, 51*(4), 1080–1087. Available from https://doi.org/10.1161/HYPERTENSIONAHA.107.104968.

Singleton, V. L. (1987). Oxygen with phenols and related reactions in musts, wines and model systems: observations and practical implications. *American Journal of Enology and Viticulture, 38*, 69–76.

Sluik, D., Bezemer, R., Sierksma, A., & Feskens, E. (2016). Alcoholic beverage preference and dietary habits: A systematic literature review. *Critical Reviews in Food Science and Nutrition, 56*(14), 2370–2382. Available from https://doi.org/10.1080/10408398.2013.841118.

Sluik, D., Van Lee, L., Geelen, A., & Feskens, E. J. (2014). Alcoholic beverage preference and diet in a representative Dutch population: The Dutch national food consumption survey 2007–2010. *European Journal of Clinical Nutrition, 68*(3), 287–294. Available from https://doi.org/10.1038/ejcn.2013.279.

Smyth, A., Teo, K. K., O'Donnell, M., Zhang, X., Rana, P., Leong, D. P., et al. (2015). Alcohol consumption and cardiovascular disease, cancer, injury, admission to hospital, and mortality: A prospective cohort study. *The Lancet, 386*(10007), 1945–1954. Available from https://doi.org/10.1016/S0140-6736(15)00235-4.

Snopek, L., Mlcek, J., Sochorova, L., Baron, M., Hlavacova, I., Jurikova, T., & Sochor, J. (2018). Contribution of red wine consumption to human health protection. *Molecules, 23*(7), 1–16. Available from https://doi.org/10.3390/molecules23071684.

Stege, P. W., Sombra, L. L., Messina, G., Martinez, L. D., & Silva, M. F. (2010). Determination of melatonin in wine and plant extracts by capillary electrochromatography with immobilized carboxylic multi-walled carbon nanotubes as stationary phase. *Electrophoresis, 31*(13), 2242–2248. Available from https://doi.org/10.1002/elps.200900782.

Thirunavukkarasu, M., Penumathsa, S. V., Samuel, S. M., Akita, Y., Zhan, L., Bertelli, A. A., & Maulik, N. (2008). White Wine-induced cardio protection against ischemia-reperfusion injury is mediated by life-extending Akt/FOXO3a/NFkB survival pathway. *Journal of Agriculture and Food Chemistry, 56*, 6733–6739. Available from https://doi.org/10.1021/jf801473v.

Tjønneland, A., Grønbæk, M., Stripp, C., & Overvad, K. (1999). Wine intake and diet in a random sample of 48763 danish men and women. *American Journal of Clinical Nutrition, 69*(1), 49–54. Available from https://doi.org/10.1093/ajcn/69.1.49.

Tolstrup, J. S., Heitmann, B. L., Tjønneland, A. M., Overvad, O. K., Sørensen, T. I. A., & Granbæk, M. N. (2005). The relation between drinking pattern and body mass index and waist and hip circumference. *International Journal of Obesity, 29*(5), 490–497. Available from https://doi.org/10.1038/sj.ijo.0802874.

Toth, A., Sandor, B., Papp, J., Rabai, M., Botor, D., Horvath, Z., & Czopf, L. (2014). Moderate red wine consumption improves hemorheological parameters in healthy volunteers. *Clinical Hemorheology and Microcirculation, 56*(1), 13–23. Available from https://doi.org/10.3233/CH-2012-1640.

Trichopoulou, A., & Critselis, E. (2004). Mediterranean diet and longevity. *European Journal of Cancer Prevention, 13*(5), 453–456. Available from https://doi.org/10.1097/00008469-200410000-00014.

Truelsen, T., Grønbæk, M., Schnohr, P., & Boysen, G. (1998). Intake of beer, wine, and spirits and risk of stroke: The Copenhagen City Heart Study. *Stroke, 29*(12), 2467–2472. Available from https://doi.org/10.1161/01.STR.29.12.2467.

Vauzour, D., Vafeiadou, K., Corona, G., Pollard, S. E., Tzounis, X., & Spencer, J. P. E. (2007). Champagne wine polyphenols protect primary cortical neurons against peroxynitrite-induced injury. *Journal of Agricultural and Food Chemistry, 55*(8), 2854–2860. Available from https://doi.org/10.1021/jf063304z.

Vilela, A. (2019). The importance of yeasts on fermentation quality and human health-promoting compounds. *Fermentation-Basel, 5*, 46. Available from https://doi.org/10.3390/fermentation5020046.

Vitalini, S., Gardana, C., Zanzotto, A., Fico, G., Faoro, F., Simonetti, P., & Iriti, M. (2011). From vineyard to glass: Agrochemicals enhance the melatonin and total polyphenol contents and antiradical activity of red wines. *Journal of Pineal Research, 51*(3), 278–285. Available from https://doi.org/10.1111/j.1600-079X.2011.00887.x.

Waterhouse, A. L. (2002). Wine phenolics. *Annals of the New York Academy of Sciences, 957*, 21–36. Available from https://doi.org/10.1111/j.1749-6632.2002.tb02903.x.

Whelan, A. P., Sutherland, W. H. F., McCormick, M. P., Yeoman, D. J., De Jong, S. A., & Williams, M. J. A. (2004). Effects of white and red wine on endothelial function in subjects with coronary artery disease. *Internal Medicine Journal, 34*(5), 224–228. Available from https://doi.org/10.1111/j.1444-0903.2004.00507.x.

Williams, M. J. A., Sutherland, W. H. F., Whelan, A. P., McCormick, M. P., & De Jong, S. A. (2004). Acute effect of drinking red and white wines on circulating levels of inflammation-sensitive molecules in men with coronary artery disease. *Metabolism: Clinical and Experimental, 53*(3), 318–323. Available from https://doi.org/10.1016/j.metabol.2003.10.012.

Yoo, Y. J., Saliba, A. J., MacDonald, J. B., Prenzler, P. D., & Ryan, D. (2013). A cross-cultural study of wine consumers with respect to health benefits of wine. *Food Quality and Preference, 28*(2), 531–538. Available from https://doi.org/10.1016/j.foodqual.2013.01.001.

Zhao, J., Stockwell, T., Roemer, A., Naimi, T., & Chikritzhs, T. (2017). Alcohol consumption and mortality from coronary heart disease: An updated *meta*-analysis of cohort studies. *Journal of Studies on Alcohol and Drugs, 78*(3), 375–386. Available from https://doi.org/10.15288/jsad.2017.78.375.

18

Enzyme applications in white wine

Antonio Álamo Aroca, Isabelle van Rolleghem and Remi Schneider

Oenobrands SAS, Montferrier sur Lez, France

18.1 Introduction to wine enzymes

18.1.1 Definition, structure, and action mechanism

Enzymes are mainly proteins that catalyze biochemical reactions in all living organisms (Cammack et al., 2006). Ribozymes are the only nonprotein enzymes since the catalytic property is due to RNA-specific structures that allow transcription of the genetic information to proteins through mRNA (Walter & Engelke, 2002).

As a catalyst, enzymes lower the activation energy needed for a specific reaction, and then make the reaction possible and faster under particular conditions (Fig. 18.1). The substrate is thus converted into products.

Enzymes are generally globular proteins, with a primary structure based on a chain of ±100 amino acids. The folding of this primary structure into secondary and tertiary structures (Fig. 18.2) defines the specificity of each enzyme for its substrate throughout the presence of active sites where the biochemical reaction takes place. In addition, to be active, many enzymes need cofactors such as metallic ions. For some of them, several units must be assembled in a specific quaternary structure (Tu, Meng, Luo, Turunen, & Zhang, 2015; Whitehurst & van Oort, 2010).

Since they are catalytic compounds, enzymes are not consumed during the reaction and remain active until all the substrate has been consumed or until they are denatured.

In all living organisms, enzyme synthesis and activity are highly regulated, and the composition of the medium plays a crucial role in those regulation mechanisms with, for instance, induction, feedback inhibition, or proteolytic degradation mechanisms. Many other factors can influence enzyme activity and the resultant performance. Temperature, pH, and the presence of irreversible or reversible inhibitors (competitive, noncompetitive, or uncompetitive) play a key role in the technological efficiency of enzymes (Segel, 1975). Typical temperature and pH effects on enzyme activity are illustrated in Fig. 18.3. Activity curves exhibit optimums for both pH and temperature that must be compatible with the application.

During winemaking the most important factors to take into consideration are temperature and pH. At very low temperatures, enzyme activity is decreased, and more enzymes might be needed to achieve the process within a reasonable time (e.g., skin contact maceration of aromatic white grapes at or close to 10°C). High temperatures, especially during grape thermo-treatments, can irreversibly denature enzymes and require the use of thermo-resistant enzyme formulations to avoid rapid denaturation, which normally occurs at 60°C. Low pH can be responsible for a loss in efficiency of added pectinases, as it disturbs the 3D enzyme structure and thus the active site shapes.

In addition, some enological products can also be inhibitory. Examples are products that result in high free SO_2 levels or fining products such as tannins or bentonite in particular, which can absorb enzymes, since they are proteins (Ribéreau-Gayon, Glories, Maujean, & Dubourdieu, 2000). Glucose concentration is also a key point for the use of glycosidases since they are inhibited at concentrations above 50 g/L (Gunata, Dugelay, Sapis, Baumes, & Bayonove, 1993).

© 2022 Elsevier Inc. All rights reserved.

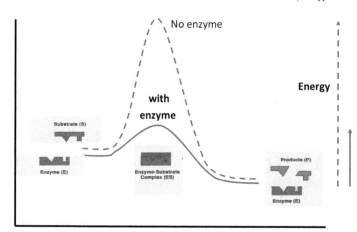

FIGURE 18.1　Schematic representation of a reaction catalyzed by an enzyme.

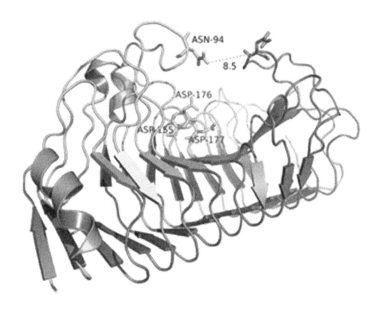

FIGURE 18.2　Example of the 3D structure of a polygalacturonase. Source: *Adapted from Tu, T., Meng, K., Luo, H., Turunen, O., Zhang, L. et al. (2015). New insights into the role of T3 loop in determining catalytic efficiency of GH28 endopolygalacturonases.*

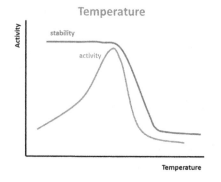

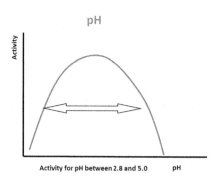

FIGURE 18.3　Typical temperature and pH impact on enzyme activity.

18.2 Enzyme origin, production, nomenclature, and regulation

18.2.1 Enzyme origin

The action of enzymes in food has been an empirical reality for centuries. Just one example of this is the α-amylases naturally produced as a result of barley germination during malting in the brewing process. Nowadays, most of the enzymes on the market are produced by microorganisms (Whitehurst & van Oort, 2010).

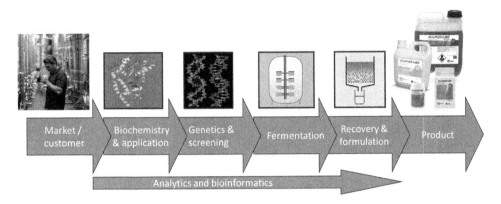

FIGURE 18.4 Scientific approach of wine enzyme development: from winemaker needs to commercial product.

This is the case for not only the wine industry, but also other food and nonfood industries such as pharma and health, detergents, fuel ethanol, textile, feed, and digestive aids.

Production processes differ depending on the enzyme desired, but are strictly controlled and regulated like other industrial and particularly food industry-related processes. AMFEP (the Association of Manufacturers and Formulators of Enzyme Products) established *Good Manufacturing Practice* in microbial food enzyme production, in order to set the fundamentals to monitor and check the identity of the production strain, guide the pure culture and operational parameters during fermentation, and assure hygienic control during enzyme recovery (AMFEP, 2020). Nowadays, the main enzyme-producing companies are in continuous search for excellence in applying scientific approaches to new enzyme development (Fig. 18.4). The process starts with the precise interpretation and evaluation of the needs of the industry (e.g., wine), followed by the application of in-house know-how on biochemistry and enzyme application, the screening for the specific strain from safe and approved microorganisms, and finally enzyme formulation and conditioning.

There are numerous origins of industrial enzymes, such as microorganisms (yeast, bacteria, and fungi), plants, fruits, and animals (Table 18.1). The majority of industrial enzymes are currently produced by fermentation with selected microorganisms.

18.2.2 Production systems

The two main systems of production are Solid State Fermentation (SSF) and Submerged Liquid Fermentation (SLF) as illustrated in Fig. 18.5.

SSF is a relatively simple technology where fungi grow on solid substrates with variable water content. Enzymes are produced by SSF using cultivation on trays with a thickness of 15−30 cm. Sterilization and inoculation of the selected strain are followed by growth in humid conditions and spraying of nutrients until the fermentation media is pressed and washed with water for enzyme recovery. SFF can be applied to all filamentous fungi such as *A. niger, A. oryzae, A. aculeatus, A. japonicus, Talaromyces emersonii,* and *Trichoderma harzianum.*

However, most of the enzymes used in the wine industry are produced using the SLF system. This process is similar to wine fermentation in stainless steel tanks, but for each fermentation a specific media under particular conditions is used to promote the optimum metabolism of the selected strain. This guarantees the desired enzyme and targeted yield. Production of enzymes with this fermentation system has some key advantages, such as ensuring constant product quality and high production yield. It also helps to obtain enzymes with the specificity required for the final application.

Wine enzymes produced by submerged fermentation come mainly from microorganisms belonging to the *Aspergillus* and *Trichoderma* genera (Table 18.2).

The production process can be divided into five different stages from strain development to packaging.

- Stages 1 and 2: the selected strain is grown to reach the fermentation stage where temperature, pH, and nutrition are carefully controlled to produce the desired enzymes.
- Stage 3: the enzyme is separated from the biomass and subjected to various filtrations up to the concentration step prior to formulation.

TABLE 18.1   Main origins of industrial enzymes.

Enzyme origin			Example of enzyme produced	
Microorganism	Yeast	• *Saccharomyces cerevisiae* • *Kluyveromyces lactis*	• Invertase • Pectinases	
	Fungi	• *Aspergillus niger* • *Trichoderma harzianum* • *Talaromyces emersonii*	• Pectinases • Cellulases • Hemicellulases • Glucanases • Glycosidases • Proteases	
	Bacteria	• *Bacillus subtilis* • *Bacillus amyloliquefaciens* • *Bacillus licheniformis* • *Bacillus stearothermophilus* • *Escherichia coli*	• Amylases • Proteases	
Plants & fruits		• Barley malt	• Amylases	
		• Papaya	• Papain	
		• Pineapple	• Bromelain	
Animal		• Egg white	• Lysozyme	
		• Pig pancreas	• Phospholipase • Amylase	
		• Calf stomach	• Chymosin	

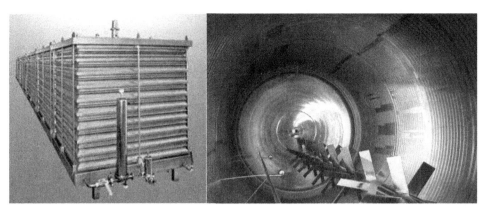

A   B

TABLE 18.2 Origin and production method of enological enzymes.

Enzymes	Origin	Production technique
Pectinases, cellulases, glucanases, xylanases	*Aspergillus niger, Aspergillus aculeatus, Trichoderma longibrachiatum*	Submerged or surface fermentation
Glycosidases	*A. niger*	Submerged or surface fermentation
Glucanases	*Trichoderma harzianum, Talaromyces emersonii*	Submerged or surface fermentation
Lysozyme	Egg white	Extraction and purification by chromatography

- Stage 4: before formulation, some producers include a purification step if necessary, while other producers focus on strain selection to produce naturally low quantities of undesired secondary enzymes like cinnamoyl esterase.
- Stage 5: wine enzymes are complex formulations of different and single fermentation products able to deliver optimal performance during the winemaking process. Up to this stage, the natural form of an enzyme is liquid and it can be packaged in this way or spray-dried depending on the final objective and application.

Normally, process-improvement enzymes are used in large volumes in big wineries and stay in their liquid form, whereas quality-enhancing enzymes are used in smaller quantities and are commercialized in dry form.

Liquid formulations are stabilized to avoid the development of microbial contamination and secure enzyme activity. Potassium chloride, glycerol, and/or ammonium sulfate are the preferred substances used for wine enzymes. The quantity of glycerol supplied by commercial liquid preparations is about 200 times lower than the glycerol content produced by yeast during alcoholic fermentation. At this dose, it has no impact (Guérin et al., 2014).

Finally, a drying and granulation process requires the use of inert support material to stabilize the enzyme. Maltodextrin is commonly used as an inert carrier for wine enzyme preparations (Fig. 18.6) (Vincent et al., 2016).

18.2.3 Nomenclature and regulation

Enzymes are usually named according to the reaction they carry out. For commodity reasons, the International Union of Biochemistry and Molecular Biology (IUBMB) and its Nomenclature Committee have proposed a classification using Enzyme Commission (EC) numbers. This classification consists of seven classes (Table 18.3), with the recent addition of EC 7 for translocases.

Within each class, enzymes are classified into three consecutive subclasses depending on their specificity. For instance, the EC tree of cellulase is:

3 Hydrolases
   3.2 Glycosylases
      3.2.1 Glycosidases, i.e., enzymes that hydrolyze *O*- and *S*-glycosyl compounds
         3.2.1.4 Cellulase

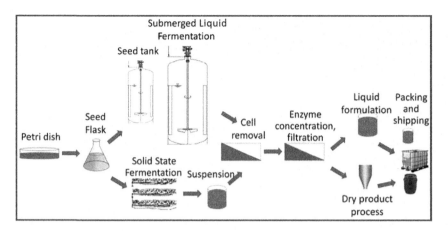

FIGURE 18.6 General diagram for the manufacturing process of enzymes. *Source: Adapted from Vincent, S., Shanahan, D., Lori, G., La Marta, J., Carrillo, R. (2016). The Generally recognized as safe (GRAS) process for industrial microbial enzymes.* Industrial Biotechnology, 12 *(5), 295–301.*

TABLE 18.3 Enzyme classification.

Class	Name	Reaction catalyzed
EC 1	Oxidoreductases	oxidation or reduction of the substrate
EC 2	Transferases	transfer of a functional group from one compound to another
EC 3	Hydrolases	cleavage of substrate into two products by hydrolysis
EC 4	Lyases	nonhydrolytic addition or removal of groups from substrates
EC 5	Isomerases	Isomerization
EC 6	Ligases	Joining two molecules with the breakdown of ATP
EC 7	Translocases	Catalyzation of ion or molecule movement across membranes

Generally speaking, enzymes are mostly processing aids that must comply with the specifications and standards of various international bodies and organizations such as the Joint FAO/WHO Expert Committee on Food Additives (JECFA), an international expert scientific committee. The Food Chemicals Codex (FCC) is a compendium of internationally recognized standards for the identity, purity, and quality of food ingredients.

Some countries have their own organizations, e.g., the Food and Drug Administration (FDA) in the United States, mandating that enzymes also need to be Generally Recognized as Safe (United States FDA GRAS Notice Inventory). Examples of other organizations include the Food Standards Australia, New Zealand (FSANZ), and the Food and Drug Regulation in Canada.

In Europe, wine has a specific regulation as determined by the OIV Codex and Code of Oenological Practices. The current regulation, (EU) 2019/934, lists all the enzymes allowed in winemaking with reference to the OIV resolutions that describe both specification and condition of use. In addition to the traditional microbiological contamination and heavy metal level limits, specification resolutions also state the allowed microorganism of origin and stabilization agents.

18.3 Enzymes in winemaking

18.3.1 Grape structure and content

In order to better understand enzyme use in white winemaking, we first need to describe the grape structure and its content (Fig. 18.7). Cell walls in plants constitute polysaccharides, glycoproteins, and lignin and determine the texture of plant tissues. In grapes, these cell walls are a strong barrier to the extraction of grape constituents. The diffusion of some cell wall constituents in must can also disturb clarification or filtration processes during winemaking.

Grape berries are composed of cells with polysaccharide-rich cell walls (Gao et al., 2019). These cells are mainly found in skin, pulp, and seeds. The skin cells represent 6–9% of the total weight, with their cell walls ensuring rigidity and protection, and are constituted mainly of polysaccharides (such as cellulose), hemicellulose, and about two-thirds of the total berry pectin content. Many compounds of interest, such as aroma and aroma precursors, are located in the skin cell, as well as part of the phenolic acids at the origin of the development of enzymatic oxidation phenomena, which lead to must browning. The pulp, with bigger cells and only one-third of the pectin content, is rich in sugar and organic acids and also contains 30–40% of the aromatic and precursor compounds, depending on the grape variety. Finally, seeds present a high concentration of harsh and bitter phenolics and tannins that are detrimental to wine quality (Gao et al., 2019).

The architectural model of primary cell walls includes different structures interconnected with each other (Carpita & Gibeaut, 1993):

- Cellulose microfibrils are linked together by a network of hemicelluloses (xyloglucan, glucomannan, and mainly xylan) supporting the structure of the wall.
- Pectic polysaccharides, composed mainly of homogalacturonan (HG), rhamnogalacturonan type I (RGI) [with side chains of arabinans (A) and arabinogalactans (AG)], and rhamnogalacturonan type II (RGII) (Figs. 18.7 and 18.8) (Fougère-Riffot, Cholet, & Bouard, 1996; Doco, Lecas, Pellerin, Brillouet, & Moutounet, 1995).
- structural wall proteins called extensins (Ducasse, 2009).

Pectin plays a key role in the hardness of the berry. They are hydrolyzed during maturation by endogenous pectinases causing berry softening and are found in the must with an average degree of methylation of 70–80% (Ribéreau-Gayon et al., 2000). Pectin gel formation and its water retention properties prevent the diffusion of aromas into the white juice before alcoholic fermentation and have a clear impact on juice extraction and clarification, as well as the filtration of juice and wine.

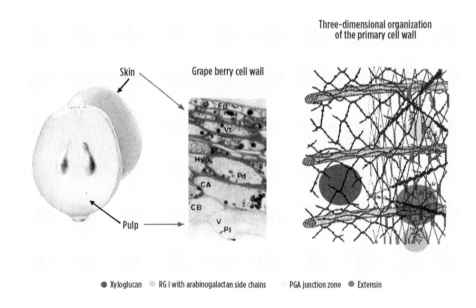

FIGURE 18.7 White grape cell wall structure. Source: *Adapted from Carpita, N.C., Gibeaut, D.M. (1993). Structural models of primary-cell walls in flowering plants — Consistency of molecular structure with the physical properties of the walls during growth.* Plant Journal, 3(1), 1–30 *and Fougère-Riffot, M., Cholet, C., Bouard, J. (1996). Evolution des parois des cellules de l'hypoderme de la baie de raisin lors de leur transformation en cellules de pulpe.* Journal International de la Vigne et du Vin, 30(2), 47–51.

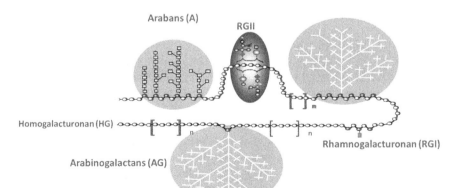

FIGURE 18.8 Schematic representation of grape berry pectins. Source: *Adapted from Doco, T., Lecas, M., Pellerin, P., Brillouet, J.M., & Moutounet, M. (1995). Les polysaccharides pectiques de la pulpe et de la pellicule de raisin. Quel devenir pendant la phase pré-fermentaire?* Revue Française d'Oenologie, 153, 16–23.

18.3.2 Endogenous enzymes in grape and juice

Grape berries contain pectinases, and more specifically, mainly pectinmethylesterase (PME) and polygalacturonase (PG) (Datunashvili, Tyrina, & Kardash, 1976; Marteau, Scheur, & Olivieri, 1963). In addition to pectinases, the most significant activity present in grapes is that of β-D-glucosidase (Crouzet & Flanzy, 1998). However, the use of exogenous industrial enzymes is needed since endogenous enzymes show limited activity at low temperature and pH and are inhibited by glucose, SO_2, red wine tannins, and alcohol (Van Rensburg & Pretorius, 2000). Enzymes are also present in must and wine, thanks to some yeasts and lactic acid bacteria. *Saccharomyces cerevisiae* is a poor secretor of glycosidases, but non-*Saccharomyces* yeasts like *Hanseniaspora*, *Kloeckera*, *Metschnikowia*, *Torulaspora*, and *Candida* spp. have been shown to produce glycosidases and esterases at more significant levels. Some strains *of Lactobacillus plantarum* produce enzymes that play a key role in wine aroma compound formation such as glycosidase, esterase, and citrate lyase (Du Toit et al., 2010). Finally, *Botrytis cinerea*, responsible for gray or noble rot on grapes, produces PME and PG in addition to laccase (Villettaz, 1984).

There are other oxidative enzymes capable of forming aldehydes and C6 alcohols responsible for wine sensory spoilage (e.g., grassy flavors), which should be limited during the early stages of the winemaking process by an appropriate pressing protocol and must clarification. The enzymes responsible for grassy flavors are acyl-hydrolase, lipoxygenase, peroxide cleavage enzyme, and alcohol dehydrogenase (Ribéreau-Gayon, Dubourdieu, Donéche, & Lonvaud, 2006).

Grapes also contain tyrosinase, which oxidizes hydroxycinnamic acids into quinones. These can condense with flavonoids, forming polymers that lead to undesirable white must browning. This mechanism will be enhanced when grape maceration is improperly conducted. The correct amount of SO_2 (30−50 ppm) and low temperatures during the prefermentative steps will limit the potential of this mechanism's detrimental effects.

In addition, grape peroxidases, active in phenol oxidation during grape maturation (Calderon, Garcia-Florneciano, Munoz, & Ros Barcelo, 1992), seem to be a minor problem in the must as they are affected by relatively low doses of SO_2.

Finally, laccase, from *B. cinerea* in rotten grapes, can compromise wine quality. Stable at must pH and resistant to SO_2, it can lead to major must oxidation. In the case of rotten grapes, maceration (skin contact) and extraction enzymes must be minimized during the prefermentative steps. Fast and efficient clarification should be the main target and, in some cases, heating up to 80°C will sufficiently denature this enzyme.

18.3.3 Exogenous enological enzymes used in white winemaking

18.3.3.1 Principal activities

18.3.3.1.1 Pectinases

Pectolytic exogenous enzyme formulations are used in white winemaking to break down grape pectin main chains, also known as polygalacturonic acid or HG (Ribéreau-Gayon et al., 2000). These enzymes are pectin lyase (PL), PME, and PG. PMEs release methanol and polygalacturonic acid (Fig. 18.9) from pectin molecules, which reduces the level of methylation and facilitates the action of PG. PG is only active on nonmethylated pectin to catalyze the degradation of polygalacturonic acid molecules into smaller oligomers (Guérin et al., 2014). Of most importance for winemaking technological efficiency is *endo*-PG, which hydrolyzes the HG inside the main chain, thereby reducing the viscosity of the must more efficiently than *exo*-PG. PL catalyzes the cleavage of glycosidic bonds and, unlike PG, has a higher affinity as the level of pectin methylation increases (Guérin et al., 2014).

18.3.3.1.2 Glycosidases

Varietal aroma compounds are present in grapes mainly as precursors. The most ubiquitous precursors are aroma glycoconjugates that can highly contribute to the typicity and quality of wine, especially white wine. One of the most documented examples is the Muscat variety where the levels of bound terpenes are very high. Other white varieties also present significant levels of aroma glycoconjugates such as Chardonnay, Viognier, and Traminer. The sugar moiety of these glycoconjugates always includes a glucose molecule, which can be bound to another sugar such as rhamnose, arabinose, or apiose.

These aroma precursors are naturally hydrolyzed during aging, but the use of exogenous glycosidases in wine can speed up this mechanism. They present a sequential hydrolysis mechanism, with diglycoside cleavage followed by the liberation of the aroma compound from the mono-glucoside. This is why not only glucosidases but also rhamnosidases, arabinosidases, and apiosidases are needed to achieve a good and efficient liberation of the aroma moiety, regardless of the sugar moiety structure (Gunata, Bitteur, Brioullet, Bayonove, & Cordonnier, 1988).

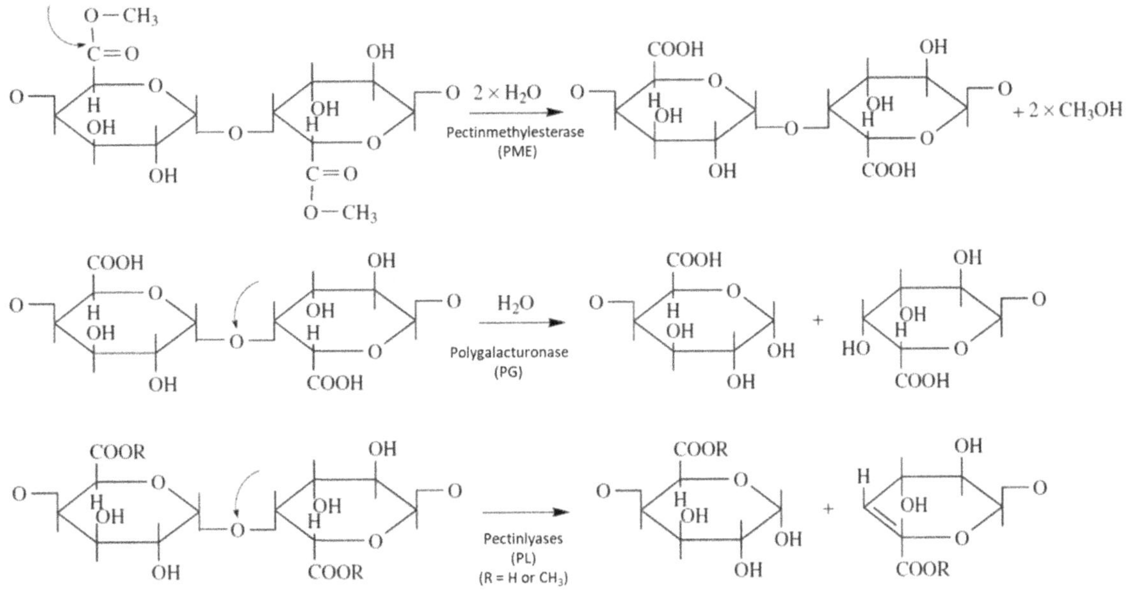

FIGURE 18.9 Mode of action of pectinases. Source: *Adapted from Ribéreau-Gayon, P., Dubourdieu, P.B., Donéche, B., Lonvaud, A. (2006). Handbook of enology volume 1. The microbiology of wine and vinifications (2nd ed.). Chichester, England: John Wiley & Sons, Ltd.*

18.3.3.1.3 β-glucanases

Exo-(1→3)-β-glucanases and (1→6)-β-D-glucanases can hydrolyze glucans originating not only from *Botrytis cinerea*, but also from yeast cell walls. They can therefore be used to facilitate pressing, clarification, and filtration of grapes and wines affected by *B. cinerea* (gray or noble rot), or during lees aging to speed up the release of compounds of interest from the yeast cells. These compounds include mannoproteins, peptides, and nucleotides, which can improve wine quality and stability.

18.3.3.1.4 Proteases (under OIV evaluation)

In wine, protein instability can lead to the appearance of haze and deposits caused by the presence of grape proteins (thaumatin-like proteins [TLPs] and chitinases). These proteins can become insoluble and aggregate to make wines appear turbid. To combat this phenomenon, the use of acid proteases, such as aspergillopepsin I, has been investigated to prevent protein instability. Heat treatment is necessary to unfold unstable proteins and make them accessible to the enzyme's action (Marangon et al., 2012).

Proteases from *A. niger* are listed in the CODEX ALIMENTARIUS specifications for Food Additives (CAC/MISC 6–2012) and are approved by the FDA "for treatment of wine and juice to reduce or remove heat labile proteins". Additionally, proteases from *A. niger* and *A. oryzae* are Generally Recognized as Safe (GRN No. 89 & 90 – United States FDA GRAS Notice Inventory). They are also approved for "use in the course of manufacture of any food" by the FSANZ; in Canada by Food and Drug Regulations for use in a range of food and beverages; and in the EU for brewing and baking, as well as cheese, fish, juice, and protein processing.

Their approval for wine application is in discussion at the OIV (aspergillopepsin I), with a compulsory heating step to achieve an efficient degradation of the unstable proteins, followed by the degradation of the protease at the end of the treatment.

18.3.3.2 Secondary activities

Fermentation with specifically selected fungi (mainly *A. niger*), to produce some of the above listed principal activities, leads to the natural production of side activities that can improve the efficiency of the enzyme preparation, or in other cases, be detrimental to wine quality.

Some of these activities will improve hydrolysis of the side chains of the pectin, known as "hairy regions," to obtain the maximum potential of the grape in terms of yield, aroma diffusion, and both optimum pressing and clarification. The activities that should be also part of the commercial enzyme formulations would be rhamnogalacturonases, arabinase, arabinosidases, and galactanases active on rhamnogalacturonan, as well as the side chains of arabinans and arabinogalactans.

Hemicelluloses, especially xylanases able to hydrolyze xylans, facilitate the diffusion of some key aroma compounds from the skin and increase must free run yield. On the contrary, cellulases present in red wine enzyme formulations to increase the extraction of color and tannins should not be present at high levels in white wine formulations. This is to avoid the release of oxidizable phenols and to limit grape skin dilaceration, which is sometimes the cause of pressing problems (Whitehurst & van Oort, 2010).

Undesired activities of contents like cinnamoyl-esterases (CE) should be naturally low in wine enzyme formulations, as they can be responsible for the formation of vinyl phenols which are detrimental to wine aroma (Morata et al., 2013). To achieve those low levels, there are two options in terms of production: either the use of *A. niger* strains selected for their low CE production or the implementation of a purification step after enzyme separation.

18.3.3.3 Lysozyme

Lysozyme is the only exogenous enzyme used in enology not produced by fermentation, but extracted from egg whites. It exhibits a bacteriolytic function and is thus able to lyse gram-positive bacteria, thereby inhibiting their growth. As lactic acid bacteria (e.g., *Oenococcus* and *Lactobacillus*) and *Pediococcus* are gram-positive, their growth can be limited and even inhibited by lysozyme addition. It is the only enzyme listed as an additive and subjected to labeling as an allergen; wines treated with lysozyme should mention "contains egg" on the label (Whitehurst & van Oort, 2010).

18.4 Enzyme application in white wine

Nowadays, winemakers have a sizeable number of commercial wine enzyme formulations available (details of the activities and technological application are presented in Table 18.4), to be used in three main scenarios: to improve the winemaking process itself, with the subsequent savings in time, labor energy, and money; to improve the quality of wine; and to solve the problems winemakers face during the prefermentative and

TABLE 18.4 Relationships between enzyme activities and technological application.

Enzyme Activities		Prefermentation				Postfermentation			
		Maceration	Pressing	Protein stabilization	Must clarification	Aging on lees	Glycosylated aroma release	Wine Clarification	Filtration
Pectinases (Principal)	Polygalacturonase (PG)	•	•		•	•		•	•
	Pectin Lyase (PL)	•	•		•	•		•	•
	Pectinmethylesterase (PME)	•	•		•	•		•	•
Pectinases (Secondary)	Rhamnogalacturonase	•	•					•	•
	Arabinofuranosidase	•	•					•	•
	Galactanase	•	•					•	•
Hemicellulases	Xylanase	•	•						
Cellulases	Endo-(1→4)-β-D-Glucanase								•
	Exo-(1→4)-β-D-Glucanase								•
Glycosidases	β-D-Glucosidase						•		
	α-L-Arabinofuranosidase						•		
	α-L-Rhamnosidase						•		
	β-D-Apiosidase						•		
Glucanases	Exo-(1→3) (1→6) β-D-Glucanases				•	•		•	•
Proteases (under OIV evaluation)	Aspergillopepsin I			•					

Adapted from Guérin, L., Pallas, S., Charrier, F., Ducasse, M.A., Ruf, J.C., Villettaz, J.C., … Pellerin, P. (2014). Enzymes en Œnologie: Production, Régulation, Applications. Institut Français de la Vigne et du Vin. Cahiers Itinéraires n° 26, 31.

fermentative steps, as well as during conservation, aging, and stabilization prior to bottling. In the following section, we will follow the applications of wine enzymes during the main white winemaking process steps.

The following sections will focus only on enzyme use (see the other dedicated chapters for technical aspects): main objectives, formulation needed, parameters to be checked to ensure enzyme technological efficiency, and some examples.

18.4.1 Prefermentative steps

During the prefermentative steps, three different enzyme applications could be required to improve the process or quality (Fig. 18.10):

- Improvement of varietal aroma extraction when skin contact is applied.
- Juice yield increase (free run and total) during pressing.
- Improvement of clarification efficiency by static settling or flotation.

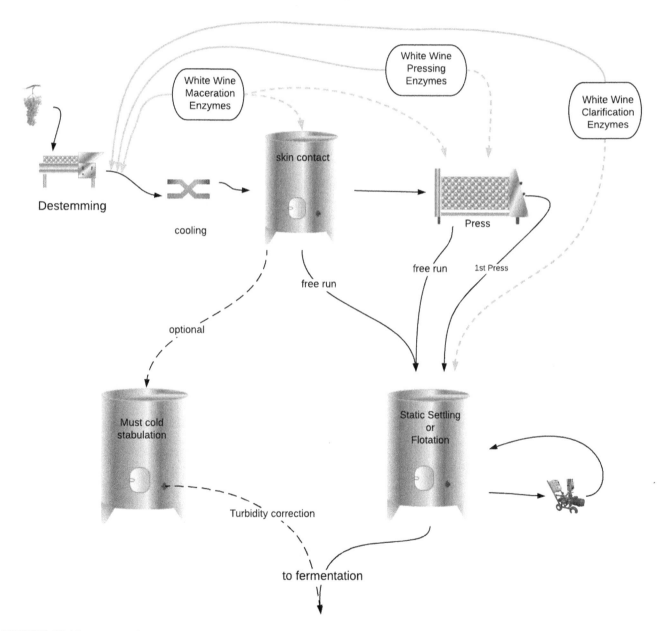

FIGURE 18.10 Main prefermentative steps in white winemaking and recommendations for enzyme application.

Other techniques, like the use of a decanter for juice extraction, will also require an enzymatic formulation to improve the yield and performance of these extraction processes.

Winemakers should follow manufacturer recommendations as enzyme dosages will vary depending on the objectives, contact time, and temperature of the grapes, juice, and wine. When used during the preformentative steps, commercial enzyme formulations should be applied to destemmed grapes as soon as possible to increase the contact time and thus their efficiency. Enzyme activities are compromised by SO_2, so the addition of each product needs to be done separately and not at the same time.

18.4.1.1 Skin contact

- Objectives: extraction of aroma and aroma precursors from the grape skins during maceration.
- Formulation needed: the formulation should include not only general pectinase activities (linear chain and side chains), but also hemicellulase activities, in order to achieve better degradation of the skin cells and favor the extraction of aroma and aroma precursors mainly located in this part of the grape berry. Cellulase can be present, but at moderate levels to avoid the extraction of phenolics that are detrimental to the must and wine quality (oxidation promotion and astringency).
- Parameters to consider: temperature and contact time are the two parameters that need to be modified to reach the desired level of extraction. Typically, skin contact maceration is performed at 5−18°C for 4−24 hours (Guérin, Charrier, Samson, & Schneider, 2013). Temperatures above 15°C for more than 12 hours of contact time can increase the extraction, but with the risk that fermentation starts during maceration.

A typical example of the improvement of aroma extraction is shown in Fig. 18.11 for Sauvignon blanc grapes (2018 harvest, Northern Italy). Skin contact for 16 hours at 12°C was applied in two different scenarios, with and without aroma extraction enzyme. The results indicate increased levels of 3-mercaptohexan-1-ol (3MH) thiol in the wine.

18.4.1.2 Juice extraction/pressing

- Objectives: increased juice yield (free run and total) and facilitation of the pressing step.
- Formulation needed: the formulation should include general pectinase activity (linear chain and side chains) and also a balanced and moderate content of hemicellulases and cellulases. This will ensure better degradation of the skin and pulp cells and favor the extraction of juice mainly located in the vacuoles of the pulp. Both hemicellulases and cellulases can be present, but at moderate levels to avoid extraction of phenolics that are detrimental to must and wine quality (e.g., oxidation promotion and astringency), and also to limit skin dilaceration that can lead to the formation of a grape mush, which is difficult to press (Whitehurst & van Oort, 2010).
- Parameters to consider: Based on the type and model of press and manufacturer recommendations, the winemaker will apply different pressing modes, depending on the grape cultivar, sanitary conditions, and winemaking objectives. Juice yield (free run and total) and pressing figures (normally in bar) to obtain the optimum yield and pressing cycle duration are the main parameters to be followed. Yield improvement can reach up to 5−15% for the first pressed juice, even when the same pressure is applied (Fig. 18.12).

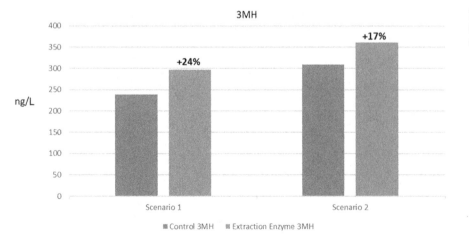

FIGURE 18.11 3MH content in a Sauvignon blanc wine when an aroma extraction enzyme is applied.

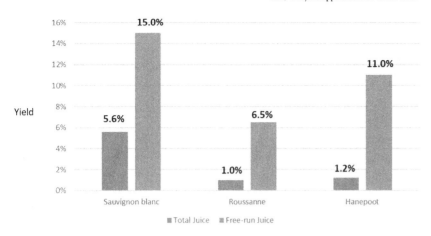

FIGURE 18.12 Increase in yield obtained with a pressing enzyme compared to a reference enzyme used during prefermentation. Sauvignon blanc and Roussanne from France (1 g/100 kg); Hanepoot from South Africa (3.3 g/100 kg).

$$Vg = \frac{|d^2 *|(\Delta\rho) * g}{18 *|\eta}$$

FIGURE 18.13 Stokes' Law, where Vg is the settling velocity of the particle (m/s), g is the gravitational field strength (m/s^2), η the viscosity of the fluid (kg·m^{-1}·s^{-1}), d the diameter of the particle (m), and $\Delta\rho$ (ρp - ρf) is the difference of particles and fluid mass densities (kg/m^3).

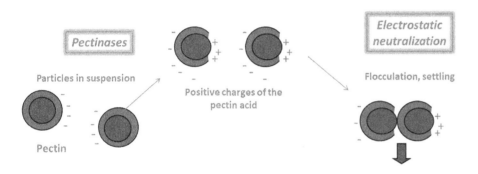

FIGURE 18.14 Electrostatic mechanism of enzymatic settling.

18.4.1.3 Juice clarification: static settling and flotation

Based on Stokes' law (Fig. 18.13), and given that juice is a viscous fluid, to increase the settling velocity of particles in suspension, we need to decrease the viscosity of the juice and increase the density and diameter of the particles.

- Objectives: to decrease the viscosity of the juice by pectin hydrolysis and increase flocculation of the particles due to the electrostatic mechanism (Fig. 18.14). Flocculation will increase the density and diameter of the particles in suspension and will also improve the settling velocity and reduce the volume of lees. Nowadays, not only in static settling but particularly in flotation, some clarification aids such as gelatin, bentonite, and silica gel are used to facilitate juice clarification, speed up the process, and maintain the stability of the lees cake over time. These will also help in aggregating the solids and improving gas bubble−particle interaction (Guérin L., 2013).
- Formulation needed: to efficiently decrease juice viscosity, we will need the combined action of the three main pectinase activities: PL, PME, and PG, especially *endo*-PG. In the case of difficult conditions such as low temperatures (below 10°C) and difficult-to-settle grape cultivars (e.g., Muscat), low-temperature resistant activities and more complex formulations are needed (Fig. 18.15). Enzyme formulations specific for flotation should present a higher content of PL, to speed up the hydrolysis of the linear pectin chain and ultimately increase the efficiency of the process.
- Parameters to consider: temperature and contact time will determine the dose, and a negative pectin test will be the main indicator of efficient settling or flotation.

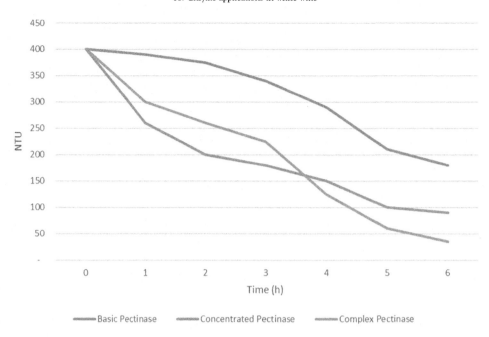

FIGURE 18.15   Turbidity evolution of a Moscatel (Argentina) when settling at 5°C with three different enzyme formulations.

- Special case when rotten grapes are present: when grapes are affected by *B. cinerea*, no skin contact should be applied and juice clarification should be carried out as rapidly as possible. In such a scenario an efficient formulation will be needed, based on pectinases' activities (linear chain and side chains) active on grape polysaccharides, and β-(1−3) (1−6) glucanases active on *B. cinerea* glucans. A glucan test, in addition to a pectin test, should be negative in order to proceed to an efficient clarification of such juices. Wine made from rotten grapes will normally be difficult to filtrate and fine, so a second dose of these enzyme formulations might be necessary.

18.4.2 Fermentative and postfermentative steps

During and after fermentation, enzymes can be applied mainly to improve wine fining, clarification, and filtration, especially when grapes are affected by *B. cinerea*. These enzymes also facilitate and improve the release of molecules such as mannoproteins (during aging on lees) and glycosylated aromas, which play an important role in the sensory profile of some grape cultivars (Fig. 18.16).

18.4.2.1 Aging on lees
- Objectives: release of compounds of interest (e.g., mannoproteins, amino acids, and nucleotides) from the yeast cytoplasm by yeast autolysis following alcoholic fermentation, for their organoleptic and stabilizing properties.
- Formulation needed: it must contain β-glucanase activity able to hydrolyze yeast cell wall glucans, thus speeding up the autolysis process. *Endo-ß-1,3-* and *endo-ß-1,6-* as well as *exo-ß-1,3-* and *exo-ß-1,6* activities are involved in this process.
- Parameters to consider: as for other enzymatic treatments, temperature and contact time are the key points. Usually, β-glucanases are poorly active below 10°C. The efficiency can be monitored by amino acid or polysaccharide analysis (Fig. 18.17), but tasting remains the best tool to decide when to rack off the wine.

18.4.2.2 Glycosylated aroma release
- Objectives: release of aroma compounds from their glycosylated precursors.
- Formulation needed: the only enzymatic formulation allowed by OIV regulation is β-glycosidase. In addition to high β-glucosidase activity, the formulations must present high levels of apiosidases, rhamnosidases, and arabinosidases in order to efficiently cleave the different forms of glycoconjugates (monoglucosides and disaccharides).

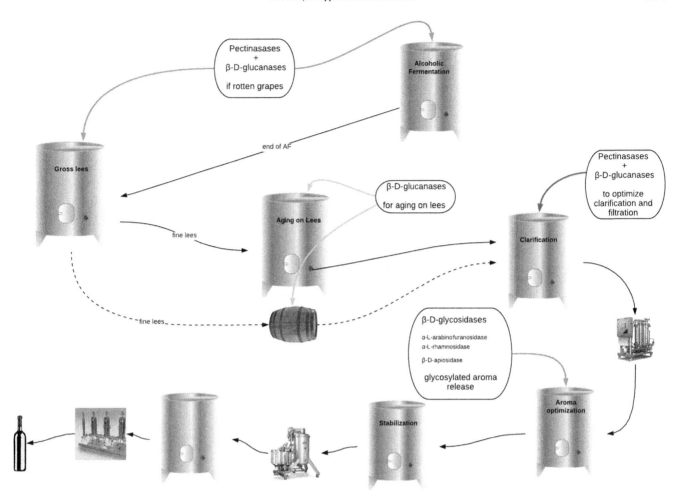

FIGURE 18.16 Fermentative and postfermentative processes in white winemaking and recommendations for enzyme applications.

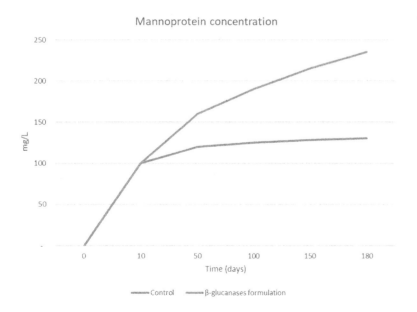

FIGURE 18.17 Mannoprotein release when a β-glucanase enzyme is added during aging on lees.

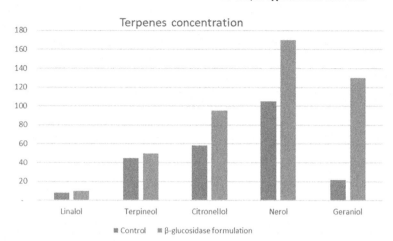

FIGURE 18.18 Effect of glycosidases on the release of terpenes (Muscat, 2 g/hL, contact time three weeks).

FIGURE 18.19 Filtrability Index (FI) and Vmax improvement in a Prosecco wine, after secondary fermentation. Filtration enzyme at 5 mL/hL, three days contact time, T$_a$ = 8°C (Italia, October 2017).

- Parameters to consider: temperature and contact time are the key points. A preliminary test should be performed at the laboratory scale to determine, for each wine, the most suitable contact time. Hydrolytic activity should stop with the addition of a small amount of bentonite (5–10 g/hL) (Fig. 18.18).

18.4.2.3 *Wine clarification, stabilization, and filtration*

- Objectives: to remove solid particles in suspension and macromolecules or aggregates in colloidal dispersion affecting wine limpidity and stability before final filtration steps and bottling.
- Formulation needed: the formulation should include general pectinase activity (linear chain and side chains) to efficiently hydrolyze grape polysaccharides still present in the wine matrix, as well as β-glucanases active on glucans affecting wine clarification and filtration.
- Parameters to consider: white wine storage is conducted at low temperatures close to 10°C, so enzyme doses are normally higher and contact time can be as long as two weeks to be efficient. Pectin and glucan tests should be negative prior to moving to clarification and filtration.

Turbidity measurements and filtration parameters such as Filterability Index (FI) and Vmax (Fig. 18.19) are key at the end of the winemaking process to succeed with final sterile filtration and bottling.

18.5 Conclusions

Industrial enzymes are mainly proteins of microbial origin, that catalyze biochemical reactions and are produced by Solid State and Submerged Liquid Fermentations of fungi such as *Aspergillus* sp. and *Trichoderma* sp. They are usually named according to the reaction they carry out, and the IUBMB and its Nomenclature Committee use EC numbers for their classification. Enzymes are mostly processing aids that must comply with the specifications and standards of various international bodies and organizations such as the OIV and its International Oenological Codex.

Grape cell walls are rich in polysaccharides such as cellulose, hemicellulose, and pectins that need to be hydrolyzed for an efficient winemaking process. Though grape berries contain pectinases, the use of exogenous industrial enzymes is needed since endogenous enzymes are affected by low temperature, and pH and are inhibited by glucose, SO_2, and alcohol.

Winemakers use commercial wine enzyme formulations to improve the winemaking process, the wine quality, and to solve problems that might arise during the journey from grapes to bottling. The principal enzyme activities used during white winemaking are those of pectinases, glycosidases, and β-glucanases, together with secondary activities such as those of rhamnogalacturonases, arabinases, arabinosidases, and galactanases active on pectin "hairy regions". Other undesired activities like CE should be naturally low in white wine enzyme formulations, as they are detrimental to wine aroma.

The future approval of acid proteases (Aspergillopepsin I) to prevent protein instability is in discussion at the OIV and will likely become an alternative to treatments with bentonite affecting wine quality and causing huge loss of wine worldwide (Schneider, 2021).

Enzyme producers, in collaboration with international bodies and organizations, will endlessly invest in new microorganism selections, improved production processes, and optimized enzyme formulations to help winemakers perform a more efficient and sustainable winemaking process, thus enhancing wine quality.

References

AMFEP (2020). https://amfep.org/about-enzymes/production/ Accessed 15.07.20.

Calderon, A. A., Garcia-Florneciano, E., Munoz, R., & Ros Barcelo, A. (1992). Gamay grapevine peroxidase: Its role in vacuolar anthocyanidin degradation. *Vitis*, *31*, 139–147.

Cammack, R., Atwood, T., Campbell, P., Parish, H., Smith, A., Vella, F., & Stirling, J. (2006). *Oxford dictionary of biochemistry and molecular biology*. Oxford: Oxford University Press.

Carpita, N. C., & Gibeaut, D. M. (1993). Structural models of primary-cell walls in flowering plants – Consistency of molecular structure with the physical properties of the walls during growth. *Plant Journal*, *3*(1), 1–30.

Crouzet, J., & Flanzy, C. (1998). *Les enzymes en œnologie. Oenologie - Fondements scientifiques et technologiques* (p. 361411) Tec & Doc Lavoisier.

Datunashvili, E. N., Tyrina, S. S., & Kardash, N. K. (1976). Some properties of grape pectin methylesterase. *Applied Biochemistry and Microbiology*, *12*, 36–40.

Doco, T., Lecas, M., Pellerin, P., Brillouet, J. M., & Moutounet, M. (1995). Les polysaccharides pectiques de la pulpe et de la pellicule de raisin. *Quel devenir pendant la phase pré-fermentaire? Revue Française d'Oenologie*, *153*, 16–23.

Du Toit, M., Engelbrecht, L., Lerm, E., & Krieger-Weber, S. (2010). Lactobacillus: The Next Generation of Malolactic Fermentation Starter Cultures - An Overview. *Food Bioprocess Technol*, *4*, 876–906.

Ducasse, M.A. (2009). Impact des enzymes de Macération sur la composition en polysaccharides et en polyphénols des Vins rouges – Etude de l'évolution de ces composés en solution modèle vin. PhD thesis, University of Montpellier II, 204 p.

Fougère-Riffot, M., Cholet, C., & Bouard, J. (1996). Evolution des parois des cellules de l'hypoderme de la baie de raisin lors de leur transformation en cellules de pulpe. *Journal International de la Vigne et du Vin*, *30*(2), 47–51.

Gao, Y., Zietsman, A. J. J., Vivier, M. A., & Moore, J. P. (2019). Deconstructing Wine Grape Cell Walls with Enzymes During Winemaking: New Insights from Glycan Microarray Technology. *Molecules (Basel, Switzerland)*, *2019*(24), 165.

Guérin, L., Charrier, F., Samson, A., & Schneider, R. (2013). Les opérations préfermentaires, Chap. 12. In R. Schneider (Ed.), *Les vins blancs de la démarche markéting à la vinification, les clés d'un pilotage réussi* (pp. 146–163). La France Agricole.

Guérin, L., Pallas, S., Charrier, F., Ducasse, M. A., Ruf, J. C., Villettaz, J. C., ... Pellerin, P. (2014). *Enzymes en Œnologie: Production, Régulation, Applications*. Institut Français de la Vigne et du Vin. Cahiers Itinéraires n°, *26*, 31.

Gunata, Z., Bitteur, S., Brioullet, J. M., Bayonove, C., & Cordonnier, R. (1988). Sequential enzymatic hydrolysis of potential aromatic glycosides from grape. *Carbohydrates. Res*, *184*, 139–149.

Gunata, Z., Dugelay, I., Sapis, J. C., Baumes, R., & Bayonove, C. (1993). Role of the enzyme in the use of the flavour potential from grape glycosides in wine making. In P. Schreier, & P. Winterhalter (Eds.), *Progress in flavour precursor studies* (pp. 219–234). Wheaton (USA): Allured.

Marangon, M., Van Sluyter, S. C., Robinson, E. M., Muhlack, R. A., Holt, H. E., Haynes, P. A., ... Waters, E. J. (2012). Degradation of white wine haze proteins by Aspergillopepsin I and II during juice flash pasteurization. *Food Chemistry*, *135*(3), 1157–1165.

Marteau, G., Scheur, J., & Olivieri, C. (1963). Le rôle des enzymes pectolytiques du raisin ou de préparations commerciales dans la clarification des jus. *Ann. Technol. Agric.*, *12*, 155–176.

Morata, A., Vejarano, R., Benito, S., Palomero, B., Gonzalez, C., Tesfaye, W., ... Suárez-Lepe, J.A., enotecUPM. Reduction of 4-ethylphenol contents in red wines using HCDC + yeasts and cinnamyl esterases. World Congress of Vine and Wine. June 2nd–7th. 2013. Bucharest.

Ribéreau-Gayon, P., Dubourdieu, P. B., Donéche, B., & Lonvaud, A. (2006). *Handbook of enology volume 1. The microbiology of wine and vinifications* (2nd Edition). Chichester, England: John Wiley & Sons, Ltd.

Ribéreau-Gayon, P., Glories, Y., Maujean, A., & Dubourdieu, D. (2000). *Handbook of enology Vol 2. The chemistry of wine stabilization and treatments* (2nd Edition). Chichester, England: John Wiley & Sons, Ltd.

Schneider R., 2021. Protease application in winemaking to achieve protein stability of the wine. Enoforum Web USA 2021. https://www.infowine.com/en/video/using_proteases_to_achieve_protein_stability_in_wines_sc_19881.htm.

Segel, I. (1975). *Enzyme Kinetics*. J. Wiley & Sons.

Tu, T., Meng, K., Luo, H., Turunen, O., Zhang, L. et al. (2015). New insights into the role of T3 loop in determining catalytic efficiency of GH28 *endo*-polygalacturonases.

Van Rensburg, P., & Pretorius, I. S. (2000). Enzymes in winemaking. Harnessing natural catalysts for efficient biotransformations — A Review. *South African Journal of Enology and Viticulture, 21*, 52—73.

Villettaz, J. C. (1984). Les enzymes en oenologie. *Bulletin de l'O.I.V., 635*, 19—29.

Vincent, S., Shanahan, D., Lori, G., La Marta, J., & Carrillo, R. (2016). The Generally recognized as safe (GRAS) process for industrial microbial enzymes. *Industrial Biotechnology, 12*(5), 295—301.

Walter, N. G., & Engelke, D. R. (2002). Ribozymes: Catalytic RNAs that cut things, make things, and do odd and useful jobs. *Biologist (London), 49*(5), 199—203.

Whitehurst, R. J., & van Oort, M. (2010). *Enzymes in Food Technology* (Second Edition). Wiley-Blackwell.

19

Near infrared for white wine analysis

A. Power[1] and D. Cozzolino[2]

[1]CREST Technology Gateway of TU Dublin, Dublin, Ireland [2]Centre for Nutrition and Food Sciences, Queensland Alliance for Agriculture and Food Innovation (QAAFI), The University of Queensland, Brisbane, QLD, Australia

19.1 Introduction

Wine is a complex alcoholic beverage and has prompted significant research in its production, maturation, and protection, particularly targeting its quality and authenticity (Margalit, 2012; Polášková, Herszage, & Ebeler, 2008; Thoukis, 1974; Waterhouse, Sacks, & Jeffery, 2016). The distinctive characteristics of a wine are defined by multiple factors including the raw material (e.g., grape composition), flavor additives, and processing (e.g., fermentation and maturation) (Margalit, 2012; Polášková et al., 2008; Thoukis, 1974; Waterhouse et al., 2016). As analytical chemistry techniques have advanced and developed, so has the knowledge of wine composition and flavor (Margalit, 2012; Polášková et al., 2008; Thoukis, 1974; Waterhouse et al., 2016). Analysts first focused on the characterization of wines, by assessing their major components, namely ethanol, organic acids, and sugars (Margalit, 2012; Polášková et al., 2008; Thoukis, 1974; Waterhouse et al., 2016). The development of chromatographic methodologies, particularly gas chromatography (GC), in the 1950s prompted a new era of wine analysis, resulting in the isolation and identification of over 1300 volatile chemicals or molecules in alcoholic beverages and the continued detection of novel compounds (Bakker & Clarke, 2011).

Despite these advances, the flavor of complex systems such as wine has yet to be fully understood and thus cannot be fully predicted (Ebeler & Thorngate, 2009; Ebeler, 2001; Gambetta et al., 2014). Particularly, as often, the perceived flavor is the result of a combination of or specific ratio of a myriad of compounds, rather than a single "impact" compound (Ebeler & Thorngate, 2009; Ebeler, 2001; Gambetta et al., 2014). Moreover, in wine, where the major products of fermentation, esters, and alcohols contribute a generic background flavor, it is often the case that trace components usually confer the distinctive facets of an individual wine (Ebeler & Thorngate, 2009; Ebeler, 2001; Gambetta et al., 2014). Both the raw ingredient (grapes) and every step involved in the production of wine contribute to its final sensory characteristics, aroma, taste, and mouthfeel (Ebeler & Thorngate, 2009; Ebeler, 2001; Gambetta et al., 2014). However, this is not the sole concern of the wine industry and the consumer. Safety and brand protection are also of high concern, as they are interlinked, where protected geographical status (PGS), protected designation of origin (PDO), protected geographical indication (PGI), and traditional specialty guaranteed (TSG) are used to characterize wine (e.g., composition and other specific properties) (Barrère, 2007; Chiodo, Casolani, & Fantini, 2011; Ditter & Brouard, 2014; Martelo-Vidal & Vazquez, 2014; Ríos-Reina et al., 2019). In turn, these also contribute to the sustainability of wine production systems, particularly regarding wine quality and the survival of the artisanal producers (Barrère, 2007; Chiodo et al., 2011; Ditter & Brouard, 2014; Martelo-Vidal & Vazquez, 2014; Ríos-Reina et al., 2019). This is coupled with a more "plugged in" and educated consumer base whose wish for greater quality and higher safety levels has resulted in a demand for more stringent analytical processes and control within the entire winemaking process (Barrère, 2007; Chiodo et al., 2011; Ditter & Brouard, 2014; Martelo-Vidal & Vazquez, 2014; Ríos-Reina et al., 2019), ultimately leading to the development of multiple analytical methodologies for determining the compositional profile of wine, both during production and postproduction (e.g., blending, storage) (Parpinello et al., 2013; Ye et al., 2014).

Another rising concern for consumers and producers alike is fraud (Bull, 2016; Muhammad & Countryman, 2019; Rutter & Bryce, 2008; Shen, 2018). Counterfeit, illegal imitations, of wine, infringe upon the legal right of a

© 2022 Elsevier Inc. All rights reserved.

product's manufacturer to protect their product, brand image, and the associated profit; this is also coupled with the intent to defraud the consumer (Bull, 2016; Muhammad & Countryman, 2019; Rutter & Bryce, 2008; Shen, 2018). The phenomenon of dealing in fraudulent materials is not new; amphorae of wine discovered in France and determined to be from around 27 BC were observed to be marked with counterfeit seals in an attempt to pass them off as the more expensive Roman import rather than a presumably local wine (Bull, 2016; Muhammad & Countryman, 2019; Rutter & Bryce, 2008; Shen, 2018).

These issues have resulted in a significant concern for the global food industry were both the production and sale of counterfeit goods was highlighted by the International Trademark Association that estimated that more than $460 billion American doillars worth of counterfeit goods were trade in 2016 (Bull, 2016; Muhammad & Countryman, 2019; Rutter & Bryce, 2008; Shen, 2018). Moreover, the scope is extensive, with few manufactured goods escaping counterfeiters' attention. Today, counterfeit consumer goods, ranging from sports gear to car parts to foodstuffs such as alcohol, have all been subject to illegal markets; moreover, while these activities are no doubt too lucrative, the very nature of this illicit and covert enterprise makes evaluating the exact value difficult (Bull, 2016; Muhammad & Countryman, 2019; Rutter & Bryce, 2008; Shen, 2018). For example, in 2016 the International Trademark Association estimated that the pirate market was worth US$460 billion; this could be a conservative figure when considering the losses suffered by legitimate business (Bull, 2016; Muhammad & Countryman, 2019; Rutter & Bryce, 2008; Shen, 2018). The development and advent of new technologies have only exacerbated the issue, reducing the cost and time required to produce fake goods and easing their dissemination, e.g., increased consumer access via the internet. This has contributed and continues to contribute to the growth of illegal markets and the profits of those associated (Bull, 2016; Muhammad & Countryman, 2019; Rutter & Bryce, 2008; Shen, 2018).

Counterfeiting offers criminals a high-income revenue stream that is comparably low risk and has significantly lesser penalties on conviction than alternative illegal activities such as drug trafficking. Rudi Kurniawan was sentenced to 10 years following his conviction for wine and mail fraud after his decade-long career of selling fraudulent vintages, including the record-breaking sale of a single wine in 2006 for US$24.7 million (Cumming, 2016). Thus the authentication of food products is a priority for consumers, the food industry, and authorities, not just as a means of preventing economic fraud, and its negative impact on both consumers and industry stakeholders, but for assuring public health and safety, as incorrectly declared or undeclared food constituents represent significant risk to health risk to consumers (Bikoff et al., 2015; Kotelnikova, 2017; Spink & Moyer, 2011).

These concerns have prompted bodies such as the International Organisation of Vine and Wine (OIV) to define a series of methodologies to standardize the analysis of grapes and wines for scientific, legal, and practical purposes. These methods are widely recognized as robust and accurate; however, it is also true that they are often costly, time intensive, destructive, and require expert (highly trained) users, which can hinder their usefulness in application throughout the enological process, denying vintners of true online quality control. This has resulted in considerable research devoted to the development of new methodologies capable of maintaining high levels of robustness and precision in quality analysis while also being capable of doing so in a manner that is rapid, noninvasive, cost effective, and user friendly.

Spectroscopy (Bao et al., 2014; dos Santos et al., 2018; Giovenzana et al., 2013; Martelo-Vidal, Domínguez-Agis, & Vázquez, 2013), particularly near infrared (NIR) coupled with novel statistical (multivariate) analysis, has emerged as a potential solution that can be successfully applied to a wide range of processes within the wine industry, enhancing its ability to analytically monitor the winemaking process in its entirety, from grape to bottle, and to assure safety and quality to the consumer (Cozzolino et al., 2012; Cozzolino, McCarthy, & Bartowsky, 2012; González-Caballero et al., 2012; Hill et al., 2013).

This chapter presents and discusses some of the advances in the field of NIR spectroscopy as well as some of the recent applications of this technique in the analysis of white wine. The potential promise of chemometric methods combined with this technique to develop new applications and some of the limitations that still dominate the uptake of this technology by the wine industry will be also discussed.

19.2 NIR spectroscopy analysis

The electromagnetic spectrum between the visible (VIS) and the microwave wavelengths is the infrared (IR) region. When IR radiation interacts with an analyte, be it solid, liquid, or gaseous, some of the incoming IR radiation is absorbed at specific wavelengths that are in turn specific to the sample, resulting in a "fingerprint" or spectrum (Cozzolino, 2009; Cozzolino et al., 2006a, 2009). Discovered in 1800 by Friedrich Wilhelm Herschel, NIR spectroscopy is based on the absorption of electromagnetic radiation over the wavelength range between 780 and

2500 nm. NIR spectroscopy assesses organic chemical structures that contain O−H, N−H, and C−H bonds, with the majority of the bands in the NIR being combination or overtone bands of the IR region's fundamental absorption region. Consequently, these absorptions are the consequence of specific compounds or molecules that are present in the food matrix. In more complex matrices such as food or wine, which contain an interlink of molecules and properties derived from their interrelations, band or peak broadening occurs in addition to the multiple bands. This results in NIR spectra characterized with broad bands rather than with individual peaks (Shen et al., 2012).

When comparing samples, often the NIR spectra between samples seems to very similar and dominated mainly by water as indicated by the absorption of O−H overtones and combination tones at 760 nm, 970 nm, 1450 nm and 1940 nm. Consequently, spectral peaks in the mid-infrared (MIR) frequencies are often sharper and better resolved than those of the NIR range. NIR spectra still contain the higher overtones of the O−H, N−H, C−H, and S−H bands from the MIR region but are weaker than the fundamental frequencies observed in the MIR (Huang et al., 2008; Nicolai et al., 2007; Osborne, 2006). Moreover, combination bands such as the CO stretch and NH bonds associated with proteins might result in a congested NIR spectrum with strongly overlapping bands. This is why NIR has yet to be exploited to the extent of other vibrational spectroscopic methods, as NIR spectra' characteristic overlap and complexity make quantification and interpretation of data difficult (Cynkar, Cozzolino, & Dambergs, 2009). Recent advances in instrumentation and computer processing have been utilized by analysts to better elucidate NIR spectra. Specifically, advancements in multivariate analysis and chemometrics have provided researchers the means to decipher and model compositional characteristics in complex matrices such as foodstuffs (Cozzolino et al., 2006b; Cynkar et al., 2009; Fernández-Novales et al., 2008; Fernández-Novales et al., 2009a, 2009b; Smyth et al., 2008) that are not readily detected by traditional targeted chemical analysis. Fig. 19.1 shows the typical NIR spectra of white wine samples analyzed using a 2-mm path length cuvette.

19.3 Multivariate analysis of chemometrics

The term chemometrics has been used to describe a number of statistical methodologies including exploratory data analysis, pattern recognition, and statistical experimental design to provide analysts with the most complete and relevant information from data sets (Chapman et al., 2019). Unlike traditional statistical approaches, which focus on a single variable, chemometrics considers multiple variables concurrently; moreover, it accounts for collinearity between variables: the variation in one variable or a group of variables, in terms of covariation with other variables (Chapman et al., 2019).

Whereas classical statistics is reductionist in its effort to understand the factors that are dominant in the observed effects, chemometrics' use of multivariate approaches in which all variables are simultaneously considered allow the method to model the data and thus predict or recognize patterns within it (Chapman et al., 2019; Cozzolino et al., 2011; Gishen, Dambergs, & Cozzolino, 2005). Ultimately, this allows the analyst to better distinguish between information and data, which is key within measurements where large amounts of results are generated (Chapman et al., 2019; Cozzolino et al., 2011; Gishen et al., 2005). This is a major strength of the technique, particularly when it is paired with spectroscopic methods (Cozzolino et al., 2011; Gishen et al., 2005) such as NIR, where the chemical and

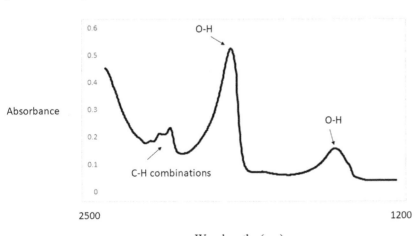

FIGURE 19.1 Typical near-infrared spectra of white wine samples scanned using a 2-mm path length cuvette.

physical basis of spectroscopic transitions within resultant spectra are not a consequence of a single species or a very isolated set of energies. Rather, the spectral data are highly correlated between measurement channels and thus from one chemical species to the next. This correlation can mean that the information provided appears, but because chemometrics accounts for such correlation, the methods exploit the correlation's similarity or redundancy, increasing its overall precision, allowing for the efficient extraction of unique information from the spectra (Cozzolino et al., 2005; Cozzolino, Smyth, & Gishen, 2003; Gishen, Cozzolino, & Dambergs, 2006).

Principal component analysis (PCA), partial least squares (PLS) regression (Cozzolino et al., 2007), principal component regression (PCR), discriminant analysis (DA), and artificial neural network (ANN) (Cozzolino et al., 2006b; Yu et al., 2008) are the most commonly applied multivariate data analysis techniques to NIR spectra. In most applications, PCA is used to investigate the structure of the data set, allowing the analyst to visualize outliers or atypical samples in both the spectral and analytical data. PLS, PCR, or multiple linear regression (MLR) is used to develop calibration models exploited by chemometrics. However, it cannot be understated that when applying any of these data-modeling techniques, it is paramount that an optimum number of variables or components is nominated. Too many variables as compared to the samples might determine that the model is overfitted as there is excessive redundancy in the X-variables (wavelengths); (Cozzolino et al., 2006b; González-Caballero et al., 2011) that is, the model is too dependent on the data set and will provide poor prediction results. Alternatively, too few variables will result in underfitting and a model not large enough to capture the full variability within the data (Cozzolino, 2009; Cozzolino et al., 2005, 2006a, 2009; Cozzolino et al., 2003; Páscoa et al., 2020).

19.4 Application of NIR and chemometrics for white wine analysis

Véstia and coworkers (Véstia et al., 2019) reported the use of FT-NIR combined with multivariate analysis to develop a robust prediction model to quantify the calcium content in white wine samples ($n = 98$) and must samples ($n = 60$) from Alentejo. The authors reported that the results of this study confirmed the potential of FT-NIR spectroscopy as a rapid nondestructive alternative to the standard atomic absorption (AA) spectrochemical method to predict calcium content in grape must and base wine (Véstia et al., 2019).

Work by Molla and coworkers (Molla et al., 2017) investigated the effect of measurement parameters such as temperature and path length during the analysis of white wine samples using NIR, UV, and VIS spectroscopy. The effect of these parameters was tested on the ability of the method to characterize and/or predict the chemical compositions of wines and their subsequent effect on the performance of the calibration models (Molla et al., 2017). The authors highlighted the importance of considering the path length and its influence on the standard error of the models based on UV, VIS, and NIR spectroscopy to predict phenolic compounds (Molla et al., 2017). The authors reported that the path length (e.g., 0.1 and 1 mm) did not induce any statistically significant differences ($p > 0.05$) in the ability of the models to measure the phenolic content in the wine samples analyzed (Molla et al., 2017). Additionally, no differences in the spectra of the samples measured using the two path lengths based on the PCA score plot were reported (Molla et al., 2017).

Ripoll and collaborators (Ripoll, Vazquez, & Vilanova, 2017), investigated the potential of UV-VIS-NIR spectroscopy coupled with chemometrics to develop a rapid method for the determination of volatile compounds in white wines from "Albariño" grapes sourced from the DO Rias Baixas (Spain). The authors determined that the technique, when paired with PLS regression models, reliably predicted a number of volatiles including E-2-hexenal, 1-hexanol, (Z)-2-hexanol, benzaldehyde, phenylethanal, cis pyran linalool oxide, and 2-phenylethanol, with a coefficient of correlation greater than 0.85 for these compounds (Ripoll et al., 2017). The authors indicated that their study highlighted the potential of UV-VIS-NIR spectroscopy as a rapid means for online in-process assessment for determining volatile compounds in white "Albariño" musts (Ripoll et al., 2017).

Cayuela and colleagues (Cayuela, Puertas, & Cantos-Villar, 2017) also investigated the combination of chemometrics with VIS-NIR spectroscopy to predict quality markers in white wine, particularly those compounds associated with sensory attributes. The authors reported that VIS-NIR spectroscopy was able to predict some of the most characteristic sensory attributes of wines including color and flavor intensity, astringency, and persistency, with validation correlation coefficients of approximately 0.9 in most cases (Cayuela et al., 2017). The authors also determined that the technique was useful for predicting critical defects such as oxidation, ethyl acetate, and acetic acid (Cayuela et al., 2017). The authors speculate that the primary potential use of this technique is as a means for contrasting or confirming the assessment of wine for the positive and negative sensory attributes determined by the sensory analysis of tasting panels, which would be of particular benefit in cases of discrepancies in the assessments of a sensory panel (Cayuela et al., 2017).

Fernandez-Novales and coworkers (Fernández-Novales et al., 2019) reported on the combination VIS-NIR spectroscopy to estimate the concentration of individual amino acids and total soluble solids (TSS) in intact grape berries in order to ascertain quality prior to the enological process. These authors utilized PLS regression to develop calibration models and reported a coefficient of determination in prediction of approximately 0.60 for asparagine (standard error of performance [SEP]: 0.45 mg N/L), tyrosine (SEP: 0.33 mg N/L), and proline (SEP: 17.5 mg N/L) using the 570−1000 nm wavelength range (Fernández-Novales et al., 2019). Similar results were obtained for lysine (SEP: 0.44 mg N/l), tyrosine (SEP: 0.26 mg N/l), and proline (SEP: 15.54 mg N/l) using the 1100−2100 nm wavelength range (Fernández-Novales et al., 2019). The authors also reported the development of models for TSS with a coefficient of determination of approximately 0.90 (SEP = 1.60ǫ Brix, regression point displacement [RPD] = 3.79) (Fernández-Novales et al., 2019). It was concluded that the contactless, nondestructive spectroscopy method could be an alternative to providing information about grape amino acid composition in intact grapes (Fernández-Novales et al., 2019).

Jara-Palacios and colleagues (Jara-Palacios et al., 2016) combined chemometrics with NIR hyperspectral imaging to predict 27 individual phenolic compounds in grape marc after freeze-drying. PLS regression models successfully predict phenolic compounds with coefficients of regression higher than 0.98 (Jara-Palacios et al., 2016). Although the method does show potential and is an attractive alternative due to its speed and simplicity, the authors suggest that it is not a substitute for conventional chemical analysis but rather more useful as a screening tool to assess the chemical composition of grape marc for researchers and for deciding the destination of this byproduct in wineries (Jara-Palacios et al., 2016).

In a similar study, Beghi and collaborators (Beghi et al., 2017) investigated the applicability of VIS and NIR spectroscopy combined with multivariate analysis to rapidly assess the phytosanitary status of grape berries. Classification analysis models were developed using PLS-DA regression. The results provided classification accuracy rates between 89.8% and 94.0% (Beghi et al., 2017). This study demonstrated the capability of the VIS-NIR system to provide useful online information about wine grape phytosanitary status for better management of the vinification process (Beghi et al., 2017).

In a study on 159 must samples, Giovenzana and collaborators (Giovenzana et al., 2018) investigated the utilization of VIS and NIR spectroscopy in both reflectance and transmittance modes as a means of evaluating and objectively quantifying the phytosanitary infection status of musts at the grape receiving area to support wineries (Giovenzana et al., 2018). The authors reported that the percentage of samples correctly classified using PLS-DA ranged between 52.5% to 90.4%, and between 68.4% to 84.3% using either reflectance or transmittance modes, respectively (Giovenzana et al., 2018). The authors highlighted that the methodology could be incorporated inline, inside the analysis tank at the grape receiving area, allowing for rapid screening during the winemaking process (Giovenzana et al., 2018).

A recent study by dos Santos et al. (2017) reported the combination of NIR with other vibrational (MIR and Raman) spectroscopic methods combined with PLS-DA to develop a classification method to discriminate between wines from four Portuguese geographic origins. The authors reported correct prediction in 87.7% of the wines; however, it was highlighted that the NIR method was not as useful as the other spectroscopic techniques used (dos Santos et al., 2017). Overall, the authors concluded that for the consolidation of robust calibration models capable of discriminating wines from different locations, it would be necessary not only to include a larger number of samples, representatively identifying their origins, but also to include additional types of wines (namely red, rose, and sparkling), to attest the suitability of the technique (dos Santos et al., 2017).

Guidetti and coworkers (Giovenzana et al., 2015) reported the development and testing of a simple optical prototype for rapid estimation of the ripening parameters of white grape for Franciacorta wine directly in the field. The device obtained reflectance spectra at four wavelengths, namely 630, 690, 750, and 850 nm, in the VIS and NIR range (Giovenzana et al., 2015). Chemometric analyses were utilized to mine the maximum useful information from optical data. In this study, PCA provided a preliminary evaluation of the data (Giovenzana et al., 2015). Correlations between the optical data matrix and ripening parameters [total soluble solids (TSS) content; titratable acidity (TA)] were developed using PLS and MLR methods (Giovenzana et al., 2015). Classification analyses were also performed with the aim of discriminating ripe and unripe grape samples (Giovenzana et al., 2015). The authors reported that PCA, MLR, and classification models clearly demonstrated the effectiveness of the simplified system in separating samples among different sampling dates and in discriminating ripe from unripe samples (Giovenzana et al., 2015). In addition, the authors highlighted that the prototype could be further optimized for augmented in-field applications (e.g., ripeness evaluation, chemicals, and physical properties prediction or shelf-life analysis) or for different applications (Giovenzana et al., 2015), ultimately providing producers with an inexpensive, portable, and user-friendly screening tool (Giovenzana et al., 2015).

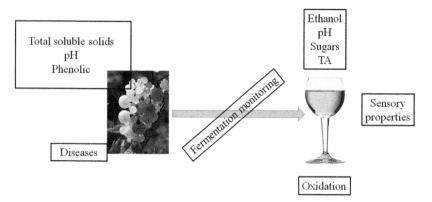

FIGURE 19.2 Applications of near-infrared spectroscopy for the analysis of white wine, from the vineyard to the consumer.

Fig. 19.2 shows the current and potential applications of NIR spectroscopy for the analysis of white wine, from the vineyard to the consumer.

19.5 Summary

Exploring, extracting, and describing NIR data derived from the analysis of grape and wine samples have increased exponentially in recent decades as highlighted by the number of applications described in the literature. The increase in the number of these applications is associated with advances in instrumentation and multivariate data manipulation techniques that have become easily available in the industry. However, one of the largest challenges facing the wider application of NIR spectroscopy as an analytical tool in the winery is related to the difficulties in interpreting the complex spectra obtained during the application of this technique.

NIR spectroscopy has been reported to successfully being explored either solely or in combination with other vibrational spectroscopy methods to measure the chemical composition and properties of grapes and wines, as well as to assess product quality and for production control, particularly as the NIR spectrum can provide analysts with a global signature of composition (fingerprint), which, combined with chemometric techniques, can be used to elucidate particular compositional characteristics not easily detected by traditional targeted chemical analysis. The primary advantages of such analysis compared to traditional chemical and chromatographic techniques are the speed and the user-friendly and noninvasive nature of the approach that makes this technology easy to use in routine operations.

Moreover, the potential for miniaturization, portability, cost-effectiveness, and reduction in analysis time and financial requirements, coupled with the "green" nature of the technology, has positioned NIR spectroscopy as a very attractive technique with a bright future in analyzing grapes at the vineyard, monitoring processes (e.g., fermentation), and assessing end-product quality.

However, some issues persist that hold back the wide application of this technology in the wine industry. One critical point is related to the lack of formal education and training in both NIR spectroscopy and chemometrics. This issue still hinders the widespread acceptance of this technology as a tool for the analysis of complex products like grapes and wines. The industry still relies on the in-house training provided by the companies that supply the instruments.

Moreover, the lack of readily accessible complete spectroscopic databases (e.g., databases containing chemical compounds of importance to wine production), coupled with our still-limited understanding of many of the molecular processes that take place during the production of wine, or the difficulties raised from the combination of these methods when big data sets are generated are still important issues that need to be resolved. A few companies know or are aware of the standard error of the lab or reference method; however, this is still the ruler that is used to evaluate the calibration models generated by the application of NIR spectroscopy.

Therefore the continued development and application of NIR spectroscopy will require novel approaches in such a manner in which analysis and interpretation of the data generated are conducted. The adoption of this approach as a modern analytical technique will determine a bright future during wine analysis. In particular, this approach allows moving away from data being a discreet range of numbers reflecting what has happened, to a more holistic and dynamic space where big data (NIR spectra) combined with data mining and modeling will improve the analysis of wine. In addition, this approach will help to fully understand the complex intra- and interrelationships and processes that result from the production of white wines of high quality. However, research and industry need to move away from the *"black box"* approach.

References

Bakker, J., & Clarke, R. J. (2011). *Wine: Favour chemistry*. John Wiley & Sons.

Bao, Y., et al. (2014). Measurement of soluble solid contents and pH of white vinegars using VIS/NIR spectroscopy and least squares support vector machine. *Food and Bioprocess Technology, 7*(1), 54−61.

Barrère, C. (2007). The genesis, evolution and crisis of an institution: The protected designation of origin in wine markets. *Journal of Institutional Economics, 3*(2), 165−181.

Beghi, R., et al. (2017). Rapid evaluation of grape phytosanitary status directly at the check point station entering the winery by using visible/ near infrared spectroscopy. *Journal of Food Engineering, 204*, 46−54.

Bikoff, J. L., et al. (2015). Fake it'til we make it: Regulating dangerous counterfeit goods. *Journal of Intellectual Property Law & Practice, 10*(4), 246−254.

Bull, T. (2016). *The grape war of China: Wine fraud and how science is fighting back. Art crime* (pp. 41−56). Springer.

Cayuela, J. A., Puertas, B., & Cantos-Villar, E. (2017). Assessing wine sensory attributes using Vis/NIR. *European Food Research and Technology, 243*(6), 941−953.

Chapman, J., et al. (2019). Spectroscopic approaches for rapid beer and wine analysis. *Current Opinion in Food Science*.

Chiodo, E., Casolani, N., & Fantini, A. (2011). Regulatory policies and consumers perception of wines with protected designation of origin: Aconjoint experiment. Ancona - 122nd EAAE Seminar "Evidence-Based Agricultural and Rural Policy Making",1−14 pp.

Cozzolino, D., et al. (2005). Relationship between sensory analysis and near infrared spectroscopy in Australian Riesling and Chardonnay wines. *Analytica Chimica Acta, 539*(1−2), 341−348.

Cozzolino, D., et al. (2006a). Analysis of grapes and wine by near infrared spectroscopy. *Journal of Near Infrared Spectroscopy, 14*(5), 279−289.

Cozzolino, D., et al. (2006b). Combining mass spectrometry based electronic nose, visible−near infrared spectroscopy and chemometrics to assess the sensory properties of Australian Riesling wines. *Analytica Chimica Acta, 563*(1−2), 319−324.

Cozzolino, D., et al. (2007). Effect of temperature variation on the visible and near infrared spectra of wine and the consequences on the partial least square calibrations developed to measure chemical composition. *Analytica Chimica Acta, 588*(2), 224−230.

Cozzolino, D., et al. (2009). A brief introduction to multivariate methods in grape and wine analysis. *International Journal of Wine Research, 1*(1), 123−130.

Cozzolino, D. (2009). Near infrared spectroscopy in natural products analysis. *Planta Medica, 75*(07), 746−756.

Cozzolino, D., et al. (2011). Technical solutions for analysis of grape juice, must, and wine: the role of infrared spectroscopy and chemometrics. *Analytical and Bioanalytical Chemistry, 401*(5), 1475−1484.

Cozzolino, D., et al. (2012). Varietal differentiation of grape juice based on the analysis of near-and mid-infrared spectral data. *Food Analytical Methods, 5*(3), 381−387.

Cozzolino, D., McCarthy, J., & Bartowsky, E. (2012). Comparison of near infrared and mid infrared spectroscopy to discriminate between wines produced by different *Oenococcus oeni* strains after malolactic fermentation: A feasibility study. *Food Control, 26*(1), 81−87.

Cozzolino, D., Smyth, H. E., & Gishen, M. (2003). Feasibility study on the use of visible and near-infrared spectroscopy together with chemometrics to discriminate between commercial white wines of different varietal origins. *Journal of Agricultural and Food Chemistry, 51*(26), 7703−7708.

Cumming, E. (2016). The great wine fraud. The Observer. Available at: https://www.theguardian.com/global/2016/sep/11/the-great-wine-fraud-a-vintage-swindle

Cynkar, W., Cozzolino, D., & Dambergs, R. G. (2009). The effect of sample storage and homogenisation techniques on the chemical composition and near infrared spectra of white grapes. *Food Research International, 42*(5−6), 653−658.

Ditter, J.-G., & Brouard, J. (2014). The competitiveness of French protected designation of origin wines: A theoretical analysis of the role of proximity. *Journal of Wine Research, 25*(1), 5−18.

Ebeler, S. E. (2001). Analytical chemistry: Unlocking the secrets of wine flavor. *Food Reviews International, 17*(1), 45−64.

Ebeler, S. E., & Thorngate, J. H. (2009). Wine chemistry and flavor: Looking into the crystal glass. *Journal of Agricultural and Food Chemistry, 57*(18), 8098−8108.

Fernández-Novales, J., et al. (2008). A feasibility study on the use of a miniature fiber optic NIR spectrometer for the prediction of volumic mass and reducing sugars in white wine fermentations. *Journal of Food Engineering, 89*(3), 325−329.

Fernández-Novales, J., et al. (2009a). Assessment of quality parameters in grapes during ripening using a miniature fiber-optic near-infrared spectrometer. *International Journal of Food Sciences and Nutrition, 60*(Suppl. 7), 265−277.

Fernández-Novales, J., et al. (2009b). Shortwave-near infrared spectroscopy for determination of reducing sugar content during grape ripening, winemaking, and aging of white and red wines. *Food Research International, 42*(2), 285−291.

Fernández-Novales, J., et al. (2019). Assessment of amino acids and total soluble solids in intact grape berries using contactless Vis and NIR spectroscopy during ripening. *Talanta, 199*, 244−253.

Gambetta, J. M., et al. (2014). Factors influencing the aroma composition of Chardonnay wines. *Journal of Agricultural and fFod Chemistry, 62*(28), 6512−6534.

Giovenzana, V., et al. (2013). Quick quality evaluation of Chilean grapes by a portable vis/NIR device. *Acta Horticulturae, 978*, 93−100.

Giovenzana, V., et al. (2015). Testing of a simplified LED based vis/NIR system for rapid ripeness evaluation of white grape (Vitis vinifera L.) for Franciacorta wine. *Talanta, 144*, 584−591.

Giovenzana, V., et al. (2018). Comparison of two immersion probes coupled with visible/near infrared spectroscopy to assess the must infection at the grape receiving area. *Computers and Electronics in Agriculture, 146*, 86−92.

Gishen, M., Cozzolino, D., & Dambergs, R. G. (2006). The analysis of grapes, wine, and other alcoholic beverages by infrared spectroscopy. *Handbook of Vibrational Spectroscopy*.

Gishen, M., Dambergs, R., & Cozzolino, D. (2005). Grape and wine analysis-enhancing the power of spectroscopy with chemometrics. A review of some applications in the Australian wine industry. *Australian Journal of Grape and Wine Research, 11*(3), 296−305.

González-Caballero, V., et al. (2011). Optimization of NIR spectral data management for quality control of grape bunches during on-vine ripening. *Sensors, 11*(6), 6109−6124.

González-Caballero, V., et al. (2012). On-vine monitoring of grape ripening using near-infrared spectroscopy. *Food Analytical Methods, 5*(6), 1377−1385.

Hill, G., et al. (2013). Near and mid-infrared spectroscopy for the quantification of botrytis bunch rot in white wine grapes. *Journal of Near Infrared Spectroscopy, 21*(6), 467−475.

Huang, H., et al. (2008). Near infrared spectroscopy for on/in-line monitoring of quality in foods and beverages: A review. *Journal of Food Engineering, 87*(3), 303−313.

Jara-Palacios, M., et al. (2016). Determination of phenolic substances of seeds, skins and stems from white grape marc by near-infrared hyperspectral imaging. *Australian Journal of Grape and Wine Research, 22*(1), 11−15.

Kotelnikova, Z. (2017). Explaining counterfeit alcohol purchases in Russia. *Alcoholism: Clinical and Experimental Research, 41*(4), 810−819.

Margalit, Y. (2012). *Concepts in wine chemistry*. Board and Bench Publishing.

Martelo-Vidal, M. J., & Vazquez, M. (2014). Rapid authentication of white wines. Part 1: Classification by designation of origin. *Agro Food Industry Hi-Tech, 25*(5), 44−48.

Martelo-Vidal, M., Domínguez-Agis, F., & Vázquez, M. (2013). Ultraviolet/visible/near-infrared spectral analysis and chemometric tools for the discrimination of wines between subzones inside a controlled designation of origin: A case study of R ías B aixas. *Australian Journal of Grape and Wine Research, 19*(1), 62−67.

Molla, N., et al. (2017). The effect of path length on the measurement accuracies of wine chemical parameters by UV, visible, and near-infrared spectroscopy. *Food Analytical Methods, 10*(5), 1156−1163.

Muhammad, A., & Countryman, A. M. (2019). In vino 'no' veritas: Impacts of fraud on wine imports in China. *Australian Journal of Agricultural and Resource Economics, 63*(4), 742−758.

Nicolai, B. M., et al. (2007). Nondestructive measurement of fruit and vegetable quality by means of NIR spectroscopy: A review. *Postharvest Biology and Technology, 46*(2), 99−118.

Osborne, B.G. (2006). Near-infrared spectroscopy in food analysis. Encyclopedia of analytical chemistry: Applications, theory and instrumentation. John Wiley & Sons, Ltd. https://doi.org/10.1002/9780470027318.

Parpinello, G. P., et al. (2013). Relationship between sensory and NIR spectroscopy in consumer preference of table grape (cv Italia). *Postharvest Biology and Technology, 83*, 47−53.

Páscoa, R. N., et al. (2020). The application of near infrared spectroscopy to wine analysis: An innovative approach using lyophilization to remove water bands interference. *Talanta, 214*, 120852.

Polášková, P., Herszage, J., & Ebeler, S. E. (2008). Wine flavor: chemistry in a glass. *Chemical Society Reviews, 37*(11), 2478−2489.

Ríos-Reina, R., et al. (2019). Data fusion approaches in spectroscopic characterization and classification of PDO wine vinegars. *Talanta, 198*, 560−572.

Ripoll, G., Vazquez, M., & Vilanova, M. (2017). Ultraviolet−visible-near infrared spectroscopy for rapid determination of volatile compounds in white grapes during ripening. *Ciência e Técnica Vitivinícola, 32*(1), 53−61.

Rutter, J., & Bryce, J. (2008). The consumption of counterfeit goods: Here be pirates? *Sociology, 42*(6), 1146−1164.

dos Santos, C. A. T., et al. (2017). Merging vibrational spectroscopic data for wine classification according to the geographic origin. *Food Research International, 102*, 504−510.

dos Santos, C. A. T., et al. (2018). Raman spectroscopy for wine analyses: A comparison with near and mid infrared spectroscopy. *Talanta, 186*, 306−314.

Shen, A. (2018). 'Being affluent, one drinks wine': Wine counterfeiting in Mainland China. *International Journal for Crime, Justice and Social Democracy, 7*(4), 16−32.

Shen, F., et al. (2012). Discrimination between Shaoxing wines and other Chinese rice wines by near-infrared spectroscopy and chemometrics. *Food and Bioprocess Technology, 5*(2), 786−795.

Smyth, H., et al. (2008). Near infrared spectroscopy as a rapid tool to measure volatile aroma compounds in Riesling wine: Possibilities and limits. *Analytical and Bioanalytical Chemistry, 390*(7), 1911−1916.

Spink, J., & Moyer, D. C. (2011). Defining the public health threat of food fraud. *Journal of food Science, 76*(9), R157−R163.

Thoukis, G. (1974). *Chemistry of wine stabilization: A review*. ACS Publications.

Véstia, J., et al. (2019). Predicting calcium in grape must and base wine by FT-NIR spectroscopy. *Food Chemistry, 276*, 71−76.

Waterhouse, A. L., Sacks, G. L., & Jeffery, D. W. (2016). *Understanding wine chemistry*. John Wiley & Sons.

Ye, M., et al. (2014). Application of FT-NIR spectroscopy to apple wine for rapid simultaneous determination of soluble solids content, pH, total acidity, and total ester content. *Food and Bioprocess Technology, 7*(10), 3055−3062.

Yu, H., et al. (2008). Prediction of enological parameters and discrimination of rice wine age using least-squares support vector machines and near infrared spectroscopy. *Journal of agricultural and Food Chemistry, 56*(2), 307−313.

Aging on lees

Piergiorgio Comuzzo, L. Iacumin and Sabrina Voce

Department of Agricultural, Food, Environmental and Animal Sciences (Di4A), University of Udine, Udine, Italy

20.1 Introduction

Aging on lees (AOL) or *sur lies* is a quite olden winemaking technology. In its former and traditional application, it consisted of the fermentation of wines in wood barrels and their subsequent aging onto the fermentation lees (Charpentier & Feuillat, 1993; Ribéreau-Gayon et al., 2006). It is a fact that racking was not a frequent practice in the past; in 1860, Redding (1860) reported that, during the first year after harvest, wines were decanted only twice, in spring and autumn, and that some wines "of a generous quality" remained on their lees for 3 or 4 years. Hence it can be inferred that AOL was a quite common procedure and that prolonged *sur lies* aging was a suitable practice for obtaining high-quality wines.

AOL has become particularly widespread in certain winemaking regions, such as Burgundy, since the beginning of the last century (Ribéreau-Gayon et al., 2006). In the 1990s, white wines aged on lees became popular in different European and non-European countries, and, probably due to the high management costs generally associated with this technology, several variations and modifications of the traditional practice have been developed. In its conventional application, AOL requires a great amount of work (e.g., for resuspending the lees, known as *bâtonnage*, or topping the barrels), continuous monitoring, and long aging time, during which wines cannot be sold, becoming an immobilized asset for the winery and increasing production costs. Moreover, such prolonged contact with fermentation lees makes the wine prone to microbial contaminations (especially if sulfur dioxide is used in low doses) because of the presence of high amounts of nutrients and viable microorganisms in lees themselves. In order to decrease this risk and to reduce production costs, AOL is currently managed not only in wood barrels but also in larger cooperage vessels, or in stainless steel vats (Boulton, Bisson, & Kunkee, 1996), eventually combined with micro oxygenation or with the addition of oak chips. In these conditions, the aging time is shorter and management becomes simpler, making AOL sustainable even for medium-price wines.

During AOL, yeasts undergo autolysis, and different substances are released from lees into the wine (Charpentier & Feuillat, 1993), provoking deep modifications in wine composition and sensory characters. To further reduce the aging time, thus accelerating autolysis, different technologies have been proposed; β-glucanase enzymes, ultrasounds, high-pressure technologies, or pulsed electric fields can be used for processing wine at the beginning of AOL or, more frequently, lees after their separation from wine. In this latter case, treated lees can be redistributed in different vats or barrels, accelerating significantly the rate of the *sur lies* aging process.

Finally, AOL is an important technique also for the production of sparkling wines. In this case, both the base wine and the final product obtained after refermentation can undergo a more or less prolonged contact with lees. This does not happen only for the sparkling wines obtained with the traditional method (e.g., Champagne, Cava, Franciacorta); a more or less extended maturation over lees has become a quite common practice also after the refermentation of certain sparkling wines produced in autoclave with the Martinotti–Charmat method (Voce et al., 2019).

This chapter focuses on all the abovementioned aspects, discussing them from the scientific point of view, while also considering the main technological and practical implications connected with their application to white wines.

White Wine Technology.
DOI: https://doi.org/10.1016/B978-0-12-823497-6.00008-9

© 2022 Elsevier Inc. All rights reserved.

20.2 Definition of wine lees

The Regulation (EU) 1308/2013 (Commission Regulation (EU) 1308/2013, 2013) defines lees as the residue that accumulates in vessels containing wine after fermentation, during storage or after the authorized treatments, as well as the residue obtained after the filtration or centrifugation of such products. European Regulation authorizes the addition of fresh, sound, and undiluted lees to dry wines, at a quantity not exceeding 5% of the volume of product treated; such lees must contain yeasts and come from recent vinification of dry wines (Commission Delegated Regulation (EU) 934/2019, 2019).

Fornairon-Bonnefond, Camarasa, Moutounet, and Salmon (2001) classified lees into three typologies: clear lees (CL), fatty lees (FL), and dry lees (DL). The first (CL) is the deposit obtained after racking or fining; CL are not dry and still contain a certain amount of wine. They are further divided into crude or virgin lees (the coarse fraction of the sediment) and fine lees (the upper and finest part of the deposit), depending on how they stratify at the bottom of the vessels during settling. FL contain a lower amount of liquid (wine) and they are obtained after draining, filtration, or centrifugation of CL. Finally, DL are the wet lees cake formed after certain winery treatments (e.g., rotary vacuum filtration); they do not contain any extractable residue of wine.

Not all of these three lees typologies (CL, FL, and DL) are used for managing AOL; indeed, only CL fractions (normally obtained without the use of fining agents) are suitable. Generally, wine aging can be managed on total lees, if no racking is carried out at the end of alcoholic fermentation, or on fine lees, if the crude lees fraction is eliminated before *sur lies* aging (Fornairon-Bonnefond et al., 2001).

20.3 Main constituents of wine lees and their modifications during aging

20.3.1 Microbiological composition of lees

Fresh lees are constituted of microbial cells (35%−45%), tartaric salts (25%−35%), and other organic residues (30%−40%) such as plant cell debris coming from grape (Fornairon-Bonnefond & Salmon, 2003; Fornairon-Bonnefond et al., 2001; Salmon et al., 2000); they constitute 2%−4% of the volume of the wine on average (Fornairon-Bonnefond et al., 2001). Microorganisms represent a significant part of fermentation lees; they are mainly yeast cells produced during alcoholic fermentation, that settle at the bottom of the tanks at the end of the fermentative process (Salmon et al., 2000). The amount of viable yeasts in fresh lees may account for up to 6−7 Log CFU/mL (Table 20.1); this concentration generally decreases during aging (Fornairon-Bonnefond et al., 2001), due to the autolytic process, determining a decrease of the lees dry weight (Leroy, Charpentier, Duteurtre, Feuillat, & Charpentier, 1990).

Lactic acid bacteria (LAB), particularly *Oenococcus* spp., are also well represented in fresh unsulfured lees (Table 20.1). It is generally reported that the presence of fermentation lees may activate the development of

TABLE 20.1 Viable microorganisms detected in two different lees samples obtained from Pinot gris.

	Microbial load (Log CFU/mL)	
	Fresh lees[a]	Aged lees[b]
Total yeasts	6.7 ± 0.1	1.8 ± 0.3
Saccharomyces spp.	6.7 ± 0.1	1.9 ± 0.1
Non-*Saccharomyces* spp.	5.3 ± 0.2	1.3 ± 0.0
Lactic acid bacteria	2.8 ± 0.1	3.9 ± 0.1
Oenococcus spp.	3.5 ± 0.1	n.d.[c]
Brettanomyces spp.	n.d.[d]	n.d.[d]

[a]Collected at the end of the alcoholic fermentation.
[b]Collected after 6 months of sur lies aging.
[c]Not detectable (<1.0 CFU/mL).
[d]Not detectable (<10.0 CFU/mL).
Modified from Carrano, G. (2016). Gestione dell'élevage sur lies mediante applicazione di tecnologie basate sull'alta pressione (Master Thesis). University of Udine.

malolactic fermentation (MLF) (Caridi, 2006; Jackson, 2008), so that *Oenococcus oeni* may grow during AOL (Patynowski, Jiranek, & Markides, 2002). It can be detected in amounts of 2−4 Log CFU/mL in nonsulfited white wines kept 1 month over lees (Carrano, 2016); this concentration may increase to 5−7 Log CFU/mL after 6 months of aging if MLF occurs spontaneously (Chinni, 2018).

So far, the effects of the microbial composition of lees on the modifications occurring in wine during AOL have been poorly investigated. In fact, up till the early 2000s, all the studies related to AOL and autolysis were focused on the genus *Saccharomyces* and the contribution of non-*Saccharomyces* spp. and LAB (which are also present in lees in nonnegligible concentrations) has been scarcely considered. More recently, different non-*Saccharomyces* strains have been characterized, demonstrating interesting features for their potential use during *sur lies* aging of wines (Domizio, Liu, Bisson, & Barile, 2017; Loira, Morata, Palomero, González, & Suárez-Lepe, 2018; Palomero, Morata, Benito, Calderón, & Suárez-Lepe, 2009; Romani et al., 2018).

The residues of yeast cells, which remain in the medium after cell death, also represent an important part of yeast lees; due to their high adsorption capacity, they play a role in the fixation of volatile thiols, medium-chain fatty acids, volatile phenols, phenolic substances released from wood, and polyphenols and aroma compounds (Fornairon-Bonnefond et al., 2001).

20.3.2 Chemical composition of lees

From a chemical point of view, lees are a very rich substrate. Certain components produced during alcoholic fermentation are concentrated in fresh lees and several compounds are released during AOL because of yeast autolysis. Lees chemical composition and modifications during AOL have been reviewed by Fornairon-Bonnefond et al. (2001). The most important lees components, from an enological point of view, are reported in Table 20.2. Their relative abundance is extremely variable, depending on several factors such as wine variety, microbial composition (e.g., fermenting yeast strain), aging conditions and length.

20.3.2.1 Polysaccharides

The main polysaccharides in yeast lees are glycoproteins. They are mostly composed of mannose and (to a minor extent) glucose, covalently linked to proteins (Llaubères, Dubourdieu, & Villettaz, 1987). The structure of yeast mannoproteins was studied extensively in the 1980s to 1990s (Klis, 1994; Llaubères et al., 1987); they are highly glycosylated proteoglycans, accounting for 40% of the total mass of yeast cell wall by-products (Li & Karboune, 2019). Carbohydrates represent the major component of their molecular structure, with a mannan/protein ratio that varies from 2.5 to 181 w/w (Li & Karboune, 2019). Their molecular weight is also extremely variable, ranging from a few dozen to several hundred kDa (Li & Karboune, 2019; Moine Leoux, 1996; Ribéreau-Gayon et al., 2006).

The concentration of such polysaccharides in fresh lees (separated from the wine) may reach several grams per liter (Carrano, 2016). During alcoholic fermentation and *sur lies* aging, they are released in wine in amounts that depend on different technological factors, such as the yeast strain, the degree of must clarification, the aging temperature, and the extent of agitation (Feuillat, 2003; Llaubères et al., 1987). In wine, their concentration varies from 150−200 mg/L to several hundred and increases during *sur lies* aging (Llaubères, Dubourdieu, & Villettaz, 1987; Ribéreau-Gayon et al., 2006).

Most of the studies about yeast polysaccharides have been carried out on the genus *Saccharomyces* (*Saccharomyces cerevisiae* or *Saccharomyces bayanus*). However, other yeasts may give a relevant contribution in defining the polysaccharide profile of wines aged on the lees. In particular, *Saccharomycodes ludwigii*, *Schizosaccharomyces pombe* (Palomero et al., 2009), and *Schizosaccharomyces japonicus* (Domizio et al., 2017; Romani et al., 2018) were found to release very high levels of polysaccharides. In the case of *Schizosaccharomyces*, such polysaccharides have been described as galactomannoproteins (Domizio et al., 2017; Palomero et al., 2009) and may be produced in concentrations 3−7 times higher than those released by *S. cerevisiae* (Domizio et al., 2017).

Yeast polysaccharides have been characterized for a wide series of properties, such as their ability to improve wine colloidal stability, interact with wine aroma compounds to modulate their volatility and perception, adsorb toxic compounds, and interact with polyphenols to improve mouthfeel and reduce astringency, as well as for their antioxidant and emulsifying capacity (Table 20.2).

TABLE 20.2 Major components of yeast lees, their concentration, and effects on wine.

	Indicative concentration in yeast lees	Type of molecules	Indicative concentration in wines aged on lees	Effects on wine	References
Polysaccharides	Up to several g/L	Mannoproteins, Galactomannoproteins (*Schizosaccharomyces* spp.), glucans	From 150–200 to several hundred mg/L	Improved tartaric stability	Leger, Charpentier, and Feuillat, (1993); Ledoux, Perrin, Paladin, and Dubourdieu (1997)
				Improved protein stability	Waters, Wallace, Tate, and Williams (1993); Moine-Ledoux and Dubourdieu, 1999
				Improved foamability in sparkling wines	(Vincenzi, Crapisi, & Curioni, 2014)
				Interaction with wine aroma compounds	Lubbers, Voilley, Feuillat, and Charpentier, (1994); Comuzzo et al. (2011)
				Interaction with phenolic compounds	Feuillat (2003)
				Increased mouthfeel	Alexandre & Guilloux-Benatier (2006)
				Adsorption of medium-chain fatty acids	(Guilloux-Benatier & Feuillat, 1991)
				Adsorption of Ochratoxin A	Caridi (2006)
				Velum formation in flor wines	Feuillat (2003); Caridi (2006)
				Antioxidant capacity	Gallardo-Chacón, Vichi, Urpí, López-Tamames, and Buxaderas (2010); Jaehrig, Rohn, Kroh, Fleischer, and Kurz (2007); Tirelli, Fracassetti, and De Noni (2010)
				Emulsifying properties	(Li and Karboune, 2019)
Nitrogen compounds	11.5–42 mg/g dry weight as total nitrogen (Fornairon-Bonnefond et al., 2001)	Proteins	From 0.4 to 2 mg/L (as protein nitrogen in sparkling wines) (Moreno-Arribas et al., 1996)	Improved foamability in sparkling wines	Moreno-Arribas, Pueyo, Nieto, Martín-Álvarez, and Polo, 2000)
		Peptides	From 17 to 90 mg/L (as peptide nitrogen in sparkling wines) (Moreno-Arribas et al., 1996)	Stimulation or inhibition of malolactic fermentation	Alexandre et al. (2001); Remize, Augagneur, Guilloux-Benatier, and Guzzo, (2005); (Rizk et al., 2016)
				Surfactant properties	Moreno-Arribas, Pueyo, and Polo, (1996)
				Sweet–bitter taste	Moreno-Arribas et al. (1996); Alexandre and Guilloux-Benatier (2006)

(*Continued*)

TABLE 20.2 (*Continued*)

	Indicative concentration in yeast lees	Type of molecules	Indicative concentration in wines aged on lees	Effects on wine	References
		Amino acids	From 39 to approx. 230 mg/L (as amino nitrogen in sparkling wines) (Moreno-Arribas et al., 1996)	Aroma precursors	Alexandre and Guilloux-Benatier (2006); Jackson (2008)
				Stimulation of alcoholic and malolactic fermentation	(Bell & Henschke, 2005)
Lipids	5.40 ± 0.46% in lees from Sherry wine (Gómez et al., 2004)	Triglycerides, glycolipids, phospholipids, free fatty acids and their methyl esters, sterols, and sterol esters	Sterol esters and triacylglycerols: 1−5 mg/L	Aroma precursors	Pueyo, Martínez-Rodríguez, Polo, Santa-María, and Bartolomé (2000)
			Free fatty acids, sterols, diacylglycerols, and monoacylglycerols: 40−600 μg/L	Foam stability in sparkling wine	Gallart, Lopez-tamames, Suberbiola, and Buxaderas (2002)
			Data collected in model wine, pH 3.00, 10% v/v ethanol; 10 g/L commercial yeast suspension (Pueyo et al., 2000)	Survival factors for yeasts during refermentation	(Ribéreau-Gayon et al., 2006)
				Involved in oxygen consumption capacity of lees	Salmon, Fornairon-Bonnefond, Mazauric, and Moutounet (2000)
Nucleotides and nucleosides	0−390 mg/100 g of yeast biomass dry weight (Charpentier et al., 2005)	5′-, 3′-, and 2′-ribonucleotides	0−3 mg/L in aged Champagne wine, depending on aging time (Charpentier et al., 2005)	Flavor enhancers	Charpentier et al. (2005)
Glutathione and other thiol molecules	GSH: Variable conc. depending on yeast strain, aging time, and conditions;	Glutathione, protein fractions of yeast cell wall containing cysteinyl residues	GSH: 0−11 mg/L depending on yeast strain, aging time, and conditions (Kritzinger, Bauer, & Du Toit, 2013)	Antioxidant capacity	Gallardo-Chacón et al. (2010)
	Cysteinylated proteins: 0.73−1.4 mmol/100 g in commercial yeast autolysates (Tirelli et al., 2010); 0.10−0.89 nmol GSH/million cells in lees, depending on aging time (Gallardo-Chacón et al., 2010)				
Adsorbed polyphenols	87−594 ng gallic acid/million cells (Gallardo-Chacón et al., 2010)	Wine polyphenolic molecules adsorbed onto the lees surface during fermentation and aging		Antioxidant capacity	Gallardo-Chacón et al. (2010)

20.3.2.2 Nitrogen compounds

Nitrogen compounds represent another important group of substances, released from yeast cells during fermentation and autolysis. Total nitrogen content in lees may range from 11.5 to 42 mg/g of dry weight (Fornairon-Bonnefond et al., 2001), including proteins, peptides, and amino acids (AA) (Alexandre & Guilloux-Benatier, 2006). During autolysis, protein nitrogen decreases in lees because of the presence of proteolytic activities in the biomass (Leroy et al., 1990); this determines the release of peptides and AA in the medium (Alexandre et al., 2001; Charpentier & Feuillat, 1993).

In sparkling wines, the release of AA during AOL occurs in two different ways (Alexandre & Guilloux-Benatier, 2006; Alexandre et al., 2001); at the end of the refermentation process, amino nitrogen increases because AA are excreted out of the cell by passive exsorption. The release of AA due to the proteolytic degradation of yeast proteins and peptides starts later, after 3−9 months, depending on base wine composition, yeast strain, and aging conditions (Alexandre & Guilloux-Benatier, 2006). Martínez-Rodríguez, Carrascosa, Martín-Álvarez, Moreno-Arribas, and Polo (2002) described a more complex scenario for the release of soluble nitrogen during the industrial production of sparkling wines, which was summarized in four steps: (i) during refermentation (first 40 days after *tirage*), AA are consumed and peptides are released because of protein/polypeptide breakdown; (ii) between 40 and 90 days after *tirage*, proteins are released and peptides are degraded to AA; (iii) 90−270 days after *tirage*, enzymatic activities promote the further release of proteins and peptides; (iv) after approximately 270 days of aging, the concentration of AA decreases.

Basing on such indications, the release of AA from lees into the wine is subjected to fluctuations during aging, which depend on the equilibrium between their liberation during AOL and their microbial transformation (decarboxylation, deamination), which leads to the formation of different wine compounds (Martínez-Lapuente et al., 2018). Additionally, MLF can modify the nitrogen composition of wines during AOL, because *O. oeni* is able to hydrolyze peptides and glycosylated proteins (Remize et al., 2005). Consequently, peptide concentration also fluctuates during AOL, reaching a maximum after 12−15 months of aging and then decreasing (Alexandre & Guilloux-Benatier, 2006); such molecules are transformed over time due to the enzymatic activities of yeasts and their molecular weight becomes lower as the aging time increases (Martínez-Rodríguez & Polo, 2000; Moreno-Arribas et al., 1996). These indications highlight that the concentration and nature of nitrogen compounds in wines aged on the lees depend on several factors, such as the yeast strain and its proteolytic activity, the aging time and the storage conditions, as well as the initial composition of the medium, and the occurrence of fermentation processes (e.g., refermentation, MLF).

Nitrogen compounds are reported to have different effects on wine composition, such as their role in stimulating refermentation and MLF, their positive action on the foam of sparkling wines, and their potential impact on wine sensory characters (Table 20.2).

20.3.2.3 Lipids

Lipid fraction in wine lees is associated with the membrane of yeast cells. In *S. cerevisiae*, lipids represent 3%−15% of the cell dry weight (Fornairon-Bonnefond et al., 2001). Their concentration in lees has been poorly investigated. Analyzing lees from Sherry wine, Gómez and coworkers found a lipid content in the range of 5.40 ± 0.46% of lees dry weight, including free fatty acids and their methyl esters, triglycerides, glycolipids, phospholipids, sterols, and sterol esters (Gómez, Igartuburu, Pando, Rodríguez Luis, & Mourente, 2004). Such lipids can be released in wine during yeast autolysis (Fornairon-Bonnefond et al., 2001), even if most of them tend to remain in the cell envelopes (Babayan & Bezrukov, 1985; Ferrari, Meunier, & Feuillat, 1987). According to Pueyo, Martínez-Rodríguez, Polo, Santa-María, and Bartolomé (2000), the release of such molecules in model wine system increased at the beginning of accelerated autolysis, to decrease after the second day for the probable effect of hydrolytic activities in the biomass; the molecules found in the autolysates were triacylglycerols (TAG), 1,3-diacylglycerols (DAG), 2-monoacylglycerols (MAG), sterols (S) and their esters (SE), and free fatty acids (FFA); phospholipids (PL) were not detected (Pueyo et al., 2000).

During AOL, the ratio between polar (phospholipids) and nonpolar lipids (TAG) in yeast changes. Piton, Charpentier, and Troton (1988) observed the formation of lipidic vesicles inside the autolyzing yeast cells; such vesicles expanded during aging as they accumulated TAG; contrary phospholipids content decreased, suggesting a transformation of PL in TAG, with DAG as the intermediate substrate. In such conditions, some lipidic molecules can be excreted into the wine (Piton et al., 1988). FFA are some of the most studied molecules among the lipidic components of yeast lees; their content in wines aged *sur lies* varies from a few dozens to 280 µg/L (Ferrari & Feuillat, 1988). The most representative FFA in yeast lees are palmitic (C16), stearic (C18), oleic (C18:1), and linoleic (C18:2) acids (Ferrari & Feuillat, 1988; Gómez et al., 2004), but minor amounts of octanoic and decanoic acids have also been reported (Fornairon-Bonnefond et al., 2001).

The most representative sterol in yeast lees is ergosterol (37−176 µg/g dry weight), followed by squalene, lanosterol, and ergost-7-en(3β)-ol (Fornairon-Bonnefond & Salmon, 2003). The release of sterols in wines *sur lies* seems negligible (Fornairon-Bonnefond et al., 2001). In model wine subjected to accelerated autolysis, Pueyo and coworkers detected less than 100 μg/L of sterols, with a peak of 150 μg/L at day 2; contrarily, sterol esters ranged between 2 and 5 mg/L during the whole 12 days of aging (Pueyo et al., 2000).

Lipids have several important functions in wines aged on lees (Table 20.2); they are important survival factors for yeasts during refermentation (Ribéreau-Gayon et al., 2006), they are involved in the foam stability of sparkling wine (Gallart et al., 2002) as they are also aroma precursors, leading to the formation of esters, aldehydes, and ketones (Pueyo et al., 2000). Furthermore, they are connected with the great capacity of fresh lees to consume oxygen; such capacity decreases during aging because of the peroxidation of certain lipidic fractions (Salmon et al., 2000).

20.3.2.4 *Other components of yeast lees*

Besides polysaccharides, nitrogen compounds, and lipids, other minor lees components play a role in determining the compositional and sensory characteristics of wines aged *sur lies.*

Nucleotides and nucleosides are released during yeast autolysis because of the degradation of nucleic acids by DNase and RNase enzymatic activities (Alexandre & Guilloux-Benatier, 2006; Leroy et al., 1990; Zhao & Fleet, 2003). Charpentier et al. (2005) detected different monophosphate nucleotides in Champagne wines; their relative abundance was dependent on aging time. In particular, 5'-nucleotides were the only compounds detected in the first 2 years of aging; 5'-UMP was the most abundant, with a peak of concentration in the samples aged for 2 years. A slower increase in the first years of aging was observed in 2'- and 3'-nucleotides, reaching a maximum after 9 years. The average concentration of the different molecules was estimated to be in the range between 30 and 1500 μg/L, depending on the aging time, with a maximum overall amount of approx. 3 mg/L found in a wine sample aged on yeasts for 9 years (Charpentier et al., 2005). Nucleotides and nucleosides are well known to be powerful flavor enhancers (Nagodawithana, 1992). Even if they are detected at a concentration below their sensory threshold (Charpentier et al., 2005), they can have an impact on the sensory properties of wines aged *sur lies* (Alexandre & Guilloux-Benatier, 2006; Charpentier et al., 2005) because of the synergistic effects they can elicit in food systems, as well as in presence of glutamate, another natural component of autolyzed yeasts and yeast-derived products (Charpentier et al., 2005; Nagodawithana, 1992).

Glutathione (GSH) and thiol compounds represent other significant components of yeast lees. In *S. cerevisiae*, GSH represents 0.1%−1% of dry cell weight (Bachhawat et al., 2009). The presence of such tripeptide in yeast lees is important because of its antioxidant capacity (Kritzinger et al., 2013). According to Lavigne, Pons, and Dubourdieu (2007), AOL allows the preservation of wine glutathione content because of the capacity of lees themselves to scavenge oxygen, limiting GSH oxidation. However, GSH concentration tends to decrease during AOL; Kritzinger and colleagues showed that the levels of the tripeptide in white wines aged over lees gradually diminish after alcoholic fermentation, becoming negligible (0.8−1.2 mg/L) after 12 months of aging (Kritzinger et al., 2013). This reduction of GSH might be connected with the formation of disulfide bridges with the cysteinyl residues of yeast cell wall mannoproteins (Vasserot, Steinmetz, & Jeandet, 2003), with the sorption of GSH by lees (Kritzinger et al., 2013), or with the capacity of the tripeptide to react with quinones (Ugliano et al., 2011). Glutathione is a powerful wine antioxidant (Sonni, Clark, Prenzler, Riponi, & Scollary, 2011; Sonni, Moore, et al., 2011) and its presence contributes to exert a protective effect on wine aroma compounds, especially volatile thiols, terpenes, and esters (Kritzinger et al., 2013).

Besides glutathione, other thiol molecules can be found in yeast lees; in particular, certain protein fractions of yeast cell wall containing cysteinyl residues have been hypothesized to play a role in the antioxidant capacity of wine lees (Gallardo-Chacón et al., 2010; Jaehrig et al., 2007; Tirelli et al., 2010). According to Gaillardo-Chacón and coworkers, cell wall thiols accounted for $25 \pm 11\%$ of the total lees surface antioxidant activity estimated using DPPH assay; this antioxidant potential decreases during the aging time due to the diminution of such thiol functions (Gallardo-Chacón et al., 2010). Besides their antioxidant capacity, the presence of thiol compounds in lees may reduce the development of reduction sulfur off-flavors in wines because of the fixation of such compounds via disulfide bridges on the cysteinyl residues of the cell wall mannoproteins (Vasserot et al., 2003).

Finally, yeast lees contain adsorbed polyphenols and aroma compounds coming from grape and must fermentation. The former group of molecules is involved in the antioxidant capacity of fresh lees (Gallardo-Chacón et al., 2010). Volatile compounds in lees and wines aged *sur lies* represent a complex and wide subject; some details will be given in the following sections.

20.4 Yeast autolysis and autophagy

Yeast autolysis is the hydrolysis of intracellular biopolymers under the effect of hydrolytic enzymes associated with cell death and the consequent formation of low-molecular-weight products (Babayan & Bezrukov, 1985). This process becomes enologically important when such products are released from the cell in the external medium. As discussed above, most of the research carried out on autolysis in enology was aimed at analyzing the compounds liberated in the wine, emphasizing proteolysis and the solubilization of polysaccharides, nucleic acids, and lipids. Actually, these phenomena correspond to the final step of cellular apoptosis, in which two main distinct phases, which are connected to one another, play a significant role: autophagy and proper autolysis.

20.4.1 Autophagy in yeasts

Autophagy is a key step in the maintenance of cellular homeostasis, in which the biosynthetic processes are as important as catabolic reactions. The mechanism of autophagy in *S. cerevisiae* has been investigated by numerous authors since this microorganism is used as a model system to understand the more complex (but similar) mechanisms of the eukaryotic cells of multicellular organisms (Almeida et al., 2007; Budovskaya, Stephan, Reggiori, Klionsky, & Herman, 2004; Reggiori & Klionsky, 2013). In particular, autophagic dysfunctions have been linked to a large number of diseases, such as cancer onset, diabetes, myopathies, certain neurodegenerative diseases, heart and liver diseases.

Autophagy may take place under winemaking conditions, contributing to the release of yeast compounds into the wine (Cebollero & Gonzalez, 2006; Cebollero, Carrascosa, & Gonzalez, 2005). When *Saccharomyces* cells suffer nutrient starvation, they enter into a nondividing resting state (stationary phase), characterized by low metabolic activity, increased resistance to environmental stresses, and modification of gene expression pattern (Budovskaya et al., 2004). Gene expression changes include the activation of genes involved in long-term survival and autophagy, which corresponds to a selection of processes for protein and macromolecules breakdown and degradation of organelles. Such processes operate by diverse mechanisms, with regard to the machinery involved, the nature of the substrate, and the site of sequestration (Brown, Hung, Dunton, & Chiang, 2010; Hoffman & Chiang, 1996; Hung, Brown, Wolfe, Liu, & Chiang, 2004).

One of the most probable mechanisms of autophagy consists of the invagination and scission of the vacuole membrane with the production of intravacuolar vesicles that undergo degradation, without any damage to the integrity of the vacuole (Müller et al., 2000; Sattler & Mayer, 2000). Other observations report that small portions of nuclear membrane and nucleoplasm protrude into the vacuole lumen, thanks to the interaction of specific proteins, leading to the formation of a vacuolar vesicle that is subjected to hydrolysis (Dunn et al., 2005; Krick et al., 2008; Roberts et al., 2003). Other cell membrane structures, such as endoplasmic reticulum, Golgi apparatus, and plasma membrane, may also be involved in the formation of autophagosomes, which are responsible for delivering proteins and other cytoplasmatic constituents to the vacuole for degradation (Geng, Nair, Yasumura-Yorimitsu, & Klionsky, 2010; Klionsky et al., 2003; Lynch-Day et al., 2010; Reggiori & Klionsky, 2013; Shintani & Klionsky, 2004; Taylor, Chen, Chou, Patel, & Jin, 2012).

Autophagy regulation in yeasts depends primarily on nutrient withdrawal. There is a complex system of interactions among the regulatory mechanisms which modulate the process: nitrogen-dependent regulation, glucose depletion, AA and phosphate starvation, mitophagy, pexophagy, transcriptional control, inositols (Reggiori & Klionsky, 2013), and sphingolipids (Yamagata, Obara, & Kihara, 2011).

20.4.2 Autolysis of yeast cell

After the degradation of proteins, macromolecules, and organelles as determined by autophagy, yeast autolysis is the lytic event of the cell, which allows the release of cytoplasmic (AA, peptides, fatty acids, and nucleotides) and cell wall (mannoproteins) compounds into the wine. It is an irreversible process initiated by intracellular enzymes. Conversely, with respect to autophagy (which starts at the beginning of the stationary phase), autolysis begins at the final stage of the stationary phase, when cell death occurs (Babayan & Bezrukov, 1985). As reported by Babayan et al. (1981), autolysis can be described using the following four steps: (1) degradation of endostructures and discharging of vacuolar proteases in the cytoplasm; (2) activation of vacuolar proteases by degradation of their specific cytoplasmic inhibitors; (3) hydrolysis of the intracellular polymeric compounds and their accumulation in the proximity to the cell wall; (4) release of the hydrolyzed products which are degraded to a molecular mass compatible to the size of the pores in the cell wall. Optimal conditions for autolysis are 45°C and pH 5, which are very different from the typical wine environment (pH 3–4, 15°C, ethanol 12% v/v) (Charpentier & Feuillat, 1993; Connew, 1998; Todd, Fleet, & Henschke, 2000). This makes natural autolysis a slow process.

Autolysis also determines the degradation of cell wall structures. The main steps can be summarized as follows: (1) glucanases hydrolyze glucans, allowing the release of mannoproteins (Klis, Mol, Hellingwerf, & Brul,

2002); (2) glucans are also released, thanks to the residual activity of solubilized glucanases and glucanases bound to the cell wall (Alexandre & Guilloux-Benatier, 2006; Lombardi, De Leonardis, Lustrato, Testa, & Iorizzo, 2015); (3) finally, the protein fraction of mannoproteins is degraded by proteases. These events can be followed using transmission electron microscopy (TEM). Healthy cells appear round and velvety, with smooth surfaces; when autolysis occurs, they appear wrinkled, wizened, and compressed. The cell wall remains unbroken during the entire process, due to a tangled structure of fibers formed by residual β-1,3-glucan associated with proteins (Schiavone et al., 2017; Tudela et al., 2012). The autolytic capacity is strain dependent (Perpetuini et al., 2016); thus strain selection can be an optimal solution to improving wine quality (Martínez-Rodríguez, Carrascosa, & Polo, 2001; Nunez, Carrascosa, González, Polo, & Martínez-Rodríguez, 2005).

20.5 Technological management of aging on lees

20.5.1 Aging on lees in still wines

20.5.1.1 Aging in wood barrels

During the aging of still wines, the contact of wine with lees may last several months, averaging from 2−3 to 15 (Fornairon-Bonnefond et al., 2001). As already mentioned, the traditional approach practiced to manage AOL consists of barrel fermentation followed by *sur lies* aging in the same container (without racking), according to the winemaking style used in Burgundy for Chardonnay wines (Ribéreau-Gayon et al., 2006). Alcoholic fermentation, however, may be also carried out in stainless steel vats, in which case the transfer of wine in wood barrels is only done in its final stages or after its completion (e.g., after the first racking). Thus AOL can be accomplished on total lees or on fine lees. Barrel aging on total lees results in a faster/higher release of polysaccharides (Ribéreau-Gayon et al., 2006) and minor color development (Chatonnet, Dubourdieu, & Boidron, 1992) with respect to the same wine aged in tank on fine lees. Moreover, AOL on total lees determines a lower sensitivity toward oxidative pinking (Dubourdieu, 1995; Ribéreau-Gayon et al., 2006). Contrarily, practical experiences suggest that aging on total lees may increase the risk of the development of sulfur off-flavors, especially for certain wine typologies susceptible to producing high amounts of crude lees; for such varieties, the winemaker shall take care of must fining while also preferring a more gentle approach during grape processing.

AOL in wood barrels is the result of a complex equilibrium among the characteristics of the wine itself, the contribution of lees, the impact of wood and their interaction, representing one of the most intriguing technical challenges for winemakers, because of the number of technological variables that can be combined when designing a given wine. For instance, different results may be achieved depending on wood composition (e.g., wood species, grain, permeability to oxygen), barrel size (e.g., *barrique*, *tonneaux*, or larger cooperage containers), barrel age (e.g., new barrels, 1- or 2-year-old vessels), and wood manufacturing (e.g., toasting level). Wood does not only release phenolic and aroma compounds in wine; a fundamental feature of wood casks is related to their capacity to provide natural microoxygenation of wine. During AOL, the oxygen permeability of wood is counterbalanced by the capacity of lees to scavenge oxygen, so that in the presence of lees, white wines are less susceptible to oxidation. At the same time, the slight and gradual oxygen transfer through the wood avoids the development of reducing off-flavors in the lees deposit (Fornairon-Bonnefond, Camarasa, Moutounet, & Salmon, 2001; Ribéreau-Gayon et al., 2006). The resuspension of the lees sediment (*bâtonnage*) is a fundamental operation from this point of view, because it homogenizes the redox potential inside the barrels, limiting the formation of gradients between the upper part of the wine (close to the ullage) and the lees layer (Jackson, 2008; Ribéreau-Gayon et al., 2006). Practically speaking, the frequency of stirring varies depending on the susceptibility of the wine to undergo oxidation or reduction during aging. Fornairon-Bonnefond et al. (2001) report (as an example) one stirring operation per week, but certain varieties very susceptible to producing sulfur off-odors may require an increased *bâtonnage* frequency, up to once or even twice per day in the first weeks of aging. The sensibility of the winemaker is fundamental to setting up a correct sequence of stirring operations by continuous sensory monitoring of barrels during aging. Stirring also accelerates the release of polysaccharides and AA from the lees into the wine (Feuillat, 2003).

MLF normally occurs in still wines aged on lees because of the stimulating effect of the lees themselves on the development of LAB. If MLF is desired, sulfur dioxide levels should be kept low (e.g., no sulfiting after alcoholic fermentation), barrels topped off, and temperature set at 16°C−18°C (Ribéreau-Gayon et al., 2006). Obviously, these conditions may favor the development of unwanted microbial strains, so it would be recommended to complete MLF before AOL, eventually by coinoculating malolactic bacteria with the selected yeast strain used for alcoholic fermentation. An immediate sulfiting or lysozyme addition after MLF may reduce further risks of microbial pollution.

Fining treatments are generally not carried out on wines during AOL. For instance, the use of bentonite is not recommended, because the presence of the fining agent in the lees may damage the organoleptic quality of the wines (Ribéreau-Gayon et al., 2006). Bentonite fining (if needed) may be carried out on must before alcoholic fermentation, or after aging, even if AOL generally increases wine colloidal stability (Ledoux et al., 1997; Lubbers et al., 1993; Moine-Ledoux & Dubourdieu, 1999; Waters et al., 1993), reducing the stabilization treatments required.

20.5.1.2 Aging in stainless steel vats

When AOL is managed in stainless steel, the variables to be taken into account are the larger size of the containers and their lower permeability to oxygen.

As a general rule, as the volume of the vessels increases, the surface/volume ratio decreases and the contact surface of lees per unit volume of wine becomes lower. In addition, in large stainless steel vats, an efficient and homogeneous resuspension of lees (normally carried out by injecting nitrogen from the bottom valve or providing vats themselves with tank mixer agitators) is more difficult than in barrels. This provokes a slower release of compounds from the lees, a limited oxygen dissolution in the tank, and a higher risk of the development of reductive off-odors during AOL (Lavigne & Dubourdieu, 1996; Ribéreau-Gayon et al., 2006). This phenomenon is emphasized by the higher pressure exerted on the lees in large-volume containers, which provokes the compacting of the sediment and stimulates the production of hydrogen sulfide and other organic thiols (Jackson, 2008). High levels of sulfur dioxide represent a further problem from this point of view because of the residual sulfite reductase activity in yeasts (Ribéreau-Gayon et al., 2006).

Different strategies have been suggested to reduce this problem. They are mostly based on combining AOL with microoxygenation, as well as with the addition of oak chips or staves, for simulating the natural permeability of wood or for miming the effects of wood-soluble substances (Del Barrio-Galán, Pérez-Magariño, & Ortega-Heras, 2011; Rodrigues, Ricardo-Da-Silva, Lucas, & Laureano, 2016; Sartini, Arfelli, Fabiani, & Piva, 2007). However, the number of papers produced about the use of these practices in white wine aging is limited and further investigations are necessary for understanding their impact on different wine varieties and winemaking conditions.

An interesting approach for reducing the development of reduction off-odors during AOL in stainless steel tanks is the one suggested by Lavigne (1995). It consists of the separation of lees after sulfiting and their storage in barrels for approximately 1 month. During this time, lees lose their sulfite reductase activity and consequently their reductive potential, so that they can be reincorporated into the wine without bringing sulfur off-odors; in addition, they are able to adsorb thiol compounds (e.g., methanethiol) eventually present in the wine itself (Ribéreau-Gayon et al., 2006). This practice may have other advantages. For instance, the separation of lees and their subsequent reincorporation in wine also allows to potentially reduce the time needed for AOL, by applying different technologies suitable for accelerating autolysis; these aspects will be further discussed in the following paragraphs.

20.5.2 Aging on lees in sparkling wines

20.5.2.1 Classic method

In sparkling wine technology, AOL is traditionally connected with the classic method (refermentation and aging in bottle). Generally, in sparkling wine production, AOL can be carried out on base wines (before refermentation) or after the *prise de mousse*.

In the production of Cava, the wine remains in contact with lees for at least 9 months, with a variable aging time, which depends on the kind of product desired. Francioli et al. reported a wide range of aging times, up to 7 years (Francioli, Torrens, Riu-Aumatell, López-Tamames, & Buxaderas, 2003). For Champagne, aging ranges from a minimum of 15 months (for nonvintage Champagne) to 3 years minimum for vintage Champagne (*millesimé*); however, such wines may stay on their lees for up to 8 years, or even longer for certain special *cuvées* (Ribéreau-Gayon et al., 2006). For the Italian Franciacorta the minimum length required is 18 months for "Franciacorta" DOCG and up to 60 months for "Franciacorta riserva" ("Disciplinare di Produzione dei vini a Denominazione di Origine Controllata e Garantita "Franciacorta,"" 2020).

During AOL, MLF may occur in bottled sparkling wines. During refermentation and aging, it is difficult to control MLF, and this may provoke an increased risk of microbial pollution and, in certain cases, a loss of freshness of the wine (Ribéreau-Gayon et al., 2006). In Champagne, when MLF is desired, it is generally carried out on base wine and managed by sequential inoculation or coinoculation (Kemp, Alexandre, Robillard, & Marchal, 2015). Biological deacidification of the base wine is suggested especially in products with high acidities. Contrarily, to reduce the risk of MLF during bottle aging, suitable levels of free and total sulfur dioxide are recommended; MLF becomes very difficult when the total SO_2 level is kept at 80–100 mg/L (Ribéreau-Gayon et al., 2006). Lysozyme (Gerbaux, Villa, Monamy, & Bertrand, 1997) or microfiltration (Ribéreau-Gayon et al., 2006) of base wine can also be helpful for controlling MLF during sparkling wine aging.

20.5.2.2 Martinotti–Charmat method

Differently, with respect to the traditional method, in which a minimum period of AOL is mandatory for wine appellation, the products obtained by refermentation in autoclave are not traditionally subjected to AOL. In fact, the Martinotti–Charmat method (MCM) has been introduced as a refermentation method suitable for obtaining sparkling wines with high freshness and intense floral and fruity notes, as well as for exalting the varietal characteristics of aromatic grapevine varieties (e.g., Muscat, Riesling, Traminer). Typical products obtained by MCM are Asti spumante and Prosecco.

In MCM, secondary fermentation in pressurized tanks lasts for about 30–40 days. Normally, after this time, wine is subjected to cross-flow filtration or centrifugation, and then bottled in isobaric conditions. Certain products, however, are subjected to a more or less prolonged AOL, which may vary from a few to several months (even 10 or more). During aging, lees are resuspended in order to favor a higher contact surface with wine that, in autoclave, is lower compared to that found in bottle. Examples of products obtained through this "extended MCM" approach are certain Prosecco and Ribolla Gialla sparkling wines from northeastern Italy. Autoclaves are normally equipped with heating/cooling systems, useful for providing suitable temperatures for refermentation and aging, as well as for allowing tartaric stabilization. According to some producers, a 6-month aging period seems appropriate for obtaining a good balance between freshness (characteristic of young sparkling wines) and a higher body and structure, finer *perlage*, and higher aroma complexity (typical features of aged sparkling wines).

20.5.2.3 Other applications of aging on lees in sparkling wine production

Besides than in bottle or autoclave, in sparkling wine production, AOL can be managed also storing base wine for a more or less prolonged time on primary fermentation lees. AOL of base wine increased the concentration of soluble colloids during the manufacturing of Trento DOC sparkling wine (Contessotto, 1999). However, this practice requires that a given number of vats be dedicated to this practice, potentially reducing the number of containers available in the winery during harvest time. For this reason, in the case of MCM, some winemakers prefer to extend the aging period of sparkling wines, rather than to age base wines, with the advantage of having the products available for marketing in the same vintage.

Finally, a peculiar method of applying AOL in sparkling wine production is the olden method for manufacturing Prosecco. It consists of storing base wine in vat over lees for the whole winter. In spring, the sugar level is eventually corrected and the base wine is bottled and subjected to refermentation, inoculating selected yeasts or even managed spontaneously. The sparkling wine obtained is kept on its lees for at least 3 months, until consumption, and it is sold without riddling and disgorging (Giotto, 2018). The idea of a spontaneous, authentic, and sugar-free product leads to a good appreciation of these peculiar products by the consumers.

20.6 Impact of AOL on wine aroma and sensory characteristics

The interaction of lees with wine volatile compounds is a complex subject. In general, AOL improves the organoleptic characters of certain wines, attenuating wood aroma, promoting the formation of compounds with floral and fruity odor, and adsorbing wine volatile compounds, aroma precursors, and off-odors (Gambetta, Bastian, Cozzolino, & Jeffery, 2014).

Various molecules, including esters, terpenols, lactones, higher alcohols, aldehydes, and ketones, may be released during yeast autolysis (Alexandre & Guilloux-Benatier, 2006; Liberatore, Pati, Del Nobile, & La Notte, 2010). The effect of AOL on wine aroma composition varies depending on aging time; for instance, in young (sparkling) wines, acetates, ethyl, and isoamyl esters prevail (Alexandre & Guilloux-Benatier, 2006), whereas vitispiranes, 1,6,6-trimethyl-1,2-dihydronaphthalene (TDN), carbonyl compounds, furans, diethyl succinate, and ethyl lactate increase with aging time. This enhances aroma complexity, modifying wine sensory perception with toasty and yeast-like notes (Kemp et al., 2015; Torrens, Rlu-Aumatell, Vichi, López-Tamames, & Buxaderas, 2010). For this reason, AOL is generally considered unsuitable for the production of fresh, varietal aromatic wines.

In sparkling wines, the occurrence or the absence of AOL (depending on the refermentation method) also affects aroma composition. Voce, Battistutta, Tat, Sivilotti, and Comuzzo (2020) found that the refermentation of Ribolla Gialla by MCM (short aging or no aging) led to higher amounts of terpenes and ethyl, isoamyl, and acetic esters, conferring higher freshness and perception of floral, green apple, and citrus notes. Contrarily, samples refermented by the classic method (and aged *sur lies* for a longer time) had a greater aroma complexity, characterized by higher concentrations of free fatty acids, higher alcohols, aging esters (e.g., ethyl lactate), lactones, and carbonyl compounds (Fig. 20.1).

AOL also determines sensory modifications concerning wine tasting attributes; the macromolecules released in wine during autolysis (e.g., mannoproteins) contribute to increasing the body and mouthfeel properties of *sur lies* wines (Alexandre & Guilloux-Benatier, 2006). Yeast macromolecules are also able to modulate the volatility and the sensory perception of certain wine compounds (Lubbers et al., 1994), depending on their chemical characteristics, wine temperature, and pH (Comuzzo et al., 2011). Finally, AOL also affects the foam properties of sparkling wines. According to Culbert et al., the products obtained by the classic method are characterized by a higher foam volume and stability (Culbert et al., 2017). Foam properties change during AOL, decreasing during the first 12 months of aging to increase again after longer aging periods (e.g., 18 months) (Moreno-Arribas et al., 2000).

20.7 Microbiological aspects connected with aging on lees

From a microbiological point of view, AOL leads to an increase in the nutrients available for spoiling yeasts and bacteria. One of the major microbiological problems during AOL is the development of yeasts of genus *Brettanomyces/Dekkera* because of their ability to produce several undesirable compounds that can seriously affect the sensory quality of wine. The presence of volatile phenols, particularly 4-ethylphenol and 4-ethylguaiacol, results in a horse-sweat odor or "animal" flavor, olfactorily perceptible even at low concentrations (Ribéreau-Gayon et al., 2006). The risk of increased growth of *Brettanomyces/Dekkera*, and consequently the production of such unpleasant flavors, is real during wine storage (Fornairon-Bonnefond et al., 2001; Renouf & Lonvaud-Funel, 2004). Guilloux-Benatier, Chassagne, Alexandre, Charpentier, and Feuillat (2001) demonstrated that *Brettanomyces* was able to grow in synthetic media containing yeast lees treated with β-glucanase, although the increase in extracellular glucose did not affect its concentration. In such conditions, the authors observed an increment of *Brettanomyces* counts of 5 Log CFU/mL in 5 days; on the other hand, volatile phenol contents were 3–5 times lower than in the control wines, suggesting that yeast lees were able to adsorb these compounds by fixation onto the cell walls. This hypothesis was also stressed by Palomero, Ntanos, Morata, Benito, and Suárez-Lepe (2011), who confirmed that the use of lees can lead to the removal of unwanted volatile compounds, thanks to their adsorbent capacity.

The adsorbent capacity of lees is also connected with the stimulation of MLF. In fact, medium-chain fatty acids and other toxic metabolites produced by yeasts can be adsorbed and eliminated from the growing medium, while the release of small quantities of glucose and AA can stimulate the growth of *O. oeni* or other LAB species (Lerm, Engelbrecht, & du Toit, 2010; Sumby, Bartle, Grbin, & Jiranek, 2019). On the other hand, the increment of free AA in the environment can also affect the content of biogenic amines. AA are the precursors of biogenic amines and ethyl carbamate, and under uncontrolled winemaking conditions, their increment may result in an increased concentration of these molecules (Smit & du Toit, 2013). Spontaneous MLF is particularly exposed to uncontrolled potential spoilage microbes, including decarboxylase-positive LAB, and in the presence of lees the risk of biogenic amine contamination is higher (Smit & du Toit, 2013). In general, numerous studies have revealed that the frequency of biogenic amine incidence in wines aged *sur lies* was higher during the first 4–6 months of aging. This is connected to the presence of vitamins and nitrogen compounds in the medium, as well as to the fact that sulfur dioxide is not able to completely stop all the biochemical reactions and enzymatic activities after MLF (Alcaide-Hidalgo, Moreno-Arribas, Martín-Álvarez, & Polo, 2007; Hernández-Orte et al., 2008; Marques, Leitão, & San Romão, 2008; Smit & du Toit, 2013). However, in some cases a decrease in the concentration of biogenic amines during long aging could be observed, supposedly due to amine oxidase activities of specific bacterial strains, which transform these unwanted molecules into aldehydes, hydrogen peroxide, and ammonia (Leuschner, Heidel, & Hammes, 1998).

20.8 Technologies for accelerating aging on lees

As already mentioned, natural autolysis is a slow process, because of the low cell death rate and enzymatic activities determined by the low aging temperatures (Torresi, Frangipane, Garzillo, Massantini, & Contini, 2014). For this reason, several months are required to observe perceptible effects in wines. Different strategies have been suggested to accelerate AOL.

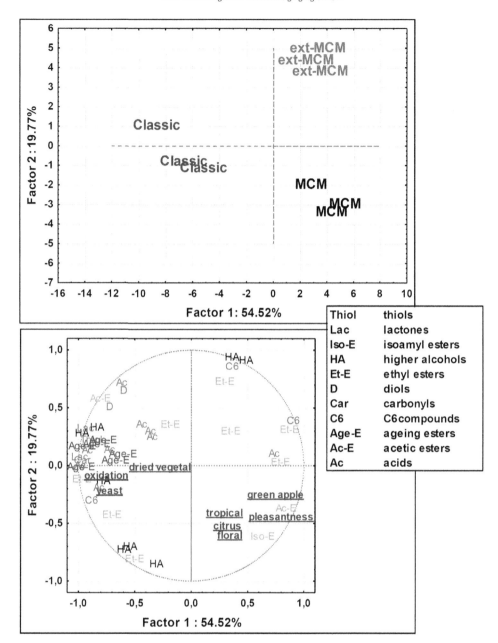

FIGURE 20.1 Comparison of aroma and sensory profile of sparkling wines obtained by different refermentation methods and aging times. *Classic*: Classic method, refermentation and aging in bottle over lees for 11 months; *MCM*: Martinotti-Charmat method, refermentation (40 days) and immediate bottling (no aging on lees); *ext-MCM*: extended Martinotti-Charmat method, refermentation (40 days) and aging over lees in autoclave for 4 months. Source: © *Modified from Voce, S., Battistutta, F., Tat, L., Sivilotti, P., & Comuzzo. (2020). Composizione chimica ed evoluzione del profilo aromatico di spumanti ribolla gialla prodotti con diversi metodi di rifermentazione.* Infowine, 3, 1–9.

20.8.1 Use of enzymes

The use of β-glucanase enzymes is the most common method available. Nowadays, a number of commercial enzyme preparations are marketed; such products are mixtures of β-(1→3)- and β-(1→6)-glucanases (International Organization of Vine and Wine, 2018) able to degrade the β-glucans of the yeast cell wall, thus improving cell wall degradation (Torresi et al., 2014). Actually, such preparations do not provoke a complete cell breakdown, but they lead to the decrease in cell wall thickness (Fig. 20.2), as well as the formation of wrinkles and folds on the cell surface (Alexandre & Guilloux-Benatier, 2006), accelerating the release of polysaccharides and other substances (Morata, Palomero, Loira, & Suárez-Lepe, 2018).

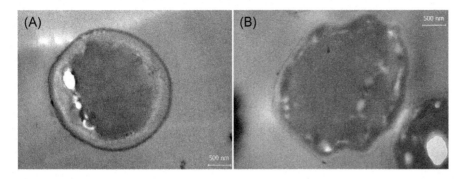

FIGURE 20.2 *Saccharomyces cerevisiae* cells before (A) and after (B) treatment with β-glucanase enzymes (1 g/L). Transmission electron microscopy (TEM), magnification 19,000 ×.

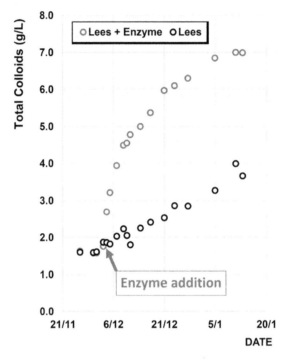

FIGURE 20.3   Effect of β-glucanase addition on the release of glucidic colloids from wine lees. Source: *Modified from Contessotto, A. (1999). Indagine sulle possibili applicazioni dei preparati ad azione glucanasica nella produzione di vini spumanti secondo il metodo classico (Master Thesis). University of Udine.*

Palomero and colleagues observed that, when β-glucanases were added to a model wine containing different *Saccharomyces* strains, autolysis was complete in 2 or 3 weeks, while conventional autolysis required at least 5 months (Palomero, Morata, Benito, Gonzalez, & Suarez Lepe, 2007). Studying sparkling wines, Torresi et al. also confirmed these results: enzymes were able to accelerate the aging of the products analyzed; the effects of the treatment were strain dependent (Torresi et al., 2014).

The addition of β-glucanase enzymes can be managed also as suggested by Contessotto (1999). Fresh lees are separated from wine after alcoholic fermentation, treated with enzymes, and then reincorporated into the wine itself before refermentation (sparkling wines) or aging (still wines); in such conditions, the author observed a significantly faster release of polysaccharides over a period of 2 months (Fig. 20.3) (Contessotto, 1999).

Enzyme preparations may also have an impact on wine aroma and sensory characteristics: wines treated with β-glucanases were characterized by higher concentrations of total volatiles, particularly ethyl esters, 2-phenylethanol, hexanol, and *trans*-3-hexenol (Masino, Montevecchi, Arfelli, & Antonelli, 2008). In sparkling wines, enzymatic treatment increased the aging characteristics of the products, conferring yeast and dough notes, as well as a more sweet taste (Rodriguez-Nogales, Fernández-Fernández, & Vila-Crespo, 2012).

20.8.2 Emerging technologies for accelerating aging on lees

Besides enzymes, several new preservation technologies have demonstrated their potential role in accelerating autolysis and AOL.

Ultrasounds (US) are high-frequency mechanical waves (20−100 kHz) able to produce compression waves in liquid food products (Morata et al., 2017). When applied to a fluid, US determine the formation and collapse (cavitation) of high-energy microbubbles, generating a localized increase of pressure (up to 50 MPa) and temperature (higher than 5500°C) (Jiranek, Grbin, Yap, Barnes, & Bates, 2008). This feature makes US an efficient technology to disrupt microbial cells, promoting the depolymerization of the cell envelope and the release of mannoproteins and polysaccharides (Morata et al., 2018). Like the other emerging techniques discussed in this section, US is generally considered a non-thermal technology; however, if the treatment is carried out for a relatively long time (higher than 5 minutes), sample temperature can increase above 60°C (Morata et al., 2017).

Concerning their potential effect on yeast autolysis, Liu et al. (2016) observed a significant acceleration in polysaccharide release during 2 weeks of aging of a model wine containing lees from two different yeast strains; the solubilization of macromolecules was even more intense in the presence of inert abrasive particles (e.g., glass beads). Similar results were obtained in other papers: US considerably increased the release of polysaccharides during AOL, accelerating autolysis in both *Saccharomyces* and non-*Saccharomyces* strains (Cacciola, Batllò, Ferraretto, Vincenzi, & Celotti, 2013; del Fresno et al., 2018, 2019; Kulkarni et al., 2015). Consequently, the sonication of lees may allow a reduction in sulfur dioxide levels in wine, allowing to shorten the aging period (del Fresno et al., 2019).

High pressurization technologies have been used for a number of decades for promoting cell disruption as well as for extracting proteins or other compounds from yeasts (Follows, Hetherington, Dunnill, & Lilly, 1971; Hetherington, Follows, Dunniil, & Lilly, 1971; Middelberg, 1995). High-pressure techniques can operate in static or dynamic (in-flow) conditions. The latter allows the continuous processing of the lees suspension, significantly reducing the treatment time, which is more suitable for processing the volumes normally found at the winery scale (Comuzzo & Calligaris, 2019). High-pressure homogenization consists of forcing a fluid to pass through a homogenization valve with a suitable geometry; the suspended particles are subjected to intense mechanical forces (cavitation, shear, turbulence) and elongation stresses, becoming twisted, deformed, and disrupted (Comuzzo & Calligaris, 2019). When the pressure applied is up to 150−200 MPa, we talk about high-pressure homogenization (HPH); when it is increased up to 300−400 MPa, we are in the field of ultra-high-pressure homogenization (UHPH) (Dumay et al., 2013).

HPH is able to promote yeast autolysis increasing the release of glucidic colloids, proteins, and AA (Comuzzo et al., 2015) proportionally to the pressure applied (50−150 MPa) and the number of passes of the yeast suspension in the homogenizer (1−10) (Comuzzo et al., 2015, 2017). Patrignani and colleagues also reported the positive effects of HPH (90 MPa) on the modulation of the autolytic phenomena of *tirage* starter cultures for sparkling wines (Patrignani et al., 2013). Finally, Chinni (2018) found that the HPH treatment of fresh lees (60 MPa, 1 pass) determined a significant increase of glucidic colloids in wines after 6 months of aging, without affecting their filterability; HPH determined a lower concentration of C6 alcohols and higher amounts of fermentative esters in the wines aged *sur lies* (Chinni, 2018).

Pulsed electric fields (PEF) also demonstrated a positive effect in accelerating yeast autolysis. PEF is based on the application of short pulses (micro- to milliseconds) of high-voltage electric current (10−80 kV/cm) to food products positioned between two electrodes (Maged & Amer Eissa, 2012); these conditions produce the formation of pores on cell membranes (electroporation), causing modifications in cell permeability, up to the permanent cell breakdown (Tsong, 1991).

Contrary to HPH, which provokes the fragmentation of the cell, producing debris (Siddiqi, Titchener-Hooker, & Shamlou, 1997), PEF does not determine initially any evident modification in the cell wall structure (Fig. 20.4).

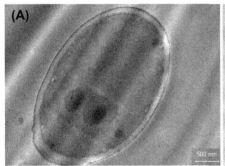 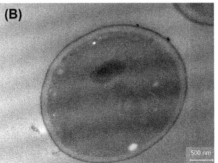

FIGURE 20.4 *Saccharomyces cerevisiae* cells before (A) and after (B) PEF treatment (25 kJ/L). Transmission electron microscopy (TEM), magnification 19000 ×.

However, a faster release of mannose in the extracellular environment was observed over 21 days, when a yeast suspension was treated by PEF (Martínez et al., 2018). Similar results were also found by (Martínez, Delso, Maza, Álvarez, & Raso, 2019) studying the evolution *sur lies* of a Chardonnay wine; the release of mannoproteins during AOL was faster in PEF-treated samples than in control. Two treatments of 5 and 10 kV/cm for 75 μs allowed to reach the maximum release of mannoproteins in 60 and 30 days, respectively. The same mannoprotein concentration in control wines was reached after 6 months (Martínez et al., 2019).

Acknowledgments

The authors are grateful to Bruno and Stefano Pittaro (Pitars Snc, San Martino al Tagliamento, PN, Italy) and to Umberto Marchiori (Marchiori Wines, Farra di Soligo, TV, Italy) for their practical indications concerning the paragraph "Aging on lees in sparkling wines."

References

Alcaide-Hidalgo, J. M., Moreno-Arribas, M. V., Martín-Álvarez, P. J., & Polo, M. C. (2007). Influence of malolactic fermentation, postfermentative treatments and ageing with lees on nitrogen compounds of red wines. *Food Chemistry, 103*(2), 572−581. Available from https://doi.org/10.1016/j.foodchem.2006.09.002.

Alexandre, H., & Guilloux-Benatier, M. (2006). Yeast autolysis in sparkling wine − A review. *Australian Journal of Grape and Wine Research, 12*(2), 119−127. Available from https://doi.org/10.1111/j.1755-0238.2006.tb00051.x.

Alexandre, H., Heintz, D., Chassagne, D., Guilloux-Benatier, M., Charpentier, C., & Feuillat, M. (2001). Protease A activity and nitrogen fractions released during alcoholic fermentation and autolysis in enological conditions. *Journal of Industrial Microbiology and Biotechnology, 26*, 235−240. Available from https://doi.org/10.1038/sj.jim.7000119.

Almeida, B., Buttner, S., Ohlmeier, S., Silva, A., Mesquita, A., Sampaio-Marques, B., ... Ludovico, P. (2007). NO-mediated apoptosis in yeast. *Journal of Cell Science, 120*(18), 3279−3288. Available from https://doi.org/10.1242/jcs.010926.

Babayan, T. L., & Bezrukov, M. G. (1985). Autolysis in yeasts. *Acta Biotechnologica, 5*(2), 129−136. Available from https://doi.org/10.1002/abio.370050205.

Babayan, T. L., Bezrukov, M. G., Latov, V. K., Belikov, V. M., Belavtseva, E. M., & Titova, E. F. (1981). Induced autolysis of *Saccharomyces cerevisiae*: Morphological effects, rheological effects, and dynamics of accumulation of extracellular hydrolysis products. *Current Microbiology, 5*(3), 163−168. Available from https://doi.org/10.1007/BF01578522.

Bachhawat, A. K., Ganguli, D., Kaur, J., Kasturia, N., Thakur, A., Kaur, H., ... Yadav, A. (2009). *Glutathione production in yeast. Yeast biotechnology: Diversity and applications* (pp. 259−280). India: Springer Netherlands. Available from https://doi.org/10.1007/s00217-015-2566-z-4_13.

Bell, S.-J., & Henschke, P. A. (2005). Implications of nitrogen nutrition for grapes, fermentation and wine. *Australian Journal of Grape and Wine Research, 11*, 242−295. Available from https://doi.org/10.1111/j.1755-0238.2005.tb00028.x.

Boulton, R. B., Bisson, L. F., & Kunkee, R. E. (1996). *Principles and practices of winemaking*. New York: Chapman & Hall.

Brown, C. R., Hung, G. C., Dunton, D., & Chiang, H. L. (2010). The TOR complex 1 is distributed in endosomes and in retrograde vesicles that form from the vacuole membrane and plays an important role in the vacuole import and degradation pathway. *Journal of Biological Chemistry, 285*(30), 23359−23370. Available from https://doi.org/10.1074/jbc.M109.075143.

Budovskaya, Y. V., Stephan, J. S., Reggiori, F., Klionsky, D. J., & Herman, P. K. (2004). The Ras/cAMP-dependent protein kinase signaling pathway regulates an early step of the autophagy process in *Saccharomyces cerevisiae. Journal of Biological Chemistry, 279*(20), 20663−20671. Available from https://doi.org/10.1074/jbc.M400272200.

Cacciola, V., Batllò, I. F., Ferraretto, P., Vincenzi, S., & Celotti, E. (2013). Study of the ultrasound effects on yeast lees lysis in winemaking. *European Food Research and Technology, 236*(2), 311−317. Available from https://doi.org/10.1007/s00217-012-1893-6.

Caridi, A. (2006). Enological functions of parietal yeast mannoproteins. *Antonie van Leeuwenhoek, 89*, 417−422. Available from https://doi.org/10.1007/s10482-005-9050-x.

Carrano, G. (2016). *Gestione dell'élevage sur lies mediante applicazione di tecnologie basate sull'alta pressione* (Master Thesis). University of Udine.

Cebollero, E., Carrascosa, A. V., & Gonzalez, R. (2005). Evidence for yeast autophagy during simulation of sparkling wine aging: A reappraisal of the mechanism of yeast autolysis in wine. *Biotechnology Progress, 21*(2), 614−616. Available from https://doi.org/10.1021/bp049708y.

Cebollero, E., & Gonzalez, R. (2006). Induction of autophagy by second-fermentation yeasts during elaboration of sparkling wines. *Applied and Environmental Microbiology, 72*(6), 4121−4127. Available from https://doi.org/10.1128/AEM.02920-05.

Charpentier, C., Aussenac, J., Charpentier, M., Prome, J. C., Duteurtre, B., & Feuillat, M. (2005). Release of nucleotides and nucleosides during yeast autolysis: Kinetics and potential impact on flavor. *Journal of Agricultural and Food Chemistry, 53*(8), 3000−3007. Available from https://doi.org/10.1021/jf040334y.

Charpentier, C., & Feuillat, M. (1993). Yeast autolysis. In G. H. Fleet (Ed.), *Wine microbiology and biotechnology* (pp. 225−242). Chur, Switzerland: Harwood Academic Publishers.

Chatonnet, P., Dubourdieu, D., & Boidron, J. N. (1992). Incidence of fermentation and aging conditions of dry white wines in barrels on their composition in substances yielded by oak wood. *Sciences Des Aliments, 12*(4), 665−685.

Chinni, G. (2018). *Trattamento delle fecce mediante tecnologie basate sulle alte pressioni ed effetti sull'evoluzione di vini conservati sur lies* (Master Thesis). University of Udine.

Commission Delegated Regulation (EU) 934/2019. (2019). Supplementing regulation (EU) No 1308/2013 of the European parliament and of the council as regards wine-growing areas where the alcoholic strength may be increased, authorised oenological practices and restrictions applicable to the production and conservation of grapevine products, the minimum percentage of alcohol for by-products and their disposal, and publication of OIV files. *Official Journal of the European Union, L, 149*, 1−52.

Commission Regulation (EU) 1308/2013. (2013). Establishing a common organisation of the markets in agricultural products and repealing council regulations (EEC) No 922/72, (EEC) No 234/79, (EC) No 1037/2001 and (EC) No 1234/2007. *Official Journal of the European Union, L, 347*, 671–854

Comuzzo, P., & Calligaris, S. (2019). Potential applications of high pressure homogenization in winemaking: A review. *Beverages, 5*(3), 56. Available from https://doi.org/10.3390/beverages5030056.

Comuzzo, P., Calligaris, S., Iacumin, L., Ginaldi, F., Palacios Paz, A. E., & Zironi, R. (2015). Potential of high pressure homogenization to induce autolysis of wine yeasts. *Food Chemistry, 185*, 340–348. Available from https://doi.org/10.1016/j.foodchem.2015.03.129.

Comuzzo, P., Calligaris, S., Iacumin, L., Ginaldi, F., Voce, S., & Zironi, R. (2017). Application of multipass high pressure homogenization under variable temperature regimes to induce autolysis of wine yeasts. *Food Chemistry, 224*, 105–113. Available from https://doi.org/10.1016/j.foodchem.2016.12.038.

Comuzzo, P., Tat, L., Fenzi, D., Brotto, L., Battistutta, F., & Zironi, R. (2011). Interactions between yeast autolysates and volatile compounds in wine and model solution. *Food Chemistry, 127*(2), 473–480. Available from https://doi.org/10.1016/j.foodchem.2011.01.026.

Connew, S. J. (1998). Yeast autolysis: A review of current research. *Australian and New Zealand Wine Industry Journal, 13*, 31–64.

Contessotto, A. (1999). *Indagine sulle possibili applicazioni dei preparati ad azione glucanasica nella produzione di vini spumanti secondo il metodo classico* (Master Thesis). University of Udine.

Culbert, J. A., McRae, J. M., Condé, B. C., Schmidtke, L. M., Nicholson, E. L., Smith, P. A., . . . Wilkinson, K. L. (2017). Influence of production method on the chemical composition, foaming properties, and quality of australian carbonated and sparkling white wines. *Journal of Agricultural and Food Chemistry, 65*(7), 1378–1386. Available from https://doi.org/10.1021/acs.jafc.6b05678.

Del Barrio-Galán, R., Pérez-Magariño, S., & Ortega-Heras, M. (2011). Techniques for improving or replacing ageing on lees of oak aged red wines: The effects on polysaccharides and the phenolic composition. *Food Chemistry, 127*(2), 528–540. Available from https://doi.org/10.1016/j.foodchem.2011.01.035.

del Fresno, J. M., Loira, I., Morata, A., González, C., Suárez-Lepe, J. A., & Cuerda, R. (2018). Application of ultrasound to improve lees ageing processes in red wines. *Food Chemistry, 157*–163. Available from https://doi.org/10.1016/j.foodchem.2018.04.041.

del Fresno, J. M., Morata, A., Escott, C., Loira, I., Cuerda, R., & Suárez-Lepe, J. A. (2019). Sonication of yeast biomasses to improve the ageing on lees technique in red wines. *Molecules, 635*. Available from https://doi.org/10.3390/molecules24030635.

Disciplinare di Produzione dei vini a Denominazione di Origine Controllata e Garantita "Franciacorta." (2020). Retrieved July 13, 2020, from https://www.franciacorta.net/it/vino/disciplinare/.

Domizio, P., Liu, Y., Bisson, L. F., & Barile, D. (2017). Cell wall polysaccharides released during the alcoholic fermentation by *Schizosaccharomyces pombe* and *S. japonicus*: Quantification and characterization. *Food Microbiology, 61*, 136–149. Available from https://doi.org/10.1016/j.fm.2016.08.010.

Dubourdieu, D. (1995). Intérêts œnologiques et risques associés à l'élevage des vins blancs sur lies en barriques. *Revue Française d'Œnologie, 155*, 30–35.

Dumay, E., Chevalier-Lucia, D., Picart-Palmade, L., Benzaria, A., Gràcia-Julià, A., & Blayo, C. (2013). Technological aspects and potential applications of (ultra) high-pressure homogenisation. *Trends in Food Science and Technology, 31*(1), 13–26. Available from https://doi.org/10.1016/j.tifs.2012.03.005.

Dunn, W. A., Cregg, J. M., Kiel, J. A. K. W., van der Klei, I. J., Oku, M., Sakai, Y., . . . Veenhuis, M. (2005). Pexophagy: The selective autophagy of peroxisomes. *Autophagy, 1*(2), 75–83. Available from https://doi.org/10.4161/auto.1.2.1737.

Ferrari, G., & Feuillat, M. (1988). L'elevage sur lie desvins blancs de Bourgogne 1. Etude des composes azotes, des acides gras et analyse sensorielle des vins. *Vitis, 27*, 183–197.

Ferrari, G., Meunier, J. M., & Feuillat, M. (1987). Dosage des acides gras totaux du vin et des levures de vinification. *Sciences Des Aliments, 1*, 61–76.

Feuillat, M. (2003). Yeast macromolecules: Origin, composition, and enological interest. *American Journal of Enology and Viticulture, 54*(3), 211–213.

Follows, M., Hetherington, P. J., Dunnill, P., & Lilly, M. D. (1971). Release of enzymes from bakers' yeast by disruption in an industrial homogenizer. *Biotechnology and Bioengineering, 13*(4), 549–560. Available from https://doi.org/10.1002/bit.260130408.

Fornairon-Bonnefond, C., Camarasa, C., Moutounet, M., & Salmon, J. M. (2001). État des connaissances scientifiques actuelles sur le phénomène d'autolyse des levures et l'élevage des vins sur lies. *Journal International des Sciences de la Vigne et du Vin, 35*(2), 57–78. Available from https://doi.org/10.20870/oeno-one.2001.35.2.990.

Fornairon-Bonnefond, C., & Salmon, J. M. (2003). Impact of oxygen consumption by yeast lees on the autolysis phenomenon during simulation of wine aging on lees. *Journal of Agricultural and Food Chemistry, 51*(9), 2584–2590. Available from https://doi.org/10.1021/jf0259819.

Francioli, S., Torrens, J., Riu-Aumatell, M., López-Tamames, E., & Buxaderas, S. (2003). Volatile compounds by SPME-GC as age markers of sparkling wines. *American Journal of Enology and Viticulture, 54*(3), 158–162.

Gallardo-Chacón, J. J., Vichi, S., Urpí, P., López-Tamames, E., & Buxaderas, S. (2010). Antioxidant activity of lees cell surface during sparkling wine sur lie aging. *International Journal of Food Microbiology, 143*(1–2), 48–53. Available from https://doi.org/10.1016/j.ijfoodmicro.2010.07.027.

Gallart, M., Lopez-tamames, E., Suberbiola, G., & Buxaderas, S. (2002). Influence of fatty acids on wine foaming. *Journal of Agricultural and Food Chemistry, 50*(24), 7042–7045. Available from https://doi.org/10.1021/jf0204452.

Gambetta, J. M., Bastian, S. E. P., Cozzolino, D., & Jeffery, D. W. (2014). Factors influencing the aroma composition of chardonnay wines. *Journal of Agricultural and Food Chemistry, 62*(28), 6512–6534. Available from https://doi.org/10.1021/jf501945s.

Geng, J., Nair, U., Yasumura-Yorimitsu, K., & Klionsky, D. J. (2010). Post-golgi sec proteins are required for autophagy in *Saccharomyces cerevisiae*. *Molecular Biology of the Cell, 21*(13), 2257–2269. Available from https://doi.org/10.1091/mbc.E09-11-0969.

Gerbaux, V., Villa, A., Monamy, C., & Bertrand, A. (1997). Use of lysozyme to inhibit malolactic fermentation and to stabilize wine after malolactic fermentation. *American Journal of Enology and Viticulture, 48*(1), 49–54.

Giotto, S. (2018). *Caratterizzazione chimico-fisica e microbiologica di vino Prosecco prodotto con metodo tradizionale* (Master Thesis). University of Udine.

Gómez, M. E., Igartuburu, J. M., Pando, E., Rodríguez Luis, F., & Mourente, G. (2004). Lipid composition of lees from Sherry wine. *Journal of Agricultural and Food Chemistry, 52*(15), 4791–4794. Available from https://doi.org/10.1021/jf030499r.

Guilloux-Benatier, M., Chassagne, D., Alexandre, H., Charpentier, C., & Feuillat, M. (2001). Influence of yeast autolysis after alcoholic fermentation on the development of *Brettanomyces/Dekkera* in wine. *Journal International Des Sciences de La Vigne et Du Vin*, *35*(3), 157–164. Available from https://doi.org/10.20870/oeno-one.2001.35.3.1701.

Guilloux-Benatier, M., & Feuillat, M. (1991). Utilisation d'adjuvants d'origine levurienne pour améliorer l'ensemencement des vins en bactéries sélectionnées. *Revue Française d'Œnologie*, *132*, 51–55.

Hernández-Orte, P., Lapeña, A. C., Peña-Gallego, A., Astrain, J., Baron, C., Pardo, I., ... Ferreira, V. (2008). Biogenic amine determination in wine fermented in oak barrels: Factors affecting formation. *Food Research International*, *41*(7), 697–706. Available from https://doi.org/10.1016/j.foodres.2008.05.002.

Hetherington, P. J., Follows, M., Dunniil, P., & Lilly, M. D. (1971). Release of protein from baker's yeast (*Saccharomyces cerevisiae*) by disruption in an industrial homogeniser. *Transactions of the Institution of Chemical Engineers*, *49*, 142–148.

Hoffman, M., & Chiang, H. L. (1996). Isolation of degradation-deficient mutants defective in the targeting of fructose-1,6-bisphosphatase into the vacuole for degradation in *Saccharomyces cerevisiae*. *Genetics*, *143*(4), 1555–1566.

Hung, G. C., Brown, C. R., Wolfe, A. B., Liu, J., & Chiang, H. L. (2004). Degradation of the gluconeogenic enzymes fructose-1,6-bisphosphatase and malate dehydrogenase is mediated by distinct proteolytic pathways and signaling events. *Journal of Biological Chemistry*, *279*(47), 49138–49150. Available from https://doi.org/10.1074/jbc.M404544200.

International Organization of Vine and Wine, O. I. V. (2018). *International Oenological Codex*. Paris: O.I.V.

Jackson, R. S. (2008). *Wine science. Principles and applications* (3rd ed.). San Diego: Academic Press.

Jaehrig, S. C., Rohn, S., Kroh, L. W., Fleischer, L. G., & Kurz, T. (2007). In vitro potential antioxidant activity of $(1 \rightarrow 3),(1 \rightarrow 6)$-β-D- glucan and protein fractions from *Saccharomyces cerevisiae* cell walls. *Journal of Agricultural and Food Chemistry*, *55*(12), 4710–4716. Available from https://doi.org/10.1021/jf063209q.

Jiranek, V., Grbin, P., Yap, A., Barnes, M., & Bates, D. (2008). High power ultrasonics as a novel tool offering new opportunities for managing wine microbiology. *Biotechnology Letters*, *30*, 1–6. Available from https://doi.org/10.1007/s10529-007-9518-z.

Kemp, B., Alexandre, H., Robillard, B., & Marchal, R. (2015). Effect of production phase on bottle-fermented sparkling wine quality. *Journal of Agricultural and Food Chemistry*, *63*(1), 19–38. Available from https://doi.org/10.1021/jf504268u.

Klionsky, D. J., Cregg, J. M., Dunn, W. A., Emr, S. D., Sakai, Y., Sandoval, I. V., ... Ohsumi, Y. (2003). A unified nomenclature for yeast autophagy-related genes. *Developmental Cell*, *5*(4), 539–545. Available from https://doi.org/10.1016/S1534-5807(03)00296-X.

Klis, F. M. (1994). Review: Cell wall assembly in yeast. *Yeast*, *10*(7), 851–869. Available from https://doi.org/10.1002/yea.320100702.

Klis, F. M., Mol, P., Hellingwerf, K., & Brul, S. (2002). Dynamics of cell wall structure in *Saccharomyces cerevisiae*. *FEMS Microbiology Reviews*, *26*(3), 239–256. Available from https://doi.org/10.1016/S0168-6445(02)00087-6.

Krick, R., Muehe, Y., Prick, T., Bremer, S., Schlotterhose, P., Eskelinen, E. L., ... Thumm, M. (2008). Piecemeal microautophagy of the nucleus requires the core macroautophagy genes. *Molecular Biology of the Cell*, *19*(10), 4492–4505. Available from https://doi.org/10.1091/mbc.E08-04-0363.

Kritzinger, E. C., Bauer, F. F., & Du Toit, W. J. (2013). Influence of yeast strain, extended lees contact and nitrogen supplementation on glutathione concentration in wine. *Australian Journal of Grape and Wine Research*, *19*(2), 161–170. Available from https://doi.org/10.1111/ajgw.12025.

Kulkarni, P., Loira, I., Morata, A., Tesfaye, W., Gonzalez, C., & Suarez-Lepe, J. A. (2015). Use of non-*Saccharomyces* yeast strains coupled with ultrasound treatment as a novel technique to accelerate ageing on lees of red wines and its repercussion in sensorial parameters. *LWT - Food Science and Technology*, *64*, 1255–1262.

Lavigne, V. (1995). Interprétation et prévention des défauts olfactifs de réduction lors de l'élevage sur lies totales. *Revue Française d'Œnologie*, *155*, 36–39.

Lavigne, V., & Dubourdieu, D. (1996). Mise en évidence et interprétation de l'aptitude des lies à éliminer certains thiols volatils du vin. *Journal International des Sciences de la Vigne et du Vin*, *30*(4), 201–206. Retrieved from http://www.jisvv.com.

Lavigne, V., Pons, A., & Dubourdieu, D. (2007). Assay of glutathione in must and wines using capillary electrophoresis and laser-induced fluorescence detection. Changes in concentration in dry white wines during alcoholic fermentation and aging. *Journal of Chromatography A*, *1139*(1), 130–135. Available from https://doi.org/10.1016/j.chroma.2006.10.083.

Ledoux, M., Perrin, V., Paladin, A., & Dubourdieu, I. (1997). Premier résultats de stabilisation tartrique des vins par addition de mannoprotéines purifiées (Mannostab™). *Journal International des Sciences de la Vigne et du Vin*, *31*(1), 23–31.

Lerm, E., Engelbrecht, L., & du Toit, M. (2010). Malolactic fermentation: The ABC's of MLF. *South African Journal of Enology and Viticulture*, *31*(2), 186–212.

Leroy, M. J., Charpentier, M., Duteurtre, B., Feuillat., & Charpentier, C. (1990). Yeast autolysis during champagne aging. *American Journal of Enology and Viticulture*, *1*, 21–28.

Leuschner, R. G., Heidel, M., & Hammes, W. P. (1998). Histamine and tyramine degradation by food fermenting microorganisms. *International Journal of Food Microbiology*, *39*(1–2), 1–10. Available from https://doi.org/10.1016/S0168-1605(97)00109-8.

Li, J., & Karboune, S. (2019). Characterization of the composition and the techno-functional properties of mannoproteins from *Saccharomyces cerevisiae* yeast cell walls. *Food Chemistry*, *297*, 124867. Available from https://doi.org/10.1016/j.foodchem.2019.05.141.

Liberatore, M. T., Pati, S., Del Nobile, M. A., & La Notte, E. (2010). Aroma quality improvement of Chardonnay white wine by fermentation and ageing in barrique on lees. *Food Research International*, *43*(4), 996–1002. Available from https://doi.org/10.1016/j.foodres.2010.01.007.

Liu, L., Loira, I., Morata, A., Suárez-Lepe, J. A., González, C., & Rauhut, D. (2016). Shortening the ageing on lees process in wines by using ultrasound and microwave treatments both combined with stirring and abrasion techniques. *European Food Research and Technology*, *242*, 559–569. Available from https://doi.org/10.1007/s00217-015-2566-z.

Llaubères, R. M., Dubourdieu, D., & Villettaz, J. -C. (1987). Exocellular polysaccharides from *Saccharomyces* in wine. *Journal of the Science of Food and Agriculture*, *41*(3), 277–286. Available from https://doi.org/10.1002/jsfa.2740410310.

Loira, I., Morata, A., Palomero, F., González, C., & Suárez-Lepe, J. A. (2018). *Schizosaccharomyces pombe*: A promising biotechnology for modulating wine composition. *Fermentation*, *4*(3). Available from https://doi.org/10.3390/fermentation4030070.

Lombardi, S. J., De Leonardis, A., Lustrato, G., Testa, B., & Iorizzo, M. (2015). Yeast autolysis in sparkling wine aging: Use of killer and sensitive *Saccharomyces cerevisiae* strains in co-culture. *Recent Patents on Biotechnology*, *9*(3), 223–230. Retrieved from http://www.benthamdirect.org/pages/all_b_bypublication.php.

Lubbers, S., Leger, B., Charpentier, C., & Feuillat, M. (1993). Effet colloide-protecteur d'extraits de parois de levures sur la stabilité tartrique d'une solution modèle. *Journal International des Sciences de la Vigne et du Vin*, 27(1), 13–22.

Lubbers, S., Voilley, A., Feuillat, M., & Charpentier, C. (1994). Influence of mannoproteins from yeast on the aroma intensity of a model wine. *LWT − Food Science and Technology*, 27(2), 108–114. Available from https://doi.org/10.1006/fstl.1994.1025.

Lynch-Day, M. A., Bhandari, D., Menon, S., Huang, J., Caid, H., Bartholomew, C. R., ... Klionsky, D. J. (2010). Trs85 directs a Ypt1 GEF, TRAPPIII, to the phagophore to promote autophagy. *Proceedings of the National Academy of Sciences of the United States of America*, 107(17), 7811–7816. Available from https://doi.org/10.1073/pnas.1000063107.

Maged, E. A. M., & Amer Eissa, A. H. (2012). Pulsed Electric Fields for Food Processing Technology. In A. H. Amer Eissa (Ed.), *Structure and function of food engineering* (pp. 275–306). London: InTechOpen.

Marques, A. P., Leitão, M. C., & San Romão, M. V. (2008). Biogenic amines in wines: Influence of oenological factors. *Food Chemistry*, 107(2), 853–860. Available from https://doi.org/10.1016/j.foodchem.2007.09.004.

Martínez, J. M., Delso, C., Aguilar, D., Cebrián, G., Álvarez, I., & Raso, J. (2018). Factors influencing autolysis of *Saccharomyces cerevisiae* cells induced by pulsed electric fields. *Food Microbiology*, 73, 67–72. Available from https://doi.org/10.1016/j.fm.2017.12.008.

Martínez, J. M., Delso, C., Maza, M. A., Álvarez, I., & Raso, J. (2019). Pulsed electric fields accelerate release of mannoproteins from *Saccharomyces cerevisiae* during aging on the lees of Chardonnay wine. *Food Research International*, 116, 795–801. Available from https://doi.org/10.1016/j.foodres.2018.09.013.

Martínez-Lapuente, L., Apolinar-Valiente, R., Guadalupe, Z., Ayestarán, B., Pérez-Magariño, S., Williams, P., & Doco, T. (2018). Polysaccharides, oligosaccharides and nitrogenous compounds change during the ageing of Tempranillo and Verdejo sparkling wines. *Journal of the Science of Food and Agriculture*, 98(1), 291–303. Available from https://doi.org/10.1002/jsfa.8470.

Martínez-Rodríguez, A. J., Carrascosa, A. V., Martín-Álvarez, P. J., Moreno-Arribas, V., & Polo, M. C. (2002). Influence of the yeast strain on the changes of the amino acids, peptides and proteins during sparkling wine production by the traditional method. *Journal of Industrial Microbiology and Biotechnology*, 29(6), 314–322. Available from https://doi.org/10.1038/sj.jim.7000323.

Martínez-Rodríguez, A. J., & Polo, M. C. (2000). Characterization of the nitrogen compounds released during yeast autolysis in a model wine system. *Journal of Agricultural and Food Chemistry*, 48(4), 1081–1085. Available from https://doi.org/10.1021/jf991047a.

Martínez-Rodríguez, A. J., Carrascosa, A. V., & Polo, M. C. (2001). Release of nitrogen compounds to the extracellular medium by three strains of *Saccharomyces cerevisiae* during induced autolysis in a model wine system. *International Journal of Food Microbiology*, 68, 155–160. Available from https://doi.org/10.1016/s0168-1605(01)00486-x.

Masino, F., Montevecchi, G., Arfelli, G., & Antonelli, A. (2008). Evaluation of the combined effects of enzymatic treatment and aging on lees on the aroma of wine from Bombino bianco grapes. *Journal of Agricultural and Food Chemistry*, 56(20), 9495–9501. Available from https://doi.org/10.1021/jf8015893.

Middelberg, A. P. J. (1995). Process-scale disruption of microorganisms. *Biotechnology Advances*, 13(3), 491–551. Available from https://doi.org/10.1016/0734-9750(95)02007-P.

Moine Leoux, V. (1996). *Recherches sur le rôle des mannoprotéines de levure vis-à-vis de la stabilisation protéique et tartrique des vins* (PhD Thesis). University of Bordeaux II.

Moine-Ledoux, V., & Dubourdieu, D. (1999). An invertase fragment responsible for improving the protein stability of dry white wines. *Journal of the Science of Food and Agriculture*, 79(4), 537–543. Available from https://doi.org/10.1002/(SICI)1097-0010(19990315)79:4 < 537::AID-JSFA214 > 3.0.CO;2-B.

Morata, A., Loira, I., Vejarano, R., González, C., Callejo, M. J., & Suárez-Lepe, J. A. (2017). Emerging preservation technologies in grapes for winemaking. *Trends in Food Science & Technology*, 67, 36–43. Available from https://doi.org/10.1016/j.tifs.2017.06.014.

Morata, A., Palomero, F., Loira, I., & Suárez-Lepe, J. A. (2018). *New trends in aging on lees. Red wine technology* (pp. 163–176). Spain: Elsevier. Available from https://doi.org/10.1016/B978-0-12-814399-5.00011-6.

Moreno-Arribas, V., Pueyo, E., Nieto, F. J., Martín-Álvarez, P. J., & Polo, M. C. (2000). Influence of the polysaccharides and the nitrogen compounds on foaming properties of sparkling wines. *Food Chemistry*, 70(3), 309–317. Available from https://doi.org/10.1016/S0308-8146(00)00088-1.

Moreno-Arribas, V., Pueyo, E., & Polo, M. C. (1996). Peptides in musts and wines. Changes during the manufacture of cavas (sparkling wines). *Journal of Agricultural and Food Chemistry*, 44(12), 3783–3788. Available from https://doi.org/10.1021/jf960307a.

Müller, O., Sattler, T., Flötenmeyer, M., Schwarz, H., Plattner, H., & Mayer, A. (2000). Autophagic tubes: Vacuolar invaginations involved in lateral membrane sorting and inverse vesicle budding. *Journal of Cell Biology*, 151(3), 519–528. Available from https://doi.org/10.1083/jcb.151.3.519.

Nagodawithana, T. (1992). Yeast-derived flavors and flavor enhancers and their probable mode of action. *Food Technology*, 46(11), 138–144.

Nunez, Y. P., Carrascosa, A. V., González, R., Polo, M. C., & Martínez-Rodríguez, A. J. (2005). Effect of accelerated autolysis of yeast on the composition and foaming properties of sparkling wines elaborated by a champenoise method. *Journal of Agricultural and Food Chemistry*, 53(18), 7232–7237. Available from https://doi.org/10.1021/jf050191v.

Palomero, F., Morata, A., Benito, S., Calderón, F., & Suárez-Lepe, J. A. (2009). New genera of yeasts for over-lees aging of red wine. *Food Chemistry*, 112(2), 432–441. Available from https://doi.org/10.1016/j.foodchem.2008.05.098.

Palomero, F., Morata, A., Benito, S., Gonzalez, M., & Suarez Lepe, J. A. (2007). Conventional and enzyme-assisted autolysis during ageing over lees in red wines: Influence on the release of polysaccharides from yeast cell walls and on wine monomeric anthocyanin content. *Food Chemistry*, 105(2), 838–846. Available from https://doi.org/10.1016/j.foodchem.2007.01.062.

Palomero, F., Ntanos, K., Morata, A., Benito, S., & Suárez-Lepe, J. A. (2011). Reduction of wine 4-ethylphenol concentration using lyophilised yeast as a bioadsorbent: Influence on anthocyanin content and chromatic variables. *European Food Research and Technology*, 232(6), 971–977. Available from https://doi.org/10.1007/s00217-011-1470-4.

Patrignani, F., Ndagijimana, M., Vernocchi, P., Gianotti, A., Riponi, C., Gardini, F., & Lanciotti, R. (2013). High-pressure homogenization to modify yeast performance for sparkling wine production according to traditional methods. *American Journal of Enology and Viticulture*, 64(2), 258–267. Available from https://doi.org/10.5344/ajev.2012.12096.

Patynowski, R. J., Jiranek, V., & Markides, A. J. (2002). Yeast viability during fermentation and sur lie ageing of a defined medium and subsequent growth of *Oenococcus oeni*. *Australian Journal of Grape and Wine Research*, 8(1), 62–69.

Perpetuini, G., Di Gianvito, P., Arfelli, G., Schirone, M., Corsetti, A., Tofalo, R., & Suzzi, G. (2016). Biodiversity of autolytic ability in flocculent *Saccharomyces cerevisiae* strains suitable for traditional sparkling wine fermentation. *Yeast*, *33*(7), 303–312. Available from https://doi.org/10.1002/yea.3151.

Piton, F., Charpentier, M., & Troton, D. (1988). Cell wall and lipid changes in *Saccharomyces cerevisiae* during aging of Champagne wine. *American Journal of Enology and Viticulture*, *39*, 221–226.

Pueyo, E., Martínez-Rodríguez, A., Polo, M. C., Santa-María, G., & Bartolomé, B. (2000). Release of lipids during yeast autolysis in a model wine system. *Journal of Agricultural and Food Chemistry*, *48*(1), 116–122. Available from https://doi.org/10.1021/jf990036e.

Redding, C. (1860). *A history and description of modern wines*. London: Henry G. Bohn.

Reggiori, F., & Klionsky, D. J. (2013). Autophagic processes in yeast: Mechanism, machinery and regulation. *Genetics*, *194*(2), 341–361. Available from https://doi.org/10.1534/genetics.112.149013.

Remize, F., Augagneur, Y., Guilloux-Benatier, M., & Guzzo, J. (2005). Effect of nitrogen limitation and nature of the feed upon *Oenococcus oeni* metabolism and extracellular protein production. *Journal of Applied Microbiology*, *98*(3), 652–661. Available from https://doi.org/10.1111/j.1365-2672.2004.02494.x.

Renouf, V., & Lonvaud-Funel, A. (2004). Les soutirages sont des étapes clefs de la stabilisation microbiologique des vins. *Journal International des Sciences de la Vigne et du Vin*, *38*(4), 219–224. Available from https://doi.org/10.20870/oeno-one.2004.38.4.914.

Ribéreau-Gayon, P., Dubourdieu, D., Doneche, B., & Lonvaud, A. (2006). *Handbook of enology. The microbiology of wine and vinifications*, (2nd (ed.)). (Vol. I). New York: John Wiley & Sons.

Ribéreau-Gayon, P., Glories, Y., Maujean, A., & Dubourdieu, D. (2006). *Handbook of enology. The chemistry of wine stabilization and treatments*, (2nd (ed.)). (Vol. II). New York: John Wiley & Sons.

Rizk, Z., El Rayess, Y., Ghanem, C., Mathieu, F., Taillandier, P., & Nehme, N. (2016). Impact of inhibitory peptides released by Saccharomyces cerevisiae BDX on the malolactic fermentation performed by *Oenococcus oeni* Vitilactic F. *International Journal of Food Microbiology*, *233*, 90–96. Available from https://doi.org/10.1016/j.ijfoodmicro.2016.06.018.

Roberts, P., Moshitch-Moshkovitz, S., Kvam, E., O'Toole, E., Winey, M., & Goldfarb, D. S. (2003). Piecemeal microautophagy of nucleus in *Saccharomyces cerevisiae*. *Molecular Biology of the Cell*, *14*(1), 129–141. Available from https://doi.org/10.1091/mbc.E02-08-0483.

Rodrigues, A., Ricardo-Da-Silva, J. M., Lucas, C., & Laureano, O. (2016). Characterization of mannoproteins during white wine (*Vitis vinifera* L. cv. ENCRUZADO) ageing on lees with stirring in oak wood barrels and in a stainless steel tank with oak staves. *OENO One*, *46*(4), 321–329. Available from https://doi.org/10.20870/oeno-one.2012.46.4.1527.

Rodriguez-Nogales, J. M., Fernández-Fernández, E., & Vila-Crespo, J. (2012). Effect of the addition of β-glucanases and commercial yeast preparations on the chemical and sensorial characteristics of traditional sparkling wine. *European Food Research and Technology*, *235*(4), 729–744. Available from https://doi.org/10.1007/s00217-019-03334-9.

Romani, C., Lencioni, L., Gobbi, M., Mannazzu, I., Ciani, M., & Domizio, P. (2018). *Schizosaccharomyces japonicus*: A polysaccharide-overproducing yeast to be used in winemaking. *Fermentation*, *4*(1). Available from https://doi.org/10.3390/fermentation4010014.

Salmon, J. M., Fornairon-Bonnefond, C., Mazauric, J. P., & Moutounet, M. (2000). Oxygen consumption by wine lees: Impact on lees integrity during wine ageing. *Food Chemistry*, *71*(4), 519–528. Available from https://doi.org/10.1016/S0308-8146(00)00204-1.

Sartini, E., Arfelli, G., Fabiani, A., & Piva, A. (2007). Influence of chips, lees and micro-oxygenation during aging on the phenolic composition of a red Sangiovese wine. *Food Chemistry*, *104*(4), 1599–1604.

Sattler, T., & Mayer, A. (2000). Cell-free reconstitution of microautophagic vacuole invagination and vesicle formation. *Journal of Cell Biology*, *151*(3), 529–538. Available from https://doi.org/10.1083/jcb.151.3.529.

Schiavone, M., Déjean, S., Sieczkowski, N., Castex, M., Dague, E., & François, J. M. (2017). Integration of biochemical, biophysical and transcriptomics data for investigating the structural and nanomechanical properties of the yeast cell wall. *Frontiers in Microbiology*, *8*, 1806. Available from https://doi.org/10.3389/fmicb.2017.01806.

Shintani, T., & Klionsky, D. J. (2004). Cargo proteins facilitate the formation of transport vesicles in the cytoplasm to vacuole targeting pathway. *Journal of Biological Chemistry*, *279*(29), 29889–29894. Available from https://doi.org/10.1074/jbc.M404399200.

Siddiqi, S. F., Titchener-Hooker, N. J., & Shamlou, P. A. (1997). High pressure disruption of yeast cells: The use of scale down operations for the prediction of protein release and cell debris size distribution. *Biotechnology and Bioengineering*, *55*(4), 642–649. Available from https://doi.org/10.1002/(SICI)1097-0290(19970820)55:4 < 642::AID-BIT6 > 3.0.CO;2-H.

Smit, A. Y., & du Toit, M. (2013). Evaluating the influence of malolactic fermentation inoculation practices and ageing on lees on biogenic amine production in wine. *Food and Bioprocess Technology*, *6*(1), 198–206. Available from https://doi.org/10.1007/s11947-011-0702-8.

Sonni, F., Clark, A. C., Prenzler, P. D., Riponi, C., & Scollary, G. R. (2011). Antioxidant action of glutathione and the ascorbic acid/glutathione pair in a model white wine. *Journal of Agricultural and Food Chemistry*, *59*(8), 3940–3949. Available from https://doi.org/10.1021/jf104575w.

Sonni, F., Moore, E. G., Clark, A. C., Chinnici, F., Riponi, C., & Scollary, G. R. (2011). Impact of glutathione on the formation of methylmethine-and carboxymethine-bridged (+)-catechin dimers in a model wine system. *Journal of Agricultural and Food Chemistry*, *59*(13), 7410–7418. Available from https://doi.org/10.1021/jf200968x.

Sumby, K. M., Bartle, L., Grbin, P. R., & Jiranek, V. (2019). Measures to improve wine malolactic fermentation. *Applied Microbiology and Biotechnology*, *103*(5), 2033–2051. Available from https://doi.org/10.1007/s00253-018-09608-8.

Taylor, R., Chen, P. H., Chou, C. C., Patel, J., & Jin, S. V. (2012). KCS1 deletion in *Saccharomyces cerevisiae* leads to a defect in translocation of autophagic proteins and reduces autophagosome formation. *Autophagy*, *8*(9), 1300–1311. Available from https://doi.org/10.4161/auto.20681.

Tirelli, A., Fracassetti, D., & De Noni, I. (2010). Determination of reduced cysteine in oenological cell wall fractions of *Saccharomyces cerevisiae*. *Journal of Agricultural and Food Chemistry*, *58*(8), 4565–4570. Available from https://doi.org/10.1021/jf904047u.

Todd, B. E. N., Fleet, G. H., & Henschke, P. A. (2000). Promotion of autolysis through the interaction of killer and sensitive yeasts: Potential application in sparkling wine production. *American Journal of Enology and Viticulture*, *51*(1), 65–72.

Torrens, J., Rlu-Aumatell, M., Vichi, S., López-Tamames, E., & Buxaderas, S. (2010). Assessment of volatlle and sensory profiles between base and sparkling wines. *Journal of Agricultural and Food Chemistry*, *58*(4), 2455–2461. Available from https://doi.org/10.1021/jf9035518.

Torresi, S., Frangipane, M. T., Garzillo, A. M. V., Massantini, R., & Contini, M. (2014). Effects of a β-glucanase enzymatic preparation on yeast lysis during aging of traditional sparkling wines. *Food Research International*, *55*, 83–92. Available from https://doi.org/10.1016/j.foodres.2013.10.034.

Tsong, T. Y. (1991). Electroporation of cell membranes. *Biophysical Journal*, *60*(2), 297–306. Available from https://doi.org/10.1016/S0006-3495(91)82054-9.

Tudela, R., Gallardo-Chacón, J. J., Rius, N., López-Tamames, E., & Buxaderas, S. (2012). Ultrastructural changes of sparkling wine lees during long-term aging in real enological conditions. *FEMS Yeast Research*, *12*(4), 466–476. Available from https://doi.org/10.1111/j.1567-1364.2012.00800.x.

Ugliano, M., Kwiatkowski, M., Vidal, S., Capone, D., Siebert, T., Dieval, J. B., . . . Waters, E. J. (2011). Evolution of 3-mercaptohexanol, hydrogen sulfide, and methyl mercaptan during bottle storage of Sauvignon blanc wines. Effect of glutathione, copper, oxygen exposure, and closure-derived oxygen. *Journal of Agricultural and Food Chemistry*, *59*(6), 2564–2572. Available from https://doi.org/10.1021/jf1043585.

Vasserot, Y., Steinmetz, V., & Jeandet, P. (2003). Study of thiol consumption by yeast lees. *Antonie van Leeuwenhoek, International Journal of General and Molecular Microbiology*, *83*(3), 201–207. Available from https://doi.org/10.1023/A:1023305130233.

Vincenzi, S., Crapisi, A., & Curioni, A. (2014). Foamability of Prosecco wine: Cooperative effects of high molecular weight glycocompounds and wine PR-proteins. *Food Hydrocolloids*, *34*, 202–207. Available from https://doi.org/10.1016/j.foodhyd.2012.09.016.

Voce, S., Battistutta, F., Tat, L., Sivilotti, P., & Comuzzo. (2020). Composizione chimica ed evoluzione del profilo aromatico di spumanti ribolla gialla prodotti con diversi metodi di rifermentazione. *Infowine*, *3*, 1–9.

Voce, S., Škrab, D., Vrhovsek, U., Battistutta, F., Comuzzo, P., & Sivilotti, P. (2019). Compositional characterization of commercial sparkling wines from cv. Ribolla Gialla produced in Friuli Venezia Giulia. *European Food Research and Technology*, *245*, 2279–2292. Available from https://doi.org/10.1007/s00217-019-03334-9.

Waters, E. J., Wallace, W., Tate, M. E., & Williams, P. J. (1993). Isolation and partial characterization of a natural haze protective factor from wine. *Journal of Agricultural and Food Chemistry*, *41*(5), 724–730. Available from https://doi.org/10.1021/jf00029a009.

Yamagata, M., Obara, K., & Kihara, A. (2011). Sphingolipid synthesis is involved in autophagy in *Saccharomyces cerevisiae*. *Biochemical and Biophysical Research Communications*, *410*, 786–791787.

Zhao, J., & Fleet, G. H. (2003). Degradation of DNA during the autolysis of *Saccharomyces cerevisiae*. *Journal of Industrial Microbiology and Biotechnology*, *30*(3), 175–182. Available from https://doi.org/10.1007/s10295-003-0028-2.

21

Barrel aging of white wines

Fernando Zamora

Department of Biochemistry and Biotechnology, Faculty of Oenology, University Rovira i Virgili, Tarragona, Spain

21.1 Brief historical introduction

Clay pots are probably the earliest containers used in winemaking. There is evidence that clay vessels were already used for making and stocking wine from a long time ago. Archeological excavations in the South Georgia region of Kvemo Kartli uncovered the presence of grape seeds in clay pots, called *Kvevris*, dating back to the 6th millennium BC (Glonti, 2010). Phoenicians, Greeks, and Romans expanded the use of clay vessels (amphorae) for the wine trade, and it was not until much later that wooden containers began to be used for the storage and transport of wine. Pliny the Elder attributes the Gauls to be the first to use barrels to store and transport beer and other goods (Work, 2014). The higher resistance of wooden casks promoted their gradual introduction to winemaking and storage, although clay vessels also remained in use. Clay and wooden vessels practically monopolized the production and storage of wine for centuries until new materials gradually displaced their use. Gradually, concrete first, and then metal containers, especially stainless steel, somewhat later displaced clay and wood as the main tank materials for wine. However, wooden barrels continued to be used, especially for aging red wines, due to their positive effects on the organoleptic quality and physical—chemical stability (Zamora, 2019).

In the particular case of white wines, stainless steel almost completely displaced the other materials, thanks to its many advantages such as ease in cleaning and disinfecting, better thermal control of fermentations, and great durability. Nevertheless, the use of wooden barrels was maintained for fermenting and/or aging the best white wines in some traditional regions such as Burgundy, Bordeaux, or Rioja (Dubourdieu, 1992; Feuillat, 1992; Jackson, 2009) and nowadays is a very common practice to elaborate premium white wines around the world (Sapies, Delteil, Feuillat, & Guilloux-Benatier, 1998; Ribéreau-Gayon, Glories, Maujean, & Dubourdieu, 2006; Schneider & Dubourdieu, 2014).

The contact with wood enriches the wine with several volatile substances that improve the intensity and complexity of its aroma Boidron, Chatonnet, & Pons, 1988; Chatonnet, 1992; Navarro et al., 2018). In addition, wood also releases phenolic compounds that contribute to texture sensations of the wine, increasing some sensory attributes such as body and mouthfeel (Quinn & Singleton, 1985; Michel et al., 2011; Navarro, Kontoudakis, Giordanengo et al., 2016; Navarro, Kontoudakis, Gómez-Alonso et al., 2016). Moreover, if the aging process is performed in contact with the lees, wine benefits from the lees autolysis process (Feuillat, Freyssinet, & Charpentier, 1989; Chatonnet, Dubourdieu, & Boidron, 1991), being enriched in several substances released from the cell degradation (Feuillat & Charpentier, 1982; Fornairon-Bonnefond, Camarasa, Moutounet, & Salmon, 2002). Consequently, aging white wines in barrels has a lot of advantages, and for that reason, this practice is increasingly applied for the production of premium white wines.

21.2 The main tree species used in cooperage and what they contribute to wine

Historically, several tree species have been used in cooperage for barrel manufacturing. It should be noted that the barrels were used to store and transport many products, both solids (grains, sugar, fruits, salted meats and fishes, etc.) and liquids (beer, cider, water, olive oil, whale oil, crude oil, etc.) and, naturally, wine (Pitte,

White Wine Technology.
DOI: https://doi.org/10.1016/B978-0-12-823497-6.00019-3

© 2022 Elsevier Inc. All rights reserved.

2018). Even now, the actual benchmark unit for crude oil trade is a barrel (American Oil & Gas Historical Society. 2016), which is equal to 42 gallons (about 158 L). However, the actual cooperage uses mainly French oaks (*Quercus petraea* and *Quercus robur*), American oak (*Quercus alba*), and, in a quite smaller proportion, other trees such as chestnut (*Castanea sativa*) or acacia (*Robinia pseudoacacia*).

The different substances released by oak into the wine during barrel aging may be naturally present in the original oak wood or may derive from other wood substances during the toasting process (Chira & Teissedre, 2013). Many volatile compounds from a wide range of chemical families have been described (Cadahía, Fernández de Simón, & Jalocha, 2003), but a few of them are significant with regard to their impact on the sensory characteristics of wines. Table 21.1 synthesizes the main substances released from different types of wood into wine and their sensory descriptors. This table also shows the influence of the toasting level of the staves on these parameters.

In terms of sensory impact, the main volatile substances released by oak wood are phenolic aldehydes and ketones, volatile phenols, furanic compounds, and β-methyl-γ-octalactones (Chatonnet, 1992; Spillman, Sefton, & Gawel, 2004).

Phenolic aldehydes and ketones contribute to the characteristic vanilla aroma of aged wines, with vanillin being the main contributor (Prida & Chatonnet, 2010). The aroma of vanilla is one of the most characteristic and appreciated notes of wines aged in oak.

The family of volatile phenols includes principally guaiacol, methylguaiacol, ethylguaiacol, vinylguaiacol, and eugenol. Briefly, eugenol contributes with a pleasant spicy aroma of clove (Boidron et al., 1988), whereas all the others contribute with smoked/toasted notes (Chatonnet, Dubourdieu, & Boidron, 1992). The presence of eugenol is always well appreciated since it provides the wine with a note of distinction. However, the influence of the other volatile phenols on the quality of the wine depends on their concentration. It can be negligible if they are below

TABLE 21.1   Main substances released by the different types of wood used in cooperage and their organoleptic impact.

Botanical origen			Furanic compounds		Vanillin		β-Methyl-γ-octolactones			Volatile phenols		Ellagitannins	
			Sensory attribute										
			Toasted nuts		Vanilla		Coconut			Toasted/Smoked notes		Astringency, Structure, Mouthfeel	
			(µg/L)	PI	(µg/L)	PI	(µg/L)	% cis isomer	PI	(µg/L)	PI	(mg/L)	PI
French Oak	Quercus petraea	Light toast	1231[a]	bt	547[a]	++	532[a]	66%[a]	++	126[a]	+	31.2[b]	+++
		Medium Toast	3717[a]; 1167[h]	bt	621[a]; 538[f]	+++	422[a]; 676[d]; 442[f]	63%[a]; 85%[d]; 75%[f]	+	140[a]; 102[f]	++	9.0[b]; 8.0[c]; 39[g]	++
		Heavy toast	3233[a]	bt	497[a]	++	161[a]	64%[a]	bt	202[a]	+++	4.7[b]	+
	Quercus robur	Mid Toasted	890[h]	bt	185[f]	+	316[f]	42%[f]	+	110[f]	++	15.0[c]; 88[g]; 130-168[e]	++++
American Oak (Quercus alba)		Low toasted	716[a]	bt	390[a]	++	2887[a]	87%[a]	++++	172[a]	++	3.6[b]	+
		Mid Toasted	4554[a]; 3155[h]	bt	1152[a]	++++	1837[a]	90%[a]	++	191[a]	++	1.1[b]; 6.0[c]; 33[g]	+
		Heavy toasted	5688[a]	bt	568[a]	++	760[a]	90%[a]	++	257[a]	+++	0.4[b]	bt
Chestnut (Castanea sativa)		Mid Toasted	1439[d]	bt	456[d]	++	53[d]	59%[d]	bt	308[d]	+++	44[g]	+++
Acacia (Robinia pseudoacacia)		Mid Toasted	936[d]	bt	233[d]	+	ND[d]		bt	210[d]	++	NDA	
Cherry (Prunus avium)		Mid Toasted	278[d]	bt	304[d]	++	ND[d]		bt	179[d]	++	NDA	

their sensory threshold; it can be pleasant when their presence provides slight notes of toasted bread, or unpleasant when their concentrations are so high that their burnt bread notes can be perceived (Navarro et al., 2018).

The furanic compound family, which includes furfural, methylfurfural, hydroxymethylfurfural, and furfuryl alcohol, contributes to smoked and toasted nut notes (Spillman et al., 2004). However, normally furanic compounds are below their perception threshold (Navarro et al., 2018), and consequently, their contribution to wine aroma is not important. However, furfural is a precursor of furfurylthiol, a volatile substance with a very pleasant coffee aroma (Tominaga, Blanchard, Darriet, & Dubourdieu, 2000) which is particularly present in wines fermented and aged in oak barrels in contact on the lees (Blanchard, Tominaga, & Dubourdieu, 2001).

Finally, β-methyl-γ-octalactones are responsible for the coconut flavor. These substances are also known as whiskey lactones since they were first detected in Bourbon whiskey (Suomalainen & Nykaenen, 1970). Whiskey lactones are present in the form of two isomers (cis and trans). The perception threshold of the cis isomer is much lower than that of the trans isomer, making its contribution to coconut perception much more important (Abbott et al., 1995).

Wood also releases some nonvolatile substances such as phenolic acids, coumarins, and especially ellagitannins into the wine (Zhang, Cai, Duan, Reeves, & He, 2015) which contribute to wine taste and texture sensations, such as body and astringency (Glabasnia & Hofmann, 2006; Michel et al., 2011). The main ellagitannins released from oak wood to wine are castalagin, vescalagin, grandinin, and roburins A, B, C, D, and E (Jourdes, Lefeuvre, & Quideau, 2009).

All these substances are released from wood into the wine, and the proportion of them will depend on several factors such as the botanical and geographical origin of the tree, the seasoning technique, the degree of toasting, and the number of times the barrel has been used before (Navarro, Kontoudakis, Giordanengo et al., 2016; Navarro, Kontoudakis, Gómez-Alonso et al., 2016; Navarro et al., 2018). Briefly, the American oak *Q. alba* is very rich in vanilla and whiskey lactones, especially the *cis* isomer (Fernández de Simón et al., 2014; Navarro et al., 2018), and relatively poor in ellagitannins (Navarro, Kontoudakis, Giordanengo et al., 2016; Navarro, Kontoudakis, Gómez-Alonso et al., 2016; Chatonnet & Dubourdieu, 1998). In contrast, the French oak *Q. robur* is very rich in ellagitannins (Vivas, Glories, Bourgeois, & Vitry, 1996; Chatonnet & Dubourdieu, 1998) and relatively poor in volatile substances (Spillman et al., 2004). The French oak *Q. petraea* stands at an intermediate point since it is relatively rich in ellagitannins but to a lower extent than *Q. robur* (Vivas et al., 1996; Navarro, Kontoudakis, Giordanengo et al., 2016; Navarro, Kontoudakis, Gómez-Alonso et al., 2016), and it is very rich in volatile substances but in a lower proportion than *Q. alba* (Fernández de Simón et al., 2014; Navarro et al., 2018). Less is known about other woods such as chestnut or acacia, but it seems that chestnut is rich in ellagitannins whereas acacia is not, and aromatically, chestnut is close to oaks, whereas Acacia provides a completely different sensory profile (Vivas et al., 1996; Fernández de Simón et al., 2014).

Fig. 21.1 illustrates visually in the form of a spider web diagram the sensory characteristics that different woods provide to wine. This figure shows clearly that American oak provides the wine with an intense aroma of vanilla and especially coconut, while it only slightly enriches its structure. The French oak *Q. robur* increases a lot the structure of the wine but aromatizes it to a much lesser extent. The other French oak, *Q. petraea*, greatly enriches the structure of the wine, but not as much as *Q. robur*, and provides it with an intense and complex aroma. Chestnut also considerably enriches the structure of the wine but its aromatic contribution is lower than that of *Q. petraea* and *Q. alba*, and its flavor is certainly somewhat more rustic. Finally, what acacia provides to

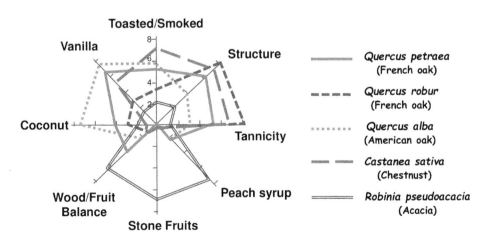

FIGURE 21.1 Spider web diagram of the sensory characteristics that different woods provides to wine.

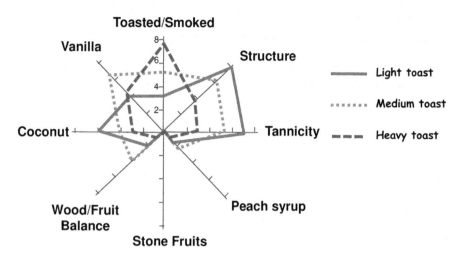

FIGURE 21.2 Spider web diagram of the sensory characteristics in function of the toasting level (Quercus petraea).

wine is completely different from that of other woods. It hardly contributes to the structure of the wine and provides it with aromas of stone fruits and peach in syrup.

Evidently, the substances that the different woods can release to the wine depend a lot on the toasting level of the staves used for making the barrels. Fig. 21.2 shows a spider web diagram of the sensory characteristics that *Q. petraea* provides to the wine in the function of the toasting level. Briefly, the higher the toasting level, the higher the vanilla, furanic compound, and volatile phenol concentrations, and the lower the ellagitannins and whiskey lactone concentrations. This means that as the toasting level increases, the structure, tannicity, and coconut notes that the wood brings to the wine decrease, while the vanilla and toasty aromas increase. In general, the other woods behave according to the level of toasting in a similar way to *Q. petraea*.

For updated reviews about the influence of the different wood trees and their toasting level during wine aging, refer to Garde-Cerdán and Ancín-Azpilicueta (2006) and Zamora (2019).

21.3 The different procedures for making white wines in barrels

Essentially, white wines can be elaborated in wood barrels using three different procedures that can exert a notable influence on the final composition of the wine and naturally on its sensory attributes. Fig. 21.3 shows the outline of the different possibilities of winemaking white wines in barrels.

The first possibility (A) is directly racking the grape juice after settling to the barrels and performing the complete alcoholic fermentation and aging inside them. In that case, the thermic control of alcoholic fermentation would be somewhat complicated since wood has great thermic inertia and it is not easy to apply effective procedures to refrigerate the fermenting must inside the barrels. However, this procedure implies that all the lees remain inside the cask, which will entail a high influence of yeast autolysis. Briefly, yeast autolysis increases the wine mouthfeel, protects wine against oxidation, and smooths the impact of the wood on the composition and sensory quality of the wine. However, the presence of all the lees may sometimes cause the appearance of a reduction taint. This procedure is called "aging with total lees" or in French "élevage sur lies totals".

The second possibility (B₁) is racking the grape juice after settling to stainless steel tanks to perform the complete alcoholic fermentation inside these vats, and once alcoholic fermentation is finished, racking the wine to the barrels. In that case, the thermic control of alcoholic fermentation will be very easy but the wine will not benefit from the presence of the lees. That means that the wine will not gain the mouthfeel associated with the release of mannoproteins and polysaccharides and will be not protected against oxidation by the lees and that the impact of wood will probably be too intense. This procedure is called "aging without lees" or in French "élevage sans lies".

Finally, a third possibility can be applied (B₂). In that case, the grape juice starts alcoholic fermentation inside stainless steel tanks and when the mid of the process is attained (around a relative density of 1040), the fermenting grape juice is racked to the barrels to finish alcoholic fermentation there. This procedure allows having good thermic control and benefits of the presence of a part of the lees. This procedure, which is called "aging with fine

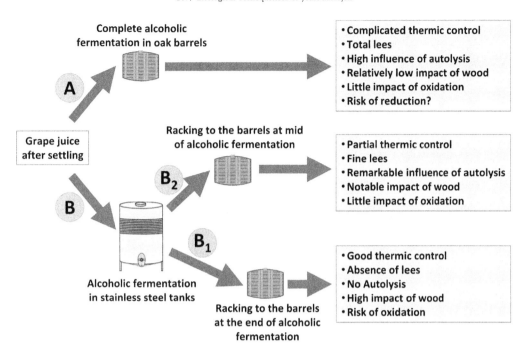

Complete alcoholic fermentation in oak barrels

- Complicated thermic control
- Total lees
- High influence of autolysis
- Relatively low impact of wood
- Little impact of oxidation
- Risk of reduction?

A

Grape juice after settling

Racking to the barrels at mid of alcoholic fermentation

- Partial thermic control
- Fine lees
- Remarkable influence of autolysis
- Notable impact of wood
- Little impact of oxidation

B₂

B

B₁

- Good thermic control
- Absence of lees
- No Autolysis
- High impact of wood
- Risk of oxidation

Alcoholic fermentation in stainless steel tanks

Racking to the barrels at the end of alcoholic fermentation

FIGURE 21.3 Outline of the different possibilities of winemaking white wines in barrels.

lees," or in French, *"élevage sur lies fines"*, is probably the most used one nowadays because it has the advantages of both the previous procedures and avoids all their drawbacks.

The organoleptic impact of each one of these procedures will be discussed in the next sections.

21.4 Enological consequences of yeast autolysis

As discussed in the previous point, the different procedures for elaborating white wines in oak barrels condition the impact of yeast autolysis a lot. But, what effects produce yeast autolysis in the wine? Fig. 21.4 synthesizes the different enological consequences of yeast autolysis.

Once alcoholic fermentation has finished, the yeast death and the autolysis process begin (Alexandre & Guilloux-Benatier, 2006). However, natural autolysis is a slow process that needs several months to achieve a real organoleptic effect. Yeast autolysis starts with a passive excretion of amino acids and other small molecules. This process, called exorption, lasts between 1 and 2 months in still wines and between 3 and 6 months in sparkling wines (Morfaux & Dupuy, 1966). Later, true autolysis begins with the degradation of membranes (cell and vacuolar), causing the progressive hydrolytic degradation of all the cell structures (Alexandre & Guilloux-Benatier, 2006), whose fragments are released into the wine. Once autolysis has begun, cell wall mannoproteins (Martínez-Lapuente, Guadalupe, Ayestarán, & Pérez-Magariño, 2015) and polysaccharides (Martínez-Lapuente et al., 2013) concentrations increase slowly. In parallel, other components of the membranes and cytoplasm such as proteins (Luguera, Moreno-Arribas, Pueyo, Bartolome, & Polo, 1998), peptides (Moreno-Arribas, Pueyo, & Polo, 1996), lipids (Pueyo, Martínez-Rodríguez, Polo, Santa-María, & Bartolomé, 2000), amino acids (Martinez-Rodriguez, Carrascosa, Martin-Alvarez, Moreno-Arribas, & Polo, 2002), and nucleotides (Charpentier et al., 2005) are also released. This process completely transforms the wine composition and therefore its sensory attributes (Kemp, Alexandre, Robillard, & Marchal, 2015).

It has been reported that mannoproteins and polysaccharides play a positive role in improving mouthfeel (Gawel, Smith, Cicerale, & Keast, 2018), and that these molecules, together with some proteins and peptides released by yeasts, can contribute to wine sweetness, smoothing the astringency of ellagitannins and the perception of acidity (Marchal, Marullo, Moine, & Dubourdieu, 2011). In addition, some peptides, amino acids, and nucleotides can contribute to the umami taste (Vilela, Ines, & Cosme, 2016) and have been described to be flavor

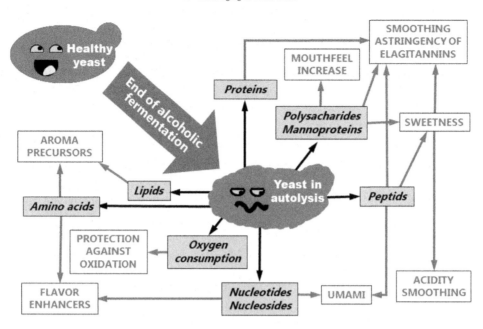

FIGURE 21.4   Oenological consequences of yeast autolysis.

enhancers. Finally, it has also been reported that lipids and amino acids are aroma precursors (Styger, Prior, & Bauer, 2011), and therefore their presence can also contribute to the aromatic complexity.

In addition, the presence of mannoproteins seems to exert a very positive effect on the wine physicochemical stability since it improves the stability of tartaric acid salts (Gerbaud, Gabas, Blounin, Pellerin, & Moutounet, 1997) and proteins (Waters, Pellerin, & Brillouet, 1994).

It has also been stated that yeast lees express antioxidant activity (Gallardo-Chacón, Vichi, Urpí, López-Tamames, & Buxaderas, 2010) and consume oxygen (Salmon, Fornairon-Bonnefond, Mazauric, & Moutounet, 2000). Consequently, the presence of lees slows down the oxidation of the wine through the consumption of oxygen. The mechanism by which the lees consume oxygen is not clear, but some authors associate it with the oxidation of membrane lipids, particularly the sterol fractions (Fornairon-Bonnefond & Salmon, 2003). However, it may also be due to the fact that lees can release glutathione (Kritzinger, Bauer, & du Toit, 2013). Regardless of the mechanism by which the lees consume oxygen, it is clear that this oxygen consumption by the lees is probably the main reason why white wines aged on lees can usually age for a far longer time than the other white wines.

In any case, yeast autolysis provides textural weightiness to the wine and increases its body. White wines aged on lees are generally richer in mouthfeel and have greater depth and flavor complexity.

21.5 Lees stirring and its importance

Once alcoholic fermentation ends, no more carbon dioxide is released, and, consequently, the lees settle at the bottom of the barrel. For this reason, it is necessary to resuspend the lees periodically in order to enhance the surface contact between wine and lees. In this way, yeast autolysis is promoted and the correct redox balance is maintained. The procedure to do this is called "lees stirring," but the French term of *bâtonnage* is used more in the world of wine. This operation was traditionally carried out using a walking stick (in French *bâton*), hence the name. Fig. 21.5 illustrates this procedure.

Nowadays, lees stirring is customary to perform using a steel rod with a paddle at the end which is introduced until reaching the bottom of the barrel where the lees are deposited. Then the steel rod is spun vigorously until the lees are completely resuspended. This operation can be done manually or with an electric drill attached to the steel rod.

Lees stirring is practiced because it provides several benefits to wine. Basically, it enhances the typical sensory attributes of aging with lees, reduces the wood impact, and, especially, it protects the wine against oxidation and reduction.

FIGURE 21.5   Lees stirring (Batônnage).

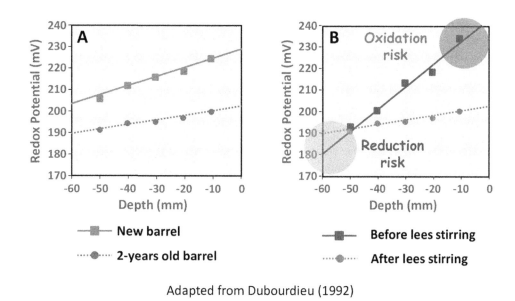

Adapted from Dubourdieu (1992)

FIGURE 21.6   Influence of the type of container and the presence of lees on the redox potential of wine. Source: *Dubourdieu, D. (1992). Vinification des vins blancs secs en barriques. In Le bois et la qualité des vins et des eaux-de-vie.* J. Inter. Sci. Vigne Vin, núm Hors de série *(pp. 137–143).*

Fig. 21.6 shows how lees stirring helps to maintain the redox balance of the wines aged *sur lies* in barrels. According to the Nernst law, redox potential depends a lot on oxygen concentration (Vivas, Zamora, & Glories, 1993). Oxygen permeates the barrel through the pores of the wood, through the joints between the staves, and through the bunghole (del Alamo-Sanza & Nevares, 2019). However, oxygen is consumed by the wine (Ferreira, Carrascon, Bueno, Ugliano, & Fernandez-Zurbano, 2015) by the ellagitannins released by the wood (Navarro, Kontoudakis, Giordanengo et al., 2016; Navarro, Kontoudakis, Gómez-Alonso et al., 2016; Pascual et al., 2017) and especially by the lees (Fornairon-Bonnefond & Salmon, 2003). Therefore the oxygen concentration and, consequently, the redox potential depend on a complicated balance between oxygen permeation and consumption. Since the lees will be at the bottom of the cask if *bâtonnage* is not performed, the oxygen in this part of the barrel will be consumed faster than that at the top. This phenomenon establishes a redox potential gradient from the top of the barrel down to the bottom (Dubourdieu, 1992). Fig. 21.6A shows how the redox potential is highest at the top and lowest at the bottom of the cask in a new barrel. This phenomenon was also found to occur in a 2-year-old barrel, but the redox values were lower and the gradient slope was less steep. This difference is because the pores of the barrels are gradually clogged over time by the precipitation of tartrates, proteins, and other colloids, which in turn gradually diminishes their capacity for oxygenation (Vivas et al., 1993).

The existence of this redox potential gradient implies that the wine at the top of the barrel has the risk of being oxidized, whereas the wine at the bottom of the cask has the risk of being reduced, as is shown in Fig. 21.6B

(Dubourdieu, 1992). Since *bâtonnage* homogenizes the wine and resuspends the lees, the redox potential gradient is damped, making the slope much less steep. Lees stirring therefore avoids wine oxidation, especially in new barrels, in which the redox potential is higher, and also protects against the appearance of reduction taint, especially in used barrels, in which the redox potential is lower.

For all these reasons, lees stirring must be applied at least twice a week.

21.6 Organoleptic consequences of the different procedures for making white wines in barrels

As mentioned in the previous sections, the different procedures for white wine winemaking in barrels exert a notable influence on the final composition of the wine and naturally on its sensory attributes. In synthesis, these differences depend mainly on whether or not the wine has fermented in the barrel and the presence or absence of the lees. If the grape juice is fermented in the barrel, the fermentative metabolism of the yeast can transform some of the substances released by wood, thus achieving a lower sensory impact (Feuillat, 1992; Fornairon-Bonnefond et al., 2002). For example, vanillin is partially reduced by yeast to vanillic alcohol and furfural to furfuryl alcohol, which significantly reduces the organoleptic impact of these substances (Chatonnet et al., 1992). In a similar way, yeast lees can absorb some volatile substances and phenolic compounds released by wood (Mazauric & Salmon, 2005; Pérez-Serradilla & Luque de Castro, 2008). All these phenomena make that the sensory wood impact of the wines fermented and aged *sur lies* is much lower than in wines that have simply been barrel-aged without the presence of the lees.

Fig. 21.7 illustrates visually in the form of a spider web diagram the sensory characteristics that different procedures for winemaking in barrels provide to white wine. This figure shows that the wines aged in barrels without the presence of lees are greatly impacted by wood since the intensities of vanilla, coconut, and toasted/smoked notes are very high. Moreover, the structure and tannicity are also very present. In contrast, the mouthfeel of these wines is relatively low since they have not benefited from yeast autolysis. In addition, some oxidation notes can be sometimes detected since they were aged without the protective effect of lees.

In contrast, the wines aged with total lees are impacted by wood to a lesser extent since the intensities of vanilla, coconut, and toasted/smoked notes, as well as the structure and tannicity, are quite lower because the lees transform and/or absorb part of the substances released by the wood. However, the mouthfeel of these wines is really very high, thanks to the fact that yeast autolysis is very present. However, sometimes reduction notes can be present since the redox potential could be too low due to the presence of total lees. In that case, lees stirring becomes more important to avoid the appearance of this taint.

Finally, white wines aged with fine lees probably have the best balance since the intensities of vanilla, coconut, and toasted/smoked notes are higher than in wines aged with total lees but lower than in wines aged without

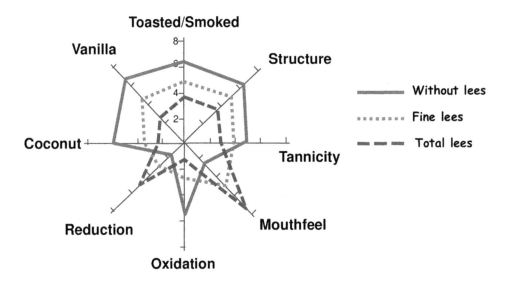

FIGURE 21.7   Spider web diagram of the sensory characteristics in function of the procedure used for winemaking.

lees. Moreover, these wines have a good mouthfeel and reasonable tannicity and are protected against oxidation since the influence of the lees is important but not so much as in the wines aged with total lees. In fact, the majority of white wines aged in barrels are elaborated with fine lees because it allows obtaining higher complexity and avoids the drawbacks that the other two procedures may originate, excess of wood and oxidation in the case of wines aged without lees or small impact of wood and reduction taint in the case of wines aged with total lees.

We must not forget that the secret of a great wine is based on its balance and complexity (Berdin & Dubourdieu, 2012). Any excess wood that imposes itself on the own nature of the wine would spoil its delicate balance. For many years, many white wines have sinned from excess wood (Robinson, 2004). Fortunately, the current market seeks subtlety more than musculature, and nowadays, most winemakers are working with these concepts.

References

Abbott, N., Puech, J. L., Bayonove, C., & Baumes, R. (1995). Determination of the aroma threshold of the cis and trans racemic forms of β-methyl-γ-octalactone by gas chromatography-sniffing analysis. *American Journal of Enology and Viticulture, 46*, 292–294.

Alexandre, H., & Guilloux-Benatier, M. (2006). Yeast autolysis in sparkling wine: A review. *Australian Journal of Grape Wine Research, 12*, 119–127.

American Oil & Gas Historical Society (2016). History of the 42-gallon oil barrel. < https://aoghs.org/transportation/history-of-the-42-gallon-oil-barrel/ > .

Berdin, G., Dubourdieu, D. (2012). Oenology and the art of winemaking: Conversations over a bottle with Denis Dubourdieu. Elytis, DL, Bordeaux, France

Boidron, J. N., Chatonnet, P., & Pons, M. (1988). Influence du bois sur certaines substances odorantes des vins. *Conn. Vigne Vin, 22*, 275–294.

Cadahía, E., Fernández de Simón, B., & Jalocha, J. (2003). Volatile compounds in Spanish, French, and American oak woods after natural seasoning and toasting. *Journal of Agriculture Food Chemistry, 51*, 5923–5932.

Chatonnet, P. (1992). Les composés aromatiques du bois de cheêne cédeés aux vins. Influence des opérations de chauffe en tonnellerie. In *Le bois et la qualité des vins et des eaux-de-vie. J. Inter. Sci. Vigne Vin*, núm Hors de série (pp. 81 – 91).

Chatonnet, P., Dubourdieu, D., & Boidron, J. N. (1991). Vinification et élevage des grands vins blancs secs en fûts de chêne: Le rôle des levures. *Rev. Œnol., 61*, 53–55.

Chatonnet, P., Dubourdieu, D., & Boidron, J. N. (1992). Incidence des conditions de fermentation et d'élevage des vins blancs secs en barriques sur leur composition en substances cédées par le bois de chêne. *Sciences des Aliments, 12*, 665–685.

Chatonnet, P., & Dubourdieu, D. (1998). Comparative study of the characteristics of American white oak (*Quercus alba*) and European oak (*Quercus petraea* and *Q. robur*) for production of barrels used in barrel ageing of wines. *Amercian Journal of Enology Viticulture, 49*, 79–85.

Blanchard, L., Tominaga, T., & Dubourdieu, D. (2001). Formation of furfurylthiol exhibiting a strong coffee aroma during oak barrel fermentation from furfural released by toasted staves. *Journal of Agriculture and Food Chemistry, 49*, 4833–4835.

Charpentier, C., Aussenac, J., Charpentier, M., Prome, J. C., Duteurtre, B., & Feuillat, M. (2005). Release of nucleotides and nucleosides during yeast autolysis: Kinetics and potential impact on flavor. *Journal of Agriculture Food Chemistry, 53*, 3000–3007.

Chira, K., & Teissedre, P. L. (2013). Relation between volatile composition, ellagitannin content and sensory perception of oak wood chips representing different toasting processes. *European Food Research Technology, 236*, 735–746.

del Alamo-Sanza, M., & Nevares, I. (2019). Oak wine barrel as an active vessel: A critical review of past and current knowledge. *Critical Reviews in Food Science and Nutrition, 58*, 2711–2726.

Dubourdieu, D. (1992). Vinification des vins blancs secs en barriques. In *Le bois et la qualité des vins et des eaux-de-vie. J. Inter. Sci. Vigne Vin*, núm Hors de série (pp. 137–143).

Fernández de Simón, B., Martín, J., Sanz, M., Cadahía, E., Esteruelas, E., & Muñoz, A. M. (2014). Volatile compounds and sensorial characterisation of red wine aged in cherry, chestnut, false acacia, ash and oak wood barrels. *Food Chemistry, 147*, 346–356.

Ferreira, V., Carrascon, V., Bueno, M., Ugliano, M., & Fernandez-Zurbano, P. (2015). Oxygen consumption by red wines. Part I: Consumption rates, relationship with chemical composition, and role of SO2. *Journal of Agriculture Food Chemistry, 63*, 10928–10937.

Feuillat, M. (1992). Vins blancs vinifiés en fûts: L'interaction du bois et des lies. *Vitis, Nov*, 103–104.

Feuillat, M., & Charpentier, C. (1982). Autolysis of yeasts in Champagne. *American Journal of Enology and Viticulture, 33*(1), 6–13.

Feuillat, M., Freyssinet, M., & Charpentier, C. (1989). L'élevage sur lies des vins blancs de Bourgogne, II. Évolution des macromolecules (polysaccharides et protéines). *Vitis, 28*, 161–176.

Fornairon-Bonnefond, C., Camarasa, C., Moutounet, M., & Salmon, J. M. (2002). New trends on yeast autolysis and wine ageing on lees: A bibliographic review. *Journal International des Sciences de la Vigne et du Vin, 36*, 49–69.

Fornairon-Bonnefond, C., & Salmon, J. M. (2003). Impact of oxygen consumption by yeast lees on the autolysis phenomenon during simulation of wine aging on lees. *Journal of Agriculture Food Chemistry, 51*, 2584–2590.

Gallardo-Chacón, J. J., Vichi, S., Urpí, P., López-Tamames, E., & Buxaderas, S. (2010). Antioxidant activity of lees cell surface during sparkling wine sur lie aging. *International Journal of Food Microbiology, 143*, 48–53.

Garde-Cerdán, T., & Ancín-Azpilicueta, C. (2006). Review of quality factors on wine ageing in oak barrels. *Trends in Food Science and Technology, 17*, 438–447.

Gawel, R., Smith, P. A., Cicerale, S., & Keast, R. (2018). The mouthfeel of white wine. *Critical Reviews in Food Science and Nutrition, 58*, 2939–2956.

Gerbaud, V., Gabas, N., Blounin, J., Pellerin, P., & Moutounet, M. (1997). Influence of wine polysaccharides and polyphenol on the crystallisation of potassium hydrogen tartrate. *Journal International des Sciences de la Vigne et du Vin, 31*, 65–83.

Glabasnia, A., & Hofmann, T. (2006). Sensory-directed identification of taste-active ellagitannins in American (*Quercus alba* L.) and European oak wood (*Quercus robur* L.) and quantitative analysis in Bourbon whiskey and oak-matured red wines. *Journal of Agriculture Food Chemistry, 54*, 3380–3390.

Glonti, T. (2010). Traditional technologies and history of Gerogian wine. *Bull. O.I.V, 83*(953-955), 335–343.

Jackson, R. S. (2009). Styles and types of wine. *Wine tasting, a professional handbook* (2nd Edition, pp. 349–386). San Diego, CA, USA: Academic Press. Elsevier.

Jourdes, M., Lefeuvre, D., & Quideau, S. (2009). C-Glycosidic ellagitannins and their influence on wine chemistry. In S. Quideau (Ed.), *Chemistry and biology of ellagitannins — An underestimated class of bioactive plant polyphenols* (pp. 320–365). Singapore: World Scientific Publishing Co.

Kemp, B., Alexandre, H., Robillard, B., & Marchal, R. (2015). Effect of production phase on bottle-fermented sparkling wine quality. *Journal of Agriculture Food Chemistry, 63*, 19–38.

Kritzinger, E. C., Bauer, F. F., & du Toit, W. J. (2013). Role of glutathione in winemaking: A review. *Journal of Agriculture Food Chemistry, 61*, 269–277.

Luguera, C., Moreno-Arribas, M. V., Pueyo, E., Bartolome, B., & Polo, M. C. (1998). Fractionation and partial characterisation of protein fractions present at different stages of the production of sparkling wines. *Food Chemistry, 63*, 465–471.

Marchal, A., Marullo, P., Moine, V., & Dubourdieu, D. (2011). Influence of yeast macromolecules on sweetness in dry wines: Role of the Saccharomyces cerevisiae protein Hsp12. *Journal of Agriculture Food Chemistry, 59*, 2004–2010.

Martínez-Lapuente, L., Guadalupe, Z., Ayestarán, B., & Ortega-Hera, M. (2013). Changes in polysaccharide composition during sparkling wine making and aging. *Journal of Agricultural and Food Chemistry, 61*, 12362–12373.

Martínez-Lapuente, L., Guadalupe, Z., Ayestarán, B., & Pérez-Magariño, S. (2015). Role of major wine constituents in the foam properties of white and rosé sparkling wines. *Food Chemistry, 174*, 330–338.

Martinez-Rodriguez, A. J., Carrascosa, A. V., Martin-Alvarez., Moreno-Arribas, M. V., & Polo, M. C. (2002). Influence of the yeast strain on the changes of the amino acids, peptides and proteins during sparkling wine production by the traditional method. *Journal of Industrial Microbiology and Biotechnology, 29*, 314–322.

Mazauric, J. P., & Salmon, J. M. (2005). Interactions between yeast lees and wine polyphenols during simulation of wine aging: I. analysis of remnant polyphenolic compounds in the resulting wines. *Journal of Agriculture Food Chemistry, 53*, 5647–5653.

Moreno-Arribas, M. C., Pueyo, E., & Polo, M. C. (1996). Peptides in musts and wines. changes during the manufacture of cavas (sparkling wines). *Journal of Agriculture Food Chemistry, 44*, 3783–3788.

Michel, J., Jourdes, M., Silva, M. A., Giordanengo, T., Mourey, N., & Teissedre, P. L. (2011). Impact of concentration of ellagitannins in oak wood on their levels and organoleptic influence in red wine. *Journal of Agriculture Food Chemistry, 59*, 5677–5683.

Morfaux, J. N., & Dupuy, P. (1966). Comparaison de l'exorption des acides aminés par une souche de Saccharomyces cerevisiae et un mutant résistant à la canavanine. *Compte Rendu de l' Académie des Sciences, 263*, 1224.

Navarro, M., Kontoudakis, N., Giordanengo, T., Gómez-Alonso, S., García-Romero, E., Fort, F., ... Zamora, F. (2016). Oxygen consumption by oak chips in a model wine solution; Influence of the botanical origin, toast level and ellagitannin content. *Food Chemistry, 199*, 822–827.

Navarro, M., Kontoudakis, N., Gómez-Alonso, S., García-Romero, E., Canals, J. M., Hermosín-Gutíerrez, I., & Zamora, F. (2016). Influence of the botanical origin and toasting level on the ellagitannin content of wines aged in new and used oak barrels. *Food Research International, 87*, 197–203.

Navarro, M., Kontoudakis, N., Gómez-Alonso, S., García-Romero, E., Canals, J. M., Hermosín-Gutíerrez, I., & Zamora, F. (2018). Influence of the volatile substances released by oak barrels into a Cabernet Sauvignon red wine and a discolored Macabeo white wine on sensory appreciation by a trained panel. *European Food Research Technology, 244*, 245–258.

Pascual, O., Vignault, A., Gombau, J., Navarro, M., Gómez-Alonso, S., García-Romero, E., ... Zamora, F. (2017). Oxygen consumption rates by different oenological tannins in a model wine solution. *Food Chemistry, 234*, 26–32.

Pérez-Serradilla, J. A., & Luque de Castro, M. D. (2008). Role of lees in wine production: A review. *Food Chemistry, 111*, 447–456.

Pitte, J. R. (2018). Tonnellerie: 2000 Ans d'histoire et de savoir-faire. *La Revue du Vin de France, 624*, 42–48.

Prida, A., & Chatonnet, P. (2010). Impact of oak-derived compounds on the olfactory perception of barrel-aged wines. *Amercian Journal of Enology Viticulture, 61*, 408–413.

Pueyo, E., Martínez-Rodríguez, A., Polo, M. C., Santa-María, G., & Bartolomé, B. (2000). Release of lipids during yeast autolysis in a model wine System. *Journal of Agriculture Food Chemistry, 48*, 116–122.

Quinn, M. K., & Singleton, V. L. (1985). Isolation and identification of ellagitannins from white oak wood and an estimation of their roles in wine. *Amercian Journal of Enology Viticulture, 36*, 148–155.

Ribéreau-Gayon, P., Glories, Y., Maujean, A., & Dubourdieu, Y. (2006). Making dry white wines in barrels. *Handbook of enology, vol 1 the microbiology of wine and vinifications* (2nd Edition, pp. 433–439). Chichester: John Wiley & sons, Ltd.

Robinson, J. (2004). Oak – Its uses and abuses. < https://www.jancisrobinson.com/articles/oak-its-uses-and-abuses >.

Salmon, J. M., Fornairon-Bonnefond, C., Mazauric, J. P., & Moutounet, M. (2000). Oxygen consumption by wine lees: Impact on lees integrity during wine ageing. *Food Chemistry, 71*, 519–528.

Sapies, J. C., Delteil, D., Feuillat, M., & Guilloux-Benatier, M. (1998). Les vins blancs secs. In C. Flanzy (Ed.), *OEnologie: Fondements scientifiques et technologiques* (pp. 718–736). Paris: Lavoisier.

Schneider, R., Dubourdieu, D. (2014). Les vins blancs de la démarche marketing à la vinification: les clés d'un pilotage reéussi. Ed. Institut français de la vigne et du vin. Editions France agricole, Paris, France.

Suomalainen, H., Nykaenen, L. (1970). Composition of whisky flavour. *Proc. Biochem., 5*, 13–18.

Spillman, P. J., Sefton, M. A., & Gawel, R. (2004). The effect of oak Wood source, location of seasoning and coopering on the composition of volatile compounds in oak-matured wines. *Australian Journal of Grape and Wine Research, 10*, 216–226.

Styger, G., Prior, B., & Bauer, F. F. (2011). Wine flavor and aroma. *Journal of Industrial Microbiology and Biotechnology, 38*, 145–1159.

Tominaga, T., Blanchard, L., Darriet, P., & Dubourdieu, D. (2000). A powerful aromatic volatile thiol, 2-furanmethanethiol, exhibiting roast coffee aroma in wines made from several Vitis vinifera grape varieties. *Journal of Agriculture Food Chemistry, 48*, 1799–1802.

Vilela, A., Ines, A., & Cosme, F. (2016). Is wine savory? Umami taste in wine. *SDRP Journal Food Science and Technology, 1*, 100−105.

Vivas, N., Glories, Y., Bourgeois, G., & Vitry, C. (1996). Les ellagitannins de bois de coeur de différentes espèces de chêne (*Quercus* sp.) et de châtaignier (*Castanea sativa* Mill.). Dosage dans les vins rouges élevés en barriques. *Journal of Science and Technology, 2*, 25−49.

Vivas, N., Zamora, F., & Glories, Y. (1993). Incidence de certains facteurs sur la consommation de l'oxygène et sur le potentiel d'oxydorreduction dans les vins. *Journal International des Sciences de la Vigne et du Vin, 27*, 23−34.

Waters, E. J., Pellerin, P., & Brillouet, J. M. (1994). A wine arabinogalactan-protein that reduces heat-induced wine protein haze. *Bioscience Biotechnology and Biochemistry, 58*, 43−48.

Work, H. H. (2014). *Wood, whiskey and barrels; A history of Barrels*. London, UK: Reaktion Books Ltd.

Zamora, F. (2019). Barrel aging; Types of wood. In A. Morata (Ed.), *Red wine technology* (pp. 125−147). London: Academic Press, Elsevier.

Zhang, B., Cai, J., Duan, C. Q., Reeves, M. J., & He, F. (2015). A review of polyphenolics in oak woods. *International Journal of Molecular Sciences, 16*, 6978−7014.

22

Use of different wood species for white wine production: wood composition and impact on wine quality

António M. Jordão[1] and Jorge M. Ricardo-da-Silva[2]

[1]Polytechnic Institute of Viseu, Agrarian Higher School, Viseu, Portugal.Chemistry Research Centre (CQ-VR) - Food and Wine Chemistry Lab, Vila Real, Portugal [2]LEAF - Linking Landscape, Environment, Agriculture and Food, Universidade de Lisboa, Instituto Superior de Agronomia, Lisboa, Portugal

22.1 Introduction

Initially, the use of wood in enology was mainly associated with the transport of wines. Already in the 7th century BC, near the Tigris river (Mesopotâmia area), palm wood was used as material for barrels manufacturing. Greeks and Romans (2000 BCE) used earthenware jars and amphorae, although these containers were fragile, heavy, and difficult to handle. Faced with this logistic problem of transporting wines from the rural areas to the consumption areas located in different Mediterranean regions, wooden containers were created and developed. Over the centuries, there were several woods used in the production of barrels and casks, with chestnut, pine, mahogany, eucalyptus, and acacia woods being the most used wood species (Taransaud, 1976; Parodi, 2000). However, oak has been one of the main wood species for this purpose, having different properties that differentiate it from the other wood species, namely high resistance, flexibility, and ease of handling, and also very low permeability. Nevertheless, it is important to note that in the mid-20th century, there was a strong abandonment of the use of wood for enological use due to the increased use of other materials, namely cement and stainless steel. From the 1990s onward, according to several authors (Vivas & Saint-Cricq de Gaulejac, 1999; Singleton, 2000), the use of wooden for barrels production, particularly from different oak wood species, reemerged rather significantly and became one of the most preferred options for wine aging. According to several authors (Carvalho, 1998; Vivas, 2005), the main oak wood species used by cooperages in the United States are *Quercus alba*, *Quercus garryana*, *Quercus macrocarpa*, and *Quercus stellata*, while in Europe, other species, namely *Quercus petraea* and *Quercus robur*, as well as species such as *Quercus cerris*, *Quercus suber*, *Quercus lyratta*, *Quercus bicollor*, and *Quercus lanuginose* are used. However, *Q. alba*, *Q. petraea*, and *Q. robur* are the main oak species used. Actually, several authors have also reported the use of other European and non-European oak species with interesting uses in enology, such as *Quercus pyrenaica* (Jordão, Ricardo-da-Silva, & Laureano, 2005, 2006; Jordão et al., 2006; Jordão, Ricardo-da-Silva, & Laureano, 2007; Gonçalves & Jordão, 2009; Fernández de Simón, Cadahía, Muiño, del Álamo, & Nevares, 2010; Gallego et al., 2012; Castro-Vázquez, Alañón, Ricardo-da-Silva, Pérez-Coello, & Laureano, 2013; Sánchez-Gómez, Ignacio, Ana, & Maria, 2018; McCallum, Lopes-Correia, & Ricardo-da-Silva, 2019), *Quercus faginea* (Fernández de Simón, Cadahía, & Jalocha, 2003; Fernández de Simón, Hernández, Cadahía, Deñas, & Estrella, 2003), *Quercus pubescens* (Jordão et al., 2019; Jordão, Lozano, & Gonzalez SanJose, 2019), and *Quercus humboldtii* from Colombia (Martínez-Gil et al., 2017; Martínez-Gil, del Álamo-Sanza, Gutiérrez-Gamboa, Moreno-Simunovic, & Nevares, 2018).

© 2022 Elsevier Inc. All rights reserved.

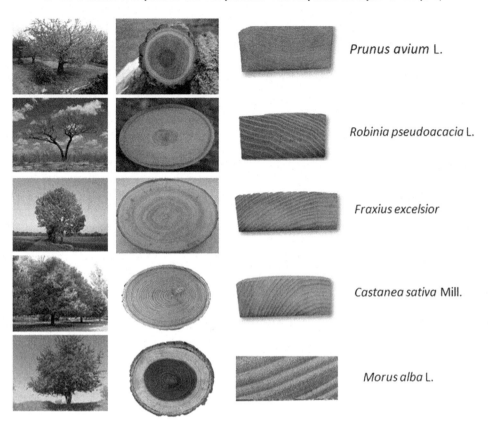

FIGURE 22.1   Some untraditional wood species (tree and cross-sections) with potential use for white wine production.

Today, oak and chestnut wood species are the only ones approved by the International Organization of Vine and Wine (OIV) for enological use (Resolution OENO 4/2005). However, in recent years, heartwoods from alternative species have been considered as possible sources of wood for wine production (Fig. 22.1), such as acacia (*Robinia pseudoacacia*), cherry (*Prunus avium* L. and *Prunus cereasus* L.), European and American ash (*Fraxinus excelsior* L. and *Fraxinus americana* L.), and mulberry (*Morus alba* L. and *Morus nigra* L.). In addition, a few studies have reported the potential use of other woods species such as *Juglans regia*, *Juniperus communis*, *Pinus heldreichii* var. *Leucodermi*s, *Prunus armeniaca*, *Fagus Syvatica*, and *Alnus glutinosa* (Gortzi, Metaxa, Mantanis, & Lalas, 2013). It is important to note that the demand for these no-oak wood species results from the fact that in recent years, the increased demand for oak wood has caused a remarkable increase in production costs (due to the limited availability of materials) and also environmental costs (due to intensive harvesting of oak trees in forests).

Actually, the main objectives of the practice of using wood are to enrich the wine with substances released by the wood, promote reactions due to contact with air diffused through the wood pores and develop certain interactive chemical reactions that take place slowly in wines, and consequently improve the wine's quality and organoleptic characteristics (Izquierdo-Cañas, Mena-Morales, & García-Romero, 2016; Cerdán & Ancín-Azpilicueta, 2006). Thus the scientific literature in recent years has contained a large number of data related to the use of oak wood in enology, particularly its impact on wine chemical and sensory characteristics. However, the great majority of studies have been carried out with different types of red wines, or by the use of oak barrels (Chatonnet, 1995; Del Álamo Sanza, Escudero, & De Castro Torío, 2004; Matějíček, Mikeš, Klejdus, Štěrbová, & Kubáň, 2005; Chatonnet, Ricardo-da-Silva, & Dubourdieu, 1997; Fernández de Simón, Cadahía, & Jalocha, 2003,; Fernández de Simón et al., 2003; Laqui-Estaña, López-Solís, Peña-Neira, Medel-Marabolí, & Obreque-Slier, 2019; McCallum et al., 2019; Prida & Chatonnet, 2010; Rasines-Perea, Jacquet, Jourdes, Quideau, & Teissedre, 2019) or oak fragments, such as chips, staves, blocks, or powders (Gonçalves & Jordão, 2009; Del Álamo Sanza, Escudero, & De Castro Torío, 2004; De Coninck, Jordão, Ricardo-da-Silva, & Laureano, 2006; Laqui-Estaña et al., 2019; Martínez-Gil, del Álamo-Sanza, Nevares, Sánchez-Gómez, & Gallego, 2020; Rubio-Bretón, Garde-Cerdán, & Martínez, 2018; Tavares, Jordão, & Ricardo-da-Silva, 2017). Also for rosé wines, the results about the potential use of different wood species, including oak, acacia, and cherry, in the fermentation and aging process were only very recently published (Nunes, Correia, Jordão, & Ricardo-da-Silva, 2020; Santos et al., 2019). In addition, other studies have also analyzed the impact of oak wood or individual oak wood components on the mechanisms of the evolution of individual red wine compounds during the aging process, to improve the knowledge of the impact of oak wood on wine composition and quality (Jordão, Ricardo-de-Silva, & Laureano, 2006, Jordão et al., 2006, 2019; Jordão, Ricardo-da-Silva, Laureano, Mullen, & Crozier, 2008; Jordão, Lozano, & González SanJosé, 2019b; Cutzach, Chatonnet, & Dubourdieu, 1999; Vivas & Glories, 1996; Vivas, Glories, Bourgeois, & Vitry, 1996). More recently, the possibility of the use of wood fragments (as already mentioned for red wines) in white wine production, including during the aging process, has also emerged. However, this option was only possible, for example, in Europe after the publication of the EEC regulation in 2006, modified by the EEC Regulamention No. 606/2009 of July 10, 2009, which modified previous rules about the use of pieces of oak for winemaking and the description and presentation of wine undergoing this treatment. It states that the pieces of oak wood used for winemaking and aging must exclusively come from the *Quercus* genus and with different toasting levels or without toasting. Thus, some works have also described the impact of the use of oak wood fragments on white wine composition and sensory profile as it is an option that is easier and cheaper to use during the white wine production (Guchu, Díaz-Maroto, Pérez-Coello, González-Viñas, & Cabezudo, 2006; Gutiérrez-Afonso, 2002; Pérez-Coello, González-Viñas, García-Romero, Cabezudo, & Sanz, 2000; Sánchez-Palomo, Alonso-Villegas, Delgado, & González-Viñas, 2017). As mentioned, the overwhelming majority of published studies regarding the use of different types of wood species in enology, with a more accentuated focus on oak wood species, have been mainly focused on red wines. Therefore there is very limited knowledge about the potential impact of the use of different alternative wood species on the white wine quality. Only in recent years have very few studies focused on this topic, particularly on the use of acacia and cherry wood species. Thus the comparative analysis of the impact of oak and no-oak wood species on white wine composition and sensory profile will be one of the main focuses of this chapter.

22.2 Composition of different wood species

It is well known that there are a great number of factors that determine the wood composition. For the different oak wood species, a great number of research works have been published about the composition of the oak woods, particularly in terms of their extractable compounds with relevance in enological terms. Among the extractable substances, ellagitannins, gallic acid, ellagic acid, various aromatic compounds, and aldehydes are the main compounds or group of compounds with enological importance. Volatile phenols and benzoic aldehydes play a very important role in the sensorial characteristics of the wines (Ibern-Gómez et al., 2001; Canas, Quaresma, Belchior, Spranger, & Bruno-de-Sousa, 2004), while hydrolyzable tannins, such as ellagitannins, are particularly important because they confer astringency and also are involved in the stabilization of pigment structures and demonstrate important antioxidant properties (Bautista-Ortin, Martinez-Cutillas, Ros-Garcia, López-Roca, & Gómez-Plaza, 2005; Escudero-Gilete, Hernanz, Galán-Lorente, Heredia, & Jara-Palacios, 2019; Fujieda, Tanaka, Suwa, Koshimizu, & Kouno, 2008; Vivas & Glories, 1996; Vivas et al., 1996). According to several authors (Vivas & Glories, 1993, 1996, Vivas et al., 1996), ellagitannins have an important role in wine oxidation processes, as they rapidly absorb dissolved oxygen and facilitate the hydroperoxidation of wine constituents and also show an important role in the condensation rate of proanthocyanidin and anthocyanins, preventing their degradation and precipitation. For oak wood species used in cooperage, in general, the concentration of these extractive compounds are affected by a large number of independent factors, namely, forest origin (Jordão, Ricardo-de-Silva, & Laureano, 2005, 2007; Chatonnet, 1995; Doussot, De Jéso, Quideau, & Pardon, 2002; Masson, Moutounet, & Puech, 1995; Puech, Feuillat, & Mosedale, 1999), species (Jordão, Ricardo-de-Silva, & Laureano, 2005, 2007; Jordão, Correia, DelCampo, & González SanJosé, 2012; Canas, Leandro, Spranger, & Belchior, 2000; Chatonnet, 1995; Doussot et al., 2002; Masson et al., 1995) heartwood age (Matricardi & Waterhouse, 1999; Viriot, Scalbert, Hervé du Penhoat, & Moutounet, 1994), and, most particularly, heat treatment that occurs in cooperage during barrel and wood fragment production (Jordão, Ricardo-de-Silva, & Laureano, 2005, 2006, 2007; Jordão et al., 2006, 2007, 2012; Cadahía, Varea, Muñoz, Fernández de Simón, & García-Vallejo, 2001; Chatonnet, 1999; Matricardi & Waterhouse, 1999). In addition, among the different oak wood species, there is a clear differentiation for the extractable substances. Thus several published works have described that European and American species have different ellagitannin concentrations, with low concentrations for American oak (*Q. alba*) compared to the European species, namely *Q. robur*, *Q. petraea*, and *Q. pyrenaica* (Feuillat et al., 1997; Jordão et al., 2007, 2012; Cadahía et al., 2001; Canas et al., 2000; Chatonnet & Dubourdieu, 1998; Vivas & Glories, 1996; Vivas et al., 1996).

Table 22.1 shows the total phenolic content and several individual phenolic compounds, particularly low-molecular-weight phenolic compounds and ellagitannins, in untoasted and toasted forms of some *Quercus* specimens, namely *Q. petraea*, *Q. alba*, and *Quercus pyrenaica*, and also cherry (*P. avium* L.) and acacia (*Robina pseudoacacia*)

TABLE 22.1   Total phenolic and individual phenolic compounds quantified in different wood chip species with potential use for white wine production.

Wood species						
	Oak wood			**Acacia**	**Cherry**	**References**
Phenolic composition	*Quercus petraea*	*Quercus alba*	*Quercus pyrenaica* Willd.	*Robinia pseudoacacia* L.	*Prunus avium* L.	
Total polyphenols[a]	1269	–	–	786	898	De Rosso, Cancian, Panighel, Dalla Vedova, and Flamini (2009)[b]
Total polyphenols[c]	61.39	61.51	65.03	51.23	46.82	Jordão, Lozano, and González SanJosé (2019b), Jordão et al. (2019a)[g]
Protocatechuic aldehyde[d]	nd	nd	nd	1.40	nd	Jordão et al. (2016)[g]
Gallic acid	1.25	nd	1.44	nd	nd	
Vanillic acid	0.14	0.33	0.128	nd	nd	
Syringic acid	0.44	0.83	0.22	nd	nd	
(+)-Catechin	nd	nd	nd	nd	18.51	
Robinetin	nd	nd	nd	118.94	nd	
Fustin[e]	nd	nd	nd	0.86	nd	
Butin[e]	nd	nd	nd	3.52	nd	
p-Coumaric acid	153.4	nd	84.3	nd	172.5	
Quercetin	5.45	5.49	2.48	nd	nd	
Naringenin	nd	nd	nd	nd	5.54	
Vescalagin[f]	19.21	6.21	25.74	nd	nd	
Castalagin[f]	24.97	5.43	32.45	nd	nd	
Ellagic acid	3.49	1.17	4.91	0.049	0.72	
Vescalagin[h]	9.87–10.6	5.41–8.24	–	–	–	Jordão et al. (2012)[i]
Castalagin[h]	12.56–19.8	2.34–3.54	–	–	–	
Ellagic acid	6.91–12.56	2.34–3.42	–	–	–	
Vescalagin[h]	12.6–17.6	1.23–5.28	11.3–14.6	–	–	Jordão, Ricardo-de-Silva, and Laureano (2007)[i]
Castalagin[h]	20.1–22.7	0.37–0.44	15.2–19.7	–	–	
Ellagic acid	2.60–4.42	1.90–3.62	3.9–20.5	–	–	
Total ellagitannins[j]	–	–	–	–	nd-0.04	Sanz et al. (2010, 2012a, 2012b, 2012c), Alañón et al. (2011)
Gallic acid[k]	70.75	–	–	9.61	nd	Soares, Garcia, Freitas, and Cabrita (2012)
Vanillic acid[k]	2.34	–	–	nd	1.86	
Syringic acid[k]	3.78	–	–	0.82	3.05	
Ellagic acid[k]	32.68	–	–	nd	nd	
Protocatechuic aldehyde[k]	nd	–	–	0.19	7.90	

[a]*values expressed as (+)-catechin equivalents.*
[b]*values expressed as mg/L and data obtained after 30 extraction days by the use of model wine solutions 12% with 60 g/L of wood chips without toasting process.*
[c]*values expressed as gallic acid equivalents.*
[d]*values expressed as protocatechuic acid equivalents.*
[e]*values expressed as naringenin equivalents.*
[f]*values expressed as ellagic acid equivalents.*
[g]*values expressed as mg/L and data obtained after 30 extraction days by the use of model wine solutions 12% with 4 g/L of wood chips.*
[h]*values expressed in mg/g of dry wood gallic acid equivalents.*
[i]*data obtained after 160 min by the use of extraction solution (water/acetone).*
[j]*values expressed as mg/g of untoasted wood and sum of castalagin; vescalagin; granidin; and A, B, C, D, and E roburins.*
[k]*values expressed in mg/L of toasted wood and data obtained after 35 extraction days by the use of model wine solutions 12% with 48 g/L of wood chips; nd – not detected.*

woods. By analysis of the data shown in this table, in general, the ranges for the most individual compounds, namely for ellagitannins in the same oak species, are broad due to factors such as the heat of the treatment (toasting process) and variability within the species. In addition, *Robinia pseudoacacia* (acacia) and also, for the majority of the studies, *P. avium* (cherry) do not contain ellagitannins in their composition (Jordão, Virginia, Correia, Ortega-Heras, & González-SanJosé, 2016; Sanz et al., 2010, 2012a, 2012b, 2012c). Other authors (Alañón et al., 2011) have detected very low quantities of the two major ellagitannins: castalagin and vescalagin. The quantities found by these authors are insignificant with respect to the usual values that are quantified in the woods used in cooperage, particularly oak woods. On the other hand, ellagitannins were found also for *Castanea sativa* Mill. (chestnut). Also, according to these authors, the ranges of the content of ellagitannins for dried chestnut are between 4.74 and 76 mg/g, while a range of values between 0.66 and 10.51 mg/g was found for toasted chestnut. The results also revealed that its heartwood has the most similar polyphenolic profile to that of oak. However, there were also some differences. For example, according to Comandini and colleagues, 1-O-galloyl castalagin, which was for the first time found in chestnut samples, might originate from the esterification of castalagin or vescalagin with a gallic acid residue (Comandini, Lerma-García, Simó-Alfonso, & Toschi, 2014). Still, in relation to ellagitannins, *Quercus frainetto* is distinguished especially from the other species by its higher content of pentosylated dimers and a monomer concentration that is similar to the rest of the *Quercus* species, which makes it the species described to have the highest content of ellagitannins with average values of 108 mg/g (Vivas & Glories, 1996; Vivas et al., 1996). For ellagic acid, this phenolic acid could be found in different wood species, including oak, acacia, cherry, and chestnut (Jordão et al., 2016; Canas, Leandro, Spranger, & Belchior, 1999). In model wine solutions, ellagic acid and other oak wood extract components contribute to a protective effect against the degradation of some wine flavanols, namely (+)-catechin and procyanidin B1, during the aging time (Jordão et al., 2008). Also, previously, other authors (Jordão, Ricardo-de-Silva, & Laureano, 2006; Jordão et al., 2006) have shown that the presence of ellagic acid played an important role against (+)-catechin oxidation in oxygen-saturated model wine solutions.

Data in the literature showed that the acacia heartwood contained high amounts of flavonoids, namely dihydrorobinetin and robinetin, at concentrations of up to 100 μmol/g. These compounds are characteristic markers of acacia wood, since they have not been detected in other woods used for cooperage, such as oak, chestnut, cherry, and mulberry (Fernández de Simón et al., 2014). The results showed in Table 22.1 also confirm this tendency. In fact, Jordão and colleagues detected several compound characteristics of acacia wood, such as robinetin, fustin, and butin, that were not detected in the oak and cherry woods (Jordão et al., 2016). In contrast, other authors (Fernández de Simón et al., 2014; Sanz et al., 2012a, 2012b, 2012c) have reported only a small amount of condensed tannins and no hydrolyzable tannins. Only very low amounts of ellagic acid was detected by Jordão et al. (2016) in acacia wood chips after 30 extraction days by the use of model wine solutions 12% with 4 g/L of wood chips, in contrast with the high values detected for the different oak wood species (Jordão et al., 2016). Cherry wood usually is characterized by an abundance of condensed tannins (procyanidin type) and also appreciable values of (+)-catechin (Table 22.1). In addition, for this wood species, some phenolic acids and their esterification products (benzoic acid, *p*-hydroxybenzoic acid, 3,4,5-trimethylphenol, *p*-coumaric acid, methylsyringate, and methylvanillate) and flavonoids, such as naringenin, aromadendrin, isosakuranetin, and taxifolin, could be used as its phenolic markers for authentication purposes (Sanz et al., 2012a, 2012b, 2012c; Springmann, Rogers, & Spiecker, 2011). Also, detected appreciable values of naringenin only for cherry woods (Jordão et al., 2016). Furthermore, it is also important to note that the detection of some condensed tannins is not exclusively from cherry wood, because previously, Pascal Chatonnet had reported the presence of very low values of procyanidins also in oak woods (Chatonnet, 1995). In other wood species with potential enological use, although with still very restricted use, such as ash wood (*F. excelsior* L.), the presence of secoiridoids, phenylethanoid glycosides, or di- and oligolignols could be a good identifiers (Sanz et al., 2012a, 2012b, 2012c). However, wines aged in contact with ash wood have not shown any specific polyphenols provided by this wood (Fernández de Simón et al., 2014), although ash has been known to have compounds not present in oak (Sanz et al., 2012a, 2012b, 2012c). In addition, although the different phenolic compounds can be used as characteristic markers to distinguish the type of wood, the polyphenolic profile variability of wine, and their evolution during aging, makes the analysis of the markers found in the wood more complex than for the woods themselves (Fernández de Simón et al., 2014). Finally, for mulberry wood, particularly for the two species that have been considered for enological use, namely *M. alba* L. and *M. nigra* L., higher polyphenolic content and antioxidant activity were obtained in wood extracts from mulberry wood in comparison with phenolic content obtained from extracts of *Q. robur* and *R. pseudoacacia* L. (Kozlovic, Jeromel, Maslov, Pollnitz, & Orlić, 2010).

Table 22.2 shows the concentration of some of the most representative volatile compounds of several wood species (oaks, acacia, and cherry) for their aromatic contribution to wine during the aging process. The majority of these compounds are formed during the drying process and especially during the toasting in cooperages.

TABLE 22.2   Volatile composition of different wood chip species with potential use for white wine production.

	Wood species					References
	Oak wood			Acacia	Cherry	
Volatile composition	*Quercus petraea*	*Quercus alba*	*Quercus pyrenaica* Willd.	*Robinia pseudoacacia* L.	*Prunus avium* L.	
Furanic aldehydes[a]	397.19	−	−	8.26	70.18	Martins, Garcia, Da Silva, and Cabrita (2012)[a]
Volatile phenols[a]	941.58	−	−	197.84	334.63	
Phenolic aldehydes[a]	1563.62	−	−	170.38	1208.42	
Phenyl ketones[a]	55.33	−	−	57.04	72.21	
Lactones[a]	14.77	−	−	nd	3.95	
Guaiacol[b]	4.46	0.91	−	0.10	0.16	Alanñón et al. (2012) Alanñón et al. (2018)
Eugenol[b]	1.05	3.44	−	0.92	0.11	
Furfural[b]	12.09	5.79	−	0.56	nd	
Vanillin[b]	45.69	70.37	−	4.70	4.68	
Trans-β-methyl-γ-octalactone[b]	2.14	1.64	−	nd	nd	
Cis-β-methyl-γ-octalactone[b]	6.12	39.37	−	nd	nd	
Guaiacol[c]	2.41	4.89	3.98	5.36	1.71	Fernández de Simón et al. (2009)
Eugenol[c]	1.83	1.29	2.12	2.36	1.50	
Furfural[c]	430	395	494	804	23.3	
5-hydroxymethylfurfural[c]	22.9	21.1	28.9	113	47.6	
5-methylfurfural[c]	35.1	38.3	56.3	94.2	31.3	
Vanillin[c]	117	102	114	77.1	68.3	
Trans-β-methyl-γ-octalactone[c]	14.6	3.36	9.77	nd	nd	
Cis-β-methyl-γ-octalactone[c]	21.1	31.8	30.0	nd	nd	
Benzaldehyde[c]	0.8	0.74	0.96	0.25	0.91	
Syringaldehyde[c]	221	226	250	272	455	
Coniferaldehyde[c]	106	96.2	174	227	145	
Sinapaldehyde[c]	263	239	439	912	804	
Guaiacol[d]	4.20×10^{-3}	12.52×10^5	4.12×10^{-3}	12.8×10^{-3}	8.22×10^{-3}	Jordão et al. (2016)[g]
Eugenol[d]	3.39×10^{-3}	59.39×10^4	12.01×10^{-3}	0.60×10^{-3}	1.29×10^{-3}	
Vanillin[d]	71.7×10^{-3}	31.5×10^6	169.4×10^{-3}	120.8×10^{-3}	1.64×10^{-3}	
Furfural[d]	134×10^{-3}	20.69×10^6	5.75×10^{-3}	13.7×10^{-3}	3.24×10^{-3}	
β-methyl-γ-octalactones[d]	119×10^{-3}	199.8×10^5	521.8×10^{-3}	16.5×10^{-3}	18.1×10^{-3}	
Ethyl cinnamate[d]	nd	nd	nd	nd	1.77×10^{-3}	
Ethyl hexanoate[d]	nd	nd	nd	nd	2.21×10^{-3}	
Benzaldehyde	11.44	12.68	7.93	nd	nd	
Coniferaldehyde[e]	1.42	1.35	0.90	nd	nd	
Syringaldehyde[f]	0.109	nd	0.11	nd	nd	

[a] *values expressed as mg/100 g toasted wood in 3-octanol equivalents and oak Quercus robur species was studied.*
[b] *values expressed as mg/g of untoasted wood.*
[c] *values expressed as mg/g of toasted wood and data obtained after 15 extraction days by the use of model wine solutions 12% with 20 g/L of wood chips.*
[d] *average peak area expressed in relative peak area in relation to internal standard and data obtained after 30 extraction days by the use of model wine solutions 12% with 4 g/L of wood chips.*
[e] *values expressed as sinapaldehyde equivalents.*
[f] *values expressed as syringic acid equivalents.*
[g] *data obtained after 30 extraction days by the use of model wine solutions 12% with 4 g/L of wood chips; nd — not detected.*

As shown in Table 22.2, oak wood species present the majority of the volatile compounds with a more representative impact on wine sensory characteristics, particularly in terms of olfactive and also in some parameters of the taste profile. In general, wood belonging to the genus *Quercus* has β-methyl-γ-octalactones (*cis* and *trans* form), usually in higher amounts in American species, such as *Q. alba*. However, it is possible to find these compounds also in European oak species. These lactones are associated with sensory descriptors like fresh oak and coconut. The *cis* isomer is a more powerful aromatic than the *trans* isomer (Jordão, Ricardo-de-Silva, & Laureano, 2005; Chatonnet & Dubourdieu, 1998; Schumacher, Alañón, Castro-Vázquez, Pérez-Coello, & Díaz-Maroto, 2013; Setzer, 2016). According to several authors, the *cis*- to *trans*-oak lactone ratio is characteristic of the wood's source, where this ratio is higher for American oak wood species than European oak species (Garde-Cerdán & Ancín-Azpilicueta, 2006; Gómez-Plaza, Pérez-Prieto, Fernández-Fernández, & López-Roca, 2004). For example, other authors have reported cis/trans ratio values of 5.7, 4.2, and 0.63 for *Q. alba*, *Q. petraea*, and *Q. pyrenaica*, respectively (Jordão, Ricardo-de-Silva, & Laureano, 2005). The same authors quantified 12 different volatile compounds, including *cis*- and *trans*-β-methyl-γ-octalactones, in toasted oak woods from *Q. alba*, *Q. petraea*, and *Q. pyrenaica* species and from different geographical origins. According to the results obtained, the highest values for β-methyl-γ-octalactones were found in *Q. alba* (22.3−23.1 mg/g of dried wood and 4.0−5.0 mg/g of dried wood, respectively, for *cis* and *trans* form), followed by *Q. petraea* (14.0−21.3 mg/g of dried wood and 5.0−6.7 mg/g of dried wood, respectively, for *cis* and *trans* form). For all of these oak species, the *cis* form was the most abundant β-methyl-γ-octalactone. However, in an opposite trend, *Q. pyrenaica* from the Guarda region showed higher values for *trans* form with 8.3 mg/g of dried wood, while for *cis* form, 5.3 mg/g of dried wood was quantified. In the same oak species but from another region (Guarda region, North of Portugal), only the *cis* form was detected and no *trans* form was found. Also, Fernández de Simón and colleagues, for the *Q. pyrenaica* species, detected more *trans* form than *cis* form (33.8 and 31.5 mg/g of seasoned wood, respectively) in untoasted wood (Fernández de Simón, Esteruelas, Muñoz, Cadáhia, & Sanz, 2009).

As can be seen in Table 22.2, in general, woods not belonging to the genus *Quercus* do not have either *cis*- or *trans*-β-methyl-γ-octalactone in their composition. However, some authors (Jordão et al., 2016; Martins et al., 2012) have detected very low values of lactones in other no-oak wood species (acacia and cherry). In addition, other authors (Caldeira, Clímaco, Bruno De Sousa, & Belchior, 2006) have also found small amounts of *cis* and *trans* forms (0.23 and 0.34 mg/g in the isomer *trans* and *cis*, respectively) in chestnut wood. De Rosso and colleagues (De Rosso et al., 2009) studied the chemical compounds realized from four different woods (oak, acacia, chestnut, and mulberry) used to make barrels for aging wines and spirits. According to these authors, acacia, chestnut, and oak released coniferaldehyde in the range of 1−10 mg/g of wood, while for mulberry wood, only very low values were detected (0.1 and 0.9 mg/g of wood). In addition, higher amounts were detected for syringaldehyde for the wood species studied, particularly for acacia, chestnut, and oak, where values higher than 10 mg/g of wood were detected. For cherry wood, only values between 1 and 10 mg/g of wood were found. Finally, benzaldehyde was only detected in cherry woods, with values between 1 and 10 mg/g of wood. These results follow the similar tendency detected by other authors (Fernández de Simón et al., 2009), where, in general, coniferaldehyde, sinapaldehyde, and syringaldehyde were detected in oak, acacia, and cherry woods (Table 22.2). However, Jordão and colleagues only detected coniferaldehyde, syringaldehyde, and benzaldehyde in oak wood species (Jordão et al., 2016). Acacia could be characterized by significant contents of benzene aldehydes; and chestnut and oak by high amounts of vanillin, eugenol and methoxyeugenol, syringaldehyde, and α-terpineol (De Rosso et al., 2009). In addition, chestnut is also characterized by high amounts of volatile compounds and fatty acids. On the contrary, according to the results also published by De Rosso and colleagues in 2009, mulberry wood shows very low amounts of volatile compounds but at the same time has high amounts of (negative) fatty acids. Vanillin, guaiacol, and eugenol are also compounds with an important impact on several sensory properties of wines. These compounds are associated with sensory descriptors like vanilla, smoky, and clove aromas, respectively. In general, vanillin is present in the majority of the different wood species with potential use in enology, namely in different oak species (*Q. alba*, *Q. pyrenaica*, *Q. petraea*, *Q. robur*, and *Q. humboldtti*), chestnut (*C. sativa*), cherry (*P. avium*), acacia (*R. pseudoacacia*), and ash (*F. americana* and *F. excelsior*). Table 22.2 shows several amounts of eugenol, guaiacol, and vanillin quantified in some of these different wood species. Some studies have reported particularly high values of vanillin in untoasted chestnut wood compared with oak wood species, with values ranging from 6.40 to 70.37 mg/g of wood for oak species, and an average value of 80.90 mg/g of wood for chestnut. However, for the other wood species, such as acacia and cherry, much lower values are generally detected (Fernández de Simón et al., 2009; Alarcón et al., 2018 et al., 2018; Alañón, Castro-Vázquez, Díaz-Maroto, & Pérez-Coello, 2012). According to several authors, the guaiacol levels in medium-toasted ash woods are much higher than those detected in the other toasted woods, including those normally found in oak species; so a more pronounced smoke character can be found when using toasted ash wood in aging wines (Martínez-Gil et al., 2018). Also, according to the same authors, the concentrations of eugenol found in *Q. pyrenaica* are very high, especially in light-toasted woods, much more so than in those normally found in traditional oaks. Thus when using this wood species, it will be possible to have more spicy wines, especially with notes of clove. However, other studies (Fernández de Simón et al., 2009; Jordão et al., 2016) have demonstrated similar values (or even less) of eugenol found among *Q. pyrenaica* and the other toasted oak wood species, and even those other wood species such as acacia, cherry, and ash. Regarding this point, Fernández de Simón et al. (2009) reported average values of eugenol to be between 1.50 and 3.21 mg/g of toasted wood in different no-oak wood species (cherry, acacia, chestnut, and ash).

After toasting, furfural, 5-hydroxymethylfurfural, and 5-methylfurfural are some of the volatile compounds formed as a result of hemicellulose thermodegradation during the toasting process. These compounds formed are associated with several sensory descriptors such as caramel, toasted, and bitter almonds. In general, these compounds are present in the different wood species used for wine aging, and in almost all species, is one of the higher compounds than that of the other volatile compounds (Table 22.2). Thus *Quercus humboldtti*, for toasted wood, shows a higher concentration of 5-methylfurfural and lower concentrations for furfural and 5-hydroxymethylfurfural compared to other oak wood species, namely *Q. petraea* and *Q. alba* (Martínez-Gil et al., 2018). In addition, other authors (Martins et al., 2012) have reported very low amounts of furanic aldehydes (which includes furfural, 5-methylfurfural, 5-hydroxymethylfurfural, and maltol) in acacia and cherry toasted woods in comparison with toasted oak and chestnut woods. Even for untoasted woods, some authors (Alañón et al., 2012; Alarcón et al., 2018) have reported very low values for furfural and 5-methylfurfural in acacia woods, while for untoasted cherry, none of these compounds were detected. Finally, not only the chemical composition of the different wood species is important for wine aging. Thus also, the physical properties of the wood species play an important role in the suitability of the different woods for cooperage and enological use. In 2014, Spanish authors Acuña, Gonzalez, De La Fuente, and Moya (2014) studied the permeability of different wood species, and according to these authors the impregnation and evaporation losses of wine are strongly dependent on the wood species. In this case, the impregnation and evaporation losses were found to decrease significantly in the species *C. sativa*, *R. pseudoacacia*, *Q. alba*, and *P. avium*. In addition, they also detected that toasted *F. excelsior* has much higher total losses than the other woods tested. Thus irrespective of the chemical composition of this wood species, this type of wood is not recommended for manufacturing wood staves for winemaking. Also, according to several authors (Cadahía, Fernández de Simón, Poveda, & Sanz, 2008), staves made from *Q. pyrenaica* for wine aging were found to have characteristics similar to those of *Q. robur*. Previously, other authors (Canas, Belchior, Mateus, Spranger, & Bruno-de-Sousa, 2002) have reported the results of the kinetics of the impregnation/evaporation of wines and brandies by the use of oak wood (*Q. robur* L.) and chestnut wood (*C. sativa* Mill.). According to these authors, there is a strong influence of the intraspecific variability of wood (differences between trees and in heartwood age) that overlaps the effect of the other factors such as botanical species and toasting level. Thus the differences between the two wood species studied were not evident, as reported by other authors (Feuillat, Perrin, & Keller, 1994; Vivas, 2000).

22.3 Different wood species used in white wine production

Compared to red wines, there are only a few studies on the use and the effect of the interaction between wood and white wines, particularly for table wines. Therefore there is limited knowledge on the impact of alternative woods to oak wood on white wine quality during the aging process. In fact, in the past decades, the white wine market was determined by the consumer of young varietal or blended wines, which should be consumed within a short period after bottling (even sometimes in the same year or in the next vintage year) to avoid the loss of aroma, freshness, and fruity characters as a consequence of the potential detrimental and negative changes (for example, oxidation process) of the compounds from varietal and fermentative origin. However, in recent years, the option to use wood, particularly oak species, during the winemaking and aging process of white wines was taken into account with the winemakers' aim of obtaining white wines with potential new sensory profiles. This tendency started fundamentally through the occasional use of oak barrels, mainly during alcoholic and/or malolactic fermentation or by aging on lees, including the use of the *bâtonnage* technique (Jiménez Moreno & Ancín Azpilicueta, 2007; Chatonnet, Dubourdieu, & Boidron, 1992; Liberatore, Pati, Nobile, & Notte, 2010; Lukić, Jedrejčić, Ganić, Staver, & Peršurić, 2015; Nunes et al., 2017). In addition, also recently, fermentation and/or aging in wood barrels or wood chips by the use of acacia or cherry woods have been also implemented to improve the quality of white wines (Alarcón et al., 2018; Délia, Jordão, & Ricardo-da-Silva, 2017). These innovations have contributed to increasing the variety of white wines in contact with different wood species, thus adapting the market to the new sensory profiles of the consumer and the demands of the different international markets. However, it is important to look for the best combination between the different wood types, namely the chemical and physical properties, aging process, and grape variety characteristics, in order to obtain white wines with good quality and potential new sensory sensations. At this point, one of the main objectives will be to use the different woods without masking the primary and secondary aromas specific to each grape variety and protect the white wines against oxidative processes that could have a negative impact on the wine characteristics. In addition, for the majority of the options where white wine wood contact is made, in general, a prior prefermentative maceration is suggested. This point is important because there are high amounts of phenols and primary aromas extracted from solid parts of the grapes, which will be important for the future stability of the white wine during the wood aging, including upon lees contact (Darias-Martín, Rodríguez, Díaz, & Lamuela-Raventós, 2000; Liberatore et al., 2010; Lukić et al., 2015).

Concerning the use of different white grapes varieties to ferment the must or age the wine in contact with different types of woods, Chardonnay is the most widely studied grape variety, particularly with the use of oak (Guchu et al., 2006; Herrero et al., 2016; Liberatore et al., 2010; Lukić et al., 2015; Rieger, 2013; Towey & Waterhouse, 1996). However, studies on other white grape varieties can be found, where different wood species were used for white wine aging, such as Verdejo (Rodríguez-Nogales, Fernández-Fernández, & Vila-Crespo, 2009), Sauvignon Blanc (Herjavec, Jeromel, Da Silva, Orlic, & Redzepovic, 2007; Herrero et al., 2016), White Listán (Gutiérrez-Afonso, 2002), Muscatel (Aleixandre et al., 2003), and Picapoll (Ibern-Gómez et al., 2001). Nevertheless, Malvazia, White Grenache, Sauvignon Blanc, Encruzado, Pinot Blanc, Loureio, and Alvarinho have also been studied in other works by comparative analysis between the use of oak, acacia, or cherry woods during white wine aging (Fernández de Simón et al., 2014; Jordão, Pina, Montalbano, Correia, & Ricardo-da-Silva, 2018; Del Galdo, Correia, Jordão, & Ricardo-da-Silva, 2019; Délia et al., 2017; Rieger, 2013). On the other hand, according to other authors (Gutiérrez-Afonso, 2002) the use of wood, in particular oak, appears to hold out special promise for neutral grape varieties with weak floral or fruity aromas.

22.3.1 Impact on white wines chemical composition

The knowledge of the chemical composition of the different wood species, especially the content of ellagitannins and volatile compounds, already covered in the previous point, is of fundamental importance to understand the composition of the wines fermented and/or aged in contact with the different woods. Ibern-Gómez and colleagues examined the differences in phenolic compounds in white wines from Chardonnay and Picapoll varieties that had been fermented in oak barrels as compared to stainless steel vats (Ibern-Gómez et al., 2001). Thus the presence of characteristic oak-wood phenols, such as coniferaldehyde, sinapinaldehyde, syringaldehyde (aromatic aldehydes), scopoletin (a coumarin), and several volatile compounds such as 4-ethyl-guaiacol, 4-vinylphenol, eugenol, and β-methyl-γ-octalactone, was also observed exclusively in the white wines fermented in oak wood. Thus this is an example of how wood and its characteristics will definitely mark the composition of white wines. Recently, the effect of wood chips from acacia and cherry on the aging process during 28 aging days of Encruzado white wines was evaluated in comparison with the conventional chips from French (*Q. petraea*) and American (*Q. alba*) oak woods (Délia et al., 2017). Several results obtained by these authors are shown in Fig. 22.2.

During the brief aging time studied, the use of wood chips was found to induce an increase in the polyphenol content due to the transfer of phenols from the wood to wine. This tendency was particularly evident for the white wines aged with acacia chips, which exhibited the highest total phenolic content (342.94 mg/L) in comparison with the remaining wines aged with cherry (328.59 mg/L) and oak (319.86 mg/L) woods after 28 aging days. For flavonoid and nonflavonoid phenols, a similar tendency was also detected for the white wines aged in contact with acacia wood chips. According to these authors, the high values of flavonoid and nonflavonoid compounds quantified in wines aged especially with acacia wood chips correspond to a potentially higher extraction of individual phenolic compounds, such as, for example, gallic acid and ellagic acid (Délia et al., 2017). Previously, a similar tendency was also reported by other authors (Sanz et al., 2011), where seasoned acacia wood showed high concentrations of flavonoid compounds; this could explain the difference found. Furthermore, several other authors (De Rosso et al., 2009; Sanz et al., 2011) have also described a particular richness of some phenolic compounds in acacia woods and consequently in wines aged in contact with this wood species. In this regard, Soares and colleagues reported, for acacia extracts produced in model wine solutions,

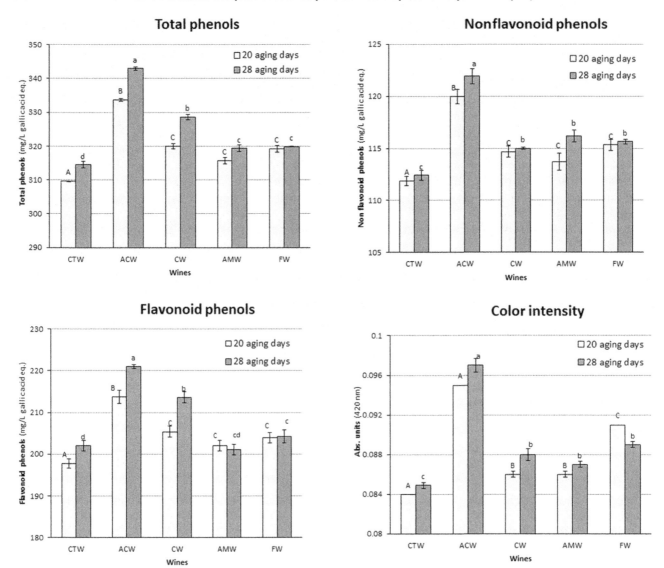

FIGURE 22.2 General phenolic composition and color intensity of Encruzado white wine aged in contact with different wood chip species after 20 and 28 aging days. *ACW*, wine aged with acacia wood chips; *AMW*, wine aged with American wood chips; *CW*, wine aged with cherry wood chips; *CTW*, control wine aged without any wood chips; *FW*, wine aged with French wood chips. *Values with the same letters for each aging day are not significantly different (Tukey test, $P < 0.05$), capital letters for 20 aging days and small letters for 28 aging days.

high values of absorbance measures at 280 nm and also total polyphenolic content (Soares et al., 2012). However, according to the results reported by other authors (Psarra, Gortzi, & Makris, 2015), who studied the extraction of polyphenols of wood chips from oak, acacia, and cherry woods into model wine systems over a period of 20 days, a more pronounced enrichment of model wines in polyphenolic substances was detected in the systems treated with oak chips. These results could be explained by the wood structure and homogeneity of polyphenolic composition as well as the fact that chips from oak wood contain significantly higher amounts of total extractable polyphenols as compared with acacia and cherry wood chips. Several authors have evaluated different phenolic markers in different varietal wines, with the majority of them from red varieties but one also using white wine produced from the white Grenache variety, aged with untoasted acacia, cherry, and chestnut wood chips (1.5 g/L) over 2 months (Fernández de Simón et al., 2014). The white wine aged with cherry wood chips showed several specific flavonoid phenols. Among them, aromadendrin, taxifolin, isosakuranetin, eriodictyol, and naringenin (compound also detected specifically in cherry wood, see Table 22.1) were highlighted. So according to these authors, some of these phenolic compounds could be proposed as possible phenolic markers for wines aged in contact with cherry wood for authenticity purposes. Also, *p*-hydroxybenzoic acid and (+)-catechin showed significantly higher concentrations in white wine aged in contact with cherry wood chips. However, the high levels of (+)-catechin found by these authors disagree with the lowest levels found by other researchers (De Rosso et al., 2009) in wines aged in cherry barrels compared to those aged in oak, acacia, chestnut, and mulberry barrels, to which the authors relate the higher oxidation levels in cherry barrels. Also, according to Fernández de Simón and colleagues, valoneic acid dilactone could be proposed as a possible phenolic marker of aging wines in contact with chestnut wood for authenticity purposes, because it was only able to be detected in these wines and during all aging times (Fernández de Simón et al., 2014).

Finally, for white wines aged in contact with acacia wood, several no-anthocyanic phenolic compounds were found in higher concentrations, including dihydrorobinetin, robinetin, and 2,4-dihydroxybenzaldehyde. Lastly, the polyphenolic profile obtained for wines aged in contact with ash wood didn't show specific polyphenols provided by this wood species. However, the results obtained by other researchers (Loupassaki, Abouzer, Basalekou, Fyssarakis, & Makris, 2016) denote that the fact that no *cis*-β-methyloctalactone was detected in white wine samples aged in acacia barrels is reasonable, since there have never been reports regarding its occurrence in the *R. pseudoacacia* species. Also related to the phenolic composition of the different wood species, and their impact on white wine composition, tannicity is a parameter that stands for the expression of the astringency perception of a wine, namely, the capacity that some phenolic compounds such as tannins have to interact with proteins, stimulating the astringent character of the wine in taste. In general, this parameter is not considered for white wines aged in contact with different woods, but it may be important to understand the role that the phenolic composition of wines may play in some sensory properties (particularly in terms of astringency). In recent years, two research works have published several data about this parameter that are applicable to white wines. Thus Délia and colleagues have reported that except for the white wines aged in contact with cherry wood chips, all wines aged in contact with oak (French and American species) and acacia wood chips showed a tendency toward an increase in tannin power values (Délia et al., 2017). This increase was particularly significant for the wine aged in contact with French oak wood chips (from 12.79 to 20.13 NTU/mL, respectively, after 20 and 28 aging days) with respect to the other white wines. According to these authors, the high phenolic composition, and especially the high ellagitannin content that characterizes French oak heartwood (Jordão, Ricardo-de-Silva, & Laureano, 2007; Jordão et al., 2012; Chira & Teissedre, 2014), may explain the high tanning power quantified in white wine aged in contact with French oak wood chips and consequently the high level of interactions between these phenolic compounds and saliva proteins inducing high levels of wine astringency. Conversely, the tanning power results for white wines aged with acacia and American oak samples were not statistically significantly different from the control white wine. More recently, Del Galdo and colleagues have described that after 60 aging days the use of blends containing cherry wood with American or French oak wood chips and also the blend containing the two oak wood chip species showed the highest tannicity values: 24.75, 22.08, and 18.88 NTU/ML, respectively (Del Galdo et al., 2019). Probably, the high phenolic composition, which particularly characterizes French oak heartwood previously mentioned previously and also the presence of (+)-catechin and procyanidins quantified in cherry wood chips (Jordão et al., 2016; Zhang, Cai, Duan, Reeves, & He, 2015), could explain the high tannicity quantified in white wine aged in contact with these blends, especially after 60 aging days. Several works have detected changes in white wine color intensity along the course of the aging time with wood contact, particularly by the use of oak wood barrels (Recamales, Sayago, González-Miret, & Hermanz, 2006; Nunes et al., 2017) and also by the use of different wood chip species (Del Galdo et al., 2019; Délia et al., 2017). The interaction between several wood compounds and wine phenolic composition induces the formation of a diversity of new compounds which affect the initial white wine color by an increase in yellow color (Vivas et al., 2008). On the other hand, the yellow color increase of aged white wines could also correspond to a higher potential extraction of phenolic compounds such as gallic acid, ellagic acid, and hydrolyzable tannins from wood to the wines during the aging process. This impact is usually particularly high for the wines aged in contact with oak wood, because these wood species generally have more levels of phenolic compounds, such as hydrolyzable tannins, than, for example, cherry heartwood (Jordão et al., 2016; Sanz et al., 2010). In fact, this tendency is also evident in the results obtained by other authors and is shown in Fig. 22.2 (Délia et al., 2017). In that case, it seems that white wines aged with acacia chips showed the highest color intensity compared to those wines aged with cherry and American or French oak wood chips. Indeed, comparing the values obtained by these authors for color difference (DE) between control white wine and wines aged in contact with the other three wood chip species, only the white wine aged with acacia wood chips showed values higher than 2 CIELAB units, specifically 2.32 units, followed by the white wines aged with American (1.35 units) and French oak chips (0.95 units), and finally by the wines aged with cherry wood chips (0.97 units), indicating that the color difference could be detected by human eyes (Spagna et al., 1996). These results could have been more clear if the wood chip concentration used had been higher (in this case, it was 0.5 g/L). Still referring to the color of white wines aged in contact with different wood species, recently, Del Galdo and colleagues studied the potential use of different wood chip blends containing oak and cherry woods (Del Galdo et al., 2019). The first clear result was that for the same aging time, the control wine showed, in general, the lowest values for color intensity. In addition, the results obtained showed a color intensity increase in all white wines aged in contact with oak wood chips (alone or in blends) than that observed for wine aged only with cherry wood chips.

One of the problems resulting from the aging of white wines is related to the oxidation mechanisms. The excessive oxidation process can induce white wine degradation, in terms of color, with the appearance of excessively yellowish-brown colors, as well as in terms of some other sensory characteristics such as aromatic degradation. The contact of white wines with wood by means of chips implies a decrease in the browning potential index due to the release of different phenolic compounds from wood to wines. Salacha and colleagues reported strong evidence that white wine browning development might be predominantly associated with flavanol compounds and, to a lesser extent, with total phenolic oxidation (Salacha, Kallithraka, & Tzourou, 2008). For these authors, total phenolic and flavanol concentrations in white wines correlated significantly with the browning rate constants of $r^2 = 0.79$ and $r^2 = 0.84$, respectively. Previously, Es-Safi and colleagues also demonstrated an oxidant process, where flavanols are converted into yellow xanthylium pigments, contributing also to white wine browning (Es-Safi, Le Guernevé, Fulcrand, Cheynier, & Moutounet, 2000). In addition, phenomena such as the formation of phenolic polymers, the precipitation of oxidized phenols, and the antioxidant properties of some phenolic compounds lead to increases in the stability of white wines to oxidation. Taking into account the potential effect of the use of different wood species on white wine browning development, there is very restricted knowledge on this topic. Only recently, two works were published about this subject. Thus in 2019 several researchers reported that for the same aging time, in general, white wines aged in contact with cherry wood chips showed the highest values of browning potential index. A similar tendency was also detected for the wines aged with cherry wood chips blended with American oak wood chips (Del Galdo et al., 2019). This tendency was also confirmed by the results previously reported by other authors (Délia et al., 2017), where white wines aged with cherry and acacia wood chips maintained high browning potential values during the aging time compared to wines aged in contact with oak wood chips. This fact was in consonance with the low content of oxidizable polyphenols that characterized particularly the cherry heartwood compared to phenolic content of oak wood (Alañón et al., 2011; De Rosso et al., 2009). This point, it's also supported by the analysis of the data shown in Table 22.1.

22.3.2 Impact on white wines sensory profile

Wines, particularly aged wines, show distinctive sensorial features that they acquire during the aging process. This is obviously true for the different red wine styles, namely table wines and liqueur wines such as Port wines. However, in recent years, winemakers have begun to produce white wines with potential new profiles, and consequently, the use of different woods for wine aging constituted an increasingly valid and widespread option. Through the aging process, physical, chemical, and biological transformations take place, improving the wine's stability and modifying the sensorial characteristics. As a consequence of these factors, an increase of the aroma complexity and different taste sensations occurs as a result of the contact between white wine and wood. The gentle microoxigenation that occurs through the pores of woods during white wine aging in barrels induces oxidation reactions, changes in color, and a reduction in astringency. In addition, the complexity increment of aroma and taste is due to the extraction of certain compounds from the wood to the white wine (Chatonnet et al., 1992; Nunes et al., 2017). Thus the different wood characteristics, namely their composition, play a key role in the sensory characteristics of white wines.

By far, different oak wood species (particularly Q. alba from North America and Q. petraea of different geographical origins in Europe) have been the most common woods used to carry out white wine aging. From different oak wood species, different volatile and nonvolatile compounds are released during the aging process (Tables 22.1 and 22.2). Recently, a few authors (Herrero et al., 2016) have related volatile compounds extracted from oak wood with different toasting levels with the sensory profile of varietal white wines from Chardonnay and Sauvignon Blanc varieties. According to these authors, eugenol, guaiacol, 4-methylguaiacol, vanillin, furfuryl alcohol, and furfural were positively related to the aroma quality perceived by experts of Sauvignon Blanc wines, while for Chardonnay wines the highest aroma scores were positively correlated with 4-vinylguaiacol and isoeugenol and negatively correlated with the presence of lactones and 4-vinylphenol in wines. Also, by the use of Chardonnay wines aged in barrels, previously, other authors (Herjavec et al., 2007; Spillman, Sefton, & Gawel, 2004) evaluated the use of oak wood on sensory profiles. The first authors confirmed the role of some oak wood-extractable compounds in Chardonnay wines aged in new oak barrels, such as the strong correlation between the volatile components of the wood produced during the toasting process and the "smoky" aroma. On the other hand, the authors of the second study reported a positive influence of the use of new Croatian oak barrels (made from *Q. petraea* and *Q. robur* oak wood) during the alcoholic fermentation on the sensory properties of white wines produced from Chardonnay and Sauvignon Blanc varieties in comparison with those fermented in steel tanks. In addition, the use of oak barrels or oak chips for white wine production also plays an important role in the sensory profile of wines. Therefore using wines from the white grape variety of Listan Blanco, Spanish researcher Gutiérrez-Afonso (2002) studied the effect of wood, in the form of oak chips (4 and 8 g/L) or in barrels made of American or French oak, on the sensory characteristics during the fermentation process. The results reported by this author showed that American oak chips produced a greater intensity of coconut and vanilla and a greater degree of bitterness and astringency than barrels. It was also observed that differences in the variety of oak used were more noticeable in oak chips than in barrels. In fact, the use of oak chips or barrels could have very different effects on the sensory characteristics of white wines. Thus other authors have reported a higher positive impact of the use of oak chips than barrels on the sensory characteristics of white wines (Sánchez-Palomo et al., 2017). According to these authors, Verdejo white wines aged in the presence of oak chips showed the highest concentration of oak lactones and furanic compounds, while Pérez-Coello and colleagues, using Airén wines, showed that the use of oak chips of American and French origin displayed the best results, with an improvement of 2 points in their acceptability rating on a scale of 9 (Pérez-Coello et al., 2000). In fact, not all white wines are suitable for aging in oak barrels since the microdiffusion of oxygen through wood pores could oxidize the wine with the consequent increment of off-flavors (for example, honey-like or cooked vegetables, besides the browning the color increase).

Also, different oak wood species and origins play an important role in the white wine sensory profile. Thus Ferreras and colleagues, using white wines made with the Treixadura variety, which is a typical variety from Galicia and Northern Portugal, obtained interesting results between the sensory profile of these wines aged for 3 months in new oak barrels of American and French origin. The sensory results indicated that the wine aged in French (Allier) oak was better scored than that aged in American oak, as it provided subtle and delicate nuances of wood, which did not mask the fruity aroma of the Treixadura wine but contributed to its aromatic complexity. On the other hand, white wine aged in the American oak barrel had the strongest oak aroma and flavor that totally masked the original aroma present in the wine and the oak attributes were more aggressive (Ferreras, Fernández, & Falqué, 2002). Additionally, other authors (Guchu et al., 2006) studied the use of toasted and untoasted chips of oak woods from different geographical provenances (American and Hungary oaks) in Chardonnay wines (4 g/L during 25 days) and found from the sensorial analyses of wines that the effect of the toasting of oak chips on wine characteristics was greater than the type of oak used. More recently, other researchers studied different fermentation (*sur lie* and classical fermentation) and aging conditions by the use of Slovenian and French oak barrels (both made from *Q. petarea* and *Q. robur* species) on lees and without lees in Chardonnay wine production. According to the results obtained, white wines produced by classical fermentation and aging in French oak barrels showed better results for sensory evaluation (especially for the use of low toasting level) than wine samples in the Slovenian oak barrels (Obradović, Mesić, Ravančić, Mijowska, & Svitlica, 2017).

In the past years, several works have described the potential sensorial quality improvement of white wines by contact with alternative woods to oak. Young and colleagues carried out very interesting research in 2020, which involved the aging of Chardonnay wines in contact with several different wood species, namely Matai, Feijoca, Cherry beech, Silver beech, Manuka, Macrocarpa, Pohutukawa, Radiata pine, Totara, Kahikatea, and Rimu, from New Zealand in comparison with oak wood from *Q. alba* over 2 weeks and with a toasted chip concentration of 5 g/L. According to the results obtained by these authors, only the Chardonnay wine aged in contact with Macrocarpa wood showed similar flavors to oak white wine. For the other wood species, unattractive sensorial features were also detected as a consequence of the use of these different wood species, which can be described as earthy, sappy, resin, paint stripper, or pencil sharpenings (Young, Kaushal, Robertson, Burns, & Nunns, 2010).

A comparative analysis of the use of oak (*Q. alba* and *Q. petraea*) and acacia (*R. pseudoacacia*) wood barrels on the sensory profile of white wines from two *Vitis vinifera* Cretan native varieties (Villana and Dafni) over 9 months was studied by Loupassaki and colleagues (Loupassaki et al., 2016). The sensory evaluation results indicated that wines that received aging in barrels made of *Q. alba* had higher average scores, exhibiting a more intense aromatic profile with notes associated with "oak" aromatic notes. On the other hand, according to these authors, white wines aged with acacia barrels showed an aromatic profile that was rather more balanced. In fact, several works are analyzing the potential use of acacia wood for white wine aging, in particular, its impact on sensory properties. Thus several researchers, after 12 months of aging, reported that Malvazija wines aged in acacia barrels showed higher finer textured, natural sweetness of the fruit with more pronounced vanilla and spicy characters than the wines aged in French and Croatian oak barrels (Kozlovic et al., 2010). In addition, the wines from the acacia barrels had more sweetness and honey taste with less pronounced oak flavor duration and were also the best, as evaluated by the panelists. In fact, the interesting results obtained by the use of acacia wood for white wine aging have therefore been mentioned in several studies. Other authors (Alarcón et al., 2018), for example, studied the use of acacia barrels and chips on the sensorial properties of Chardonnay wines. In that case, the results obtained point out that wines aged in acacia barrels may have new, clearly perceptible sensory descriptors associated with nutty, honey, and toasted notes. The responsible compounds for these aroma descriptors seem to be three volatile compounds, namely 2-acetyl pyrazine, 2-acetyl-3-methylpyrazine, and 2-acetylthiazole, which were previously identified by other authors (Culleré et al., 2013) as distinctive compounds detected in acacia wood. On the other hand, spicy or clove notes seemed not to be characteristics of the wines aged with acacia wood due to their low content of volatile phenols such as eugenol or guaiacol (Alarcón et al., 2018). Other works published (Délia et al., 2017) put out evidence of the positive impact of acacia wood chips on the sensory profile of aged white wines. These authors reported the results for their study, in which the sensorial profile of Encruzado white wines aged in contact with chips from cherry (*P. avium*), acacia (*R. pseudoacacia*), and two oak species (*Q. petraea* and *Q. alba*) had been investigated over 28 aging days. According to the results obtained, the differences were significantly related to the persistency and astringency taste descriptors. The significantly highest persistency was obtained for the white wines aged in contact with acacia and French oak wood chips in relation to the other white wines aged in contact with cherry and American oak wood chips and the control wine. For the aroma descriptors, the panel test did not detect significant differences among the white wines. However, the results for global appreciation showed better scores for white wines aged with acacia chips than for those aged with oak and cherry wood chips and the control wine.

Regarding the use of cherry wood for white wine aging, the use of this wood species, particularly from *P. avium*, is very restricted, and very few studies have been carried out about the potential impact on the sensory profile of white wines. In fact, cherry has been studied for the last 10 years in order to know about the potential effect during the aging only for red and rosé wines (Nunes et al., 2020; Santos et al., 2019; Sanz et al., 2010; Tavares et al., 2017), and also wine vinegars (Chinnici, Natali, Bellachioma, Versari, & Riponi, 2015). However, the few results published for white wines open some perspectives and increase the knowledge about the use of cherry wood for this type of wines. Délia and colleagues reported interesting results, where the white wine aged in contact with cherry chips showed similar overall appreciation scores to those obtained for the wines aged with different oak wood species (Délia et al., 2017). Also, very recently, Del Galdo and colleagues studied the use of different blends of toasted oak and cherry wood chips (0.5 g/L) on the evolution of several sensory parameters over 60 aging days of a white wine from the Encruzado grape variety (Del Galdo et al., 2019). The results obtained for sensory global appreciation at three data points (15, 30, and 60 aging days) are shown in Fig. 22.3.

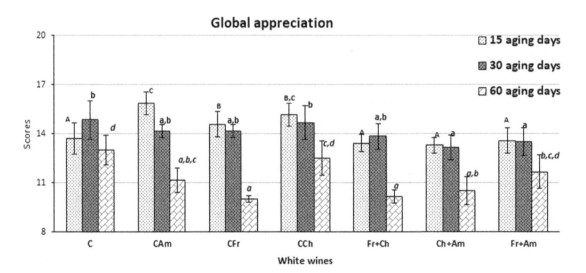

FIGURE 22.3 Evolution of sensory global appreciation scores over 60 aging days of a white wine from the Encruzado grape variety in contact with blends of oak and cherry wood chips. *C*, control wine; *CAm*, wine with American oak chips; *CFr*, wine with French oak chips; *CCh*, wine with cherry chips; *Fr + Ch*, wine with a blend of French oak and cherry chips; *Ch + Am*, wine with a blend of American oak and cherry chips; *Fr + Am*, wine with a blend of French and American oak chips; data points for same aging time showing the same letter are not significantly different (*P* < 0.05).

These results put out evidence for a significant decrease in scores attributed by the panel during the aging time considered, although the white wine aged in contact with cherry wood chips alone and the control wine showed a less pronounced decrease. In fact, after 60 aging days, these two white wines showed the significantly best overall appreciation scores from the panel test. In addition, these results led to the conclusion that contact for 60 aging days with the different wood chips studied will eventually to be excessive for the overall appreciation of the white wines. In addition, it was also clear that the use of blends containing cherry and oak wood chips did not lead to better results, concerning the wine global appreciation.

Finally, several authors (Jordão et al., 2018) have evaluated also the impact of toasted oak and cherry wood chips on the sensorial characteristics of several varietal white wines, made from Alvarinho, Viosinho, Loureiro, and Sauvignon Blanc, over 30 aging days. The results obtained were very variable depending on the grape variety. However, different tendencies were detected, namely that the impact of the use of cherry wood chips was particularly perceived for the white wine produced from the Viosinho grape variety, especially due to an increase in the fruity aroma, while the most significant differences were related to the wine aroma descriptors particularly for the Alvarinho (spicy, almond, and woody descriptors) and Loureiro (almond descriptor) white wines aged in contact with oak wood chips. In this study, it was made clear that the impact of the different wood chip species used on the sensory characteristics of the white wines will be strongly dependent on the varietal wine characteristics, namely in terms of the grapevine variety used.

22.4 Final remarks

A great number of different chemical compounds with diverse chemical structures and properties have been identified and quantified in several wood species, particularly for oak wood species. Recently, with the use of modern instrumental analysis, the quality of the study of the wood components from different botanical species has been getting better, allowing for an increasing characterization of the woods and the impact that different factors can have on their composition. All of these aspects have allowed us to better understand the impact of the use of wood on the quality of wines, particularly during their aging, either through the use of barrels or alternative wood products. Actually, oak and chestnut wood species are the only ones recommended by the OIV for enological use. Thus these wood species are traditionally used during the production and aging process of wines and spirits through the use of barrels. However, also according to the OIV the possibility of the use of wood fragments must exclusively come from the *Quercus* genus. In recent years, the increased demand for oak wood has caused a remarkable increase in production and environmental costs. Thus heartwood from alternative species has been considered as a possible source of wood for the wine production. The knowledge of the impact of the use of no-oak species on the quality of wines is still recent, particularly in terms of its use in white wines. Nevertheless, the use of different botanical wood species may be an option for winemakers to produce wines with potential new profiles. Consequently, further research will be necessary to improve the knowledge about the potential impact of the use of oak and particularly non-oak wood species on white wine quality.

References

Acuña, L., Gonzalez, D., De La Fuente, J., & Moya, L. (2014). Influence of toasting treatment on permeability of six wood species for enological use. *Holzforschung, 68*(4), 447–454. Available from https://doi.org/10.1515/hf-2013-0044.

Alañón, M. E., Castro-Vázquez, L., Díaz-Maroto, M. C., Hermosín-Gutiérrez, I., Gordon, M. H., & Pérez-Coello, M. S. (2011). Antioxidant capacity and phenolic composition of different woods used in cooperage. *Food Chemistry, 129*(4), 1584–1590. Available from https://doi.org/10.1016/j.foodchem.2011.06.013.

Alañón, M. E., Castro-Vázquez, L., Díaz-Maroto, M. C., & Pérez-Coello, M. S. (2012). Aromatic potential of *Castanea sativa* Mill. compared to *Quercus* species to be used in cooperage. *Food Chemistry, 130*(4), 875–881. Available from https://doi.org/10.1016/j.foodchem.2011.07.111.

Alarcón, M. E., Marchante, L., Alarcón, M., Díaz-Maroto, I. J., Pérez-Coello, S., & Díaz-Maroto, M. C. (2018). Fingerprints of acacia aging treatments by barrels or chips based on volatile profile, sensorial properties, and multivariate analysis. *Journal of the Science of Food and Agriculture, 98*(15), 5795–5806. Available from https://doi.org/10.1002/jsfa.9129.

Aleixandre, J. L., Padilla, A. I., Navarro, L. L., Suria, A., García, M. J., & Álvarez, I. (2003). Optimization of making barrel-fermented dry muscatel wines. *Journal of Agricultural and Food Chemistry, 51*(7), 1889–1893. Available from https://doi.org/10.1021/jf020605 + .

Bautista-Ortin, A. B., Martinez-Cutillas, A., Ros-Garcia, J. M., López-Roca, J. M., & Gómez-Plaza, E. (2005). Improving colour extraction and stability in red wines: The use of maceration enzymes and enological tannins. *International Journal of Food Science and Technology, 40*, 867–878. Available from https://doi.org/10.1111/j.1365-2621.2005.01014.x.

Cadahía, E., Fernández de Simón, B., Poveda, P., & Sanz, M. (2008). *Using Quercus pyrenaica Willd. of Castile in aging wines: Comparison with French and American oak*. Instituto Nacional de Investigación y Tecnología Agraria y Alimentaria.

Cadahía, E., Varea, S., Muñoz, L., Fernández de Simón, B., & García-Vallejo, M. C. (2001). Evolution of ellagitannins in Spanish, French, and American oak woods during natural seasoning and toasting. *Journal of Agricultural and Food Chemistry*, 49(8), 3677–3684. Available from https://doi.org/10.1021/jf010288r.

Caldeira, I., Clímaco, M. C., Bruno De Sousa, R., & Belchior, A. P. (2006). Volatile composition of oak and chestnut woods used in brandy ageing: Modification induced by heat treatment. *Journal of Food Engineering*, 76(2), 202–211. Available from https://doi.org/10.1016/j.jfoodeng.2005.05.008.

Canas, S., Belchior, A. P., Mateus, A. M., Spranger, M. I., & Bruno-de-Sousa, R. (2002). Kinetics of impregnation/evaporation and release of phenolic compounds from wood to brandy in experimental model. *Ciência e Técnica Vitivinicola*, 17, 1–14.

Canas, S., Leandro, M. C., Spranger, M. I., & Belchior, A. P. (1999). Low molecular weight organic compounds of chestnut wood (*Castanea sativa* L.) and corresponding aged brandies. *Journal of Agricultural and Food Chemistry*, 47(12), 5023–5030. Available from https://doi.org/10.1021/jf9900480.

Canas, S., Leandro, M. C., Spranger, M. I., & Belchior, A. P. (2000). Influence of botanical species and geographical origin on the content of low molecular weight phenolic compounds of woods used in Portuguese cooperage. *Holzforschung*, 54(3), 255–261. Available from https://doi.org/10.1515/HF.2000.043.

Canas, S., Quaresma, H., Belchior, A. P., Spranger, M. I., & Bruno-de-Sousa, R. (2004). Evaluation of wine brandies authenticity by the relationships between benzoic and cinnamic aldehydes and between furanic aldehydes. *Ciência e Técnica Vitivinícola*, 19, 13–27.

Carvalho, A. (1998). Identificação anatómica e caracterização física e mecânica das madeiras utilizadas no fabrico de quartolas para produção de aguardentes velhas de qualidade -Denominação Lourinhã. *Ciência e Técnica Vitivinícola*, 13, 71–105.

Castro-Vázquez, L., Alañón, M. E., Ricardo-da-Silva, J. M., Pérez-Coello, M. S., & Laureano, O. (2013). Evaluation of Portuguese and Spanish *Quercus pyrenaica* and *Castanea sativa* species used in cooperage as natural source of phenolic compounds. *European Food Research and Technology*, 237(3), 367–375. Available from https://doi.org/10.1007/s00217-013-1999-5.

Cerdán, T. G., & Ancín-Azpilicueta, C. (2006). Effect of oak barrel type on the volatile composition of wine: Storage time optimization. *LWT - Food Science and Technology*, 39(3), 199–205. Available from https://doi.org/10.1016/j.lwt.2005.01.009.

Chatonnet, P. (1995). *Influence des Procédés de Tonnellerie et des Conditions d'Élevage sur la Composition et la Qualité des Vins Élevés en Fûts de Chêne*. France: Thèse doctorat de l'Université de Bordeaux II. (p. 268).

Chatonnet, P. (1999). Discrimination and control of toasting intensity and quality of oak wood barrels. *American Journal of Enology and Viticulture*, 50(4), 479–494.

Chatonnet, P., & Dubourdieu, D. (1998). Comparative study of the characteristics of American white oak (*Quercus alba*) and European oak (*Quercus petraea* and *Q. robur*) for production of barrels used in barrel aging of wines. *American Journal of Enology and Viticulture*, 49(1), 79–85.

Chatonnet, P., Dubourdieu, D., & Boidron, J. N. (1992). Incidence des conditions de fermentation et d'élevage des vins blanc secs en barriques sur leur composition en substances cédèes par le bois de chêne. *Sciences des Aliments*, 12, 665–685.

Chatonnet, P., Ricardo-da-Silva, J. M., & Dubourdieu, D. (1997). Influence de l'utilisation de barriques en chêne sessile européen (*Quercus petraea*) ou en chêne blanc américain (*Quercus alba*) sur la composition et la qualité des vins rouges. *Revue Française d'Oenologie*, 165, 44–48.

Chinnici, F., Natali, N., Bellachioma, A., Versari, A., & Riponi, C. (2015). Changes in phenolic composition of red wines aged in cherry wood. *LWT - Food Science and Technology*, 60(2), 977–984. Available from https://doi.org/10.1016/j.lwt.2014.10.029.

Chira, K., & Teissedre, P. L. (2014). Chemical and sensory evaluation of wine matured in oak barrel: effect of oak species involved and toasting process. *European Food Research and Technology*, 240(3), 533–547. Available from https://doi.org/10.1007/s00217-014-2352-3.

Comandini, P., Lerma-García, M. J., Simó-Alfonso, E. F., & Toschi, T. G. (2014). Tannin analysis of chestnut bark samples (*Castanea sativa* Mill.) by HPLC-DAD-MS. *Food Chemistry*, 157, 290–295. Available from https://doi.org/10.1016/j.foodchem.2014.02.003.

Culleré, L., Fernández de Simón, B., Cadahía, E., Ferreira, V., Hernández-Orte, P., & Cacho, J. (2013). Characterization by gas chromatography-olfactometry of the most odor-active compounds in extracts prepared from acacia, chestnut, cherry, ash and oak woods. *LWT - Food Science and Technology*, 53(1), 240–248. Available from https://doi.org/10.1016/j.lwt.2013.02.010.

Cutzach, I., Chatonnet, P., & Dubourdieu, D. (1999). Study of the formation mechanisms of some volatile compounds during the aging of sweet fortified wines. *Journal of Agricultural and Food Chemistry*, 47(7), 2837–2846. Available from https://doi.org/10.1021/jf981224s.

Darias-Martín, J. J., Rodríguez, O., Díaz, E., & Lamuela-Raventós, R. M. (2000). Effect of skin contact on the antioxidant phenolics in white wine. *Food Chemistry*, 71(4), 483–487. Available from https://doi.org/10.1016/S0308-8146(00)00177-1.

De Coninck, G., Jordão, A. M., Ricardo-da-Silva, J. M., & Laureano, O. (2006). Evolution of phenolic composition and sensory properties in red wine aged in contact with Portuguese and French oak wood chips. *Journal International Des Sciences de La Vigne et Du Vin*, 40(1), 25–34.

De Rosso, M., Cancian, D., Panighel, A., Dalla Vedova, A., & Flamini, R. (2009). Chemical compounds released from five different woods used to make barrels for aging wines and spirits: Volatile compounds and polyphenols. *Wood Science and Technology*, 43(5–6), 375–385. Available from https://doi.org/10.1007/s00226-008-0211-8.

Del Álamo Sanza, M., Escudero, J. A. F., & De Castro Torío, R. (2004). Changes in phenolic compounds and colour parameters of red wine aged with oak chips and in oak barrels. *Food Science and Technology International*, 10(4), 233–241. Available from https://doi.org/10.1177/1082013204046095.

Del Galdo, V., Correia, A. C., Jordão, A. M., & Ricardo-da-Silva, J. M. (2019). Blends of wood chips from oak and cherry: impact on the general phenolic parameters and sensory profile of a white wine during the aging process. *Vitis*, 58, 159–169.

Délia, L., Jordão, A. M., & Ricardo-da-Silva, J. M. (2017). Einfluss von holzchips unterschiedlicher baumarten (Eiche, akazie und kirsche) auf die qualität von weiß-weinen der sorte "encruzado" bei einsatz über eine kurze periode des weinausbaus. *Mitteilungen Klosterneuburg*, 67(2), 84–96. Available from https://www.mitt-klosterneuburg.com/.

Doussot, F., De Jéso, B., Quideau, S., & Pardon, P. (2002). Extractives content in cooperage oak wood during natural seasoning and toasting; influence of tree species, geographic location, and single-tree effects. *Journal of Agricultural and Food Chemistry*, 50(21), 5955–5961. Available from https://doi.org/10.1021/jf020494e.

Escudero-Gilete, M. L., Hernanz, D., Galán-Lorente, C., Heredia, F. J., & Jara-Palacios, M. J. (2019). Potential of cooperage byproducts rich in ellagitannins to improve the antioxidant activity and color expression of red wine anthocyanins. *Foods*, 8(8). Available from https://doi.org/10.3390/foods8080336.

Es-Safi, N. E., Le Guernevé, C., Fulcrand, H., Cheynier, V., & Moutounet, M. (2000). Xanthylium salts formation involved in wine colour changes. *International Journal of Food Science and Technology*, 35(1), 63–74. Available from https://doi.org/10.1046/j.1365-2621.2000.00339.x.

Fernández de Simón, B., Cadahía, E., & Jalocha, J. (2003). Volatile compounds in a Spanish red wine aged in barrels made of Spanish, French, and American oak wood. *Journal of Agricultural and Food Chemistry*, 51(26), 7671–7678. Available from https://doi.org/10.1021/jf030287u.

Fernández de Simón, B., Hernández, T., Cadahía, E., Dueñas, M., & Estrella, I. (2003). Phenolic compounds in a Spanish red wine aged in barrels made of Spanish, French and American oak wood. *European Food Research and Technology*, 216(2), 150–156. Available from https://doi.org/10.1007/s00217-002-0637-4.

Fernández de Simón, B., Cadahía, E., Muiño, I., del Álamo, M., & Nevares, I. (2010). Volatile composition of toasted oak chips and staves and of red wine aged with them. *American Journal of Enology and Viticulture*, 61(2), 157–165. Available from http://ajevonline.org/cgi/reprint/61/2/157.

Fernández de Simón, B., Esteruelas, E., Muñoz, A. M., Cadáhia, E., & Sanz, M. (2009). Volatile compounds in acacia, chestnut, cherry, ash, and oak woods, with a view to their use in cooperage. *Journal of Agricultural and Food Chemistry*, 57(8), 3217–3227. Available from https://doi.org/10.1021/jf803463h.

Fernández de Simón, B., Sanz, M., Cadahía, E., Martínez, J., Esteruelas, E., & Muñoz, A. M. (2014). Polyphenolic compounds as chemical markers of wine ageing in contact with cherry, chestnut, false acacia, ash and oak wood. *Food Chemistry*, 143, 66–76. Available from https://doi.org/10.1016/j.foodchem.2013.07.096.

Ferreras, D., Fernández, E., & Falqué, E. (2002). Effects of oak wood on the aromatic composition of *Vitis vinifera* L. var. Treixadura wines. *Food Science and Technology International*, 8, 343–349. Available from https://doi.org/10.1106/108201302031655.

Feuillat, F., Moio, L., Guichard, E., Marinov, M., Fournier, N., & Puech, J. L. (1997). Variation in the concentration of ellagitannins and *cis*- and *trans*-β-methyl-γ-octalactone extracted from oak wood (*Quercus robur* L., *Quercus petraea* Liebl.) under model wine cask conditions. *American Journal of Enology and Viticulture*, 48(4), 509–515.

Feuillat, F., Perrin, J. R., & Keller, R. (1994). Simulation expérimentale de "l'interface tonneau": Mesure des cinétiques d'imprégnation du liquide dans le bois et d'évaporation de surface. *Journal of International des Sciences de la Vignet et du Vin*, 28, 227–245.

Fujieda, M., Tanaka, T., Suwa, Y., Koshimizu, S., & Kouno, I. (2008). Isolation and structure of whiskey polyphenols produced by oxidation of oak wood ellagitannins. *Journal of Agricultural and Food Chemistry*, 56(16), 7305–7310. Available from https://doi.org/10.1021/jf8012713.

Gallego, L., Del Álamo, M., Nevares, I., Fernández, J. A., De Simón, B. F., & Cadáhia, E. (2012). Phenolic compounds and sensorial characterization of wines aged with alternative to barrel products made of Spanish oak wood (*Quercus pyrenaica* Willd.). *Food Science and Technology International*, 18(2), 151–165. Available from https://doi.org/10.1177/1082013211427782.

Garde-Cerdán, T., & Ancín-Azpilicueta, C. (2006). Review of quality factors on wine ageing in oak barrels. *Trends in Food Science and Technology*, 17(8), 438–447. Available from https://doi.org/10.1016/j.tifs.2006.01.008.

Gómez-Plaza, E., Pérez-Prieto, L. J., Fernández-Fernández, J. I., & López-Roca, J. M. (2004). The effect of successive uses of oak barrels on the extraction of oak-related volatile compounds from wine. *International Journal of Food Science and Technology*, 39(10), 1069–1078. Available from https://doi.org/10.1111/j.1365-2621.2004.00890.x.

Gonçalves, F. J., & Jordão, A. M. (2009). Changes in antioxidant activity and the proanthocyanidin fraction of red wine aged in contact with Portuguese (*Quercus Pyrenaica* Willd.) and American (*Quercus alba* L.) oak wood chips. *Italian Journal of Food Science*, 21(1), 51–64.

Gortzi, O., Metaxa, X., Mantanis, G., & Lalas, S. (2013). Effect of artificial ageing using different wood chips on the antioxidant activity, resveratrol and catechin concentration, sensory properties and colour of two Greek red wines. *Food Chemistry*, 141(3), 2887–2895. Available from https://doi.org/10.1016/j.foodchem.2013.05.051.

Guchu, E., Díaz-Maroto, M. C., Pérez-Coello, M. S., González-Viñas, M. A., & Cabezudo, I. M. D. (2006). Volatile composition and sensory characteristics of Chardonnay wines treated with American and Hungarian oak chips. *Food Chemistry*, 99, 350–359. Available from https://doi.org/10.1016/j.foodchem.2005.07.050.

Gutiérrez-Afonso, V. L. (2002). Sensory descriptive analysis between white wines fermented with oak chips and in barrels. *Journal of Food Science*, 67(6), 2415–2419. Available from https://doi.org/10.1111/j.1365-2621.2002.tb09563.x.

Herjavec, S., Jeromel, A., Da Silva, A., Orlic, S., & Redzepovic, S. (2007). The quality of white wines fermented in Croatian oak barrels. *Food Chemistry*, 100(1), 124–128. Available from https://doi.org/10.1016/j.foodchem.2005.09.034.

Herrero, P., Sáenz-Navajas, M. P., Avizcuri, J. M., Culleré, L., Balda, P., Antón, E. C., ... Escudero, A. (2016). Study of Chardonnay and Sauvignon blanc wines from D.O.Ca Rioja (Spain) aged in different French oak wood barrels: Chemical and aroma quality aspects. *Food Research International*, 89, 227–236. Available from https://doi.org/10.1016/j.foodres.2016.08.002.

Ibern-Gómez, M., Andrés-Lacueva, C., Lamuela-Raventós, R. M., Lao-Luque, C., Buxaderas, S., & De la Torre-Boronat, M. C. (2001). Differences in phenolic profile between oak wood and stainless steel fermentation in white wines. *American Journal of Enology and Viticulture*, 52(2), 159–164.

Izquierdo-Cañas, P. M., Mena-Morales, A., & García-Romero, E. (2016). Malolactic fermentation before or during wine aging in barrels. *LWT - Food Science and Technology*, 66, 468–474. Available from https://doi.org/10.1016/j.lwt.2015.11.003.

Jiménez Moreno, N., & Ancín Azpilicueta, C. (2007). Binding of oak volatile compounds by wine lees during simulation of wine ageing. *LWT - Food Science and Technology*, 40(4), 619–624. Available from https://doi.org/10.1016/j.lwt.2006.02.027.

Jordão, A. M., Correia, A. C., DelCampo, R., & González SanJosé, M. L. (2012). Antioxidant capacity, scavenger activity, and ellagitannins content from commercial oak pieces used in winemaking. *European Food Research and Technology*, 235(5), 817–825. Available from https://doi.org/10.1007/s00217-012-1803-y.

Jordão, A. M., Costa, F., Fontes, L., Correia, A. C., Miljić, U., Puškaš, V., ... Cosme, F. (2019a). Impact of the contact time of different oak wood chips on red wine phenolic composition evolution after bottling. *BIO Web of Conferences*, 15, 02019. Available from https://doi.org/10.1051/bioconf/20191502019.

Jordão, A. M., Lozano, V., & González-SanJosé, M. L. (2019b). Influence of different wood chip extracts species on color changes and anthocyanin content in synthetic wine solutions. *Foods*, 8(7). Available from https://doi.org/10.3390/foods8070254.

Jordão, A.M., Pina, A., Montalbano, I., Correia, A.C., & Ricardo-da-Silva, J.M. (2018). Sensory profile of varietal white wines submitted to a short period of aging in contact with oak and cherry wood chips. In *Book of abstracts of forty-first world congress of vine and wine* (p. 979).

Jordão, A. M., Ricardo-da-Silva, J. M., & Laureano, O. (2005). Comparison of volatile composition of cooperage oak wood of different origins (*Quercus pyrenaica* vs *Quercus alba* and *Quercus petraea*). *Mitteilungen Klosterneuburg, 55*, 31−40.

Jordão, A. M., Ricardo-da-Silva, J. M., & Laureano, O. (2006a). Effect of oak constituents and oxygen on the evolution of malvidin-3-glucoside and (+)-catechin in model wine. *American Journal of Enology and Viticulture, 57*(3), 377−381.

Jordão, A. M., Ricardo-da-Silva, J. M., Laureano, O., Adams, A., Demyttenaere, J., Verhé, R., & De Kimpe, N. (2006b). Volatile composition analysis by solid-phase microextraction applied to oak wood used in cooperage (*Quercus pyrenaica* and *Quercus petraea*): Effect of botanical species and toasting process. *Journal of Wood Science, 52*(6), 514−521. Available from https://doi.org/10.1007/s10086-005-0796-6.

Jordão, A. M., Ricardo-da-Silva, J. M., & Laureano, O. (2007). Ellagitannins from Portuguese oak wood (*Quercus pyrenaica* Willd.) used in cooperage: Influence of geographical origin, coarseness of the grain and toasting level. *Holzforschung, 61*(2), 155−160. Available from https://doi.org/10.1515/HF.2007.028.

Jordão, A. M., Ricardo-da-Silva, J. M., Laureano, O., Mullen, W., & Crozier, A. (2008). Effect of ellagitannins, ellagic acid and volatile compounds from oak wood on the (+)-catechin, procyanidin B1 and malvidin-3-glucoside content of model wines. *Australian Journal of Grape and Wine Research, 14*(3), 260−270. Available from https://doi.org/10.1111/j.1755-0238.2008.00029.x.

Jordão, A. M., Virginia, L., Correia, A. C., Ortega-Heras, M., & González-SanJosé, M. L. (2016). Comparative analysis of volatile and phenolic composition of alternative wood chips from cherry, acacia and oak for potential use in enology. *BIO Web of Conferences, 7*, 02012. Available from https://doi.org/10.1051/bioconf/20160702012.

Kozlovic, G., Jeromel, A., Maslov, L., Pollnitz, A., & Orlić, S. (2010). Use of acacia barrique barrels − Influence on the quality of Malvazija from Istria wines. *Food Chemistry, 120*(3), 698−702. Available from https://doi.org/10.1016/j.foodchem.2009.10.065.

Laqui-Estaña, J., López-Solís, R., Peña-Neira, Á., Medel-Marabolí, M., & Obreque-Slier, E. (2019). Wines in contact with oak wood: The impact of the variety (Carménère and Cabernet Sauvignon), format (barrels, chips and staves), and aging time on the phenolic composition. *Journal of the Science of Food and Agriculture, 99*(1), 436−448. Available from https://doi.org/10.1002/jsfa.9205.

Liberatore, M. T., Pati, S., Nobile, M. A. D., & Notte, E. L. (2010). Aroma quality improvement of Chardonnay white wine by fermentation and ageing in barrique on lees. *Food Research International, 43*(4), 996−1002. Available from https://doi.org/10.1016/j.foodres.2010.01.007.

Loupassaki, S., Abouzer, M., Basalekou, M., Fyssarakis, I., & Makris, D. P. (2016). Evolution pattern of wood-related volatiles during traditional and artificial ageing of commercial red and white wines: Association with sensory analysis. *International Food Research Journal, 23*(4), 1459−1465. Available from http://www.ifrj.upm.edu.my.

Lukić, I., Jedrejčić, N., Ganić, K. K., Staver, M., & Peršurić, D. (2015). Phenolic and aroma composition of white wines produced by prolonged maceration and maturation in wooden barrels. *Food Technology and Biotechnology, 53*(4), 407−418. Available from https://doi.org/10.17113/tb.53.04.15.4144.

Martínez-Gil, A. M., Cadahía, E., de Simón, B. F., Gutiérrez-Gamboa, G., Nevares, I., & del Álamo-Sanza, M. (2017). *Quercus humboldtii* (Colombian Oak): Characterisation of wood phenolic composition with respect to traditional oak wood used in oenology. *Ciencia e Tecnica Vitivinicola, 32*(2), 93−101. Available from https://doi.org/10.1051/ctv/20173202093.

Martínez-Gil, A. M., del Álamo-Sanza, M., Gutiérrez-Gamboa, G., Moreno-Simunovic, Y., & Nevares, I. (2018). Volatile composition and sensory characteristics of Carménère wines macerating with Colombian (*Quercus humboldtii*) oak chips compared to wines macerated with American (*Q. alba*) and European (*Q. petraea*) oak chips. *Food Chemistry, 266*, 90−100. Available from https://doi.org/10.1016/j.foodchem.2018.05.123.

Martínez-Gil, A. M., del Álamo-Sanza, M., Nevares, I., Sánchez-Gómez, R., & Gallego, L. (2020). Effect of size, seasoning and toasting level of *Quercus pyrenaica* Willd. wood on wine phenolic composition during maturation process with micro-oxygenation. *Food Research International, 128*. Available from https://doi.org/10.1016/j.foodres.2019.108703.

Martins, N., Garcia, R., Da Silva, M. G., & Cabrita, M. J. (2012). Volatile compounds from oak, cherry, chestnut and acacia chips: Influence of toasting level. *Ciencia e Tecnica Vitivinicola, 27*(1), 49−57. Available from http://www.scielo.gpeari.mctes.pt/pdf/ctv/v27n1/v27n1a05.pdf.

Masson, G., Moutounet, M., & Puech, J. L. (1995). Ellagitannin content of oak wood as a function of species and of sampling position in the tree. *American Journal of Enology and Viticulture, 46*(2), 262−268.

Matějíček, D., Mikeš, O., Klejdus, B., Štěrbová, D., & Kubáň, V. (2005). Changes in contents of phenolic compounds during maturing of barrique red wines. *Food Chemistry, 90*(4), 791−800. Available from https://doi.org/10.1016/j.foodchem.2004.05.057.

Matricardi, L., & Waterhouse, A. L. (1999). Influence of toasting technique on color and ellagitannins of oak wood in barrel making. *American Journal of Enology and Viticulture, 50*(4), 519−526.

McCallum, M. J., Lopes-Correia, T., & Ricardo-da-Silva, J. M. (2019). Chemical evaluation of Carcavelos fortified wine aged in Portuguese (*Quercus pyrenaica*) and French (*Quercus robur*) oak barrels at medium and high toast. *Oeno One, 53*(3), 561−572. Available from https://doi.org/10.20870/oeno-one.2019.53.3.2284.

Nunes, I., Correia, A. C., Jordão, A. M., & Ricardo-da-Silva, J. M. (2020). Use of oak and cherry wood chips during alcoholic fermentation and the maturation process of rosé wines: Impact on phenolic composition and sensory profile. *Molecules (Basel, Switzerland), 25*, 1236. Available from https://doi.org/10.3390/molecules25051236.

Nunes, P., Muxagata, S., Correia, A. C., Nunes, F. M., Cosme, F., & Jordão, A. M. (2017). Effect of oak wood barrel capacity and utilization time on phenolic and sensorial profile evolution of an Encruzado white wine. *Journal of the Science of Food and Agriculture, 97*(14), 4847−4856. Available from https://doi.org/10.1002/jsfa.8355.

Obradović, V., Mesić, J., Ravančić, M., Mijowska, K., & Svitlica, B. (2017). Chemical and sensory properties of Chardonnay wines produced in different oak barrels. *International Journal of Nutrition and Food Engineering, 126*(6), 413−418.

Parodi, G. (2000). A proposito di barriques. *Vignevini, 3*, 77−83.

Pérez-Coello, M. S., González-Viñas, M. A., García-Romero, E., Cabezudo, M. D., & Sanz, J. (2000). Chemical and sensory changes in white wines fermented in the presence of oak chips. *International Journal of Food Science and Technology, 35*(1), 23−32. Available from https://doi.org/10.1046/j.1365-2621.2000.00337.x.

Prida, A., & Chatonnet, P. (2010). Impact of oak-derived compounds on the olfactory perception of barrel-aged wines. *American Journal of Enology and Viticulture, 61*(3), 408−413. Available from http://ajevonline.org/cgi/reprint/61/3/408.

Psarra, C., Gortzi, O., & Makris, D. P. (2015). Kinetics of polyphenol extraction from wood chips in wine model solutions: Effect of chip amount and botanical species. *Journal of the Institute of Brewing, 121*, 207−212. Available from https://doi.org/10.1002/jib.212.

Puech, J. L., Feuillat, F., & Mosedale, J. R. (1999). The tannins of oak heartwood: Structure, properties, and their influence on wine flavor. *American Journal of Enology and Viticulture, 50*(4), 469−478.

Rasines-Perea, Z., Jacquet, R., Jourdes, M., Quideau, S., & Teissedre, P. L. (2019). Ellagitannins and flavano-ellagitannins: Red wines tendency in different areas, barrel origin and ageing time in barrel and bottle. *Biomolecules, 9*(8). Available from https://doi.org/10.3390/biom9080316.

Recamales, A. F., Sayago, A., González-Miret, M. L., & Hermanz, D. (2006). The effect of time and storage conditions on the phenolic composition and color of white wine. *Food Research International, 39*, 220−229. Available from https://doi.org/10.1016/j.foodres.2005.07.009.

Rieger, T. (2013). Acacia barrels of alternative to oak. *Vineyard and Winery Management Mac-Apri*, 58−62.

Rodríguez-Nogales, J. M., Fernández-Fernández, E., & Vila-Crespo, J. (2009). Characterisation and classification of Spanish Verdejo young white wines by volatile and sensory analysis with chemometric tools. *Journal of the Science of Food and Agriculture, 89*(11), 1927−1935. Available from https://doi.org/10.1002/jsfa.3674.

Rubio-Bretón, P., Garde-Cerdán, T., & Martínez, J. (2018). Use of oak fragments during the aging of red wines. Effect on the phenolic, aromatic, and sensory composition of wines as a function of the contact time with the wood. *Beverages, 4*, 102.

Salacha, M. I., Kallithraka, S., & Tzourou, I. (2008). Browning of white wines: Correlation with antioxidant characteristics, total polyphenolic composition and flavanol content. *International Journal of Food Science and Technology, 43*(6), 1073−1077. Available from https://doi.org/10.1111/j.1365-2621.2007.01567.x.

Sánchez-Gómez, R., Ignacio, N., Ana, M.-G., & del Álamo-Sanza, M. (2018). Oxygen consumption by red wines under different micro-oxygenation strategies and *Q. Pyrenaica* chips. Effects on color and phenolic characteristics. *Beverages, 4*, 69. Available from https://doi.org/10.3390/beverages4030069.

Sánchez-Palomo, E., Alonso-Villegas, R., Delgado, J. A., & González-Viñas, M. A. (2017). Improvement of Verdejo white wines by contact with oak chips at different winemaking stages. *LWT - Food Science and Technology, 79*, 111−118. Available from https://doi.org/10.1016/j.lwt.2016.12.045.

Santos, F., Correia, A. C., Ortega-Heras, M., García-Lomillo, J., González-SanJosé, M. L., Jordão, A. M., & Ricardo-da-Silva, J. M. (2019). Acacia, cherry and oak wood chips used for a short aging period of rosé wines: Effects on general phenolic parameters, volatile composition and sensory profile. *Journal of the Science of Food and Agriculture, 99*(7), 3588−3603. Available from https://doi.org/10.1002/jsfa.9580.

Sanz, M., Cadahía, E., Esteruelas, E., Muñoz, A. M., Fernández de Simón, B., Hernández, T., & Estrella, I. (2010). Phenolic compounds in cherry (*Prunus avium*) heartwood with a view to their use in cooperage. *Journal of Agricultural and Food Chemistry, 58*(8), 4907−4914. Available from https://doi.org/10.1021/jf100236v.

Sanz, M., Fernández de Simón, B., Cadahía, E., Esteruelas, E., Muñoz, A. M., Hernández, M. T., & Estrella, I. (2012a). Polyphenolic profile as a useful tool to identify the wood used in wine aging. *Analytica Chimica Acta, 732*, 33−45. Available from https://doi.org/10.1016/j.ac.2011.12.012.

Sanz, M., Fernández de Simón, B., Cadahía, E., Esteruelas, E., Muñoz, A. M., Hernández, T., & Pinto, E. (2012b). LC-DAD/ESI-MS/MS study of phenolic compounds in ash (*Fraxinus excelsior* L. and *F. americana* L.) heartwood. Effect of toasting intensity at cooperage. *Journal of Mass Spectrometry, 47*(7), 905−918. Available from https://doi.org/10.1002/jms.3040.

Sanz, M., Fernández de Simón, B., Esteruelas, E., Muñoz, A. M., Cadahía, E., Hernández, M. T., & Martinez, J. (2012c). Polyphenols in red wine aged in acacia (*Robinia pseudoacacia*) and oak (*Quercus petraea*) wood barrels. *Analytica Chimica Acta, 732*, 83−90. Available from https://doi.org/10.1016/j.ac.2012.01.061.

Sanz, M., Fernández de Simón, B., Esteruelas, E., Muñoz, A. M., Cadahía, E., Hernández, T., & Pinto, E. (2011). Effect of toasting intensity at cooperage on phenolic compounds in Acacia (*Robinia pseudoacacia*) heartwood. *Journal of Agricultural and Food Chemistry, 59*(7), 3135−3145. Available from https://doi.org/10.1021/jf1042932.

Schumacher, R., Alañón, M. E., Castro-Vázquez, L., Pérez-Coello, M. S., & Díaz-Maroto, M. C. (2013). Evaluation of oak chips treatment on volatile composition and sensory characteristics of Merlot wine. *Journal of Food Quality, 36*(1), 1−9. Available from https://doi.org/10.1111/jfq.12012.

Setzer, W. (2016). Volatile components of oak and cherry wood chips used in aging of beer, wine and spirits. *American Journal of Essential Oil Natural Products, 4*(2), 37−40.

Singleton, V. L. (2000). Le Stockage des vins en barriques: Utilisation et variables significatives. *Journal Des Sciences et Techniques de La Tonnellerie, 6*, 1−25.

Soares, B., Garcia, R., Freitas, A. M. C., & Cabrita, M. J. (2012). Phenolic compounds released from oak, cherry, chestnut and robinia chips into a synthetic wine: Influence of toasting level. *Ciência e Técnica Vitivinicola, 27*(1), 17−26. Available from http://www.scielo.gpeari.mctes.pt/pdf/ctv/v27n1/v27n1a02.pdf.

Spagna, G., Pifferi, P. G., Rangoni, C., Mattivi, F., Nicolini, G., & Palmonari, R. (1996). The stabilization of white wines by adsorption of phenolic compounds on chitin and chitosan. *Food Research International, 29*(3−4), 241−248. Available from https://doi.org/10.1016/0963-9969(96)00025-7.

Spillman, P. J., Sefton, M. A., & Gawel, R. (2004). The contribution of volatile compounds derived during oak barrel maturation to the aroma of a Chardonnay and Cabernet Sauvignon wine. *Australian Journal of Grape and Wine Research, 10*(3), 227−235.

Springmann, S., Rogers, R., & Spiecker, H. (2011). Impact of artificial pruning on growth and secondary shoot development of wild cherry (*Prunus avium* L.). *Forest Ecology and Management, 261*(3), 764−769. Available from https://doi.org/10.1016/j.foreco.2010.12.007.

Taransaud, J. (1976). *Le livre de la tonnellerie*. Paris: La roue à livres diffusion.

Tavares, M., Jordão, A. M., & Ricardo-da-Silva, J. M. (2017). Impact of cherry, acacia and oak chips on red wine phenolic parameters and sensory profile. *Oeno One, 51*(3), 329−342. Available from https://doi.org/10.20870/oeno-one.2017.51.4.1832.

Towey, J. P., & Waterhouse, A. L. (1996). The extraction of volatile compounds from French and American oak barrels in chardonnay during three successive vintages. *American Journal of Enology and Viticulture, 47*(2), 163−172.

Viriot, C., Scalbert, A., Hervé du Penhoat, C. L. M., & Moutounet, M. (1994). Ellagitannins in woods of sessile oak and sweet chestnut dimerization and hydrolysis during wood ageing. *Phytochemistry, 36*(5), 1253−1260. Available from https://doi.org/10.1016/S0031-9422(00)89647-8.

Vivas, N. (2000). Apports récents à la connaissance du chêne de tonnellerie et à lélevage des vins rouges en barrique. *Bull. OIV, 827*, 79–108.

Vivas, N. (2005). *Manual De Tonelería: Destinado A Usuarios De Toneles.*

Vivas, N., Bourden Nonier, M. F., Absalon, C., Lizama Abad, V., Jamet, F., de Gaulejac, N. V., & Fouquet, E. (2008). Formation of flavanol-aldehyde adducts in barrel-aged white wine-possible contribution of these products to colour. *South African Journal of Enology and Viticulture, 29*(2), 98–108. Available from http://www.sasev.org/journal/sajev-articles/volume-29-2/Vivas%20et%20al%20pp%2098-108.pdf.

Vivas, N., & Glories, Y. (1993). Les phénomènes d'oxydoréduction liés à l'élevage en barriques des vins rouges: Aspects technologiques. *Revue Francaise d'Oenologie, 33*, 33–38.

Vivas, N., & Glories, Y. (1996). Role of oak wood ellagitannins in the oxidation process of red wines during aging. *American Journal of Enology and Viticulture, 47*, 103–107.

Vivas, N., Glories, Y., Bourgeois, G., & Vitry, C. (1996). The heartwood ellagitannins of different oaks (*Quercus* Sp.) and chestnut species (*Castanea sativa* Mill.). Quantity analysis of red wines aging in barrels. *Journal Des Sciences et Techniques de La Tonnellerie, 2*, 25–75.

Vivas, N., & Saint-Cricq de Gaulejac, N. (1999). The useful lifespan of new barrels abd risk realted to the use of old barrels. *Australian and New Zeland Wine Industry, 14*, 37–45.

Young, O. A., Kaushal, M., Robertson, J. D., Burns, H., & Nunns, S. J. (2010). Use of species other than oak to flavor wine: An exploratory survey. *Journal of Food Science, 75*.

Zhang, B., Cai, J., Duan, C. Q., Reeves, M. J., & He, F. (2015). A review of polyphenolics in oak woods. *International Journal of Molecular Sciences, 16*(4), 6978–7014. Available from https://doi.org/10.3390/ijms16046978.

23

Impacts of phenolics and prefermentation antioxidant additions on wine aroma

Paul Kilmartin

School of Chemical Sciences, University of Auckland, Auckland, New Zealand

23.1 Introduction

Many of the effects of oxygen on white wine aroma profiles are known to involve polyphenol oxidation. These occur both in finished wines, through gradual chemical oxidation processes, and rapidly in grape juices soon after harvest through the action of polyphenol oxidase (PPO) enzymes. Losses of fruity aromas and browning of color are problems associated with the oxidation of white wines.

In this chapter, polyphenol oxidation processes will be introduced, with a focus on the production of reactive quinones. The role of quinones in reacting with sulfur-containing aroma compounds will then be highlighted. A discussion on the effects of phenolic oxidation processes on white wine aromas follows, along with the use of antioxidants to mediate the effects of oxygen in wines, including sulfites, ascorbic acid, and glutathione. After this, issues related to must oxidation following grape harvesting are presented, including an important role for white wine aroma outcomes through the use of antioxidant additions at harvest.

23.2 Polyphenol oxidation

23.2.1 Wine oxidation linked to polyphenol content

An important early observation in relation to the susceptibility of wine to oxidize was that the oxidation of ethanol to acetaldehyde occurs through the coupled oxidation of readily oxidizable polyphenols (Wildenradt & Singleton, 1974). In the case of white wines, the polyphenols include hydroxycinnamic acids such as caftaric acid in the juice and free caffeic acid in wine, and a smaller contribution from flavanols such as catechin, which are more prominent in red wines. Both caffeic acid and catechin possess an easily oxidizable catechol moiety, with two phenolic groups situated next to each other on an aromatic ring. The oxidation of catechol-containing phenolic compounds then leads to the formation of reactive quinones (Fig. 23.1) (Danilewicz, 2016).

Without polyphenols present, ethanol and tartaric acid do not react appreciably with oxygen. Model experiments demonstrated that polyphenol oxidation proceeds more rapidly at a higher pH due to the presence of the phenolate ion form (Singleton, 1987). Only a small percentage of phenolate ions are expected at wine pH, but these may still be of importance in determining rates of oxidation, given that higher pH wines are generally more easily oxidized. Model solution experiments also showed that reactions of quinones with other polyphenols led to the formation of new polyphenol oligomers that contributed to wine browning by absorbing light at higher wavelengths. Through the regeneration of the catechol group, more units of oxygen were taken up, more so than would be expected given the number of phenolic groups originally present (Singleton, 1987).

It is also not surprising that one of the main methods of determining the total content of polyphenol compounds in beverages and foods involves an oxidizing agent, the Folin–Ciocalteu (FC) reagent

White Wine Technology.
DOI: https://doi.org/10.1016/B978-0-12-823497-6.00029-6

© 2022 Elsevier Inc. All rights reserved.

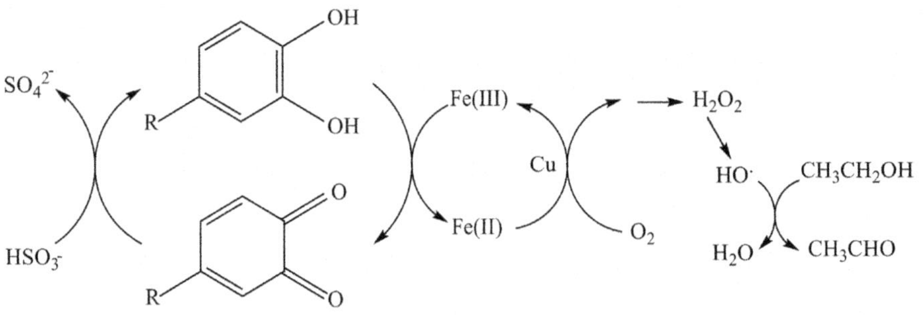

FIGURE 23.1 Polyphenol oxidation process for a representative catechol-containing phenolic compound.

(Singleton, Orthofer, & Lamuela-Raventos, 1999). The main interferents in the FC measurement of total phenolics, particularly important for white wines with a lower total phenolic content, comes from free (but not bound) SO_2, glutathione, and ascorbic acid. This interference is ascribed to their role in the rapid regeneration of quinones formed by the FC reagent, to allow further phenolic oxidation to occur. This property will be examined further below regarding the role these compounds have as important grape and wine antioxidant additives.

White wines typically contain 200–500 mg/L of total phenolics, which is five- to 10-fold lower than typical red wines. Red wines include more skin and seed-derived flavonoid tannins along with colored anthocyanins. While the content of flavonoids is low in white wines, increased concentrations of flavonoids arise with prolonged skin contact and with higher pressure applied during the pressing of juices (Maggu, Winz, Kilmartin, Trought, & Nicolau, 2007). Trials with Chardonnay showed a greater browning capacity through the addition of 200 mg/L of catechin to the wine (Singleton & Kramling, 1976). The flavonoid content in white wines has shown a greater link to browning than the concentrations of hydroxycinnamic acids (Singleton, 1987). In this context, a preference was shown for the oxidized caffeic acid quinone to couple with catechin ahead of caffeic acid, generating further condensation products (V. Cheynier, Basire, & Rigaud, 1989).

At the same time, the identification of the browning compounds formed during white wine oxidation has been difficult to determine. A model process is provided by the role of glyoxylic acid, which arises from the oxidation of tartaric acid, in cross-linking catechin molecules, with the formation of yellow-colored xanthylium pigments (George, Clark, Prenzler, & Scollary, 2006). Metals such as iron and copper are needed as catalysts for this process, and glyoxylic acid is also produced rapidly with light exposure (Grant-Preece, Schmidtke, Barril, & Clark, 2017). However, the inhibitory effects of caffeic acid on the accumulation of these pigments raise questions about the role of xanthylium pigments in white wine oxidation.

23.2.2 Oxygen and its activation in wine

Wine can take up to 6 mL/L (8.6 mg/L) of O_2 at room temperature when saturated with air, while more oxygen will dissolve at a lower temperature. Pumping of grape must and wines produces higher concentrations of O_2, while slower rates of O_2 ingress occur for wines in barrels. In addition to the reaction of O_2 with phenolics, uptake of O_2 by the yeast lees during aging can assist in minimizing unwanted effects of oxidation. An even lower level of O_2 will persist during bottle storage depending upon the O_2 permeability of the bottle closure.

The oxidizing power of O_2 needs to be mediated by both polyphenols and catalytic metals, as the triplet spin state of O_2 limits the type of molecules that can react directly with O_2 (Danilewicz, 2003). In the process, oxygen forms a number of reactive species, including hydrogen peroxide (H_2O_2) and the hydroxyl radical (OH·) (Fig. 23.1). Iron species play a role in complexing O_2 and promoting the oxidation process, with an important mediation occurring as iron species shuttle between Fe(II) and Fe(III) oxidation states, to facilitate the oxidation of polyphenols. In the later stages of oxidation, Fe(II) ions convert H_2O_2 into the very reactive OH· in the well-known Fenton reaction. The strong OH· oxidant will react with the first organic compound that it encounters and form a range of aldehydes and ketones (Waterhouse & Laurie, 2006). Given the abundance of ethanol in wine, the production of acetaldehyde is a major outcome.

23.2.3 Roles of catalytic metals in polyphenol oxidation

Copper has a further role in facilitating the redox cycling of iron and increasing the rate of polyphenol oxidation (Danilewicz, 2007). The role of metals in wine oxidation has been confirmed experimentally through the removal of Fe and Cu ions using ion exchangers, and the outcome was a lower rate of browning in white wines (Benitez, Castro, & Barroso, 2002). Higher levels of copper have also been found to increase oxygen consumption rates in wines (Ferreira, Carrascon, Bueno, Ugliano, & Fernandez-Zurbano, 2015). The role of antioxidants such as sulfite to lessen the effects of oxidation will be discussed below. However, the effect of reducing quinones back to the original catechol-containing phenolic (Fig. 23.1) can promote the overall phenolic reaction processes and uptake of O_2, by regenerating the initial substrate of oxidation (Danilewicz, 2013).

23.3 Oxidation effects on aroma compounds

A number of effects on wine aromas can be related to phenolic oxidation processes, including the removal of "reduced" compounds containing sulfur, other aroma losses, and the production of new aldehydic or "oxidized" aromas (Du Toit, Marais, Pretorius, & Du Toit, 2006). At the same time, some sulfur-containing thiol compounds make a positive contribution to wine aromas, and the loss of these compounds during wine oxidation has been recognized. A classic example in this regard is the varietal thiol 3-mercaptohexanol (3MH), which contributes grapefruit/passion fruit aromas to wines such as Sauvignon blanc (Tominaga, Murat, & Dubourdieu, 1998). The term "varietal" is applied to distinguish these thiols from unwanted reductive thiol compounds, although their presence in nearly all wines raises questions about the use of this term.

The removal of sulfur-containing aromas can be undertaken using copper fining, although the release of these same compounds later following bottle storage is a cause for concern. Increases in the concentrations of free H_2S and methanethiol were observed during wine storage without O_2 present, despite little change occurring in the concentration of total H_2S (Franco-Luesma & Ferreira, 2016). These sulfur-containing compounds can also be removed by racking the wine or other oxygenation treatments. Removal of H_2S by oxidative processes is an effective way to remove sulfidic off-odors in wines, and may even involve copper complexes (Kreitman, Danilewicz, Jeffery, & Elias, 2016).

The mechanism of removal of such reductive compounds has been ascribed to the addition of the sulfur-containing compound to quinones formed through polyphenol oxidation (Fig. 23.2). This has been demonstrated with the case of 3MH, which is able to react readily with polyphenol quinones (Blanchard, Darriet, & Dubourdieu, 2004) and consume O_2 in the process (Danilewicz, Seccombe, & Whelan, 2008). Further confirmation of this mechanism, and the competition between sulfur-containing aroma compounds and reducing agents such as SO_2 for the quinones, have also been reported (Nikolantonaki et al., 2012).

The formation of "oxidized" aromas in wine can also be related to polyphenol oxidation processes, but in a more indirect manner. The production of reactive oxygen species, leading to the creation of very reactive OH·, is an outcome of polyphenol oxidation (Fig. 23.1). In addition to the creation of acetaldehyde from ethanol, further aldehydes can be formed, which have an impact on wine aroma and are known to develop during wine aging. These include phenylacetaldehyde and methional (3-(methylthio)propionaldehyde), responsible for wine aromas described as "honey-like" and "farm-feed," respectively (A. C. S. Ferreira, de Pinho, Rodrigues, & Hogg, 2002). Further oxidized aroma compounds include sotolon (4,5-dimethyl-3-hydroxy-2(5-H)-furanone) and the compound responsible for the "kerosene" note in added Riesling, TDN (1,1,6-trimethyl-1,2-dihydronaphthalene).

Further trials with Sauvignon blanc wines have shown the direct effects of oxygen inputs on the appearance of "oxidized" versus "reductive" aromas. With no oxygen added, a "reductive" character appeared in wines, while fruitier wines were obtained after a single oxygen dose; repetitive oxidation led to the formation of acetaldehyde, along with some loss in varietal thiols (Coetzee, Van Wyngaard, Šuklje, Ferreira, & du Toit, 2016). In a closure trial on Sauvignon blanc wines, wines stored under the most airtight conditions showed high levels of H_2S, while the wines were more oxidized when closures that were highly permeable to oxygen were employed (Lopes et al., 2009). A further effect of polyphenols on wine aroma occurs through the retention of aroma compounds in wine through noncovalent interactions that affect aroma compound volatility (Aronson & Ebeler, 2004). The perception of aroma compounds associated with white wines, including varietal thiols and a representative ester, was

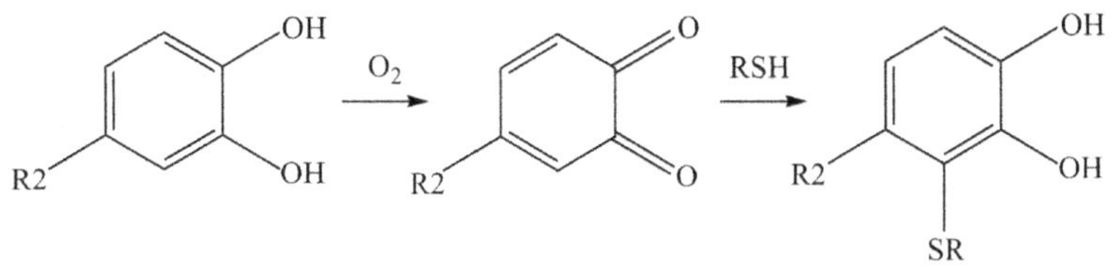

FIGURE 23.2 Mechanism for the addition of a sulfur-containing compound (RSH) to a quinone formed through the oxidation of a catechol-containing phenolic compound.

suppressed by the addition of common wine phenolics (Lund, Winz, Gardner, & Kilmartin, 2009). As phenolics change during wine oxidation and maturation, the effects of these on aroma compound volatility are an area that could be examined in future sensory studies.

23.4 Influence of wine antioxidants

The exposure of musts and wines to O_2 is one consideration in looking to limit the effects of oxidation in white wines. Winemakers can also turn to antioxidant additives to limit oxidation processes. In addition to the wide use of sulfites (SO_2), further important antioxidant additives include ascorbic acid (vitamin C) and glutathione, both of which occur naturally in the grape. Each of these additives has different mechanisms of oxidation prevention, which will be outlined in the following sections.

23.4.1 Sulfur dioxide

Additions of sulfur dioxide can occur at various stages in the winemaking process and are seen as a largely indispensable tool for modern winemaking. At the same time, limits in application rates and concentrations in wines have been applied by national and international regulations, in part due to health issues identified with the consumption of sulfites. Limits can be placed on both the level of free SO_2, the form that has an antimicrobial action and is most active as an antioxidant, and the total SO_2 content. Total SO_2 includes sulfites bound to other wine compounds that can be potentially released back into the wine.

SO_2 is notable in antioxidant terms, not for a direct reaction with O_2, but for its role in inhibiting PPO activity (Singleton, Salgues, Zaya, & Trousdale, 1985), and for scavenging hydrogen peroxide generated during polyphenol oxidation (Danilewicz, 2003). In model solutions without polyphenols present, SO_2 can be oxidized by O_2 with catalytic metals present, while oxidation of ethanol also occurs leading to the formation of acetaldehyde-bound SO_2 (Danilewicz et al., 2008; Danilewicz, 2007). As the reaction of O_2 with catechol-containing polyphenols is much faster, minimal ethanol oxidation was observed in the presence of such polyphenols, which thus serve to inhibit the autoxidation of SO_2.

A further antioxidant role for SO_2 involves the rapid reduction of oxidized polyphenol quinones (Fig. 23.1) (Cheynier et al., 1989). This action was seen in tests with the FC assay, where little response was provided by SO_2 on its own, but an enhancement for the response of catechin with SO_2 present was obtained (Saucier & Waterhouse, 1999). The controlled formation of quinones at inert electrodes also provides a suitable approach to study the interaction of antioxidants with oxidized phenolics (Makhotkina & Kilmartin, 2009). Current due to the oxidation of catechol-containing compounds increases when sulfite or glutathione are added, both to standard solutions and to white wines. A reverse current peak that is due to the reduction of the quinone back to the original polyphenol at the electrode is also suppressed in the presence of these antioxidants, showing how rapidly the quinone is removed from wines by free, but not bound, SO_2.

23.4.2 Glutathione

Glutathione can be present at concentrations in the range of 50–100 mg/kg in crushed grapes. As a sulfur-containing compound, glutathione can also react with quinones in an analogous manner to that described above for the reaction with sulfur-containing aroma compounds (Fig. 23.2). The product formed when the phenolic substrate is caftaric acid, the major hydroxycinnamic acid in grape juice, is S-glutathionyl caftaric acid, known also as the grape reaction product (GRP). Through its reaction with quinones, glutathione effectively limits the browning of musts (Singleton et al., 1985). Due to this property, the ratio of glutathione to caftaric acid has been proposed as an import measure with regard to browning susceptibility (Cheynier & Van Hulst, 1988), while the ratio of GRP to remaining caftaric acid can be considered as a measure of the extent of enzymatic oxidation that a must has experienced (Singleton, 1987).

The concentration of glutathione in juices and wines is lower when these have been subject to more oxidative conditions and also depends upon the extent of grape pressing (Maggu et al., 2007). Likewise, additions of SO_2 and ascorbic acid can assist in maintaining higher concentrations of glutathione in pressed grape juice (Du Toit, Lisjak, Stander, & Prevoo, 2007). On the other hand, changes in concentrations of caftaric acid and S-glutathionyl caftaric acid are relatively minor during fermentation and wine storage. The protective role of glutathione in wines is similar to SO_2 in its reaction with oxidized polyphenol quinones. The protection extends to the varietal aroma compound 3MH, where a protective effect was established given that glutathione will react with quinones in preference to 3MH (Blanchard et al., 2004; Dubourdieu, Moine-Ledoux, Lavigne-Cruege, Blanchard, & Tominaga, 2000). This faster reaction rate has been confirmed in model studies that have established the rapid reaction of glutathione, alongside SO_2 and ascorbic acid, with quinones (Nikolantonaki & Waterhouse, 2012).

For many years, glutathione was not an allowed additive for use in winemaking. However, the case has been made that aroma compounds such as the varietal thiols can be effectively preserved in young wines and oxidation slowed, leading to the approval of additions of glutathione up to 20 mg/L for both musts and wines (RESOLUTION OIV-OENO 445-2015). More recent bottling trials have supported this approach, where added glutathione was lost fairly soon after bottling in the presence of oxygen, and its presence led to the more rapid removal of O_2 but with little effect on SO_2 consumption (Panero, Motto, Petrozziello, Guaita, & Bosso, 2015). The browning of sparkling wine and the formation of acetaldehyde during storage was also decreased through the addition of glutathione (Webber et al., 2017).

23.4.3 Ascorbic acid

Ascorbic acid is found in grapes, but the levels of it are diminished rapidly when the must is exposed to O_2 (Singleton, 1987). As it has a lower reduction potential than most wine polyphenols, ascorbic acid is able to react directly with O_2 and, in the process, convert O_2 into reactive oxygen species, such as H_2O_2, in a similar manner to polyphenol oxidation (Fig. 23.1). A direct reaction with O_2 is not typically observed under wine conditions with SO_2 or glutathione. Ascorbic can legally be added to musts and wines and provides an alternative to sulfites for antioxidant protection of wines (Barril, Rutledge, Scollary, & Clark, 2016). However, pro-oxidative effects have at times been noted, particularly when ascorbic acid is applied in the absence of SO_2. The build-up of reactive oxygen species is suspected in this case, which can be effectively removed so long as adequate free sulfites are also present.

Ascorbic acid additions have been made within bottle closure and storage trials. In a 3-year trial on Chardonnay and Riesling, the addition of ascorbic acid generally produced less oxidized and more fruity aromas, or at least no detrimental effects on the wines (Skouroumounis et al., 2005). Ascorbic acid added to Riesling wines also led to positive sensory outcomes and retention of sulfites; however, some increase in browning was seen for wines that had more O_2 present (Morozova, Schmidt, & Schwack, 2015).

23.5 Enzymatic oxidation of musts

23.5.1 Harvesting methods and juice oxidation

The extent of juice oxidation is connected to the method of grape harvest, and how grapes and juices are handled postharvest, including pressing prior to fermentation. Machine harvesting has been widely adopted in many countries and allows grapes to be harvested with lower labor costs and with fruit at their optimum. However, the inclusion of unwanted material along with grapes is a concern, including leaves and stems, which can be readily removed during hand picking. The oxidation that occurs from berry damage and greater exposure to air during harvest and transport to the winery can be a problem. Enzymatic oxidation is further promoted by the maceration of the grapes, which provides greater access of enzymes to their substrates, in this case, PPO and phenolic compounds. Increased enzymatic activity and juice browning, typically measured as an increase in absorbance at 420 nm, has been reported in past studies on mechanical harvesting, along with the development of off-flavors (Noble, Ough, & Kasimatis, 1975; Pocock & Waters, 1998). The extended skin contact that results from mechanical harvesting may also lead to increases in pH, total nitrogen and haze-forming proteins, and a decrease in organic acids (Ough, Berg, Coffelt, & Cooke, 1971).

A different observation has been made with the late harvesting of grapes for the production of icewines (Kilmartin, Reynolds, Pagay, Nurgel, & Johnson, 2007). The wines can have very low concentrations of polyphenols and thus limited potential for the formation of quinones and removal of aroma compounds and browning. A further outcome of picking frozen grapes is freeze-concentration of aroma compounds and their precursors, leading to a higher aroma intensity in wines.

23.5.2 Polyphenol oxidase and hyperoxygenation

PPO acts on phenolics such as the major hydroxycinnamic acid in grape juices, caftaric acid. The process occurs very rapidly with O_2 available but can be restricted through the application of SO_2 to the juice (Singleton, 1987). As indicated above, the quinones formed through the PPO enzyme activity combine with glutathione, other polyphenols, and various sulfur-containing aroma compounds (Cheynier et al., 1989). PPO enzymes may remain active in some wines for a short period of time, but only when SO_2 is absent (Traverso-Rueda & Singleton, 1973).

A different strategy to limit subsequent wine oxidation is to promote full oxidation of the must through PPO activity. The technique of hyperoxidation leads to precipitates of brown oxidized phenolic compounds, which can be removed prior to fermentation (Schneider, 1998). While this leads to wines with little propensity for oxidative browning, due to the removal of the main substrates of oxidation, modifications to wine aroma have been noted. For example, changes in concentrations of volatile compounds with Chardonnay were very significant following must hyperoxygenation (Cejudo-Bastante, Hermosín-Gutiérrez, Castro-Vázquez, & Pérez-Coello, 2011).

23.5.3 Changes in C6 aroma compounds postharvest

Harvesting operations are of particular importance to the family of C6 aromas compounds, given that these are largely formed by the action of lipoxygenase and O_2 on fatty acids derived from grape lipids postharvest. Enzymes are involved in the conversion of linoleic acid to "leaf aldehyde" (E)-2-hexenal and (Z)-3-hexenal. Further reduction steps produce the associated hexenols, along with hexanol, which is often the major C6 compound found in wines (Joslin & Ough, 1978). After alcoholic fermentation, most of the (E)-2-hexenal or (E)-2-hexenol is consumed and converted to other compounds. Green aromas are associated with these compounds, such as "leaf alcohol" ((Z)-3-hexenol) with a grassy odor, while the green character is less marked with hexanol (Benkwitz et al., 2012). There have been further indications that the green and herbaceous characters of these compounds are linked to a synergistic effect with methoxypyrazines (Escudero, Campo, Fariña, Cacho, & Ferreira, 2007).

The effect of the harvesting technique, and the difference in aroma compound concentration in wines made with machine-harvested or hand-picked fruit, is quite minor for most aroma classes, including esters, higher alcohols, terpenes, and methoxypyrazines (Kilmartin, Makhotkina, Araujo, & Homer, 2015). This is not the case with the C6 compounds, where much higher concentrations of (Z)-3-hexenol and hexanol were seen in wines made from machine-harvested compared to hand-picked grapes, and for wines made from more highly pressed juices (Herbst-Johnstone et al., 2013). The only esters that increased under these conditions were (Z)-3-hexenol acetate and hexyl acetate, which are formed during fermentation from the respective C6 alcohols. The concentration of hexanol has been observed to be higher in juices that have been more exposed to oxidative conditions (Kilmartin et al., 2015).

23.5.4 Varietal thiol formation

The C6 compounds are important also as precursors to the varietal thiols 3MH and 3-mercaptohexyl acetate (3MHA). These thiol compounds have diverse aroma descriptors, such as passion fruit, grapefruit, and tropical, through to greener notes such as tomato stalk, boxwood, and sweaty armpit. This diversity is reflected in the range of responses and perceived intensities encountered when training Descriptive Analysis sensory panels using these compounds as aroma standards (Kilmartin et al., 2015). Higher concentrations of 3MH and 3MHA, by threefold or more, have been obtained with machine harvesting, or with destemming and crushing of hand-picked grapes, compared to whole-bunch pressing of the same fruit (Jouanneau, 2011).

One formation mechanism occurs via glutathione and cysteine conjugates, labeled as Glut-3MH and Cys-3MH, respectively, which can be cleaved to yield 3MH during yeast fermentation and form 3MHA through a further acetylation step (Peyrot Des Gachons, Tominaga, & Dubourdieu, 2002; Tominaga et al., 1998). The varietal thiol 4-mercapto-4-methylpentan-2-one (4MMP) is formed by an analogous pathway. Higher levels of Glut-3MH and Cys-3MH were seen to occur as a result of long-distance trucking of machine-harvested grapes, with lower levels formed from hand-picked grapes (Capone & Jeffery, 2011). However, lower levels of 3MH have been observed in wines made from more heavily pressed juices, despite these same juices containing the highest concentrations of Glut-3MH and Cys-3MH (Allen et al., 2011). 3MH can also be formed during alcoholic fermentation via the direct addition of H_2S to (E)-2-hexenal (R. Schneider, Charrier, Razungles, & Baumes, 2006), or to (E)-2-hexenol (Harsch et al., 2013). The addition of 0.1 mg/L NaSH early in fermentation, equivalent to 42 µg/L H_2S, increased the 3MH concentration from around 1600 to 3200 ng/L (Harsch et al., 2013). Even larger amounts of H_2S have been observed to form during fermentation (Ugliano, Kolouchova, & Henschke, 2011), a good part of which is removed by CO_2 purging during fermentation, showing the viability of this mechanism for 3MH formation, particularly under reductive winemaking conditions.

One source of H_2S prefermentation comes from the reduction of elemental sulfur residues that are present on grape skins and leaves, and which originate from fertilizer and fungicidal applications. The use of foliar sprays that include sulfur has been shown to increase 3MH in wines (Geffroy et al., 2017; Lacroux et al., 2008). An improvement in vine and must nitrogen status, leading to the greater formation of 3MH conjugates, was viewed as the cause of the increase in varietal thiols. Controlled additions of elemental sulfur postharvest have shown considerable increases in 3MH and 3MHA in Sauvignon blanc wine trials (Araujo et al., 2017). The persistence of higher concentrations of H_2S in some of the trial wines and links established between elemental sulfur and its conversion to H_2S during fermentation (Kwasniewski, Sacks, & Wilcox, 2014) support the mechanism where H_2S can combine with certain C6 compounds early in fermentation to form higher levels of 3MH (Harsch et al., 2013).

23.6 Applications of antioxidants at harvest

23.6.1 Sulfur dioxide

Sulfur dioxide (SO_2) additions at harvest, particularly with machine-harvested grapes, have an established place in minimizing quality loss, in part through limiting the growth of unwanted yeast and bacteria (Benedict, Morris, Fleming, & McCaskill, 1973; Morris, Cawthon, & Fleming, 1979). A second major effect of added SO_2 is lessening juice oxidation brought about by the action of PPO through enzyme inhibition and removal of polyphenols quinones before they react with further juice compounds. This is particularly important when grapes are harvested at higher midday temperatures. In the context of varietal thiol formation, juice oxidation has been shown to be critical to the potential of wines to develop aroma profiles associated with 3MH and 3MHA. In one study containing different Sauvignon blanc grapes from different vineyard sites in Marlborough, New Zealand, and at different points in prefermentation winery operations, particular juices were found to produce lower-than-expected varietal thiol concentrations (Allen et al., 2011). There was no correlation between 3MH formation in the wine and the concentrations of the conjugates Cys-3MH and Glut-3MH in the juices. Instead, the extent of juice oxidation, provided by the absorbance at 420 nm, was most important, and whenever this value exceeded 0.2 units, pointing to a relatively brown juice, the concentration of 3MH in the finished wine stayed below 1000 ng/L. This is a quite modest result for this style of wine, where values well in excess of 10,000 ng/L are often seen (Benkwitz et al., 2012).

Among the tools available to the winemaker to limit oxidation processes, antioxidant additions are among the easiest to apply. To determine the effectiveness of sulfites to maintain varietal thiol aroma potential in juices, a series of antioxidant harvesting trials were applied as a follow-up to the above study (Kilmartin et al., 2015). In this context, the application of sulfites at harvest with Marlborough Sauvignon has historically been quite varied, with rates ranging from no application at all, through to additions in the range of 60–90 ppm of SO_2.

In controlled studies, an addition of 120 ppm of SO_2 was found to maximize the production of 3MH (and 3MHA) in several trials. Relative concentrations of 3MH and 3MHA for 120 ppm of SO_2, alongside 0 and 60 ppm additions, are presented in Fig. 23.3. This figure shows averages for wines made from mechanically

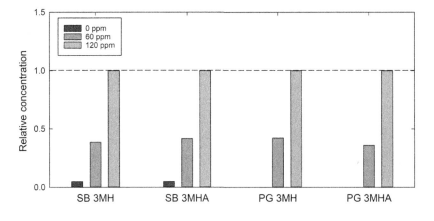

FIGURE 23.3 Relative concentrations of varietal thiols in Sauvignon blanc (SB) and Pinot gris (PG) wines produced with three different sulfite additions at harvest, and from three different sites for each variety. A value of 1 is ascribed to the wines from treatments with 120 ppm SO_2 additions. For each wine aroma compound the bars read from left to right for 0, 60 and 120 ppm additions of SO_2.

harvested fruit from three Sauvignon blanc sites, as previously published (Makhotkina, Herbst-Johnstone, Logan, Du Toit, & Kilmartin, 2013), and from hand-picked grapes that were then vigorously macerated from three Pinot gris sites, taken from a more recent unpublished trial at the University of Auckland. The results are presented in relative terms versus the concentrations of the varietal thiols found with 120-ppm SO_2 additions. The ultimate varietal thiol formation potential varied considerably between juices. The maximum values for 3MH (and 3MHA in brackets) for the wines with 120-ppm SO_2 additions were 1044 (76), 1444 (333), and 5367 (820) ng/L for the three Sauvignon blanc wines; and 2743 (947), 6260 (1672), and 12,111 (2401) ng/L for the three Pinot gris wines. These values represent a range of thiol concentrations, as typically seen in Sauvignon blanc wines, and are all well above the perception thresholds of 60 ng/L for 3MH and 4 ng/L for 3MHA (Benkwitz et al., 2012). With 60-ppm SO_2 additions, the 3MH, and 3MHA concentrations were 36%−42% of the 120-ppm addition values, and among the polyphenols present, higher GRP was observed at the expense of caftaric acid, pointing to the reaction of glutathione with caftaric acid. With no added sulfites, low levels of 3MH were seen with the Sauvignon blanc wines (23−250 ng/L), while in the Pinot gris wines, 3MH could not be detected in any of the wines.

In addition to showing the importance of antioxidant protection of the juice to varietal thiol formation, these trials have shown the ability of juices other than Sauvignon blanc to produce elevated levels of 3MH and 3MHA. There is understandably some reluctance to applying high SO_2 additions to Pinot gris fruit, given problems with pinking that can arise. However, a wine that combines traditional Pinot gris characters, with an added thiol dimension, can be produced with adequate protection against oxidation during harvest. It appears that maximum amounts of 3MH and 3MHA can be formed so long as sufficient "genuine" free SO_2 (>10 mg/L, with 420 nm absorbance < 0.1 units) can be maintained in the juices prior to fermentation. The term "genuine" has been added here, in recognition that assays for SO_2, such as the aspiration method, may include weakly bound sulfites in the "free SO_2" value, which do not limit juice oxidation. Extreme additions of SO_2 should nevertheless be avoided, given that significant delays in the onset of fermentation can occur (Makhotkina et al., 2013).

In addition to the amount of SO_2 applied, the timing of applications and the sulfite formulation are further choices available to the winemaker. Concentrated sulfite solutions can be added to the grapes in gondolas or trucks, often dispensed from two-liter plastic bottles. A drip feed of the sulfite solution on the harvester can also be made and would provide the most immediate protection. The sprinkling of the powder form of potassium metabisulfite (PMS) on top of grapes in a truckload can also be used. However, questions remain about how evenly the SO_2 spreads throughout the mass of grapes, as juice with less free SO_2 will have a lower thiol potential.

As to the timing of the application, the need to produce a good pool of C6 compounds in the juice from lipoxygenase activity would suggest that some delay in SO_2 application might assist in increasing the 3MH and 3MHA aroma potential. On the contrary, in a trial in which applications were made at various time points from as soon as feasible after mechanical harvesting through to a 2-hour delay, the 3MH and 3MHA concentrations declined progressively with the delay time (Fig. 23.4). For a 2-hour delay in application, the levels of the 3MH and 3MHA were lowered by half (Kilmartin et al., 2015). Very few differences were noted with the other wine aroma compounds analyzed, with some small changes noted with the C6 compounds present in the wines (Fig. 23.4).

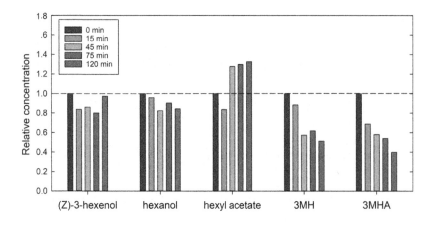

FIGURE 23.4 Relative concentrations of varietal thiols and C6 compounds in Sauvignon blanc wines produced at different time points for the addition of 50 ppm SO_2. A value of 1 is ascribed to the treatment with immediate SO_2 additions. For each aroma compound the bars read from left to right for 0 through to 120 min of delay prior to the antioxidant additions.

23.6.2 Glutathione and ascorbic acid

Further antioxidants that can potentially be applied to grape juices postharvest include ascorbic acid and glutathione, both of which occur naturally in grapes. Ascorbic acid has been used widely in New Zealand for this purpose, but the option to consider glutathione has only arisen recently following the approval of additions up to 20 mg/L of reduced glutathione to musts and wines by the OIV (RESOLUTION OIV-OENO 445-2015). The addition of glutathione immediately prior to fermentation has been considered, but in one trial, concentrations of 3MH were typically a few percent lower following a 67 mg/L addition of glutathione (Patel et al., 2010).

To avoid adding higher amounts of SO_2, both for health reasons and to avoid delays at the start of fermentation, further harvesting trials have looked at additions of 100 ppm of either ascorbic acid or glutathione alongside a "suboptimal" application of SO_2 at 30 ppm (Fig. 23.5).

Increases in 3MH and 3MHA concentrations of 50%−150% were observed in trials on Sauvignon blanc with both antioxidants (Makhotkina et al., 2014; Rosales del Prado, 2014). Ascorbic acid and glutathione could limit the formation and/or persistence of oxidized polyphenol quinones that would otherwise react with H_2S both prior to fermentation and in its early stages. This protection would allow 3MH to form by the reaction of H_2S with juice C6 compounds more readily (Harsch et al., 2013). In the second trial, 4MMP concentrations were also monitored, and a greater increase was seen with the addition of glutathione at harvest, but less of an effect with ascorbic acid (Makhotkina et al., 2014). In this case, the added glutathione could have increased the formation of the 4MMP glutathione conjugate as a precursor, with subsequent release of 4MMP during fermentation.

23.7 Concluding remarks

Controlling oxidation remains of prime importance for the production of quality white wines. The extents of exposure to O_2 at all stages of winemaking, from harvesting to storage in bottles, are choices available to the winemaker, along with the use of antioxidant additives. Wines can be readily directed to extremes that are more "reductive" through the exclusion of O_2, through to "oxidative" (aldehydic) from an oversupply of O_2. Further approaches to limit oxidation can be applied, such as the placement of wine on yeast lees to lower O_2 levels, fining of phenolics, and removal of catalytic metals with ion exchangers. In each case, full consideration needs to be given to the further effects of these procedures on wine quality.

Oxidation control is central to decisions on the extent of tropical/green aromas in white wines, associated with the varietal thiols 3MH and 3MHA to suit the needs of different consumer markets. A thiol-driven wine can be obtained through vigorous fruit maceration coupled with adequate antioxidant protection, from harvest to fermentation. The presence of residues of elemental sulfur can also elevate levels of varietal thiols, so long as care is taken to limit reductive aromas, including the removal of H_2S once 3MH and 3MHA have formed.

Conversely, to obtain wines with lower levels of 3MH and 3MHA, sulfite applications at harvest can be kept low, and a sufficiently long holding time maintained between the last application of elemental sulfur in the vineyard and harvesting date. Under these conditions, the remaining aroma compounds, which are minimally affected by grape handling and oxidation status, can have their full impact. It can be argued that this provides a better representation of the varietal expression of the grape, which is important for varieties such as Pinot gris and Chardonnay, and even with Sauvignon blanc while acknowledging the commercial success of the high

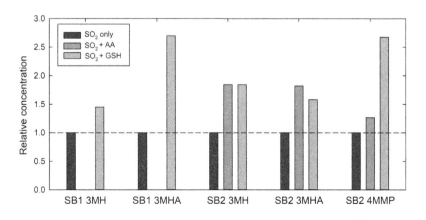

FIGURE 23.5 Relative concentrations of varietal thiols in wines from two Sauvignon blanc trials, each with grapes from three different sites. All juices received 30-ppm SO_2, while further treatments received 100 ppm of either ascorbic acid (AA) or glutathione (GSH). A value of 1 is ascribed to the treatment with only SO_2 added. For each trial wine the bars read from left to right as SO_2 only, SO_2 + AA, SO_2 + GSH.

thiol-driven style. Intermediate styles with a moderate "varietal" thiol contribution can also be considered, including blending of wines made with different harvest/oxidation strategies.

References

Allen, T., Herbst-Johnstone, M., Girault, M., Butler, P., Logan, G., Jouanneau, S., ... Kilmartin, P. A. (2011). Influence of grape-harvesting steps on varietal thiol aromas in Sauvignon blanc wines. *Journal of Agricultural and Food Chemistry, 59*(19), 10641−10650. Available from https://doi.org/10.1021/jf2018676.

Araujo, L. D., Vannevel, S., Buica, A., Callerot, S., Fedrizzi, B., Kilmartin, P. A., & du Toit, W. J. (2017). Indications of the prominent role of elemental sulfur in the formation of the varietal thiol 3-mercaptohexanol in Sauvignon blanc wine. *Food Research International (Ottawa, Ont.), 98*, 79−86. Available from https://doi.org/10.1016/j.foodres.2016.12.023.

Aronson, J., & Ebeler, S. E. (2004). Effect of polyphenol compounds on the headspace volatility of flavors. *American Journal of Enology and Viticulture, 55*, 13−21.

Barril, C., Rutledge, D. N., Scollary, G. R., & Clark, A. C. (2016). Ascorbic acid and white wine production: A reivew of beneficial versus detrimental impacts. *Australian Journal of Grape and Wine Research, 22*, 169−181.

Benedict, R. H., Morris, J. R., Fleming, J. W., & McCaskill, D. R. (1973). Effects of temperature on quality of mechanically harvested "Concord" grapes. Arkansas farm. *Research, 22*, 2.

Benitez, P., Castro, R., & Barroso, G. C. (2002). Removal of iron, copper and manganese from white wines through ion exchange techniques: Effects on their organoleptic characteristics and susceptibility to browning. *Analytica Chimica Acta, 458*(1), 197−202.

Benkwitz, F., Tominaga, T., Kilmartin, P. A., Lund, C., Wohlers, M., & Nicolau, L. (2012). Identifying the chemical composition related to the distinct aroma characteristics of New Zealand Sauvignon blanc wines. *American Journal of Enology and Viticulture, 63*, 62−72.

Blanchard, L., Darriet, P., & Dubourdieu, D. (2004). Reactivity of 3-mercaptohexanol in red wine: Impact of oxygen, phenolic fractions, and sulfur dioxide. *American Journal of Enology and Viticulture, 55*, 115−120.

Capone, D. L., & Jeffery, D. W. (2011). Effects of transporting and processing Sauvignon blanc grapes on 3-mercaptohexan-1-ol precursor concentrations. *Journal of Agricultural and Food Chemistry, 59*(9), 4659−4667. Available from https://doi.org/10.1021/jf200119z.

Cejudo-Bastante, M. J., Hermosín-Gutiérrez, I., Castro-Vázquez, L. I., & Pérez-Coello, M. S. (2011). Hyperoxygenation and bottle storage of Chardonnay white wines: Effects on color-related phenolics, volatile composition, and sensory characteristics. *Journal of Agricultural and Food Chemistry, 59*(8), 4171−4182. Available from https://doi.org/10.1021/jf104744q.

Cheynier, V., Basire, N., & Rigaud, J. (1989). Mechanism of trans-caffeoyltartaric acid and catechin oxidation in model solutions containing grape polyphenoloxidase. *Journal of Agricultural and Food Chemistry, 37*, 1069−1071.

Cheynier, V. F., & Van Hulst, M. W. J. (1988). Oxidation of trans-caftaric acid and 2-Sglutathionylcaftaric acid in model solutions. *Journal of Agricultural and Food Chemistry, 36*, 10−15.

Coetzee, C., Van Wyngaard, E., Šuklje, K., Ferreira, A. C. S., & du Toit, W. J. (2016). Chemical and sensory study on the evolution of aromatic and nonaromatic compounds during the progressive oxidative storage of a Sauvignon blanc wine. *Journal of Agricultural and Food Chemistry, 64*(42), 7979−7993. Available from https://doi.org/10.1021/acs.jafc.6b02174.

Danilewicz, J. C. (2003). Review of reaction mechanisms of oxygen and proposed intermediate reduction products in wine: Central role of iron and copper. *American Journal of Enology and Viticulture, 54*, 73−85.

Danilewicz, J. C. (2007). Interaction of sulfur dioxide, polyphenols, and oxygen in a winemodel system: central role of iron and copper. *American Journal of Enology and Viticulture, 58*, 53−60.

Danilewicz, J. C. (2013). Reactions involving iron in mediating catechol oxidation in model wine. *American Journal of Enology and Viticulture, 64*, 316−324.

Danilewicz, J. C. (2016). Fe(II):Fe(III) ratio and redox status of white wines. *American Journal of Enology and Viticulture, 67*, 146−152.

Danilewicz, J. C., Seccombe, J. T., & Whelan, J. (2008). Mechanism of interaction of polyphenols, oxygen, and sulfur dioxide in model wine and wine. *American Journal of Enology and Viticulture, 59*, 128−136.

Du Toit, W. J., Marais, J., Pretorius, I. S., & Du Toit, M. (2006). Oxygen in must and wine: A review. *South African Journal of Enology and Viticulture, 27*, 76−94.

Du Toit, W. J., Lisjak, K., Stander, M., & Prevoo, D. (2007). Using LC-MSMS to assess glutathione levels in South African white grape juices and wines made with different levels of oxygen. *Journal of Agricultural and Food Chemistry, 55*(8), 2765−2769.

Dubourdieu, D., Moine-Ledoux, V., Lavigne-Cruege, V., Blanchard, L., & Tominaga, T. (2000). Recent advances in white wine aging: the key role of the lees. In J. M. Rantz (Ed.), *Presented at the ASEV 50th Anniversary Annual Meeting*. Seattle, WA: American Society for Enology and Viticulture.

Escudero, A., Campo, E., Fariña, L., Cacho, J., & Ferreira, V. (2007). Analytical characterization of the aroma of five premium red wines. Insights into the role of odor families and the concept of fruitiness of wines. *Journal of Agricultural and Food Chemistry, 55*(11), 4501−4510.

Ferreira, A. C. S., de Pinho, P. G., Rodrigues, P., & Hogg, T. (2002). Kinetics of oxidative degradation of white wines and how they are affected by selected technological parameters. *Journal of Agricultural and Food Chemistry, 50*(21), 5919−5924.

Ferreira, V., Carrascon, V., Bueno, M., Ugliano, M., & Fernandez-Zurbano, P. (2015). Oxygen consumption by red wines. Part I: Consumption rates, relationship with chemical composition, and role of SO_2. *Journal of Agricultural and Food Chemistry, 63*(51), 10928−10937. Available from https://doi.org/10.1021/acs.jafc.5b02988.

Franco-Luesma, E., & Ferreira, V. (2016). Formation and release of H2S, methanethiol, and dimethylsulfide during the anoxic storage of wines at room temperature. *Journal of Agricultural and Food Chemistry, 64*(32), 6317−6326. Available from https://doi.org/10.1021/acs.jafc.6b01638.

Geffroy, O., Charrier, F., Poupault, P., Schneider, R., Lopez, R., Gontier, L., & Dufourcq, T. (2017). Boosting varietal thiols in white and rose wines through foliar nitrogen and sulfur spraying. In *Presented at the proceedings of the sixteenth Australian wine industry technical conference, 2016*. Adelaide, Australia.

George, N., Clark, A. C., Prenzler, P. D., & Scollary, G. R. (2006). Factors influencing the production and stability of xanthylium cation pigments in a model white wine system. *Australian Journal of Grape and Wine Research, 12*, 57−68.

Grant-Preece, P., Schmidtke, L. M., Barril, C., & Clark, A. C. (2017). Photoproduction of glyoxylic acid in model wine: Impact of sulfur dioxide, caffeic acid, pH and temperature. *Food Chemistry, 215*, 292−300. Available from https://doi.org/10.1016/j.foodchem.2016.07.131.

Harsch, M. J., Benkwitz, F., Frost, A., Colonna-Ceccaldi, B., Gardner, R. C., & Salmon, J.-M. (2013). New precursor of 3-mercaptohexan-1-ol in grape juice: Thiol-forming potential and kinetics during early stages of must fermentation. *Journal of Agricultural and Food Chemistry, 61*(15), 3703−3713. Available from https://doi.org/10.1021/jf3048753.

Herbst-Johnstone, M., Araujo, L. D., Allen, T. A., Logan, G., Nicolau, L., & Kilmartin, P. A. (2013). Effects of mechanical harvesting on Sauvignon blanc aroma. *Acta Horticulturae, 978*, 179−186.

Joslin, W. S., & Ough, C. S. (1978). Cause and fate of certain C6 compounds formed enzymatically in macerated grape leaves during harvest and wine fermentation. *American Journal of Enology and Viticulture, 29*, 11−17.

Jouanneau, S. (2011). *Survey of aroma compounds in Marlborough Sauvignon blanc wines. Regionality and small scale winemaking* (PhD Thesis). Auckland, New Zealand: The University of Auckland.

Kilmartin, P. A., Makhotkina, O., Araujo, L. D., & Homer, J. A. (2015). Influence of antioxidant additions at harvest on Sauvignon blanc wine aroma. *ACS Symposium Series, 1203*, 217−227.

Kilmartin, P. A., Reynolds, A. G., Pagay, V., Nurgel, C., & Johnson, R. (2007). Polyphenol content and browning of Canadian icewines. *Journal of Food Agriculture and Environment, 5*, 52−57.

Kreitman, G. Y., Danilewicz, J. C., Jeffery, D. W., & Elias, R. J. (2016). Reaction mechanisms of metals with hydrogen sulfide and thiols in model wine. Part 1: Copper-catalyzed oxidation. *Journal of Agricultural and Food Chemistry, 64*(20), 4095−4104. Available from https://doi.org/10.1021/acs.jafc.6b00641.

Kwasniewski, M. T., Sacks, G. L., & Wilcox, W. F. (2014). Persistence of elemental sulfur spray residue on grapes during ripening and vinification. *American Journal of Enology and Viticulture, 65*, 453−462.

Lacroux, F., Tregoat, O., Van Leeuwen, C., Pons, A., Tominaga, T., Lavigne-Cruege, V., & Dubourdieu, D. (2008). Effect of foliar nitrogen and sulphur application on aromatic expression of *Vitis vinifera* L. cv. Sauvignon blanc. *Journal International Des Sciences de La Vigne et Du Vin, 42*, 125−132.

Lopes, P., Silva, M. A., Pons, A., Tominaga, T., Lavigne, V., Saucier, C., ... Dubourdieu, D. (2009). Impact of oxygen dissolved at bottling and transmitted through closures on the composition and sensory properties of a Sauvignon blanc. wine during bottle storage. *Journal of Agricultural and Food Chemistry, 57*(21), 10261−10270. Available from https://doi.org/10.1021/jf9023257.

Lund, C. M., Winz, R., Gardner, R. C., & Kilmartin, P. A. (2009). Effect of polyphenols on the perception of key aroma compounds from Sauvignon blanc. wine. *Australian Journal of Grape and Wine Research, 15*, 18−26.

Maggu, M., Winz, R., Kilmartin, P. A., Trought, M. C. T., & Nicolau, L. (2007). Effect of skin contact and pressure on the composition of Sauvignon blanc must. *Journal of Agricultural and Food Chemistry, 55*(25), 10281−10288.

Makhotkina, O., Araujo, L. D., Olejar, K., Herbst-Johnstone, M., Fedrizzi, B., & Kilmartin, P. A. (2014). Aroma impact of ascorbic acid and glutathione additions to Sauvignon blanc at harvest to supplement sulfur dioxide. *American Journal of Enology and Viticulture, 65*, 388−393.

Makhotkina, O., Herbst-Johnstone, M., Logan, G., Du Toit, W., & Kilmartin, P. A. (2013). Influence of sulfur dioxide additions at harvest on polyphenols, C6-compounds and varietal thiols in Sauvignon blanc. *American Journal of Enology and Viticulture, 64*, 203−213.

Makhotkina, O., & Kilmartin, P. A. (2009). Uncovering the influence of antioxidants on polyphenol oxidation in wines using an electrochemical method: Cyclic voltammetry. *Journal of Electroanalytical Chemistry, 633*, 165−174.

Morozova, K., Schmidt, O., & Schwack, W. (2015). Effect of headspace volume, ascorbic acid and sulphur dioxide on oxidative status and sensory profile of Riesling wine. *European Food Research and Technology, 240*, 205−221.

Morris, J. R., Cawthon, D. L., & Fleming, J. W. (1979). Effects of temperature and SO2 addition on quality and postharvest behavior of mechanically-harvested juice grapes in Arkansas. *Journal of the American Society for Horticultural Science, 104*, 166−169.

Nikolantonaki, M., Jourdes, M., Shinoda, K., Teissedre, P.-L., Quideau, S., & Darriet, P. (2012). Identification of adducts between an odoriferous volatile thiol and oxidized grape phenolic compounds: Kinetic study of adduct formation under chemical and enzymatic oxidation conditions. *Journal of Agricultural and Food Chemistry, 60*(10), 2647−2656. Available from https://doi.org/10.1021/jf204295s.

Nikolantonaki, M., & Waterhouse, A. L. (2012). A method to quantify quinone reaction rates with wine relevant nucleophiles: a key to the understanding of oxidative loss of varietal thiols. *Journal of Agricultural and Food Chemistry, 60*(34), 8484−8491. Available from https://doi.org/10.1021/jf302017j.

Noble, A. C., Ough, C. S., & Kasimatis, A. N. (1975). Effect of leaf content and mechanical harvest on wine "quality.". *American Journal of Enology and Viticulture, 26*, 158−163.

Ough, C. S., Berg, H. W., Coffelt, R. J., & Cooke, G. M. (1971). The effect on wine quality of simulated mechanical harvest and gondola transport of grapes. *American Journal of Enology and Viticulture, 22*, 65−70.

Panero, L., Motto, S., Petrozziello, M., Guaita, M., & Bosso, A. (2015). Effect of SO2, reduced glutathione and ellagitannins on the shelf life of bottled white wines. *European Food Research and Technology, 240*, 345−356.

Patel, P., Herbst-Johnstone, M., Lee, S. A., Gardner, R. C., Weaver, R., Nicolau, L., & Kilmartin, P. A. (2010). Influence of juice pressing conditions on polyphenols, antioxidants, and varietal aroma of Sauvignon blanc microferments. *Journal of Agricultural and Food Chemistry, 58*(12), 7280−7288. Available from https://doi.org/10.1021/jf100200e.

Peyrot Des Gachons, C., Tominaga, T., & Dubourdieu, D. (2002). Sulfur aroma precursor present in S-glutathione conjugate form: identification of S-3-(hexan-1-ol)-glutathione in must from *Vitis vinifera* L. cv. Sauvignon blanc. *Journal of Agricultural and Food Chemistry, 50*(14), 4076−4079.

Pocock, K. F., & Waters, E. J. (1998). The effect of mechanical harvesting and transport of grapes, and juice oxidation, on the protein stability of wines. *Australian Journal of Grape and Wine Research, 4*, 136−139.

Rosales del Prado, D. (2014). *Effect of glutathione and inactive yeast additions to Sauvignon blanc at harvest on wine aroma* (MSc Thesis). Auckland, New Zealand: The University of Auckland.

Saucier, C. T., & Waterhouse, A. L. (1999). Synergetic activity of catechin and other antioxidants. *Journal of Agricultural and Food Chemistry, 47*(11), 4491−4494.

Schneider, R., Charrier, F., Razungles, A., & Baumes, R. (2006). Evidence for an alternative biogenetic pathway leading to 3-mercaptohexanol and 4-mercapto-4-methylpentan-2-one in wines. *Analytica Chimica Acta, 563*, 58−64.

Schneider, V. (1998). Must hyperoxidation: A review. *American Journal of Enology and Viticulture, 49*, 65−73.

Singleton, V. L. (1987). Oxygen with phenols and related reactions in musts, wines, and model systems: Observations and practical implications. *American Journal of Enology and Viticulture, 38*, 69−77.

Singleton, V. L., & Kramling, T. E. (1976). Browning of white wines and an accelerated test for browning capacity. *American Journal of Enology and Viticulture, 27*, 157−160.

Singleton, V. L., Orthofer, R., & Lamuela-Raventos, R. M. (1999). Analysis of total phenols and other oxidation substrates and antioxidants by means of Folin−Ciocalteu reagent. *Methods in Enzymology, 299*, 152−178.

Singleton, V. L., Salgues, M., Zaya, J., & Trousdale, E. (1985). Caftaric acid disappearance and conversion to products of enzymic oxidation in grape must and wine. *American Journal of Enology and Viticulture, 36*, 50−56.

Skouroumounis, G. K., Kwiatkowski, M. J., Francis, I. L., Oakey, H., Capone, D. L., Peng, Z., . . . Waters, E. J. (2005). The influence of ascorbic acid on the composition, colour and flavour properties of a Riesling and a wooded Chardonnay wine during five years' storage. *Australian Journal of Grape and Wine Research, 11*, 355−368.

Tominaga, T., Murat, M.-L., & Dubourdieu, D. (1998). Development of a method for analysing the volatile thiols involved in the characteristic aroma of wines made from *Vitis vinifera* L. Cv. Sauvignon Blanc. *Journal of Agricultural and Food Chemistry, 46*, 1044−1048.

Traverso-Rueda, S., & Singleton, V. L. (1973). Catecholase activity in grape juice and its implications in winemaking. *American Journal of Enology and Viticulture, 24*, 103−109.

Ugliano, M., Kolouchova, R., & Henschke, P. A. (2011). Occurrence of hydrogen sulfide in wine and in fermentation: Influence of yeast strain and supplementation of yeast available nitrogen. *Journal of Industrial Microbiology & Biotechnology, 38*(3), 423−429. Available from https://doi.org/10.1007/s10295-010-0786-6.

Waterhouse, A. L., & Laurie, V. F. (2006). Oxidation of wine phenolics: A critical evaluation and hypotheses. *American Journal of Enology and Viticulture, 57*, 306−313.

Webber, V., Dutra, S. V., Spinelli, F. R., Carnieli, G. J., Cardozo, A., & Vanderlinde, R. (2017). Effect of glutathione during bottle storage of sparkling wine. *Food Chemistry, 216*, 254−259. Available from https://doi.org/10.1016/j.foodchem.2016.08.042.

Wildenradt, H. L., & Singleton, V. L. (1974). The production of aldehydes as a result of oxidation of polyphenolic compounds and its relation to wine aging. *American Journal of Viticulture and Enology, 25*, 119−126.

24

A glance into the aroma of white wine

K. Chen and J. Li

College of Food Science & Nutritional Engineering, China Agricultural University, Beijing, P. R. China

24.1 Introduction

Aroma is one of the important sensory characters of white wine. More than 1300 kinds of substances were determined to make up the aroma profile of white wine. The content, properties, and balance of volatile compounds can jointly influence the white wine quality. In general, the volatile compounds in white wine are mainly affected by grape varieties, ecological conditions, ripening quality, and fermentation technology. Especially, the volatile compounds directly originating from grapes play an important role in the typicality of grape varieties and the style of the winemaking area. Moreover, the volatile compounds developed in fermentation can supply and modify the varietal aroma. The developments of aroma in grapes and the winemaking process are very complex and have been studied for many years. Although the mechanism of development of aroma in wine is still not very clear, with the continuous development and innovation of instrument analysis and other techniques, the research of grape and wine aroma has been made remarkable progress (Chen, Wen, Ma, Wen, & Li, 2019; Rapp, 1998). The purpose of this review is to provide a quick scan of the major groups of volatile compounds in white wine by summarizing the relevant aroma studies and elaborating on the impact factors on the evaluation of aroma profile.

24.2 Source of volatile compounds in white wine

At present, more than 800 kinds of volatile compounds in white wine are mainly composed of alcohols, aldehydes, ketones, acids, esters, and other components (Ebeler, 2001). Their contents range from gram per liter to nanogram per liter or are even lower than the limit of detection (Mateo & Jiménez, 2000). There are various sources of volatile compounds in white wine, which are not only directly from grape berries but also from different stages of the winemaking process, which include the volatile compounds produced by raw materials harvesting, crushing, pressing, alcohol fermentation (AF), malolactic acid fermentation, wine aging, and bottle storage.

24.2.1 Varietal aroma

The volatile compounds directly derived from grapefruit are called varietal aroma and determine the varietal typicality and original style of white wine (Fig. 24.1) (Silva Ferreira, Monteiro, Oliveira, & Guedes de Pinho, 2008). Terpenes have a strong aroma and low sensory thresholds (Mateo & Jiménez, 2000). They are the typical aroma of the *Muscat* grape and its wines. According to the content of terpenoids in white grapefruit, European varieties are classified into three categories: intensely flavored *Muscat* (total free monoterpene concentration is higher than 6 mg/L), non-*Muscat* but aromatic varieties with total free monoterpene concentration of 1−4 mg/L, and neutral varieties, which are not dependent upon monoterpenes for their flavor (Table 24.1) (Mateo & Jiménez, 2000). Terpenes are secondary metabolites synthesized by acetyl-CoA in plants (Fischer, 2007). They exist in grapefruit in free forms or tasteless glycosidic forms, and the highest concentration of terpenes is found

© 2022 Elsevier Inc. All rights reserved.

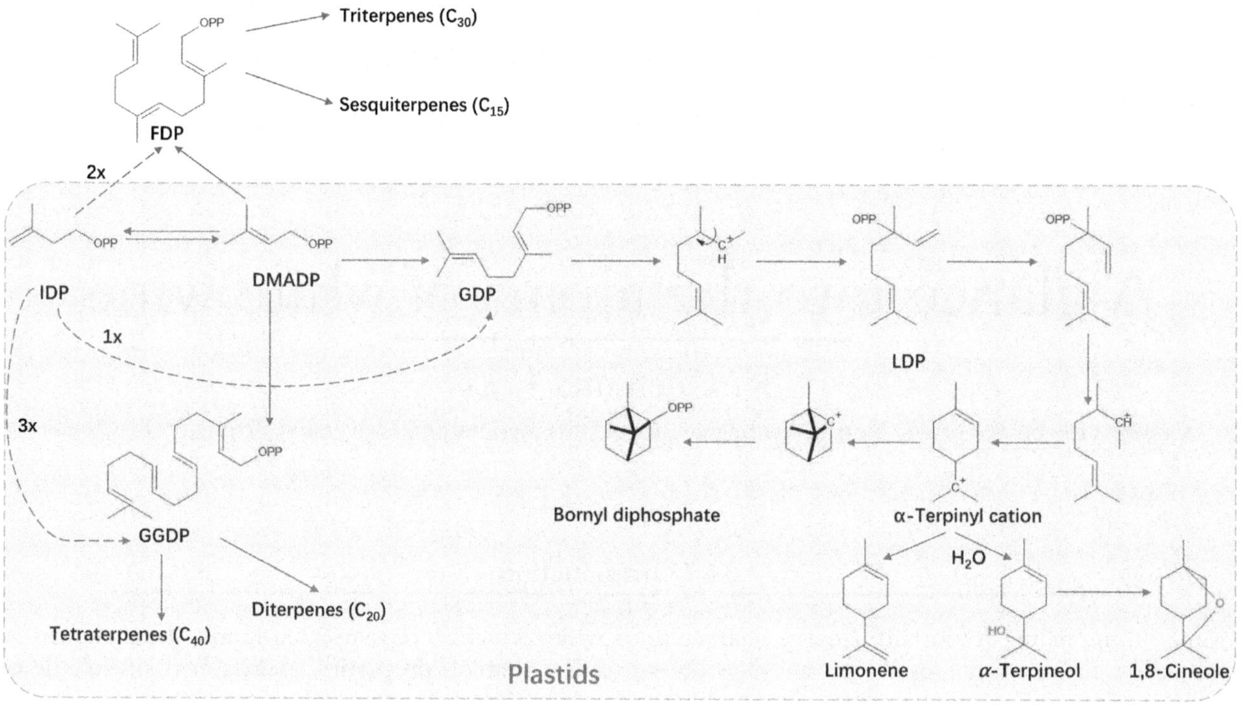

FIGURE 24.1   **Biosynthesis of representative monoterpenes from GDP in grape.** *FDP*, farnesyl diphosphate; *GGDP*, geranylgeranyl diphosphate; *GDP*, geranyl diphosphate; *LDP*, linalyl diphosphate; *DMADP*, dimethylallyl diphosphate; *IDP*, isopentenyl diphosphate. *Source: From* © *Bohlmann, J., Meyer-Gauen, G., & Croteau, R. (1998). Plant terpenoid synthases: Molecular biology and phylogenetic analysis. Proceedings of the National Academy of Sciences, 95(8), 4126–4133. https://doi.org/10.1073/pnas.95.8.4126.*

TABLE 24.1   Classification of some white varieties based on monoterpene contents.

Muscat varieties	Non-*Muscat* aromatic varieties	Neutral varieties
Canada Muscat	Traminer	Bacchus
Gewurztraminer	Huxel	Chardonnay
Muscat of Alexandria	Kerner	Chasselas
Muscat de Frontignan	Morio-Muskat	Chenin blanc
Muscato Bianco del Piemonte	Muller-Thurgau	Clairette
Muscat Ottonel	Riesling	Nobling
Moscato Italiano	Schurebe	Rkatsiteli
	Siegerebe	Sauvignon blanc
	Sylvaner	Semillon
	Wurzer	Sultana
	Italian Riesling	Trebbiano
		Verdelho
		Viognier
		Vidal blanc

in the grape skin (Cabrita, Costa Freitas, Laureano, & Di Stefano, 2006). Although it was found that some terpenes can also be synthesized by microorganisms, a few studies have reported on the natural production of terpenes by *Saccharomyces cerevisiae*. More than 70 kinds of terpenes have been identified in grapes, which are mainly composed of terpenols, esters, aldehydes, acids, monomers, polymers, and glycosides. Among grapes and wines, linalool, geraniol, nerol, citronellol, and α-terpineol are the most common compounds with *Muscat* aroma, among which geraniol and linalool are regarded as the most fragrant (Mateo & Jiménez, 2000).

24.2.2 C_{13}-norisoprenoids

Carotenoids are a type of tetraterpenoids that accumulate in the plastids of leaves, flowers, and fruits, where they contribute to red, orange, and yellow coloration, respectively (Franco, Peinado, Medina, & Moreno, 2004). Additionally, carotenoids are the precursors of apocarotenoids, visual and signaling molecules in retinal (Chen et al., 2019). Carotenoids and the terpenes containing 40 carbon atoms are oxidatively degraded to the series of derivatives of 9, 10, 11, or 13 carbon atoms (Rodríguez-Bustamante & Sánchez, 2007). Among these compounds, C_{13}-norisoprenoids derivatives with 13 carbon atoms have interesting flavor characteristics (Fig. 24.2). For example, *Riesling* wine can produce kerosene odor as cold climate is suitable for the development of 1,2-dihydro-1,1,6-trimethylnaphthalene (TDN) (Bindon, Dry, & Loveys, 2007; Schwab, Davidovich-Rikanati, & Lewinsohn, 2008), which contributes the kerosene odor, and its perception threshold is 20 μg/L (Gunata, Bayonove, Baumes, & Cordonnier, 1985; Khairallah, Reynolds, & Bowen, 2016). Moreover, the most common type of C_{13}-norisoprenoids derivatives is the oxidation of C_{13}-norisoprenoids derivatives at the C_7 or C_9 position (Yuan & Qian, 2016). β-Damascenone, which has complex floral, tropical fruit or boiled apple aroma (D. Tieman et al., 2006). Its perception threshold is very low in water (3−4 ng/L), 40−50 ng/L in model liquor solution, and 5000 ng/L in wine (Rapp, 1998). The substance of β-damascenone was originally identified in *Riesling* and *Scheurebe* grape juices (Lashbrooke, Young, Dockrall, Vasanth, & Vivier, 2013). β-Damascenone is directly derived from grapefruit, most of which exist in a tasteless glycoside binding state, and is not metabolized by yeast during fermentation (Silva Ferreira et al., 2008). Like β-damascenone, β-ionone is also present in all grape varieties and especially has a high concentration in the Muscat variety. β-Ionone has violet flavor. The sensory thresholds in water and diluted simulated liquor solutions are 120 ng/L and 800 ng/L, respectively, and the perception thresholds in liquor are 1500 ng/L (Fan, Xu, Jiang, & Li, 2010).

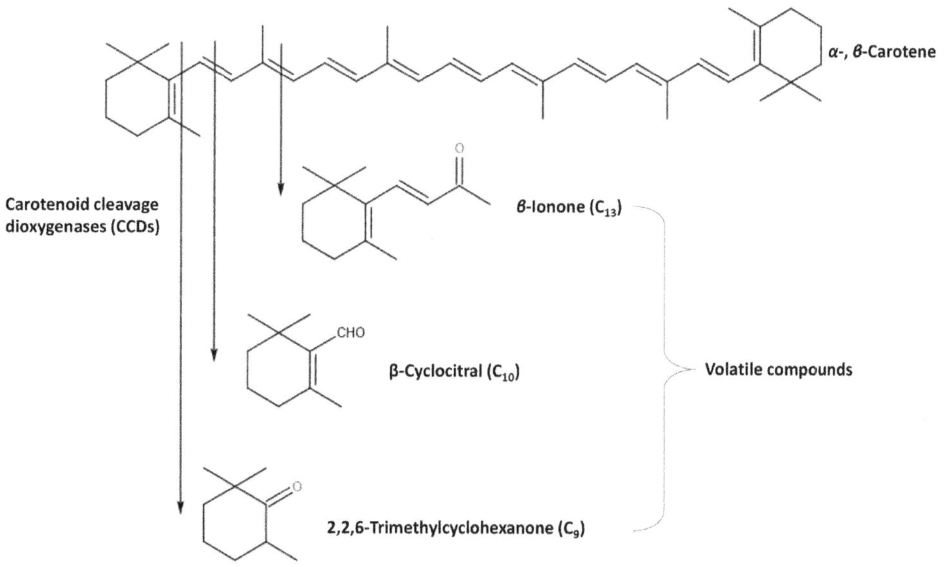

FIGURE 24.2 **Biosynthetic route for C_{13}-norisoprenoids and other aroma compounds in plants.** *ABA,* abscisic acid; *FPP,* farnesyl phosphate. Source: *From © E. Rodríguez-Bustamante & S. Sánchez (2007) Microbial Production of C13-Norisoprenoids and Other Aroma Compounds via Carotenoid Cleavage,* Critical Reviews in Microbiology, 33(3), 211−230. https://doi.org/10.1080/10408410701473306.

24.2.3 Methoxypyrazine

Methoxypyrazine, a nitrogen-containing heterocyclic compound formed by amino acid metabolism, which is typically characterized by green pepper flavor. 2-Methoxy-3-isobutylpyrazine is mainly found in the fruit stem of *Sauvignon blanc* (Hashizume & Samuta, 1997). In addition, 2-methoxy-3-isopentylpyrazine, 2-methoxy-3-isopropylpyrazine, 2-methoxy-3-sec-butyl pyrazine, and 2-methoxy-3-isopropylpyrazine were also identified in many grape varieties and wines (*Pinot Noir, Gewürztraminer, Chardonnay, Riesling*, etc.) (Roujou de Boubee, Van Leeuwen, & Dubourdieu, 2000). Moreover, studies have found changes in 2-methoxy-3-isobutylpyrazine concentration as the ripening of the grapes is affected by the environmental and cultural conditions (soil, climate, training system, etc.). In Bordeaux, the strong smell of green pepper indicates the poor ripeness of grape berry. In white wine, the smell of green pepper normally indicates aroma defect to some extent (Roujou de Boubee et al., 2000).

24.2.4 Volatile thiols

Most of sulfur containing compounds in wine are usually related to flavor defects, only semi-volatile thiols can largely contribute to particular aroma of Sauvignon Blanc wine and other white wines, for example, 4-mercapto-4-methylpentan-2-one (4MMP) was identified with a strong fragrance of boxwood and exotic fruit aroma (Fig. 24.3) (Roland, Schneider, Razungles, & Cavelier, 2011). Semi-volatile thiols are also highly aroma active compounds at very low concentrations, perception threshold of 4MMP and 3-mercaptohexyl acetate (3MHA) in model solution is 0.8 and 4.2 ng/L, respectively (Khairallah et al., 2016). 4MMP is the characteristic aroma compound in *Sauvignon blanc wine*, and it can affect aroma profile if the content of 4MMP exceeds 40 ng/L. In addition, 3MHA, 4-mercapto-4-methylpentan-2-ol (4MMPOH), 3-mercaptohexan-1-ol (3MH), and 3-mercapto-3-methylbutan-1-ol (3MMB) were also detected in white wine. It was reported that *Riesling* wine contains higher amounts of 4MMP and 3MH (Khairallah et al., 2016).

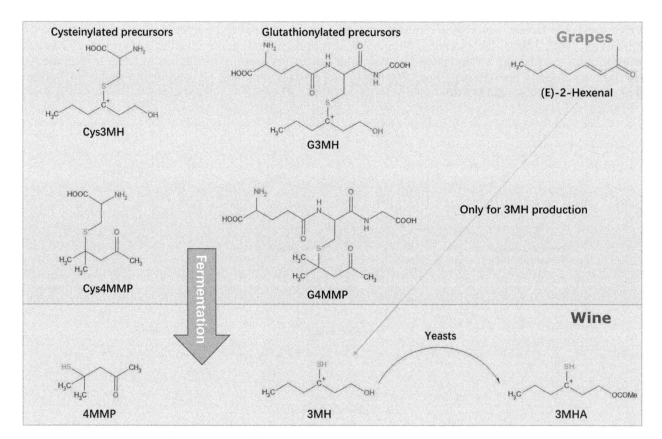

FIGURE 24.3 Biogenesis pathways for volatile thiols, 4MMP, 3MH, and 3MHA in AF. *4MMP*, 4-mercapto-4-methylpentan-2-one; *3MHA*, 3-mercaptohexyl acetate; *3MH*, 3-mercaptohexan-1-ol. Source: *From © Roland, A., Schneider, R., Razungles, A., & Cavelier, F. (2011). Varietal thiols in wine: Discovery, analysis and applications. Chemical Reviews, 111(11), 7355–7376. https://doi.org/10.1021/cr100205b.*

Some studies showed 3MHA found in poplar, grape skin, and passion fruit, with a threshold of 4 ng/L, and *Sauvignon blanc* wine could contain hundreds of ng/L (Kaewtathip & Charoenrein, 2012; Lukić, Lotti, & Vrhovsek, 2017). The threshold value of 3MH was 60 ng/L. 4MMPOH has a citrus aroma, and its content is more than 55 ng/L in wine, which is lower than the threshold (Tominaga, Baltenweck-Guyot, Peyrot des Gachons, & Dubourdieu, 2000). 3MMB has cooked leek flavor, and its content in wine is normally higher than the threshold (1500 ng/L). 3MHA is produced by the acetylation of 3MH in the process of AF (Pinu, Jouanneau, Nicolau, Gardner, & Villas-Boas, 2012). It was commonly higher in young wines and can be hydrolyzed to 3MH during aging, which results in concentration reduction. Most of the volatile thiols are bounded with cysteine by -S- bound form in grape berry. During AF, these precursors release the volatile sulfides due to the activity of β-lyase, which is mainly produced by yeasts (Makhotkina, Herbst-Johnstone, Logan, du Toit, & Kilmartin, 2013).

24.2.5 Volatile compounds produced during fermentation (fermentation aroma)

The main volatile compounds such as higher alcohols, esters, and acids are produced in fermentation (Ribéreau-Gayon, Dubourdieu, Donèche, & Lonvaud, 2006). Based on the concentration level, the volatile compounds formed in the fermentation can be the most important aroma components, especially for white wine. Higher alcohols in white wine are the products of amino acid or sugar metabolism by yeasts in AF (Fig. 24.4),

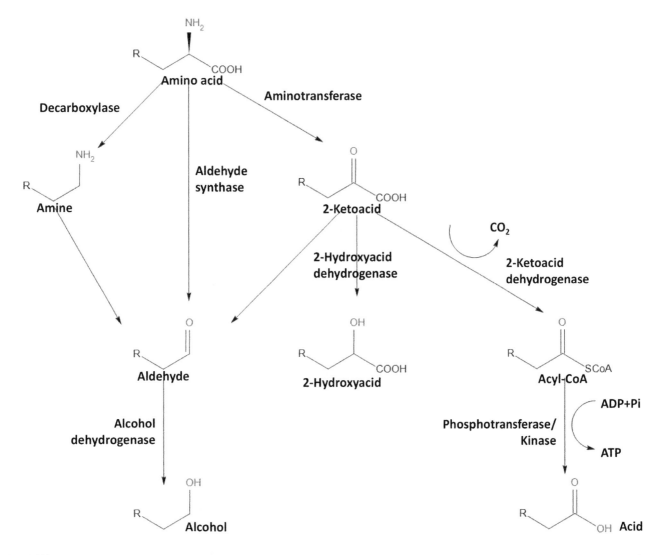

FIGURE 24.4 Biosynthetic routes for amino acid degradation to volatile compounds in yeasts. Source: *From © El Hadi, M. A. M., Zhang, F.-J., Wu, F.-F., Zhou, C.-H., & Tao, J. (2013). Advances in fruit aroma volatile research. Molecules, 18(7), 8200−8229. https://doi.org/10.3390/molecules18078200.*

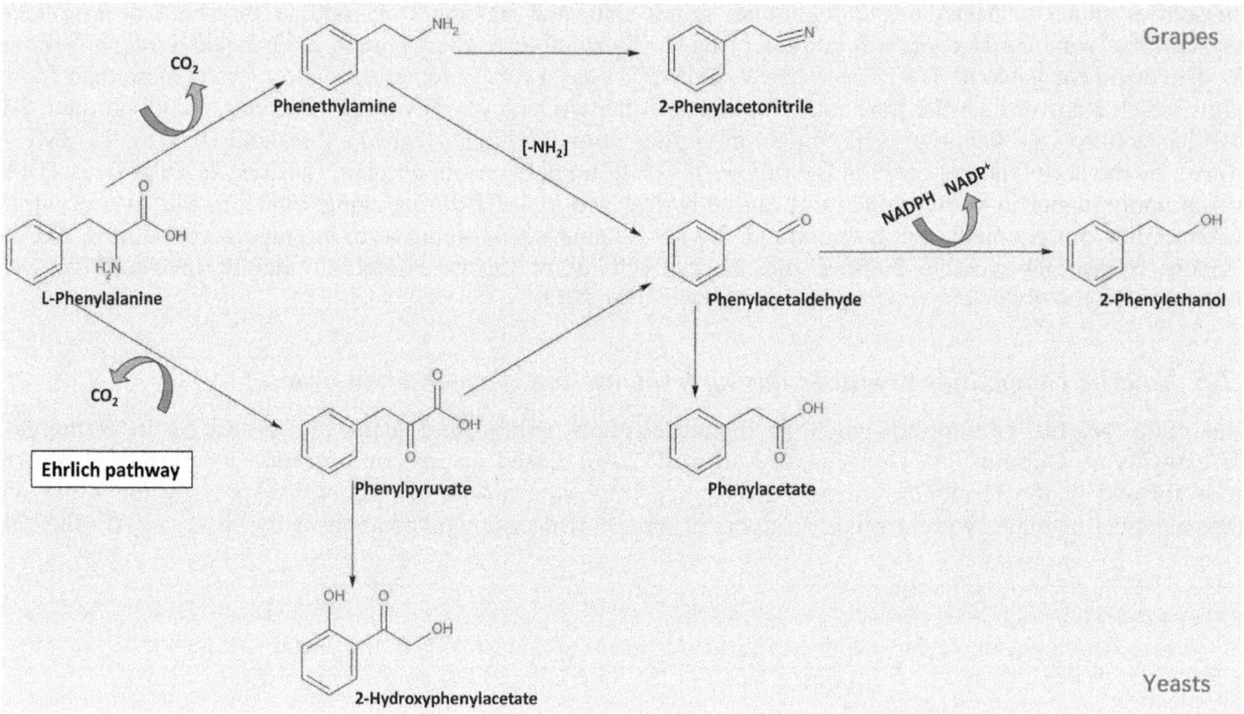

FIGURE 24.5　Synthesis pathways for the production of the volatile compounds phenylacetaldehyde and 2-phenylethanol in grapes and yeasts. Source: From © Tieman, D., Taylor, M., Schauer, N., Fernie, A. R., Hanson, A. D., & Klee, H. J. (2006). Tomato aromatic amino acid decarboxylases participate in synthesis of the flavor volatiles 2-phenylethanol and 2-phenylacetaldehyde. Proceedings of the National Academy of Sciences of the United States of America, 103(21), 8287–8292. https://doi.org/10.1073/pnas.0602469103; Gu, Y., Ma, J., Zhu, Y., & Xu, P. (2020). Refactoring Ehrlich pathway for high-yield 2-phenylethanol production in Yarrowia lipolytica. ACS Synthetic Biology, 9(3), 623–633. https://doi.org/10.1021/acssynbio.9b00468.

and the esters with fruity aroma can form in both AF and malolactic fermentation (MLF) (Chen, Han, Li, & Sheng, 2016). During AF, acetyl-CoA and ethanol are used to synthesize fatty acid ethyl ester by the catalysis of esterase. Acetyl-CoA and higher alcohols are used to synthesize acetates, which possess a faster synthesis speed than that of hydrolysis (Chen et al., 2018). The acids in wine are generally from the cleavage of higher fatty acids or the oxidation of alcohols and aldehydes. In addition, the tasteless glycosidic bound precursors of terpenes can be enzymolysis or acidolysis to the free form, releasing fruity and flowery aroma. Moreover, mercaptan, which is a cysteine precursor, can be hydrolyzed by yeast cysteine lyase to volatile thiols with a strong flavor of boxwood and broccoli (Roland et al., 2011; Silva Ferreira et al., 2008).

On the other hand, the origin of phenylalanine-derivatives in wine is the grape and largely improved with yeasts metabolism by the Ehrlich pathway during AF. This is especially true for three key aroma compounds, namely phenylacetaldehyde, 2-phenylethanol, and phenylacetate, which are derived from phenylalanine by the removal of the carboxyl and amino groups (Fig. 24.5); they contribute high fruity and flowery attributes on sensory characters of white wine. They play important roles in the biological functions of the plants. Particularly, 2-phenylethanol presents in flower reproductive structures exhibits antimicrobial and protective properties, such as acting as a potent insect attractant for pollination and attracting predatory insects (Tieman et al., 2006; Tieman, Loucas, Kim, Clark, & Klee, 2007). A study about icewine aroma showed exterior stress, off-vine treatment could be positive for the development of 2-phenylethanol (rose odor, threshold of 1000 μg/L), and late harvest significantly contributed to the accumulation of 2-phenylethanol in Vidal blanc (Chen et al., 2019).

24.2.6 Volatile compounds produced during aging (aging aroma)

The volatile compounds produced during wine aging are the new chemicals that mainly originate from oak or have different synthesis pathways, such as oak lactone (produced during aging in oak barrel), volatile phenols and phenolics (pyrolysis products of oak lignin), and furans (Maillard reaction products of cellulose and hemicellulose during aging in oak barrel) (Fenoll, Manso, Hellín, Ruiz, & Flores, 2009). The formation of potassium hydrogen tartrate can affect the material exchanges between oak barrels and wine (Villamor & Ross, 2013). Alcohol and pH affect the extraction of volatile compounds in oak barrels (Alcarde, Souza, & Bortoletto, 2014). The alcohol level is conducive to the extraction of volatile compounds from oak barrels. Sulfur dioxide (SO_2) can be combined with carbonyl compounds in wine. Likely, the concentration of SO_2 in wine affects the extraction of volatile compounds in oak barrels as well. It was shown that SO_2 can combine with carbonyl groups to form hydroxymethylfurfural, vanillin, butyraldehyde, and coniferaldehydes and delay releasing of free forms in wine (Caldeira, Anjos, Portal, Belchior, & Canas, 2010). Moreover, not all phenolic compounds containing carbonyl groups can combine with SO_2, except furfural and 5-methylfurfural, which potentially result from the weak carbonyl activity (Khairallah et al., 2016).

The extraction rate of furaldehyde from oak barrel to wine is faster than its conversion rate. In the early stage of aging, their concentration in wine is gradually increasing; after 6 months of aging, the concentration of furaldehyde reaches the maximum, and then the rate of conversion to corresponding alcohols will exceed the rate of extraction from oak barrels to wine, resulting in a significant decrease in the concentration of these furaldehydes in wine (Rodríguez Madrera, Blanco Gomis, & Mangas Alonso, 2003). Phenolics will also be transformed into corresponding alcohols by bioreduction, and the content will reach the highest after aging for 10–12 months, which is similar to the evolution of furanal. In volatile phenolics, the threshold of vanillin is low, which can give wine vanilla flavor and contribute much to aged wine aroma. The content of oak lactone in wine also changes with aging time as well (Caldeira et al., 2010; Caldeira, Belchior, & Canas, 2013). Phenolic alcohols are very stable, they are less affected by microbial and biochemical transformation, and the content in wine mainly depends on their extraction rate (Tsakiris, Kallithraka, & Kourkoutas, 2014). The concentration of 4-methylguaiacol reached the maximum after 3 months, while the concentration of guaiacol reached the maximum after 9 months (Caldeira et al., 2016). After that, the concentration of guaiacol in wine increased slowly and tended to be stable (Alcarde et al., 2014). Phenolic alcohols give wine toasted flavor, while the concentration of phenolic alcohols in wine is generally lower than the thresholds. However, in complex wine systems, these compounds may still affect the aroma quality of wine through promotion, synergism, or inhibition (Rodríguez Madrera et al., 2003). The aroma of wine evolves every minute during aging since compounds extracted from oak barrels slowly dissolve in wine. In addition, during wine aging, the esters produced from fermentation are continuously hydrolyzed in acid condition; glycosidic bound terpenes in acid condition are hydrolytically released and carry out the intermolecular conversion of analogs; a large amount of carotenoids, the precursors of norisoprenoids, are also constantly degraded (Cabrita et al., 2006; Lan et al., 2016). Therefore aging volatile compounds not only include the extraction of new products but also largely change the aroma profile from the fermentation. The contribution of oak barrels to the aging of white wine is complex. They not only give the wine some oak ingredients but also seek the balance and coordination between various volatiles (Caldeira et al., 2010).

24.3 Factors that influence aroma profile of white wine

Generally, the aroma of white wine firstly depends on the grape variety, and secondly on the natural conditions (climate, soil, cultivations, canopy management, diseases and pests, etc.) and vinification methods (Asproudi, Petrozziello, Cavalletto, & Guidoni, 2016). Particularly, under the premise of advanced technological conditions, grape varieties play a decisive role in the composition of white wine aroma. Only under suitable ecological conditions, such as climate and soil, can good grape varieties fully perform their flavor characteristics (Xu et al., 2015). Furthermore, aging and storage management also determine the final aroma characteristics of white wine.

24.3.1 Varietal differences

Different grape varieties possess different kinds of sugar, acid, ester, phenol, and other compounds. The aroma profile is also different among grape varieties, which determines the quality of grape varieties. Climate, soil, and cultivation methods are the natural factors that can determine the aroma profile. Therefore the variety is the main source of the differences among the wines, which makes the wines show their unique aroma profiles (Xu et al., 2015).

At the ripening stage of grape berry, the accumulations of free and glycosidic bound terpenes start from the veraison (Asproudi et al., 2016). These monoterpenes continually accumulate until the grapes overripen. The condition of the vineyard, especially the temperature, is highly related to this change. The water supply of grapes also affects the formation of this kind of volatile compound. In particular, the categories of terpenes in grape are scarcely affected by the local climate of producing area (Allen et al., 2011). The same variety planted in a different region shows a similar composition of terpenes (Park, Morrison, Adams, & Noble, 1991). It was found that in cold Germany and warm South Africa, the composition and categories of terpenes in Riesling were quite similar, but the content of terpenes varied greatly, and the results of sensory taste were significantly different (Rapp, 1990). Lower temperature would be conducive for the accumulation of terpenes. At present, the main aroma of grapefruits identified are monoterpenes, norisoprenoids and their derivatives, methoxypyrazines, mercaptans, and so on. Moreover, some grape varieties have specific volatile compounds, such as methyl phthalate, which is considered to be the characteristic substance of "foxy" in American grape and garden grape.

The synthesis of C_{13}-norisoprenoids is similar to that of monoterpenes, and the decrease in carotenoids follows the change of pigment, which is related to the increase in the concentration of C_{13}-norisoprenoids derivatives (TDN, vitispirane, etc.), which are mainly in the form of glycosidic bound. These changes require the action of related enzymes in grapes, which start to degrade in the form of carotenoid oxidation, and then proceed according to the glycoside mechanism (Schwab et al., 2008). During the ripening stage, irradiation could induce the decomposition of carotenoids in grapes, and the content of glycosidic C_{13}-norisoprenoids derivatives increased, which was typical in *Riesling* and *Syrah* grapes. Sometimes, during the aging period, *Riesling* wine will produce too much kerosene odor, which is related to the extremely high temperature during the ripening period. Although hot weather is suitable for sugar accumulation, wine quality could be negatively affected to some extent. Moreover, temperature, soil, grape maturity, and yield can influence the content of methoxypyrazine in grapes. In the *Sauvignon blanc* wine, the highest concentration of methoxypyrazine can reach up to 100 ng/L and gradually decrease during the ripening process. In the coldest regions of Australia, *Riesling* and *Syrah* grapes show high concentrations of methoxypyrazine. In Bordeaux, the results of wines from the same winery also showed that climate significantly influences the content of methoxypyrazine (Roujou de Boubee et al., 2000). The grapes grown in well-irrigated gravel soil have a low concentration of methoxypyrazine. The grapefruit with poor maturity is rich in oxidase (affecting the color and taste of wine) and has high activity of lipoxygenase (LOX) (Bindon et al., 2007). Under aerobic conditions, the unsaturated fatty acids are rapidly oxidized to C_6 aldehydes with "green taste", and further oxidized to form C_6 alcohols (Xu et al., 2015).

24.3.2 Environmental conditions

Terroir, include geography and climate, mainly affects the aroma profile of grape and wine. The geographical factors are the location of producing area and soil conditions, while the climatic factors mainly include light, temperature, and humidity. The terroir of producing area is very important for producing high-quality wine, which embodies the principle of suitable planting. Only the producing areas with good ecological conditions can produce high-quality wines. Different soil types, development status, water-holding capacity, fertilizer-holding capacity, structure, pH value, etc. in the production area have a significant impact on the aroma profile and sensory flavor of white wines. Enhanced light can stimulate the decomposition of carotenoids, increase the photosynthetic metabolites in grape berries, and promote the accumulation of phenolic compounds. Higher temperature is conducive to increase the sugar content and reduce the acidity of berries, while lower temperature is helpful to the synthesis of pigments. Both air humidity and soil humidity play an important role in the growth of grapes: high air humidity can cause various diseases; too high or too low soil humidity results in adverse effects on grapes; excessive precipitation or irrigation can delay ripening and reduce the content of anthocyanins in grape berries.

24.3.3 Fermentation managements

In the aspect of grape resources, the quality of grape can be improved by artificial selection of good varieties and suitable production areas, control of cultivation methods and training systems, pruning, fertilization and irrigation, pest control, etc. In terms of fermentations, different methods, fermentation materials, yeast species, fermentation conditions, and pretreatment measures directly affect the aroma profiles of white wine. The main function of AF of yeast wine is to transform sugar into alcohol, but at the same time, many byproducts such as alcohol, acid, and ester will be produced. Organic acids, higher alcohols, and esters, as well as a small amount of acetaldehyde, are the major volatile compounds. Their content and composition are affected by yeast species and fermentation conditions. Different inoculation dosages and initial sugar degrees lead to different synthesis speed and content of ester of yeasts in the fermentation process; for instance, in order to resist the high osmotic pressure produced by high sugar ($>35°$ Brix), yeast will produce a higher amount of acetic acid and glycerol during

AF of icewine. Different proportions of alcohol and ester result in the changes of aroma profiles of white wine. Yeast species were also found to influence the result of MLF, which results in different aroma profiles in the white wines. Moreover, the metabolism and interaction of a series of related microorganisms develop through the whole winemaking process. The metabolites of yeasts are the key factors that affect the sensory characteristics. It is due to the yeast metabolism that different volatile compounds are produced, which cause a direct impact on white wine aroma. Boqin Zhang et al. recently found that the aroma diversity of *Vidal blanc* icewine can be largely improved with cofermentation by using indigenous *Hanseniaspora vineae* and *Metschnikowia pulcherrima* with *S. cerevisiae* (Zhang, Shen, Duan, & Yan, 2018).

In terms of wine aging, the concentration of many aromatic compounds changes during wine aging and storage. When some wines are fermented or stored in oak, β-methyl-γ-octanol lactone (also known as oak lactone or whiskey lactone) can be dissolved from the oak barrel, which gives the wine a wooden, oaky, and coconut flavor. Some studies have shown that wood can also absorb some aromatic compounds such as phenylethyl alcohol and dodecanoic acid, thus changing their concentration in wine (Ribéreau-Gayon et al., 2006; Silva Ferreira et al., 2008). Wine can be fermented and aged in stainless steel containers, but in this way, there will be a lack of compounds from oak barrel aging, such as lactones and phenolic acids, that give the wine simple sensory characteristics. In addition to the aromatic compounds extracted from oak, chemical and microbial (acetobacter) oxidation reactions also play a key role in the flavor of aged wine, which can form acetaldehyde (nutty, sherry) and acetic acid (vinegar). However, if the oxidation reaction is not well controlled, it will produce a very high concentration of acetaldehyde and acetic acid, which will bring undesirable overall sensory impact.

Grape skin carries a large number of various yeasts, but only a few of them participate in AF. Non-*Saccharomyces* can improve the aroma complexity of white wine. For example, *Brettanomyces* can affect the development of 4-ethylphenol in wine (Gerbaux, Vincent, & Bertrand, 2002). Prior to the inoculation of *S. cerevisiae*, these wild yeasts were mainly active in the must. Due to their sensitivity to ethanol, the growth potential of these yeasts decreased rapidly after 2−3 days of fermentation in must. *S. cerevisiae* and related strains can be more tolerable to ethanol and have stronger competitiveness in the medium with high ethanol content. As a result, they become the main strains until AF is completed. Other yeasts may also appear in the fermentation and aging stage. Sometimes, due to the change of fermentation parameters, they participate in fermentation and affect the aroma composition. Wilkowska et al. investigated the collateral activity of glycosidases in commercial pectinase preparations and the release of aromas from their glycosidic counterparts in cherry juice and wine. The volatile aglycones released varied depending on the strain of yeast and bacteria used. Interestingly, the bacteria-derived glycosidases showed more than three times higher activity for free terpenoic and benzenoic aglycones than the yeast glycosidases (Wilkowska & Pogorzelski, 2017). Moreover, Zoecklein et al. studied the glycosidase activity of *S. cerevisiae*. The results showed that VL1, QA23, and EC1118 had higher glycosidase activity (Zoecklein, Marcy, Williams, & Jasinski, 1997). Some volatile thiols, such as 4MMP, 3MHA, 4MMPOH, 3MH, and 3MMB, can be characteristic volatile compounds in *Sauvignon Blanc*. While *Saccharomyces bayanus var. uvarum* can synthesize volatile thiols and produce a large amount of phenethyl alcohol. Meanwhile, *S. bayanus var. uvarum* hybrid strain was used to carry out the fermentation to make *Sauvignon Blanc* wine with high contents of 4MMP and 3MH but a low amount of phenethyl acetate (Dubourdieu, Tominaga, Masneuf, Des Gachons, & Murat, 2006).

The composition of acids, alcohols, esters, and terpenes in white wine is strongly related to the grape variety. With the selection of proper yeast strains and glucosidase, the content of acid and ester can be remarkably increased and the amount of alcohols can be reduced by adopting incomplete fermentation for white winemaking. Moreover, the content of esters, acids, and terpenes in white wine can be increased by cold maceration. A proper maceration period should be controlled in order to obtain a balance among different aroma compounds. Adding pectinase in maceration not only improves pigment extraction but also increases the content of acids and influence on the content of glycosidic bound volatile compounds. Besides, the volatile compounds in wine change constantly during aging, which especially facilitates the development of esters. In order to maintain the varietal aroma, proper conditions of the aging process should be carefully selected according to the grape variety and the changes in wine composition.

24.3.4 Glycosidase treatment

Terpenes in white wine are mainly from grape berry, and they are in the form of tasteless glycosides. All grapes contain similar glycosides, but the concentration of *Muscat* type is the highest. Glycosides volatile compounds are generally more common than the free forms. Although glucosidase is contained in grape, the optimal activity of these enzymes is pH 5.0, while the pH of must is generally 3.0−3.5, so the effect of these endogenous enzymes on the aroma precursors of glycosidase in the must is limited. In the process of white winemaking exogenous glycosidase is often added at the end of AF to facilitate the release of aromatic terpenes. In addition, there are some straight or cyclic alcohols in grape and wine, such as hexanol, benzyl alcohol, phenethyl alcohol, C_{13}-norisoprenoids, phenolic acid, and volatile phenol, which also exist as glycosidic precursors, and glycosidase can release them as free forms in a short period. Moreover, β-glycosidase produced by fungi can be inhibited by glucose (e.g., Creative & AR2000); hence the addition of glycosidase treatment is only effective for dry white wines (Chen et al., 2019).

Commercial macerating enzyme kits used in white wine are generally pectinase and lignin enzymes, which contain a small amount of "side enzyme activity" (glycosidase and cinnamate esterase). During white wine fermentation, adding macerating enzymes can cause the degradation of grape skin. As a result, the *in vivo* volatile compounds, related enzymes, or aroma precursors will be released. Interactions of the nonvolatile compounds and the volatile compounds are mainly happened in the must. Antioxidation effects of polyphenols and pigments also have a protective effect on the volatile compounds of white wine. The inhibition or enhancement of polyphenols and pigments on the quality of wine aroma cannot be ignored. The protective effect on the volatile compounds can be improved as the amount of polyphenols and pigments can be highly extracted.

24.3.5 Matrix effects

Volatile components are normally a group of small molecular compounds (molecular weight is normally lower than 300). However, the nonvolatile components in matrix are often with large molecular weight (Alcarde et al., 2014). Both of them may interact with each other and the wine aroma will be affected. Therefore only the study of volatile compounds in wine cannot be used to fully understand the overall style of wine, nor to accurately predict the aroma of grapes and wines. Due to the different chemical properties (molecular size, functional group and volatility) of volatile compounds, when the interaction occurs, the chemical components of wine are combined by different chemical bonds, such as covalent bond, hydrogen bond, or the formation of inclusions. The degree of interaction can change the distribution of volatile compounds in headspace and matrix and subsequently change the volatility of free volatiles. The content of volatile compounds directly changes the sensorial impression of aroma quality. In fact, for a certain compound, its content in the headspace may increase as the content of the other compounds increases in the matrix, but it may also be accompanied by a decrease in the content of matrix component (Villamor & Ross, 2013). The content of a single compound in the headspace and the perceived intensity are significantly affected by other chemical compounds. In most cases, the change of wine aroma can cause a change of sensorial quality, but the mechanism is still unclear. More research is needed to explain the characteristics of these naturally occurring matrix effects.

Polyphenols and polysaccharides are important nonvolatile substances in grape berries and wine, which play an important role in taste (bitter and astringent) and color (Noble, 1996). Some studies have shown that these compounds can combine with aroma components, thus changing the composition concentration and quality of aroma, but their specific effect depends on the content and physical and chemical properties of aroma compounds. Ethanol also plays an important role in changing the properties of volatile compounds in solution. Generally, higher ethanol content can reduce the activity and partition coefficient of volatile compounds in wine, increase the solubility of odorant, and reduce its volatility.

In addition to interactions with matrix components, there are also interaction effects between aroma components, which also affect the whole quality of aroma. Atanasova et al. established a mathematical model for the interaction between volatile compounds by analyzing the mixed solution of volatile compounds and oak substances (Boriana et al., 2005). Escudero et al. found that demethylated isoprene and dimethyl sulfides can enhance fruity aroma, and the principle of such a combination may result from the mechanism of physical and chemical reactions (Escudero et al., 2004; Escudero, Campo, Fariña, Cacho, & Ferreira, 2007). However, a study by Hein et al. showed that the masking effect among volatile compounds could be more significant on fruity and plant aroma (Hein, Ebeler, & Heymann, 2009). In addition, the interactions of volatile compounds can largely change the aroma characteristics of the mixed solution; for example, the combination of ethyl butyrate with fruit flavor and diacetyl with butter flavor will lead to the appearance of the salty flavor of cream candy in the wine; the mix of eugenol with clove flavor and phenethyl alcohol with rose flavor will form a special carnation-like aroma (Ebeler & Thorngate, 2009). Currently, the understanding of olfactory receptors in neurophysiology will make the mechanism of the cumulative masking effect more clear, and it would be easier to predict the aroma information of complex mixtures in the future.

24.4 A fast-growing determination method for key volatile compounds in wines: gas chromatography—olfactometry

Gas chromatography-olfactometry (GC-O) technology is considered to be an effective way to classify and arrange the odor components according to the contribution of aroma for a long time. However, after a comprehensive evaluation of the principle of GC-O analysis, some works in the literature have pointed out that the application of GC-O has limitations, and it is more suitable for wine with strong flavor. In any case, GC-O combines instrument analysis and sensory description to directly determine the odor activity value, aroma description words, and odor intensity of volatile compounds, which denotes great technological progress in the field of wine aroma; the reliability of the analysis results depends on the qualification and quantitative method of extracted volatile compounds.

Meanwhile, GC-O technology has been continuously improved and matured in application. At present, its application in wine aroma analysis can be divided into three categories: (1) to test aroma concentration; (2) to evaluate odor intensity; (3) to test the frequency of using aroma descriptors. The researchers also used gas chroma-tography-olfactometry-mass spectrometry (GC-O-MS) to detect trace volatile compounds in wine. For example, Janusz et al. used GC-O to analyze bottling to store aged white wine and found that β-damascenone can be regarded as a source of a strong fruity smell and an index of aging level (Janusz et al., 2003). Therefore the detection method of volatile compounds was developed and aroma descriptions corresponding to the smell were found. Aznar et al. also analyzed the volatile compounds in aged Spanish wine by GC-O and GC-MS and detected some characteristic markers for the first time. In addition, the GC-O technique has also been applied to the analysis of differences between different wines. GC-O is used to analyze the contribution of volatile compounds and find characteristic markers, while the method of aroma extraction dilution analysis (AEDA) is usually used to obtain the flavor dilution (FD) factor of the odor components. It will be more reasonable to carry out a simultaneous study on the relations between some wine physicochemical properties and characteristic volatiles as opposed to the evaluation of the contribution of the detected volatile compounds (Aznar, López, Cacho, & Ferreira, 2003).

The sensory quantification value provides an effective method to screen the aroma descriptors of wine features. In addition to quantifying sensory analysis results, it also provides data support for potential information of sensory analysis and effectively links sensory analysis and instrumental analysis. GC-O combined with odor activity value (OAV, the ratio of the detected content of volatile compounds to the olfactory thresholds) was also applied to the study of the relative contribution of odor substances. GC-O-MS has been developed to scan the important odor substances, calculate their OAV, and evaluate the odor activity and relative odor intensity of a substance. Some studies used this method to determine the odor-active components of *Gewürztraminer* wine and litchi wine and determined 12 kinds of high odor-active components contained in both by comparison analysis. It was concluded that these substances are the key components that give the typical litchi aroma in *Gewürztraminer* wine, and then determined these 12 components according to the OAV values, where the six volatile compounds including linalool, furanone, and geraniol were screened out.

However, the above analysis method is controversial due to the lack of sensory analysis to verify the experimental results. Recently, some studies have made up for this deficiency, that is, through sensory analysis experiments to verify the real impact of high FD values or OAVs of volatile compounds on wine aroma quality. In a particular study, the interaction effect between single aroma components and the matrix effect of wine was considered (Muñoz-González, Martín-Álvarez, Moreno-Arribas, & Pozo-Bayón, 2014). In this study, GC-O-MS was used to determine the quality and quantity of the volatile compounds in wine, and the contribution values of single volatile components (FD and OAV) were analyzed. Then according to FD value or OAV value, the active components with high impact odor were selected and added into base wine (wine with neutral aroma) or simulated wine solution (according to the detected concentration) to complete the reorganization of wine aroma. The differences and similarities between the reconstructed model wine and the original wine were analyzed by triangle tasting (Escudero et al., 2007; Pineau, Barbe, Leeuwen, & Dubourdieu, 2009). Then, importance of a volatile compound can be obtained by removal or adding the high active ingredients in the prepared model wine. Sensory evaluation was carried out until the model wine with the largest or closest difference in sensory characteristics with the original wine was found as the basis for identification of the key aroma components in the wine (Ferreira et al., 2009). Most of these experiments can show OAV and FD. However, Escudero et al. also found that omitting or adding substances with high OAV values could not change the overall aroma characteristics of young white wine (Escudero et al., 2007). This phenomenon was related to the buffering ability of wine, mainly due to the high content of ethanol, ethyl acetate, and volatile phenol in wine matrix with similar aroma properties.

The results of sensory analysis and the regression model of volatile components can be established by using the chemometrics method. The correlation between a certain aroma feature and the corresponding aroma components has been widely reported regarding the evaluation of the contribution of a single aroma component to the overall aroma (Escudero et al., 2007). Ferreira et al. using partial least squares regression (PLSR) analysis to establish the aroma quality model of Spanish wine; it was found that the turbid substances were the most influential factors on the quality, followed by some aroma components with fruit fragrance, and the turbid substances could also inhibit the aroma of wine (Ferreira et al., 2009).

24.5 Conclusion

The composition and source of wine volatile compounds are very complex. The quality of wine aroma is affected by many factors, from the raw materials to the fermentation technology to the aging process. With the improvements in chemical analysis methods, new volatile compounds are found and the "aroma fingerprint" of wine is improved constantly. The research results of the influence of winemaking technology on the formation and composition of aroma will provide theoretical guidance for the production of high-quality wine with the best aroma quality by using the most appropriate technology. Wine aroma synthesis, intermediate transformation, and decomposition are complex processes. However, the mechanism of the physiological and chemical changes of volatile compounds in grape growth, fermentation, and wine aging is still unknown.

It is true that the influence of the volatile compounds that exist in single form and in mixture on the sense organs has not yet been fully explored. Therefore future studies will focus on genome and protein grouping technology and strengthen the research on the change mechanisms of volatile compounds, thus understanding the relationship between various chemical components and sensory perception and how the interaction of odor and matrix components affects the overall flavor perception of wine. Meanwhile, the fingerprint of aroma characteristics of wine in each region should be established to provide a theoretical basis for the study of aroma profiles and sensorial characteristics in the special wine region.

References

Alcarde, A. R., Souza, L. M., & Bortoletto, A. M. (2014). Formation of volatile and maturation-related congeners during the aging of sugarcane spirit in oak barrels. *Journal of the Institute of Brewing, 120*(4), 529–536. Available from https://doi.org/10.1002/jib.165.

Allen, T., Herbst-Johnstone, M., Girault, M., Butler, P., Logan, G., Jouanneau, S., & Kilmartin, P. A. (2011). Influence of grape-harvesting steps on varietal th ol aromas in sauvignon blanc wines. *Journal of Agricultural and Food Chemistry, 59*(19), 10641–10650. Available from https://doi.org/10.1021/jf2018676.

Asproudi, A., Petrozziello, M., Cavalletto, S., & Guidoni, S. (2016). Grape aroma precursors in cv. Nebbiolo as affected by vine microclimate. *Food Chemistry, 211*, 947–956. Available from https://doi.org/10.1016/j.foodchem.2016.05.070.

Aznar, M., López, R., Cacho, J., & Ferreira, V. (2003). Prediction of aged red wine aroma properties from aroma chemical composition. Partial least squares regression models. *Journal of Agricultural and Food Chemistry, 51*(9), 2700–2707. Available from https://doi.org/10.1021/jf026115z.

Bindon, K. A., Dry, P. R., & Loveys, B. R. (2007). Influence of plant water status on the production of C13- norisoprenoid precursors in *Vitis vinifera* L. cv. cabernet sauvignon grape berries. *Journal of Agricultural and Food Chemistry, 55*(11), 4493–4500. Available from https://doi.org/10.1021/jf063331p.

Boriana, A., Thierry, T. D., Claire, C., Dominique, L., Sophie, N., & Patrick, E. (2005). Perceptual interactions in odour mixtures: Odour quality in binary mixtures of woody and fruity wine odorants. *Chemical Senses* (3).

Cabrita, M. J., Costa Freitas, A. M., Laureano, O., & Di Stefano, R. (2006). Glycosidic aroma compounds of some Portuguese grape cultivars. *Journal of the Science of Food and Agriculture, 86*(6), 922–931. Available from https://doi.org/10.1002/jsfa.2439.

Caldeira, I., Anjos, O., Portal, V., Belchior, A. P., & Canas, S. (2010). Sensory and chemical modifications of wine-brandy aged with chestnut and oak wood fragments in comparison to wooden barrels. *Analytica Chimica Acta, 660*(1–2), 43–52. Available from https://doi.org/10.1016/j.ac.2009.10.059.

Caldeira, I., Belchior, A. P., & Canas, S. (2013). Effect of alternative ageing systems on the wine brandy sensory profile. *Ciencia e Tecnica Vitivinicola, 28*(1), 9–18. Available from http://www.scielo.gpeari.mctes.pt/pdf/ctv/v28n1/v28n1a02.pdf.

Caldeira, I., Santos, R., Ricardo-Da-Silva, J. M., Anjos, O., Mira, H., Belchior, A. P., & Canas, S. (2016). Kinetics of odorant compounds in wine brandies aged in different systems. *Food Chemistry, 211*, 937–946. Available from https://doi.org/10.1016/j.foodchem.2016.05.129.

Chen, K., Escott, C., Loira, I., del Fresno, J. M., Morata, A., Tesfaye, W., & Benito, S. (2018). Use of non-*Saccharomyces* yeasts and oenological tannin in red winemaking: Influence on colour, aroma and sensorial properties of young wines. *Food Microbiology, 69*, 51–63. Available from https://doi.org/10.1016/j.fm.2017.07.018.

Chen, K., Han, S. Y., Li, M., & Sheng, W. J. (2016). Use of lysozyme and oligomeric proanthocyanidin to reduce sulfur dioxide and the evolution of volatile compounds in Italian riesling ice wine during aging process: Use of lysozyme and OPC to reduce SO 2. *Journal of Food Processing & Preservation, 41*(1).

Chen, K., Wen, J., Ma, L., Wen, H., & Li, J. (2019). Dynamic changes in norisoprenoids and phenylalanine-derived volatiles in off-vine Vidal blanc grape during late harvest. *Food Chemistry, 289*, 645−656. Available from https://doi.org/10.1016/j.foodchem.2019.03.101.

Dubourdieu, D., Tominaga, T., Masneuf, I., Des Gachons, C. P., & Murat, M. L. (2006). The role of yeasts in grape flavor development during fermentation: The example of Sauvignon blanc. *American Journal of Enology and Viticulture*. France: American Society for Enology and Viticulture Retrieved from. Available from http://ajevonline.org/.

Ebeler, S. E. (2001). Analytical chemistry: Unlocking the secrets of wine flavor. *Food Reviews International, 17*(1), 45−64. Available from https://doi.org/10.1081/FRI-100000517.

Ebeler, S. E., & Thorngate, J. H. (2009). Wine chemistry and flavor: Looking into the crystal glass. *Journal of Agricultural and Food Chemistry, 57*(18), 8098−8108. Available from https://doi.org/10.1021/jf9000555.

Escudero, A., Campo, E., Fariña, L., Cacho, J., & Ferreira, V. (2007). Analytical characterization of the aroma of five premium red wines. Insights into the role of odor families and the concept of fruitiness of wines. *Journal of Agricultural and Food Chemistry, 55*(11), 4501−4510. Available from https://doi.org/10.1021/jf0636418.

Escudero, A., Gogorza, B., Melús, M. A., Ortín, N., Cacho, J., & Ferreira, V. (2004). Characterization of the aroma of a wine from Maccabeo. Key role played by compounds with low odor activity values. *Journal of Agricultural and Food Chemistry, 52*(11), 3516−3524. Available from https://doi.org/10.1021/jf035341l.

Fan, W., Xu, Y., Jiang, W., & Li, J. (2010). Identification and quantification of impact aroma compounds in 4 nonfloral vitis vinifera varieties grapes. *Journal of Food Science, 75*(1), S81−S88. Available from https://doi.org/10.1111/j.1750-3841.2009.01436.x.

Fenoll, J., Manso, A., Hellín, P., Ruiz, L., & Flores, P. (2009). Changes in the aromatic composition of the *Vitis vinifera* grape Muscat Hamburg during ripening. *Food Chemistry, 114*(2), 420−428. Available from https://doi.org/10.1016/j.foodchem.2008.09.060.

Ferreira, V., Juan, F. S., Escudero, A., Culleré, L., Fernández-Zurbano, P., Saenz-Navajas, M. P., & Cacho, J. (2009). Modeling quality of premium spanish red wines from gas chromatography-olfactometry data. *Journal of Agricultural and Food Chemistry, 57*(16), 7490−7498. Available from https://doi.org/10.1021/jf9006483.

Fischer, U. (2007). Wine aroma. *Flavours and fragrances: Chemistry, bioprocessing and sustainability* (pp. 241−267). Germany: Springer Berlin Heidelberg. Available from https://doi.org/10.1007/978-3-540-49339-6_11.

Franco, M., Peinado, R. A., Medina, M., & Moreno, J. (2004). Off-vine grape drying effect on volatile compounds and aromatic series in must from *Pedro Ximénez* grape variety. *Journal of Agricultural and Food Chemistry, 52*(12), 3905−3910. Available from https://doi.org/10.1021/jf0354949.

Gerbaux, V., Vincent, B., & Bertrand, A. (2002). Influence of maceration temperature and enzymes on the content of volatile phenols in Pinot noir wines. *American Journal of Enology and Viticulture, 53*(2), 131−137.

Gunata, Y. Z., Bayonove, C. L., Baumes, R. L., & Cordonnier, R. E. (1985). The aroma of grapes I. Extraction and determination of free and glycosidically bound fractions of some grape aroma components. *Journal of Chromatography A, 331*(C), 83−90. Available from https://doi.org/10.1016/0021-9673(85)80009-1.

Hashizume, K., & Samuta, T. (1997). Green Odorants of grape cluster stem and their ability to cause a wine stemmy flavor. *Journal of Agricultural and Food Chemistry, 45*(4), 1333−1337. Available from https://doi.org/10.1021/jf960635a.

Hein, K., Ebeler, S. E., & Heymann, H. (2009). Perception of fruity and vegetative aromas in red wine. *Journal of Sensory Studies, 24*(3), 441−455. Available from https://doi.org/10.1111/j.1745-459X.2009.00220.x.

Janusz, A., Capone, D. L., Puglisi, C. J., Perkins, M. V., Elsey, G. M., & Sefton, M. A. (2003). (E)-1-(2,3,6-Trimethylphenyl)buta-1,3-diene: A potent grape-derived odorant in wine. *Journal of Agricultural and Food Chemistry, 51*(26), 7759−7763. Available from https://doi.org/10.1021/jf0347113.

Kaewtathip, T., & Charoenrein, S. (2012). Changes in volatile aroma compounds of pineapple (*Ananas comosus*) during freezing and thawing. *International Journal of Food Science and Technology, 47*(5), 985−990. Available from https://doi.org/10.1111/j.1365-2621.2011.02931.x.

Khairallah, R., Reynolds, A. G., & Bowen, A. J. (2016). Harvest date effects on aroma compounds in aged Riesling icewines. *Journal of the Science of Food and Agriculture, 96*(13), 4398−4409. Available from https://doi.org/10.1002/jsfa.7650.

Lan, Y. B., Qian, X., Yang, Z. J., Xiang, X. F., Yang, W. X., Liu, T., & Duan, C. Q. (2016). Striking changes in volatile profiles at sub-zero temperatures during over-ripening of "Beibinghong" grapes in Northeastern China. *Food Chemistry, 212*, 172−182. Available from https://doi.org/10.1016/j.foodchem.2016.05.143.

Lashbrooke, J. G., Young, P. R., Dockrall, S. J., Vasanth, K., & Vivier, M. A. (2013). Functional characterisation of three members of the Vitis vinifera L. carotenoid cleavage dioxygenase gene family. *BMC Plant Biology, 13*(1). Available from https://doi.org/10.1186/1471-2229-13-156.

Lukić, I., Lotti, C., & Vrhovsek, U. (2017). Evolution of free and bound volatile aroma compounds and phenols during fermentation of Muscat blanc grape juice with and without skins. *Food Chemistry, 232*, 25−35. Available from https://doi.org/10.1016/j.foodchem.2017.03.166.

Makhotkina, O., Herbst-Johnstone, M., Logan, G., du Toit, W., & Kilmartin, P. A. (2013). Influence of sulfur dioxide additions at harvest on polyphenols, C6-compounds, and varietal thiols in Sauvignon blanc. *American Journal of Enology and Viticulture, 64*(2), 203−213. Available from https://doi.org/10.5344/ajev.2012.12094.

Mateo, J. J., & Jiménez, M. (2000). Monoterpenes in grape juice and wines. *Journal of Chromatography A, 881*(1−2), 557−567. Available from https://doi.org/10.1016/S0021-9673(99)01342-4.

Muñoz-González, C., Martín-Álvarez, P. J., Moreno-Arribas, M. V., & Pozo-Bayón, M. A. (2014). Impact of the nonvolatile wine matrix composition on the in vivo aroma release from wines. *Journal of Agricultural and Food Chemistry, 62*(1), 66−73. Available from https://doi.org/10.1021/jf405550y.

Noble, A. C. (1996). Taste-aroma interactions. *Trends in Food Science and Technology, 7*(12), 439−444. Available from https://doi.org/10.1016/S0924-2244(96)10044-3.

Park, S. K., Morrison, J. C., Adams, D. O., & Noble, A. C. (1991). Distribution of free and glycosidically bound monoterpenes in the skin and mesocarp of Muscat of alexandria grapes during development. *Journal of Agricultural and Food Chemistry, 39*(3), 514−518. Available from https://doi.org/10.1021/jf00003a017.

Pineau, B., Barbe, J. C., Leeuwen, C. V., & Dubourdieu, D. (2009). Examples of perceptive interactions involved in specific\Red-\ and \Blackberry\ aromas in red wines. *Journal of Agricultural and Food Chemistry, 57*(9), 3702−3708. Available from https://doi.org/10.1021/jf803325v.

Pinu, F. R., Jouanneau, S., Nicolau, L., Gardner, R. C., & Villas-Boas, S. G. (2012). Concentrations of the volatile thiol 3-mercaptohexanol in Sauvignon blanc wines: No correlation with juice precursors. *American Journal of Enology and Viticulture, 63*(3), 407−412. Available from https://doi.org/10.5344/ajev.2012.11126.

Rapp, A. (1990). Natural flavours of wine: Correlation between instrumental analysis and sensory perception. *Fresenius' Journal of Analytical Chemistry*, 337(7), 777–785. Available from https://doi.org/10.1007/BF00322252.

Rapp, A. (1998). Volatile flavour of wine: Correlation between instrumental analysis and sensory perception. *Nahrung - Food*, 42(6), 351–363, https://doi.org/10.1002/(sici)1521-3803(199812)42:06 < 351::aid-food351 > 3.3.co;2-u.

Ribéreau-Gayon, P., Dubourdieu, D., Donèche, B., & Lonvaud, A. (2006). *Handbook of enology: The microbiology of wine and vinifications* (Vol. 1).

Rodríguez-Bustamante, E., & Sánchez, S. (2007). Microbial production of C13-norisoprenoids and other aroma compounds via carotenoid cleavage. *Critical Reviews in Microbiology*, 33(3), 211–230. Available from https://doi.org/10.1080/10408410701473306.

Rodríguez Madrera, R., Blanco Gomis, D., & Mangas Alonso, J. J. (2003). Influence of distillation system, oak wood type, and aging time on composition of cider brandy in phenolic and furanic compounds. *Journal of Agricultural and Food Chemistry*, 51(27), 7969–7973. Available from https://doi.org/10.1021/jf0347618.

Roland, A., Schneider, R., Razungles, A., & Cavelier, F. (2011). Varietal thiols in wine: Discovery, analysis and applications. *Chemical Reviews*, 111(11), 7355–7376. Available from https://doi.org/10.1021/cr100205b.

Roujou de Boubee, D., Van Leeuwen, C., & Dubourdieu, D. (2000). Organoleptic impact of 2-methoxy-3-isobutylpyrazine on red Bordeaux and Loire wines. Effect of environmental conditions on concentrations in grapes during ripening. *Journal of Agricultural and Food Chemistry*, 48(10), 4830–4834. Available from https://doi.org/10.1021/jf000181o.

Schwab, W., Davidovich-Rikanati, R., & Lewinsohn, E. (2008). Biosynthesis of plant-derived flavor compounds. *Plant Journal*, 54(4), 712–732. Available from https://doi.org/10.1111/j.1365-313X.2008.03446.x.

Silva Ferreira, A. C., Monteiro, J., Oliveira, C., & Guedes de Pinho, P. (2008). Study of major aromatic compounds in port wines from carotenoid degradation. *Food Chemistry*, 110(1), 83–87. Available from https://doi.org/10.1016/j.foodchem.2008.01.069.

Tieman, D., Taylor, M., Schauer, N., Fernie, A. R., Hanson, A. D., & Klee, H. J. (2006). Tomato aromatic amino acid decarboxylases participate in synthesis of the flavor volatiles 2-phenylethanol and 2-phenylacetaldehyde. *Proceedings of the National Academy of Sciences of the United States of America*, 103(21), 8287–8292. Available from https://doi.org/10.1073/pnas.0602469103.

Tieman, D. M., Loucas, H. M., Kim, J. Y., Clark, D. G., & Klee, H. J. (2007). Tomato phenylacetaldehyde reductases catalyze the last step in the synthesis of the aroma volatile 2-phenylethanol. *Phytochemistry*, 68(21), 2660–2669. Available from https://doi.org/10.1016/j.phytochem.2007.06.005.

Tominaga, T., Baltenweck-Guyot, R., Peyrot des Gachons, C., & Dubourdieu, D. (2000). Contribution of volatile thiols to the aromas of white wines made from several *Vitis vinifera* grape varieties. *American Journal of Enology and Viticulture*, 51(2), 178–181.

Tsakiris, A., Kallithraka, S., & Kourkoutas, Y. (2014). Grape brandy production, composition and sensory evaluation. *Journal of the Science of Food and Agriculture*, 94(3), 404–414. Available from https://doi.org/10.1002/jsfa.6377.

Villamor, R. R., & Ross, C. F. (2013). Wine matrix compounds affect perception of wine aromas. *Annual Review of Food Science and Technology*, 4(1), 1–20. Available from https://doi.org/10.1146/annurev-food-030212-182707.

Wilkowska, A., & Pogorzelski, E. (2017). Aroma enhancement of cherry juice and wine using exogenous glycosidases from mould, yeast and lactic acid bacteria. *Food Chemistry*, 237, 282–289. Available from https://doi.org/10.1016/j.foodchem.2017.05.120.

Xu, X. Q., Cheng, G., Duan, L. L., Jiang, R., Pan, Q. H., Duan, C. Q., & Wang, J. (2015). Effect of training systems on fatty acids and their derived volatiles in Cabernet Sauvignon grapes and wines of the north foot of Mt. Tianshan. *Food Chemistry*, 181, 198–206. Available from https://doi.org/10.1016/j.foodchem.2015.02.082.

Yuan, F., & Qian, M. C. (2016). Development of C13-norisoprenoids, carotenoids and other volatile compounds in *Vitis vinifera* L. Cv. Pinot noir grapes. *Food Chemistry*, 192, 633–641. Available from https://doi.org/10.1016/j.foodchem.2015.07.050.

Zhang, B.-Q., Shen, J.-Y., Duan, C.-Q., & Yan, G.-L. (2018). Use of indigenous *Hanseniaspora vineae* and *Metschnikowia pulcherrima* Co-fermentation with *Saccharomyces* cerevisiae to improve the aroma diversity of Vidal Blanc icewine. *Frontiers in Microbiology*.

Zoecklein, B. W., Marcy, J. E., Williams, J. M., & Jasinski, Y. (1997). Effect of native yeasts and selected strains of *Saccharomyces cerevisiae* on glycosyl glucose, potential volatile terpenes, and selected aglycones of White Riesling (*Vitis vinifera* L.) wines. *Journal of Food Composition and Analysis*, 10(1), 55–65. Available from https://doi.org/10.1006/jfc.1996.0518.

25

Inertization and bottling

Mark Strobl

Hochschule Geisenheim University, Geisenheim, Germany

25.1 Hazards to white wines

Avoiding negative influences on the way to the customer and until the wine is consumed has to be ensured not only by the quality of the wine preparation in the winery but, moreover, by the quality of the filling process and packaging material (Fig. 25.1). All this prevents changes to the bottled wine which do not fit the consumer expectation, as especially white wines nowadays are usually ready-to-drink products; the wines do not have time for further development in the bottle. The conservation of the quality standard at the filling date is the easiest way to provide and protect the quality the consumers receive.

Avoiding microorganisms, haze, enzyme reactions, and oxidation is easier and more projectable than hoping for a positive and uniform development of the wine in each bottle. Modern technologies provide good technical methods to achieve long shelf life of beverages nowadays, which are standard for big filling plants with a short period for return on investment because millions of bottles are produced every year. Besides this, the products have become very reliable, so the standard quality expectation of retailers and consumers is very high. While disappointed consumers change the brand or even the retailer, contracts with retailers always imply quality irregularities. Quality expectations also have to be fulfilled by smaller wineries, who only fill some thousand bottles every year and have to set higher prices for less safe products if the equipment of the big filling plants is not available.

Consumers may accept off-flavors and bad taste and doubt their own taste. Visible mistakes like hazy wine particles or sediments in the wine or damaged packages or bottles are complained about more frequently. So the filling of white wine has to comply with the wine and the product-specific preservations as well as product safety for retailers and consumers.

25.1.1 Hazard of microorganisms

Bacteria and yeasts can be responsible for a change of aroma, pH, color, and haze of the wine during the shelf life of the wine. As this unwanted fermentation, especially on the way to the consumer, is not controlled, it may lead to different, nonreproducible effects in each bottle that are not wanted by the winemaker and mostly not accepted by the wine consumer. Fermentation produces haze, alcohol, and carbon dioxide. The latter might lead to exploding bottles. This is a safety risk for the retailer and the consumer that has to be avoided. Fungi, usually only present when oxygen is available, may produce toxins or may be present with corks as well as in an uncleaned periphery; this might cause 2,4,6-trichloroanisole (TCA) as an off-flavor if it gets into the wine. Cork aromas are the most frequent complaint concerning the aroma and taste of rejected wines. The purpose of filtration is to preserve the wine from visible, olfactory, or organoleptic changes occurring in the long run. Suspended materials have to be removed, and potential turbidity formers have to be unhinged (Lindemann, 2006).

The filtration of the wine, the gasses, and the air that gets in contact with the wine and the sanitation of pipes, hoses, or containers reduce the probability of contamination (Schneider, 2019). Hygiene, cleaning, sanitation, and disinfection should be standard even in the periphery of the wine production and filling. Staff training,

© 2022 Elsevier Inc. All rights reserved.

Threads and countermeasures in wine production and filling
(after fermentation)

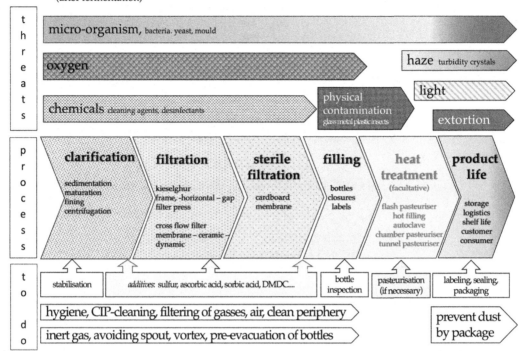

FIGURE 25.1 Overview of the process, the hazards, and the activities for safe bottled white wines.

sensitivity for dirt, development of biofilms, and potential nutrition for microorganisms that should be cleaned before biofilms are built up, which will help to keep up high sanitation standards. Biofilm adjacencies are well known to protect bacteria from harsh environmental conditions (Alexandre et al., 2016).

Pasteurization or hot filling (Krones, 2010) of sensitive wines like sweet wines or dealcoholized wines may be an expensive but secure solution, which will have an impact on the flavor and taste of the wine. Sulfur helps, and other preservative agents are expensive or have their own risks of producing an off-flavor like geraniol by sorbic acid.

Fillers are designed to be cleaned easily. Stainless steel, blank surfaces should ensure that hot water rinsing will rinse away organic material that would serve as nutrition for unwanted microorganisms. Mechanical constructions with gaps, holes, and gaskets should be avoided, but will always be present to a certain extent. Conveyers are mechanical connections of the rinsing, filling, and labeling machines with chains and gears that also transport microorganisms through equipment in both directions (Strobl, 2019). Sterility cannot be achieved, but cleaning, rinsing, and lubrication with bacteriostatic and fungistatic additives may achieve a certain equilibrium which must be treated when biofilms come up. Preventive hygiene should be standard, especially if low- or nonalcoholic wines are filled. Modern fillers are encased, and filtered air sets the filling area under excess pressure so that the air only transports germs and aerosols away from the filler with the open bottles.

25.1.2 Oxygen

Oxygen is contained in the wine at a dissolved state. It is always present in the air in the bottle and caps prior to filling. The latter two can be influenced by an appropriate filling process (Blüml, 2006). Oxygen can directly lead to the oxidation of aroma substances. Depending on time, temperature, the amount of reductive substances in the wine, and the amount of sulfur added to the wine, this may take some time. Some oxidized substances become insoluble and form cloudiness or even deposits in the wine. Lots of aerobic microorganisms have the opportunity to propagate. If residual sugars are present, a secondary fermentation changes the flavor, leads to haze, and changes the alcohol content in the bottles to unwanted amounts. The pressure of the emerging carbon dioxide may press out the corks of the bottles and may even lead to exploding bottles. A defined microoxygenation and development in the bottle of white wines are usually not possible; browning and off-flavors can best be avoided by working as oxygen free as possible.

The oxygen content in the beverage after filling is an aspect of major importance for preserving the product quality (Blüml, 2006).

Oxygen loading in the wine can occur at three points:

1. During the wine filtration process of the unfiltered wine to the filler, by contact to air in the system or even soakage of air by Venturi effects at leakages.
2. Oxygen pick-up at the filling process (Blüml, 2006).
3. Oxygen quantity in the residual air content in the bottle's headspace after filling (Blüml, 2006).

Avoiding uptake of air by fumigating tanks, filters, and bottles with nitrogen or carbon dioxide; the usage of sulfur dioxide or ascorbic acid; or the control of the gas content of the wine by membrane systems is considered before wine filling. SO_2 should be reduced for several reasons (Großmann, 2020; Giacosa et al., 2019).

25.1.3 Precipitations in the bottle

Sediments are no longer consumer expectations and are often taken for impurity and contamination of the wine. As they can be seen from the outside of the bottle, this could lead to rejection of the wine. Yeast usually removes most insoluble substances after fermentation. Maturation should be able to sediment the rest. Filtration is meant to remove all insoluble substances that could not be removed by the yeast, gravity, or centrifuges. Substances that could become insoluble in the bottle during shelf life or in the refrigerator of the consumer should be removed before filling. Temperatures colder than the consumer's refrigerator temperature should coagulate insoluble substances before filtration and be removed during the latter.

Crystallization usually takes time and can be influenced by some ions that take part in the crystallization. With high amounts of carbon dioxide in sparkling wine, tartaric acid crystals (Zoecklein, Fugelsang, Gump, & Nury, 1990) could lead to gushing. The wines should be stable or stabilized by cold or membrane stabilization. Membrane stabilization, normally used after final filtration, removes potassium or calcium directly before filling and is therefore able to be adjusted to small filling batches and changes of varieties during the filling program, while cold stabilization should be done before as it needs more time and the removal of the tartaric crystals after treatment by filtration. Stabilization of the wines by carboxymethyl cellulose is possible and should be implemented before filtration (Guise et al., 2014).

25.1.4 Chemical contamination

Pesticides from the vineyard, environmental pollutants, or cleaning agents, which might be present in the wines from harvest or production, are mostly buffered, metabolized, and removed with the yeast by sedimentation or filtration. There is a chance of detecting chemicals during wine production by checking the tanks and the batches during maturation and filtration. As the tanks after filtration and during filling usually are used more frequently, but for shorter periods, they have to be cleaned more often. Accidents with chemical residues or mixing of tanks while processing are more probable. Cross-contamination of a wine or wine product, chemical/ additive/processing aid, intermediate product, or finished product with a different starting material or product during processing and production (The concept of Good Manufacturing Practice, 2012) should be avoided. Using cleaning agents and disinfectants suitable for the winery and its employees at the point of use helps to avoid expensive and time-consuming analysis by external laboratories.

The presence of caustics, acid, and disinfection agents may be recognized by a change in pH, turbidity, and haze. A change in taste is often not recognized, as some disinfectants do not provoke this and do not form haze or color changes. This means that special laboratories with gas chromatography have to be consulted to check the separation of cleaning agents and products from time to time or if suspicious batches need to be controlled. Easy checking of peracetic acid with test stripes after the rinsing of tanks and pipes, automatic clean-in-place (CIP) cleaning, laboratory checks, and conductivity sensors in the pipes help to avoid accidents. Valves with block and bleed function avoid the merging of rinsing agents and product (Manger, 2005). Hazard Analysis Critical Control Points (HACCP) concepts help to avoid problems and sensitize the employees.

25.1.5 Physical contamination

Physical contaminations are dust, dirt in the bottle, dead insects (Fig. 25.2), macro- and microplastic, broken glass from broken or exploding bottles during filling, and glass splinters in the wine.

Usually, physical contaminations sediment during the wine production with the yeast or the finings during fermentation and maturation. The residual particles should be removed by kieselguhr or cross-flow filtration. At least at the microbiological filtration with 0.65 μm for yeast removal and 0.45 μm for removing of the bacteria, small particles bigger than that should be removed from the wine. Particles found in the wine could come from the rinsing water of the winepipes, -hoses and -tanks after the filtration or from new or even from rinsed bottles if the rinsing water was contaminated. Tap water may bring along a certain amount of microparticles (Eerkes-Medrano, Thompson, & Aldridge, 2015). Water treatment with ion exchange resin, which also consists of plastic material, may contaminate pipes, valves, gaskets (Mani et al., 2018), and rinse and pollute bottles with microplastic. Reverse osmosis - instead of ion exchanger - removes the hardness of the water and all particles from rinsing water including the salts that would be seen as spots on the bottle's surface after the rinsing and filling process.

Dust and dirt from the surroundings fall into the open bottles. Usually, new bottles are packed very well by the glass factory, but friction during transport and the opening of the shrink foil packaging could cause the open bottles to be polluted again. Especially, damaged pallets of new bottles which might even contain broken glass should be refunded by the transport insurance and discharged or returned to the supplier (The concept of Good Manufacturing Practice, 2012). Pallets that were not depleted after filling a batch should be discharged as well, especially if no rinsing machine is available to rinse out dust or insects that might have fallen into the bottle between the filling dates. For return bottles, big washing machines are necessary. The chemical treatment in big caustic baths releases lots of printing pigments of the labels that have to be removed. These have to be rinsed off with lots of clean water after the treatment. Bottle inspection and monitoring units with cameras can detect coarse contaminations in the empty bottle before the filler (Blüml, 2006). Cameras are able to detect optical deviations from clean bottles with flashlights, which produce shadows of even transparent objects and "see" objects in visible, infrared, and UV light at a speed of 50,000 bottles per hour (bph) and more (Fig. 25.3).

Reliable recognition of flawed bottles leads to rejection and the discharge of unsafe bottles by a pusher or a diverting system. Inspection machines are far more sensitive than visual controls by employees, who will get tired by the boring bottle inspection soon, even if only 1000 bph are filled. The aim is to reject faulty bottles and check the containers. This is of crucial importance not least for reasons of product liability. Course checks assure the visual appearance on the supermarket shelves (Blüml, 2006). Smaller wineries, who cannot afford an inspection machine, should rely on the original packaged glass bottles from the glass factory, especially if no rinser is used before bottling. This implies throwing away residual bottles when the wine tank is empty and new bottles are left over but would have to be kept under unsafe conditions, where dust or insects might find their way into the bottles (Fig. 25.4).

FIGURE 25.2   Insect contamination, a spider in a rejected wine bottle.

FIGURE 25.3 Bottle inspection at the glass production. Camera supervision of new bottles by flashlight.

FIGURE 25.4 Unsafe bottles not originally packed anymore. Only new bottles, originally packed, should be used at any filling plant.

25.2 Physical damage of bottles

25.2.1 Fillers

Depending on the type of filling system involved, the following filling steps can be performed:

- Evacuation;
- Flushing with ring-bowl or pure gas;
- Pressuring with ring-bowl or pure gas;
- Filling at one speed or two speeds;
- Settling and snifting (pressure release) (Blüml, 2006).

Long tube fillers release the wine about 2 cm above the bottom of the bottle. Carbon dioxide evaporates and, as it is heavier than air, forms an oxygen-free layer on the surface of the wine, which removes the air out of the bottle, thus sheltering the direct contact to the wine. Long tube fillers have less uptake of oxygen during filling than short tube fillers. If the same size bottles are always filled, the filling tubes do not have to be changed. During a filling stop, it has to be observed that wine may become touch dry on the filling pipe, and, when filling starts again, the dried matter will not be soluble again and could be seen as turbidity or flocks in the wine of the following bottles. When filling sparkling wine, these bits can be the source of gushing. It is better to finish all bottles and close them before the filler is stopped. This means some buffer capacity on the conveyers after the filler and the necessity to close all bottles after filling as soon as possible. During breaks of more than 2 m, the filler and the filling tubes should be rinsed with water to remove all organic matter, if filling stops occur.

Long tube fillers have a disadvantage if the type of bottle is changed. Usually, the tubes have to be changed as well, especially if the following bottles are different in height. Each filling tube has to be exchanged, which is time consuming. After mounting the tubes, the filler should be rinsed and sterilized again. Short tube fillers (or tubeless fillers) can be easily adjusted to other bottle forms, as long as the bottleneck of the bottles and the filling level in the bottleneck are similar. The tube of short tube fillers is not a real filling tube. It is an exhaust tube that limits the filling height as the filling is stopped when the tube is blocked by the wine and the gas in the bottle cannot leave the bottleneck anymore.

As the short tube has less surface, there are fewer problems with desiccating wine or carryover of the wine into the next bottle to be filled. As well, cleaning short tube fillers is much easier. The height the bottle has to be lifted to be pressed on the gasket is much less, so quicker filling with the same size of the filler is possible. On the other hand, contact to air in the bottle should be avoided by prerinsing the bottle with nitrogen or carbon dioxide, or even preevacuating the bottle by vacuum and rinsing it with inert gas to avoid contact or uptake of the oxygen of the air. The wine should run in a laminar flow on the inner surface of the bottle to the bottom and fill up the bottle without whirls or vortex. The gas is pressed out by the inflowing wine until the exhaust tube is blocked. The pressure equilibrium in the bottleneck stops the inflow. Pressure hammers from the ring canal may result in a higher filling level. The pressure should be adjusted and controlled over the filling process by soft opening valves, frequency-controlled pumps, and buffer tanks before the filler.

25.2.2 Filling level adjustment

Bottles that are under- or overfilled are contravention to laws (Blüml, 2006). Too much wine will cause trouble if there is tax on the wine. This might be a fiscal offense. Especially, wine bottles sealed with a cork may press out the overfilled wine out of the bottle by the cork's volume, which looks ugly and brings wine to the surface of the cork. This leads to mold, as the organic material at aerobic conditions will become nutrition for fungi. The sealing function of the cork might suffer as well. Headspace in cork-finished bottled wine should be sufficient to meet legal requirements and also be adequate so that wine expansion due to temperature fluctuations will not cause movement of the corks (The concept of Good Manufacturing Practice, 2012).

Overfilled screw-cap bottles will have no volume buffer for heat extension. This can cause leakage or breaking of the glass. Not enough wine in the bottle will disappoint the customers and is dangerous, too, as a bigger gas volume can build up pressure in the bottle when it is heated up and cause the removal of the cork or an explosion of the bottle. To be independent of pressure changes in the ring canal, electronic sensors were developed. The filling tube is constructed as an electrode with insulation at the filling level. When the wine in the bottleneck rises above the insulation, an electrical circuit between the upper and the lower part of the filling tube effected by the conductivity of the wine indicates that the bottle has reached its filling level. The signal is used to close the filling valve. The bottle is filled to the target level, independent of the pressure of the ring canal. This can be used for long tube and for short tube fillers.

Volumetric fillers are popular, where accurate filling levels have to be assured. Volumetric filling with a metering chamber or a flow meter (Blüml, 2006) ensures filling a defined volume into the bottle. Modern volumetric fillers measure the wine flow by an electromagnetic flowmeter and close the filling valve when the desired volume is reached in the bottle. This is independent of pressure drops or peaks and may easily be changed electronically when other bottle sizes shall be filled. This is also very useful when flexible containers like PET bottles, cardboard packages, or Bag in Box are filled.

25.2.3 Bursting of bottle during filling

Poor bottle quality, reduced thickness of the glass, crashes on the conveyers, scuffing, microcracks, or shards from exploding bottles harming the adjacent bottles are all reasons for potential bottle bursts. Where a breakage has occurred on the filling line, particular attention must be paid to all open adjacent bottles to ensure freedom from glass shard contamination. When breakage has occurred on the filling machine, particular attention should be paid to the bottles filled from adjacent filler heads (The concept of Good Manufacturing Practice, 2012). The broken glass might still be adhesively bound to the filling rods or gaskets and later on may fall off accidentally into one of the next bottles. As glass shards of all sizes diffuse in the area of the burst, all open bottles in this area should be emptied and the filler should be rinsed with water to rinse off all shards. Particular attention must be paid to any bottle breakages when filling sparkling wine under higher pressures (1–1.5 bar above CO_2 saturation pressure) because of the wider distribution of the shards of glass and remains of the shattered bottles.

FIGURE 25.5 Glass splinter in a rejected white wine bottle.

Any open bottles nearby should be removed and decanted (The concept of Good Manufacturing Practice, 2012). Hot filling also means more thermal stress to the inner tension of the glass in the bottles, which will result in burst bottles.

Some modern fillers run automatic bottle burst programs. If a bottle breaks while filling, this can be recognized by the filler's control unit by a drop in pressure in this filling tube or at the fixing bell of the bottle. The computer control will stop the filling of the adjacent five or six bottles. These bottles will be underfilled and can be sorted out at the filling level check. Emptying the filler carousel and rinsing the whole filler helps to avoid problems with falling shards that have landed above the open bottlenecks somewhere on the filler (Fig. 25.5). Rinsing them off avoids them descending due to gravity later when open bottles enter and leave the filler again. This should be specified in HACCP concepts for product safety.

It is recommended that routine postproduction checks also occur in all wine bottles to assess the presence of any foreign material such as glass or cork dust, for example (The concept of Good Manufacturing Practice, 2012). Filled bottle inspection can be done with an x-ray to find broken glass or filling tubes after closing the bottle even in the filled wine bottle before labeling.

25.2.4 Microbiological problems of fillers

Nowadays, fillers are made of stainless steel, and a hygienic design is applied where possible. Pipes, valves, and containers are not that difficult to be kept sterile. Rinsing with water, hot water, CIP with caustic and acid, sterilizing with hot water above 85°C, or even steaming the filler keeps the surfaces that get into contact with the wine as sterile as necessary. More and more computer-controlled independent motors have replaced big mechanical gears, but fillers still have mechanical moving parts with grease, gaps, and wear parts. Usually, the mechanical part of a filling machine is beneath, where the product is conveyed in open bottles, certainly to avoid oil or grease falling into an open bottle. When wine is running in those gaps, a biofilm will occur, which may contain wine spoilage bacteria or yeasts. Water rinsing should be done frequently, dependent on the biological status of the periphery and the sensitivity of the filled wines (residual sugars, lower alcohol or alcohol-free wines need more care). The rotating movement of a filler is able to ventilate aerosols with germs above the open bottles, and presumably, some of these germs will fall into an open bottle. Even if most of them will not survive, the survival of a selection of spoilage microorganisms is possible and can lead to product recalls.

25.2.5 Temperature during filling

As white wine should have about 1 g/L of CO_2 for its liveliness, it should be respected that the filling pressure should have a preset of 1−1.5 bar above the saturation pressure (Blüml, 2006). The colder the product, the better the CO_2 is bound and the lesser the pressure needed. Fillers with a slight underpressure to remove overfilled wine from the bottle into the ring canal of the filler by vacuum remove some CO_2 as well. These fillers should fill at temperatures as cold as possible or necessary for the desired final CO_2 content. Cold filling costs more energy and, in addition, water condensing on the outside of the bottle may cause difficulties with labeling and downstream packaging (Blüml, 2006). Problems may occur when humidity results in condensation on the bottles, especially if labeling is done immediately after bottling. Avoiding humidity in the air at bottling, labeling, and packing saves energy and leaves the wine at cellar temperature.

25.3 Closing systems

The closing of a wine bottle is usually performed by pressing a cork into the bottleneck or putting a screw cap on the thread of the bottle by torque or rolling the aluminum onto the glass thread.

25.3.1 Screw caps

Screw caps are spaced in the closure chute and passed on the capper from above. The closure is accepted by the pick-up device and fixed in position above the bottle. The capping element places the closure on the bottle, whereupon it is pressed against the bottle by spring pressure. As soon as the closure reaches its specific screw-on torque, the hysteresis clutch slips, and the closing operation is terminated (Blüml, 2006). Screw caps bring along their volume filled with air. To avoid oxygen, rinsing the caps with pure nitrogen will reduce the amount of oxygen, sealed in the bottle (Strobl, 2019). The treatment of the gas volume in the bottleneck with nitrogen or CO_2 may also decrease oxygen in the bottle so that sulfur amounts can be reduced.

25.3.2 Cork

A cork closing without fumigation or evacuation presses the bottleneck air into the wine. The air in the bottleneck containing oxygen will diminish the reductive substances, the SO_2, and oxidate aroma components. This can be minimized by evacuating the air from the bottleneck by vacuum, which makes it easier to press in the cork (Strobl, 2019). Nitrogen droppers can inject a drop of fluid nitrogen into the bottle. The nitrogen will evaporate and remove air from the bottleneck. A pressed-in cork will seal the nitrogen above the wine and build up some pressure in the bottleneck. White wine tolerates up to more than 1 g CO_2/L; rinsing with CO_2 can minimize the oxygen in the bottleneck before sealing it. As CO_2 is soluble in wine, less pressure will arise after closing the bottle compared to the nitrogen treatment.

25.3.3 Leaky closures

Leaky closures should not result from the tolerances of the bottleneck diameter at a maximum point and of cork diameter at a minimum point. The production specifications have to cover the worst-case scenario: small cork in a wide bottleneck. Bottleneck tolerances are checked at the glass production (Fig. 25.6).

Overfilled bottles might get leaky due to the wick effect, the overfilled wine in the bottle is pressed between cork and glass to the surface of the cork and can mold (Fig. 25.7).

FIGURE 25.6   Bottleneck inspection at the glassworks. The inner diameter of the bottleneck is checked because otherwise, leakage or inability to open bottles can occur.

FIGURE 25.7 Mold on the cork at an overfilled white wine bottle.

In traditional small wineries, bottles are straightened in a horizontal position for a while before labeling to look for leakages. The spilling wine stains other bottles so that these have to be washed before labeling. The other reason for keeping corked bottles in a horizontal position is the possibility of drying out of the corks and the loss of flexibility. A controlled humidity in the storage room can ensure equilibriums of evaporation and condensation of humidity in the cork.

Screw caps might not get enough pressure while forming out of the thread, so the compound mass cannot fulfill its tightening function. The compound mass is usually a safety valve function and should open at a defined pressure coming from the bottle, usually 10−15 bar. If this pressure is exceeded, the screw cap opens to release the overpressure, but it does not close anymore. This can happen while closing the bottle with very high pressure in the closing machine as well as by piling wine bottles in a vertical position with too many pallets or bottles placed on the closures of the lower wine bottles. The lower layers in such a staple might have opened compound masses by the pressure from the weight of the wine bottles above.

As untight bottles are often the reason for the soiling of the bottles and boxes, oxygen uptake, and microbiological infection, an optical check of the closures is preferred (Prabuwono et al., 2019). Ultrasonic treatment of bottles after closing will show the leakage due to the effervescence of the wine, especially in wines with higher carbon dioxide contents. A delivery control of closures and an evaluation of the quality standards should be conducted batch wise. The forces and functions of the closing machines should be checked regularly to avoid leaking wine bottles. On the other hand, the force required to open the bottles has to be checked to avoid trouble with consumers who have difficulties opening the bottle.

25.4 Pasteurization and preservatives

White wine is usually protected against most microorganisms by hygienic treatment in the wineries, sulfur, ethanol, CO_2, lack of oxygen by reductive substances, SO_2, and a low pH. High sugar content or reduced alcohol level makes wine interesting for lots of bacteria and yeasts. Uncontrolled secondary fermentations in the bottle will produce alcohol, which is certainly worst for wines that are declared alcohol free. Turbidity is an indicator of spoilage for the consumer. Pressure might press out the cork or burst the bottle. This can be prevented by pasteurizing the whole bottle. If sensitive wines are filled, further conservation methods have to be considered.

25.4.1 Preservatives

Preservatives can be ascorbic acid, which reduces oxygen in the wine but has an antibacterial effect in combination with the SO_2. Sorbic acid is similar but might lead to geranium notes as an off-flavor. Dimethyl dicarbonate (DMDC, E242, Velcorin) is exigent to dose into the wine and dangerous if set free as a concentrate. It is also quite expensive and has to be calculated against thermal treatment (Strobl, 2008).

25.4.2 Pasteurization

Heat treatment has been applied by Louis Pasteur to improve wine already in the 1860s. The temperature and the duration of the treatment are dependent on the alcohol contents of the wine, the pH, and the residual sugars and have to be defined for each wine separately in order not to overheat the product but have the microbiological effect of killing all germinable bacteria and yeasts. The alternatives are applying flash pasteurization before the filler. This is economically the best way as most of the heat energy can be recovered when the treated hot wine heats up the inflowing cold wine in a plate heat exchanger, and merely the heat losses have to be replaced. The wine temperature after treatment is a bit higher, which is good for labeling after the filling process (Strobl, 2019).

Recontamination of the wine during the filling process might spoil this effect. Applying hot filling makes sure that pasteurization of the bottle and the closure is performed with the wine itself. Depending on the ethanol contents of the wine, pH, and affinity to bacteria and yeast, the wine is heated up to 60°C and more and filled into the bottles hot. The bottle will be closed by screw caps. The cooling of the wine will occur in a slight vacuum, which would suck in a cork. The screw caps are pasteurized by the heat and steam above the wine. To reduce the thermal influence the bottles should be cooled down with water as soon as possible. For the sake of the glass, too great a difference in temperature between the cooling water and the bottle should be avoided because thermal tensions in the glass might crack the silica structure in the glass. Heat recovery of this cooling water is difficult, as lower temperatures and contamination with wine from exploding bottles make this inefficient. The fillers certainly have to be designed for hot filling.

Filling the bottle, closing it, and then pasteurizing the filled bottle after filling is a method to pasteurize the wine, the bottle and the closure in one step. This preserves the carbondioxide contents of white wines during filling. It has to be ensured that the wine in all bottles is heated up to the pasteurization temperature after the filling provess, even in the middle of the bottle and maintained for the time to achieve the pasteurization unit necessary. This can be conducted in chamber pasteurizers, a room where one or two pallets are heated up and cooled down again. Data loggers can check the pasteurization temperatures in the middle of a prepared test bottle especially in the center of the pallet, which is more vulnerable to heating up and cooling down again. Tunnel pasteurizers transport the bottles on conveyers through a rain of warm and then cold water to heat them up, keep the temperature for the time necessary for the pasteurization unit, and then cool them down again. Little heat recovery of the cooling water, used to preheat the incoming bottles, is possible. Small wineries, producing small amounts of alcohol-free or low-alcohol wines, sometimes use autoclaves on a laboratory scale to pasteurize their wines (Fig. 25.8) or just stack bottles in vats with hot water to achieve the pasteurization effect.

Heating up the wine should always avoid the presence of oxygen before, as the temperature will immediately oxidate the aromas. Also, stabilization prior to pasteurization has to be adjusted by treatment with bentonite, silica gel, polyvinylpolypyrrolidone (PVPP), enzymes, or other finings, which may be necessary as the heat may form insoluble components that can be observed as haze or turbidity. Tests before applying such a treatment should be conducted on a laboratory scale.

FIGURE 25.8   Autoclave for pasteurizing low-alcohol wines.

25.5 Labeling

Wrong declaration because of confounded tanks or labels may lead to tax problems and fiscal offense. Especially, wines with alcohol-free labels that accidentally contain alcoholic wine because of a mix-up of tanks or labels can cause big trouble. Quality checks and plausibility checks during production by computers or employees can diminish such faults. Label controls with cameras can compare the label on the conveyers after filling with the data of the filling tanks and the conductivity of the wine in the filler. Wrong labeling is sometimes more expensive than the wine itself, as withdrawal from the market can really harm the wineries. HACCP concepts should also focus on accurate labeling, not only for tax reasons, but also to prevent consumers from the effects of too much Vol% alcohol in the botteled wine, if less Vol% or none alcohol is indicated on the label. . Laboratory checks, staff instructions and education, computer control, plausibility checks, on-the-spot quality checks for color, conductivity, and the correct alcohol content within the fault tolerance should be verified and labelled.

25.6 Shelf life

Wine is known as a stable product with a long shelf-life; no "best before" has to be printed on the products. Modern wines do not always get better during aging. The bottles collect dust on their way to the consumer and should be protected by foils and boxes. Light is a form of energy and alters the aroma components. Light protection is provided by the colored glass: the darker the better. Wines in white glass bottles suffer most from sunlight or artificial light in the supermarkets. Light-tight boxes avoid the impact of light.

25.7 Extortion

Extortioners poison products or just pretend to do so. The product, customer, consumer, and brand can be harmed due to them. Withdrawal of poisoned products from the market can ruin whole enterprises. The sealing of the bottles and boxes should always undoubtedly show the original sealing. The wine closure has to be sealed by caps, shrink foils, and a sealed box to display possible manipulation. If extortion occurs, the police have to be informed at once; they have more experience in this subject than commonly known to the public. The concerned retailers and consumers have to be alarmed. Good traceability of the products helps to avoid comprehensive product recall and public notice. Machine-readable codes and logistic documentation ensure minimizing a certain extent of the problem.

25.8 Effectivity of bottling white wine

In most wineries, bottling is the most expensive and labor-intensive step. The machinery is expensive and sensitive and well-trained staff is necessary. Big plants bottle around the clock with 30,000 bph and more on several filling lines with the highest quality and safety levels. Arranging the bottling, the wine, packaging material, storage and logistics, and readiness for delivery is usually not only the last chance to avoid flawed products but also to make mistakes. More and more small wineries concentrate on winemaking and bring their wines to service providers instead of paying for the building and the machinery and the craftsmen for just a few days per year. The predominantly unused machinery also ages and stagnates while technical standards develop. Just as winemaking is a job for professionals, so is wine bottling, especially because this process is the last chance to avoid quality issues before the wine reaches the consumer.

25.9 Future tasks in wine bottling

Food production is normally divided into the agricultural part, the food production part, and the packaging and logistics part serving the retailers. Efficient, big brands set the benchmarks and establish quality demands of the retailers' and the consumers' expectations. Although wine still bases on traditional farm production and custom-made wines, the demands are set especially for the final product. Future wine bottling plants will need fully stabilized wines, as haze, precipitations, or changes in colors or taste will not be accepted by the winemakers, retailers, and wine consumers. The impact of avoiding microorganisms and minimizing oxygen content will be the challenge on the product side, especially if low-alcohol wines or nonalcohol wines gain a greater market share. Filling and closing mean sealing safe bottles that must not do any harm to the customer. The wine has to be at its best brightness, color, flavor, and taste when it finally reaches the consumer after the logistic period and shelf life. Traceability will help to shorten processes and to gain knowledge about the final wine drinkers. It also helps at recall campaigns, if necessary. Consumer standards and expectations proceed toward perfection. Big brands need huge capacities just filling one sort of bottles, bottling as effectively as possible. On the other hand, small amounts of wines and special bottles, labels, and packages will keep the product wine interesting. The same requirements for small batches have to be fulfilled as it is standard for big brands. Flexible bottling plants with interesting alternatives to the standard wine bottle will be needed with high, up-to-date bottling quality. New labeling methods, like direct print on the bottle's surface, are able to print each bottle individually at a speed of 60,000 bph (Reynolds, 2018). Individual bottles with custom-made contents will be available, ordered via the internet, and delivered in small batches. As well, big batches will cater to the lower-price segment in the worldwide competition. The sealing of bottles has to be a barrier against physical damage, dirt, light, oxygen uptake, losses, and even extortion. Automatization will increase and filling plants will become more expensive; they will need more bottles to fill to be run efficiently. This will result in service filling centers as service providers that are specialized in the filling part of the wine production instead of each vineyard filling its own wines on sometimes very old, hardly used filling machines operated by employees who run the machines just once or twice a month in comparison to professional wine filling staff, doing nothing else but bottling. Besides energy consumption, the recycling of the materials used for filling, transport, and efficient distribution of the wine will always need improvement.

References

Alexandre, B., Coelho, C., Briandet, R., Canette, A., Gougeon, R., Alexandre, H., . . . Weidmann, S. (2016). Effect of biofilm formation by oenococcus oeni on malolactic fermentation and the release of aromatic compounds in wine. *Frontier in Microbiology*, 7. Available from https://doi.org/10.3389/fmicb.2016.00613.

Blüml, S. (2006). 12. Filling. In (1st ed.H.-M. Eßlinger (Ed.), *Handbook of brewing* (Vol. 1, pp 275–320). Weinheim: Wiley-VCH. Available from https://doi.org/10.1002/9783527623488.ch12.

Eerkes-Medrano, D., Thompson, R. C., & Aldridge, D. C. (2015). *Microplastics in freshwater systems: A review of the emerging threats, identification of knowledge gaps and prioritisation of research needs. Water research* (pp. 63–82). NX Amsterdam, The Netherlands: Elsevier BV. Available from https://doi.org/10.1016/j.watres.2015.02.012.

Giacosa, S., Río Segade, S., Cagnasso, E., Caudana, A., Rolle, L., & Gerbi, V. (2019). *Chapter 21 - SO$_2$ in wines: Rational use and possible alternatives*, In *Science direct*, (1st ed., Vol. 1, pp. 309–321). Amsterdam: Elsevier. Available from https://doi.org/10.1016/C2017-0-01326-5.

Großmann, M. (2020). SO$_2$ senken. *Der Deutsche Weinbau*, 07(07), 32–35.

Guise, R., Filipe-Ribeiro, L., Nascimento, D., Bessa, O., Nunes, F. M., & Cosme, F. (2014). Comparison between different types of carboxylmethylcellulose and other oenological additives used for white wine tartaric stabilization. *Food Chemistry*, 156, 250–257. Available from https://doi.org/10.1016/j.foodchem.2014.01.081, Epub 2014 Feb 5.

Krones. (2010). *Compliments of the season*. Neutraublingen: Krones. Retrieved from https://www.krones.cn/en/products/references/compliments-of-the-season.php.

Lindemann, B. (2006). Filtration and stabilisation. In (1st ed.H. M. Eßlinger (Ed.), *Handbook of brewing* (Vol. 1, pp. 225–234). Weinheim: Wiley-VCH. Available from https://doi.org/10.1002/9783527623488.

Manger, H.J. (2005). Brauerei Forum, 11.

Mani, T., Blarer, P., Storck, F. R., Pitroff, M., Wernicke, T., & Burkhardt-Holm, P. (2018). Repeated detection of polystyrene microbeads in the Lower Rhine river. In C. Sonne, & E. Zeng (Eds.), *Environmental pollution* (pp. 634–641). Amsterdam: Elsevier. Available from https://doi.org/10.1016/j.envpol.2018.11.036.

Prabuwono, A. S., Usino, W., Yazdi, L., Basori, A. H., Bramantoro, A., Syamsuddin, I., . . . Allehaibi, K. H. S. (2019). Automated visual inspection for bottle caps using fuzzy logic. *TEM Journal*, 8(1), 107–112. Available from https://doi.org/10.18421/TEM81-15.

Reynolds, P. (2018). Direct digital printing on rigid containers. *Packaging World*, 04. Retrieved from https://www.packworld.com/home/article/13374742/direct-digital-printing-on-rigid-containers.

Schneider, I. (2019). A high-performance sheet solution for TAB. *Beer Brewing International*, 02, 12–15.

Strobl, M. (2008). Haltbar machung von Getränken. In (3rd ed., H. W. Back (Ed.), *Getränke* (Vol. 5, pp 275–321). Hamburg: Behr's Verlag. Retrieved from https://www.behrs.de/titel/mikrobiologie-der-lebensmittel/216?p = 35.

Strobl, M. (2019). 22 Red wine bottling and packaging. In A. Morata (Ed.), *Red wine technology* (Vol. 1, pp. 323–339). London, San Diego, Cambridge, Oxford: Elsevier. Retrieved from https://www.elsevier.com/books/red-wine-technology/morata/978-0-12-814399-5.

The concept of Good Manufacturing Practice. (2012). *The code of good manufacturing practice for the Australian grape and wine industry* (2nd ed., p. 33). Glen Osmond, SA 5064 Australia: The Australian Wine Research Institute, ISBN:0958683999.

Zoecklein, B. W., Fugelsang, K. C., Gump, B. H., & Nury, F. S. (1990). *Chapter 13 tartaric acid and its salts. Production wine analysis* (pp. 289–315). Boston, MA: Springer. Available from https://doi.org/10.1007/978-1-4615-8146-8_13.

White winemaking in cold climates

Belinda Kemp[1,2], Andreea Botezatu[3], Hannah Charnock[1], Debra Inglis[1,2], Richard Marchal[4], Gary Pickering[1,2,5], Fei Yang[2] and James Willwerth[1,2]

[1]Department of Biological Sciences, Faculty of Mathematics and Science, Brock University, St. Catharines, ON, Canada [2]Cool Climate Oenology & Viticulture Institute (CCOVI), Brock University, St. Catharines, ON, Canada [3]The Horticulture Department, Texas A&M AgriLife Extension Service, Texas A&M University, College Station, TX, United States [4]Laboratoire d'Oenologie de Chimie Appliquée, Unité de Recherche Vignes et Vins de Champagne (URVVC), Reims, France [5]National Wine and Grape Industry Centre, Charles Sturt University, Wagga Wagga, NSW, Australia

26.1 Introduction

The choice of white grape varieties, clones, and rootstocks planted in cool/cold climate wine regions plays a major role in the winemaking decisions and techniques used to produce cool/cold climate wines. Traditionally, cool/cold climate wines have been associated with higher acid and lower sugar concentrations at harvest than warm climate wines, resulting in juice and/or wine adjustments at harvest (Zhang et al., 2015). Depending on regional wine production legislation adjustments can include sugar addition to juice, deacidification for sparkling and still wines in cold years, and acidulation in hotter growing seasons for wine styles like Icewines. Current acid reduction methods for juice or wine can involve chemical methods that use alkali salt to react with organic acid in the juice (Li et al., 2019). Physical deacidification methods include cryogenic freezing, electrodialysis, organic extraction, and anion-exchange resin (Li et al., 2019). Biological acid reduction methods also exist for reducing malic acid content and are commercially available including the bacteria *Oenococcus oeni*, (formerly referred to as *Leuconostoc oenos*), and *Lactobacillus plantarum* (Krieger-Weber, Heras, & Suarez, 2020). Non-*Saccharomyces* yeast *Schizosaccharomyces pombe* has been known for some time to also decrease malic acid (Vilela, 2019).

The seasonal weather variation influenced by climate change, specifically extreme weather events, are factors that influence grape chemical composition (acidity and sugar levels) (Neethling, Petitjean, Quénol, & Barbeau, 2017). However, once harvest dates have been determined, and the grapes picked (manually or mechanically), the processing and winemaking techniques used by the winemaker are dependent upon the grape variety and wine style required. Depending on the winemaking goals for each grape variety, different pressing cycles can be used for juice extraction (Roman et al., 2020). The juice fractions in commercial wineries are usually divided and managed differently, depending on the chemical composition of the juice fraction. The decision is based on pH levels and potassium concentrations, varietal aromas, or phenolic compounds that influence wine composition and sensory properties (Roman et al., 2020). Crushing and destemming, skin contact in the press after crushing, or whole-bunch pressing can be used according to the aim of the winemaker. Yeast choice (inoculated or uninoculated), choice of fermentation vessel (stainless steel tank, oak barrel, concrete eggs), fermentation temperature, juice contact with grape skins and seeds, the type of yeast nutrient product added to the juice/fermentation, malolactic fermentation (MLF), and the use of oak for aging wines, all impact the aroma profile, flavor, and appearance of white wine. For example, low temperatures (10°C−15°C) increase the production of volatile aroma compounds (esters, acetates, medium-chained fatty acids) in white wines. However, the presence and concentration of aroma compounds in white wines are also highly influenced by the amount of aroma precursors located in the grape skin and pulp (Zhang et al., 2015). Additional influences on white wine appearance, aroma profile,

339

© 2022 Elsevier Inc. All rights reserved.

and flavor include oxygen management before or during fermentation (Day, Schmidt, Pearson, Kolouchova, & Smith, 2019), aging wine in contact with yeast lees (Sancho-Galán, Amores-Arrocha, Jiménez-Cantizano, & Palacios, 2020) fining for clarification and stabilization, and filtration. In this chapter, we discuss further factors that influence the appearance, aroma profile, and flavor of white wines produced in cool/cold climates. These include grape varieties (clones and rootstocks), varietal aroma (specifically methoxypyrazines), winemaking techniques (use of grape skin contact, protein haze removal), and wine style specific to cold climate wine regions (Icewine).

26.2 *Vitis vinifera* varieties, clones, and rootstocks for cold climates

Cool or cold climate viticulture regions require proper grapevine selection and management to ensure quality grape and wine production. Matching cultivar, clone, and rootstock to the vineyard site is one of the most critical decisions that must be made. In cool climates, the most limiting factors include winter survival and adequate fruit maturity to allow for consistent production and quality of grapes. The colder the region, the more critical it is to have proper grapevine selection as well as sound management practices to ensure winter survival, good production, and levels of grape maturity.

Grapevine suitability is strongly associated with environmental conditions. For cool and cold climates, ideal cultivars have some key attributes: adaptability to a range of soils and climates, good to excellent cold tolerance, and ability to mature fruit in shorter growing seasons and in cooler conditions (i.e., lower heat units). To obtain optimal terroir expression, grapes should mature under cool conditions at the end of the growing season (Van Leeuwen & Seguin, 2006). Therefore cultivar selection should not only be based on their tolerance to the extremely low temperatures for that region but also match the growing season conditions to ensure high wine quality.

In general, white cultivars are often more desirable than red cultivars in colder regions because they can be more fruit forward at lower maturity levels and lack green flavors. It is also more widely acceptable for white wines to be produced with some residual sugar to balance higher acidity found in grapes produced in colder regions. There are wide ranges of cultivars used for white wine production that include a large number of hybrid and *V. vinifera* cultivars. Some common *V. vinifera* cultivars grown in cool climate areas are described in Table 26.1.

TABLE 26.1   Common cool-climate *Vitis vinifera* white cultivars and their general viticultural characteristics.

Cultivar	Maturity	Annual freezing tolerance (relative to other cool-climate *V. vinifera*)[a]	Maturity groupings and biologically effective degree days to ripeness[b]
Auxerrois	Early season	Very good	Group 2 (1080 Edays)
Bacchus	Early season	Very good	Group 2 (1080 Edays)
Chardonnay	Mid season	Very good	Group 3 (1140 Edays)
Chasselas	Early season	Good	Group 2 (1080 Edays)
Ehrenfelser	Mid season	Poor	Group 3 (1140 Edays)
Gewürztraminer	Mid-late season	Moderate	Group 3 (1140 Edays)
Grüner Veltliner	Late season	Good	Group 4 (1200 Edays)
Müller-Thurgau	Mid season	Poor	Group 2 (1080 Edays)
Muscat Ottonel	Early season	Poor	Group 2 (1080 Edays)
Optima	Early season	Very good	Group 1 (1020 Edays)
Ortega	Early season	Very good	Group 1 (1020 Edays)
Pinot Blanc	Mid season	Moderate	Group 3 (1140 Edays)
Pinot gris	Mid season	Good	Group 3 (1140 Edays)
Riesling	Late season	Very good	Group 4 (1200 Edays)
Scheurebe	Early season	Moderate	Group 3 (1140 Edays)
Siegerrebe	Early season	Moderate	Group 2 (1080 Edays)

[a]*Dami, Li, and Zhang (2016), CCOVI VineAlert website (http://www.ccovi.ca/vine-alert).*
[b]*Gladstones (1992).*

TABLE 26.2 Characteristics of common cool-climate rootstocks.

Rootstock	Parentage	Ease of propagation	Phylloxera protection	Nematode resistance	Calcareous soil adaptation	Soil Recommendation	Scion vigor
Riparia Gloire de Montpellier	*V. riparia*	High	High	Low	Low	Deep, moist, fertile soils	Low
Couderc 3309 (3309C)	*V. riparia x V. rupestris*	High	High	Low	Low	Deep, well-drained, moist soils	Low to medium
Millardet et de Grasset 101−14 (101−14 Mtg.)	*V. riparia x V. rupestris*	High	High	Medium	Low	Deep, moist, soils	Medium
Selektion Oppenheim 4 (SO4)	*V. riparia x V. berlandieri*	High	High	Medium	Medium	Light, well-drained, low fertile soils	Medium
Teleki 5C (5C)	*V. riparia x V. berlandieri*	High	High	Medium	Medium	Moist, heavy soils	Medium
Kober 5BB (5BB)	*V. riparia x V. berlandieri*	High	High	Low-Medium	Medium	Moist, heavy soils	Medium

Modified from Cousins, P. (2005). Evolution, Genetics and Breeding: Viticultural Applications of the origins of our rootstocks. In: Cousins, P. & Striegler, R.K. (Eds.),
Proceedings of the 2005 rootstock symposium (pp. 1−7). Osage Beach, MO: Southwest Missouri State University; Shaffer, R., Sampaio, T.L., Pinkerton, J., & Vasconcelos,
M.C. (2004). Grapevine rootstocks for Oregon vineyards. Corvallis, OR: Oregon State University Extension.

There are many clonal selections for important white *V. vinifera* cultivars such as Chardonnay and Riesling, which have been grown for centuries in cool climate regions, whereas other cultivars have very few clones. Clones have slight variations in traits that can make them more desirable to grow based on their growth habit (Anderson, Smith, Williams, & Wolpert, 2008), production capacity (Wolpert, Kasimatis, & Weber, 1994), cluster morphology (Castagnoli & Vasconcelos, 2006), or cold tolerance (Hébert-Haché, Inglis, Kemp, & Willwerth, 2020). Some clones also have desirable traits for winemaking with respect to the timing of maturity, sugar:acid ratio, or unique aroma compounds. One example is the Chardonnay ENTAV clone 809, which is considered a "Musqué clone" with higher monoterpene content that has fruity, floral, and Muscat characteristics compared to more neutral-flavored clones (Duchêne et al., 2009). Clonal selection can be used to help improve wine quality or achieve a certain wine style, but knowledge of clonal performance is still lacking in many viticulture regions. It should be noted that while proper clonal selection is an important criterion, it is not as critical as grape variety or rootstock selection.

Many cool climate regions are located in areas with a presence of phylloxera, so the use of rootstocks is essential since *V. vinifera* cultivars are not tolerant to this pest. In addition to phylloxera tolerance, grafting scions to rootstocks can improve tolerance for many abiotic stresses (Corso & Bonghi, 2014), biotic pests (Naegele, Cousins, & Daane, 2020), or to mitigate vine vigor (Migicovsky et al., 2019). Rootstocks used in cool climate regions generally require excellent cold tolerance and adaptability to a wide range of soil conditions including varying levels of soil moisture. The cold hardiness of a rootstock does not translate to the scion grafted to it (Cousins, 2005; Hébert-Haché et al., 2020). However, rootstock mortality can occur if a cold tender rootstock is selected, so hardiness is still an important consideration where winters are cold. In many cases, rootstocks used in cool/cold climate regions have *Vitis riparia* in the parentage, which provides superior cold tolerance (Table 26.2). Some regions require very specific rootstock selection due to a particular soil attribute, such as areas with very calcareous soils or arid conditions. However, some very common rootstocks are typically used in cool climate regions across the world. Some of these commonly used rootstocks are described in Table 26.2 (Shaffer, Sampaio, Pinkerton, & Vasconcelos, 2004).

26.3 Methoxypyrazines in white wines

26.3.1 Sources and significance of methoxypyrazines

Alkyl-methoxypyrazines (MPs) are aromatic compounds found in various species of fruits, spices, vegetables, and nuts. They have also been identified in several grape cultivars and their wines. MPs can have a positive impact on the aroma profile of some common varietal wines and their blends, particularly Sauvignon blanc and Cabernet Sauvignon, but they can also have a detrimental effect at higher concentrations where they can contribute undesirable "green"

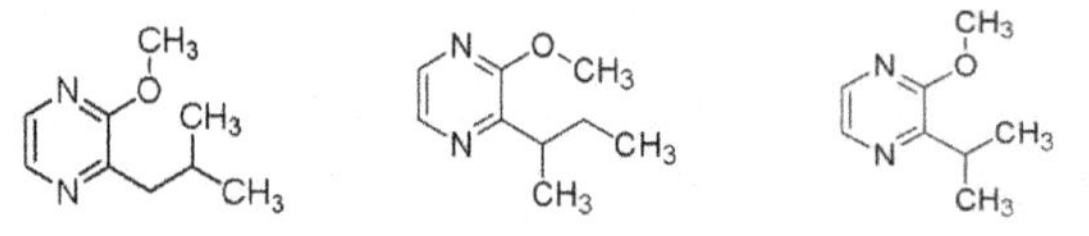

FIGURE 26.1   The most prevalent methoxypyrazines in white wine.

characters (Allen, Lacey, Harris, & Brown, 1991). While the grape is the usual source of MPs in wine, they can also be sourced from *Coccinellidae* beetles found in vineyards if they are incorporated in with the grapes at harvest, in which case the resulting unpleasant characters are known as ladybug taint (LBT) (Botezatu et al., 2012). Three MPs are consistently reported in wines across a range of grape varieties, namely isobutyl methoxypyrazine (IBMP), isopropyl methoxypyrazine (IPMP), and sec-butyl methoxypyrazine (SBMP), while two others, namely 3-methyl-2-methoxypyrazine (EMP) and 2,5-dimethyl-3-methoxypyrazine (DMMP), have been tentatively reported in both grapes and wine (Allen & Lacey, 1993; Botezatu et al., 2016) (Fig. 26.1).

Human detection thresholds for MPs are very low: 0.3−2 ng/L and 16 ng/L for IPMP and IBMP in wine, respectively (Pickering, Karthik, Inglis, Sears, & Ker, 2007, Sala, Busto, Guasch, & Zamora, 2004), and 1−2 ng/L for SBMP in water (Sala et al., 2004), making them potent odorants at the concentrations found naturally in many wines. Noteworthy is the large variation in the sensitivity of individuals to MPs. For instance, detection thresholds for IPMP in a Gewürztraminer varietal wine ranged from 0.3 to 95 ng/L (Pickering et al., 2007); the taster with the lowest threshold was over 350 times more sensitive than the individual with the highest threshold. Each MP may contribute subtly different but important aroma qualities to white wine. For instance, IBMP is typically described as "green pepper/bell pepper-like," whereas IPMP is also characterized as "peanut and potato-like" (Allen et al., 1991). The specific contribution of SBMP is less clear, but likely includes related earthy and herbaceous dimensions (Botezatu, Kotseridis, Inglis, & Pickering, 2013), and reported SBMP levels in wine to tend to be lower than those for IBMP and IPMP (Alberts, Stander, Paul, & De Villiers, 2009; Botezatu, Kemp, & Pickering, 2016). However, possible additive effects have been proposed, whereby individual MPs present at sub-threshold levels may work synergistically to influence the aroma profile of the wine (Botezatu et al., 2016).

MPs in white wines have been reported in several grape varieties, including Sauvignon blanc, Semillon, Chardonnay, and Riesling (Allen et al., 1991; Botezatu et al., 2016), most of which are extensively grown in cool climate winemaking regions (Fig. 26.2). Sauvignon blanc is of particular interest, as its characteristic aroma and flavor profile is primarily defined by just two classes of odorants: MPs, which provide the grassy/herbaceous notes, and volatile thiols (particularly 3-mercapto-hexanol [3MH] and 3-mercapto-hexylacetate [3MHA]), which contribute the tropical fruit attributes (Benkwitz et al., 2012). Sauvignon blanc wines grown in cooler regions tend to have higher levels of MPs compared to those from warmer regions (Alberts et al., 2009), which correspond to greater vegetative and grassy characters. In more complex examples popular with many consumers, these notes are typically balanced with tropical/fruity aromas of gooseberry, grapefruit, guava, or passionfruit generated by the thiols (Lund et al., 2009). Thus in whites such as Sauvignon blanc, Semillon, and their blends, MPs are key odorants that contribute to the typicity of the styles, but a fine balance needs to be achieved with the other aromatic components to produce high-quality wines.

26.3.1.1 Managing methoxypyrazines in cool climate whites

The concentration of MPs in grapes and wine is influenced by a range of viticultural, climatological, and enological factors, including leaf removal (Allen & Lacey, 1993), sunlight exposure Allen et al. (1993), grape ripening, climate, maceration duration (Kotseridis, Anocibar Beloqui, Bayonove, Baumes, & Bertrand, 1999), and the yeast strain used (Swiegers et al., 2009).

Aside from cultivar, light exposure and climate likely have the largest effect on both the synthesis and degradation of MPs in grapes (Sidhu, Lund, Kotseridis, & Saucier, 2015). Exposing clusters to sunlight can result in reduced MP concentration in the berries, while high vine vigor leads to elevated levels (Marais et al., 1999). The timing of leaf removal to improve light exposure appears particularly important, with earlier (after berry-set but pre-*véraison*) interventions being most effective, leading to IBMP reductions of up to 60% in both white and red

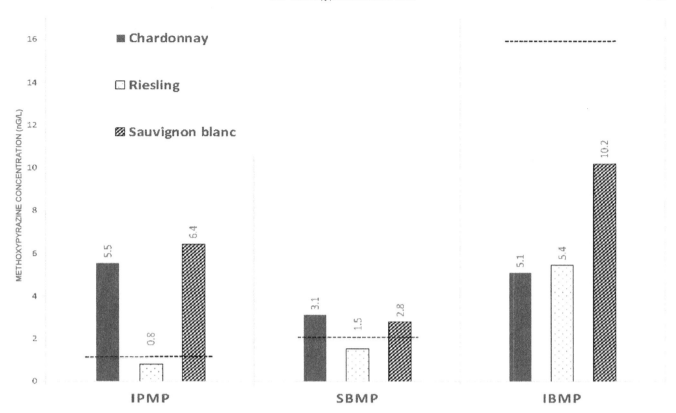

FIGURE 26.2 Average concentrations (ng/L) of 3-isopropyl 2-methoxypyrazine (IPMP), 3-sec-butyl-2-methoxypyrazine (SBMP), and 3-isobutyl 2 methoxypyrazine (IBMP) in wines from three white cultivars. The dotted horizontal line indicates the human detection threshold for each compound.

cultivars (Gregan & Jordan, 2016; Marais et al., 1999). The pruning system can also affect MP concentrations, with minimal pruning resulting in lower MP levels in one study (Allen & Lacey, 1993).

Achieving an optimum balance in white wine between MPs, thiols, and other key odorants, including esters and terpenes, should be a primary goal from an enological perspective. Cool climates tend to favor the accumulation of higher MP concentrations in grapes, so winemaking techniques in these regions are more likely to be focused on reducing MP levels in juice and wine. Harvesting the fruit late helps, as MP content reduces with increasing ripeness. However, other interventions are often needed, especially in years where the grapes have not achieved full physiological ripeness.

Juice clarification significantly reduces MP concentrations, and clarification time also plays a role, with a longer settling period (48 hours) being more effective than a shorter one (24 h) (Kotseridis et al., 2008). However, bentonite does not appear to be any more effective at reducing MPs than natural settling of the juice (Kotseridis et al., 2008). Given the distribution of MPs in the grape clusters (more than 98% are located in the stems and skins), destemming and gentle pressing are advisable when dealing with grapes with elevated MP levels (Roujou de Boubée, Cumsille, Pons, & Dubourdieu, 2002). Conversely, if an increase in MPs is desired, including some stems in the press as well as short skin macerations are options. Trade-offs to be considered are the maturity of the stems themselves, and the potential for harshness from phenolic components from both the stems and skins.

Yeast strain choice may be another tool for dealing with high MP levels. Unfortunately, there is no evidence to date that commercial yeast strains can degrade MPs during fermentation. On the contrary, one study reported a significant *increase* in IPMP concentration in wines fermented with the Lalvin BM45™ *Saccharomyces cerevisiae* yeast (Pickering et al., 2008), suggesting that it should not be used in juices with elevated MP levels. In the same study, a decrease in green aroma and flavor notes was reported in wines produced with Lalvin D21™, likely due to a masking effect from the increased fruitiness produced by this strain.

Another option includes using yeast strains that increase thiol levels, helping to balance the MP-derived characters. Swiegers et al. (2009) evaluated several *S. cerevisiae* strains for fermenting Sauvignon blanc wines and reported that Anchor Wine Yeast VIN7™ produced wines that were the most preferred sensorially by a panel of wine industry professionals, whereas wine vinified with Anchor Wine Yeast VIN13™ had the highest concentration of thiols. In the study, the authors recommend the use of thiol-converting strains rather than thiol-releasing ones, as the conversion of 3MH to 3MHA leads to improved sensory profiles due to the lower detection threshold of 3MHA. An alternative strategy is to consider coinoculation with *Pichia kluyveri* (e.g., Viniflora FrootZen), which, when used with specific *S. cerevisiae* strains, can lead to marked increases in 3MHA (Anfang, Brajkovich, & Goddard, 2009).

Various fining options have been evaluated for their capacity to lower MP levels in wine, with limited success (Pickering et al., 2006). A modest IPMP reduction (11%) was noted in activated charcoal-treated wines, which corresponded to reduced intensity of asparagus flavor. Other MP-related attributes were not altered by the charcoal treatment. Other, less conventional materials show potential with silicone-treatment of both musts (Ryona, Reinhardt, & Sacks, 2012) and wine (Botezatu & Pickering, 2015), yielding large reductions in several MPs, and importantly, with minimal change to nontarget volatile compounds (Botezatu et al., 2016; Ryona et al., 2012). Treatment with a biodegradable polylactic acid-based polymer also led to similar reductions in wine with elevated levels of IBMP, SBMP, and IPMP (Botezatu et al., 2016). Practical integration of these polymers into the winemaking process needs to be further investigated, considering their significant promise.

Winemakers can also influence MP composition through decisions on packaging and closure types. Using Tetra Pak cartons led to significantly lower MP concentrations across an 18-month storage trial compared to glass bottles; however, rapid loss of free sulfur dioxide (SO_2) was observed with Tetra Pak (Blake et al., 2009) MPs than for wine closed with natural cork or screwcap, especially later in the storage trial, most likely due to sorptive effects.

MPs are a class of volatile compounds especially relevant to cool-climate winemaking. While important and desirable components of the aroma signature of Sauvignon and Semillon-based wines, they also detract from quality at elevated concentrations, particularly in wines produced from other white cultivars. Cool-climate wine-growers have significant opportunity to influence MP content in their white wines, primarily through viticultural practices, although several enological techniques can also influence MP composition.

26.4 White winemaking with skin contact

Skin contact white wine production is defined as a prefermentative maceration where grape skins and solids are in contact with the juice before or during fermentation, following a typical red winemaking technique (Fig. 26.3). As with all winemaking processes, utilizing skin and seeds in contact with white grape juice and wine can be modified by winemakers according to the flavor/appearance they require. This strategy aims to enhance white wine quality by liberating characteristic components of the grape variety, which are concentrated in the berry skins and seeds. Flavor components, including glycosidically bound aroma-precursors, phenolic compounds, and other constituents, are infused into the wine during the maceration process, thereby enhancing varietal typicity (Jackson, 1996). Considerable variation arises from differences in vintage, region, grape cultivar, contact time, temperature, and conditions (i.e., SO_2 levels, enzyme activity), which affect the composition, flavor, and stability of the finished wine. Involuntary skin contact may also occur due to fruit damage during harvest or transport.

The duration of the skin contact (weeks, months, or even years) is responsible for the color of the wine, which can range from white to the orange/amber color derived from phenolic compounds. This expanding global wine category (orange/amber) has seen a surge in popularity over the past decade. However, the practice's origins can be traced back 8000 years to the Caucasus region in the modern-day Republic of Georgia where "kvevri," large egg-shaped earthenware vessels, ferment and age white wine on skins for up to 6 months (Glonti, 2010). In addition to Georgia's historical ties, only two other global wine-producing regions have established regulatory criteria for the production of skin-contact wines. South Africa introduced the "skin macerated white wines" category in 2015, and the Canadian province of Ontario included the "skin-fermented white" classification under the Vintner's Quality Alliance of Ontario (VQAO) in 2016 (Lorteau, 2018). The United States lists "amber wine" as an official designation; however, this does not include further details as to the production techniques or methods.

Aromatic varieties, including Riesling, Gewürztraminer, and Muscat, contain high levels of free and glycosidically bound compounds in their skins, which are potentially important contributors to wine flavor. For this reason, these varieties stand to benefit substantially from this production technique. Other varieties including Chardonnay, Sauvignon blanc, and Chenin blanc have also been investigated (Cheynier et al., 1989; Ferreira et al., 1995; Ough, et al., 1969). Conjugates of various aroma-active compounds, including monoterpenes, C_{13}-norisoprenoids, benzene derivatives, and aliphatic alcohols, can be partially hydrolyzed by acids or enzymes during fermentation to yield their odor-active

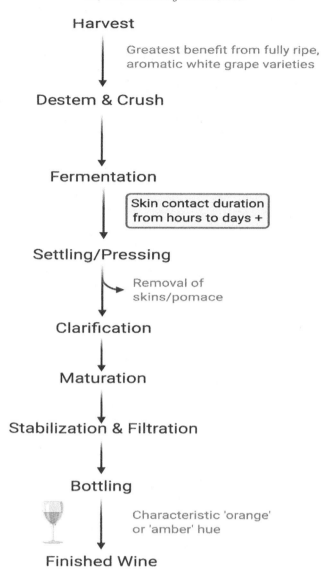

FIGURE 26.3 Flowchart depicting winemaking with juice/wine in contact with skins and seeds.

counterparts. In an evaluation of maceration periods for Muscat Blanc à Petits Grains, the most prolonged treatment (23 h) at 18°C produced a significant increase in both free and glycosidically bound aroma compounds, along with an enhanced sensory perception of the body, likely due to phenol levels (Sánchez Palomo, Pérez-Coello, Díaz-Maroto, González Viñas, & Cabezudo, 2006). Due to the low activity of grape and yeast glycosides under fermentation conditions, exogenous glycosidic and pectolytic enzymes may be a useful tool to supplement the liberation of volatile compounds and aid in the proliferation of varietal characteristics (Cabaroglu et al., 2003).

In Sauvignon blanc wine production, skin contact treatments increased the levels of varietal linked sulfur-flavor precursors, S-cysteine compounds, by up to 50% (Peyrot des Gachons et al., 2002). This study also localized the 3-mercaptohexan-1-ol (P-3MH) precursor, responsible for grapefruit and passionfruit aromas, to the grape skin where it was measured at an eightfold higher concentration than in the equivalent juice. Maceration (19 h, 18°C) led to a 50% increase in the extraction of P-3MH in the juice prior to pressing. It is of note that grape ripeness also plays a factor in skin maceration as inadequate maturity, or disease, will lead to the extraction of undesirable green and herbaceous characteristics (Sokolowsky, Rosenberger, & Fischer, 2013). C_6 aldehydes and alcohols are known contributors of herbaceous aromas, including methyl-2 propanol, n-butanol, and ethyl butanoate. During prefermentation maceration, fatty acid precursors to C_6 compounds are enzymatically degraded, leading to an increase in C_6 species during the early hours of skin contact, followed by a subsequent decrease due to adsorption onto grape solids (Ferreira et al., 1995). Pressing rather than settling following skin contact will thus elevate undesirable herbaceous qualities.

The simultaneous extraction of phenolic components with aromatic compounds from the skins can negatively influence quality parameters, including astringency, bitterness, and oxidative browning precursors. A study of six white grape varieties showed that increased skin contact treatment (24, 48, and 120 h) led to an increase in total phenolic compounds with an increasing astringency rating across all varieties (Singleton, Sieberhagen, Wet, & Wyk, 1975). However, the degree to which phenolic compounds contribute to astringency is different in white wine compared to red wine due to the absence of anthocyanins in white grapes (Oberholster et al., 2009; Townshend et al., 2019). Although anthocyanins do not directly impart astringency, they can increase the solubility of tannins, which carry astringent properties (Jackson, 1996; Townshend et al., 2019). Other classes of phenolic compounds, including hydroxycinnamates and flavanols, influence wine quality and increase with both maceration time and temperature. Hydroxycinnamates are the most abundant class of phenolic compounds found in skin contact white wine and are associated with oxidation and browning precursors (Darias-Martín et al., 2000). Ramey, Bertrand, Ough, Singleton, and Sanders (1986) identified temperatures below 10°C as effective in limiting phenol (especially flavonoid extraction), thereby reducing adverse sensory qualities and color instability. This study also identified temperature control below 15°C minimized the levels of heat-unstable proteins and reduced the requirement for bentonite additions. Oxygen levels are also associated with phenol composition. In a study on maceration conditions of Sauvignon blanc, oxidatively produced wines were lower in polyphenol and flavonoid concentrations due to oxidative polymerization and precipitation (Marais, 1998). The researchers also demonstrated that low-temperature maceration (0°C) for 15 h under reductive conditions positively affected wine quality parameters and presented an increased fruity/ester aromatic intensity.

Decreased acidity in wines in contact with white grape skins and seeds is associated with high potassium levels liberated from skins and seeds. Potassium is the most abundant inorganic ion in wine, and increased concentrations lead to the precipitation of potassium bitartrate, therefore reducing tartaric acid levels (Boulton, 1980). Consequently, as total acidity (TA g/L) declines, pH increases, although alcohol levels and volatile acidity remain relatively unchanged (Ferreira et al., 1995; Ough et al., 1969). The extent of pH increase appears to depend partly on the grape variety and differences in the growing region, possibly linked to varying levels of potassium found in the soil and climate variation. Temperature and contact time also have a significant impact on acidity (Table 26.3).

26.5 Protein stability in white wines

Consumers expect a white wine to be clear, i.e. free of haze and without visible particles. If a haze appears in the bottle, particles form a deposit and will be suspended in the wine when the bottle is handled. The particles, essentially composed of insolubilized proteins, are then visible in the glass. Even if this haze does not affect the olfactory and gustatory perceptions (Peng, Pocock, Waters, Francis, & Williams, 1997), it is undeniably considered as a defect and depreciates the organoleptic image that a consumer will keep of a wine. It is therefore important to clarify wine, but also to ensure that its clarity remains stable after bottling, during transport, and throughout the wine's storage.

Scientific studies devoted to protein precipitation and the estimation of the haze risk have been published for 90 years (Ribéreau-Gayon, 1932). There has also been a renewed interest in protein haze over the last 20 years, since the development of mass analytical techniques and the publication of the grapevine genome in 2007. Continued research demonstrates that the subject remains fully topical in 2020, likely due to the number of parameters contributing to the appearance of turbidity in bottles. To this end, the studies dealing with protein insolubilization in white wines each deal with a particular, precise point, allowing a better understanding of the phenomenon, the conditions in which it appears, the compounds involved, and the mechanisms.

In parallel to these studies, which focus on the compounds involved in protein haze and their characteristics, other studies are looking at how to deal with protein haze.

At the industrial scale, as mentioned in the skin contact in the white winemaking section, treatment with bentonite remains the reference for eliminating the haze risk in white wine, whatever the grape variety, the country, and the climate. In fact, it has been widely demonstrated that treating a must or a white wine with bentonite stabilizes any wine from a colloidal point of view. However, clay particles do not only remove proteins from the must or wine. Various studies have established the loss of color, aromas, and foamability in sparkling wines, and the volume of lees generated along with the fact that some bentonite clays contain a significant amount of heavy metals. For these reasons, many winemakers now wish to abandon bentonite treatment (Theron, Bely, & Divol, 2018). It is in these technical and scientific contexts that technological alternatives to bentonite are studied. However, knowledge of thermo-unstable proteins in wine is needed in order to understand how to treat them (Waters, Hayasaka, Tattersall, Adams, & Williams,

TABLE 26.3 Impact of skin-contact winemaking treatments on acidity, pH, and potassium levels of white wines made from cool-climate grape varieties.

Grape variety	Treatment[a]		pH	TA (g/L)	Tartaric acid (g/L)	Potassium (mg/L)	References
	Time (h)	Temp. (°C)					
Chardonnay	0	15	3.06	6.6	3.5	761	Cheynier et al. (1989)
	4	15	3.09	6.4	2.9	702	
	0	20	3.07	5.4	4.4	710	Ferreira et al. (1995)
	16	20	3.25	4.8	3.6	860	
	0	15	3.35	8.91	3.07	670	
	6	15	3.46	8.32	2.70	767	
	12	15	3.41	8.22	2.73	830	
	24	15	3.54	8.15	2.52	950	
	0	21.1	3.27	0.58	0.24	560	Ough et al. (1969)
	48	21.1	3.51	0.51	0.13	780	
Sauvignon blanc	0	21.1	3.39	0.57	0.19	820	
	48	21.1	3.86	0.41	0.12	1570	
Pinot blanc	0	21.1	3.46	0.62	0.11	1130	
	48	21.1	3.86	0.60	0.12	2020	
Riesling	0	21.1	3.15	0.69	0.24	688	
	48	21.1	3.35	0.67	0.17	860	
Gewürztraminer	0	21.1	3.24	0.59	0.20	616	
	48	21.1	3.62	0.48	0.11	670	
Chenin blanc	0	21.1	3.35	0.58	0.11	800	
	48	21.1	3.43	0.67	0.07	907	
	0	14.4	3.51	0.77	n.d.	700	
	4	14.4	3.61	0.77	n.d.	920	
	12	14.4	3.69	0.77	n.d.	920	
	0	18.3	3.51	0.79	n.d.	790	
	4	18.3	3.60	0.76	n.d.	850	
	12	18.3	3.60	0.74	n.d.	1120	
	0	35	3.51	0.81	n.d.	850	
	4	35	3.60	0.84	n.d.	1020	
	12	35	3.80	0.77	n.d.	1320	

[a]Control treatment: skin contact time = 0 h.

1998). The two groups of proteins capable of colloidal haze are grape berry proteins: the thaumatin-like proteins (TLPs) and chitinases (Marangon et al., 2014; Marangon, Sauvage, Waters, & Vernhet, 2011; Pocock, Hayasaka, McCarthy, & Waters, 2000). As soon as the grapes begin ripening, TLPs and chitinases are the majority proteins, and this remains the same at harvest time (Pocock et al., 2000). This increase in thermo-unstable grape proteins throughout ripening has been confirmed by published studies (Tattersall et al., 2001; Tian et al., 2019).

At maturity, a must can contain from 30 to 250 mg/L of total proteins depending on the grape variety (Pocock et al., 2000). A study dedicated to Muscat has shown that the accumulation (quantity) and expression (nature) of the major proteins in the grape are highly variable, for the same vineyard plot, from one year to the next (Monteiro, Piçarra-Pereira, Teixeira, Loureiro, & Ferreira, 2003). Environmental conditions play a determining role in the composition and concentration of the berry's proteins. Finally, the overripeness of the grapes is accompanied by a clear increase in the protein content (TLPs and chitinases) of the juices, due in particular to the loss of water from the grapes (Pocock et al., 2000). Harvest conditions can also influence the unstable protein content of the musts. The protein content of juices from mechanical harvesting is higher than in juices from manual harvesting. This difference can be explained by the presence of proteins released from the berry skin during the maceration phase. This release of proteins from the skin is increased by the transport of a mechanical harvest over long distances.

The protein composition of a must can vary considerably, depending on many factors that include the grape variety, the maturity of the harvest, the methods of picking and transport to the winery, and the vintage. Additionally, the soil the vine grows in, the rootstock, vineyard management, and the yield per hectare, as well as the presence of other wine compounds such as sulfates (Marangon et al., 2011), characteristics such as the pH of the wine (Lambri, Dordoni, Giribaldi, Riva Violetta, & Giuffrida, 2013), and the ionic strength (Marangon et al., 2011), all play a role in the risk of haze. However, there is still a lack of data to estimate the importance of these different parameters in relation to the risk of haze.

To act on these thermo-unstable proteins and prevent haze risk in white wine, there are several technical possibilities, the effectiveness of which varies depending on the composition of the must or the wine. Among these techniques, four are subtractive. Their principle can be summarized as follows:

- Modify the colloidal environment of the wine, with yeast proteins mannoproteins (Dupin et al., 2000; Dupin et al., 2000), or yeast protein extracts (Fernandes, Neto, Centeno, De Fátima Teixeira, & Gomes, 2015), to prevent plant proteins from flocculation. In this case, the proteins remain present in the wine in a soluble state.
- Degrade the endogenous grape proteins with enzymes. Fragments of plant proteins, i.e., oligopeptides that are no longer able to form particles by flocculation, will then remain in the wine. The enzymes that have been tested include proteases from *Saccharomyces* (Dizy & Bisson, 2000), *Aspergillus niger* (Marangon et al., 2012), and *Botrytis cinerea*. Proteases from plants have also been tested (Benucci et al., 2015). Eliminate proteins from wine via materials such as zirconium oxide (Lucchetta et al., 2013; Marangon et al., 2011) or charged nanoparticles (Mierczynska-Vasilev, Boyer, Vasilev, & Smith, 2017) that adsorb proteins. The adsorbent-protein complexes are then removed from the wine, as in the case of bentonite treatment.
- Remove proteins from the wine by retaining them on a membrane by ultrafiltration (Flores et al., 1990).
- Remove proteins from the wine by using a fining treatment using exogenous macromolecules, i.e., an enological product acting as a flocculating agent such as carrageenans (Ratnayake et al., 2019) or chitosan (Colangelo et al., 2018). The flocculating agent-protein complexes are eliminated from the wine after sedimentation.

Heat the must or wine to induce thermal denaturation of the proteins, which then flocculate and form sediment (Marangon et al., 2012). Protein insolubilization is induced in the tank to prevent it from appearing in the bottle.

Among these different technical options, the use of *Aspergillus* proteases is a particularly interesting option in view of the results obtained in France and Australia, a country (along with New Zealand) where aspergillopepsins are already authorized in enology to avoid protein haze in white wines.

26.6 Icewine production and factors impacting the aroma profile

Icewine is a sweet dessert wine of critical importance to the Canadian wine industry and for international trade, valued annually at CAN \$100 million (VQA Ontario). It is made from grapes naturally frozen on the vine, which are then pressed, trapping ice within the skins, releasing juice twice as concentrated in solutes as table wine juice (Kontkanen et al., 2004; Inglis & Pickering, 2013). Concentrated solutes (sugars and salts) place wine yeast under hyperosmotic stress, leading to reduced cell growth, altered metabolism, 10-fold higher acetic acid, twofold higher glycerol, fermentation difficulties, and, sometimes, organoleptic faults in the wine (Heit et al., 2018; Kontkanen et al., 2004; Nurgel, Pickering, & Inglis, 2004; Pigeau & Inglis 2005; Inglis & Pickering, 2013; Yang et al., 2017). These fermentation challenges are becoming increasingly relevant in common table wine production due to increased sugar in fruit from global warming (Noti, Vaudano, Pessione, & Garcia-Moruno, 2015). Icewines are characterized by intense, complex aromas of apricot, peach, pear, dried fruit, honey, and butterscotch, to name a few, with these characteristics varying based on variety and hang time on the vine

(Bowen & Reynolds 2015; Bowen, Reynolds, & Lesschaeve, 2016; Inglis & Pickering, 2013). When Ontario's Inniskillin winery won the Grand Prix d'Honneur at Vinexpo in 1991 for their 1989 Vidal Icewine, the prestigious award helped place Canada on the international wine map (Inglis & Pickering, 2013).

Icewine is produced across Canada coast to coast in British Columbia, Ontario, Quebec, and Nova Scotia, with 90% of production occurring in the cool climate of Ontario (Grape Growers of Ontario Annual Report, 2019). Strict regulations are set by the Vintners Quality Alliance (VQA) for production. All blocks dedicated for Icewine production must be registered each year with the VQA by November 15 with the location, variety, acreage, and estimated tonnage recorded. In Canada, Vidal is the most common grape used for Icewine production (73%), followed by Riesling (12%) and Cabernet Franc (10%), with additional *Vitis vinifera* varieties used in small amounts (VQA Ontario). The temperature must be at or below $-8°C$ to harvest and press the frozen fruit to release the concentrated juice. Grapes are harvested at $-10°C$ to $-12°C$ to ensure that the pressed juice attains a concentration of 35° Brix prior to fermentation in order to be classified as Icewine, with any individual press fraction never below 32° Brix. Often Icewine juices are in the 38–40° Brix range, with 300–500 mg N/L to support the yeast during fermentation (Inglis & Pickering, 2013). The weight of the harvested, frozen grapes is recorded and compared to the volume of pressed juice as part of the regulation process to ensure that volumes agree with the tonnage processed. The fermented Icewine must have final alcohol between 7% and 14.9% v/v, residual sugar of at least 125 g/L, and a maximum allowable volatile acidity of 2.1 g/L.

The average concentration of basic chemical compositions in Canadian Icewines has been reported for 51 wines: total residual sugar 215 g/L; ethanol 10.2% v/v; titratable acidity 10.7 g/L; pH 3.4; acetic acid 1.30 g/L; ethyl acetate 240 mg/L; glycerol 12.4 g/L; and iso-amyl alcohol 67 mg/L (Nurgel et al., 2004). Typically, the titratable acidity measurement increases in Icewine over that found in juice by 2 g/L due to yeast production of acetic and succinic acid (Pigeau et al., 2007). However, these acids are much weaker organic acids and do not provide the same acid perception as tartaric or malic acid. Therefore often Icewines are further acidulated with tartaric acid to balance the high level of residual sugar in the wine (Inglis & Pickering, 2013).

Grape hang time has also been found to affect the organoleptic profile of Icewine (Bowen & Reynolds, 2015; Bowen et al., 2016; Khairallah, Reynolds, & Bowen, 2016). Early harvest of Riesling Icewine grapes in November increased some aliphatic compounds and esters, many of which have higher thresholds that are difficult for consumers to perceive, whereas a later harvest in January substantially intensified several highly odor-active compounds (Khairallah et al., 2016). A similar study using both Vidal and Riesling Icewine grapes found that most aroma compounds were at the highest concentration in the latest harvest period; 16 of 24 for Vidal harvested in February and 17 of 23 for Riesling harvested in January, including ethyl isobutyrate, ethyl 3-methylbutyrate, 1-hexanol, 1-octen-3-ol, 1-octanol, cis-rose oxide, nerol oxide, ethyl benzoate, ethyl phenylacetate, γ-nonalactone, and β-damascenone (Bowen & Reynolds, 2015). Interestingly, Icewines made from grapes harvested later did not show new unique compounds but rather showed higher concentrations of compounds and increased intensities for most aroma and flavor characteristics (Bowen & Reynolds, 2015; Bowen et al., 2016; Khairallah et al., 2016). Vidal Icewines made from a late harvest date in February were associated with most of the odor-active compounds such as terpenes and norisoprenoids, which were positively correlated to dried fruit/raisin, honey, and nut flavors, whereas these grapes harvested in mid-December showed most esters negatively correlated to tropical fruit, sherry, peach, and honey aromas, and 4-vinylguaiacol inversely correlated to the honey aroma, caramelized flavor, and bitter taste (Bowen et al., 2016). Similarly, Riesling Icewines made from a late harvest in January were also associated with dried fruit/raisin aroma, nut aroma, and nut and sherry flavor, whereas Icewine made from these same grapes harvested earlier in mid-December was associated more with lychee and citrus aromas and tropical fruit, tangerine, citrus, and floral flavors and was correlated to four esters, namely ethyl butyrate, ethyl hexanoate, ethyl octanoate, and ethyl cinnamate, and 4-vinylguaiacol (Bowen et al., 2016).

Saccharomyces cerevisiae wine yeast display properties (Jimenez-Marti et al., 2011a, 2011b) and DNA sequences (Hauser et al., 2001) distinct from well-studied *S. cerevisiae* laboratory yeast, and these differences impact yeast performance during fermentation and organoleptic properties of wine (Pretorius, 2000). Studies on Icewine made from *S. cerevisiae* wine species and spontaneous fermentation have demonstrated that different yeast species produce distinct compounds or higher concentration of respective compounds, but the yeast effect also depended on the vintage and cultivar (Chen et al., 2020). For example, *Saccharomyces* showed a positive correlation with ethanol, ethyl acetate, and isoamylol, and autochthonous *Hanseniaspora* and *Candida* were found negatively correlated with hexanol (Chen et al., 2020). Commercial strains have been evaluated to reduce undesired volatile acidity associated with Icewine fermentation (Kontkanen et al., 2004; Inglis, 2008) and to understand its production (Heit et al., 2018; Pigeau & Inglis 2005; Pigeau & Inglis 2007; Yang et al., 2017). *S. bayanus* and its hybrid with

S. cerevisiae have displayed low volatile acidity production in Icewine fermentations (Bellon, Yang, Day, Inglis, & Chambers, 2015). A 53% reduced production in volatile acidity was also observed in a mixed inoculation of *Torulaspora delbrueckii* with *S. cerevisiae* in must with 360 g/L sugar (Bely, Stoeckle, Masneuf-Pomarède, & Dubourdieu, 2008). Cofermentation of *Hanseniaspora vineae* and *Metschnikowia pulcherrima* with *S. cerevisiae* demonstrated improved aromatic diversity of Icewine produced, and the sequential inoculation of *H. vineae* with *S. cerevisiae* displayed higher levels of volatiles associated with fruity, flowery, and sweet attributes (Zhang, Shen, Duan, & Yan, 2018).

Canada has become a global leader in Icewine production over the past 30 years since its commercial-scale production in the late 1980s. Protecting its authenticity as well as overcoming the challenges of global warming with shorter harvest windows during the winter months continue to drive innovation for this wine style in Canada.

26.7 Conclusion

White wines from cool/cold climates have been described as more aromatic, lighter in body, and higher in acidity than those from warmer regions. However, white winemaking in cool/cold climate regions can present many challenges, but the choice of cold-tolerant grape varieties, clones, and rootstocks can mitigate these. White wine styles suited to the grape varieties and climatic conditions, and practices that reduce MPs are now commonplace. Adapting wine style to the cool/cold climate, as is the case with Icewine in Canada, has proved to be highly successful. The old and new white winemaking methods converging, such as the juice/wine being in contact with grape skins and seeds, is also evident in the range of white wine styles incipient in regions previously thought to be too cold for wine production. The range of techniques available to the winemaker continues to expand, from commercial bacteria and non-*Saccharomyces*/*Saccharomyces* non-*cerevisiae* yeasts to fermentation vessels and equipment. It is likely that in the future, varieties new to regions will begin to be planted and evaluated, and further winemaking techniques will emerge.

References

Alberts, P., Stander, M. A., Paul, S. O., & De Villiers, A. (2009). Survey of 3-alkyl-2-methoxypyrazine content of South African Sauvignon blanc wines using a novel LC-APCI-MS/MS method. *Journal of Agricultural and Food Chemistry*, 57(20), 9347–9355. Available from https://doi.org/10.1021/jf9026475.

Allen, M., Lacey, M., Harris, R., & Brown, W. V. (1991). Contribution of methoxypyrazines to Sauvignon blanc wine aroma. *American Journal of Enology and Viticulture*, 42(2), 109–112.

Allen, M. S., & Lacey, M. J. (1993). Methoxypyrazine grape flavor: Influence of climate, cultivar and viticulture. *Wein-Wissenschaft*, 48.

Anderson, M. M., Smith, R. J., Williams, M. A., & Wolpert, J. A. (2008). Viticultural evaluation of French and California Chardonnay clones grown for production of sparkling wine. *American Journal of Enology and Viticulture*, 59(1), 73–77.

Anfang, N., Brajkovich, M., & Goddard, M. R. (2009). Co-fermentation with *Pichia kluyveri* increases varietal thiol concentrations in Sauvignon blanc. *Australian Journal of Grape and Wine Research*, 15(1), 1–8. Available from https://doi.org/10.1111/j.1755-0238.2008.00031.x.

Bellon, J. R., Yang, F., Day, M. P., Inglis, D. L., & Chambers, P. J. (2015). Designing and creating *Saccharomyces* interspecific hybrids for improved, industry relevant, phenotypes. *Applied Microbiology and Biotechnology*, 99(20), 8597–8609. Available from https://doi.org/10.1007/s00253-015-6737-4.

Bely, M., Stoeckle, P., Masneuf-Pomarède, I., & Dubourdieu, D. (2008). Impact of mixed *Torulaspora delbrueckii-Saccharomyces cerevisiae* culture on high-sugar fermentation. *International Journal of Food Microbiology*, 122(3), 312–320. Available from https://doi.org/10.1016/j.ijfoodmicro.2007.12.023.

Benkwitz, F., Tominaga, T., Kilmartin, P. A., Lund, C., Wohlers, M., & Nicolau, L. (2012). Identifying the chemical composition related to the distinct aroma characteristics of New Zealand Sauvignon blanc wines. *American Journal of Enology and Viticulture*, 63(1), 62–72. Available from https://doi.org/10.5344/ajev.2011.10074.

Benucci, I., Esti, M., & Liburdi, K. (2015). Effect of wine inhibitors on the proteolytic activity of papain from Carica papaya. L. latex. *Biotechnol Prog.*, 31(1), 48–54.

Blake, A., Kotseridis, Y., Brindle, I. D., Inglis, D., Sears, M., & Pickering, G. J. (2009). Effect of closure and packaging type on 3-Alkyl-2-methoxypyrazines and other impact odorants of Riesling and Cabernet franc wines. *Journal of Agricultural and Food Chemistry*, 57(11), 4680–4690. Available from https://doi.org/10.1021/jf803720k.

Botezatu, A., Kemp, B. S., & Pickering, G. J. (2016). Chemical and sensory evaluation of silicone and polylactic acid-based remedial treatments for elevated methoxypyrazine levels in wine. *Molecules*, 21(9). Available from https://doi.org/10.3390/molecules21091238.

Botezatu, A., & Pickering, G. J. (2015). Application of plastic polymers in remediating wine with elevated alkyl-methoxypyrazine levels. *Food Additives and Contaminants – Part A Chemistry, Analysis, Control, Exposure and Risk Assessment*, 32(7), 1199–1206. Available from https://doi.org/10.1080/19440049.2015.1028106.

Botezatu, A. I., Kotseridis, Y., Inglis, D., & Pickering, G. J. (2013). Occurrence and contribution of alkyl methoxypyrazines in wine tainted by *Harmonia axyridis* and *Coccinella septempunctata*. *Journal of the Science of Food and Agriculture*, 93(4), 803–810. Available from https://doi.org/10.1002/jsfa.5800.

Botezatu, A., & Pickering, G. J. (2012). Determination of Ortho- and Retronasal Detection Thresholds and Odor Impact of 2,5-Dimethyl-3-Methoxypyrazine in Wine. *J. of Food Sci.*, 77(11), S394–S398.

Boulton, R. (1980). The general relationship between potassium, sodium and pH in grape juice and wine. *Am. J. Enol & Vitic*, 31(2), 2–6.

Bowen, A. J., & Reynolds, A. G. (2015). Aroma compounds in Ontario Vidal and Riesling Icewines. I. Effects of harvest date. *Food Res. Int.*, 76, 540–549.

Bowen, A. J., Reynolds, A. G., & Lesschaeve, I. (2016). Harvest date and crop level influence sensory and chemical profiles of Ontario Vidal blanc and Riesling Icewines. *Food Research International*, 89, 591–603. Available from https://doi.org/10.1016/j.foodres.2016.09.005.

Cabaroglu, T., Selli, S., Canbas, A., Lepoutre, J. P., & Günata, Z. (2003). Wine flavor enhancement through the use of exogenous fungal glycosidases. *Enz & Micro Tech.*, 33(5), 581–587.

Castagnoli, S. P., & Vasconcelos, M. C. (2006). Field performance of 20 "Pinot noir" clones in the Willamette Valley of Oregon. *HortTechnology*, 16(1), 153–161.

Chen, Y., Zhang, W., Yi, H., Wang, B., Xiao, J., Zhou, X., & Shi, X. (2020). Microbial community composition and its role in volatile compound formation during the spontaneous fermentation of ice wine made from Vidal grapes. *Process Biochemistry*, 92, 365–377. Available from https://doi.org/10.1016/j.procbio.2020.01.027.

Cheynier, V., Rigaud, J., Souquet, J.-M. M., Barillere, J. M., & Moutounet, M. (1989). Effect of pomace contact and hyperoxidation on the phenolic composition and quality of Grenache and Chardonnay wines. *Am. J. Enol. & Vitic.*, 40(1), 36–42.

Colangelo, D., Torchio, F., De Faveri, D. M., & Lambri, M. (2018). The use of chitosan as alternative to bentonite for wine fining: Effects on heat-stability, proteins, organic acids, colour, and volatile compounds in an aromatic white wine. *Food Chem.*, 264, 301–309.

Corso, M., & Bonghi, C. (2014). Grapevine rootstock effects on abiotic stress tolerance. *Plant Science Today*, 108–113. Available from https://doi.org/10.14719/pst.2014.1.3.64.

Cousins, P. (2005). Evolution, Genetics and Breeding: Viticultural Applications of the origins of our rootstocks. In P. Cousins, & R. K. Striegler (Eds.), *Proceedings of the 2005 rootstock symposium* (pp. 1–7). Osage Beach, MO: Southwest Missouri State University.

Crandles, M., Cliff, M., & van Vuuren, J. J. (2004). Impact of Yeast Strain on the Production of Acetic Acid, Glycerol, and the Sensory Attributes of Icewine. *Am. J. Enol. Vitic.*, 55(4), 371–378.

Dami, I. E., Li, S., & Zhang, Y. (2016). Evaluation of primary bud freezing tolerance of twenty-three winegrape cultivars new to the eastern united States. *American Journal Enology and Viticulture*, 67, 139–145.

Darias-Martín, J. J., Rodríguez, O., Díaz, E., & Lamuela-Raventós, R. M. (2000). Effect of skin contact on the antioxidant phenolics in white wine. *Food Chem*, 71(4), 483–487.

Day, M. P., Schmidt, S. A., Pearson, W., Kolouchova, R., & Smith, P. A. (2019). Effect of passive oxygen exposure during pressing and handling on the chemical and sensory attributes of Chardonnay wine. *Australian Journal of Grape and Wine Research*, 25(2), 185–200. Available from https://doi.org/10.1111/ajgw.12384.

Dizy, M., & Bisson, L. F. (2000). Proteolytic activity of yeast strains during grape juice fermentation. *Am. J. Enol. Vitic.*, 51, 155–167.

Duchêne, E., Legras, J. L., Karst, F., Merdinoglu, D., Claudel, P., Jaegli, N., & Pelsy, F. (2009). Variation of linalool and geraniol content within two pairs of aromatic and nonaromatic grapevine clones. *Australian Journal of Grape and Wine Research*, 15(2), 120–130. Available from https://doi.org/10.1111/j.1755-0238.2008.00039.x.

Dupin, I. V. S., McKinnon, B. M., Ryan, C., Boulay, M., Markides, A. J., Jones, G. P., . . . Waters, E. J. (2000). Saccharomyces cerevisiae mannoproteins that protect wine from protein haze: Their release during fermentation and lees contact and a proposal for their mechanism of action. *J. Ag. & Food Chem.*, 48, 3098–3105.

Fernandes, J. P., Neto, R., Centeno, F., De Fátima Teixeira, M., & Gomes, A. C. (2015). Unveiling the potential of novel yeast protein extracts in white wines clarification and stabilization. *Frontiers in Chemistry*, 3. Available from https://doi.org/10.3389/fchem.2015.00020.

Ferreira, B., Hory, C., Bard, M. H., Taisant, C., Olsson, A., & Le Fur, Y. (1995). Effects of skin contact and settling on the level of the C18:2, C18:3 fatty acids and C6 compounds in burgundy chardonnay musts and wines. *Food Quality and Preference*, 6(1), 35–41. Available from https://doi.org/10.1016/0950-3293(94)P4210-W.

Flores, J. H., Heatherbell, D. A., & McDaniel, M. R. (1990). Ultrafiltration of wine: effect of ultrafiltration on white Riesling and Gewürztraminer wine composition and stability. *Am. J. Enol & Vitic.*, 41(3), 207–214.

Gladstones, J. (1992). *Viticulture and Environment*. Adelaide, Australia: Wine- titles.

Glonti, T. (2010). Traditional technologies and history of Georgian wine. *Bulletin of the International Office of Vine and Wine (OIV)*, 335–343.

Grape Growers of Ontario Annual Report. (2019). Available online: http://www.grapegrowersofontario.com/sites/default/files/flipbook/2019%20Annual%20Report/2019AnnualReport/index.html?r = 62 (accessed on 16 July 2020).

Gregan, S. M., & Jordan, B. (2016). Methoxypyrazine accumulation and O-Methyltransferase gene expression in Sauvignon blanc grapes: The role of leaf removal, light exposure, and berry development. *J. Ag. & Food Chem.*, 64(11), 2200–2208.

Hashizume, K., & Samuta, T. (1997). Green Odorants of grape cluster stem and their ability to cause a wine stemmy flavor. *Journal of Agricultural and Food Chemistry*, 45(4), 1333–1337. Available from https://doi.org/10.1021/jf960635a.

Hauser, N. C., Fellenberg, K., Gil, R., Bastuck, S., Hoheisel, J. D., & Pérez-Ortín, J. E. (2001). Whole genome analysis of a wine yeast strain. *Comp. Funct Genom.*, 2, 69–79.

Hébert-Haché, A., Inglis, D., Kemp, B., & Willwerth, J. (2020). Clone and rootstock interactions influence the cold hardiness of *Vitis vinifera* cvs. Riesling and Sauvignon blanc. *American Journal Enology and Viticulture*, 67, 139–145.

Inglis, D.L., & Pickering, G.P. (2013). Vintning on thin ice – The making of Canada's iconic dessert wine. the World of Niagara Wine.

Jackson, R.S. (1996). Fermentation – Maceration (skin contact). In *Wine science – Principles and applications*, 3rd edition (pp. 336–339).

Heit, C., Martin, S. J., Yang, F., & Inglis, D. L. (2018). Osmoadaptation of wine yeast (Saccharomyces cerevisiae) during Icewine fermentation leads to high levels of acetic acid. *J. Appl. Microbiol.*, 124, 1506–1520.

Jimenez-Marti, E., Gomar-Alba, M., Palacios, A., Ortiz-Julien, A., & del Olmo, M. (2011a). Towards an understanding of the adaptation of wine yeasts to must: relevance of the osmotic stress response. *Appl Microbiol. Biotechnol.*, 89, 1551–1561.

Khairallah, R., Reynolds, A. G., & Bowen, A. J. (2016). Harvest date effects on aroma compounds in aged Riesling icewines. *Journal of the Science of Food and Agriculture, 96*(13), 4398−4409. Available from https://doi.org/10.1002/jsfa.7650.

Kontkanen, D., Inglis, D. L., Pickering, G. J., & Reynolds, A. (2004). Effect of yeast inoculation rate, acclimatization, and nutrient addition on Icewine fermentation. *Am. J. Enol. & Vitic., 55,* 363−370.

Kotseridis, Y., Anocibar Beloqui, A., Bayonove, C. L., Baumes, R. L., & Bertrand, A. (1999). Effets de certains facteurs viticoles et Œnologiques sur les teneurs en 2-méthoxy-3-isobutylpyrazine. *Journal International des Sciences de la Vigne et du Vin, 33*(1), 19−23. Available from https://doi.org/10.20870/oeno-one.1999.33.1.1040.

Kotseridis, Y. S., Spink, M., Brindle, I. D., Blake, A. J., Sears, M., Chen, X., ... Pickering, G. J. (2008). Quantitative analysis of 3-alkyl-2-methoxypyrazines in juice and wine using stable isotope labelled internal standard assay. *Journal of Chromatography A, 1190*(1−2), 294−301. Available from https://doi.org/10.1016/j.chroma.2008.02.088.

Krieger-Weber, S., Heras, J. M., & Suarez, C. (2020). *Lactobacillus plantarum*, a new biological tool to control malolactic fermentation: A review and an outlook. *Beverages, 6*(2), 23. Available from https://doi.org/10.3390/beverages6020023.

Lambri, M., Dordoni, R., Giribaldi, M., Riva Violetta, M., & Giuffrida, M. G. (2013). Effect of pH on the protein profile and heat stability of an Italian white wine. *Food Research International, 54*(2), 1781−1786. Available from https://doi.org/10.1016/j.foodres.2013.09.038.

Li, N., Wei, Y., Li, X., Wang, J., Zhou, J., & Wang, J. (2019). Optimization of deacidification for concentrated grape juice. *Food Science and Nutrition, 7*(6), 2050−2058. Available from https://doi.org/10.1002/fsn3.1037.

Lorteau, S. (2018). A comparative legal analysis of skin-contact wine definitions in Ontario and South Africa. *Journal of Wine Research, 29*(4), 265−277. Available from https://doi.org/10.1080/09571264.2018.1532881.

Lucchetta, M., Pocock, K. F., Waters, E. J., & Marangon, M. (2013). Use of zirconium dioxide during fermentation as an alternative to protein fining with bentonite for white wines. *Am. J. Enol. Vitic., 64,* 400−404.

Lund, C. M., Thompson, M. K., Benkwitz, F., Wohler, M. W., Triggs, C. M., Gardner, R., & Nicolau, L. (2009). New Zealand sauvignon blanc distinct flavor characteristics: Sensory, chemical, and consumer aspects. *American Journal of Enology and Viticulture, 60*(1), 1−12.

Marais, J. (1998). Effect of grape temperature, oxidation and skin contact on Sauvignon blanc juice and wine composition and wine quality. *S. Afr. J. Wine Res, 19*(1), 10−16.

Marais, J., Hunter, J. J., & Haasbroek, P. D. (1999). Effect of canopy microclimate, season and region on Sauvignon blanc grape composition and wine quality. *S. Afr. J. Wine Res., 20*(1), 19−30.

Marangon, M., Sauvage, F-X., Waters, E. J., & Vernhet, A. (2011). Effects of ionic strength and sulfate upon thermal aggregation of grape chitinases and thaumatin-like proteins in a model system. *J. Ag & Food Chem., 59,* 2652−2662.

Marangon, M., Van Sluyter, S. C., Robinson, E. M. C., Muhlack, R. A., Holt, H. E., Haynes, P. A., ... Waters, E. J. (2012). Degradation of white wine haze proteins by Aspergillopepsin I and II during juice flash pasteurization. *Food Chem., 135,* 1157−1165.

Marangon, M., Van Sluyter, S. C., Waters, E. J., & Menz, R. I. (2014). Structure of haze forming proteins in white wines: Vitis vinifera thaumatin-like proteins. *PLoS One, 9,* e113757.

Mierczynska-Vasilev, A., Boyer, P., Vasilev, K., & Smith, P. A. (2017). A novel technology for the rapid, selective, magnetic removal of pathogenesis-related proteins from wines. *Food Chemistry, 232,* 508−514. Available from https://doi.org/10.1016/j.foodchem.2017.04.050.

Migicovsky, Z., Cousins, P., Jordan, L.M., Myles, S., Striegler, R.K., Verdegaal, P., & Chitwood, D.H. (2019). Rootstock choice can dramatically affect grapevine growth. bioRxiv.

Monteiro, S., Piçarra-Pereira, M. A., Teixeira, A. R., Loureiro, V. B., & Ferreira, R. B. (2003). Environmental conditions during vegetative growth determine the major proteins that accumulate in mature grapes. *Journal of Agricultural and Food Chemistry, 51*(14), 4046−4053. Available from https://doi.org/10.1021/jf020456v.

Naegele, R. P., Cousins, P., & Daane, K. M. (2020). Identification of *Vitis* cultivars, rootstocks, and species expressing resistance to a *planococcus* mealybug. *Insects, 11*(2). Available from https://doi.org/10.3390/insects11020086.

Neethling, E., Petitjean, T., Quénol, H., & Barbeau, G. (2017). Assessing local climate vulnerability and winegrowers' adaptive processes in the context of climate change. *Mitigation and Adaptation Strategies for Global Change, 22*(5), 777−803. Available from https://doi.org/10.1007/s11027-015-9698-0.

Noti, O., Vaudano, E., Pessione, E., & Garcia-Moruno, E. (2015). Short-term response of different *Saccharomyces cerevisiae* strains to hyperosmotic stress caused by inoculation in grape must: RT-qPCR study and metabolite analysis. *Food Microbiology, 52,* 49−58. Available from https://doi.org/10.1016/j.fm.2015.06.011.

Nurgel, C., Pickering, G. J., & Inglis, D. L. (2004). Sensory and chemical characteristics of Canadian Icewines. *Journal of the Science of Food and Agriculture, 84*(13), 1675−1684. Available from https://doi.org/10.1002/jsfa.1860.

Oberholster, A., Francis, I. L., Iland, P. G., & Waters, E. J. (2009). Mouthfeel of white wines made with and without pomace contact and added anthocyanins. *Aus J. Grape & Wine Res., 15*(1), 59−69.

Ough, C. S., Berg, H. W., & Amerine, M. A. (1969). Substances extracted during skin contact with white musts. I-II. *Am. J. Vitic. & Enol., 20*(2), 93−107.

Peng, Z., Pocock, K. F., Waters, E. J., Francis, I. L., & Williams, P. J. (1997). Taste properties of grape (*Vitis vinifera*) pathogenesis-related proteins isolated from wine. *Journal of Agricultural and Food Chemistry, 45*(12), 4639−4643. Available from https://doi.org/10.1021/jf970194a.

Peyrot des Gachons, C., Tominga, T., & Dubourdieu, D. (2002). Research Note: Localization of S-cysteine conjugates in the berry: Effect of skin contact on aromatic potential of *Vitis vinifera* L. cv. Sauvignon blanc must. *American Journal of Viticulture & Enology, 53*(2), 2000−2002.

Pickering, G. J., Karthik, A., Inglis, D., Sears, M., & Ker, K. (2007). Determination of ortho- and retronasal detection thresholds for 2-isopropyl-3-methoxypyrazine in wine. *Journal of Food Science, 72*(7), S468−S472. Available from https://doi.org/10.1111/j.1750-3841.2007.00439.x.

Pickering, G. J., Lin, J., Reynolds, A., Soleas, G., & Riesen, R. (2006). The evaluation of remedial treatments for wine affected by Harmonia axyridis. *Int. J. Food Sci. & Tech., 41,* 77−86.

Pickering, G. J., Spink, M., Kotseridis, Y., Inglis, D., Brindle, I. D., Sears, M., & Beh, A. L. (2008). Yeast strain affects 3-isopropyl-2-methoxypyrazine concentration and sensory profile in Cabernet Sauvignon wine. *Australian Journal of Grape and Wine Research, 14*(3), 230−237. Available from https://doi.org/10.1111/j.1755-0238.2008.00026.x.

Pigeau, G. M., & Inglis, D. L. (2005). Upregulation of ALD3 and GPD1 in S. cerevisiae during Icewine fermentation. *J. Appl. Microbiol., 99,* 112−125.

Pigeau, G. M., & Inglis, D. L. (2007). Response of wine yeast (S. cerevisiae) aldehyde dehydrogenase to acetaldehyde stress during Icewine fermentation. *J. Appl. Microbiol., 103,* 1576−1586.

Pocock, K. F., Hayasaka, Y., McCarthy, M. G., & Waters, E. J. (2000). Thaumatin-like proteins and chitinases, the haze-forming proteins of wine, accumulate during ripening of grape (*Vitis vinifera*) berries and drought stress does not affect the final levels per berry at maturity. *Journal of Agricultural and Food Chemistry, 48*(5), 1637−1643. Available from https://doi.org/10.1021/jf9905626.

Pretorius, I. S. (2000). Tailoring wine yeast for the new millennium: Novel approaches to the ancient art of winemaking. *Yeast, 16*(8), 675−729. Available from https://doi.org/10.1002/1097-0061(20000615)16:8 < 675::AID-YEA585 > 3.0.CO;2-B.

Ramey, D., Bertrand, A., Ough, C. S., Singleton, V. L., & Sanders, E. (1986). Effects of skin contact temperature on Chardonnay must and wine composition. *American Journal of Enology and Viticulture, 37*(2), 99−106.

Ratnayake, S., Stockdale, V., Grafton, S., Munro, P., Robinson, A. L., Pearson, W., & Bacic, A. (2019). Carrageenans as heat stabilisers of white wine. *Australian Journal of Grape and Wine Research, 25*(4), 439−450. Available from https://doi.org/10.1111/ajgw.12411.

Ribéreau-Gayon, J. (1932). Les matières albuminoïdes des vins blancs. *Ann. Falsif. Fraudes, 25,* 518−524.

Roman, T., Nardin, T., Trenti, G., Barnaba, C., Nicolini, G., & Larcher, R. (2020). Press fractioning of grape juice: A first step to manage potential atypical aging development during winemaking. *American Journal of Enology and Viticulture, 71*(1), 17−25. Available from https://doi.org/10.5344/ajev.2019.19030.

Roujou de Boubée, D., Cumsille, A. M., Pons, M., & Dubourdieu, D. (2002). Location of 2-methoxy-3-isobutylpyrazine in Cabernet Sauvignon grape bunches and its extractability during vinification. *American Journal of Enology and Viticulture, 53*(1), 1−5.

Ryona, I., Reinhardt, J., & Sacks, G. L. (2012). Treatment of grape juice or must with silicone reduces 3-alkyl-2-methoxypyrazine concentrations in resulting wines without altering fermentation volatiles. *Food Research International, 47*(1), 70−79. Available from https://doi.org/10.1016/j.foodres.2012.01.012.

Sala, C., Busto, O., Guasch, J., & Zamora, F. (2004). Influence of vine training and sunlight exposure on the 3-alkyl-2-methoxypyrazines content in musts and wines from the *Vitis vinifera* variety Cabernet Sauvignon. *Journal of Agricultural and Food Chemistry, 52*(11), 3492−3497. Available from https://doi.org/10.1021/jf049927z.

Sánchez Palomo, E., Pérez-Coello, M. S., Díaz-Maroto, M. C., González Viñas, M. A., & Cabezudo, M. D. (2006). Contribution of free and glycosidically-bound volatile compounds to the aroma of muscat \a petit grains\ wines and effect of skin contact. *Food Chemistry, 95*(2), 279−289. Available from https://doi.org/10.1016/j.foodchem.2005.01.012.

Sancho-Galán, P., Amores-Arrocha, A., Jiménez-Cantizano, A., & Palacios, V. (2020). Physicochemical and nutritional characterization of winemaking lees: A new food ingredient. *Agronomy, 996.* Available from https://doi.org/10.3390/agronomy10070996.

Shaffer, R., Sampaio, T. L., Pinkerton, J., & Vasconcelos, M. C. (2004). *Grapevine rootstocks for Oregon vineyards.* Corvallis, OR: Oregon State University Extension.

Sidhu, D., Lund, J., Kotseridis, Y., & Saucier, C. (2015). Methoxypyrazine analysis and influence of viticultural and enological procedures on their levels in grapes, musts, and wines. *Crit. Rev. Food Sci & Nutri., 55*(4), 485−502.

Singleton, V. L., Sieberhagen, H., Wet., & Wyk. (1975). Composition and sensory qualities of wines prepared from white grapes by fermentation with and without grape solids. *American Journal of Enology and Viticulture, 26*(2), 62−69.

Sokolowsky, M., Rosenberger, A., & Fischer, U. (2013). Sensory impact of skin contact on white wines characterized by descriptive analysis, time-intensity analysis and temporal dominance of sensations analysis. *Food Quality and Preference, 39,* 1−13. Available from https://doi.org/10.1016/j.foodqual.2014.07.002.

Swiegers, J. H., Kievit, R. L., Siebert, T., Lattey, K. A., Bramley, B. R., Francis, I. L., & Pretorius, I. S. (2009). The influence of yeast on the aroma of Sauvignon Blanc wine. *Food Microbiology, 26*(2), 204−211. Available from https://doi.org/10.1016/j.fm.2008.08.004.

Theron, L.W., Bely, M., & Divol, B. (2018). Pour un vin blanc sans bentonite. De nouvelles options contre la casse blanche. Revue des Oenologues.

Tattersall, D. B., Pocock, K. F., Hayasaka, Y., Adams, K., van Heeswijck, R., Waters, E. J., & Høj, P. B. (2001). Pathogenesis related proteins − their accumulation in grapes during berry growth and their involvement in white wine heat instability. Current knowledge and future perspectives in relation to winemaking practices. In Roubelakis-Angelakis KA (ed.), Molecular Biology and Biotechnology of the Grapevine. *Kluwer Academic, Dordrecht, the Netherlands,* 183−201.

Townshend, E. R., Harrison, R., & Tian, B. (2019). The phenolic composition of orange wine - effects of skin contact and sulfur dioxide addition on white wine tannin. *Wine & Viticulture Journal. Autumn,* 25−32.

Van Leeuwen, C., & Seguin, G. (2006). The concept of terroir in viticulture. *Journal of Wine Research, 17*(1), 1−10. Available from https://doi.org/10.1080/09571260600633135.

Vilela, A. (2019). Use of nonconventional yeasts for modulating wine acidity. *Fermentation, 5*(1), 27. Available from https://doi.org/10.3390/fermentation5010027.

Waters, E. J., Hayasaka, Y., Tattersall, D. B., Adams, K. S., & Williams, P. J. (1998). Sequence analysis of grape (*Vitis vinifera*) berry chitinases that cause haze formation in wines. *Journal of Agricultural and Food Chemistry, 46*(12), 4950−4957. Available from https://doi.org/10.1021/jf980421o.

Wolpert, J. A., Kasimatis, A. N., & Weber, E. (1994). Field performance of six Chardonnay clones in the Napa Valley. *American Journal of Enology and Viticulture, 45*(4), 393−400. Retrieved from http://ajevonline.org/.

Yang, F., Heit, C., & Inglis, D. L. (2017). Cytosolic redox status of wine yeast (Saccharomyces cerevisiae) under hyperosmotic stress during Icewine fermentation. *Fermentation, 3,* 61.

Zhang, B., Shen, J., Duan, C., & Yan, G. (2018). Use of Indigenous *Hanseniaspora vineae* and *Metschnikowia pulcherrima* co-fermentation with *Saccharomyces cerevisiae* to improve the aroma diversity of Vidal Blanc Icewine. *Frontiers in Microbioliogy, 9*, 2303. Available from https://doi.org/10.3389/fmicb.2018.02303.

Zhang, S., Petersen, M. A., Liu, J., Toldam-Andersen, T. B., Ebeler, S. E., & Hopfer, H. (2015). Influence of prefermentation treatments on wine volatile and sensory profile of the new disease tolerant cultivar Solaris. *Molecules, 20*(12), 21609–21625. Available from https://doi.org/10.3390/molecules201219791.

27

White winemaking in cold regions with short maturity periods in Northwest China

Tengzhen Ma[1,2], Shunyu Han[1,2,3], Bo Zhang[1,2], Faisal Eudes Sam[1,2], Wei Li[1,2] and Junxiang Zhang[3,4]

[1]College of Food Science and Engineering, Gansu Agricultural University, Lanzhou, P.R. China [2]Gansu Key Laboratory of Viticulture and Enology, Lanzhou, P.R. China [3]China Wine Industry Technology Institute, Yinchuan, P.R. China [4]Wine School, Ningxia University, Yinchuan, P.R. China

27.1 Introduction

White wine is generally made from white grape varieties that are usually pressed to get must by using a pneumatic presser. Only high-quality must is used in the making of white wine and it should be clarified before fermentation. The clarified must is often fermented at low temperatures, usually, around 16°C−18°C for a period of 20−30 days, during which time the must sugar is converted into alcohol to make wine. In China, the consumption of white wine is much lower than that of red wine, hence white wine production only accounts for 12% of the total wine production compared to red wine. Chinese wine production regions can be roughly divided into 11 major zones, with the majority located in Northern China. Due to the cold temperatures in winter, vines in Northern China are usually buried during winter and unearthed in spring to survive the cold and dry winter. This practice is commonly known as buried viticulture. Only a few of the major grape varieties, such as Chardonnay, Riesling, Italian Riesling, Vidal, Semillon, Ugni blanc, and Longyan grape (a Chinese variety), are now widely planted and used in the production of white wines. This chapter reviews the chemical composition of grapes and wines, including sugars, organic acids, phenolic compounds, and volatile compounds. Furthermore, winemaking technologies and some innovations in white wine making in China's cold regions have been described.

27.2 White wine market in China

In China, wine is known as '红酒' (Hóngjiǔ) to a huge portion of Chinese people. Regarding white wine patronage in China, only a few number of people purchase white wine (Chinese Wine data Analysis Report in 2017), while it shares nearly the same amount of consumption compared to red wine in Western countries. In 2015, Wade Shepard, a reporter from South China Morning Post (SCMP) published an article 'Why Pale Wine Waved the White Flag in China.' In that article, he explained that the reason why white wine failed in China was because of Chinese culture. In China, red color symbolizes fortune, luck, and wealth, while white represents the color of death in folk tradition. Another reason why white wine showed less popularity is generally because white wine contains much less phenolic content than red wine, which reflects major health properties from wine. Also, white wine is usually served chilled, and most Chinese consumers do not like to drink such cold and sour alcoholic beverages although the flavor is remarkable.

From the Chinese Wine Data Analysis Report in 2017 written by the China Alcoholic Drinks Association and Exact information, it was pointed out that for consumers to buy wine based on color, red wine accounted for the

© 2022 Elsevier Inc. All rights reserved.

The proportion of consumers buying different types of wine

Red Wine ▪ White wine ▪ Sparkling wine ▪ Rose wine

FIGURE 27.1   The proportion of consumers buying different types of wine. *Source: Hong Kong Trade Development Council.*

highest proportion at 96%, while white wine only accounted for 40% (Fig. 27.1). This also demonstrated that red wine shows much more popularity than the other types of wine in Chinese market.

27.3 General climatic and agronomic conditions of cold regions in China

According to the traditional way, from the ecological similarity of planting area and scale of division, Chinese wine production regions can be roughly divided into 11 major zones, including the Northeast, Beijing—Tianjin—Hebei (also called Jing-Jin-Ji), Shandong (also called Jiaodong Peninsula), Old Course of the Yellow River, Loess Plateau, Inner Mongolia, Eastern Region of Helan Mountains, Hexi Corridor, Southwest Alpine, Xinjiang, and others (Fig. 27.2). Recently, a study based on the meteorological data of 2294 stations in 30 years (from 1982 to 2011), and the digital elevation model (DEM) data with a resolution of 3″ (90 m), produced both climatic and varietal regionalization map of wine grapes grown in China (Wang et al., 2017). In that study, the suitable cultivation areas for grapes in China were divided into 4 districts and 12 subdistricts (Table 27.1). The subregions were further divided according to the different active accumulated temperatures of different matured grape varieties. The wine grape varieties regionalization map of the 12 subdistricts was also categorized according to the climatic characteristics and distribution range of each subdistrict (Fig. 27.3).

As can be seen from Fig. 27.3, Gansu Hexi Corridor and the Eastern region of Ningxia Helan Mountains are the two important cold regions with short maturity periods in Northwest China, thus most data in this chapter are based on these regions.

Gansu Hexi Corridor is located on longitude 93° 99′—104° 43′ East and latitude 36° 46′ − 40° 12′ North, while the Eastern region of the Helan Mountains is located on longitude 105° 45′—106° 47′ East and latitude 37° 43′ − 39° 23′ North, with both regions showing similar climatic conditions (Table 27.2). The Qilian and Helan Mountains provide a natural barrier which holds back the Gobi Desert, reduces the harmful effects of frost, stops cold air, strong winds sand storms, as well as increase accumulated temperature. More so, the snow water from the Qilian Mountain and Yellow River (the Mother River in China) flows through the Helan Mountains which also provides water for irrigation. The major features of climate conditions in these areas are similar, with long sunshine duration, strong solar radiation, low precipitation, low atmospheric humidity, and a fairly wide range of daily temperatures. However, compared to some of the famous regions all over the world, these regions are dry and hot in summer, but cold and cool in spring and autumn (Fig. 27.4).

27.4 Buried viticulture

Although global warming has evolved over the years, the extreme minimum temperature in winter in Gansu and Ningxia is usually below −20°C. For almost all the grape varieties, the fatal weakness is not only the extremely low temperature but also the great quantities of water evaporation caused by extreme droughts in spring and winter, which is also called 'drought-freezing'. Therefore, in order to protect grapes against cold and drought conditions during winter, measures must be taken, and after years of experience and studies, the best way is to bury the vine under soil (Fig. 27.5). Before the soil is frozen, the vines are taken down off the trellis after

FIGURE 27.2 Chinese wine production regions.

Northeast

Beijing-Tianjin-Hebei

Shandong

Old Course of the Yellow River

Loess Plateau

Inner Mongolia

Eastern Region of Helan Mountain

Hexi Corridor

Southwest Alpine

Xinjiang

Others

TABLE 27.1   The division standards of wine grape climatic regionalization.

Climatic region	I Cool-warm region	II Moderate-warm region	III Warm region	IV Thermal region
	160 days ≤ FFP ≤ 180 days	180 days ≤ FFP ≤ 200 days	200 days ≤ FFP ≤ 220 days	FFP > 220 days
1.0 ≤ DI ≤ 1.6	1 Cool-warm with subhumid region	4 Moderate-warm with subhumid region	7 Warm with subhumid region	10 Thermal with subhumid region
1.6 < DI ≤ 3.5	2 Cool-warm with semiarid region	5 Moderate-warm with semiarid region	8 Warm with semiarid region	11 Thermal with semiarid region
DI > 3.5	3 Cool-warm with arid region	6 Moderate-warm with arid region	9 Warm with arid region	12 Thermal with arid region

Note: DI: Dryness index; *FFP*: frost-free period.

pruning, buried into the soil (more than 30 cm underground), and left in the soil until the coming spring in the following year. This measure is inconceivable in the rest of the world famous regions, and burying the vines with soil is not welcomed by growers, but it is a very important and inevitable work for all the vineyards in Northern China.

As the majority of Chinese wine grape regions are located in soil-burying zones, this special viticulture management makes the soil surface exposed for nearly half a year. During the cold winter and the dry windy spring in Northern China, the exposed soil is mostly prone to erosion and sandstorm in viticulture regions (Wang et al., 2015). As a result, a new viticultural method was developed (Wang et al., 2015). The method requires the pruning of shoots at the early stages of winter after which the shoots are hung on the wires until next spring. This plays a

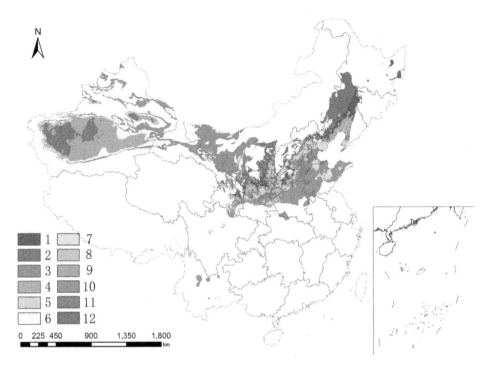

FIGURE 27.3   The climatic regionalization map of wine grape in China.

TABLE 27.2   The terroir of Gansu Hexi Corridor and the Eastern Region of Ningxia Helan mountains.

Production regions	Mountain	Irrigation water	Average elevation	Accumulated temperature (≥10°C)	Temperature difference between day and night	Annual precipitation	Sunshine duration	Annual frost-free period	Major soil type
Gansu Hexi Corridor	Qilian Mountains	Snow water	1500 m	≥ 3200°C	Around 15°C	37.3–230 mm	/	≥ 160 days	Gravel and sandy soil
Eastern region of Ningxia Helan Mountains	Helan Mountains	Yellow River	1100 m	3400°C–3800°C	12°C–15°C	150–240 mm	1700–2000 h	160–180 days	Sierozem and eolian sandy soil

FIGURE 27.4   The vineyard in Gansu (A) and Ningxia (B) regions.

FIGURE 27.5 Vines with and without shoots buried in the soil during winter.

protective role as a kind of windbreak in reducing the wind speed, lowering sandstorm pollution (Fig. 27.5), as well as puting off the high intensity of labor force until unearthing process in the following year (Fig. 27.6).

The burial time starts from the lowest temperature (≤0°C) at night to the freezing of the soil, and it usually takes place in the middle of October to early November in the Gansu and Ningxia regions. In general, young vines of 1−2 years are buried earlier than old vines of more than 3 years (which are cold-resistant). If the vines are buried too early, they are liable to have bud rot due to high temperature and low humidity in the soil. However, when buried too late the branches become vulnerable to early frost and the frozen soil is usually difficult to be used in covering the vines. Both manual and mechanical methods are used to complete buried viticulture. The approach of manual burying involves removing the pruned and sprayed branches from the trellis, straightening them out in the same direction, and pressing them down in the soil slowly. The branches can also be bundled by using a thin rope and then buried in the immobilized soil. In this process, excessive pressure or bending is forbidden to avoid the breakage of the branches.

The vines are unearthed next spring, between April and May, before the bud expansion period, and the best time should be decided according to the local climate and the phenology of the cultivar. Both artificial and mechanical methods are used to complete the unearthing, and this work should be carefully performed to prevent damage to branches and buds. In order to remove the soil more quickly in spring, different methods and cover materials including film mulching, industrial cotton, straw mattress, and plastic are used (Duan et al., 2019). In addition, different types of vine-burying machines (for burying of vines) and cleaning machines (for digging up vines) have also been developed (Duan et al., 2019).

27.5 Grape varieties in the cold regions of China

According to the International Organisation of Vine and Wine (OIV, 2017), the total grape planting area of China was 852,900 ha in 2017, and for wine grape about 85,290 ha, which occupied nearly 10% of the total grape

FIGURE 27.6   Vines unearthed from the soil in the next spring.

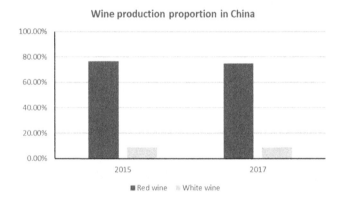

FIGURE 27.7   Wine production proportion in China. Source: *China alcoholic drinks association.*

planting area. Other interesting data were on the quantity of wine production in 2017 (Fig. 27.7), where red wine dominated at 74.82%, while white wine only showed 8.62%. A similar observation was made in 2015 with red wine dominating at 76.38%, while white wine only recorded 8.86%.

Although the history of grape planting in the Gansu and Ningxia regions is dated back 2400 years from the Han dynasty following the famous Silk Road, modern viticulture began in the 1980s where different varieties of wine grape were introduced from France, Georgia, and Italy. Currently, the Chinese Agricultural Academy of Sciences has three national grape germplasm resource nurseriesincluding Zhengzhou Fruit Research Institute, Institute of Special Animal and Plant Sciences, and Pomology Institute in the Shanxi Academy of Agricultural Sciences as well as other several local nurseries (Duan et al., 2019). Thousands of grape germplasms have been identified and conserved, but the major varieties for industrial winemaking are Chardonnay, Riesling, Italian Riesling, Vidal, Semillon, and Ugni blanc. The characteristics of some of the varieties are listed below.

Chardonnay, which originated from France, is very adaptable and can be grown in different soils. Chardonnay has become one of the most widely planted grape varieties. It is a mid-early variety with medium yield and compared to other white varieties, Chardonnay is more resistant to the late white rot. This variety can be used to make white wines and sparkling wines, with a pronounced aroma of typical acacia flora, and when fermented and aged in oak barrels, it usually generates notes of toast, vanilla, and dried fruits (Zhan et al., 2015).

Riesling originated from Germany and is a mid-early variety with a medium yield. Compared to Germany, Riesling grown in Gansu and Ningxia does not show characteristics of high acidity and mineral aromas but possesses high ripeness, deep flavor with blossom perfume, aromas of tropical fruits and sweet mellow, roundness, and good mouth-feel (Zhan et al., 2015). In Gansu and Ningxia regions, Riesling can be made into dry or semi-dry wine.

Italian Riesling originated from Italy, and it is a mid-early variety with medium yield. This variety has thin skin and is sensitive to late white rot (Zhan et al., 2015). Italian Riesling can be made into dry and semi-dry wines with intense flowery and fruity aromas. Furthermore, it is also used to make high-quality icewine in Gansu Qilian winery.

Vidal is a native grape variety in France but it is widely grown in Canada to make icewine. Because of its strong cold tolerance, it is also widely planted in Northern China. Vidal is a mid-late maturity variety with high yields. It is resistant to mildew, has thick skin, and it is not easy to get crinkled, thus it is suitable for making noble wine and icewine (Zhan et al., 2015).

Ugni blanc originated from the Mediterranean area and it is a late harvest variety with a high yield. The alcohol content of Ugni blanc wine is always lower than other varieties, with very high acidity and freshness, but with a poor fruity aroma, and it is mostly used to make high-quality brandy (Zhan et al., 2015). A traditional white grape variety in China known as Longyan, has the potential to be utilized as both a table grape and a wine grape. As a late-harvested variety, the Longyan grape has been widely cultivated in Beijing-Tianjin-Heibei, Shandong, and Loess Plateau regions for the development of wine characterized by a green to yellow color, fresh fruity flavor, and good taste.

Recently, a company named Summerland vineyard located in Hexi Corridor (Gansu Province) introduced 43 varieties (including 76 cultivars) from France and Italy and after 8 years of testing, they found some varieties such as Chardonnay, Riesling, Semillon, Malvasia, Arneis, Sauvignon Rytos, Muscat Yellow, Muscat White, and Soreli could be planted in this cold region with short maturity periods.

27.6 Chemical composition of grapes and must

The quality of grapes is mainly composed of primary metabolites including sugars and organic acids, as well as secondary metabolites such as phenolic compounds and volatile compounds (Pavloušek & Kumšta, 2011). Among white grape varieties, the quality mainly depends on the balance between ideal sugar−acid ratio, moderate pH, and varietal characteristics.

27.6.1 Sugars and organic acids compounds

With the ripening of grapes, the accumulated sugars in berries are mainly glucose and fructose, and their content increases gradually as the grapes ripen (Kuhn et al., 2014). The accumulation of sugar in grape berries contrasts with changes in organic acids.

As the growth of grape berries progresses, the sugar content and composition changes, with the sugar content showing a double S-shaped growth pattern (Davies & Robinson, 1996). The first rapid growth stage of grape berries after fruiting is due to the increase of cell number and cell expansion. In most varieties, the first rapid growth phase is followed by a lag phase, during which there is little or no growth. The second growth stage after the lag stage is the coloration stage, which is the last stage of significant accumulation of sugars (Zhu et al., 2019).

Grape berries are mainly thought to consist of high concentrations of fermentable sugars accompanied by complex polyphenols that are associated with the color and taste of wine (Conde et al., 2007). The organic acid composition mainly reflects the metabolism of tartaric acid and malic acid in the process of development and maturation and has an important influence on grapes in winemaking (Preiner et al., 2013). Therefore, the quality of white grape varieties should be focused on the desired sugar−acid ratio, with a moderate combination of pH, and individual varietal characteristics (Muñoz et al., 2011).

The sugar and acid contents of some white grape varieties before harvest from 2016−2019 in Mogao winery in Gansu Hexi Corridor are shown in Table 27.3.

TABLE 27.3 Sugar and acid contents of white varieties in Gansu's Mogao winery.

Variety	Total sugar (g/L)			Total acid (g/L)		
	2016	2017	2019	2016	2017	2019
Italian Riesling	212	218	210	6.5	6.0	6.3
Pinot blanc	232	235	233	7.5	6.2	7.6
Riesling	228	208	215	6.7	6.4	6.8
Chardonnay	236	224	238	6.8	8.3	6.0
Rkatsiteli	207	210	195	8.2	7.5	7.6
Ugni blanc	167	166	172	10.4	11.8	10.6

The sensory quality of wine grapes is primarily determined by the soluble solids, total acids, organic acids, and other components, as well as the balance between these parameters, which is important in the wine sensory profile (Liu et al., 2006). Organic acids can balance the taste of wine grapes, but high acidity will have a negative impact on the taste (Volschenk et al., 2006). The pH is also related to the acidity of grapes and wine. Although pH is often indirectly proportional to acid concentration, there is no direct link between pH and titratable acidity or pH and total acidity.

For white wine grape varieties, flavonoids and other compounds derived from phenyl propane metabolism constitute the main phenolic compounds, which are very important quality attributes in fruits (Hadi et al., 2013). During the ripening stage of grapes, volatile compounds, which are the main secondary metabolites in wine grapes also contribute to the flavor and aroma characteristics desired by consumers (Adel, 2008). Grape volatile compounds have been reviewed separately in this chapter.

27.6.2 Phenolic compounds of grapes and wine

Grape polyphenols are monomeric and polymeric molecules located in the skins as anthocyanins, flavan-3-ols, flavonols, dihydroflavonols, hydroxycinnamoyl, tartaric acids, hydroxybenzoic acids, and hydroxystilbenes, while in the seeds, polyphenols include flavan-3-ols and gallic acid. In the grape must, polyphenols are found as hydroxycinnamoyl tartaric acids (Ivanova et al., 2011). Usually, the content of total polyphenols in white grape varieties is lower compared to that of red grape varieties, since they do not synthesize anthocyanins and tannins.

Phenolic compounds are highly diverse in different varieties and growing environments. Berry phenols contribute to the quality of wine and have beneficial effects on many aspects of human health (Corder et al., 2006). Phenolic compounds are made up of one or more aromatic benzene rings with one or more hydroxyl groups and can exist as free molecules (glycoside ligands), conjugates (glycosides), oligomers, or polymers (Crozier et al., 2006). An example of a simple aromatic phenolic compound is phenolic acid, which contains two distinct constituent carbon frames, namely hydroxyl cinnamic acid and hydroxybenzoic acid with basic skeletons C6-C3 and C6-C1, respectively (Andrés et al., 2010). The total phenol content of white wine is also related to the antioxidant activity. White wine total phenol content is substantially lower than that of red and rose wines hence has the lowest antioxidant activity.

The quality of white wines can also be affected by different treatments. Lin et al. (2020) discovered that a drying process with a 30% water loss rate increased the content and total amount of phenolic compounds as well as the antioxidant properties of wine samples. Additionally, the wine's brightness, yellow tone, and aroma were enhanced. Liang et al. (2019) also investigated the effect of adding glutathione (GSH) to dry white wine. The color of wine without GSH addition darkened (reduction in L* and increase in a* and b*), whereas the change was slow in wine with GSH addition. Furthermore, there was a decrease in total phenol concentration and phenol oxidation, which were both typically linked to the wine color change.

27.6.3 Volatile compounds of grapes and wine

Volatile compounds play an important role in grape and wine quality by influencing aroma and varietal characteristics. The type and concentration of volatile compounds are determined by the grape variety, vineyard condition, soil, weather, agronomic practices, and winemaking methods (Alem et al., 2018). Terpenes (including monoterpenes and sesquiterpenes) norisoprenoids, methoxypyrazines, C6 alcohols and aldehydes and thiols are the most important aroma compounds in white grapes, some of which exist in the form of free or glycosylated molecules. The glycosylated precursors can be released during berry ripening, fermentation, or aging process.

Headspace solid-phase microextraction (HS-SPME) combined with gas chromatography-mass spectrometry (GC−MS) was used to detect the difference of volatile compounds in Italian Riesling, Chardonnay, and Riesling grapes (Sun et al., 2014). The results showed that terpenes and norisoprenoids were the most important compounds in white grapes although there are variations among different varieties. More so, Italian Riesling showed the highest concentration in terpenes and total volatile compounds, especially β-pinene, which was 2−6 times higher than the other varieties, while the content of 4-terpineol in Chardonnay was only half the amount of that in Italian Riesling and Riesling. Aroma compounds such as p-cymene were lower in Riesling grapes, while the contents of cis-3-hexen-1-ol, 2-ethyl-1-hexanol, and 2-dihydronaphthalene (TDN) were significantly higher than the other two varieties.

Cui et al. (2012) used HS-SPME-GC-MS to analyze the aroma compounds of Muscat Hamburg (2010 vintage) dry white wines made with the same brewing technology from three different wine regions in China: Tianjin, Heibei, and Shandong. A total of 58 volatile compounds were identified, with 23 of them quantified. Wines from the Tianjin region were found to have larger concentrations of terpenes and esters, as well as lower levels of higher alcohols, which contributed to the wine's floral and fruity aromas and sweet, musky flavors (Cui et al., 2012).

27.7 Wine composition as influenced by winemaking technology

White wine is generally made from white grape varieties, while some of them are made from light-colored red grapes such as Pinot Noir in Champagne. However, some white wines can also be made by strong-colored red varieties. In August of 2019, one Cabernet Sauvignon white wine made by Chateau Changyu Moser XV (located in the Eastern region of the Helan Mountains) got the Decanter World Wine Awards with up to 90 points. Although the wine still showed a light pink color, it is a successful case of making white wine from red grape varieties. In this circumstance, enologists press the grape with whole bunches, whereas some white varieties are pressed to obtain must after destemming Fig. 27.8. In recent years, cold maceration has become more and more popular in white wine production. In this technology, destemmed and crushed grapes including pulp, juice, skins, and seeds are put together at 4-10°C for 6-24 hours to extract aroma precursors and phenolic compounds from skins. A study proved that skin contact at low temperature can improve the quality of 'Guankou' white wine (Wang, Chen, et al., 2017).

When the influence of juice extraction processes on the mouthfeel and taste of white wine and their relationship to wine composition were investigated, it was discovered that the amount and type of interaction of juice with skins affected both the total phenolic concentration and phenolic composition in the wine (Gawel, Smith, & Cicerale, 2018). More so, the wine's pH strongly influenced perceived viscosity, astringency drying, and acidity. Despite a five-fold variation in total phenolics among wines, differences in bitter taste were small. Another study explored the use of cold treatment for Ecolly grapes (a new white variety in China) to enhance wine aroma (Li et al., 2016). The results showed that low-temperature treatments at either −8°C or −20°C can significantly improve the quality of wine aroma by increasing the aroma intensity to some extent.

In white wine production, the must need to be clarified by sedimentation, centrifugation, or clarification to get a clear juice before fermentation. However, if the must is over clarified, it will also influence the fermentation rate. In addition, in order to keep more aroma compounds, white wine is usually fermented at lower temperatures between 10°C and 20°C, although this method needs more time to finish the fermentation.

Jin et al. (2017) evaluated the effect of fermentation temperature (from 15°C to 25°C) on the aroma quality of Chardonnay wine. The results showed that aroma quality was affected significantly by fermentation temperature. From the study, the intensity of fruit aroma (green apple, pineapple, banana, lemon, pear, etc.) first increased and then decreased subsequently with increase in temperature. Higher fermentation temperatures resulted in more sweet fruity aromas, but also in unpleasant odors, whereas lower temperatures resulted in a lesser concentration of fermentative volatiles. Aromatic esters and higher alcohols also increased at first and subsequently decreased with increasing temperature, which was attributed to the shortening duration of fermentation (Jin, Li, & Liu, 2017).

Another important method to enhance the aroma character of white wine is barrel fermentation. In this case, not only the flavor compounds are extracted, but the barrel itself could also provide enough oxygen for yeast. For some wines with high acidity, malic acid fermentation (MLF) could be used to decrease the acidity and increase the aroma complexity. Chardonnay is a variety suited for MLF, which often results in a light creamy aroma and structure, however MLF is rarely used in Riesling wine because enologists always want to keep the wine's fresh acidity.

FIGURE 27.8 Grape harvesting and sorting in Gansu Hexi Corridor.

Ma et al. (2020) evaluated the influence of clarification treatments on volatile composition and aromatic attributes of Italian Riesling icewine samples. Italian Riesling icewines from the Hexi Corridor region of China were clarified by using fining agents such as bentonite and soybean protein, membrane filtration, and centrifugation. The clarity, physicochemical indices, volatile compounds, and sensory attributes of treated wines were investigated. Both the fining agents and mechanical clarification treatments increased the transmittance and decreased the color intensity of icewine samples. Bentonite fining agent significantly influenced the total sugar content, total acidity, and volatile acidity. Volatile compounds analysis revealed that both the types and concentrations of volatile compounds of wine samples decreased. Membrane filtration treatment had the greatest influence on volatile compounds, whereas soybean protein fining agent had a much lower impact (Ma et al., 2020).

To explore the correspondence between the evolution of volatile compounds and the development of sensory characteristics during wine aging in bottle, Liu et al. (2017) studied the various volatile compounds and sensory quality in Chardonnay wines during 18 months of bottle storage. From the findings, volatile compounds presented three evolution patterns along with bottle aging. The first trend showed higher accumulation of volatile compounds after 9 months of post-bottling, while the second trend displayed a constant decrease in volatile compounds concentration during the aging process. The third trend was characterized by a high concentration of volatile compounds at the middle of wine bottle storage followed by a continuous decrease. Most volatile compounds including esters, higher alcohols, and terpenes were found in the third trend. Floral and fruity odors in wine samples gradually declined during the 18 months of bottle storage, whereas most of the aroma components significantly reduced after 9-month bottle of aging (Liu, Zhang, & Lan, 2017).

27.8 Innovations in winemaking in cold regions

As reported earlier, Chinese consumers do not like white wines as much as red wines, hence enologists have to develop other products using white grapes. Recently, sparkling wines, low-alcohol wines, icewines, fortified

TABLE 27.4 Innovations in cold-region winemaking in China.

Product	Treatment	Variety	Impacts on wine composition	References
Sweet wine	Dehydration	Chardonnay	The alcohol, residual sugar, and acidity content increased. Phenols, antioxidant activity, brightness, yellow tone, aroma, and taste also improved.	Wang et al. (2020)
Sweet wine	Lianbai, concentrated by heating in earthen pot	Longyan	The contents of alcohols, esters, acids, aldehydes, and terpenes in lianbai-3 wine were all higher than those in the other two lianbai wines, while the total odor activity value, fruity aroma, and plant aroma were the highest.	Dong et al. (2019)
Icewine	Strain selection and cofermentation	Vidal	Indigenous *Saccharomyces* strains were selected and Non-*Saccharomyces cerevisiae* was used to improve the aroma quality and diversity of icewine products.	Shen et al. (2020)
Low-alcohol wine	Low temperature and nitrogen consumption technology	Muscat Hamburg	Alcohol content was easier to control, and the wine body was stable.	Lv et al. (2016)
Sparkling wine	Different fining agent	Chardonnay	Soy protein as fining agent before fermentation had higher degree of clarity, foam stability, and aroma quality. Soy protein could be used as an effective clarifying agent in sparkling wine production.	Ma et al. (2018)
Brandy	Ultra-high pressure technique	Ugni blanc	Ultra-high pressure technique had a great application prospect in the aging of esters in brandy.	Yan et al. (2020)

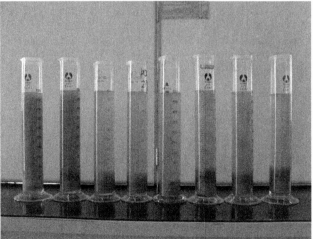

FIGURE 27.9 Semillon grape and must clarification in Qilian winery in Gansu Hexi Corridor.

wines, brandy, and sweet wines (from off-vine or late harvest grapes or dehydration treatments) have been produced by white grapes in China.(Table 27.4) (Fig. 27.9).

Funding: This work was supported by the National Key Research and Development Project (Item No. 2019YFD1002500), and the horticulturists, enologists, and sommeliers training and monograph writing in Gansu Hexi Corridor (Grant No. Ganshangcaiwufa 2017-466).

Acknowledgment

We would like to express our gratitude to professor Bi Yang from Gansu Agricultural University for his support in writing this book chapter, Huo Xingsan from China Alcoholic Drinks Associationand Liang Liang from Gansu Alcohol Administration for providing wine data in China, Hao Yan from the Institute of Fruit and Floriculture Research, Gansu Academy of Agricultural Science (Lanzhou, China) for providing information on buried viticultural, Wang Lei from China Wine Industry Technology Institute (Yinchuan, China), Chen Yanxiong from Qilian winery, Niu Yulin from Mogao winery, Li Jia and Guo Zhijun from Zixuan winery for providing the figures and tables.

References

Adel, A. K. (2008). Flavor quality of fruits and vegetables. *Journal of the Science of Food and Agriculture, 88*(11), 1863–1868.

Alem, H., Rigou, P., Schneider, R., et al. (2018). Impact of agronomic practices on grape aroma composition: A review. *Journal of the Science of Food and Agriculture, 99*(3), 975–985.

Andrés, L. C., Medina, R. A., Llorach, R., et al. (2010). Phenolic compounds: Chemistry and occurrence in fruits and vegetables. *Fruit and Vegetable Phytochemicals: Chemistry, Nutritional Value and Stability, 2010*, 53–80.

Conde, C., Silva, P., Fontes, N., et al. (2007). Biochemical changes throughout grape berry development and fruit and wine quality. *Food, 1*(1), 1–22.

Corder, R., Mullen, W., Khan, N. Q., et al. (2006). Red wine procyanidins and vascular health. *Nature, 444*(7119), 566–566.

Crozier, A., Jaganath, I. B., & Clifford, M. N. (2006). *Phenols, polyphenols and tannins: An overview. Plant Secondary Metabolites: Occurrence* (pp. 1–10). Blackwell Publishing Ltd.

Cui, Y., Wang, W., Zhang, F. Q., et al. (2012). Comparative analysis of the aroma compounds of Muscat Hamburg dry white wines from different wine regions in China. *Advanced Materials Research, 554*, 1581–1584.

Davies, C., & Robinson, S. P. (1996). Sugar accumulation in grape berries cloning of two putative vacuolar invertase cDNAs and their expression in grapevine tissues. *Plant Physiology, 111*(1), 275–283.

Dong, Z. G., Liu, Z. H., Wang, Y., et al. (2019). Comparison of aroma components of wine brewed by Longyan grapes under three kinds of lianbai concentration processes. *China brewing, 38*(12), 48–52. Available from https://doi.org/10.11882/j.issn.0254-5071.2019.12.010.

Duan, C. Q., Liu, C. H., Liu, F. Z., et al. (2019). Fruit scientific research in New China in the past 70 years: Grape. *Journal of Fruit Science, 36* (10), 1292–1301. Available from https://doi.org/10.13925/j.cnki.gsxb.Z05.

Gawel, Richard, Smith, Paul, Cicerale, Sara, et al. (2018). Critical Reviews in Food Science and Nutrition. *The Mouthfeel of White Wine, 58*(17), 2939–2957. Available from https://doi.org/10.1080/10408398.2017.1346584.

Hadi, M., Zhang, F. J., Wu, F. F., et al. (2013). Advances in fruit aroma volatile research. *Molecules (Basel, Switzerland), 18*(7), 8200–8229.

Ivanova, V., Stefova, M., Vojnoski, B., et al. (2011). Identification of polyphenolic compounds in red and white grape varieties grown in R. Macedonia and changes of their content during ripening. *Food Research International, 44*(9), 2851–2860.

Jin, Guojie, Li, Aihua, Liu, Hao, et al. (2017). Influence of Fermentation Temperature on Aroma Quality of Dry Chardonnay Wine. *Journal of Chinese Institute of Food Science and Technology, 17*(10), 134–144. Available from https://doi.org/10.16429/j.1009-7848.2017.10.019.

Kuhn, N., Guan, L., Dai, Z. W., et al. (2014). Berry ripening: Recently heard through the grapevine. *Journal of Experimental Botany, 65*(16), 4543–4559.

Li, N. N., Wang, H., Tang, G. D., et al. (2016). Effect of cold treatment of trapes on aroma components of Ecolly dry white wine. *Food Science, 37*(13), 71–76. Available from https://doi.org/10.7506/spkx1002-6630-201613013.

Liang, X. F., Wang, H. R., Song, C. H., et al. (2019). Effects of glutathione addition on dry white wine. *Liquor-Making Science & Technology, 02*, 30–35. Available from https://doi.org/10.13746/j.njkj.2018261.

Liu, D., Zhang, Y., Lan, Y. B., et al. (2017). Evolution of Volatile Compounds and Sensory in Bottled White Wines and Their Correspondence. *Journal of Chinese Institute of Food Science and Technology, 17*(4), 228–240. Available from https://doi.org/10.16429/j.1009-7848.2017.04.028.

Liu, H. F., Wu, B. H., Fan, P. G., et al. (2006). Sugar and acid concentrations in 98 grape cultivars analyzed by principal component analysis. *Journal of the Science of Food and Agriculture, 86*(10), 1526–1536.

Lv, W., Tang, M. H., & Cui, Y. (2016). Application of low temperature and nitrogen consumption technology in brewing of low-alcohol wine. *Liqupr Making Science and technology, 10*, 56–60. Available from https://doi.org/10.13746/j.njkj.2016215.

Ma, T. Z., Gong, P. F., Lu, R. R., et al. (2020). Effect of different clarification treatments on the volatile composition and aromatic attributes of "Italian Riesling" Ice wine. *Molecules (Basel, Switzerland), 25*(11), 2657.

Ma, T. Z., Lu, R. R., Hao, N., et al. (2018). Quality of sparkling wines as influenced by different fining agents added pre-fermentation. *Food and Fermentation Sciences & Technology, 54*(5), 93–100. Available from https://doi.org/10.3969/j.issn.1674-506X.2018.05-020.

Muñoz, R. P., Robledo, P., Manríquez, D., et al. (2011). Characterization of sugars and organic acids in commercial varieties of table grapes. *Chilean journal of agricultural research, 71*(3), 452.

Pavloušek, P., & Kumšta, M. (2011). Profiling of primary metabolites in grapes of interspecific grapevine varieties: Sugars and organic acids. *Czech Journal of Food Sciences, 29*(4), 361–372.

Preiner, D., Tupajić, P., Kontić, J. K., et al. (2013). Organic acids profiles of the most important Dalmatian native grapevine (*V. vinifera* L.) cultivars. *Journal of food composition and analysis, 32*(2), 162–168.

Shen, J. Y., Zhang, B. Q., Duan, C. Q., et al. (2020). Selection of indigenous Saccharomyces cerevisiae strains from spontaneous fermentation of Vidal icewine in Huanren region and evaluation of their oenological properties. *Food Science*, *41*(2), 148–157. Available from https://doi.org/10.7506/spkx1002-6630-20180903-022.

Sun, S. S., Fan, W. L., Xu, Y., et al. (2014). Comparison of volatile components in three white grape varieties. *Food and Fermentation Industries*, *40*(5), 193–198. Available from https://doi.org/10.13995/j.cnki.11-1802/ts.2014.05.006.

Volschenk, H., Van, V., & Viljoen, B. M. (2006). Malic acid in wine: Origin, function and metabolism during vinification. *S. Afr. J. Enol. Vitic.*, *27*(2), 123–136.

Wang, S., Li, H., & Wang, H. (2015). Wind erosion prevention effect of suspending shoots on wires after winter pruning in soil-burying zones over-wintering. *Transactions of the Chinese Society of Agricultural Engineering*, *31*(12), 206–212. Available from https://doi.org/10.11975/j.issn.1002-6819.2015.12.028.

Wang, L., Li, H., & Wang, H. (2017). Climatic regionalization of grape in China Ⅱ: Wine grape varieties regionalization (in Chinese). *Chin Sci Bull*, *62*, 1539–1554. Available from https://doi.org/10.1360/N972017-00249.

Wang, S. S., Chen, H. M., Dong, Z., et al. (2017). Effects of different winemaking technologies on the quality of dry white wine produced from "Guankou" grape variety. *Food Science*, *38*(21), 138–145. Available from https://doi.org/10.7506/spkx1002-6630-201721022.

Wang, L., Zhao, P., Liu, Y., et al. (2020). The effect of dehydration treatment on Chardonnay wine. *Food and Fermentation Industries*, *46*(7), 83–88. Available from https://doi.org/10.13995/j.cnki.11-1802/ts.023134.

Yan, H. K., Zhang, B., Qi, L., et al. (2020). Process optimization of accelerated ageing of esters in brandy by ultra-high pressure technique. *Food and Fermentation Industries*, *46*(13), 152–159.

Zhan, J. C., & Li, D. M. (2015). *Wine grape varieties* (pp. 84–108). Beijing, China: China Agricultural University Press.

Zhu, J., Génard, M., Poni, S., et al. (2019). Modelling grape growth in relation to whole-plant carbon and water fluxes. *Journal of Experimental Botany*, *70*(9), 2505–2521.

Further reading

Jiang, B., & Zhang, Z. (2010). Volatile compounds of young wines from Cabernet Sauvignon, Cabernet Gernischet and Chardonnay varieties grown in the Loess Plateau Region of China. *Molecules (Basel, Switzerland)*, *15*(12), 9184–9196.

Neta, E. C., Johanningsmeier, S., & McFeeters, R. (2007). The chemistry and physiology of sour taste – a review. *Journal of Food Science*, *72*(2), 33–38.

28

Dealcoholization of white wines

Matthias Schmitt and Monika Christmann

Institute of Oenologie, Hochschule Geisenheim University, Geisenheim, Germany

28.1 Introduction

Due to the tendency for average temperatures to rise in recent years, there has been an increase in the sugar content of must and consequently in the alcohol content of wine (Alston, Fuller, Lapsley, Soleas, & Tumber, 2015; Mira de Orduña, 2010). Numerous international publications have documented this development in warmer but also cool climate wine-growing regions.

In the past, especially in cooler regions such as the German wine-growing areas, it was a standard procedure to increase the alcohol content of wine in many different ways. Grape varieties and clones have been selected according to their high potential alcohol content, the viticultural practice has been optimized in this respect as well, the yields have been reduced, and training systems have been optimized by larger canopies with more leaf area. In addition to that, vinification with selected pure yeast cultures and targeted fermentation control has a share that the wines tend to contain more and more alcohol.

Alcohol as a wine ingredient is not only very controversially discussed but also strongly regulated by law. If a wine is marketed under a protected geographical indication such as, for example, in the Rheingau region, the wine must contain at least 7% alcohol by volume. Sweet wines of the categories Beerenauslese, Trockenbeerenauslese, and Icewine are excluded from this. They must have at least 5.5 vol.% actual alcohol.

As a general rule, the alcoholic strength may be increased by up to 3 vol.% by concentration or enrichment using sucrose or rectified concentrated grape must. Since 2009, a partial reduction of alcohol by physical processes such as distillation under vacuum or by means of modern membrane processes such as osmotic distillation is also permitted under certain conditions. According to OIV Resolution OENO-394A-2012, this allows the wine to be reduced by up to 20% of its actual alcohol content.

28.2 Sensory effect of ethanol

Alcohol is, in terms of volume, besides water, the most important component in wine. Furthermore, alcohol (ethanol) is also the major volatile substance in wine. It is superior to other flavors by a factor of 105−1011 (Taylor et al., 2010).

Until some years ago, the influence of alcohol on the sensory characteristics of wine was not critically examined. In recent times, many attempts have been made to increase the alcohol content of wine in many wine-growing regions. Often there has been no critical examination of the amount of alcohol that is really necessary. Only recently has the interaction of alcohol with other wine ingredients been further investigated. Increasingly, wines with excessive alcohol content are being found. Therefore the question arises: How much alcohol does a wine need?

The influence of alcohol on the sensory characteristics of wine is controversial and very complex. Alcohol has gustatory, olfactory, and trigeminal stimulating properties (Green, 1987; Mattes & DiMeglio, 2001; Williams, 1972; Yu & Pickering, 2008).

White Wine Technology.
DOI: https://doi.org/10.1016/B978-0-12-823497-6.00028-4

© 2022 Elsevier Inc. All rights reserved.

The direct sensory influence of alcohol is sweetness, and the higher the alcohol content of the wine, the more the sweetness perception of the wine increases (Martin & Pangborn, 1970; Scinska et al., 2000). The acid perception is lowered with increasing alcohol content (Fischer & Noble, 1994). To a certain extent, alcohol produces a sensation of heat in wine, which can create a burning sensation at excessive alcohol contents (Gawel, van Sluyter, & Waters, 2007; Grainger, 2009). Alcohol gives the wine body and fullness (Amerine and Roessler 1976; Gawel et al., 2007; Grainger, 2009; Pickering, Heatherbell, Vanhanen, & Barnes, 1998). How much the alcohol increases the sensation of body and fullness depends on the wine matrix. There are scientific sources pointing out that body and fullness in wine are only slightly changed by alcohol (Gawel et al., 2007; Pickering et al., 1998). It can be assumed that, depending on the wine matrix, the alcohol has a different influence on the perception of body and fullness and that the matrix of model wines cannot be reproduced in a sufficiently complex way.

Wines with a pronounced body and fullness appear more valuable and achieve higher prices for the consumer (Fischer & Noble, 1994; Szolnoki, Hoffmann, & Herrmann, 2008). This is an important motive to optimize the alcohol content upward. However, with increasing ethanol content the bitterness of the wine increases (Arnold & Noble, 1978; Fischer, 2010; Mattes & DiMeglio, 2001; Meillon et al., 2010). Fontoin et al. (2008) could show with model wine solutions that astringency decreased with increasing alcohol levels. The volatility of the aromas in the wine is reduced by increasing alcohol content. This makes these wines appear less fruity (Christmann, 2001; Conner, Birkmyre, Paterson, & Piggott, 1998; Escudero, Campo, Fariña, Cacho, & Ferreira, 2007; Goldner, Zamora, Di Leo Lira, Gianninoto, & Bandoni, 2009). The majority of white wines belong to a product category that is fresh and fruity, not to have too much bitterness and astringency. Those wines are usually less full-bodied compared to red wine.

Wines with low alcohol content are often considered unripe, due to immature tannins, higher acidity, and less pronounced primary aromas. In contrast, wines with higher alcohol contents are generally associated with pronounced ripe aromas and softer tannins (King, Dunn, Randall, & Heymann, 2013). By partially reducing the alcohol content of wine, physical processes can be used to change this common classification of wine by alcohol content. Thus even wines with moderate alcohol contents can exhibit the typical ripe and complex aroma spectrum of wines with usually high alcohol contents.

Compared to purely aqueous solutions, ethanol creates certain convection in wine through its volatilization, which accelerates the release of aromas into the headspace. However, this is less important for the sensory evaluation of wine, since a certain liquid column is necessary for convection. In wine tasting, however, such a liquid column is not directly supporting the wine, since for retronasal perception in the oral cavity only a very thin liquid film over a very short time determines the perception of aromas (Taylor et al., 2010).

Due to its physical properties, ethanol is not only noticeable in the taste of wine. Above certain alcohol contents, so-called surface tension effects cause tears or oval fields in the glass when the wine is swiveled intensively (Taylor et al., 2010).

28.2.1 Sensory effect of partial alcohol reduction

From which intensity of alcohol reduction, by means of physical processes, the treated wine changes in terms of sensory character to the control depends to a large extent on how gently and properly the processes are applied by the operator.

Several scientific works are dealing with that issue. There are in general no significant differences between the initial wine and the treated variant when 2 vol.% in alcohol content has been reduced (Diban, Athes, Bes, & Souchon, 2008; Meillon et al., 2010; Schmitt, 2016; Urbano, 2007), even if the wine was tasted by trained panelists.

Due to that, the so-called sweet spot theory, which suggests that every wine has an alcohol content that suits it the best (Christmann, 2001; Fischer, 2010; King & Heymann, 2014; Wollan, 2005), can be clearly doubted, as a targeted alcohol reduction according to good practice is not distinguished in blind tasting from the control with differences of 2 vol.% and more.

The alcohol reduction of wine by means of physical processes is always accompanied by a certain loss of volatile aromatic substances (Christmann, 1988; Diban et al., 2008; Fedrizzi et al., 2014; King & Heymann, 2014; Schmitt, 2016).

When a partial alcohol reduction is applied to white wine to an extent of more than 2 vol.%, several sensory attributes, which are directly linked to ethanol, are influenced.

The major change is linked to a loss of fullness and body (Schmitt, 2016). Significantly reduced alcohol contents can be buffered in terms of the perceived reduction in body and fullness by leaving residual sugar, adding

CO_2 or tannins. Furthermore, an increased body sensation based on mannoproteins by so-called *"sur lie"* aging or direct addition could help to overcome the influence of the treatment for white wines. While reducing the alcohol content, the sweetness sensation of white wines is slightly changed.

Even with the gentle application of alcohol reduction, the fruitiness of white wine is reduced according to the extent of the treatment. On the one hand, this goes back to the oxidation resulting from pumping the wine several times, and on the other hand, due to separating alcohol fractions including also further volatiles (Longo et al., 2017).

Bitterness is reduced when the alcohol content is lowered (King & Heymann, 2014). As many white wines suffer from increased bitterness in warm and dry conditions, targeted alcohol can help to support or replace the reduction of phenols by fining or oxidation.

It becomes clear in what complex way alcohol reduction influences the sensory characteristics of wine. A targeted treatment by more than 2 vol.% under gentle conditions might help to create a clearly different white wine style. Meillon et al. (2010) showed with red wine that the variant with 3 vol.% less alcohol was rated higher in harmony and balance compared to the control (Meillon et al., 2010). Tastings with Riesling reduced by 4 vol.% showed equal preferences for the control and the severely reduced sample (Schmitt, 2016). But this was not the case for a similar test with Pinot Blanc and Müller-Thurgau. Here the severe alcohol reduction did get significantly lower preferences compared to the untreated control (Schmitt, 2016).

28.3 Different approaches for alcohol reduction

28.3.1 Viticultural strategies

Excessive alcohol contents in white wines can be tackled in various ways. The earliest point to do so is in viticulture.

As supposedly the simplest and most economical solution, higher yields through targeted pruning could help to achieve lower sugar contents at harvest, as the leaf area to fruit ratio is lower. This strategy is just suitable in areas or for products where there are no strict yield limitations. Excessive yields bear the hazard of severe plant stress, especially in dry conditions. Increased phenol contents will be the case that leads to more bitter white wines. A reduced foliage area by massive shortening of the shoots can also help to change the leaf-fruit ratio and thus to store less sugar in the berries (Vasconcelos & Castagnoli, 2000).

However, if the weather is not as good as expected toward the end of the vegetation phase, especially in cool climate areas, there is a lack of ripeness.

Earlier harvest date is often mentioned as a solution to avoid excessive alcohol content, especially in very warm vintages; however, there is a clear discrepancy between the sugar content of the berries and the aroma ripeness (Mira de Orduña, 2010). The grapes become really aromatic only at very high sugar contents. This is an important criterion for the production of fruity white wines. An earlier harvest with the aim of moderate alcohol content is therefore accompanied by a deficit in aromas. Moreover, such wines are more prone to atypical aging (ATA) notes (Schneider, 2014).

This phenomenon is often described by sensations of mothballs, furniture varnish, acacia blossom, wet wool, and dirty dishrag and goes back to 2-aminoacetophenone and some other compounds generated by degradation of indole-3-acetic acid as the source of that off-flavor (Schneider, 2014).

Furthermore, not fully ripe grape seeds have higher contents of bitter and astringent phenols. Therefore grape processing of unripe grape material should be very gentle to avoid unnecessary maceration processes.

28.3.2 Microbiological approaches

As an alternative approach to direct alcohol reduction by physical processes, yeasts with reduced alcohol yield are conceivable (Contreras et al., 2014; Duchêne, Huard, Dumas, Schneider, & Merdinoglu, 2010; Röcker, Strub, Ebert, & Grossmann, 2016). Appropriate yeasts on the market are able to reduce the alcohol content by 1−2 vol.% It is questionable which substances are produced by the yeast instead of alcohol and whether this is beneficial to the quality of the wines. As, in most cases, excessively high sugar contents are going along with a lack of acidity, the metabolism toward more acidity would be ideal as an alternative side product of the yeast.

So far not yet permitted for wine production in most countries, but potentially very interesting, is the prefermentative use of the enzyme glucose oxidase (GOX). GOX is able to convert glucose into gluconic acid under aerobic conditions. After the aeration phase of about 30 h, the must is fermented anaerobically under normal conditions. This allows wines to be produced with about 2% less alcohol by volume. In terms of sensory characteristics, the wines produced in this way are clearly characterized by high acidity and are less typical compared to the common style of a specific region (Röcker, Schmitt, Pasch, Ebert, & Grossmann, 2016).

The tendency of increasing sugar content also increasingly leads to problems to stop the alcoholic fermentation completely. High sugar contents in must mean high osmotic stress for the yeast, which can delay the start of fermentation. This often promotes the appearance of undesirable microorganisms, which in turn lead to wine defects such as an increase in volatile acidity or the formation of ethyl acetate (Ribéreau-Gayon, 2007).

The complete metabolism of the sugar by the yeast to produce dry wines is further complicated by high alcohol contents toward the end of the fermentation and any desired malolactic fermentation is also made more difficult by the stress factor alcohol (Bisson, 1999). Even though the industry has selected numerous apparently very alcohol-tolerant yeasts and bacterial strains, the problem of stuck and sluggish fermentation is still an issue worldwide.

Partial alcohol reduction with physical processes is a promising new solution to tackle fermentation problems. As alcohol, one of the major factors causing stuck/sluggish fermentation can be removed. By removing alcohol from such a wine that could not ferment to dryness, a restart of the fermentation is clearly promoted.

Furthermore, this is a very interesting strategy to bring wines for sparkling wine production to an appropriate alcohol level of below 12 vol.%.

28.3.3 Alcohol reduction by physical methods

Based on physical methods, there are several technologies available to reduce the alcohol content in wine. They are either based on distillation under vacuum or on membrane processes.

28.3.3.1 Vacuum distillation

Distillation is a thermal separation procedure where liquids are vaporized and this vapor is then condensed again (Kolb, 2004). Distillation is a separation process for substances according to their relative volatility.

The technology of distilling under vacuum is not new. The first commercial plants were installed at the beginning of the 20th century.

Vacuum distillation achieves lower boiling points by creating a constant vacuum in the column. By lowering the pressure inside the unit, the volatility of the components is increased, and consequently, the boiling point of ethanol is reduced. Due to that the thermal energy needed to boil the ethanol is reduced. This minimizes the thermal load on the components of the liquid to be dealcoholized. The alcohol reduction of wine can take place at about 26°C−35°C (Diban et al., 2008).

During distillation the alcohol content of the liquid to be boiled decreases. Consequently, the alcohol content of the vapors continues to decrease as well (Kolb, 2004). To increase the alcohol content in the distillate fraction, the rising vapors are amplified in the distillation column. This is done by allowing the rising vapors to flow through the so-called trays of the column against an inflowing liquid. The vapor is enriched with volatile components such as ethanol, whereas the incoming liquid is enriched with high boiling components from the vapor. Depending on the application, the columns have different numbers of reinforcing trays. This process is called countercurrent distillation or rectification. It is cheaper and less expensive than multiple repetitions of the single-stage distillation (Ignatowitz, 2015). While using three so-called bubble trays, the alcohol content of the rising steam can theoretically be increased up to 83.5 vol.%. In practice, however, this value is never reached because the alcohol content in the starting liquid decreases during distillation. Furthermore, the heat and mass transfer on the respective bottoms is not always ideal (Kolb, 2004).

In general, the alcohol content in the distillate can be increased up to a content of 97.2 vol.%. Then a so-called azeotrope is formed. With an aqueous alcohol solution of 97.2 vol.%, the boiling temperature at normal pressure is 78.15°C and thus below the usual boiling point of ethanol. Therefore this is called a negative boiling point (Kolb, 2004). Since the rising vapors from this mixture have the same composition as the starting liquid, the amplification factor is 1.0, and thus no further concentration is possible.

In industrial plants for vacuum rectification, no further temperature reduction can be detected during evaporation if the pressure is reduced below 1 mbar. This is due to the pressure losses caused by the flow in the pipes between the distillation column and the condenser (Bethge, 1996). Spirit fractions of high alcohol content reduce the volume loss during dealcoholization.

In order to reduce the aroma losses caused by the alcohol reduction, the distillate is fed into the so-called aroma washout in countercurrent to the alcohol-free wine after the rectification. During this process, a part of the aromas from the distillate is transferred back into the alcohol-free wine.

Due to mild processing temperatures and the short treatment duration, there is no thermal product damage indicated by the formation of hydroxymethylfurfural (HMF).

Vacuum rectification is a continuous process that can be operated on systems with a capacity of 300 L/h or more. Common industrial plants are designed for flow rates of 1000–5000 L/h. Such rectification columns are installed on-site and are not mobile.

Usually, the alcohol content is reduced to values below 4 g/L (<0.5 vol.%). Afterward, dealcoholized subfraction is blended back to the original wine to achieve the targeted alcohol content. The alcohol-free fraction has a very high susceptibility to microbiological spoilage due to the lack of alcohol as a biocide and the nearly complete loss of all SO_2 in wine. Consequently, the alcohol-free fraction should be blended as soon as possible with the initial wine to acchieve more microbiological stability due to the presence of ethanol. Further product damage, especially for white wines can occur due to oxidation. The rectification process goes along with certain oxygen uptake in addtion to that the transport of the wine to the rectification column and back to the winery mean a certain oxygen uptake. Modern gasmanagement technologies and minimizing headspace during transport could help to reduce unwanted oxygen uptake.

28.3.3.2 Spinning cone column

A special form of vacuum rectification is the spinning cone column (SCC). The SCC is used in the food and beverage industry for aroma separation and aroma recovery (Christmann, 2001).

Unlike conventional columns for vacuum rectification, no static internals to increase the alcohol content are installed. Within the cylinder of the spinning cone, pairs of a fixed and a movable cone are installed. Due to this setup, the trickling wine forms a thin film. On the underside of the movable cones, there are ribs that swirl the rising vapors and thus enable an intensive exchange between the wine and the so-called stripping phase.

The special design of the SCC helps to overcome the disadvantage of conventional columns for vacuum rectification. The mass transfer in the co-line is reduced by the application of the vacuum because, instead of a turbulent flow, there is only a laminar flow of the boiling gases. This general disadvantage of distillation under vacuum is balanced by mounting rotating inserts in the column. The trickling liquid is transformed into a thin liquid film by its rotation. This liquid film is on average less than 1 -mm thick (Langrish, Makarytchev, Fletcher, & Prince, 2003). This results in a very efficient interaction between vapors and wine, which reduces the necessary residence time in the column. In addition, the design of the SCC allows working with viscous or high solids slurries, unlike columns for vacuum rectification (Christmann, 2001).

The SCC has been discussed controversially in some wine-related publications in the past. Unlike vacuum distillation, the wine passes the SCC twice. The first passage separates the wine from a fraction of very volatile aromas and 2–3 vol.% of the alcohol content. This fraction, rich in very volatile aromas, can be compared to a prerun fraction from a normal rectification process. It has between 50 and 60 vol.% of alcohol.

In a second step, the wine passes through the SCC column again, and the alcohol content is reduced severely. Afterward, the prerun fraction can be blended with the severely alcohol-reduced wine to buffer aroma losses. This practice allows to reduce the aroma loss a bit, but the recovered aromas are not necessarily just positive aromas. Some very volatile components, like higher alcohols, can appear glue-like and disturbing if they are in very high concentrations. Furthermore, very volatile aroma components like isoamyl acetate are not necessarily an indicator of great white wine quality. It mainly indicates that a wine was fermented slowly by cool fermentation temperatures.

28.3.3.3 Osmotic distillation

Osmotic distillation is a process in which two liquids are separated by a microporous, nonwetting, hydrophobic membrane. Both liquids are passed along this membrane, and none of the liquids penetrates the membrane pores. The volatile components, which are dissolved in the particular liquids, can pass the membrane by evaporating and permeating through the pores of the membrane. In the other liquid on the opposite side of the membrane, these substances then go into solution. Due to the hydrophobic nature of the membrane, water cannot permeate the pores of the membrane. Consequently, ions, colloids, and macromolecules, as they do not evaporate and diffuse through the membrane, are completely retained.

Osmotic distillation is an isothermal membrane separation process at atmospheric pressure. The driving factor for substance flow is the vapor pressure difference of a volatile component between the two sides of the membrane. The volatile components permeate from the side of the membrane with the higher vapor pressure to the side with the lower vapor pressure until equilibrium is reached (Curcio & Drioli, 2005). In osmotic distillation for the partial alcohol reduction of wine, water is used as a stripping medium. Besides possible losses of volatile aroma components, the ethanol flux is of essential interest. For short treatment times and low membrane surface need, the ethanol flux should be optimized.

The ethanol flux is influenced by a number of factors. Varavuth, Jiraratananon, and Atchariyawut (2009) showed that higher velocities of the feed and strip media increase the alcohol transfer through the membrane. Furthermore, the temperature has an influence. With rising temperatures, the flux of volatile components increases. With respect to the white wine quality the treatment temperature should not exceed a constant temperature of 25°C to avoid thermal quality damages.

In the English-language literature, there are several synonyms for osmotic distillation like membrane distillation, transmembrane distillation, capillary distillation, or pervaporation (Lawson & Lloyd, 1997). Other sources also speak of isothermal membrane distillation (Hogan, Canning, Peterson, Johnson, & Michaels, 1998).

For successful osmotic distillation, it is important that both sides of the membrane are sufficiently hydrophobic, the pores of the membrane are not soaked, and no water can penetrate the membrane by capillary forces (Lawson & Lloyd, 1997).

To maintain the initial hydrophobic nature of the membrane, careful cleaning with appropriate agents and storage under dry conditions is essential. The use of common membrane cleaning agents containing surfactants irreversibly destroys the hydrophobic property. The same applies to storage in a damp state.

The gas and vapor passage through the membrane pores is achieved by diffusion. The permeation of volatile molecules through the air space of the membrane pores can be described by Knudsen and Fick diffusion, depending on the pore radius (Lawson & Lloyd, 1997).

When versatile osmotic distillation plants are used for the partial alcohol reduction of wine, the treatment takes place relatively simply through batch processing. It is recommended to reduce a partial quantity of about 4 vol.% in alcohol content and to blend it with the starting wine to the targeted alcohol content. This prevents, among other things, treatment beyond the permissible range. Before use the membrane contactors should be tested for their sensory neutral status. Either off-flavors are carried over by adsorption and subsequently washed out of a contacted wine to another (Schaefer, 2014), or false tones could be produced in the membrane itself due to improper cleaning and storage, thus contaminating wines by the treatment.

Various literature references suggest that a simultaneous water transfer takes place between both sides of the membrane. The higher the process temperature, the higher the water transfer (Varavuth et al., 2009). If the so-called strip water is degassed before treatment to avoid undesired gas input into the wine, the water transfer is also increased (Hogan et al., 1998). If the wine temperature is higher than the water temperature, the water transfer also increases (Baker, 2002). In their work, Varavuth et al. (2009) quantify the water transfer to be up to $3 \, \text{L/m}^2 \, \text{h}^{-1}$. However, if the hydrophobic properties of the membrane are damaged by improper cleaning and storage, it can be assumed that the water transfer increases. If there is a vapor transfer through the membrane, it can be assumed that the water vapor is composed relatively more of light oxygen atoms. The issue is water addition to wine is controversially discussed worldwide. Most countries do not allow a targeted water addition. Only technically unavoidable amounts of water, such as for rehydration of yeasts, are permitted.

Therefore numerous sophisticated analytical methods, such as checking the isotopic relation of O16/18, have been developed to detect the addition of water.

Even if relatively small amounts of water are added to the wine, osmotic distillation for alcohol reduction could simulate significantly higher water contents in the wine.

The technique of osmotic distillation is widely used in various industries. It can be used for the aeration and degassing of liquids as well as for alcohol reduction of beer and wine (Lawson & Lloyd, 1997).

The use of hydrophobic membranes offers a further interesting application in winemaking: the targeted removal and addition of gas, often described as gas management. The so-called gas management with such membranes delivers the opportunity for targeted removal of oxygen prior to bottling (Claude, Marc, & Günter, 2011). So this gives the chance to avoid negative influences on the white wine quality such as oxidation of aroma and browning. Furthermore, this helps to lower the SO_2 demand and to create a higher homogeneity in bigger lots of wine that are bottled.

Furthermore, the CO_2 content can be adjusted inline just before the wine enters the bottling line (Claude et al., 2011). This application offers a very precise and accurate way to increase the CO_2 content of white wine to make it taste fresher and livelier. Depending on the variety and consumer demand, 1.2–1.5 g/L are often added in this way.

28.3.3.4 Coupling of nanofiltration or reverse osmosis with further treatments

Reverse osmosis is a process for the concentration of liquids with low solid content. The transport through the membrane takes place by diffusion through a semipermeable membrane (Petersen, 1993). In reverse osmosis, a pressure must be applied that exceeds the osmotic pressure of the solution to be concentrated. The separation of different components is based on retention in terms of molecule size and the solution diffusion mechanism. Reverse osmosis was originally developed for water treatment and desalination, but a number of other applications are possible in the beverage and food industry.

For the concentration of must, reverse osmosis is applied without any further process step (Zürn, 1979). Other applications in winemaking based on reverse osmosis, however, require a second process. For this subsequent step, various other processes are used depending on the application target. If reverse osmosis is used for alcohol reduction of wine, a permeate solution is separated in the first step. This aqueous solution contains, besides alcohol, only some volatile aroma components. During the second step, alcohol is separated from this fraction.

This can be achieved either by a further membrane process like osmotic distillation or by distillation. The so-called diafiltration, i.e., replacing the permeate solution of reverse osmosis with water, would also be possible. However, wine regulations in most wine-producing countries worldwide do not allow this step, as targeted water addition is forbidden.

Bui, Dick, Moulin, and Galzy (1986) describe a different approach to alcohol reduction. In experiments, they combine two treatments with different reverse osmosis membranes. In the first step, a permeate solution with an alcohol content of about 6 vol.% is separated from the wine. In the second step, this permeate solution is reduced in alcohol content with a membrane less permeable to alcohol. In this step, a permeate solution is obtained that has an alcohol content of only 2 vol.% This fraction is blended with the retentate of the first treatment step and an alcohol-reduced wine is obtained (Bui et al., 1986).

Until today, this approach has not been established in practice, nor is there an equipment manufacturer who follows this approach. Most probably, the increased production and operation costs compared to other approaches are the reason for this. The coupling of two reverse osmosis treatments is to be estimated with relatively high energy costs. In addition, the performance of alcohol separation per m^2 membrane area is quite low, whereby the plant costs are increased due to the increased required membrane area.

Nanofiltration is a membrane process similar to reverse osmosis. The definition by Porter describes the separation limit of the membranes to be between 100 and 500 Dalton (Porter, 1989). The pore size is around 0.001 μm, and the usual working pressure is between 10 and 30 bar, and for some applications, even up to 40 bar (Porter, 1989).

In comparison to reverse osmosis, nanofiltration is applied with lower pressure and produces more permeate per unit of the membrane surface. This is achieved due to the membrane structure and the pore size of the membranes. However, this means that other wine components permeate to a higher extent through the nanofiltration membrane. In the second step of the treatment, the alcohol reduction of the permeate solution is consequently going along with higher losses or damages of desired aroma substances.

There are further applications for reverse osmosis in winemaking, such as the reduction of volatile acidity. During this process, the reverse osmosis permeate solution is treated with ion exchange resin. Consequently, acetic acid can be removed in a targeted way, whereas the other organic acids in wine are not reduced. Furthermore, off-flavors arising from rotten grape material are not removed by this treatment (Christmann, 2001).

Other off-flavors such as smoke taint, resulting from grape exposure to smoke from forest fires, can be removed by coupling reverse osmosis and treatment with adsorbent resins in the second step of the process. This significantly reduces volatile phenols such as guaiacol and 4-methylguaiacol (Fudge, Ristic, Wollan, & Wilkinson, 2011).

Membrane processes, in contrast to distillation procedures, show the advantage of being mobile and having quite low requirements regarding the winery's infrastructure. Just electricity and partly softened water are necessary. In comparison to that, distillation plants require high capacities in terms of steam and cooling. Furthermore, they are not mobile. Membrane plants are suitable to reduce the alcohol content by not more than 4−5 vol.; if the wine should be reduced to a higher degree, distillation plants are more suitable.

References

Alston, J. M., Fuller, K. B., Lapsley, J. T., Soleas, G., & Tumber, K. P. (2015). Splendide mendax: False label claims about high and rising alcohol content of wine. *Journal of Wine Economics, 10*(3), 275−313. Available from https://doi.org/10.1017/jwe.2015.33.

Amerine, M. A., & Roessler, E. B. (1976). *Wines, their sensory evaluation.* WH Freeman.

Arnold, R. A., & Noble, A. C. (1978). Bitterness and astringency of grape seed phenolics in a model wine solution. *American Journal of Enology and Viticulture, 29*(3), 150−152.

Baker, R. W. (2002). Membrane technology. Encyclopedia of polymer science and technology, (Vol 3).

Bethge, D. (1996). Destillation im Fein-und Hochvakuum. *Vakuum in Forschung und Praxis, 8*(2), 84−87.

Bisson, L. F. (1999). Stuck and sluggish fermentations. *American Journal of Enology and Viticulture, 50*(1), 107.

Bui, K., Dick, R., Moulin, G., & Galzy, P. (1986). A reverse osmosis for the production of low ethanol content wine. *American Journal of Enology and Viticulture, 37*(4), 297−300.

Christmann, M. (1988). Veränderungen von Weininhaltsstoffen unter besonderer Berücksichtigung der Weinaromen bei der Entalkoholisierung von Weinen mit kombinierter Dialyse-Vakuumdestillation. Veröffentlichungen der Forschungsanstalt Geisenheim.

Christmann, M. (2001). Neue oenologische Verfahren: Detaillierte Darstellung der wichtigsten. Techniken: Meininger.

Claude, V. J., Marc, V. V., & Günter, W. (2011). Exact management of dissolved gases of wines by membrane contactor. *Bulletin de l'OIV, 84*(962−964), 179−187.

Conner, J. M., Birkmyre, L., Paterson, A., & Piggott, J. R. (1998). Headspace concentrations of ethyl esters at different alcoholic strengths. *Journal of the Science of Food and Agriculture, 77*(1), 121−126.

Contreras, A., Hidalgo, C., Henschke, P. A., Chambers, P. J., Curtin, C., & Varela, C. (2014). Evaluation of non-Saccharomyces yeasts for the reduction of alcohol content in wine. *Applied and Environmental Microbiology, 80*(5), 1670−1678. Available from https://doi.org/10.1128/AEM.03780-13.

Curcio, E., & Drioli, E. (2005). Membrane distillation and related operations—A review. *Separation and Purification Reviews, 34*(1), 35−86.

Diban, N., Athes, V., Bes, M., & Souchon, I. (2008). Ethanol and aroma compounds transfer study for partial dealcoholization of wine using membrane contactor. *Journal of membrane Science, 311*(1−2), 136−146.

Duchêne, E., Huard, F., Dumas, V., Schneider, C., & Merdinoglu, D. (2010). The challenge of adapting grapevine varieties to climate change. *Climate Research, 41*(3), 193−204. Available from https://doi.org/10.3354/cr00850.

Escudero, A., Campo, E., Fariña, L., Cacho, J., & Ferreira, V. (2007). Analytical characterization of the aroma of five premium red wines. Insights into the role of odor families and the concept of fruitiness of wines. *Journal of Agricultural and Food Chemistry, 55*(11), 4501−4510.

Fedrizzi, B., Nicolis, E., Camin, F., Bocca, E., Carbognin, C., Scholz, M., et al. (2014). Stable isotope ratios and aroma profile changes induced due to innovative wine dealcoholisation approaches. *Food and Bioprocess Technology, 7*(1), 62−70.

Fischer, U. (2010). Sensorische bedeutung des alkohols im wein. *Eine unterschätzte Einflussgröße. Das Deutsche Weinmagazin, 19*, 108−110.

Fischer, U., & Noble, A. C. (1994). The effect of ethanol, catechin concentration, and pH on sourness and bitterness of wine. *American Journal of Enology and Viticulture, 45*(1), 6−10.

Fudge, A. L., Ristic, R., Wollan, D., & Wilkinson, K. L. (2011). Amelioration of smoke taint in wine by reverse osmosis and solid phase adsorption. *Australian Journal of Grape and Wine Research, 17*(2), S41−S48.

Gawel, R., van Sluyter, S., & Waters, E. J. (2007). The effects of ethanol and glycerol on the body and other sensory characteristics of Riesling wines. *Australian Journal of Grape and Wine Research, 13*(1), 38−45.

Goldner, M. C., Zamora, M. C., Di Leo Lira, P. A. O. L. A., Gianninoto, H., & Bandoni, A. (2009). Effect of ethanol level in the perception of aroma attributes and the detection of volatile compounds in red wine. *Journal of Sensory Studies, 24*(2), 243−257.

Grainger, K. (2009). *Wine quality: tasting and selection.* John Wiley & Sons.

Green, B. G. (1987). The sensitivity of the tongue to ethanol. *Annals of the New York Academy of Sciences, 510*(1), 315−317.

Hogan, P. A., Canning, R. P., Peterson, P. A., Johnson, R. A., & Michaels, A. S. (1998). A new option: osmotic distillation. *Chemical Engineering Progress, 94*(7), 49−61.

Ignatowitz, E. (2015). Chemietechnik. 12. In: *Auflage, Europa-Lehrmittel.*

King, E. S., Dunn., Randall, L., & Heymann, H. (2013). The influence of alcohol on the sensory perception of red wines. *Food Quality and Preference, 28*(1), 235−243.

King, E. S., & Heymann, H. (2014). The effect of reduced alcohol on the sensory profiles and consumer preferences of white wine. *Journal of Sensory Studies, 29*(1), 33−42.

Kolb, E. (2004). Spirituosentechnologie: Behr.

Langrish, T. A. G., Makarytchev, S. V., Fletcher, D. F., & Prince, R. G. H. (2003). Progress in understanding the physical processes inside spinning cone columns. *Chemical Engineering Research and Design, 81*(1), 122−130.

Lawson, K. W., & Lloyd, D. R. (1997). Membrane distillation. *Journal of Membrane Science, 124*(1), 1−25.

Longo, R., Blackman., John, W., Torley, P. J., Rogiers, S. Y., & Schmidtke, L. M. (2017). Changes in volatile composition and sensory attributes of wines during alcohol content reduction. *Journal of the Science of Food and Agriculture, 97*(1), 8−16.

Martin, S., & Pangborn, R. M. (1970). Taste interaction of ethyl alcohol with sweet, salty, sour and bitter compounds. *Journal of the Science of Food and Agriculture, 21*(12), 653−655.

Mattes, D., & DiMeglio, D. (2001). Ethanol ingestion and perception. *Physiology and Behavior, 72*, 217−229.

Meillon, S., Viala, D., Medel, M., Urbano, C., Guillot, G., & Schlich, P. (2010). Impact of partial alcohol reduction in Syrah wine on perceived complexity and temporality of sensations and link with preference. *Food Quality and Preference, 21*(7), 732−740. Available from https://doi.org/10.1016/j.foodqual.2010.06.005.

Mira de Orduña, R. (2010). Climate change associated effects on grape and wine quality and production. *Food Research International, 43*(7), 1844−1855. Available from https://doi.org/10.1016/j.foodres.2010.05.001.

Petersen, R. J. (1993). Composite reverse osmosis and nanofiltration membranes. *Journal of Membrane Science, 83*(1), 81−150. Available from https://doi.org/10.1016/0376-7388(93)80014-O.

Pickering, G. J., Heatherbell, D. A., Vanhanen, L. P., & Barnes, M. F. (1998). The effect of ethanol concentration on the temporal perception of viscosity and density in white wine. *American Journal of Enology and Viticulture, 49*(3), 306−318.

Porter, M.C. (1989). *Handbook of industrial membrane technology.* United States: N.P.

Ribéreau-Gayon, P. (2007). *Handbook of enology* (2nd ed.). Chichester: J. Wiley & Sons.

Röcker, J., Schmitt, M., Pasch, L., Ebert, K., & Grossmann, M. (2016). The use of glucose oxidase and catalase for the enzymatic reduction of the potential ethanol content in wine. *Food chemistry, 210*, 660−670. Available from https://doi.org/10.1016/j.foodchem.2016.04.093.

Röcker, J., Strub, S., Ebert, K., & Grossmann, M. (2016). Usage of different aerobic non-Saccharomyces yeasts and experimental conditions as a tool for reducing the potential ethanol content in wines. *European Food Research and Technology, 242*(12), 2051−2070. Available from https://doi.org/10.1007/s00217-016-2703-3.

Schaefer, V. (2014). Untersuchungen zum Auftreten dumpf-muffiger Fehltöne im Wein: Untersuchungen zum Auftreten, der Herkunft, Behandlung und Vermeidung sensorisch wirksamer, dumpf-muffiger Fehltöne im Wein, die durch Trauben, Weinbearbeitung, Schönung. Weinbehandlung und Abfüllen verursacht werden können: Ges. zur Förderung der Hochsch. Geisenheim.

Schmitt, M. (2016). *Teilweise Alkoholreduzierung von Wein mittels physikalischer Verfahren - Alkoholmanagement*. Geisenheim: Gesellschaft zur Förderung der Hochschule Geisenheim (Geisenheimer Berichte, Band 80).

Schneider, V. (2014). Atypical aging defect: Sensory discrimination, viticultural causes, and enological consequences. A review. *American Journal of Enology and Viticulture, 65*(3), 277–284. Available from https://doi.org/10.5344/ajev.2014.14014.

Scinska, A., Koros, E., Habrat, B., Kukwa, A., Kostowski, W., & Bienkowski, P. (2000). Bitter and sweet components of ethanol taste in humans. *Drug and Alcohol Dependence, 60*(2), 199–206. Available from https://doi.org/10.1016/S0376-8716(99)00149-0.

Szolnoki, G., Hoffmann, D., & Herrmann, R. (2008). Quantifizierung des Einflusses der äußeren Produktgestaltung auf die Geschmacksbewertung und auf die Kaufbereitschaft bei Wein mittels eines. Charakteristika-Modells: Inst. für Agrarpolitik und Marktforschung.

Taylor, A. J., Tsachaki, M., Lopez, R., Morris, C., Ferreira, V., & Wolf, B. (2010). Odorant release from alcoholic beverages. In: Neil C. Da Costa und Robert J. Cannon (Eds.), *Flavors in Noncarbonated Beverages* (pp. S. 161–175). Washington, DC: American Chemical Society (ACS Symposium Series).

Urbano, C. (2007). R-index and triangular tests to determine the perception threshold of a reduction of alcohol content in wine, 7th Urbano: C.

Varavuth, S., Jiraratananon, R., & Atchariyawut, S. (2009). Experimental study on dealcoholization of wine by osmotic distillation process. *Separation and Purification Technology, 66*(2), 313–321.

Vasconcelos, M. C., & Castagnoli, S. (2000). Leaf Canopy Structure and Vine Performance. *American Journal of Enology and Viticulture, 51*(4), 390.

Williams, A. A. (1972). Flavour effects of ethanol in alcoholic beverages.

Wollan, D. (2005). Controlling excess of alcohol in wine. *Australian and New Zealand Wine Industry Journal, 20*, 48–50.

Yu, P., & Pickering, G. J. (2008). Ethanol difference thresholds in wine and the influence of mode of evaluation and wine style. *American Journal of Enology and Viticulture, 59*(2), 146–152, Online verfügbar unter. Available from https://www.ajevonline.org/content/59/2/146.

Zürn, F. (1979). Anwendungsmöglichkeiten der Umkehrosmose (Hyperfiltration) zur Qualitätsverbesserung von Most und Wein. Aus dem Institut für Weinchemie und Getränkeforschung der Forschungsanstalt.

White wine tasting: Understanding taster responses based on flavor neuronal processing

Manuel Malfeito-Ferreira

Linking Landscape Environment Agriculture and Food Research Center (LEAF), Instituto Superior de Agronomia, Universidade de Lisboa, Lisbon, Portugal

29.1 Introduction

The tasting of foods and beverages has been an activity intimately linked to human evolution since the dawn of Humanity (McGovern, 2019). The ability to discern the different sensory properties was essential to the selection of healthy and nutritive food sources. The description of sensory characteristics has been known since the first written texts, but the development of systematic tasting methodologies only appeared by the middle of the 20th century, leading to sensory analysis being a discipline essential to the food industry (Doty, 2015). In wines, the process evolved in parallel, and the initial sensory concerns were mostly related to the prevention of off-flavors and directed to winemaking professionals (Amerine & Roessler, 1976; Ribéreau-Gayon & Peynaud, 1958).

Tasting and sensory analysis are logically intertwined, but while the former may be an occasional action, the latter requires adequate and extensive training to calibrate individual responses (Box 29.1). The differences between both definitions are not well defined and may be materially absent when tasting has the same methodological requirements as sensory analysis.

The present basic tasting methods involve the sequential evaluation of visual, smell, taste, mouthfeel, and overall assessment, with additional gradations in each of these features when necessary (Jackson, 2017; Léglise, 1984; Peynaud, 1987). The tasting protocols are widely available and deeply rooted in the mind of scholars, winemakers, sommeliers, critics, and consumers reflecting the present worldwide interest in wine consumption (Broadbent, 1979; Robinson, 2008; Schuster, 2017). In parallel, sensory science has impressively evolved in methodologies and scope, from in-house quality control to consumer research (Ares & Varela, 2018; Heymann & Ebeler, 2016). However, recent advances in neuroscience question the applicability and significance of commonly accepted methods, opening the way for alternative approaches.

Simultaneously, the definition of wine quality is not straightforward. It may be similar to the overall definition used for food commodities when wines are a product subjected to the international market rules and must "fit to purpose," i.e., must follow the sensory properties that guarantee commercial success (Norris & Lee, 2002). For these wines, the development in analytical chemistry and sensory analysis together with powerful statistical tools enabled us to define and understand the impact of winemaking options on sensory properties and consumer acceptance. The research in this field uses approaches similar to those for other food products (Civille & Oftedal, 2012; Hopfer, Nelson, Ebeler, & Heymann, 2015), but there are wines that are not tractable in the same way. The so-called "fine wines" or "vins de garde" have characteristics that are only partially captured by analytical measurements. Their evaluation depends on properties (e.g., harmony, complexity) requiring other tasting approaches. For these wines, the concept of quality goes beyond meeting the global consumer demand, being strongly reliant on broad cultural elements, and is more directed to specific niche markets (Box 29.2).

© 2022 Elsevier Inc. All rights reserved.

Charters (2003) discussed the absence of a clear definition of fine wine quality and presented several approaches by academic and professional personalities, which reflects the vagueness of the concept.

The question of quality perception is also relevant for understanding the wide variety of responses by consumers. The level of involvement, practice, and knowledge or consumption frequency is taken into account to ascertain the degree of expertise (Sáenz-Navajas et al., 2016). The boundaries between the different segments are frequently not clear, ranging from those without expertise (e.g., novice, naïve, unexperienced) to those with long years of practice (e.g., experts, experienced, winemakers, *connoisseurs*). Using the reported preferences for wines, Charters & Pettigrew (2007) distinguished between consumers with high and low involvement that broadly match the previous definition of wine quality (Box 29.3). The high involvement relied on sensory appreciation driven by holistic features (e.g., structure, concentration, complexity) that are considered paradigmatic dimensions because they reflect other factors like the origin, grape variety, or aging potential. The low involvement is explained by the lesser importance to the cognitive tasks while a medium segment may be defined as having intermediate behavior.

The understanding of the previously described concepts enables us to define broad wine styles that will be used as examples throughout the chapter (Box 29.4). The international commercial wine style corresponds to those wines characterized by a "fit to purpose" quality, evaluated in-house by sensory analytical methods that are highly appreciated by low-involved consumers. In the opposed profile, fine wines have distinct holistic

---

BOX 29.1

Sensory approaches

Sensory analysis: A scientific discipline used to evoke, measure, analyze, and interpret those responses to products that are perceived by the senses of sight, smell, touch, taste, and hearing [Institute of Food Technologists (IFT)] quoted by Heymann & Ebeler, 2016.

Wine Tasting: The act of tasting a wine with care in order to appreciate its quality; to submit it to examination by our senses, in particular, those of taste and smell; to try and understand it by discovering its various qualities and faults and putting then into words. It is to study, analyze, describe, define, judge, and classify (Jéan Ribereau-Gayon quoted by Peynaud, 1987).

BOX 29.2

Quality definitions

Wine Sensory Quality: The acceptance of the sensory characteristics by consumers who are regular users of the product category or who comprise the target market of the product. In wine, it corresponds to the development of wine styles for a targeted population (Norris & Lee, 2002).

Fine Wine Quality: "…it is the totality of its properties, that is to say the properties which render it acceptable or desirable. Quality only exists in relation to the individual and then only in as far as he has the ability to perceive it and approve it…" (Peynaud, 1987). "… Quality can be assessed by the length of time the flavour lingers in the mouth, by its richness and subtlety and by its aftertaste…" (Broadbent, 1979).

BOX 29.3

Consumer segmentation

Low-involvement consumers: Subjects that associate quality with pleasantness driven by sensory aspects like smoothness, flavor, and drinkability.

High-involvement consumers: Subjects to whom the quality perception is a cognitively driven process influenced by holistic properties of the structure, concentration, complexity, and aging potential together with cultural dimensions of origin and grape variety.

BOX 29.4

Wine styles

International commercial style: Wines characterized by a "fit to purpose" quality, evaluated by sensory analytical methods in order to be appreciated by most consumers in a globalized market.

Fine wines: Wines characterized by tasting methods assessing holistic properties in order to be appreciated by individuals understanding their cultural meaning.

---

properties appreciated by high-involved experts and *connoisseurs* that benefit from being aware of other facets related to history and culture. These wines have a strong link to the so-called classical European wines, but the definition is conceptual and not geographical.

The aim of the present chapter is to present a critical appraisal of current wine tasting methodologies and to understand what may be improved, keeping in mind how the brain processes sensory information. The purpose is not to describe the vast array of available analytical techniques but to understand and interpret the outputs. Illustrative examples of two quality concepts, sensory approaches, consumer segments, and wine styles will be used for the sake of simplicity, given the difficult distinction between both when it comes to synthesizing the wealth of present knowledge. Hopefully, the gap between sensory scientists, winemakers, other wine professionals, and consumers will be reduced, all benefiting from mutual understanding.

29.2 Neuronal processing of sensory stimuli

Tasting depends on the individual ability to detect and perceive the sensory stimuli elicited by the physical—chemical composition of foods and beverages. Sensory science ultimately aims both to replace our senses with instruments and to make us behave like instruments (Heymann & Ebeler, 2016); the concepts are intimately linked to those of analytical chemistry. The improvements in analytical chemistry continuously provide deeper insights into flavor-active molecules, and recent research has shown that it is possible, to a certain extent, to predict sensory characteristics based on electronic-mouth and electronic-nose methodologies (Haddad, Medhanie, Roth, Harel, & Sobel, 2010). However, the genetic basis for the remarkable human ability to smell has not changed since the dawn of mankind (McGann, 2017). Therefore the gap between flavor analytical chemistry and human perception has not ceased to increase, despite the improvements acquired by adequate and extensive sensory training.

The outstanding complexity of the human neuronal system has just begun to be unveiled with direct implications in the comprehension of the external sensory world. Hence, the sensory analytical methods not only should be adapted to the object of study (e.g., wine), as in any other scientific field, but should also be adapted to the "instrument" of tasting (e.g., the brain). In fact, the flavor of the wine only gets its meaning when it is interpreted by the brain. Before this step, the numerous molecules possibly detected by the most powerful chromatograph, although being responsible for the flavor, have no flavor (Chen, 2020; Small, 2012).

The description of the physiological mechanisms of sensory perception is beyond the scope of this chapter and may be found in excellent textbooks edited by neuroscientists (Murray & Wallace, 2011; Shepherd, 2017; Small & Green, 2012) and sensory researchers (Heymann & Ebeler, 2016). The focus here is to present those phenomena underlying the function of the five senses, namely vision, gustation, olfaction, touch, and hearing with direct implications for wine tasting.

29.2.1 The peculiarity of olfaction

Among all senses, olfaction plays a central role in tasting, even if foods without smell are popularly described as "tasteless." The peculiarity of the olfactory perception among the other four senses lies in the straightforward transmission of stimuli from the receptor cells in the nose to the brain section (olfactory cortex) where conscious processing takes place without passing through the so-called "thalamic relay." Briefly, odorant molecules aspired by both nostrils reach the receptor neurons in the olfactory epithelium and the electric signals are transmitted to the olfactory bulb, where a "smell image" is formed. This image is sent to the piriform cortex, in the primary olfactory cortex, where a cortical microcircuit reformats it as a content-addressable memory. Then the signals are directed to the orbitofrontal cortex (OFC), where a diverse pattern of multiple bidirectional connections has links to brain networks that regulate cognition, emotion, memory, and behavior. Here is where representations of individual components are assembled into holistic objects ("odor objects"). Therefore the signals of volatile molecules go directly to the most evolved part of the brain where higher cognitive functions that make us human exist (Shepherd, 2017).

The inputs from the eye, tongue (taste and mouthfeel), or ear cross the thalamus, at the core of the brain, before being processed in different cortical areas in the middle and back of the brain and then sent to the OFC (Fig. 29.1).

To understand the implications of the smell perception on wine tasting, it is helpful to compare the physiological mechanisms with those of vision and of emotion.

29.2.1.1 The comparison with visual images

The biological basis of odor coding is fundamentally different from the coding principles of the visual system (Barwich, 2019), but from an applied perspective the particularities of "odor objects" may be effectively explained by comparing with visual objects or images. According to Shepherd (2017), the most relevant characteristics are as follows:

i. The odor object is synthetic

The visual observation of a painting or a face is not performed by individuating the different colors or features but by first perceiving the whole. Similarly, the odor object is sensed as a whole in the olfactory cortex. Therefore the sense of smell is essentially synthetic and not analytic. Moreover, to individualize the colors of a painting it is not necessary to appreciate the whole and may distort the reality, mainly when colors are diffused as in impressionist masterpieces. In wines the effect of trying to individuate the odors is similar, especially in complex wines not dominated by simple odors.

ii. Guessing the rest and the effect of the context

Another similarity with visual images is related to "guessing the rest." A familiar face is recognized by seeing only a small part or by identifying general features when seen at distance. The efficacy of the identification depends on the context. Familiar faces may not be recognized when the person is not supposed to be met. In smelling, the recognition of initial odors may be used to anticipate the rest of the experience because odor object processing allows filling the rest of the pattern when degraded inputs arrive. Thus the odor object is seen as complete, which promotes perceptual stability when the context is familiar. Conversely, the memory based on the conscious accumulation of experiences enables the brain to imagine smells without indeed having perceived them, in a process known as mental imagery (Croijmans, Speed, Arshamian, & Majid, 2020).

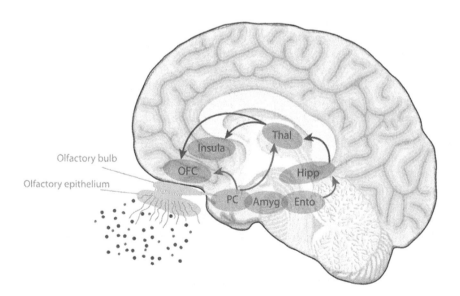

FIGURE 29.1 Schematic view of the human olfactory system. The primary and secondary olfactory cortices are represented in blue and green, respectively. Amyg, amygdala; Ento, entorhinal cortex; Hipp, hippocampus; OFC, orbitofrontal cortex; PC, piriform cortex; Thal, thalamus. *Source: From Saive et al., 2014.*

iii. The poor ability to name odors in simple aromatic mixtures

The olfactory world is constituted of thousands of volatile molecules, but there is evidence that the number of recognizable odors that contribute anything to the aroma of all foods is approximately 250 (Rochelle, Prévost, & Acree, 2018). Studies with large mixtures of odorants (formulated to be of equal potency) showed that a subject's ability to detect individual odorants in these mixtures was vanishingly small (Rochelle et al., 2018). Even in a simple mixture of three aromatic molecules, only about 5% of the trained individuals recognized all, with lower percentages for mixtures of four or five (Laing & Francis, 1989). These authors posited that odors in mixtures blend to form a new odor with few of the characteristics of the original ones, being corroborated by further research (Laing, Link, Jinks, & Hutchinson, 2002; Marshall, Laing, Jinks, & Hutchinson, 2006). Recently, Frank, Fletcher, & Hettinger (2017) added new data to support these observations, but the implications of such behavior barely seem to have refrained from the utilization of a large number of aroma descriptors in sensory analysis.

iv. The "white" smell and the aroma base

The perception of odor objects has other implications as well that illustrate smell as a synthetic sense. Weiss et al. (2012) found that different mixtures of about 30, or more, equal-intensity components that span the stimulus space smelled alike despite not having a single component in common. This singularity originated the concept of a "white" smell, similar to the visual whites (absence of color). The effect did not happen in mixtures containing fewer components or in mixtures that did not span the stimulus space.

In wines, this "white" smell sensation resembles the wine "aroma base" as hypothesized by Ferreira (2010). According to this concept, the aroma of wine is characterized by a less number of molecules (about 30) that act as a buffer, because differences in their concentrations are not reflected in different aromatic perceptions. These molecules are the same in white or red wines and only the so-called impact molecules, or families of molecules, may break the aromatic "buffer." Examples of impact volatiles are terpenes in varieties like Muscat or off-flavors. In the absence of impact molecules, wines may be described as odorless and different varietals are indistinguishable (Campo, Do, Ferreira, & Valentin, 2008). These wines just smell like wines, which tasters associate with a light sweetish and pungent, alcoholic, and somewhat fruity aroma (Ferreira, 2010). Then the nature of wine aroma limits the broad utilization of "odor activity values" (OAV) to predict the flavor of wines (Pozo-Bayón, Muñoz-González, & Esteban-Fernández, 2016), especially those regarded as fine wines (McKay & Buica, 2020).

v. Olfactory illusions

Similar to vision, smell is also subjected to illusions driven by extrinsic cues. The observation of red color in white wines added of odorless colour additives induces the use of descriptors memorized as being from reds, even in trained experts (Morrot, Brochet, & Dubourdieu, 2001). Smith & McSweeney (2019) showed distinct flavor descriptions in white wines under red light compared to white light, including effects on mouthfeel properties. Therefore flavor descriptions should be made with wines in dark glasses or using red ambient light (Heymann & Ebeler, 2016).

Verbal labels may also "fool" the smell perceptions. Herz & Von Clef (2001) showed that when subjects smelled identical pairs of five mixtures but with different verbal labels (e.g., "parmesan" and "vomit" for butyric acid) the response was that the mixtures smelled differently. These authors concluded that different semantic associations result in diverging perceptions and qualitative evaluations of the same stimulus simply by experiencing an odor with a different conceptual tag. Similarly, Romano et al. (2020) found that experienced tasters used more frequently descriptors related to organic wines (e.g., animal/undergrowth) when informed about the mode of production.

These illusions have overwhelming implications in tasting, mainly when using descriptive analysis (DA) or profiling methods, regarded as the gold standard for sensory scientists (Heymann & Ebeler, 2016). Furthermore, the use of extended lists of smells in questionnaires has a psychological effect of adding flavors (Lelièvre, Chollet, Abdi, & Valentin, 2008). In our experience, the citation of oak notes in blank wines is frequent when the effect of oak is being tested.

29.2.1.2 Olfaction as an emotional sense

The unique properties of the sense of smell led Yeshurun & Sobel (2010) to define it as an emotional sense, where the emotion is not a result of the sensory stimuli but an integral part of the perception (Table 29.1). In fact, the first spontaneous reaction to odorants is a verbal description related to pleasantness (e.g., "it is good" or "I like it"). For these authors, the odor object is the integration of its pleasantness, induced by its external state (volatile molecules) with the subjective internal state at the moment when it is sensed. The human piriform cortex

TABLE 29.1 Characteristics of the sense of olfaction similar to emotional reactions according to Yeshurun and Sobel (2010) with direct implications for tasting methodologies.

Characteristics	Implication	Illustrative references
Superior detection and discrimination of odor objects	Utilization of methods comparing samples with well-defined standards (pivots or poles)	Thuillier et al. (2015), Wilson et al. (2018), Pearson et al. (2020)
Poor verbalization of odors	Restrict descriptive analysis to highly trained individuals and a little number of descriptors	Campo et al. (2008), Brand et al. (2020)
Pleasantness is the main dimension of perception	Assessment of sensory and hedonic scores in parallel with emotional responses	Coste et al. (2018), Souza-Coutinho et al. (2020)
Memory is based on a given pleasantness	Past sensory experiences influence flavor recognition (e.g., background culture and training)	Croijmans and Majid (2016), Tempère et al. (2019b)
Influence of internal factors (satiety, mood, ambient temperature) on pleasantness	Assess tasting in different contexts (e.g., immersive ambient, during meals, domestic consumption)	Orth and Bourrain (2005), Pappalardo et al. (2019), Tempèere et al. (2019a), Mackenzie, Hannum, Forzley, Popper, and Simons (2020), Mahieu, Visalli, Thomas, & Schlich (2020)

processes emotional information even before odor perception (Schulze, Bestgen, Lech, Kuchinke, & Suchan, 2017). The implications of these features should not be neglected when tasting wine and may explain, at least in part, the recent development methodologies that are appropriate to capture the wholeness of wine tasting (Table 29.1).

29.2.2 Descriptive analysis of odors

The first two characteristics in Table 29.1 concern the ability to identify and name odors. The capacity of human detection and discrimination among different smells is remarkable, exceeding that of vision (Keller, 2017). The difficulty is identifying and naming the odors, as described previously. Then to consider odor mixtures as a single vector is consistent with a synthetic rather than analytical brain processing mechanism in olfaction (Snitz et al., 2013), particularly evidenced in recent wine flavor research (Ferreira et al., 2016). The explanation for this fact lies in the incompatibility between human conceptual semantic resources and the olfactory system, even in cultures where the identification ability is higher (Young, 2020). Hughson & Boakes (2002) found a better matching between flavors and wines with a short list (14) of wine descriptors than with a long or no list. Brand et al. (2020), in DA by frequency of citation, restricted the possible choices to four or five descriptors because large numbers added biases to the statistical analysis. However, this number is multiplied by the number of assessors, and the most frequent descriptors may be more than 20, giving the false idea that they can be sensed by one person. The same restrictions should be applied to classical DA measured by intensity scales; otherwise minor flavor active compounds may distort the results and give an incorrect idea of complexity. Thus the inclusion of triangle or paired tests, only by sample comparison, would be a rigorous strategy to validate the differences in the flavor profiles determined by DA (González-Centeno, Chira, & Teissedre, 2019).

The requirement for extensive training to standardize perceptions and vocabulary is a serious limitation to the participation of experienced wine professionals that are reluctant to commit to long training sessions. To overcome this drawback, methodologies based on comparisons to standards (e.g., pivot profile, projective mapping) have witnessed an increasing interest (Barton, Hayward, Richardson, & McSweeney, 2020; Dehlholm, Brockhoff, Meinert, Aaslyng, & Bredie, 2012; Hayward, Jantzi, Smith, & McSweeney, 2020; Jaeger, Roigard, Le Blond, Hedderley, & Giacalone, 2019; Jantzi, Hayward, Barton, Richardson, & McSweeney, 2020), which is supported by the highly efficient neuronal ability to discriminate different "odor objects."

Moreover, the difficulty in verbalizing the perception is comparable to the difficulty in describing faces or emotions, although explained by different neuronal processing (Young, 2020). The effort to name a flavor makes its recognition more difficult (Melcher & Schooler, 1996). Sensory training may diminish this difficulty but it may be an illusion when fluent vocabulary is used together with memory to describe expected flavors. In fact, flavor descriptors are metaphors, in the sense that a wine smells "like a rose" and not smells "to roses." Then fluency in wine description may just be a matter of literary practice and not of sensory acuity. Furthermore, different flavor familiarities result in different descriptors simply because no one describes objects that never came across (Williamson, Mueller-Loose, Lockshin, & Francis, 2017). For instance, insisting on training wine experts in the recognition of "stinking skunk" is senseless in countries where skunks do not exist.

Outside the laboratory rooms, the use of aroma descriptions is "out of control." In our opinion, to show long lists of descriptors (e.g., aroma-wheels, *Le Nez du vin*©), accompanied by nicely designed figures (e.g., spider graphs, fruit or plants inside glasses), gives the idea that it is possible for the individual to name them all. Moreover, it seems that the gold standard of consumer appreciation should depend on DA, when it should not. Another behavioral implication is that it makes tasting look like an exercise of guessing the odors, further extended to guessing the grapes varieties, regions, or vintages. These tasks say more about the taster than about the wine, with failure being rather frequent, especially when wines are out of the expected context. A possible technical implication would be the addition of artificial flavors to increase appreciation (Saltman et al., 2017), which is not allowed in wines by the International Organisation of Vine and Wine (OIV).

29.2.3 Descriptive analysis of emotions

The use of DA may only provide a product description without liking implications, but the inclusion of methodologies such as preference mapping is aimed at that purpose. However, despite all improvements, it is now realized that the outputs only give an idea of the complexity of food choice by the consumer (Jaeger et al., 2019). Therefore the emotional nature of olfaction has been explored by recent sensory research. Similar to DA, long lists of emotions and feelings have been produced and food-related questionnaires are now available together with emotions wheels where liking, preference, or appreciation is always underlying the responses (Desmet & Schifferstein, 2008). The sensory drivers of emotions may not correspond to the sensory drivers of liking because they depend on the type of food and consumer groups (Spinelli, Monteleone, Ares, & Jaeger, 2019) but are considered as going "beyond liking" and being more stable indicators of day-life behavior (Köster & Mojet, 2015).

The most recent achievements showing that emotions may be handled as flavor molecules comprised the use of an electronic nose tuned to human odor pleasantness estimates (Haddad et al., 2010). The apparatus generated odorant pleasantness ratings with above 80% similarity to average human ratings, and with above 90% accuracy at discriminating between categorically pleasant or unpleasant odorants. Similar results were obtained in two cultures (native Israeli and native Ethiopian) without retuning of the apparatus, suggesting that, unlike in vision and audition, in olfaction, there is a systematic predictable link between stimulus structure and stimulus pleasantness. This goes in contrast to the popular notion that odorant pleasantness is completely subjective (Haddad et al., 2010) as assumed in hedonic or affective assessments (Heymann & Ebeler, 2016), thus supporting the inclusion of emotions in tasting methodologies.

29.2.4 Flavor has an integrative sense

The integration of information from the senses of vision, gustation, olfaction, and the somatosensory and trigeminal systems, responsible for astringency, heat, and prickling sensations, results in the flavor (Heymann & Ebeler, 2016). The neural mechanisms, although with different processes, are interconnected and difficult to individualize (Lundström, Gordon, Wise, & Frasnelli, 2012). Small & Green (2012) claimed that during tasting, a "flavor object" is formed on both retronasal olfactory, gustatory, and somatosensory stimuli in a sort of perceptual *gestalt*. In fact, the poorer ability to identify components in odor-taste mixtures than in taste mixtures indicates that interactions occur between the two senses, challenging the proposal that odors and tastes are processed independently when present in complex chemosensory stimuli (Laing et al., 2002). Probably, this is the reason why orthonasal odor descriptors are more diverse than flavor descriptors in wine sensory analysis (Pearson, Schmidtke, Francis, & Blackman, 2020). In addition, the process of "oral referral," by which smells seem to originate from the mouth, is another evidence of the multisensory feature of flavor (Spence, 2016). Moreover, flavor is also affected by nonsensory cues (Piqueras-Fiszman & Spence, 2015) being regarded as a multimodal neural construct, sort of like a new sense resulting from the others (Hannum, Stegman, Fryer, & Simons, 2018).

The complexity of flavor interactions makes it less tractable by sensory analytical methods, as demonstrated by the few dedicated articles (Palczak, Blumenthal, Rogeaux, & Delarue, 2019) when compared with the profusion of research based only on orthonasal smells as if wine was only a perfume. In fact, wine tasting involves the retronasal pathway to produce the flavors that are not fully captured only by the orthonasal function and influence in-mouth properties (Goodstein, Bohlscheid, Evans, & Ross, 2014). Even the act of spitting prevents understanding phenomena occurring below the throat in the digestive tract (Sclafani & Ackroff, 2012). Likewise, the question of food pairing, also relevant to wine appreciation, increases the complexity of chemical reactions between both matrices and with the sensory receptors, resulting in a diversity of individual responses that prevent generalizations probably explaining the scarcity of dedicated research (Spence, 2020).

29.2.5 The high range of sensory sensitivities

The variation in sensory perception is driven by genetic and acquired factors due to unconscious exposure or dedicated training (Keller & Vosshall, 2016). Besides olfaction (Fig. 29.2), the other senses of taste and oral touch contribute to creating a sensory world with a high diversity of inputs. The complexity of interactions is at the chemical and colloidal levels in the wine and at perception in the brain, turning the task of individuating the weight of each of them difficult. The sensory sensitivity to oral stimuli has been studied with a wide diversity of methods encompassing the different physiological factors influencing perception (Table 29.2).

Hypothetically, a general chemosensory sensitivity could result from associations among taste, smell, and trigeminal stimulation thresholds. Though there may be a common underlying factor that determines some individual differences, a general chemical sensitivity that spans chemosensory modalities is thought to be unlikely (Lundström et al., 2012). To complicate the picture, assigning verbal descriptors to stimuli to evaluate recognition thresholds is highly dependent on familiarity and experience (Keller & Vosshall, 2016). Parr (2019) eloquently

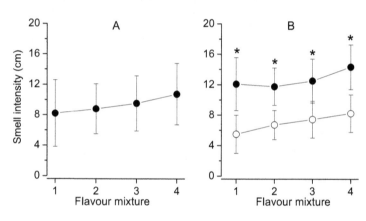

FIGURE 29.2 Smell intensity of white wine spiked with increasing flavor mixture concentrations (1, control wine; 2, control plus 0.5 mL/L; 3, control plus 1 mL/L; 4, control plus 2 mL/L) (A, all tasters; B, Low-sensitivity tasters and High-sensitivity tasters). Asterisks denote significant differences ($P < .05$) between High- (•) and Low-sensitivity (○) tasters. Intensity was measured as the distance (in cm) from the top of the glass to the nose of the taster. *Source: From Vitorino (2018).*

TABLE 29.2 The individual diversity of sensory phenotypes.

Sensory function	Sensory measurement	Sensory phenotype	References
Taste	Bitterness intensity	PROP tasters, nontasters and super-tasters	Pickering and Robert (2006)
	Various taste intensities	Hypogeusia, hypergeusia	Webb et al. (2015)
Touch	Astringency intensities	High, low, and intermediate sensitivities	Pickering et al. (2004)
Taste and Smell	Detection and recognition thresholds	High or low sensitivities, specific anosmia	Tempère et al. (2011)
Smell (ortho- and retronasal)	Suprathreshold sensitivities	High and low sensitivities	Doty (2015)
	Retronasal detection thresholds	Hydrolysis of glucosides in saliva	Parker et al. (2019)
Salivary flow	Weight of saliva	High, low salivators	Criado et al. (2019)
Thermal taste	Taste and mouthfeel intensities after heating the tongue	Thermal and nonthermal tasters	Yang et al. (2014), Small-Kelly and Pickering (2020)
Chemesthesis	Irritation sensitivities	Different sensitizations	Carstens et al. (2002)

summarized the implications of these diverse behaviors in sensory analysis: "individual differences are genetic and acquired, but are not error in the machine." The "machine" function becomes even more difficult to analyze when genetic variability in the olfactory genes also induces different pleasantness responses (Eriksson et al., 2012). Then, more than formatting sensory responses to obtain a consensus, the answer is to develop methods that accommodate these differences because they reflect "machines" that work differently, namely when different levels of expertise are involved.

29.2.6 The perceptual features of expertise in odor analysis

The difference between expertise in odor analysis usually groups consumers into naïve and experienced subjects. The experience is evaluated by the ability of perceptual grouping to assign classes and meanings to olfactory stimuli in order to form context-sensitive associations (Köster et al., 2014) which require prolonged olfactory training in a complex mixture (Croijmans & Majid, 2016). In practice, expertise is defined by the ability to separate aromas in different families and subfamilies as illustrated in aroma wheels (Noble et al., 1984) or in perfumery (Edwards, 2010). However, the fact that odors are perceived as synthetic objects seems to be in contradiction with their possible segregation. Barwich (2017, 2019) addressed this issue by postulating that olfaction affords perceptual categorization without the need to form odor objects when experts are involved, thus being different from the visual objects. Overall, this expertise builds on the analytical ability to separate sensory information and focus on salient features by experts who must overcome the synthetic nature of olfaction. Then they are able to differentiate the different smells and categorize them into families of similarity under a process called figure-ground segregation in odor perception. It is not immediate but a process of perceptual grouping that involves sequential processing of attention and iterative interpretation.

What primarily drives olfactory perception is not discrete, stable odor object formation but context-sensitive decision-making (Barwich, 2017, 2019). The function of olfaction is to recognize changes in context. This author further explains that, upon encountering olfactory information, the brain essentially has to decide on the grouping, salience, and value of olfactory input by matching this information with learned templates of odor categorization and other cues (e.g., cross-modal, verbal), all indicating the context in which a decision about the incoming olfactory information takes place. Therefore the same physical information of a stimulus can be grouped differently and matched against various cognitive templates, particularly in complex stimulus. Then different experts may describe the bouquet of wine in separate qualitative terms simply because different features in the wine appear salient to them in their perceptual evaluation, and so they pay attention to different olfactory notes in their characterization.

To conclude, experts only have a limited better sensitivity over nonexperts that is acquired by perceptual and linguistic training (Croijmans & Majid, 2016). Consumers can acquire this expertise probably with less time than previously supposed. In fact, Koenig, Coulon-Leroy, Symoneaux, Cariou, & Vigneau (2020) showed that the level of wine expertise played a minor role in creating the semantic representation of wine odors, affecting mainly the knowledge of specialized terms. This approximate semantic behavior is understandable by the familiarity with the popular fancy aroma representations but the question of how many flavors may be identified in simultaneous remains to be clarified.

The thin line between mental imagery and the imaginary.
The metaphorical process of mental imagery, where a smell is perceived even in its absence, is improved by expertise and plays an important role in odor discrimination and categorization (Croijmans et al., 2020; Tempère, de Revel, & Sicard, 2019b). Then the philosophical hypothesis of Barwich (2017, 2019) to explain expertise in odor segregation needs experimental validation because other cognitive valences (e.g., memory associations, semantic fluency, imagery) may be engaged instead of truly perceiving a wide diversity of scents. The ability to "guess the rest" is constitutive of olfaction, and the question of how much of the imagery is a fantasy is yet to be answered.

This situation is particularly evident when experts have memorized the aromatic profile of a varietal wine. Honoré-Chedozeau, Chollet, Lelièvre-Desmas, Ballester, and Valentin (2020) explained that experts activate a "wine tasting script" just like novices that are familiar with wine descriptions. For instance, the perception of volatile thiols (e.g., tropical fruit) may create the image of a typical Sauvignon Blanc and a dozen, or more, of descriptors come straightforward, all correct even if not sensed (Hughson & Boakes, 2002; Parr, Green, White, & Sherlock, 2007). The exercise is very attractive but when the context changes (e.g., the same varietal with other dominant flavors, different varietals with thiolic character), the task is probably only achieved by chance (Campo et al., 2008). The possibility of having functional magnetic resonance images (fMRI) able to distinguish imagery from imagination should solve this question because tasters are true believers of their discriminating capabilities (Honoré-Chedozeau et al., 2020).

However, this problem is only relevant for commercial wines dependent on intense sweetish flavors to be appreciated by consumers (Lezaeta, Bordeu, Agosin, Pérez-Correa, & Varela, 2018). The recognition of fine wines does not require flavor description, and so this ability of flavor segregation should not be used as a requirement to define expertise in wines (see section 29.4). Then, for the sake of fine wine recognition the different behaviors of experts and nonexperts could be better elucidated by the different ways of processing information instead of flavor identification.

29.2.7 The top-down and bottom-up processing of information

Recent research on the role of expertise in olfaction further elicited the underlying effects of extensive training in terms of functional and structural modifications in the neural processing of odors (Royet, Plailly, Saive, Veyrac, & Delon-Martin, 2013). Olfaction involves a significant amount of high-level processing, such as experience, memory, attention, expectation, and intention, in aspects beyond those described in the previous section. It is further penetrated by cognitive influences through education, knowledge, imagination, and culture (Chen et al., 2014). Therefore it is reasonable to interpret the sensory responses as in other areas of human behavior where only cognitive factors are involved (e.g., economy, psychology).

The strategies to process information to obtain knowledge are known as top-down and bottom-up. The first one dominates when the conceptual information (top) influences the sensory response (down), while the second corresponds to the cases where the sensory information (bottom) determines the conceptual response (up).

Briefly, in the presence of a new wine, the intuitive question made by the expert would be "have I tasted it before?" while the novice would ask, "what is this?" However, the same person may behave differently depending on the task to be performed, and learning is also possible for the expert. Moreover, "right" or "wrong" answers may come from both processing mechanisms. The coincidence or divergence between experts and novices are consistent with the neural processing of sensory stimuli in the OFC where the most complex human reactions take place.

The implications may be understood in wine tasting using two illustrative cases of different styles of Chardonnay wines and the definition of consumers with low and high involvement given by Charters & Pettigrew (2007):

i. Coincident preferences between experts and novices

The first white wine style corresponds to the hypothetic responses to a warm climate Chardonnay from California with oak aging (Fig. 29.3). The expert recognizes the paradigmatic dimension of oak flavors and deduces that the base wine must have been chosen among the best in the winery, originating from vineyards with the lowest yields and perfect grape balance. Hence, the final response is of high appreciation. The novice is not aware of this but is also attracted by the pleasant flavors of vanilla and buttery and smooth mouthfeel, thus eliciting high appreciation and correspondent high perceived quality. Then, due to different reasons, the result is the same and the consensus is welcome by both parties without further discussion. The expert may add several memorized descriptors that are readily acknowledged by the newcomer in a hopefully memorable experience.

ii. Divergent preferences between experts and novices

The second style corresponds to the responses caused by a cool climate Chardonnay from Burgundy with high quality standards (Fig. 29.4). For the expert, the reductive flavors and sour lingering taste are recognized as indicators of "minerality," thus meaning a paradigmatic relation to reputed wine regions. The response is obviously of high appreciation. On the contrary, the wine stinks like a skunk (or "baby diapers" or "dirty

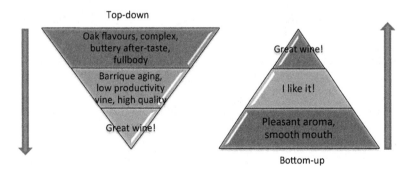

FIGURE 29.3 Hypothetical flavor information processing by an expert (left) and naïve consumers (right) tasting an oak-aged Chardonnay from a warm climate.

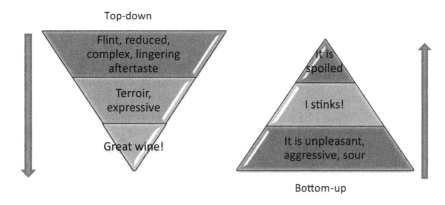

Top-down

Bottom-up

FIGURE 29.4 Hypothetical flavor information processing by an expert (left) and naïve consumers (right) tasting a 1st Cru Burgundy Chardonnay.

socks," to name a few uneducated but right guesses) and has a marked acidity to the inexperienced taster. This response is characteristic of other elicited by reduced wines as described by Lezaeta et al. (2018), who showed a similar grouping of unpleasant aromas (herbaceous, moisture, earthy, reduced) together with "mineral" with consumers tasting *Sauvignon blanc* wines. Then the answer to these perceptions can only be of displeasure, being impossible to accept that the wine may be of outstanding quality. The initial unpleasant aromas of the quality Chardonnays may explain why it does not match the expected fruity and floral descriptors as noticed by Campo et al. (2008). Then, the expert should understand this response driven by true sensory perception and explain the distinctive character of this wine style that requires time to develop. Otherwise, the experience is unforgettable for bad reasons and the expert does not manage to be recognized by the novice.

Other examples could be drawn, but in essence, the coincident responses between experts and nonexperts are related with sweetish flavors and smooth mouthfeel, consistent with the quality "fit to the purpose" of commercial style when the expert is aware that the base wine has the quality standard of fine wines. Fine wines typically induce opposite responses that must be properly explained using the holistic attributes by the expert.

Keeping in mind the present knowledge on the basics of neuronal processing of sensory information, together with the illustrative opposed examples of quality definition, consumer segmentation, and wine styles, we are now in a position to address its implications in the current sensory and tasting methodologies.

29.3 Technical tasting approaches

A wide diversity of tasting methodologies has been presented in the continuously updated editions of *Wine Tasting, a Professional Handbook* (Jackson, 2017). The focus here is to discuss the significance of the results yielded by different approaches learned in educational courses from universities and other wine professional institutions.

29.3.1 Multipoint score sheets

Classical wine tasting depends on the evaluation of analytic and synthetic (holistic) descriptors. The former include those where the wine is divided into parts related to visual, smell, taste, and somatosensory responses. The latter intend to describe sensations that are the reflection of these analytic sensory features but that cannot be fully predicted from them while capturing the whole of the wine sensory characteristics. That is the case of body, complexity, persistence, harmony, elegance, balance, memorability, and aging potential, among others, usually being placed at the end of the tasting schemes as "overall quality."

The result for each of the descriptors may be recorded in written or numerical forms so that the sum of the analytical and the synthetic data reflects the quality in terms of a score. Hence, the tasting sequence is contrary to flavor neuronal processing because the first sniff immediately elicits an emotional response, initially influenced by sight if glasses are not dark, even if not conscious. Keeping this in mind, the difficulty in developing objective tasting sheets is understandable because pleasantness reactions are continuous along the tasting sequence. In fact, one of the most difficult tasks is to separate wine appreciation from individual preferences, mainly in the synthetic properties as explained in Section 29.4.

29.3.1.1 Technical and popular scoresheets

The development of an objective tasting method has been a constant pursuit of sensory scholars to evaluate wine quality. The seminal work of the University of California, Davis, led by Maynard Amerine, produced a tasting sheet conceived to help winemakers to prevent wine defects and assess the trueness to type (Amerine & Roessler, 1976). The system comprises the rating of several attributes whose sum gives a score reflecting the wine quality. Earlier models had a maximum of 100 points, but Amerine realized that the high variability (more than 20 points) for the same wine justified diminishing the range to 20. The system is known as the 20-point Davis scoresheet and is used in industry and research (Table 29.3). Contemporaneous tasting approaches developed in Bordeaux University were essentially the same (Ribéreau-Gayon & Peynaud, 1958), although the outputs were not numeric and less prone to statistical analysis.

Later, the OIV also developed a tasting sheet. The updated version is required for the recognition of international wine challenges by this institution (Table 29.3). In this sheet the inclusion of the adjective "harmonious" and "positive" introduces additional sources of subjectivity across smell and taste evaluations.

These sheets may be increased by describing the attributes of color and aspect, smell, taste, mouthfeel, and other synthetic parameters that reflect the overall quality (WSET, 2011). The adaptation of the descriptors is also necessary for evaluating other wine types like sparkling and sweet wines, but they are essentially equal to those used for still wines. This type of approach has also been so widely adopted by the popular wine press that it seems to have originated from wine critics but 100-point schemes have now been improved for nearly 100 years (Amerine & Roessler, 1976).

The system apparently fulfills the requirement for grading wine quality in the sense of "fit to purpose" for the international wine market. By giving a score, it seems that the quality may be rated by a precise mathematical system, with increasing power enhanced by a larger range, as shown by recent web applications going up to 1000 points (https://www.wine-lister.com, assessed July 30, 2020). However, the increase in the range is illusive because the results of the quality evaluation are the same (Parr, Green, & Geoffrey White, 2006). In fact, Amerine, supported by statistical analysis, stated that the system could well work with a score of 10. A contemporaneous, recognized professional expert, Michel Broadbent, only used 5 (Broadbent, 1979).

However, the essential limitation of these sheets is related to their subjective nature as reflected by several biases that ultimately result from using them for a purpose different from the original. What begun by being an objective tasting sheet for winery application soon became a tool for hedonic appreciation that measures individual preferences. Hence, the score should be understood as a numerical metaphor, i.e., "this wine under my view has an outstanding quality given by a score of 18 out of 20." With time, the tasting sheets were adapted to a role to which they were not meant and presently their development is more dominated by divulgation than by science, increasing the gap between both worlds. In fact, Amerine warned out that it should not be used by, or to inform, consumers without specific training.

TABLE 29.3 UC Davis and OIV scoresheets.

UC Davis		OIV		
Appearance	0–2	Visual	Limpidity	1–5
Color	0–2		Aspect other than limpidity	2–10
Aroma	0–4	Nose	Genuineness	2–6
Aroma flaws	0–2		Positive intensity	2–8
Sugar, dryness	1 (Weak), 2 (Appropriate)		Quality	8–16
Body and mouthfeel	1 (Absent), 2 (Good)	Taste	Genuineness	2–6
Flavor, balance	1–2		Positive intensity	2–8
Astringency	0–2 (reds and rosés)		Harmonious persistence	4–8
	2 (whites)		Quality	10–22
Overall appreciation	0–2	Harmony	Overall judgement	7–11
Total	3–20	Total		40–100

29.3.2 Scoresheets as style preference indicators

The high susceptibility of these scoresheets to subjective responses makes them an appropriate tool to understand individual preferences. This may be exemplified by two case studies, with rather different taster's cohorts, but where the final output was essentially the same illustrating the present dominant commercial wine style.

i. Two expert panels using the Davis 20-point scorecard (Grohmann, Peña, & Joy, 2018) reported the score given by tasters from Montreal and Okanagan to red wines from different countries and styles. The Okanagan panel was composed of winemakers, sensory scientists, and WSET level 3 professionals. The Montreal cohort gathered sommeliers, wine journalists, and consultants. The results showed that the best-rated wines were different. For the Okanagan cohort, among the preferred wines was the Apothic Red with noticeable sweetness, described as a "fruit bomb," while this wine was the least liked by the Montreal cohort. Then, using the same tasting sheet and the same wine, the outputs were opposite in terms of wine styles. The analysis of the panel composition strongly indicates that preference was mediated by different educational or learning points of view. Then both panels had opposite concepts of wine quality, with one directed toward the international commercial style, the other toward a style consistent with the definition of fine wines.

ii. Large panels in an international challenge using the OIV tasting sheet

Mundus Vini is one of the most reputed international challenges that use the OIV scorecard. First observations showed that gold-awarded wines were dominated by high ethanol content and sugar levels higher than the limit for dry wines ($>4\,g/L$), even considering the modulation of dryness definition by acidity adopted by the OIV (Loureiro, Brasil, & Malfeito-Ferreira, 2016). Later, the competition added sensory and quality descriptors to the tasting sheet, which made it possible to get further insight into the nature of quality evaluation.

The data were retrieved from the internet corresponded to five different challenge editions with different tasting panels. Table 29.4 shows the descriptors that explained the higher appreciation of Great Gold awards when compared with the Gold ones (Malfeito-Ferreira, Diako, & Ross, 2019). Then it seems that despite all individual preferences supposed to be found among numerous tasting panels, a kind of global quality profile may be summarized from all challenge editions. This profile corresponds to the international commercial style driven by sweetish flavors and a full and smooth mouthfeel. Wine styles closer to the concept of fine wines did not appear to have been recognized. Overall, these observations were further evidenced by Guld, Nyitrainé Sárdy, Gere, & Rácz (2020) and Ailer, Valšíková, Jedlička, Mankovecký, & Baro (2020) using the same 100-point OIV scheme, in Hungary and Slovakia, respectively.

29.3.3 Indications for good tasting practices and challenge organization

The previous observations justify a deeper analysis to understand why fine wines are not recognized in these challenges. Of course, this may be the result of overall lesser appreciation for these flavor profiles by the judges, but we argue that the methodology also contributes to their relative devaluation.

Tasting manuals list a series of instructions to minimize tasting biases, mainly in cases of analytical sensory studies (Lesschaeve & Noble, 2010). Some of these are difficult to implement in large competitions and may be understood as a trade-off between method soundness and organization practicability. Others could be easily implemented to improve the scientific significance of the results and the recognition of a wider range of perceived quality, as described below.

i. Tasting organization (sequence and number of wines)

The tasting begins with the aspect and smell that both induce an unconscious emotional response that drives the other reactions. Intensely flavored and full-bodied wines are favored overwhelming less exuberant wines. The high number of wines in the tasting flights thus favors those more intense in the nose and in the mouth, and those that quickly show their qualities once poured in the glass. Wines with less flavor and less body that require more than 2—3 minutes to open are penalized by this type of procedure.

These shortcomings could be minimized by limiting the number of samples in each flight to be at most four or five according to the good tasting practices elaborated by Heymann & Ebeler (2016). The impact of potent wines was also analyzed by King, Dunn, & Heymann (2013), who advised a separation into two groups at 14% (v/v) ethanol.

TABLE 29.4 Log odd ratios of significant predictors of white and red wine award of Grand Gold medal (Malfeito-Ferreira et al., 2019).

Wine type	Predictor	β-value
White	Complex	1.17
	Exotic Fruit	0.24
	Body	0.12
	Astringency	−0.04
	Acidity	−0.16
Red	Complex	0.71
	Potential	0.35
	Spicy	0.11
	Body	0.10
	Dried fruit	0.09
	Red berries	−0.17
	Green vegetative	−0.39

ii. Division in flavor styles

Wines are basically grouped according to color (white, rosé, and red), stillness (sparkling or non-sparkling), or sweetness (dry, naturally sweet, or fortified). However, this separation does not reflect very different wine styles that may be present in each of the categories, particularly in the dry group where minute amounts of added concentrated juice have a significant effect in smoothening the palate of sour and astringent wines.

The quality should be evaluated according to the standard in each class, just as dogs' races are separated in the respective competitions. In fact, Amerine & Roessler (1976) constantly warned that these sheets are to be used for wines within the same style. The same evaluation of an oaky Chardonnay or a dry Riesling does not give an idea of the different qualities within the style. The appearance of an increasing number of whites obtained with maceration (e.g., orange or amber wines) further justifies stylistic groupings, especially because astringency is a devaluating factor.

iii. Score variability and judge inconsistency

The public availability of ratings given to wines in competitions and by the wine press provides a large wealth of data that has been handled mostly by scholars with an economical background. With the present global quality, a wine is seldom given a score of less than 85/100 and what looked like to be a wide range leads to a rather small range of variation (Bodington & Malfeito-Ferreira, 2017). Further, the high variability in averages might make the extreme scores be placed in a gold level while the other may be below silver, being the results statistically equal. The awards are based on mathematical averages of all scores for a given wine and ties are solved by increasing the number of decimal places, which is not a statistically sound process, although it is easily understood by the competitors and consumers (Bodington & Malfeito-Ferreira, 2019).

The general conclusion is that the competition ratings should not be relied upon as a measure of quality because of poor judges' performance (Cliff & King, 1997, 1999; Scaman, Dou, Cliff, Yuksel, & King, 2001; Ashton, 2017; Hodgson, 2008; Honoré-Chedozeau, Ballester, Chatelet, & Lempereur, 2015; Honoré-Chedozeau et al., 2020) and lack of reliable statistical data treatment (Ashenfelter & Quandt, 2012). The observation that these ratings may be poor science is not precise, simply because they are not science but numerical metaphors.

As a corollary of these limitations, Alexandre & Ballester (2019) concluded that the probability of medals being awarded to a wine is due to chance and so the odds of getting a medal increases with the number of submitted wines by a producer, a region, or a country.

iv. The standards of global quality

Despite all the previously mentioned drawbacks, some of them easily solved and others related to the subjective nature of evaluation, the main qualitative bias is related to the valorization of the commercial wine style that, by receiving the highest awards, may be regarded as the prototype of superior quality (Malfeito-Ferreira et al., 2019). The awarded wines have characteristics consistent with consumer liking positively driven by sweetness and negatively related to bitterness and sourness (Rodrigues & Parr, 2019). In addition, sugar levels also increase the sensation of the wine body (Sáenz-Navajas et al., 2012), and so the valorization of fullness favors its addition at bottling (e.g. under the form of rectified concentrated grape must). Winemakers should be aware of these implications when having to decide the winemaking options.

In conclusion, the awards or the scores in wine press are indicators of wine quality "fit to the purpose" of meeting the international market demands. They are excellent means for brand divulgation in a global wine market reflecting the most fashionable preferences (Herbst & Arnim, 2009), but the biases are too much to be used as a reliable tool to assess fine wine quality. Then, which should be the indicators of fine wine quality?

29.4 The evaluation of fine wines

The absence of an objective definition for fine wine quality has been addressed by Charters (2003), but the two expert views described in Box 29.3 suffice to understand that fine wines require another sort of approach to be properly appreciated. This approach should take into consideration the wine as a whole, as the brain immediately does upon the first look, before dividing it into parts as conventional sensory analysis instructs. This is now the time to address the question of holistic or synthetic properties.

29.4.1 Synthetic properties

The sensory elicited properties that reflect the overall fine wine quality have several denominations, besides holistic or synthetic (Ashton, 2017). From an esthetic perspective, these properties are known as emergent (Burnham & Skilleas, 2012), while abstract, high-order, or umbrella descriptors are also commonly mentioned (Parr et al., 2020). Lezaeta et al. (2018) coined these attributes as extrinsic, which may be confused with extrinsic cues such as label, bottle, or price.

In opposition to analytical descriptors, these synthetic properties are characterized as follows:

i. are not a measure of individual molecules or cannot be obtained by the sum of several molecules;
ii. depend on the chemical nature of the wine but cannot be predicted by wine physico-chemical analysis;
iii. are not individual sensations, or groups of sensations, but to a certain extent are relations among sensations.

The difficulty in capturing their essence by instruments perhaps explains why published research is mostly owed to teams including psychologists, sociologists, philosophers, or economists. However, the ultimate goal is the same for all and an effort should be made to set future shared research. Otherwise, others more oriented to trade or critics may take the lead without a scientific background, as noted by Marks (2015) and Parr et al. (2020).

The list of synthetic properties is long, including long-accepted terms of harmony, intensity, transparency, structure, texture, finesse, body, richness, complexity, elegance, depth, gracefulness, delicacy, typicality, memorability, or aging potential. The choice is mostly idiosyncratic. For instance, texture is not mentioned by Jackson (2017), being a synonym of mouthfeel for Cheynier & Sarni-Manchado (2010), and is included in a sensory mouthfeel wheel (Gawel, Smith, Cicerale, & Keast, 2018). The choice may also fluctuate according to the popular trend of the moment as for "minerality" that should be considered as a synthetic property given the vagueness of the concept. Instead, Parr, Maltman, Easton, & Ballester (2018) advise the use of only sensory descriptors (e.g., gunflint, salty, acidity) for "mineral" wines, but the concept is out of control in the popular wine press. Despite all these ill-defined features, their crucial significance to define fine wines justify the efforts to set adequate tasting methods that require sensory, perceptual, conceptual, memory judgment, and language skills to be properly addressed (Cheynier & Sarni-Manchado, 2010; Parr et al., 2020).

The abstract concepts of wine quality tend to be only associated with fine wines. However, the observation of expert responses shows that the concept is not clear because several descriptors may be more indicative of the quality of the commercial wine style. The case reported by Parr et al. (2020) is illustrative of this behavior. The fine quality of New Zealand Pinot Noir was associated with complexity and varietal typicality. However, the additional character of flavor attractiveness casts some doubts on the ability to discern complexity. Similarly, "expressiveness" appeared to depend on potent flavor and full-body, which are not appropriate indicators of harmony or elegance. Moreover, brownish color may be a cue influencing smell, taste, and mouthfeel properties, diminishing product expectations. This concept of fine wine quality, biased to the commercial style, was not observed in a French cohort (Parr, Ballester, Peyron, Grose, & Valentin, 2015; Parr et al., 2020).

Overall, these results show the weight of educational factors on wine quality evaluation and liking, although Spence & Wang (2019) considered these effects mediated by learning to be not so impressive when compared to other food and beverages. These authors hypothesized that the complex nature of wine explains, at least in part, this behavior. Moreover, Smith (2019) posited that fine wine tasting is a hard task requiring years of practice to improve perception and categorization of the various wine sensory facets. However, the expertise, as measured by qualitative categorization, constitutes a necessary but not a sufficient element in training the aesthetic appraisal of odors or flavors because the categorization is not primarily an aesthetic activity (Barwich, 2017). Moreover, the olfaction as an emotional sense has also implications in pleasantness and pleasure (Chatterjee, 2014). Then the cortical domains of pleasure and aesthetic evaluation are coincident, probably explaining the difficulty in separating one from the other. This cognitive task is very hard and we argue that here lies the core of expertise in recognizing fine wines.

29.4.2 Preferences and aesthetic assessment

Olfaction has been considered out of the traditional philosophical aesthetic considerations, contrarily to vision and audition, because of its subjective and inherently affective character that allegedly reflects only personal preferences, without a sufficient cognitive dimension (Barwich, 2017). However, recent advancements in neuroscience have shown that the strict distinction between perception and cognition is no longer valid (Keller, 2017), debunking the myth that olfaction is an unsophisticated sense (Barwich, 2017). Indeed, Skov (2019) showed that esthetic objects (e.g., art, faces, landscapes) are appreciated by the OFC using the same neurobiological mechanisms such as food, drinks, or money, coining the process as a "common currency" to deal with pleasure and reward.

Coincidently, philosophers Burnham & Skilleas (2012) also argued that wine may be regarded as an aesthetic object and enumerated a series of characteristics common to others in the fields of literature, painting, music, or landscapes (Table 29.5). Therefore the rules of aesthetic evaluation may and should be applied to wine.

Burnham & Skilleas (2012) postulated that to have an esthetic judgment the assessors must have competence either to create and to justify it. Thus, according to these authors, the nature of competence, or expertise, includes the following:

  i. Cultural competence (e.g., knowledge of wine types, regions, winemaking techniques);
 ii. Experiential competence (e.g., tasting expertise improved by practice);
iii. Aesthetic competence (e.g., perception of synthetic/emergent properties);
 iv. Communication intelligibility (e.g., understand and explain consumer reactions).

Sackris (2019) posited that experts or nonexperts are unable to separate enjoyment from aesthetic evaluation but, in our opinion, that is precisely why the level of competence may be assessed by the ability to distinguish personal preferences from the aesthetic evaluation. The observation that some characteristics show an emotional attachment (e.g., attractive fruit aromatics) indicates that liking, appreciation, or preference, mediates the definition of fine wine quality. In other words, the object of attraction can only be defined by highly esteemed descriptors. This is particularly obvious when wine preferences are subjected to trends that may blur the perception of aesthetic properties. For example, powerful full-bodied red wines dominated by oak presently seem to give place, at least in the trendy popular press, to light reds with noticeable sourness. These wines may be more persistent and complex but the question is if these properties are recognized or if what is recognized is their higher

TABLE 29.5 Experienced attributes of aesthetic objects present in wine.

Attributes of aesthetic objects	Definition	Examples
Vagueness	Presence of characteristics that are difficult to assess	Odor or flavor description
		Cross-modal interactions between senses
		Variation in individual perception
Moving target	Evolve with time	Aging potential
		Develop in the glass
		Change with food or the occasion
Richness	Valorization of properties beyond a physical object	Cultural, historical, and social influences

Adapted from Burnham, D., & Skilleas, O. (2012). The aesthethics of wines.

aggressiveness, exactly as the potency was. These salient features act as cues that are easily recognizable: the brain perceives the sign and, by top-down regulation, associates high quality to these heightened sensations, possibly without recognizing the synthetic ones. The object of preference changed but the lack of aesthetic competence persisted probably mediated by top-down mechanisms.

29.4.2.1 The top-down induced bias: the halo and familiarity effects

The cognitive bias induced by top-down processes may be present in either experts or novices, as illustrated by the effects of "halo" (Apaolaza, Hartmann, Echebarria, & Barrutia, 2017) and familiarity (Priilaid, Human, Pitcher, Smith, & Varkel, 2017). In terms of flavor processing, subjects feel the same taste or provide descriptions adapted to the preference, but the perceived pleasantness is heightened, consistent with the emotional nature of the olfaction (Spence, 2020). These pleasurable effects are similar to those induced by food, sex, or money, which have been shown to engage brain areas related to pleasure and memory (Chatterjee, 2014).

A remarkable example of the halo effect has been observed in the evaluation of organic wines by experienced faculty students (Romano et al., 2020). These authors tested the same organic wines in three sessions with the information of being conventional, organic, or of unknown origin. The wines were clean or faulty and were properly described as such by the students. The correct recognition of off-flavors was regarded as the cue to organic label but, surprisingly, defective wines were more valorized and appreciated. The emotional nature of the appreciation was evidenced by the overwhelming citation of "unpleasant" in the last session when information on the mode of production was absent. Thus experienced tasters, when they had no cues, behaved like novices and reacted emotionally. This case shows that the ideological halo effect enabled the distinction between preference from aesthetic judgment, which has not been considered likely to occur in wines (Boncinellia, Casinia, Continia, Gerinia, & Scozzafava, 2016; Sackris, 2019). This behavior contradicts all teaching directed to flavor purity implemented since the beginning of wine higher education in Davis and Bordeaux Universities.

The effect of familiarity, neophobia, or ethnocentrism may be explained by the same neuronal processing. The question "where have I tasted this wine" is a sign that the brain is looking for similar flavor patterns that induce pleasantness coincident with the expectations (Boncinellia et al., 2016). Within familiar wines, even naïve subjects use top-down processing which must be coincident with the inputs coming from the senses (bottom-up). Unfamiliar wines may show the opposite halo effect ("horn effect") because the sensory unpleasant cues induced such behavior (Antunes, 2018).

29.4.2.2 Widening the range of perceived quality

When the concept of quality is limited to wines that are "fit to purpose" of pleasing the global market, it is hard to accept that unpleasant wines may be regarded as "fine wines." Therefore training is essential to open the range of perceived quality of unfamiliar wines by improving the perceptual ability of synthetic properties (Wang & Prešern, 2018). These authors, even if the perception improvement has been questioned due to methodological flaws (Capehart, 2019), showed that, with time, older wines became more appreciated. The less experienced subjects showed an increased preference for higher acidity and alcohol and decreased preference for oak flavors. Similarly, this change in liking toward fine wines has also been observed by Tempère et al. (2019a), in experiments performed on consecutive days in a domestic context. These authors attributed the shift to the boring effect of the commercial-style wines dominated by ethanol, full-body, and smooth mouthfeel. Thus it is possible to broaden the range of perceived quality by simply tasting a larger range of wine styles but, from an educational perspective, it would be interesting to have a tasting approach that could evidence this behavior.

29.5 Understanding wine styles based on emotional tasting responses

In wines, research has shown the different associations between flavors and emotional responses supporting the efforts put on increasing consumer acceptance (Mora, Urdaneta, & Chaya, 2018; Niimi, Danner, & Bastian, 2019). The positive emotions (e.g., happiness, relaxed, well-being, desire, romantic) are triggered by sweetish flavors (e.g., fruity, floral, vanilla, chocolate) (Ristic et al., 2019), reflecting that wine is no different from other foods. Contrarily, bitter, salty, sour, or barnyard induced sadness, fear, disgust, and irritation (Ristic et al., 2019). Thus it is expectable that the winemaker should make wines that elicit smell and taste sensations consonant with the overall sweetish flavor to assure the fitness of the commercial wine style. If not, under a psychological effect known as cognitive dissonance, consumers strongly reject the product (Civille & Oftedal, 2012).

However, another possibility is to use the emotions elicited by wines as an educational tool particularly adequate to consumers without previous practice or knowledge nor wishing to spend time in extensive learning programs. Here the cognitive dissonance may be explored in the case of fine wines where the unpleasant experience is striking. Then by adequate communication, the professional explains the reasons for the dissonance and the experience becomes memorable (Coste, Sousa, & Malfeito-Ferreira, 2018; Souza-Coutinho, Brasil, Souza, Sousa, & Malfeito-Ferreira, 2020).

29.5.1 The emotional wine tasting sheet

The utilization of emotional responses together with analytical and synthetic descriptors was first proposed by Loureiro et al. (2016) as a means of broadening the range of perceived quality by consumers in a short number of sessions. The objective is accomplished by tasting blind two wines of clearly opposite wine styles using two well-defined wines as standards and profiting from the efficient ability of the brain to compare samples, similar to the methods using wine pivots or poles. In Table 29.6, the underlying sensory descriptors of the emotional responses are listed and defined.

TABLE 29.6 Emotion-based tasting sheet and attribute description (Loureiro et al., 2016).

	Attribute and score range	Short description
Nose	Initial Impression – Distaste (1) to Attraction (5)	Emotion: wine's appeal after the first smell
	Intensity – Weak (1) to Strong (5)	Evaluate intensity as the distance between the nose and the glass top when the smell begins to be sensed
	Elegance – Cloying (1) to Subtle (5)	"Cloying/ostensive" – feeling after imagining smelling and drinking the wine every day for a fortnight "Subtle" – opposite of cloying and associated with wines that have a delicate smell
	Complexity – Easy (1) to Difficult to describe (5)	"Easy" – odors easily identified, tasters agree with one or two descriptors "Difficult" – absence of dominating main odor, several descriptors arise from different tasters
	Expectations for the mouth – Low (1) to High (5)	Emotion: expectations for the mouth assessment that were created by the smell
Mouth	Relation to smell – Disappointing (1) to Surprisingly good (5)	Emotion: response to the expectations raised by the olfactory assessment
	Thermal sensation – Cool (1) to Hot (2)	Tactile sensation of heat revealed by the wines when tasted at the same temperature
	Creaminess[a] – Dry (1) to Jammy (5)	Tactile sensation of texture
	Fullness[b] – Light (1) to Full-bodied (5)	Tactile sensation of wine body
	Harshness[b] – Smooth (1) to Abrasive (5)	Tactile sensation of roughness
	Persistence – Short (1) to Long-lasting (5)	Duration of the sensations in mouth and retronasal pathway
	Prevailing flavors – Sweet (S), Salty (S), Harsh/Bitter (H/B), Acid (Ac)	Identification of the dominating taste and flavors
Final nose	Evolution of the fragrance in the glass – Unchanged (1) to Fully developed (5)	Changes of the smell in the glass during time
	Duration of the fragrance in the glass – Short (1) to Very Prolonged (5)	Time of smell permanence in the glass
Overall	Disagreeable (1) to Exciting (5)	Emotion: overall response to the tasted wine
Visual	Color – White (W), Rosé (P), Red (R)	Just record the color
	Appearance – Clear (C), Cloudy (CI), Murky (M)	Just record the appearance
	Condition – Young (Y), Developed (D), Tired (T)	Just record the condition

[a]Only for free-run white wines.
[b]Only for red wines or whites with maceration.

The wines must be very well chosen with clear differences; otherwise the experience is not obvious. The commercial wines have gold medals and the fine wines should be from a reputed region like Burgundy, at least at 1st Cru level, with initial reduced character.

The application of improved versions of this scheme has been further tested in controlled experiments showing that consumers without formal training could distinguish white wines with the two opposite styles (Fig. 29.5). Fine wines were correctly considered more complex, persistent, and developing with time. Therefore naïve subjects showed the perceptual ability to recognize these synthetic properties simply by comparing them with a standard.

The wines W2 and W5 were distant from the ideal wine that can be explained by their associated unpleasant emotions (Fig. 29.6). The responses are in agreement with the notion of quality as conceived by consumers, but when the labels are disclosed, the surprise effect is remarkable. Then the awareness that the supposed poor-quality wines, even noted as faulty, were in fact fine wines of indisputable quality and higher price makes the experience memorable. Several sessions may be held with other wines of the same opposite styles to set the ideas, but the learning is prompt, as explained by the psychological effect of "cognitive dissonance." According to these results, the question of fine wine recognition is not of perceptual nature but of cognition, and so the task is not hard, nor does it require years of practice and expertise to segregate fine wine sensory properties.

29.5.2 A proposal for wine systematic tasting at the winery level

The previous approach is primarily dedicated to consumer education, but it may be the first step in a wine systematic tasting approach to be applied at the winery level to support viticulture and winemaking decisions. The perception of the synthetic variables is made by comparison with a pivot given that a holistic evaluation is specially fit to this tasting possibility (Pearson et al., 2020). The pivot may be replaced by a memorized standard acquired by practice but fine-tuning benefits from actual comparisons. These tastings at the winery level should be done knowing the samples (not blind) because it improves the judgment (Priilaid, 2007).

The systematic approach includes the parameters considered most important for the definition of the style and consequent market implications: quality concept, grape growing and wine manipulation, complexity and harmonious persistence, aging potential, sense of place, gastronomic potential, and price segmentation.

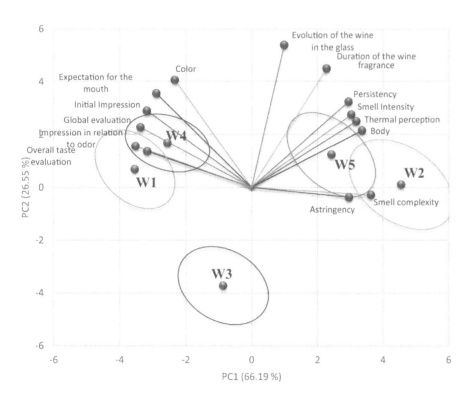

FIGURE 29.5 Two-dimensional map made from principal component analysis (PCA) of the wine descriptor scores (red circles) using the emotional tasting sheet by consumer panel. W1 and W4 represented the international commercial style with intense fruity and floral flavors. W2 and W5 represented the concept of fine wines, with W5 being a Chardonnay Burgundy 1er Cru. W3 was a popular wine with 11% (v/v) ethanol, fizziness, and 11 g/L sugar. *Source: From Souza-Coutinho et al. (2020).*

29.5.2.1 Quality concept

The first option begins with the decision of the wine style as defined by the previous emotional tasting sheet. The winemaker must imagine behaving as the target consumer to predict wine acceptance. The utilization of emotional responses enables the winemaker to understand the consumer responses and more effectively adapt the wine to fit market demands. For niche wines with a cultural background (e.g., clay jar, historic or heroic wines) the output must be according to the required profile.

The expression of winemaker aesthetic preferences may find a place here to produce an "author wine." The concept of the wine defines the grape growing and winemaking options.

29.5.2.2 Grape growing and wine manipulation

The viticulture decisions on the level of grape growing and ripening are determinant to obtain the expected wine style. For example, excessive productions, irrigation, or overripening are not likely to produce fine wines but are required for large-scale wines. Likewise, grape health is also a determining factor, and the technological options should be decided accordingly.

The manipulation or modulation of wine sensory properties is at the core of winemaking. Without proper technology, all wines would turn into vinegar except for sweet wines with more than 16% (v/v). Hence, manipulation may range from a minimalist approach to extreme intervention when wines are modified to meet consumer preferences. Table 29.7 describes the general effect of the most frequent options. Most of these techniques are intended to increase pleasant responses due to sweeter flavors and smoother mouthfeel and have been extensively supported by analytical and sensory research using state-of-the-art methodologies.

The increasing appreciation of the so-called "mineral" wines justifies modulating wine acidity accordingly in regions of warmer climates. The effect of fermenting acids (lactic and succinic acids) seems to have also the property of increasing the saltiness perception in addition to the expected acidifying effect. Therefore an increasing interest of the research community in addressing the production of wines with these characteristics is expected.

The manipulation should also be in agreement with the target price. For instance, the utilization of oak derivatives, type of oak (e.g., American or French), or the size of barriques should be decided accordingly. Likewise, the sparkling wine method (e.g., bottle fermentation or Charmat) has a direct influence on the final bottle cost.

29.5.2.3 Complexity and harmonious persistence

The complexity and harmonious persistence are the most relevant perceived synthetic properties that assess the fine wine quality. The definition of complexity should take into account the holistic perception of flavors as

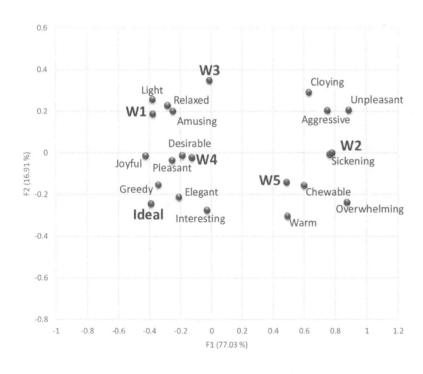

FIGURE 29.6   PCA plot of wines (red circles) in relation to the Ideal wine according to emotional responses. W1 and W4 represented the international commercial style with intense fruity and floral flavors. W2 and W5 represented the concept of fine wines, with W5 being a Chardonnay Burgundy 1er Cru. W3 was a popular wine with 11% (v/v) ethanol, fizziness, and 11 g/L sugar. *Source: From Souza-Coutinho et al. (2020).*

TABLE 29.7 Technological options to modulate wine sensory characteristics and improve appreciation.

Emotional responses	Sensory effect	Technological options	Illustrative references
Increase aroma pleasantness	Increase "sweetish" flavors	Reductive vinification	Antonelli et al. (2010)
		Fermenting at low temperatures	Torija et al. (2003), Beltran, Novo, Guillamón, Mas, and Rozès (2008)
		Utilization of enhanced aroma producing yeasts	Van Wyk, Grossmann, Wendland, Von Wallbrunn, and Pretorius (2019)
		Utilization of enzymes	Claus and Mojsov (2018)
		Blends with aromatic varieties	Ganić et al. (2003)
		Addition of wine flavor extracts	Lezaeta et al. (2018)
		Utilization of oak	Liberatore et al. (2010), González-Centeno et al. (2019)
Decrease in-mouth aggressiveness	Increase "sweetish" flavors and/or "smooth" mouthfeel	Prefermentative maceration	Cejudo-Bastante, Castro-Vázquez, Hermosín-Gutiérrez, & Pérez-Coello (2011), Darias-Martín, Rodríguez, Díaz, and Lamuela-Raventós (2000)
		Utilization of oak	Liberatore et al. (2010) ,González-Centeno et al. (2019)
		Elevage sur lies or addition of mannoproteins	Gawel et al. (2016)
		Malolactic fermentation	Lasik-Kurdys et al. (2018)
	Increase sweetness perception	Addition of concentrated juice	Saliba, Wragg, and Richardson (2009), Blackman et al. (2010), Sena-Esteves et al. (2018)
	Increase fizziness	Addition of carbon dioxide	Gawel, Schulkin, Smith, Espinase, and McRae (2020)
		Secondary sparkling wine fermentation	Culbert et al. (2017)
Increase the surprise effect	Increase persistence	Acidification with different acids	Ruivo (2018)
		Utilization of acidifying yeasts and nonthermal grape treatment	Morata et al. (2020)
	Increase complexity	Bottle aging	Karbowiak et al. (2019), Pati et al. (2020)

"odor objects." Spence & Wang (2019) and Palczak et al. (2019) presented the diversity of definitions on the subject and, among them, the difficulty to identify the aromas could be applied to complex wines. In our experience, consumers understand that more complex wines are harder to describe (Souza-Coutinho et al., 2020). In our opinion, associating complexity with the detection of a flavor molecule, or a group of molecules, may reduce the holistic perception. For instance, regarding off-flavors, the detection of the faulty molecule may be taken as a synonym for spoilage precluding the perception of complexity (Malfeito-Ferreira, 2018). The neuronal ability to "fill the rest" when a molecule is sensed, sparing the energy engaged for further attention, may be the explanation for this behavior. Another hint of complexity is the development of new flavors after pouring the wine into the glass. By comparing with a simple fruit-forward wine that remains unchanged, the task is easily accomplished.

The mouth persistence is also relatively easy to explain by the time the sensation lingers after swallowing or spitting. The harmony and elegance are understood as the balance among all sensations. However, these properties are not so easy to be noticed by nonexperts, and usually, complex and persistent wines are also harmonious and elegant. Here, the comparison with attractive full-bodied wines with short finish is very helpful to recognize the difference.

29.5.2.4 Aging potential

The aging potential is a major characteristic of fine wines, which are also referred to as "vins de garde." The ability to age is linked to wine physico-chemical composition and may be predicted by sensory analysis

(Langlois, Ballester, Campo, Dacremont, & Peyro, 2010). Some molecules are considered as markers of aging potential (Aleixandre-Tudo & du Toit, 2020; Pons, Lavigne, Darriet, & Dubourdieu, 2013), and a chemometric approach has been proposed for red wines (Aleixandre-Tudo & du Toit, 2020). In white wines, the problem of "atypical wine aging" is related to the accumulation of molecules that induce a character of old wines in young ones (Schneider, 2014). The present tendency to reduce sulfite levels justifies further research on the topic (Pati, Crupi, Savastano, Benucci, & Esti, 2020).

Typically, wines that age well are from odorless varieties and high fixed acidity. Bottle aging develops a bouquet and mouthfeel smoothens, probably because of the "mellowed" flavor character where aging markers (e.g., sotolon, 2-aminoacetophenone) are harmoniously incorporated in the bouquet. Certain indicator molecules may be detected [e.g., those responsible for walnut in Chardonnay (Gros et al., 2017)] but the issue is to perceive the odor object of a fine old wine.

Winemakers appreciating fruit-forward wines should recognize this character and do not associate the lack of fruitiness with oxidation. The question of recognizing the type is also relevant to understand the aging ability, as exemplified by oxidative notes in Sherry wines and volatile acidity in old fortified or naturally sweet wines.

29.5.2.5 Sense of place

The concepts of localness, sense of place, or *terroir* presume that distinction in wine flavors may be detected according to the origin. The notion is vividly discussed in the wine press and winemakers should understand the myths associated with the terroir topic (Matthews, 2015). In its strictest definition, wine is distinguished only by factors depending on soil characteristics, but it can have a broad sense of typicality for a region or wine type (Burnham & Skilleas, 2012). These two extreme cases may be understood using Chardonnay from Burgundy, where two neighboring vine plots provide different wines, or a blend of wines representing the vast Australian country or the multiple Douro valley terroirs.

The terroir concept depends on producing a wine with aesthetic value (Burnham & Skilleas, 2012), being questionable if it should be claimed for dry wines obtained from overmatured grapes in deep uniform soils with excessive irrigation (Van Leeuwen & Seguin, 2006).

Another case where the strict terroir concept does not apply is related to highly reputed wines, like Port, Madeira, Sherry, or Champagne. The case of noble-rot wines (e.g., Tokay, Sauternes) or ice wines even challenges the requirement for sound grapes in conventional winemaking. The technological options (e.g., fortification, aging under pellicle, second fermentation, processing frozen grapes) are indispensable to modify the base wine into the required character, including blending from different vintages. These manipulations induce a change in the base wine character stronger than those required to meet the commercial market demands for dry still wines. However, they fulfill all the criteria for aesthetic appreciation and are an example of long-lasting luxury products in the wine business. The question may be circumvented by introducing the human factor in the definition of the terroir concept (Charters, 2006).

29.5.2.6 Gastronomic potential

Food pairing has hardly been addressed in wine sensory research (Spence, 2020). The question of food pairing should also be in the mind of the winemaker to understand the full range of wine aptitudes. There are wines that do not require food to be appreciated and others that only show their qualities under appropriate culinary preparations (Harrington, 2008). In particular, those that have apparent "deficiencies" (e.g., bitterness, sourness, excessive astringency, less body) benefit from food accompaniment. The factors affecting this parameter have been studied less and involve numerous variables with strong cultural habituation (Eschevins, Giboreau, Julien, & Dacremont, 2019; Kustos, Heymann, Jeffery, Goodman, & Bastian, 2020).

29.5.2.7 The price

A record that ends with the target price of the wine may illustrate the systematization of the previously described parameters. Commercial wines logically follow the market price, and exclusive wines are in the range of luxury products (Thach, Charters, & Cogan-Marie, 2018). The question is how to persuade opinion leaders of their uniqueness (Humphreys & Carpenter, 2018). Here begins a new challenge with which winemakers are being increasingly faced that is likely to be successfully accomplished with a holistic perception of all factors involved in wine production (Fig. 29.7).

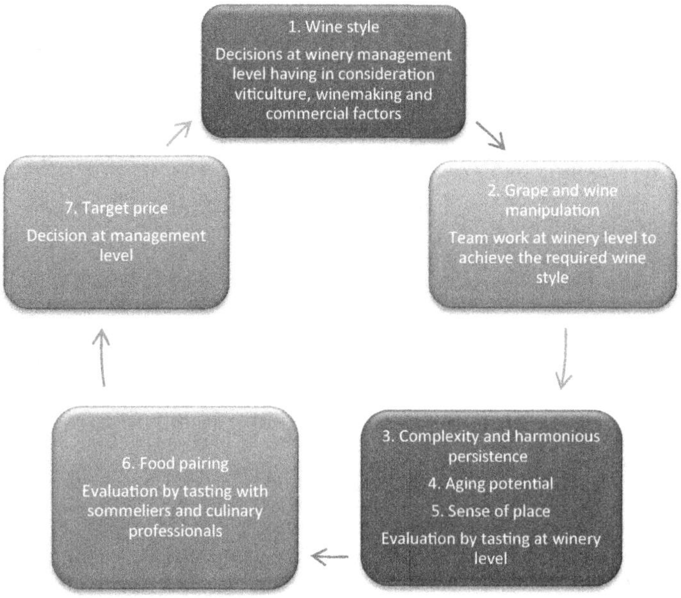

FIGURE 29.7   A holistic approach to wine production.

29.6 Final considerations

The evolution of wine tasting and sensory analysis has been remarkable in recent years, but the gap between the scientific achievements and their implications on wine understanding outside scholars' world has apparently increased. The consequences derive mainly from the use of methods that require extensive training and are not adequate to communicate the different aspects of the wine quality to the consumer. The issue of commercial wine quality directed to fit the purpose of the global wine market has been deeply and successfully addressed. However, when it comes to fine wines, the analytical sensory approaches have shortcomings that need to be tackled by the scientific community at the risk that wine evaluation would no longer require science. The evolution of sensory methods benefiting from the expertise of winemakers and *connoisseurs* is most welcome and should thus be stimulated, together with emotion-based methods adapted to consumers' education.

The holistic approach to wine tasting is also required for wine science given that its multiple facets depend on a broad range of disciplines. The extraordinary developments in each of the scientific fields have the drawback of increasing specialization without a global approach that may capture the wine *gestalt*, a German word in psychology where understanding the whole must precede the analysis of the parts (Millar, 2019). These pitfalls have been addressed by Wilk (2012), concerned with the actual limits of food studies that demand a more synthetic and empirical, i.e., holistic, approach. The remarks raised by this author are particularly valid for wine science, where all the scholars involved in Enology ("the science of wine") and other fields (e.g., psychology, sociology, history, geography, anthropology) would benefit from a close interaction leading to mutual understanding. Otherwise, the progress in new approaches to tackle the worldwide interest in wine (Tiefenbacher & Townsend, 2020) barely requires those directly involved in the central core of Enology. This interdisciplinary approach could be the body of *Enosophy* ("the knowledge of wine"), a new discipline embracing all the areas related to wine and reflecting its increasing significance in the present globalized world.

References

Ailer, Š., Valšíková, M., Jedlička, J., Mankovecký, J., & Baro, M. (2020). Influence of sugar and ethanol content and color of wines on the sensory evaluation: From wine competition "Nemčiňany" wine days. *In Slovak Republic (2013–2016)*. Erwerbs-Obstbau. Available from https://doi.org/10.1007/s10341−020−00486-x.

Aleixandre-Tudo, J. L., & du Toit, W. (2020). A chemometric approach to the evaluation of the ageing ability of red wines. *Chemometrics and Intelligent Laboratory Systems, 203*. Available from https://doi.org/10.1016/j.chemolab.2020.104067.

Alexandre, H., & Ballester, J. (2019). *Petit traité de mythologie œnologique*. Les Éditions Du Net. St. Ouen.

Amerine, M., & Roessler, E. (1976). *Wines, their sensory evaluation*. Freeman Press.

Antunes, F. (2018). *Consumer preference for warm or cool climate wine styles is dependent on emotional responses and familiarity*. American Association on Intellectual and Developmental Disabilities (AAIDD).

Antonelli, A., Arfelli, G., Masino, F., & Sartini, E. (2010). Comparison of traditional and reductive winemaking: Influence on some fixed components and sensorial characteristics. *European Food Research and Technology*, 231, 85−91. Available from https://doi.org/10.1007/s00217-010-1250-6.

Apaolaza, V., Hartmann, P., Echebarria, C., & Barrutia, J. M. (2017). Organic label's halo effect on sensory and hedonic experience of wine: A pilot study. *Journal of Sensory Studies*, 32(1). Available from https://doi.org/10.1111/joss.12243.

Ares, A., & Varela, P. (2018). Methods in consumer research, (1st (ed.)). (Vol. 1 and 2). Cambridge: Woodhead Pub.

Ashenfelter, O., & Quandt, R. (2012). Analyzing a wine tasting statistically. *Chance*, 12, 16−20.

Ashton, R. (2017). Dimensions of expertise in wine evaluation. *Journal of Wine Economics*, 59−83. Available from https://doi.org/10.1017/jwe.2016.27.

Barton, A., Hayward, L., Richardson, C. D., & McSweeney, M. B. (2020). Use of different panellists (experienced, trained, consumers and experts) and the projective mapping task to evaluate white wine. *Food Quality and Preference*, 83. Available from https://doi.org/10.1016/j.foodqual.2020.103900.

Barwich, A. S. (2017). Up the nose of the beholder? Aesthetic perception in olfaction as a decision-making process. *New Ideas in Psychology*, 47, 157−165. Available from https://doi.org/10.1016/j.newideapsych.2017.03.013.

Beltran, G., Novo, M., Guillamón, J. M., Mas, A., & Rozès, N. (2008). Effect of fermentation temperature and culture media on the yeast lipid composition and wine volatile compounds. *International Journal of Food Microbiology*, 121, 169−177. Available from https://doi.org/10.1016/j.ijfoodmicro.2007.11.030.

Barwich, A. S. (2019). A Critique of olfactory objects. *Frontiers in Psychology*, 10. Available from https://doi.org/10.3389/fpsyg.2019.01337.

Blackman, J., Saliba, A., & Schmidtke, L. (2010). Sweetness acceptance of novices, experienced consumers and winemakers in hunter valley semillon wines. *Food Quality and Preference*, 21, 679−683. Available from https://doi.org/10.1016/j.foodqual.2010.05.001.

Bodington, J., & Malfeito-Ferreira, M. (2017). The 2016 wines of Portugal challenge: General implications of more than 8400 wine-score observations. *Journal of Wine Research*, 28(4), 313−325. Available from https://doi.org/10.1080/09571264.2017.1392291.

Bodington, J., & Malfeito-Ferreira, M. (2019). Should ties be broken in commercial wine competitions? when yes, what method is practical and defensible? *Journal of Wine Economics*, 14(3), 298−308. Available from https://doi.org/10.1017/jwe.2019.35.

Boncinellia, F., Casinia, L., Continia, C., Gerinia, F., & Scozzafava, G. (2016). The consumer loves typicality but prefers the international wine. *Agriculture and Agricultural Science Procedia*, 8, 236−242.

Brand, J., Valentin, D., Kidd, M., Vivier, M. A., Næs, T., & Nieuwoudt, H. H. (2020). Comparison of pivot profile© to frequency of attribute citation: Analysis of complex products with trained assessors. *Food Quality and Preference*, 84. Available from https://doi.org/10.1016/j.foodqual.2020.103921.

Broadbent, M. (1979). Pocket guide to winetasting.

Burnham, D., & Skilleas, O. (2012). The aesthethics of wines.

Campo, E., Do, B., Ferreira, V., & Valentin, D. (2008). Aroma properties of young Spanish monovarietal white wines: A study using sorting task, list of terms and frequency of citation. *Australian Journal of Grape and Wine Research*, 14(2), 104−115. Available from https://doi.org/10.1111/j.1755-0238.2008.00010.x.

Capehart, K. W. (2019). Does blind tasting work? Another look. *Journal of Wine Economics*, 14(3), 309−320. Available from https://doi.org/10.1017/jwe.2019.25.

Charters. (2003). *Perceptions of wine quality*. Edith Cowan Univeristy.

Charters, S. (2006). *Wine & society: The social and cultural context of a drink*. Routledge.

Carstens, E., Carstens, M., Dessirier, J., O'Mahony, M., Simons, C. T., Sudo, M., & Sudo, S. (2002). It hurts so good: Oral irritation by spices and carbonated drinks and the underlying neural mechanisms. *Food Quality and Preference*, 13(7−8), 431−443. Available from https://doi.org/10.1016/S0950-3293(01)00067-2.

Cejudo-Bastante, M. J., Castro-Vázquez, L., Hermosín-Gutiérrez, I., & Pérez-Coello, M. S. (2011). Combined effects of prefermentative skin maceration and oxygen addition of must on color-related phenolics, volatile composition, and sensory characteristics of airén white wine. *Journal of Agricultural and Food Chemistry*, 59, 12171−12182. Available from https://doi.org/10.1021/jf202679y.

Charters, S., & Pettigrew, S. (2007). The dimensions of wine quality. *Food Quality and Preference*, 18(7), 997−1007. Available from https://doi.org/10.1016/j.foodqual.2007.04.003.

Chatterjee, A. (2014). *The aesthetic brain: How we evolved to desire beauty and enjoy art*. USA: Oxford University Press.

Chen, C. F., Zou, D. J., Altomare, C. G., Xu, L., Greer, C. A., & Firestein, S. J. (2014). Nonsensory target-dependent organization of piriform cortex. *Proceedings of the National Academy of Sciences of the United States of America*, 111(47), 16931−16936. Available from https://doi.org/10.1073/pnas.1411266111.

Chen, J. (2020). It is important to differentiate sensory property from the material property. *Trends in Food Science and Technology*, 96, 268−270. Available from https://doi.org/10.1016/j.tifs.2019.12.014.

Cheynier, V., & Sarni-Manchado, P. (2010). Wine taste and mouthfeel. *In Managing wine quality: viticulture and wine quality* (pp. 29−72). France: Elsevier Inc. Available from https://doi.org/10.1533/9781845699284.1.30.

Civille, G. V., & Oftedal, K. N. (2012). Sensory evaluation techniques − Make good for you taste good. *Physiology and Behavior*, 107(4), 598−605. Available from https://doi.org/10.1016/j.physbeh.2012.04.015.

Claus, H., & Mojsov, K. (2018). Enzymes for wine fermentation: Current and perspective applications. *Fermentation*, 4(3). Available from https://doi.org/10.3390/fermentation4030052.

Cliff, M., & King, M. (1997). The evaluation of judges at wine competitions: The application of eggshell plots. *Journal of Wine Research*, 8(2), 75−80.

Cliff, M., & King, M. (1999). Use of principal component analysis for the evaluation of judge performance at wine competitions. *Journal of Wine Research*, 10(1), 25−32.

Coste, A., Sousa, P., & Malfeito-Ferreira, M. (2018). Wine tasting based on emotional responses: An expedite approach to distinguish between warm and cool climate dry red wine styles. *Food Research International*, 106, 11−21. Available from https://doi.org/10.1016/j.foodres.2017.12.039.

Criado, C., Chaya, C., Fernández-Ruíz, V., Álvarez, M. D., Herranz, B., & Pozo-Bayón, M. Á. (2019). Effect of saliva composition and flow on inter-individual differences in the temporal perception of retronasal aroma during wine tasting. *Food Research International, 126*. Available from https://doi.org/10.1016/j.foodres.2019.108677.

Croijmans, I., & Majid, A. (2016). Not all flavor expertise is equal: The language of wine and coffee experts. *PLoS One, 11*(6). Available from https://doi.org/10.1371/journal.pone.0155845.

Croijmans, I., Speed, L. J., Arshamian, A., & Majid, A. (2020). Expertise Shapes Multimodal Imagery for Wine. *Cognitive Science, 44*(5). Available from https://doi.org/10.1111/cogs.12842.

Culbert, J. A., Ristic, R., Ovington, L. A., Saliba, A. J., & Wilkinson, K. L. (2017). Influence of production method on the sensory profile and consumer acceptance of Australian sparkling white wine styles. *Australian Journal of Grape and Wine Research, 23*, 170−178. 10.1111/ajgw.12277.

Darias-Martín, J. J., Rodríguez, O., Díaz, E., & Lamuela-Raventós, R. M. (2000). Effect of skin contact on the antioxidant phenolics in white wine. *Food Chemistry, 71*, 483−487. Available from https://doi.org/10.1016/S0308-8146(00)00177-1.

Dehlholm, C., Brockhoff, P. B., Meinert, L., Aaslyng, M. D., & Bredie, W. L. P. (2012). Rapid descriptive sensory methods - Comparison of Free Multiple Sorting, Partial Napping, Napping, Flash Profiling and conventional profiling. *Food Quality and Preference, 26*(2), 267−277. Available from https://doi.org/10.1016/j.foodqual.2012.02.012.

Desmet, P. M. A., & Schifferstein, H. N. J. (2008). Sources of positive and negative emotions in food experience. *Appetite, 50*(2−3), 290−301. Available from https://doi.org/10.1016/j.appet.2007.08.003.

Doty, R. (2015). Introduction and Historical Perspective (Ch. 1). In: Doty, R. (ed.) Handbook of Olfaction and Gustation, pp 1−36. John Wiley & Sons, Inc., New Jersey. Available from https://doi.org/10.1002/97811189717582015.

Edwards, M. (2010). *Fragrances of the world: Parfums du Monde.*

Eriksson, N., Wu, S., Do, C. B., Kiefer, A. K., Tung, J. Y., & Mountain, J. L. (2012). A genetic variant near olfactory receptor genes influences cilantro preference. *Flavour, 1*, 22. Available from https://doi.org/10.1186/2044-7248-1-22.

Eschevins, A., Giboreau, A., Julien, P., & Dacremont, C. (2019). From expert knowledge and sensory science to a general model of food and beverage pairing with wine and beer. *International Journal of Gastronomy and Food Science, 17*. Available from https://doi.org/10.1016/j.ijgfs.2019.100144.

Ferreira, V. (2010). Volatile aroma compounds and wine sensory attributes. *In Managing wine quality: Viticulture and wine quality* (pp. 3−28). Spain: Elsevier Inc. Available from https://doi.org/10.1533/9781845699284.1.3.

Ferreira, V., Sáenz-Navajas, M. P., Campo, E., Herrero, P., de la Fuente, A., & Fernández-Zurbano, P. (2016). Sensory interactions between six common aroma vectors explain four main red wine aroma nuances. *Food Chemistry, 199*, 447−456. Available from https://doi.org/10.1016/j.foodchem.2015.12.048.

Frank, M. E., Fletcher, D. B., & Hettinger, T. P. (2017). Recognition of the component odors in mixtures. *Chemical Senses, 42*(7), 537−546. Available from https://doi.org/10.1093/chemse/bjx031.

Ganić, K. K., Staver, M., Peršurić, Đ., Banović, M., Komes, D., & Gracin, L. (2003). Influence of blending on the aroma of malvasia istriana wine. *Food Technology and Biotechnology, 41*, 305−314.

Gawel, R., Schulkin, A., Smith, P. A., Espinase, D., & McRae, J. M. (2020). Effect of dissolved carbon dioxide on the sensory properties of still white and red wines. *Australian Journal of Grape and Wine Research, 26*, 172−179. Available from https://doi.org/10.1111/ajgw.12429.

Gawel, R., Smith, P. A., Cicerale, S., & Keast, R. (2018). The mouthfeel of white wine. *Critical Reviews in Food Science and Nutrition, 58*(17), 2939−2956. Available from https://doi.org/10.1080/10408398.2017.1346584.

Gawel, R., Smith, P. A., & Waters, E. J. (2016). Influence of polysaccharides on the taste and mouthfeel of white wine. *Australian Journal of Grape and Wine Research, 22*, 350−357. 10.1111/ajgw.12222.

González-Centeno, M. R., Chira, K., & Teissedre, P. L. (2019). Use of oak wood during malolactic fermentation and ageing: Impact on chardonnay wine character. *Food Chemistry, 278*, 460−468. Available from https://doi.org/10.1016/j.foodchem.2018.11.049.

Goodstein, E. S., Bohlscheid, J. C., Evans, M., & Ross, C. F. (2014). Perception of flavor finish in model white wine: A time-intensity study. *Food Quality and Preference, 36*, 50−60. Available from https://doi.org/10.1016/j.foodqual.2014.02.012.

Grohmann, B., Peña, C., & Joy, A. (2018). Wine quality and sensory assessments: Do distinct local groups of wine experts differ? *Journal of Wine Research, 29*(4), 278−289. Available from https://doi.org/10.1080/09571264.2018.1532882.

Gros, J., Lavigne, V., Thibaud, F., Gammacurta, M., Moine, V., Dubourdieu, D., & Marchal, A. (2017). Toward a molecular understanding of the typicality of chardonnay wines: Identification of powerful aromatic compounds reminiscent of hazelnut. *Journal of Agricultural and Food Chemistry, 65*(5), 1058−1069. Available from https://doi.org/10.1021/acs.jafc.6b04516.

Guld, Z., Nyitrainé Sárdy, D., Gere, A., & Rácz, A. (2020). Comparison of sensory evaluation techniques for Hungarian wines. *Journal of Chemometrics, 34*(4). Available from https://doi.org/10.1002/cem.3219.

Haddad, R., Medhanie, A., Roth, Y., Harel, D., & Sobel, N. (2010). Predicting odor pleasantness with an electronic nose. *PLoS Computational Biology, 6*(4). Available from https://doi.org/10.1371/journal.pcbi.1000740.

Hannum, M., Stegman, M. A., Fryer, J. A., & Simons, C. T. (2018). Different olfactory percepts evoked by orthonasal and retronasal odorant delivery. *Chemical Senses, 43*(7), 515−521. Available from https://doi.org/10.1093/chemse/bjy043.

Harrington, R. (2008). *Food and wine pairing: A sensory experience.* John Wiley & Sons.

Hayward, L., Jantzi, H., Smith, A., & McSweeney, M. B. (2020). How do consumers describe cool climate wines using projective mapping and ultra-flash profile? *Food Quality and Preference, 86*. Available from https://doi.org/10.1016/j.foodqual.2020.104026.

Herbst, F., & Arnim. (2009). The role and influence of wine awards as perceived by the South African wine consumers. *Acta Commercii, 9*, 90−101.

Herz, R. S., & Von Clef, J. (2001). The influence of verbal labeling on the perception of odors: Evidence for olfactory illusions? *Perception, 30*(3), 381−391. Available from https://doi.org/10.1068/p3179.

Heymann, H., & Ebeler, S. E. (2016). Sensory and Instrumental Evaluation of Alcoholic Beverages. *Sensory and instrumental evaluation of alcoholic beverages* (pp. 1−265). Elsevier Inc. undefined: Retrieved from. Available from http://www.sciencedirect.com/science/book/9780128027271.

Hodgson, R. (2008). An examination of judge reliability at a major United States wine competition. *Journal of Wine Economics*, 105−113. Available from https://doi.org/10.1017/s1931436100001152.

Honoré-Chedozeau, C., Ballester, J., Chatelet, B., & Lempereur, V. (2015). Wine competition: From between-juries consistency to sensory perception of consumers. *Bio Web of Conferences, 5*, 03009.

Honoré-Chedozeau, C., Chollet, S., Lelièvre-Desmas, M., Ballester, J., & Valentin, D. (2020). From perceptual to conceptual categorization of wines: What is the effect of expertise? *Food Quality and Preference, 80*. Available from https://doi.org/10.1016/j.foodqual.2019.103806.

Hopfer, H., Nelson, J., Ebeler, S. E., & Heymann, H. (2015). Correlating wine quality indicators to chemical and sensory measurements. *Molecules (Basel, Switzerland), 20*(5), 8453–8483. Available from https://doi.org/10.3390/molecules20058453.

Hughson, A. L., & Boakes, R. A. (2002). The knowing nose: The role of knowledge in wine expertise. *Food Quality and Preference, 13*(7–8), 463–472. Available from https://doi.org/10.1016/S0950-3293(02)00051-4.

Humphreys, A., & Carpenter, G. S. (2018). Status Games: Market driving through social influence in the United States Wine industry. *Journal of Marketing, 82*(5), 141–159. Available from https://doi.org/10.1509/jm.16.0179.

Jackson, R. (2017). *Wine tasting: A professional handbook.* Elsevier.

Jaeger, S. R., Roigard, C. M., Le Blond, M., Hedderley, D. I., & Giacalone, D. (2019). Perceived situational appropriateness for foods and beverages: consumer segmentation and relationship with stated liking. *Food Quality and Preference, 78*. Available from https://doi.org/10.1016/j.foodqual.2019.05.001.

Jantzi, H., Hayward, L., Barton, A., Richardson, C. D., & McSweeney, M. B. (2020). Investigating the effect of extrinsic cues on consumers' evaluation of red wine using a projective mapping task. *Journal of Sensory Studies, 35*(3). Available from https://doi.org/10.1111/joss.12568.

Karbowiak, T., Crouvisier-Urion, K., Lagorce, A., Ballester, J., Geoffroy, A., Roullier-Gall, C., & Bellat, J. (2019). Wine aging: A bottleneck story. *Npj Science of Food, 3*, 10.1038/s41538-019-0045-9.

Keller, A. (2017). The distinction between perception and judgment, if there is one, is not clear and intuitive. *Behavioral and Brain Sciences, 39*.

Keller, A., & Vosshall, L. B. (2016). Olfactory perception of chemically diverse molecules. *BMC Neuroscience, 17*(1). Available from https://doi.org/10.1186/s12868-016-0287-2.

King, E. S., Dunn, R. L., & Heymann, H. (2013). The influence of alcohol on the sensory perception of red wines. *Food Quality and Preference, 28*(1), 235–243. Available from https://doi.org/10.1016/j.foodqual.2012.08.013.

Koenig, L., Coulon-Leroy, C., Symoneaux, R., Cariou, V., & Vigneau, E. (2020). Influence of expertise on semantic categorization of wine odors. *Food Quality and Preference, 83*. Available from https://doi.org/10.1016/j.foodqual.2020.103923.

Köster, E. P., & Mojet, J. (2015). From mood to food and from food to mood: A psychological perspective on the measurement of food-related emotions in consumer research. *Food Research International, 76*(2), 180–191. Available from https://doi.org/10.1016/j.foodres.2015.04.006.

Köster, E. P., Van Der Stelt, O., Nixdorf, R. R., Linschoten, M. R. I., De Wijk, R. A., & Mojet, J. (2014). Olfactory imagination and odor processing: Three same-different experiments. *Chemosensory Perception, 7*(2), 68–84. Available from https://doi.org/10.1007/s12078-014-9165-4.

Kustos, M., Heymann, H., Jeffery, D. W., Goodman, S., & Bastian, S. E. P. (2020). Intertwined: What makes food and wine pairings appropriate? *Food Research International, 136*. Available from https://doi.org/10.1016/j.foodres.2020.109463.

Laing, D., & Francis, G. (1989). The capacity of humans to identify odors in mixtures. *Physiology & Behavior*, 809–814. Available from https://doi.org/10.1016/0031-9384(89)90041-3.

Laing, D., Link, C., Jinks, A. L., & Hutchinson, I. (2002). The limited capacity of humans to identify the components of taste mixtures and taste-odour mixtures. *Perception, 31*(5), 617–635. Available from https://doi.org/10.1068/p3205.

Langlois, J., Ballester, J., Campo, E., Dacremont, C., & Peyron, D. (2010). Combining olfactory and gustatory clues in the judgment of aging potential of red wine by wine professionals. *American Journal of Enology and Viticulture, 61*, 15–22.

Lasik-Kurdys, M., Majcher, M., & Nowak, J. (2018). Effects of different techniques of malolactic fermentation induction on diacetyl metabolism and biosynthesis of selected aromatic esters in cool-climate grape wines. *Molecules, 23*, 10.3390/molecules23102549.

Léglise, M. (1984). *Une initiation à la dégustation des grands vins.* Jeanne Laffitte.

Lelièvre, M., Chollet, S., Abdi, H., & Valentin, D. (2008). What is the validity of the sorting task for describing beers? A study using trained and untrained assessors. *Food Quality and Preference, 19*(8), 697–703. Available from https://doi.org/10.1016/j.foodqual.2008.05.001.

Lesschaeve, I., & Noble, A. C. (2010). *Sensory analysis of wine. In Managing wine quality: Viticulture and wine quality* (pp. 189–217). Canada: Elsevier Inc. Available from https://doi.org/10.1533/9781845699284.2.189.

Lezaeta, A., Bordeu, E., Agosin, E., Pérez-Correa, J. R., & Varela, P. (2018). White wines aroma recovery and enrichment: Sensory-led aroma selection and consumer perception. *Food Research International, 108*, 595–603. Available from https://doi.org/10.1016/j.foodres.2018.03.044.

Liberatore, M., Pati, S., Nobile, M., & Notte, E. (2010). Aroma quality improvement of chardonnay white wine by fermentation and ageing in barrique on lees. *Food Research International, 43*, 996–1002. Available from https://doi.org/10.1016/j.foodres.2010.01.007.

Loureiro, V., Brasil, R., & Malfeito-Ferreira, M. (2016). A New wine tasting approach based on emotional responses to rapidly recognize classic European wine styles. *Beverages, 2*(1), 6. Available from https://doi.org/10.3390/beverages2010006.

Lundström, J. N., Gordon, A. R., Wise, P., & Frasnelli, J. (2012). Individual differences in the chemical senses: Is there a common sensitivity? *Chemical Senses, 37*(4), 371–378. Available from https://doi.org/10.1093/chemse/bjr114.

Mackenzie E., Hannum, S., Forzley, R., Popper, C., & Simons, T. (2020). Further validation of the engagement questionnaire (EQ): Do immersive technologies actually increase consumer engagement during wine evaluations? *Food Quality and Preference, 85*, 103966, Available from https://doi.org/10.1016/j.foodqual.2020.103966.

Mahieu, B., Visalli, M., Thomas, A., & Schlich, P. (2020). Free-comment outperformed check-all-that-apply in the sensory characterisation of wines with consumers at home. *Food Quality and Preference, 84*, 103937. Available from https://doi.org/10.1016/j.foodqual.2020.103937.

Malfeito-Ferreira, M. (2018). Two decades of "horse sweat" taint and brettanomyces yeasts in wine: Where do we stand now? *Beverages, 32*. Available from https://doi.org/10.3390/beverages4020032.

Malfeito-Ferreira, M., Diako, C., & Ross, C. F. (2019). Sensory and chemical characteristics of 'dry' wines awarded gold medals in an international wine competition. *Journal of Wine Research, 30*(3), 204–219. Available from https://doi.org/10.1080/09571264.2019.1652154.

Marks, D. (2015). Seeking the Veritas about the Vino: fine wine ratings as wine knowledge. *Journal of Wine Research, 26*(4), 319–335. Available from https://doi.org/10.1080/09571264.2015.1083953.

Marshall, K., Laing, D. G., Jinks, A. L., & Hutchinson, I. (2006). The capacity of humans to identify components in complex odor-taste mixtures. *Chemical Senses, 31*(6), 539–545. Available from https://doi.org/10.1093/chemse/bjj058.

Matthews, M. (2015). *Terroir and other myths of Winegrowing*. Univeristy of California Press.

McGann, J. P. (2017). Poor human olfaction is a 19th-century myth. *Science (New York, N.Y.)*, *356*(6338). Available from https://doi.org/10.1126/science.aam7263.

McGovern, P. E. (2019). *Ancient wine: The search for the origins of viniculture*. Princeton Univeristy.

McKay, M., & Buica, A. (2020). Factors influencing olfactory perception of selected off-flavourcausing compounds in red wine-A review. *South African Journal of Enology and Viticulture*, *41*(1), 56−71. Available from https://doi.org/10.21548/41-1-3669.

Melcher, J. M., & Schooler, J. W. (1996). The misremembrance of wines past: Verbal and perceptual expertise differentially mediate verbal overshadowing of taste memory. *Journal of Memory and Language*, *35*(2), 231−245. Available from https://doi.org/10.1006/jmla.1996.0013.

Millar, B. (2019). Smelling objects. *Synthese*, *196*(10), 4279−4303. Available from https://doi.org/10.1007/s11229-017-1657-8.

Mora, M., Urdaneta, E., & Chaya, C. (2018). Emotional response to wine: Sensory properties, age and gender as drivers of consumers' preferences. *Food Quality and Preference*, *66*, 19−28. Available from https://doi.org/10.1016/j.foodqual.2017.12.015.

Morata, A., Escott, C., Bañuelos, M. A., Loira, I., Del Fresno, J. M., González, C., & Suárez-lepe, J. A. (2020). Contribution of non-saccharomyces yeasts to wine freshness. A review. *Biomolecules*, *10*(1). Available from https://doi.org/10.3390/biom10010034.

Morrot, G., Brochet, F., & Dubourdieu, D. (2001). The color of odors. *Brain and Language*, *79*(2), 309−320. Available from https://doi.org/10.1006/brln.2001.2493.

Murray, M. M., & Wallace, M. T. (2011). The neural bases of multisensory processes. *The neural bases of multisensory processes*, 978−143981219.

Niimi, J., Danner, L., & Bastian, S. E. (2019). Wine leads us by our heart not our head: emotions and the wine consumer. *Current Opinion in Food Science*, *27*, 23−28. Available from https://doi.org/10.1016/j.cofs.2019.04.008.

Noble, A., Arnold, R., Masuda, B., Pecore, S., Schmidt, J., & Stern, P. (1984). Progress towards a standardized system of wine aroma terminology. *American Journal of Enology and Viticulture*, *35*, 107−109.

Norris, L. & Lee, T. (2002). How do flavour and quality of wine relate? Research Paper for Flavour Sense and E. J. Gallo Winery, Modesto, CA (retrieved from www.citeseerx.ist.psu.edu on the 10th July 2020).

Orth, U. R., & Bourrain, A. (2005). Ambient scent and consumer exploratory behaviour: A causal analysis. *Journal of Wine Research*, *16*(2), 137−150. 10.1080/09571260500327671.

Palczak, J., Blumenthal, D., Rogeaux, M., & Delarue, J. (2019). Sensory complexity and its influence on hedonic responses: A systematic review of applications in food and beverages. *Food Quality and Preference*, *71*, 66−75. Available from https://doi.org/10.1016/j.foodqual.2018.06.002.

Pappalardo, G., Selvaggi, R., Pecorino, B., Lee, J. Y., & Nayga, R. M. (2019). Assessing experiential augmentation of the environment in the valuation of wine: Evidence from an economic experiment in mt. etna, italy. *Psychology and Marketing*, *36*(6), 642−654. 10.1002/mar.21202.

Parker, M., Onetto, C., Hixson, J., Bilogrevic, E., Schueth, L., Pisaniello, L., & Francis, L. (2019). Factors contributing to interindividual variation in retronasal odour perception from aroma glycosides: The role of odourant sensory detection threshold, oral microbiota, and hydrolysis in saliva. *Journal of Agricultural and Food Chemistry*, 10.1021/acs.jafc.9b05450.

Parr, W., Maltman, A., Easton, S., & Ballester, J. (2018). Minerality in Wine: Towards the Reality behind the Myths. *Beverages*, *77*. Available from https://doi.org/10.3390/beverages4040077.

Parr, W. V. (2019). Demystifying wine tasting: Cognitive psychology's contribution. *Food Research International*, *124*, 230−233. Available from https://doi.org/10.1016/j.foodres.2018.03.050.

Parr, W. V., Ballester, J., Peyron, D., Grose, C., & Valentin, D. (2015). Perceived minerality in Sauvignon wines: Influence of culture and perception mode. *Food Quality and Preference*, *41*, 121−132. Available from https://doi.org/10.1016/j.foodqual.2014.12.001.

Parr, W. V., Green, J. A., & Geoffrey White, K. (2006). Wine judging, context and New Zealand Sauvignon blanc. *Revue Europeenne de Psychologie Appliquee*, *56*(4), 231−238. Available from https://doi.org/10.1016/j.erap.2005.09.011.

Parr, W. V., Green, J. A., White, K. G., & Sherlock, R. R. (2007). The distinctive flavour of New Zealand Sauvignon blanc: Sensory characterisation by wine professionals. *Food Quality and Preference*, *18*(6), 849−861. Available from https://doi.org/10.1016/j.foodqual.2007.02.001.

Parr, W. V., Grose, C., Hedderley, D., Medel Maraboli, M., Masters, O., Araujo, L. D., & Valentin, D. (2020). Perception of quality and complexity in wine and their links to varietal typicality: An investigation involving Pinot noir wine and professional tasters. *Food Research International*, *137*. Available from https://doi.org/10.1016/j.foodres.2020.109423.

Pati, S., Crupi, P., Savastano, M. L., Benucci, I., & Esti, M. (2020). Evolution of phenolic and volatile compounds during bottle storage of a white wine without added sulfite. *Journal of the Science of Food and Agriculture*, *100*(2), 775−784. Available from https://doi.org/10.1002/jsfa.10084.

Pearson, W., Schmidtke, L., Francis, I. L., & Blackman, J. W. (2020). An investigation of the Pivot© Profile sensory analysis method using wine experts: Comparison with descriptive analysis and results from two expert panels. *Food Quality and Preference*, *83*. Available from https://doi.org/10.1016/j.foodqual.2019.103858.

Peynaud, E. (1987). *The taste of wine: The art and science of wine appreciation*. San Francisco: The Wine Appreciation Guild Ltd.

Pickering, G. J., & Robert, G. (2006). Perception of mouthfeel sensations elicited by red wine are associated with sensitivity to 6-N-propylthiouracil. *Journal of Sensory Studies*, *21*(3), 249−265. 10.1111/j.1745-459X.2006.00065.x.

Pickering, G. J., Simunkova, K., & DiBattista, D. (2004). Intensity of taste and astringency sensations elicited by red wines is associated with sensitivity to PROP (6-n-propylthiouracil). *Food Quality and Preference*, *15*(2), 147−154. 10.1016/S0950-3293(03)00053-3.

Piqueras-Fiszman, B., & Spence, C. (2015). Sensory expectations based on product-extrinsic food cues: An interdisciplinary review of the empirical evidence and theoretical accounts. *Food Quality and Preference*, *40*, 165−179. Available from https://doi.org/10.1016/j.foodqual.2014.09.013.

Pons, A., Lavigne, V., Darriet, P., & Dubourdieu, D. (2013). Role of 3-methyl-2,4-nonanedione in the flavor of aged red wines. *Journal of Agricultural and Food Chemistry*, *61*(30), 7373−7380. Available from https://doi.org/10.1021/jf400348h.

Pozo-Bayón, M. Á., Muñoz-González, C., & Esteban-Fernández, A. (2016). *Wine preference and wine aroma perception. Wine safety, consumer preference, and human health* (pp. 139−162). Spain: Springer International Publishing. Available from https://doi.org/10.1007/978−3−319−24514-0_7.

Priilaid, D. (2007). The placebo of place: Terroir effects in the blind and sighted quality assessments of South African varietal wines. *Journal of Wine Research*, *18*(2), 87−106. Available from https://doi.org/10.1080/09571260701660862.

Priilaid, D., Human, G., Pitcher, K., Smith, T., & Varkel, C. (2017). Are consumers' quality perceptions influenced by brand familiarity, brand exposure and brand knowledge? Results from a wine tasting experiment. *South African Journal of Business Management*, *48*(2), 45–54. Available from https://doi.org/10.4102/sajbm.v48i2.27.

Ribéreau-Gayon, J., & Peynaud, E. (1958). *Analyse et contrôle des vins (Ch. 2)* (2nd ed., p. 262). Paris: Librairie Polytechnique CH. Béranger.

Ristic, R., Danner, L., Johnson, T. E., Meiselman, H. L., Hoek, A. C., Jiranek, V., & Bastian, S. E. P. (2019). Wine-related aromas for different seasons and occasions: Hedonic and emotional responses of wine consumers from Australia, UK and USA. *Food Quality and Preference*, *71*, 250–260. Available from https://doi.org/10.1016/j.foodqual.2018.07.011.

Robinson, J. (2008). *How to taste wine: A guide to enjoying wine*. Simon & Schuster.

Rochelle, M. M., Prévost, G. J., & Acree, T. E. (2018). Computing odor images. *Journal of Agricultural and Food Chemistry*, *66*(10), 2219–2225. Available from https://doi.org/10.1021/acs.jafc.6b05573.

Rodrigues, H., & Parr, W. V. (2019). Contribution of cross-cultural studies to understanding wine appreciation: A review. *Food Research International*, *115*, 251–258. Available from https://doi.org/10.1016/j.foodres.2018.09.008.

Romano, M., Chandra, M., Harutunyan, M., Savian, T., Villegas, C., Minim, V., & Malfeito-Ferreira, M. (2020). Off-flavours and unpleasantness are cues for the recognition and valorization of organic wines by experienced tasters. *Foods*, *9*(1). Available from https://doi.org/10.3390/foods9010105.

Royet, J. P., Plailly, J., Saive, A. L., Veyrac, A., & Delon-Martin, C. (2013). The impact of expertise in olfaction. *Frontiers in Psychology*, *4*. Available from https://doi.org/10.3389/fpsyg.2013.00928.

Ruivo, I. (2018). Sensory and preference evaluation of the addition of organic acids to white wines. Mater thesis. In *Viticulture and Enology*, University of Porto.

Sackris, D. C. (2019). What Jancis Robinson didn't know may have helped her. *Erkenntnis*, *84*(4), 805–822. Available from https://doi.org/10.1007/s10670-018-9981-z.

Sáenz-Navajas, M. P., Ballester, J., Fernández-Zurbano, P., Ferreira, V., Peyron, D., & Valentin, D. (2016). Wine quality perception: A sensory point of view. In *Wine safety, consumer preference, and human health* (pp. 119–138). Spain: Springer International Publishing. Available from https://doi.org/10.1007/978-3-319-24514-0_6.

Sáenz-Navajas, M. P., Campo, E., Avizcuri, J. M., Valentin, D., Fernández-Zurbano, P., & Ferreira, V. (2012). Contribution of non-volatile and aroma fractions to in-mouth sensory properties of red wines: Wine reconstitution strategies and sensory sorting task. *Analytica Chimica Acta*, *732*, 64–72. Available from https://doi.org/10.1016/j.ac.2011.12.042.

Saive, A.-L., Royet, J.-P., & Plailly, J. (2014). A review on the neural bases of episodic odour memory: from laboratory-based to autobiographical approaches. *Frontiers in Behavioral Neuroscience*, *8*, 240. = 10.3389/fnbeh.2014.00240.

Saliba, A. J., Wragg, K., & Richardson, P. (2009). Sweet taste preference and personality traits using a white wine. *Food Quality and Preference*, *20*, 572–575. Available from https://doi.org/10.1016/j.foodqual.2009.05.009.

Saltman, Y., Johnson, T. E., Wilkinson, K. L., Ristic, R., Norris, L. M., & Bastian, S. E. P. (2017). Natural flavor additives influence the sensory perception and consumer liking of Australian chardonnay and shiraz wines. *American Journal of Enology and Viticulture*, *68*(2), 243–251. Available from https://doi.org/10.5344/ajev.2016.16057.

Scaman, C. H., Dou, J., Cliff, M. A., Yuksel, D., & King, M. C. (2001). Evaluation of wine competition judge performance using principal component similarity analysis. *Journal of Sensory Studies*, *16*(3), 287–300. Available from https://doi.org/10.1111/j.1745-459X.2001.tb00302.x.

Schneider, V. (2014). Atypical aging defect: Sensory discrimination, Viticultural causes, and enological consequences. a review. *American Journal of Enology and Viticulture*, *65*(3), 277–284. Available from https://doi.org/10.5344/ajev.2014.14014.

Schulze, P., Bestgen, A. K., Lech, R. K., Kuchinke, L., & Suchan, B. (2017). Preprocessing of emotional visual information in the human piriform cortex. *Scientific Reports*, *7*(1). Available from https://doi.org/10.1038/s41598-017-09295-x.

Schuster, M. (2017). *Essential winetasting: The complete practical winetasting*. .. ISBN13: 9781845330200.

Sclafani, A., & Ackroff, K. (2012). Role of gut nutrient sensing in stimulating appetite and conditioning food preferences. *American Journal of Physiology — Regulatory Integrative and Comparative Physiology*, *302*, R1119–R1133. Available from https://doi.org/10.1152/ajpregu.00038.2012.

Sena-Esteves, M., Mota, M., & Malfeito-Ferreira, M. (2018). Patterns of sweetness preference in red wine according to consumer characterization. *Food Research International*, *106*, 38–44.

Shepherd, G. (2017). Neuroenology, how the brain creates the taste of wine. *Flavour*, *4*, 19. Available from https://doi.org/10.1186/s13411-014-0030-9.

Skov, M. (2019). Aesthetic appreciation: The view from neuroimaging. *Empirical Studies of the Arts*, *37*(2), 220–248. Available from https://doi.org/10.1177/0276237419839257.

Small, D. M. (2012). Flavor is in the brain. *Physiology and Behavior*, *107*(4), 540–552. Available from https://doi.org/10.1016/j.physbeh.2012.04.011.

Small, D. M., & Green, B. G. (2012). A proposed model of a flavor modality. In *The neural bases of multisensory processes*. Boca Raton, FL: CRC Press/Taylor & Francis.

Small-Kelly, S., & Pickering, G. (2020). Variation in orosensory responsiveness to alcoholic beverages and their Constituents—The role of the thermal taste phenotype. *Chemosensory Perception*, *13*, 45–58. Available from https://doi.org/10.1007/s12078-019-09266-8.

Smith, A. M., & McSweeney, M. B. (2019). Partial projective mapping and ultra-flash profile with and without red light: A case study with white wine. *Journal of Sensory Studies*, *34*(5). Available from https://doi.org/10.1111/joss.12528.

Smith, B. C. (2019). Getting more out of wine: Wine experts, wine apps and sensory science. *Current Opinion in Food Science*, *27*, 123–129. Available from https://doi.org/10.1016/j.cofs.2019.10.007.

Snitz, K., Yablonka, A., Weiss, T., Frumin, I., Khan, R. M., & Sobel, N. (2013). Predicting odor perceptual similarity from odor structure. *PLoS Computational Biology*, *9*(9). Available from https://doi.org/10.1371/journal.pcbi.1003184.

Souza-Coutinho, M., Brasil, R., Souza, C., Sousa, P., & Malfeito-Ferreira, M. (2020). Consumers associate high-quality (fine)wines with complexity, persistence, and unpleasant emotional responses. *Foods*, *9*(4). Available from https://doi.org/10.3390/foods9040452.

Spence, C. (2016). Oral referral: On the mislocalization of odours to the mouth. *Food Quality and Preference, 50*, 117—128. Available from https://doi.org/10.1016/j.foodqual.2016.02.006.

Spence, C. (2020). Wine psychology: Basic & applied. *Cognitive Research: Principles and Implications, 5*(1). Available from https://doi.org/10.1186/s41235-020-00225-6.

Spence, C., & Wang, Q. J. (2019). Wine expertise: Perceptual learning in the chemical senses. *Current Opinion in Food Science, 27*, 49—56. Available from https://doi.org/10.1016/j.cofs.2019.05.003.

Spinelli, S., Monteleone, E., Ares, G., & Jaeger, S. R. (2019). Sensory drivers of product-elicited emotions are moderated by liking: Insights from consumer segmentation. *Food Quality and Preference, 78*. Available from https://doi.org/10.1016/j.foodqual.2019.103725.

Tempère, S., Cuzange, E., Malak, J., Bougeant, J. C., De Revel, G., & Sicard, G. (2011). The training level of experts influences their detection thresholds for key wine compounds. *Chemosensory Perception, 4*, 99—115. 10.1007/s12078-011-9090-8.

Tempère, S., de Revel, G., & Sicard, G. (2019). Impact of learning and training on wine expertise: A review. *Current Opinion in Food Science, 27*, 98—103. Available from https://doi.org/10.1016/j.cofs.2019.07.001.

Thach, L., Charters, S., & Cogan-Marie, L. (2018). Core tensions in luxury wine marketing: The case of Burgundian wineries. *International Journal of Wine Business Research, 30*(3), 343—365. Available from https://doi.org/10.1108/IJWBR-04-2017-0025.

Thuillier, B., Valentin, D., Marchal, R., & Dacremont, C. (2015). Pivot© profile: A new descriptive method based on free description. *Food Quality and Preference, 42*, 66—77. 10.1016/j.foodqual.2015.01.012.

Tiefenbacher, J. P., & Townsend, C. (2020). The semiofoodscape of wine: The changing global landscape of wine culture and the language of making, selling, and drinking wine. *Handbook of the changing world language map*, 1—44. Available from https://doi.org/10.1007/978-3-319-73400-2_213-2.

Torija, M. J., Beltran, G., Novo, M., Poblet, M., Guillamón, J. M., Mas, A., & Rozès, N. (2003). Effects of fermentation temperature and saccharomyces species on the cell fatty acid composition and presence of volatile compounds in wine. *International Journal of Food Microbiology, 85*, 127—136. 10.1016/S0168-1605(02)00506-8.

Van Leeuwen, C., & Seguin, G. (2006). The concept of terroir in viticulture. *Journal of Wine Research, 17*(1), 1—10. Available from https://doi.org/10.1080/09571260600633135.

Van Wyk, N., Grossmann, M., Wendland, J., Von Wallbrunn, C., & Pretorius, I. S. (2019). The whiff of wine yeast innovation: Strategies for enhancing aroma production by yeast during wine fermentation. *Journal of Agricultural and Food Chemistry, 67*, 13496—13505. 10.1021/acs.jafc.9b06191.

Vitorino, G. (2018). *Effect of wine flavour on the perception of wine taste and preference. Master Thesis.* Instituto Superior de Agronomia, Universidade de Lisboa.

Wang, Q. J., & Prešern, D. (2018). Does blind tasting work? Investigating the impact of training on blind tasting accuracy and wine preference. *Journal of Wine Economics, 13*(4), 375—383. Available from https://doi.org/10.1017/jwe.2018.36.

Webb, J., Bolhuis, D., Cicerale, S., Hayes, J., & Keast, R. (2015). The relationships between common measurements of taste function. *Chemical Perception, 8*, 11—18.

Weiss, T., Snitz, K., Yablonka, A., Khan, R. M., Gafsou, D., Schneidman, E., & Sobel, N. (2012). Perceptual convergence of multi-component mixtures in olfaction implies an olfactory white. *Proceedings of the National Academy of Sciences of the United States of America, 109*(49), 19959—19964. Available from https://doi.org/10.1073/pnas.1208110109.

Wilk, R. (2012). The limits of discipline: Towards interdisciplinary food studies. *Physiology and Behavior, 107*(4), 471—475. Available from https://doi.org/10.1016/j.physbeh.2012.04.023.

Williamson, P., Mueller-Loose, S., Lockshin, L., & Francis, I. (2017). More hawthorn and less dried longan: the role of information and taste on red wine consumer preferences in China. *Australian Journal of Grape and Wine Research, 24*, 113—124.

Wilson, C., Brand, J., du Toit, W., & Buica, A. (2018). Polarized projective mapping as a rapid sensory analysis method applied to South African Chenin Blanc wines. *LWT, 92*, 140—146. Available from https://doi.org/10.1016/j.lwt.2018.02.022.

WSET. (2011). *Understanding style and quality.* Wine & Spirit Education Trust.

Yang, Q., Hollowood, T., & Hort, J. (2014). Phenotypic variation in oronasal perception and the relative effects of PROP and thermal taster status. *Food Quality and Preference, 38*, 83—91. 10.1016/j.foodqual.2014.05.013.

Yeshurun, Y., & Sobel, N. (2010). An odor is not worth a thousand words: From multidimensional odors to unidimensional odor objects. *Annual Review of Psychology, 61*, 219—241. Available from https://doi.org/10.1146/annurev.psych.60.110707.163639.

Young, B. D. (2020). Smell's puzzling discrepancy: Gifted discrimination, yet pitiful identification. *Mind and Language, 35*(1), 90—114. Available from https://doi.org/10.1111/mila.12233.

Index

Printed in the United States
by Baker & Taylor Publisher Services